the hist[illegible] and philosophy of science

a reader

Also available from Bloomsbury:

The Bloomsbury Companion to the Philosophy of Science, edited by Steven French and Juha Saatsi

An Historical Introduction to the Philosophy of Mathematics: A Reader, edited by Russell Marcus and Mark McEvoy

Philosophy of Science: Key Concepts, by Steven French

the history and philosophy of science

a reader

Edited by Daniel J. McKaughan and Holly VandeWall

Bloomsbury Academic
An imprint of Bloomsbury Publishing Plc

B L O O M S B U R Y
LONDON • OXFORD • NEW YORK • NEW DELHI • SYDNEY

Bloomsbury Academic
An imprint of Bloomsbury Publishing Plc

50 Bedford Square
London
WC1B 3DP
UK

1385 Broadway
New York
NY 10018
USA

www.bloomsbury.com

BLOOMSBURY and the Diana logo are trademarks of Bloomsbury Publishing Plc

First published 2018

British Library Cataloguing-in-Publication Data
A catalogue record for this book is available from the British Library.

ISBN: HB: 978-1-4742-3273-9
PB: 978-1-4742-3272-2
ePDF: 978-1-4742-3275-3
ePub: 978-1-4742-3274-6

Library of Congress Cataloging-in-Publication Data
A catalog record for this book is available from the Library of Congress.

Cover image © Bridgeman Images

Typeset by Newgen KnowledgeWorks Pvt. Ltd., Chennai, India
Printed and bound in Great Britain

Contents

Part V Elements in Transition: Chemistry, Air, Atoms, and Heat 531

Acknowledgments

We would like to express our thanks to Colleen Coalter, senior commissioning editor at Bloomsbury, who was a delight to work with from the beginning to the end of the publication process. Drew Alexander, graduate research assistant than which none greater can be conceived, also deserves special thanks for his help in the assembly and editing of texts during the 2015–16 academic year.

We are especially grateful for and indebted to the inspiring teachers from whom we received our graduate training in the History and Philosophy of Science Program at the University of Notre Dame and to its associated institutional structures such as the John J. Reilly Center and Program of Liberal Studies. In particular, we express our gratitude to the many ways in which the following faculty invested in our education in the history and philosophy of science: Matthew Ashley, Katherine Brading, Michael Crowe, Jim Cushing, Robert Goulding, Gary Gutting, Chris Hamlin, Anja Jauernig, Lynn Joy, Don Howard, Edward Manier, Vaughn McKim, Ernan McMullin, Gerald McKenny, Phil Mirowski, Lenny Moss, Janet Kourany, Phillip Sloan, Jole Shackelford, Kristin Shrader-Frechette, and Tom Stapleford.

At Boston College we wish in particular to thank Pat Byrne, from whom we inherited a well-thought-out New Scientific Visions course and who entrusted us with bringing our own vision to it; Brian Braman, who continues to ably direct and attract students to the Perspectives Program; and our colleague Marius Stan, who we shall enjoy seeing leave his own distinctive mark on history and philosophy of science education at Boston College in the years ahead.

We hope to have produced a volume worthy both of the ND HPS brand of history and philosophy of science and the Boston College Perspectives Program.

Thanks to Phillip Sloan, Carmen Giunta, and Greg Froelich, who graciously granted permission to use their selections and translations for this volume free of charge.

We would each like to thank the other for the pleasures of sharing in the various tasks of setting up the course out of which this collection of reading grew with a friend, and to note for the record that we both shared equally in the preparation of and work on every facet of the volume. Finally, we would each like to thank our spouses for their graciousness, patience, and support through the cost of time and challenges that our level of commitment to teaching and scholarship has involved.

Copyright Permission Acknowledgments (in Order of Appearance)

Such a text would not be possible without copyright permission from a variety of sources and we are grateful to have received permission to reprint material from the following sources. Please notify the publisher regarding any mistakes or omissions that should be corrected in future reprints or subsequent editions.

Part I: The Birth of Natural Philosophy: Science and Mathematics in the Ancient Hellenistic World

Aristotle, *On the Generation of Animals*, trans. Arthur Platt Book 1:1–2, Book 2:1 (lines: 715a1–716b13, 731b19–734b28), used by permission of site administrator Robert J. Robbins, ed. www.esp.org.

Eratosthenes, *Measurement of the Earth* as described by Cleomedes in *On the Circular Motion of the Heavenly Bodies*. In: Page, T. E., Capps, E., Rouse, W. H. D., Post, L. A., & Warmington, E. H. (eds), *Selections Illustrating the History of Greek Mathematics: From Aristarchus to Pappus*. Translated by I. Thomas. Vol. 2. London; Cambridge, MA: William Heinemann Ltd; Harvard University Press, 1941, reprinted in 1951, 1957. Used by permission of Harvard University Press.

Ptolemy, *Almagest*, translated by R. Catesby Taliaferro. In: Hutchins, Robert Maynard (ed.), *Ptolemy, Copernicus, Kepler: Great Books of the Western World*. Vol. 16. Chicago: Encyclopaedia Britannica, 1952. Used by permission of Encyclopaedia Britannica.

Part II: Translation, Appropriation, and Critical Engagement: Science in the Roman and Medieval Islamic and European Worlds

Philoponus, *Commentary on Aristotle's Physics*, translated by I. E. Drabkin. In: Cohen, Morris R., & Drabkin, I. E. (eds), *A Source Book in Greek Science* (Source Books in the History of the Sciences). Cambridge: Harvard University Press, 1948, 217–23. Used by permission of Harvard University Press.

Ibn al-Haytham (Alhazen), *Perspectiva (Book of Optics)*, translated by David C. Lindberg. In: Grant, Edward, *A Source Book in Medieval Science*. Cambridge, MA: Harvard University Press, 1974. Used by permission of Harvard University Press.

Albertus Magnus (Albert the Great), *On Plants*, translated by Edward Grant. In: Grant, Edward, *A Source Book in Medieval Science*. Cambridge, MA: Harvard University Press, 1974. Used by permission of Harvard University Press.

Aquinas, *On the Motion of the Heart (De Motu Cordis)*, translated by Gregory Froelich. Used by permission of Gregory Froelich.

Oresme, Nicole (c. 1360). *Tractatus de configurationibus qualitatum et motuum (A Treatise on the Configuration of Qualities and Motions)*, Paris, translated by Marshall Clagett. In: Clagett, Marshall, *The Science of Mechanics in the Middle Ages*. Madison: The University of Wisconsin Press, 1959. Used by permission of University of Wisconsin Press.

Buridan, John (1509). *Subtilissimae Quaestiones super octo Physicorum libros Aristotelis (Question on the Eight Books of the Physics of Aristotle)*, Paris, translated by Marshall Clagett. In: Clagett, Marshall, *The Science of Mechanics in the Middle Ages*. Madison: The University of Wisconsin Press, 1959. Used by permission of University of Wisconsin Press.

Part III: Revolutions in Astronomy and Mechanics: From Copernicus to Newton

Copernicus, Nicolaus (1515). *The Commentariolus.* In: Rosen, Edward, *Three Copernican Treatises: The "Commentariolus of Copernicus," the "Letter against Werner," the "Narratio Prima of Rheticus,"* Translated with Introduction and Notes by Edward Rosen (3rd ed.). New York: Dover, 1959. Used by permission of Dover Press.

Brahe, Tycho (1598). *Astronomiæ instauratæ mechanica (Instruments for the Restoration of Astronomy)*, Wandsbek. In: *Tycho Brahe's Description of His Instruments and Scientific Work as Given in Astronomiae Instauratae Mechanica.* Translated and edited by H. Ræder, E. Strömgren, and B. Strömgren. Copenhagen: Munksgaard, 1946. Used by permission of the Danish National Academy.

Galilei, Galileo (1632). *Dialogue concerning the Two Chief World Systems.* English translation by Thomas Salusbury in 1661. *Dialogue on the Great World Systems. Galileo Galilei.* Revised and annotated by Giorgio de Santillana. Chicago: University of Chicago Press, 1953. Used by permission of University of Chicago Press.

Descartes, René (1664). *Le Monde, ou Traite de la lumiere (The World, or Treatise on the Light)*. Translation and introduction by Michael Sean Mahoney. New York: Abaris Books, 1979. Used by permission of Abaris Books.

Part IV: Investigating the Invisible: Light, Electricity, and Magnetism

Hooke, Robert (1672). *Critique of Newton's Theory of Light and Colors.* In: Thomas Birch, *The History of the Royal Society*, vol. 3. London, 1757, 10–15. Used by permission of Robert Iliffe.

Part V: Elements in Transition: Chemistry, Air, Atoms, and Heat

Paracelsus, *Of the Nature of Things* (1537). In: Linden, Stanton J. (ed.), *The Alchemy Reader: From Hermes Trismegistus to Isaac Newton*. Cambridge: Cambridge University Press, 2003, 153–62. Used by permission of Cambridge University Press.

Newton, Isaac. *The Key* (c. 1650s–1670s), translated by Betty Jo Teeter Dobbs. In: Linden, Stanton J. (ed.), *The Alchemy Reader: From Hermes Trismegistus to Isaac Newton*. Cambridge: Cambridge University Press, 2003, 244–6. Used by permission of Cambridge University Press.

Boyle, Robert (1672). *Tracts Written by the Honourable Robert Boyle, Containing New Experiments, Touching the Relation between Flame and Air, and about Explosions*. In: Leicester, Henry M. and Klickstein, Herbert S. (eds), *A Source Book in Chemistry: 1400–1900*. Cambridge: Harvard University Press, 1952, 43–7. Used by permission of Harvard University Press.

Becher, Johann Joachim (1669). *Concerning the First Principle of Metals and Stones, Which Is Called Vitreous Stone, or Stony Earth* [terra lapidia] *and Improperly Salt* from *"Acta Laboratorii Chymici Monacensis, Seu Physicae Subterranae,"* translated by Henry M. Leicester and Herbert S. Klickstein. In: Leicester, Henry M., and Klickstein, Herbert S. (eds), *A Source Book in Chemistry: 1400–1900*. Cambridge: Harvard University Press, 1952, 56–8. Used by permission of Harvard University Press.

Stahl, Georg Ernst (1697). *Foundation of the Fermentative Art* (*Zymotechnia Fundamentalis seu Fermentationis Theoria generalis*), translated by Henry M. Leicester and Herbert S. Klickstein. In: Leicester, Henry M., and Klickstein, Herbert S. (eds), *A Source Book in Chemistry: 1400–1900*. Cambridge: Harvard University Press, 1952, 59–60. Used by permission of Harvard University Press.

Stahl, Georg Ernst (1718). *Random Thoughts and Useful Concerns* (*Zufällige Gedanken und nützliche Bedencken*), translated by Henry M. Leicester and Herbert S. Klickstein. In: Leicester, Henry M., and Klickstein, Herbert S. (eds), *A Source Book in Chemistry: 1400–1900*. Cambridge: Harvard University Press, 1952, 60–1. Used by permission of Harvard University Press.

Stahl, Georg Ernst (1723). *Dogmatic and Experiential Foundations of Chemistry* (*Fundamenta chymiae dogmaticae et experimentalis*), translated by Peter Shaw. In: Leicester, Henry M., and Klickstein, Herbert S. (eds), *A Source Book in Chemistry: 1400–1900*. Cambridge: Harvard University Press, 1952, 61–3. Used by permission of Harvard University Press.

Lavoisier, Antoine-Laurent de, and Laplace, Pierre-Simon marquis de (1780). *Memoir on Heat*, translated by M. L. Gabriel. In: Gabriel, Mordecai L., and Fogel, Seymour (eds), *Great Experiments in Biology*. Englewood Cliffs, NJ: Prentice-Hall, 1955, 85–93. Used by permission of Prentice-Hall.

Parts VI: The Earth and All Its Creatures: Developments in Geology and Biology; and VII: The Emergence of Evolution: Darwin and His Interlocutors

Descartes, René (1664). *L'homme, et un Traitté de la formation du foetus (Treatise on Man and on the Formation of the Fetus [Description of the Human Body])*, ed.

Claude Clerselier. Paris: Charles Angot. Translated from original French by P. R. Sloan. Used with permission of Phillip Sloan.

Buffon, Georges-Louis Leclerc, Comte de (1749). *The Theory of the Earth.* In: *Natural History, Volume I (of X) Containing a Theory of the Earth, a General History of Man, of the Brute Creation, and of Vegetables, Mineral, &c. &c*. Translated into English from the original 1749 French *Imprimerie Royale* edition of the *Histoire naturelle* by James Smith Barr. London: H.D. Symonds, 1797. With emendations from the original French by P. R. Sloan. Used with permission of Phillip Sloan.

Buffon, Georges-Louis Leclerc, Comte de (1749). *History of Animals.* In: *Natural History, Volume II (of X) Containing a Theory of the Earth, a General History of Man, of the Brute Creation, and of Vegetables, Mineral, &c. &c*. Translated into English from the original 1749 French *Imprimerie Royale* edition of the *Histoire naturelle* by James Smith Barr. London: H.D. Symonds, 1797. With emendations from the original French by P. R. Sloan. Used with permission of Phillip Sloan.

Cuvier, Georges (1828). "General Conclusion on the Organization of the Fishes." In Cuvier, G., & Valenciennes, A. *Histoire naturelle des Poissons* (Paris: Levrault, 1828) I, 543–51. Translated into English by P. R. Sloan. Used with permission of Phillip Sloan.

Introduction

Daniel J. McKaughan and
Holly VandeWall

We are pleased to present this collection of foundational texts in all the major fields of science as it developed in Western culture from antiquity through the end of the nineteenth century. Our aim is to make accessible key scientific discoveries, observations, experiments, contributions to methodology, and theories as they are described in the works of great innovators, such as Aristotle, Euclid, Copernicus, Galileo, Descartes, Harvey, Newton, Lavoisier, Darwin, Faraday, and Maxwell. We hope that this volume will prove useful for introducing a wide audience of potential readers to a broad historical and philosophical perspective on the origin and development of sciences such as astronomy, physics, chemistry, and the life sciences.

These carefully selected original source readings are intended to serve as the core sourcebook for an advanced undergraduate or graduate-level class in the history of science and/or philosophy of science. We think that it is important for people working in these fields to be acquainted with these texts and hope that the volume will become a standard introduction to these disciplines. The collection grew out of our teaching of an intensive, year-long double-credit introduction to the history and philosophy of science and mathematics emphasizing direct engagement with primary texts at Boston College. In the fall of 2008, Pat Byrne, our department chair, commissioned us with the responsibility of revitalizing the "New Scientific Visions" course as part of the great books program (Perspectives) at Boston College. Although the field of history and philosophy of science is active and professionally well-established today, we found it surprisingly difficult to find a collection of primary source material in the history of science that didn't begin with Copernicus or end with Newton, that took the ancient world seriously and didn't forget that geology is a science, and that allowed the development of method and theory in the history of science to emerge in context. So we painstakingly began piecing one together, using existing course materials, things we had read in graduate school, and recommendations from colleagues. We have continued to bring about fifty students a year through our program. Over the years we have continued to make changes, removing things that didn't work, adding material to cover a wider and more interesting array of topics, cutting readings that were too long, and expanding those that didn't cover enough.

The readings, cut for brevity and intelligibility, are accompanied by discussion questions at the end of each reading and suggestions for further reading at the end of each section. The texts are not arranged purely chronologically but grouped into sections arranged by topic and thematic unity to help clarify the development of disciplines, methods, and the unification of theories. The section divisions also allow instructors focusing their course on the Copernican revolution or the background to Darwinian evolution to point students to just those sections relevant for the class. Other texts of this kind tend to focus more narrowly, for example, on mechanics and astronomy, neglecting the material in Parts I, IV, and V of our collection. Moreover, they tend to shift from reading scientific texts to philosophy of science texts at the end of the seventeenth century. By focusing on the most accessible selections from the primary source materials, this collection remains focused on the scientists themselves. These readings show natural philosophers having lively discussions about religion and race, politics and patronage. Real, colorful, flesh-and-blood people emerge in these pages—authors struggling to learn how to learn about the natural world. Their writings contain dissent and disdain, humor and scathing criticism, arrogance and humility—all in the context of natural philosophy.

The selections in Part I, "The Birth of Natural Philosophy: Science and Mathematics in the Ancient Hellenistic World," present the development of several branches of science in antiquity, attending to the theories of matter, motion, life, mathematics, and cosmology that shaped the Western scientific tradition.

Part II, "Translation, Appropriation, and Critical Engagement: Science in the Roman and Medieval Islamic and European Worlds," reflects both the extent to which the ancient worldview was adopted and developed in the medieval Islamic and European worlds, and the degree to which this work influenced and inspired the revolutions of the sixteenth and seventeenth centuries.

Part III, "Revolutions in Astronomy and Mechanics: From Copernicus to Newton," contains the critique of Aristotelian theory and development of methods, instruments, and techniques that led to profound, worldview shifting revolutions in astronomy and mechanics in the sixteenth and seventeenth centuries.

Part IV, "Investigating the Invisible: Light, Electricity, and Magnetism," presents controversies over the particle and wave theories of light, important milestones in investigations of the nature of electricity and magnetism, and the stunning theoretical unification of all three of these phenomena in the work of Faraday and Maxwell.

Part V, "Elements in Transition: Chemistry, Air, Atoms, and Heat," explores the shift from alchemical work to modern chemistry, including the identification of different kinds of air, important steps in the development of the periodic table of elements, the analysis of heat in terms of molecular motion rather than as a caloric fluid, and the emergence of thermodynamics.

Part VI, "The Earth and All Its Creatures: Developments in Geology and Biology," focuses on the extraordinary theoretical and experimental developments in biology, geology, and anthropology, from Harvey's discovery of the circulation of the blood through Lyell's uniformitarian geology.

Part VII, "The Emergence of Evolution: Darwin and His Interlocutors," focuses on Darwinian evolution, with selections from both the *Origin of Species* and the *Descent of Man* as well as both critical and supportive responses to Darwin's work from his contemporaries.

We envision careful reading and discussion of these texts giving rise to a rich set of historical and philosophical conversations that help students to understand key scientific ideas, to explore the social and cultural context in which they emerge, and to reflect on their significance.

Rationale: Why Should the General Reader Care about the History and Philosophy of Science?

The independent intellectual value of the history and philosophy of science needs no defense. Permit us, however, a few all too brief remarks on the potential value of cultivating a historically and philosophically informed perspective on where science came from and how it developed and how this contributes to the task of understanding and evaluating scientific material with an eye to larger socially important issues.

Science has come to exert enormous influence in our society, on its industries, government policies, educational institutions, world affairs, and our views of reality. Perhaps now, as much as in any previous era, opportunities to learn about and reflect on the origin and development of science are crucial to informed civic participation on a host of science and technology related social and moral issues our global civilization will face in the years ahead. We do not pretend, of course, that solutions to our future social challenges are to be found in a collection of texts from the history of science. Neither do these texts cover the content and methods of contemporary scientific inquiry. However, at least a general familiarity with the history and philosophy of science, not just among aspiring scientists, but among our citizens, is arguably vital to ensuring that science will be structured in ways that both allow it to continue to flourish and to be put in the service of the common public good.

We think that an exposure to and an engagement with issues related to the history and philosophy of science can contribute to scientific literacy and responsible reflection on science in a host of important ways. Chief goals for university-level courses taught in conjunction with the reader might include:

1. To carefully analyze and discuss some of the great classical and contemporary texts and figures from natural sciences such as astronomy, physics, chemistry, and the life sciences
2. To understand something of the social and cultural context in which scientific ideas emerge
3. To equip students with the tools and skills needed to read, interpret, and critically engage central problems and ideas from the history and philosophy of Western science
4. To identify and describe the roles that mathematics plays in scientific inquiry
5. To explore central themes such as the question of whether there is a scientific method (or methods) and how these developed over time
6. To cultivate the kind of background knowledge, historical perspective, and independent habits of thought about scientific activity that help to enable informed citizen participation on important questions that arise at the intersection of science and society

There is much value to be gained from approaching the study of science from the perspective of the liberal arts—asking questions about science and its place in culture that scientists themselves often don't directly ask or attempt to answer as part of their professional work, seeking to understand the sorts of questions people have asked about the natural world at various times, the assumptions they made, the methods they employed, the answers that they found satisfying, and the implications of the ideas and values embodied in their traditions.

For each of the readings, ask yourself the following sorts of questions: What ideas, activities, and values are explicit or implicit in the way the author approaches understanding the natural world? How do these activities compare to or set it apart from other disciplines such as metaphysics, theology, literature, ethics, political philosophy, or engineering? What sorts of questions are the authors asking and how do they go about attempting to answer them? By what criteria do they evaluate various proposals (e.g., mathematical rigor, accurately identifying the cause, etc.)? What motivates such inquiries? What roles does this activity play in society (e.g., mythology, utility in prediction, practical application, inquiry for its own sake, etc.), and what does this tell us about the needs, interests, and capabilities of their society? What sort of literary genre is the piece written in (e.g., a poem, a letter, a lecture, a dissertation, or a dialogue)? Who do you think is the intended audience (e.g., other experts in their specific scientific field, all natural philosophers, the public, a patron)? What role do sense experience, reason, mathematics, or other factors play in the author's epistemology or views about the sources of knowledge? Given the philosophical outlook expressed in these readings, what sorts of implications does it have for the place of inquiry into nature in the author's philosophy? Does the reading indicate anything about the state of the various disciplines or fields of inquiry in the author's culture? What areas are developed and what technologies are available in the author's society? How were they using mathematics or thinking about the nature of living beings, the stars, and our place in the universe? In what ways did great scientific achievements come about by breaking from tradition, and in what ways did they depend upon it? How did the separate sciences begin to emerge as distinct disciplines? How did mathematical advances often give rise to new scientific visions? What are the scope and limits or promises and perils of science as it is concretely instantiated at any particular moment in time? What will we make of the powers science unleashes, with potential for great good and great ill? It is our hope that those who undertake the challenging but enjoyable work of cultivating such a perspective on the history of human attempts to understand the natural world, life and the stars, matter and how things work, and what these might contribute to big questions about our place in the cosmos and where it all came from will be better positioned to think independently about how science might flourish and how best to use it.

A note about terminology: we use the terms *natural philosophy* and *science* interchangeably in the introduction and reading questions of this volume. The term *science* did not come into common use as a descriptor of investigation into natural phenomena until the late eighteenth century (though it derives from the Latin *scientia* and Greek *epistēmē* meaning knowledge, words often used in contrast to mere opinion). Terms for specific areas of study—astronomy, optics, mechanics, and anatomy, for example—existed, but someone wishing to indicate his or her interest in more than one of these areas would have termed themselves a *natural philosopher* until almost the turn of the twentieth century. Galileo angled for years for the job of "Court Philosopher" to the Medici. During his years traveling on the Beagle voyage, Darwin's crewmates called him "Philosopher." (Darwin's actual title on the voyage was *ship's naturalist*.) But science/scientist are the terms in common use today, and we use it in this volume, for both clarity and brevity, even in periods where it is not how the authors would have referred to themselves or their work.

Part I

The Birth of Natural Philosophy: Science and Mathematics in the Ancient Hellenistic World

Nature of Man (c. 400 BCE)

Hippocrates[1]

I. He who is accustomed to hear speakers discuss the nature of man beyond its relations to medicine will not find the present account of any interest. For I do not say at all that a man is air, or fire, or water, or earth, or anything else that is not an obvious constituent of a man; such accounts I leave to those that care to give them. Those, however, who give them have not in my opinion correct knowledge. For while adopting the same idea they do not give the same account. Though they add the same appendix to their idea—saying that "what is" is a unity, and that this is both unity and the all—yet they are not agreed as to its name. One of them asserts that this one and the all is air, another calls it fire, another, water, and another, earth; while each appends to his own account evidence and proofs that amount to nothing. The fact that, while adopting the same idea, they do not give the same account, shows that their knowledge too is at fault. . . .

II. Now about these men I have said enough, and I will turn to physicians. Some of them say that a man is blood, others that he is bile, a few that he is phlegm. Physicians, like the metaphysicians, all add the same appendix. For they say that a man is a unity, giving it the name that severally they wish to give it; this changes its form and its power, being constrained by the hot and the cold, and becomes sweet, bitter, white, black and so on. But in my opinion these views also are incorrect. Most physicians then maintain views like these, if not identical with them; but I hold that if man were a unity he would never feel pain, as there would be nothing from which a unity could suffer pain. And even if he were to suffer, the cure too would have to be one. But as a matter of fact cures are many.

[1] As with other texts in the Hippocratic Corpus, there is some question about the authorship of *Nature of Man*. In *History of Animals* III.3, Aristotle attributes the work to Polybus, the son-in-law of Hippocrates.

For in the body are many constituents, which, by heating, by cooling, by drying or by wetting one another contrary to nature, engender diseases; so that both the forms of diseases are many and the healing of them is manifold. But I require of him who asserts that man is blood and nothing else, to point out a man when he does not change his form or assume every quality, and to point out a time, a season of the year or a season of human life, in which obviously blood is the only constituent of man. For it is only natural that there should be one season in which blood-in-itself appears as the sole constituent. My remarks apply also to him who says that man is only phlegm, and to him who says that man is bile. I for my part will prove that what I declare to be the constituents of a man are, according to both convention and nature, always alike the same; it makes no difference whether the man be young or old, or whether the season be cold or hot. I will also bring evidence, and set forth the necessary causes why each constituent grows or decreases in the body.

III. . . .Therefore, since such is the nature both of all other things and of man, man of necessity is not one, but each of the components contributing to generation has in the body the power it contributed. Again, each component must return to its own nature when the body of a man dies, moist to moist, dry to dry, hot to hot and cold to cold. Such too is the nature of animals, and of all other things. All things are born in a like way, and all things die in a like way. For the nature of them is composed of all those things I have mentioned above, and each thing, according to what, has been said, ends in that from which it was composed. So that too is whither it departs.

IV. The body of man has in itself blood, phlegm, yellow bile and black bile; these make up the nature of his body, and through these he feels pain or enjoys health. Now he enjoys the most perfect health when these elements are duly proportioned to one another in respect of compounding, power and bulk, and when they are perfectly mingled. Pain is felt when one of these elements is in defect or excess, or is isolated in the body without being compounded with all the others. For when an element is isolated and stands by itself, not only must the place which it left become diseased, but the place where it stands in a flood must, because of the excess, cause pain and distress. In fact when more of an element flows out of the body than is necessary to get rid of superfluity, the emptying causes pain. If, on the other hand, it be to an inward part that there takes place the emptying, the shifting and the separation from other elements, the man certainly must, according to what has been said, suffer from a double pain, one in the place left, and another in the place flooded.

V. Now I promised to show that what are according to me the constituents of man remain always the same, according to both convention and nature. These constituents are, I hold, blood, phlegm, yellow bile and black bile. First I assert that the names of these according to convention are separated, and that none of them has the same name as the others; furthermore, that according to nature their essential forms are separated, phlegm being quite unlike blood, blood being quite unlike bile, bile being quite unlike phlegm. How could they be like one another, when their colours appear not alike to the sight nor does their touch seem alike to the hand? For they are not equally warm, nor cold, nor dry, nor moist. Since then they are so different from one another in essential form and in power, they cannot be one, if fire and water are not one. From the following evidence you may know that these elements are not all one, but that each of them has its own power and its own nature. If you were to give a man a medicine which withdraws phlegm, he will vomit you phlegm; if you give him one which withdraws bile, he will vomit you bile. Similarly too black bile is purged away if you give a medicine which withdraws black bile. And if you wound a man's body so as to cause a wound, blood will flow from him. And you will find all these things happen on any day and on any night, both in winter and in summer, so long as the man can draw breath in and then breathe it out again, or until he is deprived of one of the elements congenital with him. Congenital with him (how should they not be so?) are the elements already mentioned. First, so long as a man lives he manifestly has all these elements always in him; then he is born out of a human being having all these elements, and is nursed in a human being having them all, I mean those elements I have mentioned with proofs.

. . .

VII. Phlegm increases in a man in winter; for phlegm, being the coldest constituent of the body, is closest akin to winter. A proof that phlegm is very cold is that if you touch phlegm, bile and blood, you will find phlegm the coldest. And yet it is the most viscid, and after black bile requires most force for its evacuation. But things that are moved by force become hotter under the stress of the force. Yet in spite of all this, phlegm shows itself the coldest element by reason of its own nature. That winter fills the body with phlegm you can learn from the following evidence. It is in winter that the sputum and nasal discharge of men is fullest of phlegm; at this season mostly swellings become white, and diseases generally phlegmatic. And in spring too phlegm still remains strong in the body, while the blood increases. For the cold relaxes, and the rains come on, while the blood accordingly increases through the showers and the hot days. For these conditions of the year are most akin to the nature of blood, spring being moist and warm. You can learn the truth from the following facts. It is chiefly in spring and summer that men are attacked by dysenteries, and by hemorrhage from the nose, and they are then hottest and red. And in summer blood is still strong, and bile rises in the body and extends until autumn. In autumn blood becomes small in quantity, as autumn is opposed to its nature, while bile prevails in the body during the summer season and during autumn. You may learn this truth from the following facts. During this season men vomit bile without an emetic, and when they take purges the discharges are most bilious. It is plain too from fevers and from the complexions of men. But in summer phlegm is at its weakest. For the season is opposed to its nature, being dry and warm. But in autumn blood becomes least in man, for autumn is dry and begins from this point to chill him. It is black bile which in autumn is greatest and strongest. When winter comes on, bile being chilled becomes small in quantity, and phlegm increases again because of the abundance of rain and the length of the nights. All these elements then are always comprised in the body of a man, but as the year goes round they become now greater and now less, each in turn and according to its nature. For just as every year participates in every element, the hot, the cold, the dry and the moist—none in fact of these elements would last for a moment without all the things that exist in this universe, but if one were to fail all would disappear, for by reason of the same necessity all things are constructed and nourished by one another—even so, if any of these congenital elements were to fail, the man could not live. In the year sometimes the winter is most powerful, sometimes the spring, sometimes the summer and sometimes the autumn. So too in man sometimes phlegm is powerful, sometimes blood, sometimes bile, first yellow, and then what is called black bile. The clearest proof is that if you will give the same man to drink the same drug four times in the year, he will vomit, you will find, the most phlegmatic matter in the winter, the moistest in the spring, the most bilious in the summer, and the blackest in the autumn.

. . .

IX. Furthermore, one must know that diseases due to repletion are cured by evacuation, and those due to evacuation are cured by repletion; those due to exercise are cured by rest, and those due to idleness are cured by exercise. To know the whole matter, the physician must set himself against the established character of diseases, of constitutions, of seasons and of ages; he must relax what is tense and make tense what is relaxed. For in this way the diseased part would rest most, and this, in my opinion, constitutes treatment.

Translated by W. H. S. Jones

Reading and Discussion Questions

1. Hippocrates accuses both physicians and metaphysicians of being overly concerned with "unity." What does he think that the human body consists of and how does he approach questions of health?
2. How is Hippocrates thinking about the relationship between health, the seasons, and atmospheric conditions?

On the Nature of Things (c. 50 BCE)

Titus Lucretius Carus

Nothing Exists per se Except Atoms and the Void

But, now again to weave the
tale begun,
All nature, then, as self-sustained,
consists
Of twain of things: of bodies and
of void
In which they're set, and where
they're moved around.
For common instinct of our race
declares
That body of itself exists: unless
This primal faith, deep-founded, fail
us not,
Naught will there be whereunto
to appeal
On things occult when seeking
aught to prove
By reasonings of mind. Again,
without
That place and room, which we do
call the inane,
Nowhere could bodies then be
set, nor go
Hither or thither at all—as shown
before.
Besides, there's naught of which
thou canst declare
It lives disjoined from body, shut
from void—
A kind of third in nature. For
whatever
Exists must be a somewhat; and
the same,
If tangible, however fight and slight,
Will yet increase the count of
body's sum,
With its own augmentation big
or small;
But, if intangible and
powerless ever
To keep a thing from passing
through itself

On any side, 'twill be naught else but that
Which we do call the empty, the inane.
Again, whate'er exists, as of itself,
Must either act or suffer action on it,
Or else be that wherein things move and be:
Naught, saving body, acts, is acted on;
Naught but the inane can furnish room. And thus,
Beside the inane and bodies, is no third
Nature amid the number of all things—
Remainder none to fall at any time
Under our senses, nor be seized and seen
By any man through reasonings of mind.
Name o'er creation with what names thou wilt,
Thou'lt find but properties of those first twain,
Or see but accidents those twain produce.

A property is that which not at all
Can be disjoined and severed from a thing
Without a fatal dissolution: such,
Weight to the rocks, heat to the fire, and flow
To the wide waters, touch to corporal things,
Intangibility to the viewless void.
But state of slavery, pauperhood, and wealth,
Freedom, and war, and concord, and all else
Which come and go whilst nature stands the same,
We're wont, and rightly, to call accidents.
Even time exists not of itself; but sense
Reads out of things what happened long ago,
What presses now, and what shall follow after:
No man, we must admit, feels time itself,
Disjoined from motion and repose of things.
Thus, when they say there "is" the ravishment
Of Princess Helen, "is" the siege and sack
Of Trojan Town, look out, they force us not
To admit these acts existent by themselves,
Merely because those races of mankind
(Of whom these acts were accidents) long since
Irrevocable age has borne away:
For all past actions may be said to be
But accidents, in one way, of mankind,—
In other, of some region of the world.
Add, too, had been no matter, and no room
Wherein all things go on, the fire of love
Upblown by that fair form, the glowing coal
Under the Phrygian Alexander's breast,
Had ne'er enkindled that renowned strife

Of savage war, nor had the
wooden horse
Involved in flames old Pergama, by
a birth
At midnight of a brood of the
Hellenes.
And thus thou canst remark that
every act
At bottom exists not of itself,
nor is
As body is, nor has like name
with void;
But rather of sort more fitly to
be called
An accident of body, and of place
Wherein all things go on.

Character of the Atoms

Bodies, again,
Are partly primal germs of things,
and partly
Unions deriving from the primal
germs.
And those which are the primal
germs of things
No power can quench; for in the
end they conquer
By their own solidness; though
hard it be
To think that aught in things has
solid frame;
For lightnings pass, no less than
voice and shout,
Through hedging walls of houses,
and the iron
White-dazzles in the fire, and rocks
will burn
With exhalations fierce and burst
asunder.
Totters the rigid gold dissolved
in heat;
The ice of bronze melts conquered
in the flame;
Warmth and the piercing cold
through silver seep,
Since, with the cups held rightly in
the hand,
We oft feel both, as from above
is poured
The dew of waters between their
shining sides:
So true it is no solid form
is found.
But yet because true reason and
nature of things
Constrain us, come, whilst in few
verses now
I disentangle how there still exist
Bodies of solid, everlasting
frame—
The seeds of things, the primal
germs we teach,
Whence all creation around us
came to be.
First since we know a twofold
nature exists,
Of things, both twain and utterly
unlike—
Body, and place in which all
things go on—
Then each must be both for and
through itself,
And all unmixed: where'er be
empty space,
There body's not; and so where
body bides,

There not at all exists the
void inane.
Thus primal bodies are solid,
without a void.
But since there's void in all
begotten things,
All solid matter must be round
the same;
Nor, by true reason canst thou
prove aught hides
And holds a void within its
body, unless
Thou grant what holds it be a
solid. Know,
That which can hold a void of
things within
Can be naught else than matter in
union knit.
Thus matter, consisting of a
solid frame,
Hath power to be eternal, though
all else,
Though all creation, be
dissolved away.
Again, were naught of empty
and inane,
The world were then a solid; as,
without
Some certain bodies to fill the
places held,
The world that is were but a
vacant void.
And so, infallibly, alternate-wise
Body and void are still distinguished,
Since nature knows no wholly full
nor void.
There are, then, certain bodies,
possessed of power
To vary forever the empty and
the full;
And these can nor be sundered
from without
By beats and blows, nor from
within be torn
By penetration, nor be overthrown
By any assault soever through
the world—
For without void, naught can be
crushed, it seems,
Nor broken, nor severed by a cut
in twain,
Nor can it take the damp, or
seeping cold
Or piercing fire, those old
destroyers three;
But the more void within a thing,
the more
Entirely it totters at their sure assault.
Thus if first bodies be, as I have
taught,
Solid, without a void, they must
be then
Eternal; and, if matter ne'er
had been
Eternal, long ere now had all
things gone
Back into nothing utterly, and all
We see around from nothing had
been born—
But since I taught above that
naught can be
From naught created, nor the once
begotten
To naught be summoned back,
these primal germs
Must have an immortality of frame.
And into these must each thing be
resolved,
When comes its supreme hour, that
thus there be
At hand the stuff for plenishing
the world.

So primal germs have solid
singleness
Nor otherwise could they have been
conserved
Through aeons and infinity of time

For the replenishment of wasted worlds.
Once more, if nature had given a scope for things
To be forever broken more and more,
By now the bodies of matter would have been
So far reduced by breakings in old days
That from them nothing could, at season fixed,
Be born, and arrive its prime and top of life.
For, lo, each thing is quicker marred than made;
And so whate'er the long infinitude
Of days and all fore-passed time would now
By this have broken and ruined and dissolved,
That same could ne'er in all remaining time
Be builded up for plenishing the world.
But mark: infallibly a fixed bound
Remaineth stablished 'gainst their breaking down;
Since we behold each thing soever renewed,
And unto all, their seasons, after their kind,
Wherein they arrive the flower of their age.

Again, if bounds have not been set against
The breaking down of this corporeal world,
Yet must all bodies of whatever things
Have still endured from everlasting time
Unto this present, as not yet assailed
By shocks of peril. But because the same
Are, to thy thinking, of a nature frail,
It ill accords that thus they could remain
(As thus they do) through everlasting time,
Vexed through the ages (as indeed they are)
By the innumerable blows of chance.

So in our programme of creation, mark
How 'tis that, though the bodies of all stuff
Are solid to the core, we yet explain
The ways whereby some things are fashioned soft—
Air, water, earth, and fiery exhalations—
And by what force they function and go on:
The fact is founded in the void of things.
But if the primal germs themselves be soft,
Reason cannot be brought to bear to show
The ways whereby may be created these
Great crags of basalt and the during iron;
For their whole nature will profoundly lack
The first foundations of a solid frame.
But powerful in old simplicity,
Abide the solid, the primeval germs;
And by their combinations more condensed,

All objects can be tightly knit and bound
And made to show unconquerable strength.
Again, since all things kind by kind obtain
Fixed bounds of growing and conserving life;
Since Nature hath inviolably decreed
What each can do, what each can never do;
Since naught is changed, but all things so abide
That ever the variegated birds reveal
The spots or stripes peculiar to their kind,
Spring after spring: thus surely all that is
Must be composed of matter immutable.
For if the primal germs in any wise
Were open to conquest and to change, 'twould be
Uncertain also what could come to birth
And what could not, and by what law to each
Its scope prescribed, its boundary stone that clings
So deep in Time. Nor could the generations
Kind after kind so often reproduce
The nature, habits, motions, ways of life,
Of their progenitors.

And then again,
Since there is ever an extreme bounding point
Of that first body which our senses now
Cannot perceive: That bounding point indeed
Exists without all parts, a minimum
Of nature, nor was e'er a thing apart,
As of itself—nor shall hereafter be,
Since 'tis itself still parcel of another,
A first and single part, whence other parts
And others similar in order lie
In a packed phalanx, filling to the full
The nature of first body: being thus
Not self-existent, they must cleave to that
From which in nowise they can sundered be.
So primal germs have solid singleness,
Which tightly packed and closely joined cohere
By virtue of their minim particles—
No compound by mere union of the same;
But strong in their eternal singleness,
Nature, reserving them as seeds for things,
Permitteth naught of rupture or decrease.

Moreover, were there not a minimum,
The smallest bodies would have infinites,
Since then a half-of-half could still be halved,
With limitless division less and less.
Then what the difference 'twixt the sum and least?
None: for however infinite the sum,

Yet even the smallest would consist
the same
Of infinite parts. But since true
reason here
Protests, denying that the mind can
think it,
Convinced thou must confess such
things there are
As have no parts, the minimums of
nature.
And since these are, likewise
confess thou must
That primal bodies are solid and
eterne.
Again, if Nature, creatress of all
things,
Were wont to force all things to be
resolved
Unto least parts, then would she
not avail
To reproduce from out them
anything;
Because whate'er is not endowed
with parts
Cannot possess those properties
required
Of generative stuff—divers
connections,
Weights, blows, encounters,
motions, whereby things
Forevermore have being and
go on.

Translated by William Ellery Leonard

Reading and Discussion Questions

1. How are proponents of atomism thinking about the relationship between appearances and reality?
2. Are any arguments or evidence offered for the view? How might the atomists have supported their claims or defended their views against criticism?

Timaeus (c. 360 BCE)

Plato

CRITIAS: Let me proceed to explain to you, Socrates, the order in which we have arranged our entertainment. Our intention is, that Timaeus, who is the most of an astronomer amongst us, and has made the nature of the universe his special study, should speak first, beginning with the generation of the world and going down to the creation of man; next, I am to receive the men whom he has created of whom some will have profited by the excellent education which you have given them; and then, in accordance with the tale of Solon, and equally with his law, we will bring them into court and make them citizens, as if they were those very Athenians whom the sacred Egyptian record has recovered from oblivion, and thenceforward we will speak of them as Athenians and fellow-citizens.

SOCRATES: I see that I shall receive in my turn a perfect and splendid feast of reason. And now, Timaeus, you, I suppose, should speak next, after duly calling upon the Gods.

TIMAEUS: All men, Socrates, who have any degree of right feeling, at the beginning of every enterprise, whether small or great, always call upon God. . . . I am asking a question which has to be asked at the beginning of an enquiry about anything—was the world, I say, always in existence and without beginning? or created, and had it a beginning? Created, I reply, being visible and tangible and having a body, and therefore sensible; and all sensible things are apprehended by opinion and sense and are in a process of creation and created. Now that which is created must, as we affirm, of necessity be created by a cause. But the father and maker of all this universe is past finding out; and even if we found him, to tell of him to all men would be impossible. And there is still a question to be asked about him: Which of the patterns had the artificer in view when he made the world—the pattern of the unchangeable, or of that which is created? If the world be indeed fair and the artificer good, it is manifest that he must have looked to that which is eternal; but if what cannot be said without blasphemy is true, then to the created pattern. Everyone will see that he must have looked to, the eternal; for the world is the fairest of creations and he is the best of causes. And having been created in this way, the world has been framed in the likeness of that which is apprehended by reason and mind and is unchangeable, and must therefore of necessity, if this is admitted, be a copy of something. Now it is all-important that the beginning of everything should be according to nature. And in speaking of the copy and the original we may assume that words are akin to the matter which they describe; when they relate to the lasting and permanent and intelligible, they ought to be lasting and unalterable, and, as far as their nature allows, irrefutable and immovable—nothing less. But when they express only the copy or likeness and not the eternal things themselves, they need only be likely and analogous to the real words. As being is to becoming, so is truth to belief. If then, Socrates, amid the many opinions about the gods and the generation of the universe, we are not able to give notions which are altogether and in every respect exact and consistent with one another, do not be surprised. Enough, if we adduce probabilities as likely as any others; for we must remember that I who am the speaker, and you who are the judges, are only mortal men, and we ought to accept the tale which is probable and enquire no further.

SOCRATES: Excellent, Timaeus; and we will do precisely as you bid us. The prelude is charming, and is already accepted by us—may we beg of you to proceed to the strain?

TIMAEUS: Let me tell you then why the creator made this world of generation. He was good, and the good can never have any jealousy of anything. And being free from jealousy, he desired that all things should be as like himself as they could be. This is in the truest sense the origin of creation and of the world, as we shall do well in believing on the testimony of wise men: God desired that all things should be good and nothing bad, so far as this was attainable. Wherefore also finding the whole visible sphere not at rest, but moving in an irregular and disorderly fashion, out of disorder he brought order, considering that this was in every way better than the other. Now the deeds of the best could never be or have been other than the fairest; and the creator, reflecting on the things which are by nature visible, found that no unintelligent creature taken as a whole was fairer than the intelligent taken as a whole; and that intelligence could not be present in anything which was devoid of soul. For which reason, when he was framing the universe, he put intelligence in soul, and soul in body, that he might be the creator of a work which was by nature fairest and best. Wherefore, using the language of probability, we may say that the world became a living creature truly endowed with soul and intelligence by the providence of God.

This being supposed, let us proceed to the next stage: ... Are we right in saying that there is one world, or that they are many and infinite? There must be one only, if the created copy is to accord with the original. For that which includes all other intelligible creatures cannot have a second or companion; in that case there would be need of another living being which would include both, and of which they would be parts, and the likeness would be more truly said to resemble not them, but that other which included them. In order then that the world might be solitary, like the perfect animal, the creator made not two worlds or an infinite number of them; but there is and ever will be one only-begotten and created heaven.

Now that which is created is of necessity corporeal, and also visible and tangible. And nothing is visible where there is no fire, or tangible which has no solidity, and nothing is solid without earth. Wherefore also God in the beginning of creation made the body of the universe to consist of fire and earth. But two things cannot be rightly put together without a third; there must be some bond of union between them. And the fairest bond is that which makes the most complete fusion of itself and the things which it combines; and proportion is best adapted to effect such a union. For whenever in any three numbers, whether cube or square, there is a mean, which is to the last term what the first term is to it; and again, when the mean is to the first term as the last term is to the mean—then the mean becoming first and last, and the first and last both becoming means, they will all of them of necessity come to be the same, and having become the same with one another will be all one. If the universal frame had been created a surface only and having no depth, a single mean would have sufficed to bind together itself and the other terms; but now, as the world must be solid, and solid bodies are always compacted not by one mean but by two, God placed water and air in the mean between fire and earth, and made them to have the same proportion so far as was possible (as fire is to air so is air to water, and as air is to water so is water to earth); and thus he bound and put together a visible and tangible heaven. And for these reasons, and out of such elements which are in number four, the body of the world was created, and it was harmonized by proportion, and therefore has the spirit of friendship; and having been reconciled to itself, it was indissoluble by the hand of any other than the framer.

Now the creation took up the whole of each of the four elements; for the Creator compounded the world out of all the fire and all the water and all the air and all the earth, leaving no part of any of them nor any power of them outside. His intention was, in the first place, that the animal[1] should be as far as possible a perfect whole and of perfect parts: secondly, that it should be one, leaving no remnants out of which another such world might be created: and also that it should be free from old age and unaffected by disease. Considering

[1] Other translations use "Living Being" instead of animal.

that if heat and cold and other powerful forces which unite bodies surround and attack them from without when they are unprepared, they decompose them, and by bringing diseases and old age upon them, make them waste away—for this cause and on these grounds he made the world one whole, having every part entire, and being therefore perfect and not liable to old age and disease. And he gave to the world the figure which was suitable and also natural. Now to the animal which was to comprehend all animals, that figure was suitable which comprehends within itself all other figures. Wherefore he made the world in the form of a globe, round as from a lathe, having its extremes in every direction equidistant from the centre, the most perfect and the most like itself of all figures; for he considered that the like is infinitely fairer than the unlike. This he finished off, making the surface smooth all around for many reasons; in the first place, because the living being had no need of eyes when there was nothing remaining outside him to be seen; nor of ears when there was nothing to be heard; and there was no surrounding atmosphere to be breathed; nor would there have been any use of organs by the help of which he might receive his food or get rid of what he had already digested, since there was nothing which went from him or came into him: for there was nothing beside him. Of design he was created thus, his own waste providing his own food, and all that he did or suffered taking place in and by himself. For the Creator conceived that a being which was self-sufficient would be far more excellent than one which lacked anything; and, as he had no need to take anything or defend himself against any one, the Creator did not think it necessary to bestow upon him hands: nor had he any need of feet, nor of the whole apparatus of walking; but the movement suited to his spherical form was assigned to him, being of all the seven that which is most appropriate to mind and intelligence; and he was made to move in the same manner and on the same spot, within his own limits revolving in a circle. All the other six motions were taken away from him, and he was made not to partake of their deviations. And as this circular movement required no feet, the universe was created without legs and without feet.

Such was the whole plan of the eternal God about the god that was to be, to whom for this reason he gave a body, smooth and even, having a surface in every direction equidistant from the centre, a body entire and perfect, and formed out of perfect bodies. And in the centre he put the soul, which he diffused throughout the body, making it also to be the exterior environment of it; and he made the universe a circle moving in a circle, one and solitary, yet by reason of its excellence able to converse with itself, and needing no other friendship or acquaintance. Having these purposes in view he created the world a blessed god.

...

When the father creator saw the creature which he had made moving and living, the created image of the eternal gods, he rejoiced, and in his joy determined to make the copy still more like the original; and as this was eternal, he sought to make the universe eternal, so far as might be. Now the nature of the ideal being was everlasting, but to bestow this attribute in its fullness upon a creature was impossible. Wherefore he resolved to have a moving image of eternity, and when he set in order the heaven, he made this image eternal but moving according to number, while eternity itself rests in unity; and this image we call time. For there were no days and nights and months and years before the heaven was created, but when he constructed the heaven he created them also. They are all parts of time, and the past and future are created species of time, which we unconsciously but wrongly transfer to the eternal essence; for we say that he "was," he "is," he "will be," but the truth is that "is" alone is properly attributed to him, and that "was" and "will be" only to be spoken of becoming in time, for they are motions, but that which is immovably the same cannot become older or younger by time, nor ever did or has become, or hereafter will be, older or younger, nor is subject at all to any of those states which affect moving and sensible things and of which generation is the cause. These are the forms of time, which imitates eternity and revolves according to a law of number. Moreover, when we say that what has become is become and what becomes is becoming, and that what will become is about to become and that the non-existent is non-existent—all these are inaccurate modes of expression. But perhaps this whole subject will be more suitably discussed on some other occasion.

Time, then, and the heaven came into being at the same instant in order that, having been created together, if ever there was to be a dissolution of them, they might be dissolved together. It was framed after the pattern of the eternal nature, that it might resemble this as far as was possible; for the pattern exists from eternity, and the created heaven has been, and is, and will be, in all time. Such was the mind and thought of God in the creation of time. The sun and moon and five other stars, which are called the planets, were created by him in order to distinguish and preserve the numbers of time; and when he had made their several bodies, he placed them in the orbits in which the circle of the other was revolving—in seven orbits seven stars. First, there was the moon in the orbit nearest the earth, and next the sun, in the second orbit above the earth; then came the morning star and the star sacred to Hermes, moving in orbits which have an equal swiftness with the sun, but in an opposite direction; and this is the reason why the sun and Hermes and Lucifer overtake and are overtaken by each other. To enumerate the places which he assigned to the other stars, and to give all the reasons why he assigned them, although a secondary matter, would give more trouble than the primary. These things at some future time, when we are at leisure, may have the consideration which they deserve, but not at present.

Now, when all the stars which were necessary to the creation of time had attained a motion suitable to them, and had become living creatures having bodies fastened by vital chains, and learnt their appointed task, moving in the motion of the diverse, which is diagonal, and passes through and is governed by the motion of the same, they revolved, some in a larger and some in a lesser orbit—those which had the lesser orbit revolving faster, and those which had the larger more slowly. Now by reason of the motion of the same, those which revolved fastest appeared to be overtaken by those which moved slower although they really overtook them; for the motion of the same made them all turn in a spiral, and, because some went one way and some another, that which receded most slowly from the sphere of the same, which was the swiftest, appeared to follow it most nearly. That there might be some visible measure of their relative swiftness and slowness as they proceeded in their eight courses, God lighted a fire, which we now call the sun, in the second from the earth of these orbits, that it might give light to the whole of heaven, and that the animals, as many as nature intended, might participate in number, learning arithmetic from the revolution of the same and the like. Thus then, and for this reason the night and the day were created, being the period of the one most intelligent revolution. And the month is accomplished when the moon has completed her orbit and overtaken the sun, and the year when the sun has completed his own orbit. Mankind, with hardly an exception, have not remarked the periods of the other stars, and they have no name for them, and do not measure them against one another by the help of number, and hence they can scarcely be said to know that their wanderings, being infinite in number and admirable for their variety, make up time. And yet there is no difficulty in seeing that the perfect number of time fulfils the perfect year when all the eight revolutions, having their relative degrees of swiftness, are accomplished together and attain their completion at the same time, measured by the rotation of the same and equally moving. After this manner, and for these reasons, came into being such of the stars as in their heavenly progress received reversals of motion, to the end that the created heaven might imitate the eternal nature, and be as like as possible to the perfect and intelligible animal.

Thus far and until the birth of time the created universe was made in the likeness of the original, but inasmuch as all animals were not yet comprehended therein, it was still unlike. What remained, the creator then proceeded to fashion after the nature of the pattern. Now as in the ideal animal the mind perceives ideas or species of a certain nature and number, he thought that this created animal ought to have species of a like nature and number. There are four such; one of them is the heavenly race of the gods; another, the race of birds whose way is in the air; the third, the watery species; and the fourth, the pedestrian and land creatures. Of the heavenly and divine, he created the greater part out of fire, that they might be the brightest of all things and fairest to behold, and he fashioned them after the likeness of the universe in the figure of a circle, and made them follow the intelligent motion of the supreme, distributing them over the whole circumference of heaven, which was to be a true cosmos or glorious world spangled with them all over. And he gave to each of them two movements: the first, a movement on the same spot after the same manner, whereby they ever continue to think consistently the same thoughts about the same things; the second, a forward movement, in which they are controlled by the revolution of the same and the like; but by the other five motions they were unaffected, in order that each of them might attain the highest perfection. And for this reason the fixed stars were created, to be divine and eternal animals, ever-abiding and revolving after the same manner and on the same spot; and the other stars which reverse their motion and are subject to deviations of this kind, were created in the manner already described. The earth, which is our nurse, clinging around the pole which is extended through the universe, he framed to be the guardian and artificer of night and day, first and eldest of gods that are in the interior of heaven. Vain would be the attempt to tell all the figures of them circling as in dance, and their juxtapositions, and the return of them in their revolutions upon themselves, and their approximations, and to say which of these deities in their conjunctions meet, and which of them are in opposition, and in what order they get behind and before one another, and when they are severally eclipsed to our sight and again reappear, sending terrors and intimations of the future to those who cannot calculate their movements—to attempt to tell of all this without a visible representation of the heavenly system would be labor in vain. Enough on this head; and now let what we have said about the nature of the created and visible gods have an end.

To know or tell the origin of the other divinities is beyond us, and we must accept the traditions of the men of old time who affirm themselves to be the offspring of the gods—that is what they say—and they must surely have known their own ancestors. How can we doubt the word of the children of the gods? Although they give no probable or certain proofs, still, as they declare that they are speaking of what took place in their own family, we must conform to custom and believe them. In this manner, then, according to them, the genealogy of these gods is to be received and set forth.

Oceanus and Tethys were the children of Earth and Heaven, and from these sprang Phorcys and Cronos and Rhea, and all that generation; and from Cronos and Rhea sprang Zeus and Here, and all those who are said to be their brethren, and others who were the children of these.

Now, when all of them, both those who visibly appear in their revolutions as well as those other gods who are of a more retiring nature, had come into being, the creator of the universe addressed them in these words: "Gods, children of gods, who are my works, and of whom I am the artificer and father, my creations are indissoluble, if so I will. All that is bound may be undone, but only an evil being would wish to undo that which is harmonious and happy. Wherefore, since ye are but creatures, ye are not altogether immortal and indissoluble, but ye shall certainly not be dissolved, nor be liable to the fate of death, having in my will a greater and mightier bond than those with which ye were bound at the time of your birth. And now listen to my instructions: Three tribes of mortal beings remain to be created—without them the universe will be incomplete, for it will not contain every kind of animal which it ought to contain, if it is to be perfect. On the other hand, if they were created by me and received life at my hands, they would be on an equality with the gods. In order then that they may be mortal, and that this universe may be truly universal, do ye, according to your natures, betake yourselves to the formation of animals, imitating the power which was shown by me in creating you. The part of them worthy of the name immortal, which is called divine and is the guiding principle of those who are willing to follow justice and you—of that divine part I will myself sow the seed, and having made a beginning, I will hand the work over to you. And do ye then interweave the mortal with the immortal, and make and beget living creatures, and give them food, and make them to grow, and receive them again in death."

Thus he spoke, and once more into the cup in which he had previously mingled the soul of the universe he poured the remains of the elements, and mingled them in much the same manner; they were not, however, pure as before, but diluted to the second and third degree. And having made it he divided the whole mixture into souls equal in number to the stars, and assigned each soul to a star; and having there placed them as in a chariot, he showed them the nature of the universe, and declared to them the laws of destiny, according to which their first birth would be one and the same for all—no one should suffer a disadvantage at his hands; they were to be sown in the instruments of time severally adapted to them, and to come forth the most religious of animals; and as human nature was of two kinds, the superior race would hereafter be called man. Now, when they should be implanted in bodies by necessity, and be always gaining or losing some part of their bodily substance, then in the first place it would be necessary that they should all have in them one and the same faculty of sensation, arising out of irresistible impressions; in the second place, they must have love, in which pleasure and pain mingle; also fear and anger, and the feelings which are akin or opposite to them; if they conquered these they would live righteously, and if they were conquered by them, unrighteously. He who lived well during his appointed time was to return and dwell in his native star, and there he would have a blessed and congenial existence. But if he failed in attaining this, at the second birth he would pass into a woman, and if, when in that state of being, he did not desist from evil, he would continually be changed into some brute who resembled him in the evil nature which he had acquired, and would not cease from his toils and transformations until he followed the revolution of the same and the like within him, and overcame by the help of reason the turbulent and irrational mob of later accretions, made up of fire and air and water and earth, and returned to the form of his first and better state. Having given all these laws to his creatures, that he might be guiltless of future evil in any of them, the creator sowed some of them in the earth, and some in the moon, and some in the other instruments of time; and when he had sown them he committed to the younger gods the fashioning of their mortal bodies, and desired them to furnish what was still lacking to the human soul, and having made all the suitable additions, to rule over them, and to pilot the mortal animal in the best and wisest manner which they could, and avert from him all but self-inflicted evils.

When the creator had made all these ordinances he remained in his own accustomed nature, and his children heard and were obedient to their father's word, and receiving from him the immortal principle of a mortal creature, in imitation of their own creator they borrowed portions of fire, and earth, and water, and air from the world, which were hereafter to be restored—these they took and welded them together, not with the indissoluble chains by which they were themselves bound, but with little pegs too small to be visible, making up out of all the four elements each separate body, and fastening the courses of the immortal soul in a body which was in a state of perpetual influx and efflux. Now these courses, detained as in a vast river, neither overcame nor were overcome; but were hurrying and hurried to and fro, so that the whole animal was moved and progressed, irregularly however and irrationally and anyhow, in all the six directions of motion, wandering backwards and forwards, and right and left, and up and down, and in all the six directions. For great as was the advancing and retiring flood which provided nourishment, the affections produced by external contact caused still greater tumult—when the body of any one met and came into collision with some external fire, or with the solid earth or the gliding waters, or was caught in the tempest borne on the air, and the motions produced by any of these impulses were carried through the body to the soul.

. . .

The sight in my opinion is the source of the greatest benefit to us, for had we never seen the stars, and the sun, and the heaven, none of the words which we have spoken about the universe would ever have been uttered. But now the sight of day and night, and the months and the revolutions of the years, have created number, and have given us a conception of time, and the power of enquiring about the nature of the universe; and from this source we have derived philosophy, than which no greater good ever was or will be given by the gods to mortal man. This is the greatest boon of sight: and of the lesser benefits why should I speak? Even the ordinary man if he were deprived of them would bewail his loss, but in vain. This much let me say however: God invented and gave us sight to the end that we might behold the courses of intelligence in the heaven, and apply them to the courses of our own intelligence which are akin to them, the unperturbed to the perturbed; and that we, learning them and partaking of the natural truth of reason, might imitate the absolutely unerring courses of God and regulate our own vagaries. The same may be affirmed of speech and hearing: they have been given by the gods to the same end and for a like reason. For this is the principal end of speech, whereto it most contributes. Moreover, so much of music as is adapted to the sound of the voice and to the sense of hearing is granted to us for the sake of harmony; and harmony, which has motions akin to the revolutions of our souls, is not regarded by the intelligent votary of the Muses as given by them with a view to irrational pleasure, which is deemed to be the purpose of it in our day, but as meant to correct any discord which may have arisen in the courses of the soul, and to be our ally in bringing her into harmony and agreement with herself; and rhythm too was given by them for the same reason, on account of the irregular and graceless ways which prevail among mankind generally, and to help us against them.

. . .

[W]e must consider the nature of fire, and water, and air, and earth, such as they were prior to the creation of the heaven, and what was happening to them in this previous state; for no one has as yet explained the manner of their generation, but we speak of fire and the rest of them, whatever they mean, as though men knew their natures, and we maintain them to be the first principles and letters or elements of the whole, when they cannot reasonably be compared by a man of any sense even to syllables or first compounds. And let me say thus much: I will not now speak of the first principle or principles of all things, or by whatever name they are to be called, for this reason—because it is difficult to set forth my opinion according to the method of discussion which we are at present employing. Do not imagine, any more than I can bring myself to imagine, that I should be right in undertaking so great and difficult a task. Remembering what I said at first about probability, I will do my best to give as probable an explanation as any other-or rather, more probable; and I will first go back to the beginning and try to speak of each thing and of all. Once more, then, at the commencement of my discourse, I call upon God, and beg him to be our savior out of a strange and unwonted enquiry, and to bring us to the haven of probability. So now let us begin again.

. . .

At first, [the various elements] were all without reason and measure. But when the world began to get into order, fire and water and earth and air had only certain faint traces of themselves, and were altogether such as everything might be expected to be in the absence of God; this, I say, was their nature at that time, and God fashioned them by form and number. Let it be consistently maintained by us in all that we say that God made them as far as possible the fairest and best, out of things which were not fair and good. And now I will endeavor to show you the disposition and generation of them by an unaccustomed argument, which am compelled to use; but I believe that you will be able to follow me, for your education has made you familiar with the methods of science.

In the first place, then, as is evident to all, fire and earth and water and air are bodies. And every sort of body possesses solidity, and every solid must necessarily be contained in planes; and every plane rectilinear figure is composed of triangles; and all triangles are originally of two kinds, both of which are made up of one right and two acute angles; one of them has at either end of the base the half of a divided right angle, having equal sides, while in the other the right angle is divided into unequal parts, having unequal sides. These, then, proceeding by a combination of probability with demonstration, we assume to be the original elements of fire and the other bodies; but the principles which are prior to these God only knows, and he of men who is the friend God. And next we have to determine what are the four most beautiful bodies which are unlike one another, and of which some are capable of resolution into one another; for having discovered thus much, we shall know the true origin of earth and fire and of the proportionate and intermediate elements. And then we shall not be willing to allow that there are any distinct kinds of visible bodies fairer than these. Wherefore we must endeavor to construct the four forms of bodies which excel in beauty, and then we shall be able to say that we have sufficiently apprehended their nature. Now of the two triangles, the isosceles has one form only; the scalene or unequal-sided has an infinite number. Of the infinite forms we must select the most beautiful, if we are to proceed in due order, and anyone who can point out a more beautiful form than ours for the construction of these bodies, shall carry off the palm, not as an enemy, but as a friend. Now, the one which we maintain to be the most beautiful of all the many triangles (and we need not speak of the others) is that of which the double forms a third triangle which is equilateral; the reason of this would be long to tell; he who disproves what we are saying, and shows that we are mistaken, may claim a friendly victory. Then let us choose two triangles, out of which fire and the other elements have been constructed, one isosceles, the other having the square of the longer side equal to three times the square of the lesser side.

Now is the time to explain what was before obscurely said: there was an error in imagining that all the four elements might be generated by and into one another; this, I say, was an erroneous supposition, for there are generated from the triangles which we have selected four kinds—three from the one which has the sides unequal; the fourth alone is framed out of the isosceles triangle. Hence they cannot all be resolved into one another, a great number of small bodies being combined into a few large ones, or the converse. But three of them can be thus resolved and compounded, for they all spring from one, and when the greater bodies are broken up, many small bodies will spring up out of them and take their own proper figures; or, again, when many small bodies are dissolved into their triangles, if they become one, they will form one large mass of another kind. So much for their passage into one another. I have now to speak of their several kinds, and show out of what combinations of numbers each of them was formed. The first will be the simplest and smallest construction, and its element is that triangle which has its hypotenuse twice the lesser side. When two such triangles are joined at the diagonal, and this is repeated three times, and the triangles rest their diagonals and shorter sides on the same point as a centre, a single equilateral triangle is formed out of six triangles; and four equilateral triangles, if put together, make out of every three plane angles one solid angle, being that which is nearest to the most obtuse of plane angles; and out of the combination of these four angles arises the first solid form which distributes into equal and similar parts the whole circle in which it is inscribed.

The second species of solid is formed out of the same triangles, which unite as eight equilateral triangles and form one solid angle out of four plane angles, and out of six such angles the second body is completed. And the third body is made up of 120 triangular elements, forming twelve solid angles, each of them included in five plane equilateral triangles, having altogether twenty bases, each of which is an equilateral triangle. The one element [i.e., the triangle which has its hypotenuse twice the lesser side] having generated these figures, generated no more; but the isosceles triangle produced the fourth elementary figure, which is compounded of four such triangles, joining their right angles in a centre, and forming one equilateral quadrangle. Six of these united form eight solid angles, each of which is made by the combination of three plane right angles; the figure of the body thus composed is a cube, having six plane quadrangular equilateral bases. There was yet a fifth combination which God used in the delineation of the universe. . . .

Now, he who, duly reflecting on all this, enquires whether the worlds are to be regarded as indefinite or definite in number, will be of opinion that the notion of their indefiniteness is characteristic of a sadly indefinite and ignorant mind. He, however, who raises the question whether they are to be truly regarded as one or five, takes up a more reasonable position. Arguing from probabilities, I am of opinion that they are one; another, regarding the question from another point of view, will be of another mind. But, leaving this enquiry, let us proceed to distribute the elementary forms, which have now been created in idea, among the four elements.

To earth, then, let us assign the cubical form; for earth is the most immoveable of the four and the most plastic of all bodies, and that which has the most stable bases must of necessity be of such a nature. Now, of the triangles which we assumed at first, that which has two equal sides is by nature more firmly based than that which has unequal sides; and of the compound figures which are formed out of either, the plane equilateral quadrangle has necessarily, a more stable basis than the equilateral triangle, both in the whole and in the parts. Wherefore, in assigning this figure to earth, we adhere to probability; and to water we assign that one of the remaining forms which is the least moveable; and the most moveable of them to fire; and to air that which is intermediate. Also we assign the smallest body to fire, and the greatest to water, and the intermediate in size to air; and, again, the acutest body to fire, and the next in acuteness to, air, and the third to water. Of all these elements, that which has the fewest bases must necessarily be the most moveable, for it must be the acutest and most penetrating in every way, and also the lightest as being composed of the smallest number of similar particles: and the second body has similar properties in a second degree, and the third body in the third degree. Let it be agreed, then, both according to strict reason and according to probability, that the pyramid is the solid which is the original element and seed of fire; and let us assign the element which was next in the order of generation to air, and the third to water. We must imagine all these to be so small that no single particle of any of the four kinds is seen by us on account of their smallness: but when many of them are collected together their aggregates are seen. And the ratios of their numbers, motions, and other properties, everywhere God, as far as necessity allowed or gave consent, has exactly perfected, and harmonized in due proportion.

From all that we have just been saying about the elements or kinds, the most probable conclusion is as follows:—earth, when meeting with fire and dissolved by its sharpness, whether the dissolution take place in the fire itself or perhaps in some mass of air or water, is borne hither and thither, until its parts, meeting together and mutually harmonizing, again become earth; for they can never take any other form. But water, when divided by fire or by air, on reforming, may become one part fire and two parts air; and a single volume of air divided becomes two of fire. Again, when a small body of fire is contained in a larger body of air or water or earth, and both are moving, and the fire struggling is overcome and broken up, then two volumes of fire form one volume of air; and when air is overcome and cut up into small pieces, two and a half parts of air are condensed into one part of water. Let us consider the matter in another way. When one of the other elements is fastened upon by fire, and is cut by the sharpness of its angles and sides, it coalesces with the fire, and then ceases to be cut by them any longer. For no element which is one and the same with itself can be changed by or change another of the same kind and in the same state. But so long as in the process of transition the weaker is fighting against the stronger, the dissolution continues. Again, when a few small particles, enclosed in many larger ones, are in process of decomposition and extinction, they only cease from their tendency to extinction when they consent to pass into the conquering nature, and fire becomes air and air water. But if bodies of another kind go and attack them [i.e., the small particles], the latter continue to be dissolved until, being completely forced back and dispersed, they make their escape to their own kindred, or else, being overcome and assimilated to the conquering power, they remain where they are and dwell with their victors, and from being many become one. And owing to these affections, all things are changing their place, for by the motion of the receiving vessel the bulk of each class is distributed into its proper place; but those things which become unlike themselves and like other things, are hurried by the shaking into the place of the things to which they grow like.

Now all unmixed and primary bodies are produced by such causes as these. As to the subordinate species which are included in the greater kinds, they are to be attributed to the varieties in the structure of the two original triangles. For either structure did not originally produce the triangle of one size only, but some larger and some smaller, and there are as many sizes as there are species of the four elements. Hence when they are mingled with themselves and with one another there is an endless variety of them, which those who would arrive at the probable truth of nature ought duly to consider.

Translated by Benjamin Jowett

Reading and Discussion Questions

1. Describe the origin story that Timaeus tells. What sort of confidence do you think that Plato places in it—what epistemological status do you think that the author takes it to have?
2. What explanation does Timaeus offer for the observed order and regularity in the world?
3. Compare and contrast Timaeus' account to origin stories from other cultures with which you are familiar. What similarities or differences do you see, for example, between Timaeus' account and the Genesis narratives to be found in the Hebrew Bible?
4. Compare and contrast the view of matter found here in Plato with that of the atomists found in the Lucretius reading.

Philebus (c. 350 BCE)

Plato

SOCRATES:	And now, having subjected pleasure to every sort of test, let us not appear to be too sparing of mind and knowledge: let us ring their metal bravely, and see if there be unsoundness in any part, until we have found out what in them is of the purest nature; and then the truest elements both of pleasure and knowledge may be brought up for judgment.
PROTARCHUS:	Right.
SOCRATES:	Knowledge has two parts,—the one productive, and the other educational?
PROTARCHUS:	True.
SOCRATES:	And in the productive or handicraft arts, is not one part more akin to knowledge, and the other less; and may not the one part be regarded as the pure, and the other as the impure?
PROTARCHUS:	Certainly.
SOCRATES:	Let us separate the superior or dominant elements in each of them.
PROTARCHUS:	What are they, and how do you separate them?
SOCRATES:	I mean to say, that if arithmetic, mensuration, and weighing be taken away from any art, that which remains will not be much.
PROTARCHUS:	Not much, certainly.
SOCRATES:	The rest will be only conjecture, and the better use of the senses which is given by experience and practice, in addition to a certain power of guessing, which is commonly called art, and is perfected by attention and pains.
PROTARCHUS:	Nothing more, assuredly.
SOCRATES:	Music, for instance, is full of this empiricism; for sounds are harmonized, not by measure, but by skillful conjecture; the music of the flute is always trying to guess the pitch of each vibrating note,

and is therefore mixed up with much that is doubtful and has little which is certain.

PROTARCHUS: Most true.

SOCRATES: And the same will be found to hold good of medicine and husbandry and piloting and generalship.

PROTARCHUS: Very true.

SOCRATES: The art of the builder, on the other hand, which uses a number of measures and instruments, attains by their help to a greater degree of accuracy than the other arts.

PROTARCHUS: How is that?

SOCRATES: In ship-building and house-building, and in other branches of the art of carpentering, the builder has his rule, lathe, compass, line, and a most ingenious machine for straightening wood.

PROTARCHUS: Very true, Socrates.

SOCRATES: Then now let us divide the arts of which we were speaking into two kinds,—the arts which, like music, are less exact in their results, and those which, like carpentering, are more exact.

PROTARCHUS: Let us make that division.

SOCRATES: Of the latter class, the most exact of all are those which we just now spoke of as primary.

PROTARCHUS: I see that you mean arithmetic, and the kindred arts of weighing and measuring.

SOCRATES: Certainly, Protarchus; but are not these also distinguishable into two kinds?

PROTARCHUS: What are the two kinds?

SOCRATES: In the first place, arithmetic is of two kinds, one of which is popular, and the other philosophical.

PROTARCHUS: How would you distinguish them?

SOCRATES: There is a wide difference between them, Protarchus; some arithmeticians reckon unequal units; as for example, two armies, two oxen, two very large things or two very small things. The party who are opposed to them insist that every unit in ten thousand must be the same as every other unit.

PROTARCHUS: Undoubtedly there is, as you say, a great difference among the votaries of the science; and there may be reasonably supposed to be two sorts of arithmetic.

SOCRATES: And when we compare the art of mensuration which is used in building with philosophical geometry, or the art of computation which is used in trading with exact calculation, shall we say of either of the pairs that it is one or two?

PROTARCHUS: On the analogy of what has preceded, I should be of opinion that they were severally two.

SOCRATES: Right; but do you understand why I have discussed the subject?

PROTARCHUS: I think so, but I should like to be told by you.

SOCRATES: The argument has all along been seeking a parallel to pleasure, and true to that original design, has gone on to ask whether one sort of knowledge is purer than another, as one pleasure is purer than another.

PROTARCHUS: Clearly; that was the intention.

SOCRATES: And has not the argument in what has preceded, already shown that the arts have different provinces, and vary in their degrees of certainty?

PROTARCHUS: Very true.

SOCRATES: And just now did not the argument first designate a particular art by a common term, thus making us believe in the unity of that art; and then again, as if speaking of two different things, proceed to enquire whether the art as pursed by philosophers, or as pursued by non-philosophers, has more of certainty and purity?

PROTARCHUS: That is the very question which the argument is asking.

SOCRATES: And how, Protarchus, shall we answer the enquiry?

PROTARCHUS: O Socrates, we have reached a point at which the difference of clearness in different kinds of knowledge is enormous.

SOCRATES: Then the answer will be the easier.

PROTARCHUS: Certainly; and let us say in reply, that those arts into which arithmetic and mensuration enter, far surpass all others; and that of these the arts or sciences which are animated by the pure philosophic impulse are infinitely superior in accuracy and truth.

SOCRATES: Then this is your judgment; and this is the answer which, upon your authority, we will give to all masters of the art of misinterpretation?

PROTARCHUS: What answer?

SOCRATES: That there are two arts of arithmetic, and two of mensuration; and also several other arts which in like manner have this double nature, and yet only one name.

PROTARCHUS: Let us boldly return this answer to the masters of whom you speak, Socrates, and hope for good luck.

SOCRATES: We have explained what we term the most exact arts or sciences.

PROTARCHUS: Very good.

SOCRATES: And yet, Protarchus, dialectic will refuse to acknowledge us, if we do not award to her the first place.

PROTARCHUS: And pray, what is dialectic?

SOCRATES: Clearly the science which has to do with all that knowledge of which we are now speaking; for I am sure that all men who have a grain of intelligence will admit that the knowledge which has to do with being and reality, and sameness and unchangeableness, is by far the truest of all. But how would you decide this question, Protarchus?

PROTARCHUS: I have often heard Gorgias maintain, Socrates, that the art of persuasion far surpassed every other; this, as he says, is by far the best of them all, for to it all things submit, not by compulsion, but of their own free will. Now, I should not like to quarrel either with you or with him.

SOCRATES: You mean to say that you would like to desert, if you were not ashamed?

PROTARCHUS: As you please.

SOCRATES: May I not have led you into a misapprehension?

PROTARCHUS: How?

SOCRATES: Dear Protarchus, I never asked which was the greatest or best or usefullest of arts or sciences, but which had clearness and accuracy, and the greatest amount of truth, however humble and little useful an art. And as for Gorgias, if you do not deny that his art has the advantage in usefulness to mankind, he will not quarrel with you for saying that the study of which I am speaking is superior in this particular of essential truth; as in the comparison of white colours, a little whiteness, if that little be only pure, was said to be superior in truth to a great mass which is impure. And now let us give our best attention and consider well, not the comparative use or reputation of the sciences, but the power or faculty, if there be such, which the soul has of loving the truth, and of doing all things for the sake of it; let us search into the pure element of mind and intelligence, and then we shall be able to say whether the science of which I have been speaking is most likely to possess the faculty, or whether there be some other which has higher claims.

PROTARCHUS: Well, I have been considering, and I can hardly think that any other science or art has a firmer grasp of the truth than this.

SOCRATES: Do you say so because you observe that the arts in general and those engaged in them make use of opinion, and are resolutely engaged in the investigation of matters of opinion? Even he who supposes himself to be occupied with nature is really occupied with the things of this world, how created, how acting or acted upon. Is not this the sort of enquiry in which his life is spent?

PROTARCHUS: True.

SOCRATES: He is labouring, not after eternal being, but about things which are becoming, or which will or have become.

PROTARCHUS: Very true.

SOCRATES: And can we say that any of these things which neither are nor have been nor will be unchangeable, when judged by the strict rule of truth ever become certain?

PROTARCHUS: Impossible.

SOCRATES: How can anything fixed be concerned with that which has no fixedness?

PROTARCHUS: How indeed?

SOCRATES: Then mind and science when employed about such changing things do not attain the highest truth?

PROTARCHUS: I should imagine not.

SOCRATES: And now let us bid farewell, a long farewell, to you or me or Philebus or Gorgias, and urge on behalf of the argument a single point.

PROTARCHUS: What point?

SOCRATES: Let us say that the stable and pure and true and unalloyed has to do with the things which are eternal and unchangeable and unmixed, or if not, at any rate what is most akin to them has; and that all other things are to be placed in a second or inferior class.

PROTARCHUS: Very true.

SOCRATES: And of the names expressing cognition, ought not the fairest to be given to the fairest things?

PROTARCHUS: That is natural.

SOCRATES: And are not mind and wisdom the names which are to be honoured most?

PROTARCHUS: Yes.

SOCRATES: And these names may be said to have their truest and most exact application when the mind is engaged in the contemplation of true being?

PROTARCHUS: Certainly.

Translated by Benjamin Jowett

Reading and Discussion Questions

1. A distinction between pure and applied science has been influential in the history of science. In this text Socrates distinguishes between "pure" and "impure" sciences. What factors does the dialogue suggest contribute to purity and what sorts of values are expressed in this discussion?
2. What sorts of disciplines and crafts are already well developed in Plato's society?

The Republic (c. 370 BCE)

Plato

Book VI

(Socrates and Glaucon) . . .

But you see that without the addition of some other nature there is no seeing or being seen?

How do you mean?

Sight being, as I conceive, in the eyes, and he who has eyes wanting to see; colour being also present in them, still unless there be a third nature specially adapted to the purpose, the owner of the eyes will see nothing and the colours will be invisible.

Of what nature are you speaking?

Of that which you term light, I replied.

True, he said.

Noble, then, is the bond which links together sight and visibility, and great beyond other bonds by no small difference of nature; for light is their bond, and light is no ignoble thing?

Nay, he said, the reverse of ignoble.

And which, I said, of the gods in heaven would you say was the lord of this element? Whose is that light which makes the eye to see perfectly and the visible to appear?

You mean the sun, as you and all mankind say.

May not the relation of sight to this deity be described as follows?

How?

Neither sight nor the eye in which sight resides is the sun?

No.

Yet of all the organs of sense the eye is the most like the sun?

By far the most like.

And the power which the eye possesses is a sort of effluence which is dispensed from the sun?

Exactly.

Then the sun is not sight, but the author of sight who is recognised by sight?

True, he said.

And this is he whom I call the child of the good, whom the good begat in his own likeness, to be in the visible world, in relation to sight and the things of sight, what the good is in the intellectual world in relation to mind and the things of mind:

Will you be a little more explicit? he said.

Why, you know, I said, that the eyes, when a person directs them towards objects on which the light of day is no longer shining, but the moon and stars only, see dimly, and are nearly blind; they seem to have no clearness of vision in them?

Very true.

But when they are directed towards objects on which the sun shines, they see clearly and there is sight in them?

Certainly.

And the soul is like the eye: when resting upon that on which truth and being shine, the soul perceives and understands, and is radiant with intelligence; but when turned towards the twilight of becoming and perishing, then she has opinion only, and goes blinking about, and is first of one opinion and then of another, and seems to have no intelligence?

Just so.

Now, that which imparts truth to the known and the power of knowing to the knower is what I would have you term the idea of good, and this you will deem to be the cause of science, and of truth in so far as the latter becomes the subject of knowledge; beautiful too, as are both truth and knowledge, you will be right in esteeming this other nature as more beautiful than either; and, as in the previous instance, light and sight may be truly said to be like the sun, and yet not to be the sun, so in this other sphere, science and truth may be deemed to be like the good, but not the good; the good has a place of honour yet higher.

What a wonder of beauty that must be, he said, which is the author of science and truth, and yet surpasses them in beauty; for you surely cannot mean to say that pleasure is the good?

God forbid, I replied; but may I ask you to consider the image in another point of view?

In what point of view?

You would say, would you not, that the sun is not only the author of visibility in all visible things, but of generation and nourishment and growth, though he himself is not generation?

Certainly.

In like manner the good may be said to be not only the author of knowledge to all things known, but of their being and essence, and yet the good is not essence, but far exceeds essence in dignity and power.

Glaucon said, with a ludicrous earnestness: By the light of heaven, how amazing!

Yes, I said, and the exaggeration may be set down to you; for you made me utter my fancies.

And pray continue to utter them; at any rate let us hear if there is anything more to be said about the similitude of the sun.

Yes, I said, there is a great deal more.

Then omit nothing, however slight.

I will do my best, I said; but I should think that a great deal will have to be omitted.

I hope not, he said.

You have to imagine, then, that there are two ruling powers, and that one of them is set over the intellectual world, the other over the visible. I do not say heaven, lest you should fancy that I am playing upon the name ('ourhanoz, orhatoz'). May I suppose that you have this distinction of the visible and intelligible fixed in your mind?

I have.

Now take a line which has been cut into two unequal parts, and divide each of them again in the same proportion, and suppose the two main divisions to answer, one to the visible and the other to the intelligible, and then compare the subdivisions in respect of their clearness and want of clearness, and you will find that the first section in the sphere of the visible consists of images. And by images I mean, in the first place, shadows, and in the second place, reflections in water and in solid, smooth and polished bodies and the like: Do you understand?

Yes, I understand.

Imagine, now, the other section, of which this is only the resemblance, to include the animals which we see, and everything that grows or is made.

Very good.

Would you not admit that both the sections of this division have different degrees of truth, and that the copy is to the original as the sphere of opinion is to the sphere of knowledge?

Most undoubtedly.

Next proceed to consider the manner in which the sphere of the intellectual is to be divided.

In what manner?

Thus:—There are two subdivisions, in the lower of which the soul uses the figures given by the former division as images; the enquiry can only be hypothetical, and instead of going upwards to a principle descends to the other end; in the higher of the two, the soul passes out of hypotheses, and goes up to a principle which is above hypotheses, making no use of images as in the former case, but proceeding only in and through the ideas themselves.

I do not quite understand your meaning, he said.

Then I will try again; you will understand me better when I have made some preliminary remarks. You are aware that students of geometry, arithmetic, and the kindred sciences assume the odd and the even and the figures and three kinds of angles and the like in their several branches of science; these are their hypotheses, which they and everybody are supposed to know, and therefore they do not deign to give any account of them either to themselves or others; but they begin with them, and go on until they arrive at last, and in a consistent manner, at their conclusion?

Yes, he said, I know.

And do you not know also that although they make use of the visible forms and reason about them, they are thinking not of these, but of the ideals which they resemble; not of the figures which they draw, but of the absolute square and the absolute diameter, and so on—the forms which they draw or make, and which have shadows and reflections in water of their own, are converted by them into images, but they are really seeking to behold the things themselves, which can only be seen with the eye of the mind?

That is true.

And of this kind I spoke as the intelligible, although in the search after it the soul is compelled to use hypotheses; not ascending to a first principle, because she is unable to rise above the region of hypothesis, but employing the objects of which the shadows below are resemblances in their turn as images, they having in relation to the shadows and reflections of them a greater distinctness, and therefore a higher value.

I understand, he said, that you are speaking of the province of geometry and the sister arts.

And when I speak of the other division of the intelligible, you will understand me to speak of that other sort of knowledge which reason herself attains by the power of dialectic, using the hypotheses not as first principles, but only as hypotheses—that is to say, as steps and points of departure into a world which is above hypotheses, in order that she may soar beyond them to the first principle of the whole; and clinging to this and then to that which depends on this, by successive steps she descends again without the aid of any sensible object, from ideas, through ideas, and in ideas she ends.

I understand you, he replied; not perfectly, for you seem to me to be describing a task which is really tremendous; but, at any rate, I understand you to say that knowledge and being, which the science of dialectic contemplates, are clearer than the notions of the arts, as they are termed, which proceed from hypotheses only: these are also contemplated by the understanding, and not by the senses: yet, because they start from hypotheses and do not ascend to a principle, those who contemplate them appear to you not to exercise the higher reason upon them, although when a first principle is added to them they are cognizable by the higher reason. And the habit which is concerned with geometry and the cognate sciences I suppose that you would term understanding and not reason, as being intermediate between opinion and reason.

You have quite conceived my meaning, I said; and now, corresponding to these four divisions, let there be four faculties in the soul—reason answering to the highest, understanding to the second, faith (or conviction) to the third, and perception of shadows to the last—and let there be a scale of them, and let us suppose that the several faculties have clearness in the same degree that their objects have truth.

I understand, he replied, and give my assent, and accept your arrangement.

Book VII

And now, I said, let me show in a figure how far our nature is enlightened or unenlightened: behold! human beings living in an underground den, which has a mouth open towards the light and reaching all along the den; here they have been from their childhood, and have their legs and necks chained so that they cannot move, and can only see before them, being prevented by the chains from turning round their heads. Above and behind them a fire is blazing at a distance, and between the fire and the prisoners there is a raised way; and you will see, if you look, a low wall built along the way, like the screen which marionette players have in front of them, over which they show the puppets.

I see.

And do you see, I said, men passing along the wall carrying all sorts of vessels, and statues and figures of animals made of wood and stone and various materials, which appear over the wall? Some of them are talking, others silent.

You have shown me a strange image, and they are strange prisoners.

Like ourselves, I replied; and they see only their own shadows, or the shadows of one another, which the fire throws on the opposite wall of the cave?

True, he said; how could they see anything but the shadows if they were never allowed to move their heads?

And of the objects which are being carried in like manner they would only see the shadows?

Yes, he said.

And if they were able to converse with one another, would they not suppose that they were naming what was actually before them?

Very true.

And suppose further that the prison had an echo which came from the other side, would they not be sure to fancy when one of the passers-by spoke that the voice which they heard came from the passing shadow?

No question, he replied.

To them, I said, the truth would be literally nothing but the shadows of the images.

That is certain.

And now look again, and see what will naturally follow if the prisoners are released and disabused of their error. At first, when any of them is liberated and compelled suddenly to stand up and turn his neck round and walk and look towards the light, he will suffer sharp pains; the glare will distress him, and he will be unable to see the realities of which in his former state he had seen the shadows; and then conceive someone saying to him, that what he saw before was an illusion, but that now, when he is approaching nearer to being and his eye is turned towards more real existence, he has a clearer vision—what will be his reply? And you may further imagine that his instructor is pointing to the objects as they pass and requiring him to name them—will he not be perplexed? Will he not fancy that the shadows which he formerly saw are truer than the objects which are now shown to him?

Far truer.

And if he is compelled to look straight at the light, will he not have a pain in his eyes which will make him turn away to take refuge in the objects of vision which he can see, and which he will conceive to be in reality clearer than the things which are now being shown to him?

True, he said.

And suppose once more, that he is reluctantly dragged up a steep and rugged ascent, and held fast until he is forced into the presence of the sun himself, is he not likely to be pained and irritated? When he approaches the light his eyes will be dazzled, and he will not be able to see anything at all of what are now called realities.

Not all in a moment, he said.

He will require to grow accustomed to the sight of the upper world. And first he will see the shadows best, next the reflections of men and other objects in the water, and then the objects themselves; then he will gaze upon the light of the moon and the stars and the spangled heaven; and he will see the sky and the stars by night better than the sun or the light of the sun by day?

Certainly.

Last of all he will be able to see the sun, and not mere reflections of him in the water, but he will see him in his own proper place, and not in another; and he will contemplate him as he is.

Certainly.

He will then proceed to argue that this is he who gives the season and the years, and is the guardian of all that is in the visible world, and in a certain way the cause of all things which he and his fellows have been accustomed to behold?

Clearly, he said, he would first see the sun and then reason about him.

And when he remembered his old habitation, and the wisdom of the den and his fellow-prisoners, do you not suppose that he would felicitate himself on the change, and pity them?

Certainly, he would.

And if they were in the habit of conferring honours among themselves on those who were quickest to observe the passing shadows and to remark which of them went before, and which followed after, and which were together; and who were therefore best able to draw conclusions as to the future, do you think that he would care for such honours and glories, or envy the possessors of them? Would he not say with Homer,

'Better to be the poor servant of a poor master', and to endure anything, rather than think as they do and live after their manner?

Yes, he said, I think that he would rather suffer anything than entertain these false notions and live in this miserable manner.

Imagine once more, I said, such a one coming suddenly out of the sun to be replaced in his old situation; would he not be certain to have his eyes full of darkness?

To be sure, he said.

And if there were a contest, and he had to compete in measuring the shadows with the prisoners who had never moved out of the den, while his sight was still weak, and before his eyes had become steady (and the time which would be needed to acquire this new habit of sight might be very considerable), would he not be ridiculous? Men would say of him that up he went and down he came without his eyes; and that it was better not even to think of ascending; and if any one tried to loose another and lead him up to the light, let them only catch the offender, and they would put him to death.

No question, he said.

This entire allegory, I said, you may now append, dear Glaucon, to the previous argument; the prison-house is the world of sight, the light of the fire is the sun, and you will not misapprehend me if you interpret the journey upwards to be the ascent of the soul into the intellectual world according to my poor belief, which, at your desire, I have expressed—whether rightly or wrongly God knows. But, whether true or false, my opinion is that in the world of knowledge the idea of good appears last of all, and is seen only with an effort; and, when seen, is also inferred to be the universal author of all things beautiful and right, parent of light and of the lord of light in this visible world, and the immediate source of reason and truth in the intellectual; and that this is the power upon which he who would act rationally either in public or private life must have his eye fixed.

I agree, he said, as far as I am able to understand you.

Moreover, I said, you must not wonder that those who attain to this beatific vision are unwilling to descend to human affairs; for their souls are ever hastening into the upper world where they desire to dwell; which desire of theirs is very natural, if our allegory may be trusted.

Yes, very natural.

And is there anything surprising in one who passes from divine contemplations to the evil state of man, misbehaving himself in a ridiculous manner; if, while his eyes are blinking and before he has become accustomed to the surrounding darkness, he is compelled to fight in courts of law, or in other places, about the images or the shadows of images of justice, and is endeavouring to meet the conceptions of those who have never yet seen absolute justice?

Anything but surprising, he replied.

Anyone who has common sense will remember that the bewilderments of the eyes are of two kinds, and arise from two causes, either from coming out of the light or from going into the light, which is true of the mind's eye, quite as much as of the bodily eye; and he who remembers this when he sees any one whose vision is perplexed and weak, will not be too ready to laugh; he will first ask whether that soul of man has come out of the brighter life, and is unable to see because unaccustomed to the dark, or having turned from darkness to the day is dazzled by excess of light. And he will count the one happy in his condition and state of being, and he will pity the other; or, if he have a mind to laugh at the soul which comes from below into the light, there will be more reason in this than in the laugh which greets him who returns from above out of the light into the den.

That, he said, is a very just distinction.

But then, if I am right, certain professors of education must be wrong when they say that they can put a knowledge into the soul which was not there before, like sight into blind eyes.

They undoubtedly say this, he replied.

...

And all arithmetic and calculation have to do with number?

Yes.

And they appear to lead the mind towards truth?

Yes, in a very remarkable manner.

Then this is knowledge of the kind for which we are seeking, having a double use, military and philosophical; for the man of war must learn the art of number or he will not know how to array his troops, and the philosopher also, because he has to rise out of the sea of change and lay hold of true being, and therefore he must be an arithmetician.

That is true.

And our guardian is both warrior and philosopher?

Certainly.

Then this is a kind of knowledge which legislation may fitly prescribe; and we must endeavour to persuade those who are to be the principal men of our State to go and learn arithmetic, not as amateurs, but they must carry on the study until they see the nature of numbers with the mind only; nor again, like merchants or retail-traders, with a view to buying or selling, but for the sake of their military use, and of the soul herself; and because this will be the easiest way for her to pass from becoming to truth and being.

That is excellent, he said.

Yes, I said, and now having spoken of it, I must add how charming the science is! And in how many ways it conduces to our desired end, if pursued in the spirit of a philosopher, and not of a shopkeeper!

How do you mean?

I mean, as I was saying, that arithmetic has a very great and elevating effect, compelling the soul to reason about abstract number, and rebelling against the introduction of visible or tangible objects into the argument. You know how steadily the masters of the art repel and ridicule anyone who attempts to divide absolute unity when he is calculating, and if you divide, they multiply (meaning either (1) that they integrate the number because they deny the possibility of fractions; or (2) that division is regarded by them as a process of multiplication, for the fractions of one continue to be units), taking care that one shall continue one and not become lost in fractions.

That is very true.

Now, suppose a person were to say to them: O my friends, what are these wonderful numbers about which you are reasoning, in which, as you say, there is a unity such as you demand, and each unit is equal, invariable, indivisible—what would they answer?

They would answer, as I should conceive, that they were speaking of those numbers which can only be realized in thought.

Then you see that this knowledge may be truly called necessary, necessitating as it clearly does the use of the pure intelligence in the attainment of pure truth?

Yes; that is a marked characteristic of it.

And have you further observed, that those who have a natural talent for calculation are generally quick at every other kind of knowledge; and even the dull, if they have had an arithmetical training, although they may derive no other advantage from it, always become much quicker than they would otherwise have been.

Very true, he said.

And indeed, you will not easily find a more difficult study, and not many as difficult.

You will not.

And, for all these reasons, arithmetic is a kind of knowledge in which the best natures should be trained, and which must not be given up.

I agree.

Let this then be made one of our subjects of education. And next, shall we enquire whether the kindred science also concerns us?

You mean geometry?

Exactly so.

Clearly, he said, we are concerned with that part of geometry which relates to war; for in pitching a camp, or taking up a position, or closing or extending the lines of an army, or any other military maneuver, whether in actual battle or on a march, it will make all the difference whether a general is or is not a geometrician.

Yes, I said, but for that purpose a very little of either geometry or calculation will be enough; the question relates rather to the greater and more advanced part of geometry—whether that tends in any degree to make more easy the vision of the idea of good; and thither, as I was saying, all things tend which compel the soul to turn her gaze towards that place, where is the full perfection of being, which she ought, by all means, to behold.

True, he said.

Then if geometry compels us to view being, it concerns us; if becoming only, it does not concern us?

Yes, that is what we assert.

Yet anybody who has the least acquaintance with geometry will not deny that such a conception of the science is in flat contradiction to the ordinary language of geometricians.

How so?

They have in view practice only, and are always speaking, in a narrow and ridiculous manner, of squaring and extending and applying and the like—they confuse the necessities of geometry with those of daily life; whereas knowledge is the real object of the whole science.

Certainly, he said.

Then must not a further admission be made?

What admission?

That the knowledge at which geometry aims is knowledge of the eternal, and not of aught perishing and transient.

That, he replied, may be readily allowed, and is true.

Then, my noble friend, geometry will draw the soul towards truth, and create the spirit of philosophy, and raise up that which is now unhappily allowed to fall down.

Nothing will be more likely to have such an effect.

Then nothing should be more sternly laid down than that the inhabitants of your fair city should by all means learn geometry. Moreover the science has indirect effects, which are not small.

Of what kind? he said.

There are the military advantages of which you spoke, I said; and in all departments of knowledge, as experience proves, anyone who has studied geometry is infinitely quicker of apprehension than one who has not.

Yes indeed, he said, there is an infinite difference between them.

Then shall we propose this as a second branch of knowledge which our youth will study?

Let us do so, he replied.

And suppose we make astronomy the third—what do you say?

I am strongly inclined to it, he said; the observation of the seasons and of months and years is as essential to the general as it is to the farmer or sailor.

I am amused, I said, at your fear of the world, which makes you guard against the appearance of insisting upon useless studies; and I quite admit the difficulty of believing that in every man there is an eye of the soul which, when by other pursuits lost and dimmed, is by these purified and re-illumined; and is more precious far than ten thousand bodily eyes, for by it alone is truth seen. Now there are two classes of persons: one class of those who will agree with you and will take your words as a revelation; another class to whom they will be utterly unmeaning, and who will naturally deem them to be idle tales, for they see no sort of profit which is to be obtained from them. And therefore you had better decide at once with which of the two you are proposing to argue. You will very likely say with neither, and that your chief aim in carrying on the argument is your own improvement; at the same time you do not grudge to others any benefit which they may receive.

I think that I should prefer to carry on the argument mainly on my own behalf.

Then take a step backward, for we have gone wrong in the order of the sciences.

What was the mistake? he said.

After plane geometry, I said, we proceeded at once to solids in revolution, instead of taking solids in themselves; whereas after the second dimension the third, which is concerned with cubes and dimensions of depth, ought to have followed.

That is true, Socrates; but so little seems to be known as yet about these subjects.

Why, yes, I said, and for two reasons:—in the first place, no government patronises them; this leads to a want of energy in the pursuit of them, and they are difficult; in the second place, students cannot learn them unless they have a director. But then a director can hardly be found, and even if he could, as matters now stand, the students, who are very conceited, would not attend to him. That, however, would be otherwise if the whole State became the director of these studies and gave honour to them; then disciples would want to come, and there would be continuous and earnest search, and discoveries would be made; since even now, disregarded as they are by the world, and maimed of their fair proportions, and although none of their votaries can tell the use of them, still these studies force their way by their natural charm, and very likely, if they had the help of the State, they would someday emerge into light.

Yes, he said, there is a remarkable charm in them. But I do not clearly understand the change in the order. First you began with a geometry of plane surfaces?

Yes, I said.

And you placed astronomy next, and then you made a step backward?

Yes, and I have delayed you by my hurry; the ludicrous state of solid geometry, which, in natural order, should have followed, made me pass over this branch and go on to astronomy, or motion of solids.

True, he said.

Then assuming that the science now omitted would come into existence if encouraged by the State, let us go on to astronomy, which will be fourth.

The right order, he replied. And now, Socrates, as you rebuked the vulgar manner in which I praised astronomy before, my praise shall be given in your own spirit. For every one, as I think, must see that astronomy compels the soul to look upwards and leads us from this world to another.

Everyone but myself, I said; to everyone else this may be clear, but not to me.

And what then would you say?

I should rather say that those who elevate astronomy into philosophy appear to me to make us look downwards and not upwards.

What do you mean? he asked.

You, I replied, have in your mind a truly sublime conception of our knowledge of the things above. And I dare say that if a person were to throw his head back and study the fretted ceiling, you would still think that his mind was the percipient, and not his eyes. And you are very likely right, and I may be a simpleton: but, in my opinion, that knowledge only which is of being and of the unseen can make the soul look upwards, and whether a man gapes at the heavens or blinks on the ground, seeking to learn some particular of sense, I would deny that he can learn, for nothing of that sort is matter of science; his soul is looking downwards, not upwards, whether his way to knowledge is by water or by land, whether he floats, or only lies on his back.

I acknowledge, he said, the justice of your rebuke. Still, I should like to ascertain how astronomy can be learned in any manner more conducive to that knowledge of which we are speaking?

I will tell you, I said: The starry heaven which we behold is wrought upon a visible ground, and therefore, although the fairest and most perfect of visible things, must necessarily be deemed inferior far to the true motions of absolute swiftness and absolute slowness, which are relative to each other, and carry with them that which is contained in them, in the true number and in every true figure. Now, these are to be apprehended by reason and intelligence, but not by sight.

True, he replied.

The spangled heavens should be used as a pattern and with a view to that higher knowledge; their beauty is like the beauty of figures or pictures excellently wrought by the hand of Daedalus, or some other great artist, which we may chance to behold; any geometrician who saw them would appreciate the exquisiteness of their workmanship, but he would never dream of thinking that in them he could find the true equal or the true double, or the truth of any other proportion.

No, he replied, such an idea would be ridiculous.

And will not a true astronomer have the same feeling when he looks at the movements of the stars? Will he not think that heaven and the things in heaven are framed by the Creator of them in the most perfect manner? But he will never imagine that the proportions of night and day, or of both to the month, or of the month to the year, or of the stars to these and to one another, and any other things that are material and visible can also be eternal and subject to no deviation—that would be absurd; and it is equally absurd to take so much pains in investigating their exact truth.

I quite agree, though I never thought of this before.

Then, I said, in astronomy, as in geometry, we should employ problems, and let the heavens alone if we would approach the subject in the right way and so make the natural gift of reason to be of any real use.

That, he said, is a work infinitely beyond our present astronomers.

Yes, I said; and there are many other things which must also have a similar extension given to them, if our legislation is to be of any value. But can you tell me of any other suitable study?

No, he said, not without thinking.

Motion, I said, has many forms, and not one only; two of them are obvious enough even to wits no better than ours; and there are others, as I imagine, which may be left to wiser persons.

But where are the two?

There is a second, I said, which is the counterpart of the one already named.

And what may that be?

The second, I said, would seem relatively to the ears to be what the first is to the eyes; for I conceive that as the eyes are designed to look up at the stars, so are the ears to hear harmonious motions; and these are sister sciences—as the Pythagoreans say, and we, Glaucon, agree with them?

Yes, he replied.

But this, I said, is a laborious study, and therefore we had better go and learn of them; and they will tell us whether there are any other applications of these sciences. At the same time, we must not lose sight of our own higher object.

What is that?

There is a perfection which all knowledge ought to reach, and which our pupils ought also to attain, and not to fall short of, as I was saying that they did in astronomy. For in the science of harmony, as you probably know, the same thing happens. The teachers of harmony compare the sounds and consonances which are heard only, and their labour, like that of the astronomers, is in vain.

Yes, by heaven! he said; and 'tis as good as a play to hear them talking about their condensed notes, as they call them; they put their ears close alongside of the strings like persons catching a sound from their neighbour's wall—one set of them declaring that they distinguish an intermediate note and have found the least interval which should be the unit of measurement; the others insisting that the two sounds have passed into the same—either party setting their ears before their understanding.

You mean, I said, those gentlemen who tease and torture the strings and rack them on the pegs of the instrument: I might carry on the metaphor and speak after their manner of the blows which the plectrum gives, and make accusations against the strings, both of backwardness and forwardness to sound; but this would be tedious, and therefore I will only say that these are not the men, and that I am referring to the Pythagoreans, of whom I was just now proposing to enquire about harmony. For they too are in error, like the astronomers; they investigate the numbers of the harmonies which are heard, but they never attain to problems—that is to say, they never reach the natural harmonies of number, or reflect why some numbers are harmonious and others not.

That, he said, is a thing of more than mortal knowledge.

A thing, I replied, which I would rather call useful; that is, if sought after with a view to the beautiful and good; but if pursued in any other spirit, useless.

Very true, he said.

Now, when all these studies reach the point of inter-communion and connection with one another, and come to be considered in their mutual affinities, then, I think, but not till then, will the pursuit of them have a value for our objects; otherwise there is no profit in them.

I suspect so; but you are speaking, Socrates, of a vast work.

What do you mean? I said; the prelude or what? Do you not know that all this is but the prelude to the actual strain which we have to learn? For you surely would not regard the skilled mathematician as a dialectician?

Assuredly not, he said; I have hardly ever known a mathematician who was capable of reasoning.

But do you imagine that men who are unable to give and take a reason will have the knowledge which we require of them?

Neither can this be supposed.

And so, Glaucon, I said, we have at last arrived at the hymn of dialectic. This is that strain which is of the intellect only, but which the faculty of sight will nevertheless be found to imitate; for sight, as you may remember, was imagined by us after a while to behold the real animals and stars, and last of all the sun himself. And so with dialectic; when a person starts on the discovery of the absolute by the light of reason only, and without any assistance of sense, and perseveres until by pure intelligence he arrives at the perception of the absolute good, he at last finds himself at the end of the intellectual world, as in the case of sight at the end of the visible.

Exactly, he said.

Then this is the progress which you call dialectic?

True.

But the release of the prisoners from chains, and their translation from the shadows to the images and to the light, and the ascent from the underground den to the sun, while in his presence they are vainly trying to look on animals and plants and the light of the sun, but are able to perceive even with their weak eyes the images in the water (which are divine), and are the shadows of true existence (not shadows of images cast by a light of fire, which compared with the sun is only an image)—this power of elevating the highest principle in the soul to the contemplation of that which is best in existence, with which we may compare the raising of that faculty which is the very light of the body to the sight of that which is brightest in the material and visible world—this power is given, as I was saying, by all that study and pursuit of the arts which has been described.

I agree in what you are saying, he replied, which may be hard to believe, yet, from another point of view, is harder still to deny. This, however, is not a theme to be treated of in passing only, but will have to be discussed again and again. And so, whether our conclusion be true or false, let us assume all this, and proceed at once from the prelude or preamble to the chief strain (a play upon the Greek word, which means both 'law' and 'strain'), and describe that in like manner. Say, then, what is the nature and what are the divisions of dialectic, and what are the paths which lead thither; for these paths will also lead to our final rest.

Dear Glaucon, I said, you will not be able to follow me here, though I would do my best, and you should behold not an image only but the absolute truth, according to my notion. Whether what I told you would or would not have been a reality I cannot venture to say; but you would have seen something like reality; of that I am confident.

Doubtless, he replied.

But I must also remind you, that the power of dialectic alone can reveal this, and only to one who is a disciple of the previous sciences.

Of that assertion you may be as confident as of the last.

And assuredly no one will argue that there is any other method of comprehending by any regular process all true existence or of ascertaining what each thing is in its own nature; for the arts in general are concerned with the desires or opinions of men, or are cultivated with a view to production and construction, or for the preservation of such productions and constructions; and as to the mathematical sciences which, as we were saying, have some apprehension of true being—geometry and the like—they only dream about being, but never can they behold the waking reality so long as they leave the hypotheses which they use unexamined, and are unable to give an account of them. For when a man knows not his own first principle, and when the conclusion and intermediate steps are also constructed out of he knows not what, how can he imagine that such a fabric of convention can ever become science?

Impossible, he said.

Then dialectic, and dialectic alone, goes directly to the first principle and is the only science which does away with hypotheses in order to make her ground secure; the eye of the soul, which is literally buried in an outlandish slough, is by her gentle aid lifted upwards; and she uses as handmaids and helpers in the work of conversion, the sciences which we have been discussing. Custom terms them sciences, but they ought to have some other name, implying greater clearness than opinion and less clearness than science: and this, in our previous sketch, was called understanding. But why should we dispute about names when we have realities of such importance to consider?

Why indeed, he said, when any name will do which expresses the thought of the mind with clearness?

At any rate, we are satisfied, as before, to have four divisions; two for intellect and two for opinion, and to call the first division science, the second understanding, the third belief, and the fourth perception of shadows, opinion being concerned with becoming, and intellect with being; and so to make a proportion:—

As being is to becoming, so is pure intellect to opinion. And as intellect is to opinion, so is science to belief, and understanding to the perception of shadows.

But let us defer the further correlation and subdivision of the subjects of opinion and of intellect, for it will be a long enquiry, many times longer than this has been.

As far as I understand, he said, I agree.

And do you also agree, I said, in describing the dialectician as one who attains a conception of the essence of each thing? And he who does not possess and is therefore unable to impart this conception, in whatever degree he fails, may in that degree also be said to fail in intelligence? Will you admit so much?

Yes, he said; how can I deny it?

And you would say the same of the conception of the good? Until the person is able to abstract and define rationally the idea of good, and unless he can run the gauntlet of all objections, and is ready to disprove them, not by appeals to opinion, but to absolute truth, never faltering at any step of the argument—unless he can do all this, you would say that he knows neither the idea of good nor any other good; he apprehends only a shadow, if anything at all, which is given by opinion and not by science;—dreaming and slumbering in this life, before he is well awake here, he arrives at the world below, and has his final quietus.

In all that I should most certainly agree with you.

And surely you would not have the children of your ideal State, whom you are nurturing and educating—if the ideal ever becomes a reality—you would not allow the future rulers to be like posts (literally 'lines', probably the starting-point of a race-course), having no reason in them, and yet to be set in authority over the highest matters?

Certainly not.

Then you will make a law that they shall have such an education as will enable them to attain the greatest skill in asking and answering questions?

Yes, he said, you and I together will make it.

Dialectic, then, as you will agree, is the coping-stone of the sciences, and is set over them; no other science can be placed higher—the nature of knowledge can no further go?

I agree, he said.

But to whom we are to assign these studies, and in what way they are to be assigned, are questions which remain to be considered.

Yes, clearly.

You remember, I said, how the rulers were chosen before?

Certainly, he said.

The same natures must still be chosen, and the preference again given to the surest and the bravest, and, if possible, to the fairest; and, having noble and generous tempers, they should also have the natural gifts which will facilitate their education.

And what are these?

Such gifts as keenness and ready powers of acquisition; for the mind more often faints from the severity of study than from the severity of gymnastics: the toil is more entirely the mind's own, and is not shared with the body.

Very true, he replied.

Further, he of whom we are in search should have a good memory, and be an unwearied solid man who is a lover of labour in any line; or he will never be able to endure the great amount of bodily exercise and to go through all the intellectual discipline and study which we require of him.

Certainly, he said; he must have natural gifts.

The mistake at present is, that those who study philosophy have no vocation, and this, as I was before saying, is the reason why she has fallen into disrepute: her true sons should take her by the hand and not bastards.

What do you mean?

In the first place, her votary should not have a lame or halting industry—I mean, that he should not be half industrious and half idle: as, for example, when a man is a lover of gymnastic and hunting, and all other bodily exercises, but a hater rather than a lover of the labour of learning or listening or enquiring. Or the occupation to which he devotes himself may be of an opposite kind, and he may have the other sort of lameness.

Certainly, he said.

And as to truth, I said, is not a soul equally to be deemed halt and lame which hates voluntary falsehood and is extremely indignant at herself and others when they tell lies, but is patient of involuntary falsehood, and does not mind wallowing like a swinish beast in the mire of ignorance, and has no shame at being detected?

To be sure.

And, again, in respect of temperance, courage, magnificence, and every other virtue, should we not carefully distinguish between the true son and the bastard? for where there is no discernment of such qualities states and individuals unconsciously err; and the state makes a ruler, and the individual a friend, of one who, being defective in some part of virtue, is in a figure lame or a bastard.

That is very true, he said.

All these things, then, will have to be carefully considered by us; and if only those whom we introduce to this vast system of education and training are sound in body and mind, justice herself will have nothing to say against us, and we shall be the saviours of the constitution and of the State; but, if our pupils are men of another stamp, the reverse will happen, and we shall pour a still greater flood of ridicule on philosophy than she has to endure at present.

That would not be creditable.

Certainly not, I said; and yet perhaps, in thus turning jest into earnest I am equally ridiculous.

In what respect?

I had forgotten, I said, that we were not serious, and spoke with too much excitement. For when I saw philosophy so undeservedly trampled underfoot of men I could not help feeling a sort of indignation at the authors of her disgrace: and my anger made me too vehement.

Indeed! I was listening, and did not think so.

But I, who am the speaker, felt that I was. And now let me remind you that, although in our former selection we chose old men, we must not do so in this. Solon was under a delusion when he said that a man when he grows old may learn many things—for he can no more learn much than he can run much; youth is the time for any extraordinary toil.

Of course.

And, therefore, calculation and geometry and all the other elements of instruction, which are a preparation for dialectic, should be presented to the mind in childhood; not, however, under any notion of forcing our system of education.

Why not?

Because a freeman ought not to be a slave in the acquisition of knowledge of any kind. Bodily exercise, when compulsory, does no harm to the body; but knowledge which is acquired under compulsion obtains no hold on the mind.

Very true.

Then, my good friend, I said, do not use compulsion, but let early education be a sort of amusement; you will then be better able to find out the natural bent.

That is a very rational notion, he said.

Translated by Benjamin Jowett

Reading and Discussion Questions

1. Explain the significance of the divided line and cave analogies for Plato's theory of knowledge. Where on the divided line would inquiry into nature be placed? On this view, can one have knowledge of sensible objects?
2. What role does Plato envision for astronomy in the education of the guardians? What sort of place does Plato's philosophy envision for the study of the natural world and to what extent is such inquiry valued?

The Categories (c. 350 BCE)

Aristotle

6

Part 3

When one thing is predicated of another, all that which is predicable of the predicate will be predicable also of the subject. Thus, 'man' is predicated of the individual man; but 'animal' is predicated of 'man'; it will, therefore, be predicable of the individual man also: for the individual man is both 'man' and 'animal'.

If genera are different and co-ordinate, their differentiae are themselves different in kind. Take as an instance the genus 'animal' and the genus 'knowledge'. 'With feet', 'two-footed', 'winged', 'aquatic', are differentiae of 'animal'; the species of knowledge are not distinguished by the same differentiae. One species of knowledge does not differ from another in being 'two-footed'.

But where one genus is subordinate to another, there is nothing to prevent their having the same differentiae: for the greater class is predicated of the lesser, so that all the differentiae of the predicate will be differentiae also of the subject.

Part 4

Expressions which are in no way composite signify substance, quantity, quality, relation, place, time, position, state, action, or affection. To sketch my meaning roughly, examples of substance are 'man' or 'the horse', of quantity, such terms as 'two cubits long' or 'three cubits long', of quality, such attributes as 'white', 'grammatical'. 'Double', 'half', 'greater', fall under the category of relation; 'in the market place', 'in the Lyceum', under that of place; 'yesterday', 'last year', under that of time. 'Lying', 'sitting', are terms indicating position, 'shod', 'armed', state; 'to lance', 'to cauterize', action; 'to be lanced', 'to be cauterized', affection.

No one of these terms, in and by itself, involves an affirmation; it is by the combination of such terms that positive or negative statements arise. For every assertion must, as is admitted, be either true or false, whereas expressions which are not in any way composite such as 'man', 'white', 'runs', 'wins', cannot be either true or false.

Part 5

Substance, in the truest and primary and most definite sense of the word, is that which is neither predicable of a subject nor present in a subject; for instance, the individual man or horse. But in a secondary sense those things are called substances within which, as species, the primary substances are included; also those which, as genera, include the species. For instance, the individual man is included in the species 'man', and the genus to which the species belongs is 'animal'; these, therefore—that is to say, the species 'man' and the genus 'animal—are termed secondary substances.

It is plain from what has been said that both the name and the definition of the predicate must be predicable of the subject. For instance, 'man' is predicated of the individual man. Now in this case the name of the species 'man' is applied to the individual, for we use the term 'man' in describing the individual; and the definition of 'man' will also be predicated of the individual man, for the individual man is both man and animal. Thus, both the name and the definition of the species are predicable of the individual.

With regard, on the other hand, to those things which are present in a subject, it is generally the case that neither their name nor their definition is predicable of that in which they are present. Though, however, the definition is never predicable, there is nothing in certain cases to prevent the name being used. For instance, 'white' being present in a body is predicated of that in which it is present, for a body is called white: the definition, however, of the colour 'white' is never predicable of the body.

Everything except primary substances is either predicable of a primary substance or present in a primary substance. This becomes evident by reference to particular instances which occur. 'Animal' is predicated of the species 'man', therefore of the individual man, for if there were no individual man of whom it could be predicated, it could not be predicated of the species 'man' at all. Again, colour is present in body, therefore in individual bodies, for if there were no individual body in which it was present, it could not be present in body at all. Thus everything except primary substances is either predicated of primary substances, or is present in them, and if these last did not exist, it would be impossible for anything else to exist.

Of secondary substances, the species is more truly substance than the genus, being more nearly related to primary substance. For if anyone should render an account of what a primary substance is, he would render a more instructive account, and one more proper to the subject, by stating the species than by stating the genus. Thus, he would give a more instructive account of an individual man by stating that he was man than by stating that he was animal, for the former description is peculiar to the individual in a greater degree, while the latter is too general. Again, the man who gives an account of the nature of an individual tree will give a more instructive account by mentioning the species 'tree' than by mentioning the genus 'plant'.

Moreover, primary substances are most properly called substances in virtue of the fact that they are the entities which underlie everything else, and that everything else is either predicated of them or present in them. Now the same relation which subsists between primary substance and everything else subsists also between the species and the genus: for the species is to the genus as subject is to predicate, since the genus is predicated of the species, whereas the species cannot be predicated of the genus. Thus we have a second ground for asserting that the species is more truly substance than the genus.

Of species themselves, except in the case of such as are genera, no one is more truly substance than another. We should not give a more appropriate account of the individual man by stating the species to which he belonged, than we should of an individual horse by adopting the same method of definition. In the same way, of primary substances, no one is more truly substance than another; an individual man is not more truly substance than an individual ox.

It is, then, with good reason that of all that remains, when we exclude primary substances, we concede to species and genera alone the name 'secondary substance', for these alone of all the predicates convey a knowledge of primary substance. For it is by stating the species or the genus that we appropriately define any individual man; and we shall make our definition more exact by stating the former than by stating the latter. All other things that we state, such as that he is white, that he runs, and so on, are irrelevant to the definition. Thus it is just that these alone, apart from primary substances, should be called substances.

Further, primary substances are most properly so called, because they underlie and are the subjects of everything else. Now the same relation that subsists between primary substance and everything else subsists also between the species and the genus to which the primary substance belongs, on the one hand, and every attribute which is not included within these, on the other. For these are the subjects of all such. If we call an individual man 'skilled in grammar', the predicate is applicable also to the species and to the genus to which he belongs. This law holds good in all cases.

It is a common characteristic of all substance that it is never present in a subject. For primary substance is neither present in a subject nor predicated of a subject; while, with regard to secondary substances, it is clear from the following arguments (apart from others) that they are not present in a subject. For 'man' is predicated of the individual man, but is not present in any subject: for manhood is not present in the individual man. In the same way, 'animal' is also predicated of the individual man, but is not present in him. Again, when a thing is present in a subject, though the name may quite well be applied to that in which it is present, the definition cannot be applied. Yet of secondary substances, not only the name, but also the definition, applies to the subject: we should use both the definition of the species and that of the genus with reference to the individual man. Thus substance cannot be present in a subject.

Yet this is not peculiar to substance, for it is also the case that differentiae cannot be present in subjects. The characteristics 'terrestrial' and 'two-footed' are predicated of the species 'man', but not present in it. For they are not in man. Moreover, the definition of the differentia may be predicated of that of which the differentia itself is predicated. For instance, if the characteristic 'terrestrial' is predicated of the species 'man', the definition also of that characteristic may be used to form the predicate of the species 'man': for 'man' is terrestrial.

The fact that the parts of substances appear to be present in the whole, as in a subject, should not make us apprehensive lest we should have to admit that such parts are not substances: for in explaining the phrase 'being present in a subject', we stated that we meant 'otherwise than as parts in a whole'.

It is the mark of substances and of differentiae that, in all propositions of which they form the predicate, they are predicated univocally. For all such propositions have for their subject either the individual or the species. It is true that, inasmuch as primary substance is not predicable of anything, it can never form the predicate of any proposition. But of secondary substances, the species is predicated of the individual, the genus both of the species and of the individual. Similarly the differentiae are predicated of the species and of the individuals. Moreover, the definition of the species and that of the genus are applicable to the primary substance, and that of the genus to the species. For all that is predicated of the predicate will be predicated also of the subject. Similarly, the definition of the differentiae will be applicable to the species and to the individuals. But it was stated above that the word 'univocal' was applied to those things which had both name and definition in common. It is, therefore, established that in every proposition, of which either substance or a differentia forms the predicate, these are predicated univocally.

All substance appears to signify that which is individual. In the case of primary substance this is indisputably true, for the thing is a unit. In the case of secondary substances, when we speak, for instance, of 'man' or 'animal', our form of speech gives the impression that we are here also indicating that which is individual, but the impression is not strictly true; for a secondary substance is not an individual, but a class with a certain qualification; for it is not one and single as a primary substance is; the words 'man', 'animal', are predicable of more than one subject.

Yet species and genus do not merely indicate quality, like the term 'white'; 'white' indicates quality and nothing further, but species and genus determine the quality with reference to a substance: they signify substance qualitatively differentiated. The determinate qualification covers a larger field in the case of the genus that in that of the species: he who uses the word 'animal' is herein using a word of wider extension than he who uses the word 'man'.

Another mark of substance is that it has no contrary. What could be the contrary of any primary substance, such as the individual man or animal? It has none. Nor can the species or the genus have a contrary. Yet this characteristic is not peculiar to substance, but is true of many other things, such as quantity. There is nothing that forms the contrary of 'two cubits long' or of 'three cubits long', or of 'ten', or of any such term. A man may contend that 'much' is the contrary of 'little', or 'great' of 'small', but of definite quantitative terms no contrary exists.

Substance, again, does not appear to admit of variation of degree. I do not mean by this that one substance cannot be more or less truly substance than another, for it has already been stated that this is the case; but that no single substance admits of varying degrees within itself. For instance, one particular substance, 'man', cannot be more or less man either than himself at some other time or than some other man. One man cannot be more man than another, as that which is white may be more or less white than some other white object, or as that which is beautiful may be more or less beautiful than some other beautiful object. The same quality, moreover, is said to subsist in a thing in varying degrees at different times. A body, being white, is said to be whiter at one time than it was before, or, being warm, is said to be warmer or less warm than at some other time. But substance is not said to be more or less that which it is: a man is not more truly a man at one time than he was before, nor is anything, if it is substance, more or less what it is. Substance, then, does not admit of variation of degree.

The most distinctive mark of substance appears to be that, while remaining numerically one and the same, it is capable of admitting contrary qualities. From among things other than substance, we should find ourselves unable to bring forward any which possessed this mark. Thus, one and the same colour cannot be white and black. Nor can the same one action be good and bad: this law holds good with everything that is not substance. But one and the selfsame substance, while retaining its identity, is yet capable of admitting contrary qualities. The same individual person is at one time white, at another black, at one time warm, at another cold, at one time good, at another bad. This capacity is found nowhere else, though it might be maintained that a statement or opinion was an exception to the rule. The same statement, it is agreed, can be both true and false. For if the statement 'he is sitting' is true, yet, when the person in question has risen, the same statement will be false. The same applies to opinions. For if anyone thinks truly that a person is sitting, yet, when that person has risen, this same opinion, if still held, will be false. Yet although this exception may be allowed, there is, nevertheless, a difference in the manner in which the thing takes place. It is by themselves changing that substances admit contrary qualities. It is thus that that which was hot becomes cold, for it has entered into a different state. Similarly that which was white becomes black, and that which was bad good, by a process of change; and in the same way in all other cases it is by changing that substances are capable of admitting contrary qualities. But statements and opinions themselves remain unaltered in all respects: it is by the alteration in the facts of the case that the contrary quality comes to be theirs. The statement 'he is sitting' remains unaltered, but it is at one time true, at another false, according to circumstances. What has been said of statements applies also to opinions. Thus, in respect of the manner in which the thing takes place, it is the peculiar mark of substance that it should be capable of admitting contrary qualities; for it is by itself changing that it does so.

If, then, a man should make this exception and contend that statements and opinions are capable of admitting contrary qualities, his contention is unsound. For statements and opinions are said to have this capacity, not because they themselves undergo modification, but because this modification occurs in the case of something else. The truth or falsity of a statement depends on facts, and not on any power on the part of the statement itself of admitting contrary qualities. In short, there is nothing which can alter the nature of statements and opinions. As, then, no change takes place in themselves, these cannot be said to be capable of admitting contrary qualities.

But it is by reason of the modification which takes place within the substance itself that a substance is said to be capable of admitting contrary qualities; for a substance admits within itself either disease or health, whiteness or blackness. It is in this sense that it is said to be capable of admitting contrary qualities.

To sum up, it is a distinctive mark of substance, that, while remaining numerically one and the same, it is capable of admitting contrary qualities, the modification taking place through a change in the substance itself.

Let these remarks suffice on the subject of substance.

Part 14

There are six sorts of movement: generation, destruction, increase, diminution, alteration, and change of place.

It is evident in all but one case that all these sorts of movement are distinct each from each. Generation is distinct from destruction, increase and change of place from diminution, and so on. But in the case of alteration it may be argued that the process necessarily implies one or other of the other five sorts of motion. This is not true, for we may say that all affections, or nearly all, produce in us an alteration which is distinct from all other sorts of motion, for that which is affected need not suffer either increase or diminution or any of the other sorts of motion. Thus alteration is a distinct sort of motion; for, if it were not, the thing altered would not only be altered, but would forthwith necessarily suffer increase or diminution or some one of the other sorts of motion in addition; which as a matter of fact is not the case. Similarly that which was undergoing the process of increase or was subject to some other sort of motion would, if alteration were not a distinct form of motion, necessarily be subject to alteration also. But there are some things which undergo increase but yet not alteration. The square, for instance, if a gnomon is applied to it, undergoes increase but not alteration, and so it is with all other figures of this sort. Alteration and increase, therefore, are distinct.

Speaking generally, rest is the contrary of motion. But the different forms of motion have their own contraries in other forms; thus destruction is the contrary of generation, diminution of increase, rest in a place, of change of place. As for this last, change in the reverse direction would seem to be most truly its contrary; thus motion upwards is the contrary of motion downwards and vice versa.

In the case of that sort of motion which yet remains, of those that have been enumerated, it is not easy to state what is its contrary. It appears to have no contrary, unless one should define the contrary here also either as 'rest in its quality' or as 'change in the direction of the contrary quality', just as we defined the contrary of change of place either as rest in a place or as change in the reverse direction. For a thing is altered when change of quality takes place; therefore either rest in its quality or change in the direction of the contrary may be called the contrary of this qualitative form of motion. In this way becoming white is the contrary of becoming black; there is alteration in the contrary direction, since a change of a qualitative nature takes place.

Translated by E. M. Edghill

Reading and Discussion Questions

1. What is the distinction between primary and secondary substances that Aristotle is making in this reading? Why does he say that "if the primary substances did not exist it would be impossible for any of the other things to exist"?
2. How does this relate to his project of characterizing individuals into categories? How is this different from Plato's view of how we categorize?
3. What are some examples of qualities or attributes that substances can have?

Posterior Analytics (c. 350 BCE)

Aristotle

7

Book I

Part 1

All instruction given or received by way of argument proceeds from pre-existent knowledge. This becomes evident upon a survey of all the species of such instruction. The mathematical sciences and all other speculative disciplines are acquired in this way, and so are the two forms of dialectical reasoning, syllogistic and inductive; for each of these latter make use of old knowledge to impart new, the syllogism assuming an audience that accepts its premises, induction exhibiting the universal as implicit in the clearly known particular. Again, the persuasion exerted by rhetorical arguments is in principle the same, since they use either example, a kind of induction, or enthymeme, a form of syllogism.

The pre-existent knowledge required is of two kinds. In some cases admission of the fact must be assumed, in others comprehension of the meaning of the term used, and sometimes both assumptions are essential. Thus, we assume that every predicate can be either truly affirmed or truly denied of any subject, and that 'triangle' means so and so; as regards 'unit' we have to make the double assumption of the meaning of the word and the existence of the thing. The reason is that these several objects are not equally obvious to us. Recognition of a truth may in some cases contain as factors both previous knowledge and also knowledge acquired simultaneously with that recognition—knowledge, this latter, of the particulars actually falling under the universal and therein already virtually known. For example, the student knew beforehand that the angles of every triangle are equal to two right angles; but it was only at the actual moment at which he was being led on to recognize this as true in the instance before him that he came to know 'this figure inscribed in the semicircle' to be a triangle. For some things (viz. the singulars finally reached which are not predicable of anything else as subject) are only learnt in this way, i.e. there is here no recognition through a middle of a minor term as subject to a major. Before he was led on to recognition or before he actually drew a conclusion, we should perhaps say that in a manner he knew, in a manner not.

If he did not in an unqualified sense of the term know the existence of this triangle, how could he know without qualification that its angles were equal to two right angles? No: clearly he knows not without qualification but only in the sense that he knows universally. If this distinction is not drawn, we are faced with the dilemma in the Meno: either a man will learn nothing or what he already knows; for we cannot accept the solution which some people offer. A man is asked, 'Do you, or do you not, know that every pair is even?' He says he does know it. The questioner then produces a particular pair, of the existence, and so a fortiori of the evenness, of which he was unaware. The solution which some people offer is to assert that they do not know that every pair is even, but only that everything which they know to be a pair is even: yet what they know to be even is that of which they have demonstrated evenness, i.e. what they made the subject of their premise, viz. not merely every triangle or number which they know to be such, but any and every number or triangle without reservation. For no premise is ever couched in the form 'every number which you know to be such', or 'every rectilinear figure which you know to be such': the predicate is always construed as applicable to any and every instance of the thing. On the other hand, I imagine there is nothing to prevent a man in one sense knowing what he is learning, in another not knowing it. The strange thing would be, not if in some sense he knew what he was learning, but if he were to know it in that precise sense and manner in which he was learning it.

Part 2

We suppose ourselves to possess unqualified scientific knowledge of a thing, as opposed to knowing it in the accidental way in which the sophist knows, when we think that we know the cause on which the fact depends, as the cause of that fact and of no other, and, further, that the fact could not be other than it is. Now that scientific knowing is something of this sort is evident—witness both those who falsely claim it and those who actually possess it, since the former merely imagine themselves to be, while the latter are also actually, in the condition described. Consequently the proper object of unqualified scientific knowledge is something which cannot be other than it is.

There may be another manner of knowing as well—that will be discussed later. What I now assert is that at all events we do know by demonstration. By demonstration I mean a syllogism productive of scientific knowledge, a syllogism, that is, the grasp of which is *eo ipso* such knowledge. Assuming then that my thesis as to the nature of scientific knowing is correct, the premises of demonstrated knowledge must be true, primary, immediate, better known than and prior to the conclusion, which is further related to them as effect to cause. Unless these conditions are satisfied, the basic truths will not be 'appropriate' to the conclusion. Syllogism there may indeed be without these conditions, but such syllogism, not being productive of scientific knowledge, will not be demonstration. The premises must be true: for that which is non-existent cannot be known—we cannot know, e.g. that the diagonal of a square is commensurate with its side. The premises must be primary and indemonstrable; otherwise they will require demonstration in order to be known, since to have knowledge, if it be not accidental knowledge, of things which are demonstrable, means precisely to have a demonstration of them. The premises must be the causes of the conclusion, better known than it, and prior to it; its causes, since we possess scientific knowledge of a thing only when we know its cause; prior, in order to be causes; antecedently known, this antecedent knowledge being not our mere understanding of the meaning, but knowledge of the fact as well. Now 'prior' and 'better known' are ambiguous terms, for there is a difference between what is prior and better known in the order of being and what is prior and better known to man. I mean that objects nearer to sense are prior and better known to man; objects without qualification prior and better known are those further from sense. Now the most universal causes are furthest from sense and particular causes are nearest to sense, and they are thus exactly opposed to one another. In saying that the premises of demonstrated knowledge must be primary, I mean that they must be the 'appropriate' basic truths, for I identify primary premise and basic truth. A 'basic truth' in a demonstration is an immediate proposition. An immediate proposition is one which has no other proposition prior to it. A proposition is either part of an enunciation, i.e. it predicates a single attribute of a single subject. If a proposition is dialectical, it assumes either part indifferently; if it is demonstrative, it lays down one part to the definite exclusion of the other because that part is true. The term 'enunciation' denotes either part of a contradiction indifferently. A contradiction is an opposition which of its own nature excludes a middle. The part of a contradiction which conjoins a predicate with a subject is an affirmation; the part disjoining them is a negation. I call an immediate basic truth of syllogism a 'thesis' when, though it is not susceptible of proof by the teacher, yet ignorance of it does not constitute a total bar to progress on the part of the pupil: one which the pupil must know if he is to learn anything whatever is an axiom. I call it an axiom because there are such truths and we give them the name of axioms par excellence. If a thesis assumes one part or the other of an enunciation, i.e. asserts either the existence or the non-existence of a subject, it is a hypothesis; if it does not so assert, it is a definition. Definition is a 'thesis' or a 'laying something down', since the arithmetician lays it down that to be a unit is to be quantitatively indivisible; but it is not a hypothesis, for to define what a unit is is not the same as to affirm its existence.

Now since the required ground of our knowledge—i.e. of our conviction—of a fact is the possession of such a syllogism as we call demonstration, and the ground of the syllogism is the facts constituting its premises, we must not only know the primary premises—some if not all of them—beforehand, but know them better than the conclusion: for the cause of an attribute's inherence in a subject always itself inheres in the subject more firmly than that attribute; e.g. the cause of our loving anything is dearer to us than the object of our love. So since the primary premises are the cause of our knowledge—i.e. of our conviction—it follows that we know them better—that is, are more convinced of them—than their consequences, precisely because of our knowledge of the latter is the effect of our knowledge of the premises. Now a man cannot believe in anything more than in the things he knows, unless he has either actual knowledge of it or something better than actual knowledge. But we are faced with this paradox if a student whose belief rests on demonstration has not prior knowledge; a man must believe in some, if not in all, of the basic truths more than in the conclusion. Moreover, if a man sets out to acquire the scientific knowledge that comes through demonstration, he must not only have a better knowledge of the basic truths and a firmer conviction of them than of the connexion which is being demonstrated: more than this, nothing must be more certain or better known to him than these basic truths in their character as contradicting the fundamental premises which lead to the opposed and erroneous conclusion. For indeed the conviction of pure science must be unshakable.

Part 13

Knowledge of the fact differs from knowledge of the reasoned fact. To begin with, they differ within the same science and in two ways: (1) when the premises of the syllogism are not immediate (for then the proximate cause is not contained in them—a necessary condition of knowledge of the reasoned fact): (2) when the premises are immediate, but instead of the cause the better known of the two reciprocals is taken as the middle; for of two reciprocally predicable terms the one which is not the cause may quite easily be the better known and so become the middle term of the demonstration. Thus (2, a) you might prove as follows that the planets are near because they do not twinkle: let C be the planets, B not twinkling, A proximity. Then B is predicable of C; for the planets do not twinkle. But A is also predicable of B, since that which does not twinkle is near—we must take this truth as having been reached by induction or sense-perception. Therefore A is a necessary predicate of C; so that we have demonstrated that the planets are near. This syllogism, then, proves not the reasoned fact but only the fact; since they are not near because they do not twinkle, but, because they are near, do not twinkle. The major and middle of the proof, however, may be reversed, and then the demonstration will be of the reasoned fact. Thus: let C be the planets, B proximity, A not twinkling. Then B is an attribute of C, and

A—not twinkling—of B. Consequently A is predicable of C, and the syllogism proves the reasoned fact, since its middle term is the proximate cause. Another example is the inference that the moon is spherical from its manner of waxing. Thus: since that which so waxes is spherical, and since the moon so waxes, clearly the moon is spherical. Put in this form, the syllogism turns out to be proof of the fact, but if the middle and major be reversed it is proof of the reasoned fact; since the moon is not spherical because it waxes in a certain manner, but waxes in such a manner because it is spherical. (Let C be the moon, B spherical, and A waxing.) Again (b), in cases where the cause and the effect are not reciprocal and the effect is the better known, the fact is demonstrated but not the reasoned fact. This also occurs (1) when the middle falls outside the major and minor, for here too the strict cause is not given, and so the demonstration is of the fact, not of the reasoned fact. For example, the question 'Why does not a wall breathe?' might be answered, 'Because it is not an animal'; but that answer would not give the strict cause, because if not being an animal causes the absence of respiration, then being an animal should be the cause of respiration, according to the rule that if the negation of causes the non-inherence of y, the affirmation of x causes the inherence of y; e.g. if the disproportion of the hot and cold elements is the cause of ill health, their proportion is the cause of health; and conversely, if the assertion of x causes the inherence of y, the negation of x must cause y's non-inherence. But in the case given this consequence does not result; for not every animal breathes. A syllogism with this kind of cause takes place in the second figure. Thus: let A be animal, B respiration, C wall. Then A is predicable of all B (for all that breathes is animal), but of no C; and consequently B is predicable of no C; that is, the wall does not breathe. Such causes are like far-fetched explanations, which precisely consist in making the cause too remote, as in Anacharsis's account of why the Scythians have no flute-players; namely because they have no vines.

Thus, then, do the syllogism of the fact and the syllogism of the reasoned fact differ within one science and according to the position of the middle terms. But there is another way too in which the fact and the reasoned fact differ, and that is when they are investigated respectively by different sciences. This occurs in the case of problems related to one another as subordinate and superior, as when optical problems are subordinated to geometry, mechanical problems to stereometry, harmonic problems to arithmetic, the data of observation to astronomy. (Some of these sciences bear almost the same name; e.g. mathematical and nautical astronomy, mathematical and acoustical harmonics.) Here it is the business of the empirical observers to know the fact, of the mathematicians to know the reasoned fact; for the latter are in possession of the demonstrations giving the causes, and are often ignorant of the fact: just as we have often a clear insight into a universal, but through lack of observation are ignorant of some of its particular instances. These connexions have a perceptible existence though they are manifestations of forms. For the mathematical sciences concern forms: they do not demonstrate properties of a substratum, since, even though the geometrical subjects are predicable as properties of a perceptible substratum, it is not as thus predicable that the mathematician demonstrates properties of them. As optics is related to geometry, so another science is related to optics, namely the theory of the rainbow. Here knowledge of the fact is within the province of the natural philosopher, knowledge of the reasoned fact within that of the optician, either qua optician or qua mathematical optician. Many sciences not standing in this mutual relation enter into it at points; e.g. medicine and geometry: it is the physician's business to know that circular wounds heal more slowly, the geometer's to know the reason why.

Translated by G. R. G. Mure

Reading and Discussion Questions

1. What is induction? What is deduction?
2. Explain what Aristotle means by a scientific deduction and what would be involved in giving a demonstration. Explain how his discussion of why the planets do not twinkle serves as an example of Aristotelian demonstration.
3. Notice at the end of the reading what kinds of disciplines are available and well developed in his period.

On the Heavens (c. 350 BCE)

Aristotle

8

Book II

11

With regard to the shape of each star, the most reasonable view is that they are spherical. It has been shown that it is not in their nature to move themselves, and, since nature is no wanton or random creator, clearly she will have given things which possess no movement a shape particularly unadapted to movement. Such a shape is the sphere, since it possesses no instrument of movement. Clearly then their mass will have the form of a sphere. Again, what holds of one holds of all, and the evidence of our eyes shows us that the moon is spherical. For how else should the moon as it waxes and wanes show for the most part a crescent-shaped or gibbous figure, and only at one moment a half-moon? And astronomical arguments give further confirmation; for no other hypothesis accounts for the crescent shape of the sun's eclipses. One, then, of the heavenly bodies being spherical, clearly the rest will be spherical also.

12

There are two difficulties, which may very reasonably here be raised, of which we must now attempt to state the probable solution: for we regard the zeal of one whose thirst after philosophy leads him to accept even slight indications where it is very difficult to see one's way, as a proof rather of modesty than of overconfidence.

Of many such problems one of the strangest is the problem why we find the greatest number of movements in the intermediate bodies, and not, rather, in each successive body a variety of movement proportionate to its distance from the primary motion. For we should expect, since the primary body shows one motion only, that the body which is nearest to it should move with the fewest movements, say two, and the one next after that with three, or some similar arrangement. But the opposite is the case. The movements of the sun and moon are fewer than those of some of the planets. Yet these planets are farther from the centre and thus nearer to the primary body than they, as observation has itself revealed. For we have seen the moon, half-full, pass beneath the planet Mars, which vanished on its shadow side and came forth by the bright and shining part. Similar accounts of other stars are given by the Egyptians and Babylonians, whose observations have been kept for very many years past, and from whom much of our evidence about particular stars is derived. A second difficulty which may with equal justice be raised is this. Why is it that the primary motion includes such a multitude of stars that their whole array seems to defy counting, while of the other stars each one is separated off, and in no case do we find two or more attached to the same motion?

On these questions, I say, it is well that we should seek to increase our understanding, though we have but little to go upon, and are placed at so great a distance from the facts in question. Nevertheless there are certain principles on which if we base our consideration we shall not find this difficulty by any means insoluble. We may object that we have been thinking of the stars as mere bodies, and as units with a serial order indeed but entirely inanimate; but should rather conceive them as enjoying life and action. On this view the facts cease to appear surprising. For it is natural that the best-conditioned of all things should have its good without action, that which is nearest to it should achieve it by little and simple action, and that which is farther removed by a complexity of actions, just as with men's bodies one is in good condition without exercise at all, another after a short walk, while another requires running and wrestling and hard training, and there are yet others who however hard they worked themselves could never secure this good, but only some substitute for it. To succeed often or in many things is difficult. For instance, to throw ten thousand Coan* throws with the dice would be impossible, but to throw one or two is comparatively easy. In action, again, when A has to be done to get B, B to get C, and C to get D, one step or two present little difficulty, but as the series extends the difficulty grows. We must, then, think of the action of the lower stars as similar to that of animals and plants. For on our earth it is man that has the greatest variety of actions—for there are many goods that man can secure; hence his actions are various and directed to ends beyond them—while the perfectly conditioned has no need of action, since it is itself the end, and action always requires two terms, end and means. The lower animals have less variety of action than man; and plants perhaps have little action

* A "Coan" or "Chian" throw is double ones.

and of one kind only. For either they have but one attainable good (as indeed man has), or, if several, each contributes directly to their ultimate good. One thing then has and enjoys the ultimate good, other things attain to it, one immediately by few steps, another by many, while yet another does not even attempt to secure it but is satisfied to reach a point not far removed from that consummation. Thus, taking health as the end, there will be one thing that always possesses health, others that attain it, one by reducing flesh, another by running and thus reducing flesh, another by taking steps to enable himself to run, thus further increasing the number of movements, while another cannot attain health itself, but only running or reduction of flesh, so that one or other of these is for such a being the end. For while it is clearly best for any being to attain the real end, yet, if that cannot be, the nearer it is to the best the better will be its state. It is for this reason that the earth moves not at all and the bodies near to it with few movements. For they do not attain the final end, but only come as near to it as their share in the divine principle permits. But the first heaven finds it immediately with a single movement, and the bodies intermediate between the first and last heavens attain it indeed, but at the cost of a multiplicity of movement.

As to the difficulty that into the one primary motion is crowded a vast multitude of stars, while of the other stars each has been separately given special movements of its own, there is in the first place this reason for regarding the arrangement as a natural one. In thinking of the life and moving principle of the several heavens one must regard the first as far superior to the others. Such a superiority would be reasonable. For this single first motion has to move many of the divine bodies, while the numerous other motions move only one each, since each single planet moves with a variety of motions. Thus, then, nature makes matters equal and establishes a certain order, giving to the single motion many bodies and to the single body many motions. And there is a second reason why the other motions have each only one body, in that each of them except the last, i.e. that which contains the one star, is really moving many bodies. For this last sphere moves with many others, to which it is fixed, each sphere being actually a body; so that its movement will be a joint product. Each sphere, in fact, has its particular natural motion, to which the general movement is, as it were, added. But the force of any limited body is only adequate to moving a limited body.

The characteristics of the stars which move with a circular motion, in respect of substance and shape, movement and order, have now been sufficiently explained.

13

It remains to speak of the earth, of its position, of the question whether it is at rest or in motion, and of its shape.

I. As to its position there is some difference of opinion. Most people—all, in fact, who regard the whole heaven as finite—say it lies at the centre. But the Italian philosophers known as Pythagoreans take the contrary view. At the centre, they say, is fire, and the earth is one of the stars, creating night and day by its circular motion about the centre. They further construct another earth in opposition to ours to which they give the name counterearth. In all this they are not seeking for theories and causes to account for observed facts, but rather forcing their observations and trying to accommodate them to certain theories and opinions of their own. But there are many others who would agree that it is wrong to give the earth the central position, looking for confirmation rather to theory than to the facts of observation. Their view is that the most precious place befits the most precious thing: but fire, they say, is more precious than earth, and the limit than the intermediate, and the circumference and the centre are limits. Reasoning on this basis they take the view that it is not earth that lies at the centre of the sphere, but rather fire. The Pythagoreans have a further reason. They hold that the most important part of the world, which is the centre, should be most strictly guarded, and name it, or rather the fire which occupies that place, the 'Guardhouse of Zeus', as if the word 'centre' were quite unequivocal, and the centre of the mathematical figure were always the same with that of the thing or the natural centre. But it is better to conceive of the case of the whole heaven as analogous to that of animals, in which the centre of the animal and that of the body are different. For this reason they have no need to be so disturbed about the world, or to call in a guard for its centre: rather let them look for the centre in the other sense and tell us what it is like and where nature has set it. That centre will be something primary and precious; but to the mere position we should give the last place rather than the first. For the middle is what is defined, and what defines it is the limit, and that which contains or limits is more precious than that which is limited, seeing that the latter is the matter and the former the essence of the system.

II. As to the position of the earth, then, this is the view which some advance, and the views advanced concerning its rest or motion are similar. For here too there is no general agreement. All who deny that the earth lies at the centre think that it revolves about the centre, and not the earth only but, as we said before, the counter-earth as well. Some of them even consider it possible that there are several bodies so moving, which are invisible to us owing to the interposition of the earth. This, they say, accounts for the fact that eclipses of the moon are more frequent than eclipses of the sun: for in addition to the earth each of these moving bodies can obstruct it. Indeed, as in any case the surface of the earth is not actually a centre but distant from it a full hemisphere, there is no more difficulty, they think, in accounting for the observed facts on their view that we do not dwell at the centre, than on the common view that the earth is in the middle. Even as it is, there is nothing in the observations to suggest that we are removed from the centre by half the diameter of the earth. Others, again, say that the earth, which lies at the centre, is 'rolled', and thus in motion, about the axis of the whole heaven. So it stands written in the Timaeus.

III. There are similar disputes about the shape of the earth. Some think it is spherical, others that it is flat and drum-shaped. For evidence they bring the fact that, as the sun rises and sets, the part concealed by the earth shows a straight and not a curved edge, whereas if the earth were spherical the line of section would have to be circular. In this they leave out of account the great distance of the sun from the earth and the great size of the circumference, which, seen from a distance on these apparently small circles appears straight. Such an appearance ought not to make them doubt the circular shape of the earth. But they have another argument. They say that because it is at rest, the earth must necessarily have this shape. For there are many different ways in which the movement or rest of the earth has been conceived.

The difficulty must have occurred to everyone. It would indeed be a complacent mind that felt no surprise that, while a little bit of earth, let loose in mid-air moves and will not stay still, and more there is of it the faster it moves, the whole earth, free in midair, should show no movement at all. Yet here is this great weight of earth, and it is at rest. And again, from beneath one of these moving fragments of earth, before it falls, take away the earth, and it will continue its downward movement with nothing to stop it. The difficulty then, has naturally passed into a common place of philosophy; and one may well wonder that the solutions offered are not seen to involve greater absurdities than the problem itself.

By these considerations some have been led to assert that the earth below us is infinite, saying, with Xenophanes of Colophon, that it has 'pushed its roots to infinity', in order to save the trouble of seeking for the cause. Hence the sharp rebuke of Empedocles, in the words 'if the deeps of the earth are endless and endless the ample ether—such is the vain tale told by many a tongue, poured from the mouths of those who have seen but little of the whole. Others say the earth rests upon water. This, indeed, is the oldest theory that has been preserved, and is attributed to Thales of Miletus. It was supposed to stay still because it floated like wood and other similar substances, which are so constituted as to rest upon water but not upon air. As if the same account had not to be given of the water which carries the earth as of the earth itself! It is not the nature of water, any more than of earth, to stay in mid-air: it must have something to rest upon. Again, as air is lighter than water, so is water than earth: how then can they think that the naturally lighter substance lies below the heavier? Again, if the earth as a whole is capable of floating upon water, that must obviously be the case with any part of it. But observation shows that this is not the case. Any piece of earth goes to the bottom, the quicker the larger it is. These thinkers seem to push their inquiries some way into the problem, but not so far as they might. It is what we are all inclined to do, to direct our inquiry not by the matter itself, but by the views of our opponents: and even when interrogating oneself one pushes the inquiry only to the point at which one can no longer offer any opposition. Hence a good inquirer will be one who is ready in bringing forward the objections proper to the genus, and that he will be when he has gained an understanding of all the differences.

Anaximenes and Anaxagoras and Democritus give the flatness of the earth as the cause of its staying still. Thus, they say, it does not cut, but covers like a lid, the air beneath it. This seems to be the way of flat-shaped bodies: for even the wind can scarcely move them because of their power of resistance. The same immobility, they say, is produced by the flatness of the surface which the earth presents to the air which underlies it; while the air, not having room enough to change its place because it is underneath the earth, stays there in a mass, like the water in the case of the water-clock. And they adduce an amount of evidence to prove that air, when cut off and at rest, can bear a considerable weight.

Now, first, if the shape of the earth is not flat, its flatness cannot be the cause of its immobility. But in their own account it is rather the size of the earth than its flatness that causes it to remain at rest. For the reason why the air is so closely confined that it cannot find a passage, and therefore stays where it is, is its great amount: and this amount great because the body which isolates it, the earth, is very large. This result, then, will follow, even if the earth is spherical, so long as it retains its size. So far as their arguments go, the earth will still be at rest.

In general, our quarrel with those who speak of movement in this way cannot be confined to the parts; it concerns the whole universe. One must decide at the outset whether bodies have a natural movement or not, whether there is no natural but only constrained movement. Seeing, however, that we have already decided this matter to the best of our ability, we are entitled to treat our results as representing fact. Bodies, we say, which have no natural movement, have no constrained movement; and where there is no natural and no constrained movement there will be no movement at all. This is a conclusion, the necessity of which we have already decided, and we have seen further that rest also will be inconceivable, since rest, like movement, is either natural or constrained. But if there is any natural movement, constraint will not be the sole principle of motion or of rest. If, then, it is by constraint that the earth now keeps its place, the so-called 'whirling' movement by which its parts came together at the centre was also constrained. (The form of causation supposed they all borrow from observations of liquids and of air, in which the larger and heavier bodies always move to the centre of the whirl. This is thought by all those who try to generate the heavens to explain why the earth came together at the centre. They then seek a reason for its staying there; and some say, in the manner explained, that the reason is its size and flatness, others, with Empedocles, that the motion of the heavens, moving about it at a higher speed, prevents movement of the earth, as the water in a cup, when the cup is given a circular motion, though it is often underneath the bronze, is for this same reason prevented from moving with the downward movement which is natural to it.) But suppose both the 'whirl' and its flatness (the air beneath being withdrawn) cease to prevent the earth's motion, where will the earth move to then? Its movement to the centre was constrained, and its rest at the centre is due to constraint; but there must be some motion which is natural to it. Will this be upward motion or downward or what? It must have some motion; and if upward and downward motion are alike to it, and the air above the earth does not prevent upward movement, then no more could air below it prevent downward movement. For the same cause must necessarily have the same effect on the same thing.

Further, against Empedocles there is another point which might be made. When the elements were separated off by Hate, what caused the earth to keep its place? Surely the 'whirl' cannot have been then also the cause. It is absurd too not to perceive that, while the whirling movement may have been responsible for the original coming together of the art of earth at the centre, the question remains, why now do all heavy bodies move to the earth. For the whirl surely does not come near us. Why, again, does fire move upward? Not, surely, because of the whirl. But if fire is naturally such as to move in a certain direction, clearly the same may be supposed to hold of earth. Again, it cannot be the whirl which determines the heavy and the light. Rather that movement caused the pre-existent heavy and light things to go to the middle and stay on the surface respectively. Thus, before ever the whirl began, heavy and light existed; and what can have been the ground of their distinction, or the manner and direction of their natural movements? In the infinite chaos there can have been neither above nor below, and it is by these that heavy and light are determined.

It is to these causes that most writers pay attention: but there are some, Anaximander, for instance, among the ancients, who say that the earth keeps its place because of its indifference. Motion upward and downward and sideways were all, they thought, equally inappropriate to that which is set at the centre and indifferently related to every extreme point; and to move in contrary directions at the same time was impossible: so it must needs remain still. This view is ingenious but not true. The argument would prove that everything, whatever it be, which is put at the centre, must stay there. Fire, then, will rest at the centre: for the proof turns on no peculiar property of earth. But this does not follow. The observed facts about earth are not only that it remains at the centre, but also that it moves to the centre. The place to which any fragment of earth moves must necessarily be the place to which the whole moves; and in the place to which a thing naturally moves, it will naturally rest. The reason then is not in the fact that the earth is indifferently related to every extreme point: for this would apply to any body, whereas movement to the centre is peculiar to earth. Again it is absurd to look for a reason why the earth remains at the centre and not for a reason why fire remains at the extremity. If the extremity is the natural place of fire, clearly earth must also have a natural place. But suppose that the centre is not its place, and that the reason of its remaining there is this necessity of indifference—on the analogy of the hair which, it is said, however great the tension, will not break under it, if it be evenly distributed, or of the men who, though exceedingly hungry and thirsty, and both equally, yet being equidistant from food and drink, is therefore bound to stay where he is—even so, it still remains to explain why fire stays at the extremities. It is strange, too, to ask about things staying still but not about their motion—why, I mean, one thing, if nothing stops it, moves up, and another thing to the centre. Again, their statements are not true. It happens, indeed, to be the case that a thing to which movement this way and that is equally inappropriate is obliged to remain at the centre. But so far as their argument goes, instead of remaining there, it will move, only not as a mass but in fragments. For the argument applies equally to fire. Fire, if set at the centre, should stay there, like earth, since it will be indifferently related to every point on the extremity. Nevertheless it will move, as in fact it always does move when nothing stops it, away from the centre to the extremity. It will not, however, move in a mass to a single point on the circumference—the only possible result on the lines of the indifference theory—but rather each corresponding portion of fire to the corresponding part of the extremity, each fourth part, for instance, to a fourth part of the circumference. For since no body is a point, it will have parts. The expansion, when the body increased the place occupied, would be on the same principle as the contraction, in which the place was diminished. Thus, for all the indifference theory shows to the contrary, earth also would have moved in this manner away from the centre, unless the centre had been its natural place.

We have now outlined the views held as to the shape, position, and rest or movement of the earth.

14

Let us first decide the question whether the earth moves or is at rest. For, as we said, there are some who make it one of the stars, and others who, setting it at the centre, suppose it to be 'rolled' and in motion about the pole as axis. That both views are untenable will be clear if we take as our starting-point the fact that the earth's motion, whether the earth be at the centre or away from it, must needs be a constrained motion. It cannot be the movement of the earth itself. If it were, any portion of it would have this movement; but in fact every part moves in a straight line to the centre. Being, then, constrained and unnatural, the movement could not be eternal. But the order of the universe is eternal. Again, everything that moves with the circular movement, except the first sphere, is observed to be passed, and to move with more than one motion. The earth, then, also, whether it move about the centre or as stationary at it, must necessarily move with two motions. But if this were so, there would have to be passings and turnings of the fixed stars. Yet no such thing is observed. The same stars always rise and set in the same parts of the earth.

Further, the natural movement of the earth, part and whole alike, is the centre of the whole—whence the fact that it is now actually situated at the centre—but it might be questioned since both centres are the same, which centre it is that portions of earth and other heavy things move to. Is this their goal because it is the centre of the earth or because it is the centre of the whole? The goal, surely, must be the centre of the whole. For fire and other light things move to the extremity of the area which contains the centre. It happens, however, that the centre of the earth and of the whole is the same. Thus they do move to the centre of the earth, but accidentally, in virtue of the fact that the earth's centre lies at the centre of the whole. That the centre of the earth is the goal of their movement is indicated by the fact that heavy bodies moving towards the earth do not parallel but so as to make equal angles, and thus to a single centre, that of the earth. It is clear, then, that the earth must be at the centre and immovable, not only for the reasons already given, but also because heavy bodies forcibly thrown quite straight upward return to the point from which they started, even if they are thrown to an infinite distance. From these considerations then it is clear that the earth does not move and does not lie elsewhere than at the centre.

From what we have said the explanation of the earth's immobility is also apparent. If it is the nature of earth, as observation shows, to move from any point to the centre, as of fire contrariwise to move from the centre to the extremity, it is impossible that any portion of earth should move away from the centre except by constraint. For a single thing has a single movement, and a simple thing a simple: contrary movements cannot belong to the same thing, and movement away from the centre is the contrary of movement to it. If then no portion of earth can move away from the centre, obviously still less can the earth as a whole so move. For it is the nature of the whole to move to the point to which the part naturally moves. Since, then, it would require a force greater than itself to move it, it must needs stay at the centre. This view is further supported by the contributions of mathematicians to astronomy, since the observations made as the shapes change by which the order of the stars is determined, are fully accounted for on the hypothesis that the earth lies at the centre. Of the position of the earth and of the manner of its rest or movement, our discussion may here end.

Its shape must necessarily be spherical. For every portion of earth has weight until it reaches the centre, and the jostling of parts greater and smaller would bring about not a waved surface, but rather compression and convergence of part and part until the centre is reached. The process should be conceived by supposing the earth to come into being in the way that some of the natural philosophers describe. Only they attribute the downward movement to constraint, and it is better to keep to the truth and say that the reason of this motion is that a thing which possesses weight is naturally endowed with a centripetal movement. When the mixture, then, was merely potential, the things that were separated off moved similarly from every side towards the centre. Whether the parts which came together at the centre were distributed at the extremities evenly, or in some other way, makes no difference. If, on the one hand, there were a similar movement from each quarter of the extremity to the single centre, it is obvious that the resulting mass would be similar on every side. For if an equal amount is added on every side the extremity of the mass will be everywhere equidistant from its centre, i.e. the figure will be spherical. But neither will it in any way affect the argument if there is not a similar accession of concurrent fragments from every side. For the greater quantity, finding a lesser in front of it, must necessarily drive it on, both having an impulse whose goal is the centre, and the greater weight driving the lesser forward till this goal is reached. In this we have also the solution of a possible difficulty. The earth, it might be argued, is at the centre and spherical in shape: if, then, a weight many times that of the earth were added to one hemisphere, the centre of the earth and of the whole will no longer be coincident. So that either the earth will not stay still at the centre, or if it does, it will be at rest without having its centre at the place to which it is still its nature to move. Such is the difficulty. A short consideration will give us an easy answer, if we first give precision to our postulate that any body endowed with weight, of whatever size, moves towards the centre. Clearly it will not stop when its edge touches the centre. The greater quantity must prevail until the body's centre occupies the centre. For that is the goal of its impulse. Now it makes no difference whether we apply this to a clod or common fragment of earth or to the earth as a whole. The fact indicated does not depend upon degrees of size but applies universally to everything that has the centripetal impulse. Therefore earth in motion, whether in a mass or in fragments, necessarily continues to move until it occupies the centre equally every way, the less being forced to equalize itself by the greater owing to the forward drive of the impulse.

If the earth was generated, then, it must have been formed in this way, and so clearly its generation was spherical; and if it is ungenerated and has remained so always, its character must be that which the initial generation, if it had occurred, would have given it. But the spherical shape, necessitated by this argument, follows also from the fact that the motions of heavy bodies always make equal angles, and are not parallel. This would be the natural form of movement towards what is naturally spherical. Either then the earth is spherical or it is at least naturally spherical. And it is right to call anything that which nature intends it to be, and which belongs to it, rather than that which it is by constraint and contrary to nature. The evidence of the senses further corroborates this. How else would eclipses of the moon show segments shaped as we see them? As it is, the shapes which the moon itself each month shows are of every kind straight, gibbous, and concave—but in eclipses the outline is always curved: and, since it is the interposition of the earth that makes the eclipse, the form of this line will be caused by the form of the earth's surface, which is therefore spherical. Again, our observations of the stars make it evident, not only that the earth is circular, but also that it is a circle of no great size. For quite a small change of position to south or north causes a manifest alteration of the horizon. There is much change, I mean, in the stars which are overhead, and the stars seen are different, as one moves northward or southward. Indeed there are some stars seen in Egypt and in the neighbourhood of Cyprus which are not seen in the northerly regions; and stars, which in the north are never beyond the range of observation, in those regions rise and set. All of which goes to show not only that the earth is circular in shape, but also that it is a sphere of no great size: for otherwise the effect of so slight a change of place would not be quickly apparent. Hence one should not be too sure of the incredibility of the view of those who conceive that there is continuity between the parts about the pillars of Hercules and the parts about India, and that in this way the ocean is one. As further evidence in favour of this they quote the case of elephants, a species occurring in each of these extreme regions, suggesting that the common characteristic of these extremes is explained by their continuity. Also, those mathematicians who try to calculate the size of the earth's circumference arrive at the figure 400,000 stades. This indicates not only that the earth's mass is spherical in shape, but also that as compared with the stars it is not of great size.

Translated by J. L. Stocks

Reading and Discussion Questions

1. Aristotle is aware that he is the beneficiary of ideas drawn from several ongoing intellectual traditions. Give some examples and describe how he engages with the ideas of his predecessors. What sorts of attitudes, styles, or habits of thought does he display in his interaction with their ideas?
2. What is his argument for the belief that the moon is closer to us than Mars?
3. Aristotle gives a series of arguments about the shape of the earth. What does he conclude about the shape of the earth and its approximate size? Does he believe that the earth is in motion or at rest, and how does he reach this conclusion?
4. How does his argument for the position of the earth relate to the doctrine of natural place?
5. How would you use the ideal of demonstration from the *Posterior Analytics* to understand the reason why lunar eclipses occur?

Meteorology (c. 350 BCE)

Aristotle

9

Book I

Part 1

We have already discussed the first causes of nature, and all natural motion, also the stars ordered in the motion of the heavens, and the physical elements—enumerating and specifying them and showing how they change into one another—and becoming and perishing in general. There remains for consideration a part of this inquiry which all our predecessors called meteorology. It is concerned with events that are natural, though their order is less perfect than that of the first of the elements of bodies. They take place in the region nearest to the motion of the stars. Such are the Milky Way, and comets, and the movements of meteors. It studies also all the affections we may call common to air and water, and the kinds and parts of the earth and the affections of its parts. These throw light on the causes of winds and earthquakes and all the consequences the motions of these kinds and parts involve. Of these things some puzzle us, while others admit of explanation in some degree. Further, the inquiry is concerned with the falling of thunderbolts and with whirlwinds and fire-winds, and further, the recurrent affections produced in these same bodies by concretion. When the inquiry into these matters is concluded let us consider what account we can give, in accordance with the method we have followed, of animals and plants, both generally and in detail. When that has been done we may say that the whole of our original undertaking will have been carried out.

After this introduction let us begin by discussing our immediate subject.

Part 2

We have already laid down that there is one physical element which makes up the system of the bodies that move in a circle, and besides this four bodies owing their existence to the four principles, the motion of these latter bodies being of two kinds: either from the centre or to the centre. These four bodies are fire, air, water, earth. Fire occupies the highest place among them all, earth the lowest, and two elements correspond to these in their relation to one another, air being nearest to fire, water to earth. The whole world surrounding the earth, then, the affections of which are our subject, is made up of these bodies. This world necessarily has a certain continuity with the upper motions: consequently all its power and order is derived from them. (For the originating principle of all motion is the first cause. Besides, that element is eternal and its motion has no limit in space, but is always complete; whereas all these other bodies have separate regions which limit one another.) So we must treat fire and earth and the elements like them as the material causes of the events in this world (meaning by material what is subject and is affected), but must assign causality in the sense of the originating principle of motion to the influence of the eternally moving bodies.

Part 3

Let us first recall our original principles and the distinctions already drawn and then explain the 'milky way' and comets and the other phenomena akin to these.

Fire, air, water, earth, we assert, originate from one another, and each of them exists potentially in each, as all things do that can be resolved into a common and ultimate substrate.

The first difficulty is raised by what is called the air. What are we to take its nature to be in the world surrounding the earth? And what is its position relatively to the other physical elements. (For there is no question as to the relation of the bulk of the earth to the size of the bodies which exist around it, since astronomical demonstrations have by this time proved to us that it is actually far smaller than some individual stars. As for the water, it is not observed to exist collectively and separately, nor can it do so apart from that volume of it which has its seat about the earth: the sea, that is, and rivers, which we can see, and any subterranean water that may be hidden from our observation.) The question is really about that which lies between the earth and the nearest stars. Are we to consider it to be one kind of body or more than one? And if more than one, how many are there and what are the bounds of their regions?

We have already described and characterized the first element, and explained that the whole world of the upper motions is full of that body.

This is an opinion we are not alone in holding: it appears to be an old assumption and one which men have held in the past, for the word ether has long been used to denote that element. Anaxagoras, it is true, seems to me to think that the word means the same as fire. For he thought that the upper regions were full of fire, and that men referred to those regions when they spoke of ether. In the latter point he was right, for men seem to have assumed that a body that was eternally in motion was also divine in nature; and, as such a body was different from any of the terrestrial elements, they determined to call it 'ether'.

For the same opinions appear in cycles among men not once nor twice, but infinitely often.

Now there are some who maintain that not only the bodies in motion but that which contains them is pure fire, and the interval between the earth and the stars air: but if they had considered what is now satisfactorily established by mathematics, they might have given up this puerile opinion. For it is altogether childish to suppose that the moving bodies are all of them of a small size, because they so to us, looking at them from the earth.

This a matter which we have already discussed in our treatment of the upper region, but we may return to the point now.

If the intervals were full of fire and the bodies consisted of fire every one of the other elements would long ago have vanished.

However, they cannot simply be said to be full of air either; for even if there were two elements to fill the space between the earth and the heavens, the air would far exceed the quantity required to maintain its proper proportion to the other elements. For the bulk of the earth (which includes the whole volume of water) is infinitesimal in comparison with the whole world that surrounds it. Now we find that the excess in volume is not proportionately great where water dissolves into air or air into fire. Whereas the proportion between any given small quantity of water and the air that is generated from it ought to hold good between the total amount of air and the total amount of water. Nor does it make any difference if any one denies that the elements originate from one another, but asserts that they are equal in power. For on this view it is certain amounts of each that are equal in power, just as would be the case if they actually originated from one another.

So it is clear that neither air nor fire alone fills the intermediate space.

It remains to explain, after a preliminary discussion of difficulties, the relation of the two elements air and fire to the position of the first element, and the reason why the stars in the upper region impart heat to the earth and its neighbourhood. Let us first treat of the air, as we proposed, and then go on to these questions.

Since water is generated from air, and air from water, why are clouds not formed in the upper air? They ought to form there the more, the further from the earth and the colder that region is. For it is neither appreciably near to the heat of the stars, nor to the rays reflected from the earth. It is these that dissolve any formation by their heat and so prevent clouds from forming near the earth.

For clouds gather at the point where the reflected rays disperse in the infinity of space and are lost. To explain this we must suppose either that it is not all air which water is generated, or, if it is produced from all air alike, that what immediately surrounds the earth is not mere air, but a sort of vapour, and that its vaporous nature is the reason why it condenses back to water again. But if the whole of that vast region is vapour, the amount of air and of water will be disproportionately great. For the spaces left by the heavenly bodies must be filled by some element. This cannot be fire, for then all the rest would have been dried up. Consequently, what fills it must be air and the water that surrounds the whole earth—vapour being water dissolved.

After this exposition of the difficulties involved, let us go on to lay down the truth, with a view at once to what follows and to what has already been said. The upper region as far as the moon we affirm to consist of a body distinct both from fire and from air, but varying degree of purity and in kind, especially towards its limit on the side of the air, and of the world surrounding the earth. Now the circular motion of the first element and of the bodies it contains dissolves, and inflames by its motion, whatever part of the lower world is nearest to it, and so generates heat. From another point of view we may look at the motion as follows. The body that lies below the circular motion of the heavens is, in a sort, matter, and is potentially hot, cold, dry, moist, and possessed of whatever other qualities are derived from these. But it actually acquires or retains one of these in virtue of motion or rest, the cause and principle of which has already been explained. So at the centre and round it we get earth and water, the heaviest and coldest elements, by themselves; round them and contiguous with them, air and what we commonly call fire. It is not really fire, for fire is an excess of heat and a sort of ebullition; but in reality, of what we call air, the part surrounding the earth is moist and warm, because it contains both vapour and a dry exhalation from the earth. But the next part, above that, is warm and dry. For vapour is naturally moist and cold, but the exhalation warm and dry; and vapour is potentially like water, the exhalation potentially like fire. So we must take the reason why clouds are not formed in the upper region to be this: that it is filled not with mere air but rather with a sort of fire.

However, it may well be that the formation of clouds in that upper region is also prevented by the circular motion. For the air round the earth is necessarily all of it in motion, except that which is cut off inside the circumference which makes the earth a complete sphere. In the case of winds it is actually observable that they originate in marshy districts of the earth; and they do not seem to blow above the level of the highest mountains. It is the revolution of the heaven which carries the air with it and causes its circular motion, fire being continuous with the upper element and air with fire. Thus its motion is a second reason why that air is not condensed into water.

But whenever a particle of air grows heavy, the warmth in it is squeezed out into the upper region and it sinks, and other particles in turn are carried up together with the fiery exhalation. Thus the one region is always full of air and the other of fire, and each of them is perpetually in a state of change.

So much to explain why clouds are not formed and why the air is not condensed into water, and what account must be given of the space between the stars and the earth, and what is the body that fills it.

As for the heat derived from the sun, the right place for a special and scientific account of it is in the treatise about sense, since heat is an affection of sense, but we may now explain how it can be produced by the heavenly bodies which are not themselves hot.

We see that motion is able to dissolve and inflame the air; indeed, moving bodies are often actually found to melt. Now the sun's motion alone is sufficient to account for the origin of terrestrial warmth and heat. For a motion that is to have this effect must be rapid and near, and that of the stars is rapid but distant, while that of the moon is near but slow, whereas the sun's motion combines both conditions in a sufficient degree. That most heat should be generated where the sun is present is easy to understand if we consider the analogy of terrestrial phenomena, for here, too, it is the air that is nearest to a thing in rapid motion which is heated most. This is just what we should expect, as it is the nearest air that is most dissolved by the motion of a solid body.

This then is one reason why heat reaches our world. Another is that the fire surrounding the air is often scattered by the motion of the heavens and driven downwards in spite of itself.

Shooting-stars further suffix to prove that the celestial sphere is not hot or fiery: for they do not occur in that upper region but below: yet the more and the faster a thing moves, the more apt it is to take fire. Besides, the sun, which most of all the stars is considered to be hot, is really white and not fiery in colour.

Part 4

Having determined these principles let us explain the cause of the appearance in the sky of burning flames and of shooting-stars, and of 'torches', and 'goats', as some people call them. All these phenomena are one and the same thing, and are due to the same cause, the difference between them being one of degree.

The explanation of these and many other phenomena is this. When the sun warms the earth the evaporation which takes place is necessarily of two kinds, not of one only as some think. One kind is rather of the nature of vapour, the other of the nature of a windy exhalation. That which rises from the moisture contained in the earth and on its surface is vapour, while that rising from the earth itself, which is dry, is like smoke. Of these the windy exhalation, being warm, rises above the moister vapour, which is heavy and sinks below the other. Hence the world surrounding the earth is ordered as follows. First below the circular motion comes the warm and dry element, which we call fire, for there is no word fully adequate to every state of the fumid evaporation: but we must use this terminology since this element is the most inflammable of all bodies. Below this comes air. We must think of what we just called fire as being spread round the terrestrial sphere on the outside like a kind of fuel, so that a little motion often makes it burst into flame just as smoke does: for flame is the ebullition of a dry exhalation. So whenever the circular motion stirs this stuff up in any way, it catches fire at the point at which it is most inflammable. The result differs according to the disposition and quantity of the combustible material. If this is broad and long, we often see a flame burning as in a field of stubble: if it burns lengthwise only, we see what are called 'torches' and 'goats' and shooting-stars. Now when the inflammable material is longer than it is broad sometimes it seems to throw off sparks as it burns. (This happens because matter catches fire at the sides in small portions but continuously with the main body.) Then it is called a 'goat'. When this does not happen it is a 'torch'. But if the whole length of the exhalation is scattered in small parts and in many directions and in breadth and depth alike, we get what are called shooting-stars.

The cause of these shooting-stars is sometimes the motion which ignites the exhalation. At other times the air is condensed by cold and squeezes out and ejects the hot element; making their motion look more like that of a thing thrown than like a running fire. For the question might be raised whether the 'shooting' of a 'star' is the same thing as when you put an exhalation below a lamp and it lights the lower lamp from the flame above. For here too the flame passes wonderfully quickly and looks like a thing thrown, and not as if one thing after another caught fire. Or is a 'star' when it 'shoots' a single body that is thrown? Apparently both cases occur: sometimes it is like the flame from the lamp and sometimes bodies are projected by being squeezed out (like fruit stones from one's fingers) and so are seen to fall into the sea and on the dry land, both by night and by day when the sky is clear. They are thrown downwards because the condensation which propels them inclines downwards. Thunderbolts fall downwards for the same reason: their origin is never combustion but ejection under pressure, since naturally all heat tends upwards.

...

Part 9

Let us go on to treat of the region which follows next in order after his and which immediately surrounds the earth. It is the region common to water and air, and the processes attending the formation of water above take place in it. We must consider the principles and causes of all these phenomena too as before. The efficient and chief and first cause is the circle in which the sun moves. For the sun as it approaches or recedes, obviously causes dissipation and condensation and so gives rise to generation and destruction. Now the earth remains but the moisture surrounding it is made to evaporate by the sun's rays and the other heat from above, and rises. But when the heat which was raising it leaves it, in part dispersing to the higher region, in part quenched through rising so far into the upper air, then the vapour cools because its heat is gone and because the place is cold, and condenses again and turns from air into water. And after the water has formed it falls down again to the earth.

The exhalation of water is vapour: air condensing into water is cloud. Mist is what is left over when a cloud condenses into water, and is therefore rather a sign of fine weather than of rain; for mist might be called a barren cloud. So we get a circular process that follows the course of the sun. For according as the sun moves to this side or that, the moisture in this process rises or falls. We must think of it as a river flowing up and down in a circle and made up partly of air, partly of water. When the sun is near, the stream of vapour flows upwards; when it recedes, the stream of water flows down: and the order of sequence, at all events, in this process always remains the same. So if 'Oceanus' had some secret meaning in early writers, perhaps they may have meant this river that flows in a circle about the earth.

So the moisture is always raised by the heat and descends to the earth again when it gets cold. These processes and, in some cases, their varieties are distinguished by special names. When the water falls in small drops it is called a drizzle; when the drops are larger it is rain.

Part 13

Let us explain the nature of winds, and all windy vapours, also of rivers and of the sea. But here, too, we must first discuss the difficulties involved: for, as in other matters, so in this no theory has been handed down to us that the most ordinary man could not have thought of.

Some say that what is called air, when it is in motion and flows, is wind, and that this same air when it condenses again becomes cloud and water, implying that the nature of wind and water is the same. So they define wind as a motion of the air. Hence some, wishing to say a clever thing, assert that all the winds are one wind, because the air that moves is in fact all of it one and the same; they maintain that the winds appear to differ owing to the region from which the air may happen to flow on each occasion, but really do not differ at all. This is just like thinking that all rivers are one and the same river, and the ordinary unscientific view is better than a scientific theory like this. If all rivers flow from one source, and the same is true in the case of the winds, there might be some truth in this theory; but if it is no more true in the one case than in the other, this ingenious idea is plainly false. What requires investigation is this: the nature of wind and how it originates, its efficient cause and whence they derive their source; whether one ought to think of the wind as issuing from a sort of vessel and flowing until the vessel is empty, as if let out of a wineskin, or, as painters represent the winds, as drawing their source from themselves. . . .

Part 14

The same parts of the earth are not always moist or dry, but they change according as rivers come into existence and dry up. And so the relation of land to sea changes too and a place does not always remain land or sea throughout all time, but where there was dry land there comes to be sea, and where there is now sea, there one day comes to be dry land. But we must suppose these changes to follow some order and cycle. The principle and cause of these changes is that the interior of the earth grows and decays, like the bodies of plants and animals. Only in the case of these latter the process does not go on by parts, but each of them necessarily grows or decays as a whole, whereas it does go on by parts in the case of the earth. Here the causes are cold and heat, which increase and diminish on account of the sun and its course. It is owing to them that the parts of the earth come to have a different character, that some parts remain moist for a certain time, and then dry up and grow old, while other parts in their turn are filled with life and moisture. Now when places become drier the springs necessarily give out, and when this happens the rivers first decrease in size and then finally become dry; and when rivers change and disappear in one part and come into existence correspondingly in another, the sea must needs be affected.

If the sea was once pushed out by rivers and encroached upon the land anywhere, it necessarily leaves that place dry when it recedes; again, if the dry land has encroached on the sea at all by a process of silting set up by the rivers when at their full, the time must come when this place will be flooded again.

But the whole vital process of the earth takes place so gradually and in periods of time which are so immense compared with the length of our life, that these changes are not observed, and before their course can be recorded from beginning to end whole nations perish and are destroyed. Of such destructions the most utter and sudden are due to wars; but pestilence or famine cause them too. Famines, again, are either sudden and severe or else gradual. In the latter case the disappearance of a nation is not noticed because some leave the country while others remain; and this goes on until the land is unable to maintain any inhabitants at all. So a long period of time is likely to elapse from the first departure to the last, and no one remembers and the lapse of time destroys all record even before the last inhabitants have disappeared. In the same way a nation must be supposed to lose account of the time when it first settled in a land that was changing from a marshy and watery state and becoming dry. Here, too, the change is gradual and lasts a long time and men do not remember who came first, or when, or what the land was like when they came. This has been the case with Egypt. Here it is obvious that the land is continually getting drier and that the whole country is a deposit of the river Nile. But because the neighbouring peoples settled in the land gradually as the marshes dried, the lapse of time has hidden the beginning of the process. However, all the mouths of the Nile, with the single exception of that at Canopus, are obviously artificial and not natural.

And Egypt was nothing more than what is called Thebes, as Homer, too, shows, modern though he is in relation to such changes. For Thebes is the place that he mentions; which implies that Memphis did not yet exist, or at any rate was not as important as it is now. That this should be so is natural, since the lower land came to be inhabited later than that which lay higher. For the parts that lie nearer to the place where the river is depositing the silt are necessarily marshy for a longer time since the water always lies most in the newly formed land. But in time this land changes its character, and in its turn enjoys a period of prosperity. For these places dry up and come to be in good condition while the places that were formerly well-tempered some day grow excessively dry and deteriorate. This happened to the land of Argos and Mycenae in Greece. In the time of the Trojan wars the Argive land was marshy and could only support a small population, whereas the land of Mycenae was in good condition (and for this reason Mycenae was the superior). But now the opposite is the case, for the reason we have mentioned: the land of Mycenae has become completely dry and barren, while the Argive land that was formerly barren owing to the water has now become fruitful. Now the same process that has taken place in this small district must be supposed to be going on over whole countries and on a large scale.

Men whose outlook is narrow suppose the cause of such events to be change in the universe, in the sense of a coming to be of the world as a whole. Hence they say that the sea being dried up and is growing less, because this is observed to have happened in more places now than formerly. But this is only partially true. It is true that many places are now dry, that formerly were covered with water. But the opposite is true too: for if they look they will find that there are many places where the sea has invaded the land. But we must not suppose that the cause of this is that the world is in process of becoming. For it is absurd to make the universe to be in process because of small and trifling changes, when the bulk and size of the earth are surely as nothing in comparison with the whole world....

Book IV

Part 1

We have explained that the qualities that constitute the elements are four, and that their combinations determine the number of the elements to be four.

Two of the qualities, the hot and the cold, are active; two, the dry and the moist, passive. We can satisfy ourselves of this by looking at instances. In every case heat and cold determine, conjoin, and change things of the same kind and things of different kinds, moistening, drying, hardening, and softening them. Things dry and moist, on the other hand, both in isolation and when present together in the same body are the subjects of that determination and of the other affections enumerated. The account we give of the qualities when we define their character shows this too. Hot and cold we describe as active, for 'congregating' is essentially a species of 'being active': moist and dry are passive, for it is in virtue of its being acted upon in a certain way that a thing is said to be 'easy to determine' or 'difficult to determine'. So it is clear that some of the qualities are active and some passive.

. . .

Part 2

We must now describe the next kinds of processes which the qualities already mentioned set up in actually existing natural objects as matter.

Of these concoction is due to heat; its species are ripening, boiling, broiling. Inconcoction is due to cold and its species are rawness, imperfect boiling, imperfect broiling. (We must recognize that the things are not properly denoted by these words: the various classes of similar objects have no names universally applicable to them; consequently we must think of the species enumerated as being not what those words denote but something like it.) Let us say what each of them is. Concoction is a process in which the natural and proper heat of an object perfects the corresponding passive qualities, which are the proper matter of any given object. For when concoction has taken place we say that a thing has been perfected and has come to be itself. It is the proper heat of a thing that sets up this perfecting, though external influences may contribute in some degrees to its fulfilment. Baths, for instance, and other things of the kind contribute to the concoction of food, but the primary sources is the proper heat of the body. . . .

Concoction ensues whenever the matter, the moisture, is mastered. For the matter is what is what is determined by the natural heat in the object, and as long as the ratio between them exists in it a thing maintains its nature. Hence things like the liquid and solid excreta and waste-stuffs in general are signs of health, and concoction is said to have taken place in them; for they show that the proper heat has mastered the indeterminate matter.

Things that undergo a process of concoction necessarily become thicker and hotter; for the action of heat is to make things more compact, thicker, and drier. . . .

Translated by E. W. Webster

Reading and Discussion Questions

1. What are the five basic elements that Aristotle discusses in this reading? What are the properties of each of these elements, and what is their natural direction of motion?

2. How does Aristotle use the properties of these elements to explain meteors, lightning, the falling of heavy objects, and the circular motion of celestial objects?
3. Describe Aristotle's account of the water cycle, including evaporation and condensation, and his argument for why water cannot be merely flowing out of the earth.

Physics (c. 350 BCE)

Aristotle

10

Book II

Part 1

Of things that exist, some exist by nature, some from other causes.

'By nature' the animals and their parts exist, and the plants and the simple bodies (earth, fire, air, water)—for we say that these and the like exist 'by nature'.

All the things mentioned present a feature in which they differ from things which are not constituted by nature. Each of them has within itself a principle of motion and of stationariness (in respect of place, or of growth and decrease, or by way of alteration). On the other hand, a bed and a coat and anything else of that sort, qua receiving these designations i.e. in so far as they are products of art—have no innate impulse to change. But in so far as they happen to be composed of stone or of earth or of a mixture of the two, they do have such an impulse, and just to that extent which seems to indicate that nature is a source or cause of being moved and of being at rest in that to which it belongs primarily, in virtue of itself and not in virtue of a concomitant attribute.

I say 'not in virtue of a concomitant attribute', because (for instance) a man who is a doctor might cure himself. Nevertheless it is not in so far as he is a patient that he possesses the art of medicine: it merely has happened that the same man is doctor and patient—and that is why these attributes are not always found together. So it is with all other artificial products. None of them has in itself the source of its own production. But while in some cases (for instance houses and the other products of manual labour) that principle is in something else external to the thing, in others those which may cause a change in themselves in virtue of a concomitant attribute—it lies in the things themselves (but not in virtue of what they are).

'Nature' then is what has been stated. Things 'have a nature' which have a principle of this kind. Each of them is a substance; for it is a subject, and nature always implies a subject in which it inheres.

The term 'according to nature' is applied to all these things and also to the attributes which belong to them in virtue of what they are, for instance the property of fire to be carried upwards—which is not a 'nature' nor 'has a nature' but is 'by nature' or 'according to nature'.

What nature is, then, and the meaning of the terms 'by nature' and 'according to nature', has been stated. That nature exists, it would be absurd to try to prove; for it is obvious that there are many things of this kind, and to prove what is obvious by what is not is the mark of a man who is unable to distinguish what is self-evident from what is not. (This state of mind is clearly possible. A man blind from birth might reason about colours. Presumably therefore such persons must be talking about words without any thought to correspond.)

Some identify the nature or substance of a natural object with that immediate constituent of it which taken by itself is without arrangement, e.g. the wood is the 'nature' of the bed, and the bronze the 'nature' of the statue.

As an indication of this Antiphon points out that if you planted a bed and the rotting wood acquired the power of sending up a shoot, it would not be a bed that would come up, but wood—which shows that the arrangement in accordance with the rules of the art is merely an incidental attribute, whereas the real nature is the other, which, further, persists continuously through the process of making.

But if the material of each of these objects has itself the same relation to something else, say bronze (or gold) to water, bones (or wood) to earth and so on, that (they say) would be their nature and essence.

Consequently some assert earth, others fire or air or water or some or all of these, to be the nature of the things that are. For whatever any one of them supposed to have this character—whether one thing or more than one thing—this or these he declared to be the whole of substance, all else being its affections, states, or dispositions. Every such thing they held to be eternal (for it could not pass into anything else), but other things to come into being and cease to be times without number.

This then is one account of 'nature', namely that it is the immediate material substratum of things which have in themselves a principle of motion or change.

Another account is that 'nature' is the shape or form which is specified in the definition of the thing.

For the word 'nature' is applied to what is according to nature and the natural in the same way as 'art' is applied to what is artistic or a work of art. We should not say in the latter case that there is anything artistic about a thing, if it is a bed only potentially, not yet having the form of a bed; nor should we call it a work of art. The same is true of natural compounds. What is potentially flesh or bone has not yet its own 'nature', and does not exist until it receives the form specified in the definition, which we name in defining what flesh or bone is. Thus in the second sense of 'nature' it would be the shape or form (not separable except in statement) of things which have in themselves a source of motion. (The combination of the two, e.g. man, is not 'nature' but 'by nature' or 'natural'.)

The form indeed is 'nature' rather than the matter; for a thing is more properly said to be what it is when it has attained to fulfilment than when it exists potentially. Again man is born from man, but not bed from bed. That is why people say that the figure is not the nature of a bed, but the wood is—if the bed sprouted not a bed but wood would come up. But even if the figure is art, then on the same principle the shape of man is his nature. For man is born from man.

We also speak of a thing's nature as being exhibited in the process of growth by which its nature is attained. The 'nature' in this sense is not like 'doctoring', which leads not to the art of doctoring but to health. Doctoring must start from the art, not lead to it. But it is not in this way that nature (in the one sense) is related to nature (in the other). What grows qua growing grows from something into something. Into what then does it grow? Not into that from which it arose but into that to which it tends. The shape then is nature.

'Shape' and 'nature', it should be added, are in two senses. For the privation too is in a way form. But whether in unqualified coming to be there is privation, i.e. a contrary to what comes to be, we must consider later.

Part 2

We have distinguished, then, the different ways in which the term 'nature' is used.

The next point to consider is how the mathematician differs from the physicist. Obviously physical bodies contain surfaces and volumes, lines and points, and these are the subject-matter of mathematics.

Further, is astronomy different from physics or a department of it? It seems absurd that the physicist should be supposed to know the nature of sun or moon, but not to know any of their essential attributes, particularly as the writers on physics obviously do discuss their shape also and whether the earth and the world are spherical or not.

Now the mathematician, though he too treats of these things, nevertheless does not treat of them as the limits of a physical body; nor does he consider the attributes indicated as the attributes of such bodies. That is why he separates them; for in thought they are separable from motion, and it makes no difference, nor does any falsity result, if they are separated. The holders of the theory of Forms do the same, though they are not aware of it; for they separate the objects of physics, which are less separable than those of mathematics. This becomes plain if one tries to state in each of the two cases the definitions of the things and of their attributes. 'Odd' and 'even', 'straight' and 'curved', and likewise 'number', 'line', and 'figure', do not involve motion; not so 'flesh' and 'bone' and 'man'—these are defined like 'snub nose', not like 'curved'.

Similar evidence is supplied by the more physical of the branches of mathematics, such as optics, harmonics, and astronomy. These are in a way the converse of geometry. While geometry investigates physical lines but not qua physical, optics investigates mathematical lines, but qua physical, not qua mathematical.

Since 'nature' has two senses, the form and the matter, we must investigate its objects as we would the essence of snubness. That is, such things are neither independent of matter nor can be defined in terms of matter only. Here too indeed one might raise a difficulty. Since there are two natures, with which is the physicist concerned? Or should he investigate the combination of the two? But if the combination of the two, then also each severally. Does it belong then to the same or to different sciences to know each severally?

If we look at the ancients, physics would to be concerned with the matter. (It was only very slightly that Empedocles and Democritus touched on the forms and the essence.)

But if on the other hand art imitates nature, and it is the part of the same discipline to know the form and the matter up to a point (e.g. the doctor has a knowledge of health and also of bile and phlegm, in which health is realized, and the builder both of the form of the house and of the matter, namely that it is bricks and beams, and so forth): if this is so, it would be the part of physics also to know nature in both its senses.

Again, 'that for the sake of which', or the end, belongs to the same department of knowledge as the means. But the nature is the end or 'that for the sake of which'. For if a thing undergoes a continuous change and there is a stage which is last, this stage is the end or 'that for the sake of which'. (That is why the poet was carried away into making an absurd statement when he said 'he has the end for the sake of which he was born'. For not every stage that is last claims to be an end, but only that which is best.)

For the arts make their material (some simply 'make' it, others make it serviceable), and we use everything as if it was there for our sake. (We also are in a sense an end. 'That for the sake of which' has two senses: the distinction is made in our work On Philosophy.) The arts, therefore, which govern the matter and have knowledge are two, namely the art which uses the product and the art which directs the production of it. That is why the using art also is in a sense directive; but it differs in that it knows the form, whereas the art which is directive as being concerned with production knows the matter. For the helmsman knows and prescribes what sort of form a helm should have, the other from what wood it should be made and by means of what operations. In the products of art, however, we make the material with a view to the function, whereas in the products of nature the matter is there all along.

Again, matter is a relative term: to each form there corresponds a special matter. How far then must the physicist know the form or essence? Up to a point, perhaps, as the doctor must know sinew or the smith bronze (i.e. until he understands the purpose of each): and the physicist is concerned only with things whose forms are separable indeed, but do not exist apart from matter. Man is begotten by man and by the sun as well. The mode of existence and essence of the separable it is the business of the primary type of philosophy to define.

Part 3

Now that we have established these distinctions, we must proceed to consider causes, their character and number. Knowledge is the object of our inquiry, and men do not think they know a thing till they have grasped the 'why' of (which is to grasp its primary cause). So clearly we too must do this as regards both coming to be and passing away and every kind of physical change, in order that, knowing their principles, we may try to refer to these principles each of our problems.

In one sense, then, (1) that out of which a thing comes to be and which persists, is called 'cause', e.g. the bronze of the statue, the silver of the bowl, and the genera of which the bronze and the silver are species.

In another sense (2) the form or the archetype, i.e. the statement of the essence, and its genera, are called 'causes' (e.g. of the octave the relation of 2:1, and generally number), and the parts in the definition.

Again (3) the primary source of the change or coming to rest; e.g. the man who gave advice is a cause, the father is cause of the child, and generally what makes of what is made and what causes change of what is changed.

Again (4) in the sense of end or 'that for the sake of which' a thing is done, e.g. health is the cause of walking about. ('Why is he walking about?' we say. 'To be healthy', and, having said that, we think we have assigned the cause.) The same is true also of all the intermediate steps which are brought about through the action of something else as means towards the end, e.g. reduction of flesh, purging, drugs, or surgical instruments are means towards health. All these things are 'for the sake of' the end, though they differ from one another in that some are activities, others instruments.

This then perhaps exhausts the number of ways in which the term 'cause' is used.

As the word has several senses, it follows that there are several causes of the same thing not merely in virtue of a concomitant attribute), e.g. both the art of the sculptor and the bronze are causes of the statue. These are causes of the statue qua statue, not in virtue of anything else that it may be—only not in the same way, the one being the material cause, the other the cause whence the motion comes. Some things cause each other reciprocally, e.g. hard work causes fitness and vice versa, but again not in the same way, but the one as end, the other as the origin of change. Further the same thing is the cause of contrary results. For that which by its presence brings about one result is sometimes blamed for bringing about the contrary by its absence. Thus we ascribe the wreck of a ship to the absence of the pilot whose presence was the cause of its safety.

All the causes now mentioned fall into four familiar divisions. The letters are the causes of syllables, the material of artificial products, fire, &c., of bodies, the parts of the whole, and the premises of the conclusion, in the sense of 'that from which'. Of these pairs the one set are causes in the sense of substratum, e.g. the parts, the other set in the sense of essence—the whole and the combination and the form. But the seed and the doctor and the adviser, and generally the maker, are all sources whence the change or stationariness originates, while the others are causes in the sense of the end or the good of the rest; for 'that for the sake of which' means what is best and the end of the things that lead up to it. (Whether we say the 'good itself' or the 'apparent good' makes no difference.)

Such then is the number and nature of the kinds of cause.

Now the modes of causation are many, though when brought under heads they too can be reduced in number. For 'cause' is used in many senses and even within the same kind one may be prior to another (e.g. the doctor and the expert are causes of health, the relation 2:1 and number of the octave), and always what is inclusive to what is particular. Another mode of causation is the incidental and its genera, e.g. in one way 'Polyclitus', in another 'sculptor' is the cause of a statue, because 'being Polyclitus' and 'sculptor' are incidentally conjoined. Also the classes in which the incidental attribute is included; thus 'a man' could be said to be the cause of a statue or, generally, 'a living creature'. An incidental attribute too may be more or less remote, e.g. suppose that 'a pale man' or 'a musical man' were said to be the cause of the statue.

All causes, both proper and incidental, may be spoken of either as potential or as actual; e.g. the cause of a house being built is either 'house-builder' or 'house-builder building'.

Similar distinctions can be made in the things of which the causes are causes, e.g. of 'this statue' or of 'statue' or of 'image' generally, of 'this bronze' or of 'bronze' or of 'material' generally. So too with the incidental attributes. Again we may use a complex expression for either and say, e.g. neither 'Polyclitus' nor 'sculptor' but 'Polyclitus, sculptor'.

All these various uses, however, come to six in number, under each of which again the usage is twofold. Cause means either what is particular or a genus, or an incidental attribute or a genus of that, and these either as a complex or each by itself; and all six either as actual or as potential. The difference is this much, that causes which are actually at work and particular exist and cease to exist simultaneously with their effect, e.g. this healing person with this being-healed person and that house-building man with that being-built house; but this is not always true of potential causes—the house and the housebuilder do not pass away simultaneously.

In investigating the cause of each thing it is always necessary to seek what is most precise (as also in other things): thus man builds because he is a builder, and a builder builds in virtue of his art of building. This last cause then is prior: and so generally.

Further, generic effects should be assigned to generic causes, particular effects to particular causes, e.g. statue to sculptor, this statue to this sculptor; and powers are relative to possible effects, actually operating causes to things which are actually being effected.

This must suffice for our account of the number of causes and the modes of causation.

Part 7

It is clear then that there are causes, and that the number of them is what we have stated. The number is the same as that of the things comprehended under the question 'why'. The 'why' is referred ultimately either (1), in things which do not involve motion, e.g. in mathematics, to the 'what' (to the definition of 'straight line' or 'commensurable', &c.), or (2) to what initiated a motion, e.g. 'why did they go to war?—because there had been a raid'; or (3) we are inquiring 'for the sake of what?—that they may rule'; or (4), in the case of things that come into being, we are looking for the matter. The causes, therefore, are these and so many in number.

Now, the causes being four, it is the business of the physicist to know about them all, and if he refers his problems back to all of them, he will assign the 'why' in the way proper to his science—the matter, the form, the mover, 'that for the sake of which'. The last three often coincide; for the 'what' and 'that for the sake of which' are one, while the primary source of motion is the same in species as these (for man generates man), and so too, in general, are all things which cause movement by being themselves moved; and such as are not of this kind are no longer inside the province of physics, for they cause motion not by possessing motion or a source of motion in themselves, but being themselves incapable of motion. Hence there are three branches of study, one of things which are incapable of motion, the second of things in motion, but indestructible, the third of destructible things.

The question 'why', then, is answered by reference to the matter, to the form, and to the primary moving cause. For in respect of coming to be it is mostly in this last way that causes are investigated—what comes to be after what? what was the primary agent or patient?' and so at each step of the series.

Now the principles which cause motion in a physical way are two, of which one is not physical, as it has no principle of motion in itself. Of this kind is whatever causes movement, not being itself moved, such as (1) that which is completely unchangeable, the primary reality, and (2) the essence of that which is coming to be, i.e. the form; for this is the end or 'that for the sake of which'. Hence since nature is for the sake of something, we must know this cause also. We must explain the 'why' in all the senses of the term, namely, (1) that from this that will necessarily result ('from this' either without qualification or in most cases); (2) that 'this must be so if that is to be so' (as the conclusion presupposes the premises); (3) that this was the essence of the thing; and (4) because it is better thus (not without qualification, but with reference to the essential nature in each case).

Book IV

Part 6

The investigation of similar questions about the void, also, must be held to belong to the physicist—namely whether it exists or not, and how it exists or what it is—just as about place. The views taken of it involve arguments both for and against, in much the same sort of way. For those who hold that the void exists regard it as a sort of place or vessel which is supposed to be 'full' when it holds the bulk which it is capable of containing, 'void' when it is deprived of that—as if 'void' and 'full' and 'place' denoted the same thing, though the essence of the three is different.

We must begin the inquiry by putting down the account given by those who say that it exists, then the account of those who say that it does not exist, and third the current view on these questions.

Those who try to show that the void does not exist do not disprove what people really mean by it, but only their erroneous way of speaking; this is true of Anaxagoras and of those who refute the existence of the void in this way. They merely give an ingenious demonstration that air is something—by straining wine-skins and showing the resistance of the air, and by cutting it off in clepsydras. But people really mean that there is an empty interval in which there is no sensible body. They hold that everything which is in body is body and say that what has nothing in it at all is void (so what is full of air is void). It is not then the existence of air that needs to be proved, but the non-existence of an interval, different from the bodies, either separable or actual—an interval which divides the whole body so as to break its continuity, as Democritus and Leucippus hold, and many other physicists—or even perhaps as something which is outside the whole body, which remains continuous.

These people, then, have not reached even the threshold of the problem, but rather those who say that the void exists.

(1) They argue, for one thing, that change in place (i.e. locomotion and increase) would not be. For it is maintained that motion would seem not to exist, if there were no void, since what is full cannot contain anything more. If it could, and there were two bodies in the same place, it would also be true that any number of bodies could be together; for it is impossible to draw a line of division

beyond which the statement would become untrue. If this were possible, it would follow also that the smallest body would contain the greatest; for 'many a little makes a mickle': thus if many equal bodies can be together, so also can many unequal bodies.

Melissus, indeed, infers from these considerations that the All is immovable; for if it were moved there must, he says, be void, but void is not among the things that exist.

This argument, then, is one way in which they show that there is a void.

(2) They reason from the fact that some things are observed to contract and be compressed, as people say that a cask will hold the wine which formerly filled it, along with the skins into which the wine has been decanted, which implies that the compressed body contracts into the voids present in it.

Again (3) increase, too, is thought to take always by means of void, for nutriment is body, and it is impossible for two bodies to be together. A proof of this they find also in what happens to ashes, which absorb as much water as the empty vessel.

The Pythagoreans, too, (4) held that void exists and that it enters the heaven itself, which as it were inhales it, from the infinite air. Further it is the void which distinguishes the natures of things, as if it were like what separates and distinguishes the terms of a series. This holds primarily in the numbers, for the void distinguishes their nature.

These, then, and so many, are the main grounds on which people have argued for and against the existence of the void.

Part 8

Let us explain again that there is no void existing separately, as some maintain. If each of the simple bodies has a natural locomotion, e.g. fire upward and earth downward and towards the middle of the universe, it is clear that it cannot be the void that is the condition of locomotion. What, then, will the void be the condition of? It is thought to be the condition of movement in respect of place, and it is not the condition of this.

Again, if void is a sort of place deprived of body, when there is a void where will a body placed in it move to? It certainly cannot move into the whole of the void. The same argument applies as against those who think that place is something separate, into which things are carried; viz. how will what is placed in it move, or rest? Much the same argument will apply to the void as to the 'up' and 'down' in place, as is natural enough since those who maintain the existence of the void make it a place.

And in what way will things be present either in place—or in the void? For the expected result does not take place when a body is placed as a whole in a place conceived of as separate and permanent; for a part of it, unless it be placed apart, will not be in a place but in the whole. Further, if separate place does not exist, neither will void.

If people say that the void must exist, as being necessary if there is to be movement, what rather turns out to be the case, if one the matter, is the opposite, that not a single thing can be moved if there is a void; for as with those who for a like reason say the earth is at rest, so, too, in the void things must be at rest; for there is no place to which things can move more or less than to another; since the void in so far as it is void admits no difference.

The second reason is this: all movement is either compulsory or according to nature, and if there is compulsory movement there must also be natural (for compulsory movement is contrary to nature, and movement contrary to nature is posterior to that according to nature, so that if each of the natural bodies has not a natural movement, none of the other movements can exist); but how can there be natural movement if there is no difference throughout the void or the infinite? For in so far as it is infinite, there will be no up or down or middle, and in so far as it is a void, up differs no whit from down; for as there is no difference in what is nothing, there is none in the void (for the void seems to be a non-existent and a privation of being), but natural locomotion seems to be differentiated, so that the things that exist by nature must be differentiated. Either, then, nothing has a natural locomotion, or else there is no void.

Further, in point of fact things that are thrown move though that which gave them their impulse is not touching them, either by reason of mutual replacement, as some maintain, or because the air that has been pushed pushes them with a movement quicker than the natural locomotion of the projectile wherewith it moves to its proper place. But in a void none of these things can take place, nor can anything be moved save as that which is carried is moved.

Further, no one could say why a thing once set in motion should stop anywhere; for why should it stop here rather than here? So that a thing will either be at rest or must be moved ad infinitum, unless something more powerful get in its way.

Further, things are now thought to move into the void because it yields; but in a void this quality is present equally everywhere, so that things should move in all directions.

Further, the truth of what we assert is plain from the following considerations. We see the same weight or body moving faster than another for two reasons, either because there is a difference in what it moves through, as between water, air, and earth, or because, other things being equal, the moving body differs from the other owing to excess of weight or of lightness.

Now the medium causes a difference because it impedes the moving thing, most of all if it is moving in the opposite direction, but in a secondary degree even if it is at rest; and especially a medium that is not easily divided, i.e. a medium that is somewhat dense. A, then, will move through B in time G, and through D, which is thinner, in time E (if the length of B is equal to D), in proportion to the density of the hindering body. For let B be water and D air;

then by so much as air is thinner and more incorporeal than water, A will move through D faster than through B. Let the speed have the same ratio to the speed, then, that air has to water. Then if air is twice as thin, the body will traverse B in twice the time that it does D, and the time G will be twice the time E. And always, by so much as the medium is more incorporeal and less resistant and more easily divided, the faster will be the movement.

Now there is no ratio in which the void is exceeded by body, as there is no ratio of 0 to a number. For if 4 exceeds 3 by 1, and 2 by more than 1, and 1 by still more than it exceeds 2, still there is no ratio by which it exceeds 0; for that which exceeds must be divisible into the excess + that which is exceeded, so that will be what it exceeds 0 by + 0. For this reason, too, a line does not exceed a point unless it is composed of points! Similarly the void can bear no ratio to the full, and therefore neither can movement through the one to movement through the other, but if a thing moves through the thickest medium such and such a distance in such and such a time, it moves through the void with a speed beyond any ratio. For let Z be void, equal in magnitude to B and to D. Then if A is to traverse and move through it in a certain time, H, a time less than E, however, the void will bear this ratio to the full. But in a time equal to H, A will traverse the part O of A. And it will surely also traverse in that time any substance Z which exceeds air in thickness in the ratio which the time E bears to the time H. For if the body Z be as much thinner than D as E exceeds H, A, if it moves through Z, will traverse it in a time inverse to the speed of the movement, i.e. in a time equal to H. If, then, there is no body in Z, A will traverse Z still more quickly. But we supposed that its traverse of Z when Z was void occupied the time H. So that it will traverse Z in an equal time whether Z be full or void. But this is impossible. It is plain, then, that if there is a time in which it will move through any part of the void, this impossible result will follow: it will be found to traverse a certain distance, whether this be full or void, in an equal time; for there will be some body which is in the same ratio to the other body as the time is to the time.

To sum the matter up, the cause of this result is obvious, viz. that between any two movements there is a ratio (for they occupy time, and there is a ratio between any two times, so long as both are finite), but there is no ratio of void to full.

These are the consequences that result from a difference in the media; the following depend upon an excess of one moving body over another. We see that bodies which have a greater impulse either of weight or of lightness, if they are alike in other respects, move faster over an equal space, and in the ratio which their magnitudes bear to each other. Therefore they will also move through the void with this ratio of speed. But that is impossible; for why should one move faster? (In moving through plena it must be so; for the greater divides them faster by its force. For a moving thing cleaves the medium either by its shape, or by the impulse which the body that is carried along or is projected possesses.) Therefore all will possess equal velocity. But this is impossible.

It is evident from what has been said, then, that, if there is a void, a result follows which is the very opposite of the reason for which those who believe in a void set it up. They think that if movement in respect of place is to exist, the void cannot exist, separated all by itself; but this is the same as to say that place is a separate cavity; and this has already been stated to be impossible.

But even if we consider it on its own merits the so-called vacuum will be found to be really vacuous. For as, if one puts a cube in water, an amount of water equal to the cube will be displaced; so too in air; but the effect is imperceptible to sense. And indeed always in the case of any body that can be displaced, must, if it is not compressed, be displaced in the direction in which it is its nature to be displaced—always either down, if its locomotion is downwards as in the case of earth, or up, if it is fire, or in both directions—whatever be the nature of the inserted body. Now in the void this is impossible; for it is not body; the void must have penetrated the cube to a distance equal to that which this portion of void formerly occupied in the void, just as if the water or air had not been displaced by the wooden cube, but had penetrated right through it.

But the cube also has a magnitude equal to that occupied by the void; a magnitude which, if it is also hot or cold, or heavy or light, is none the less different in essence from all its attributes, even if it is not separable from them; I mean the volume of the wooden cube. So that even if it were separated from everything else and were neither heavy nor light, it will occupy an equal amount of void, and fill the same place, as the part of place or of the void equal to itself. How then will the body of the cube differ from the void or place that is equal to it? And if there can be two such things, why cannot there be any number coinciding?

This, then, is one absurd and impossible implication of the theory. It is also evident that the cube will have this same volume even if it is displaced, which is an attribute possessed by all other bodies also. Therefore if this differs in no respect from its place, why need we assume a place for bodies over and above the volume of each, if their volume be conceived of as free from attributes? It contributes nothing to the situation if there is an equal interval attached to it as well. [Further it ought to be clear by the study of moving things what sort of thing void is. But in fact it is found nowhere in the world. For air is something, though it does not seem to be so—nor, for that matter, would water, if fishes were made of iron; for the discrimination of the tangible is by touch.]

It is clear, then, from these considerations that there is no separate void.

Translated by R. P. Hardie and R. K. Gaye

Reading and Discussion Questions

1. What are Aristotle's four causes of change? What role do these play in understanding observed changes in the natural world? How do the four causes relate to his concepts of matter, form, and telos or end, and ideas about natural place? Try using Aristotle's four causes to explain the motion of Mars, the flight

of an arrow, the growth of a flower from seed, why a hot air balloon rises, and why a sheep gains weight when it eats grass.

2. Notice that Aristotle's use of "motion" is not limited to changes in place (local motion). What other kinds of change does his use of the term include?
3. Explain Aristotle's argument that the existence of a void is impossible. What role does his claim that the speed of a falling body is directly proportional to its weight and inversely proportional to the density (resistance) of the medium through which it falls play in this line of argument?

On the Soul (c. 350 BCE)

Aristotle

Book II

Part 1

Let the foregoing suffice as our account of the views concerning the soul which have been handed on by our predecessors; let us now dismiss them and make as it were a completely fresh start, endeavouring to give a precise answer to the question, 'what is soul?' i.e. to formulate the most general possible definition of it.

We are in the habit of recognizing, as one determinate kind of what is, substance, and that in several senses, (a) in the sense of matter or that which in itself is not 'a this', and (b) in the sense of form or essence, which is that precisely in virtue of which a thing is called 'a this', and thirdly (c) in the sense of that which is compounded of both (a) and (b). Now matter is potentiality, form actuality; of the latter there are two grades related to one another as e.g. knowledge to the exercise of knowledge.

Among substances are by general consent reckoned bodies and especially natural bodies; for they are the principles of all other bodies. Of natural bodies some have life in them, others not; by life we mean self-nutrition and growth (with its correlative decay). It follows that every natural body which has life in it is a substancc in the sense of a composite.

But since it is also a body of such and such a kind, viz. having life, the body cannot be soul; the body is the subject or matter, not what is attributed to it. Hence the soul must be a substance in the sense of the form of a natural body having life potentially within it. But substance is actuality, and thus soul is the actuality of a body as above characterized. Now the word actuality has two senses corresponding respectively to the possession of knowledge and the actual exercise of knowledge. It is obvious that the soul is actuality in the first sense, viz. that of knowledge as possessed, for both sleeping and waking presuppose the existence of soul, and of these waking corresponds to actual knowing, sleeping to knowledge possessed but not employed, and, in the history of the individual, knowledge comes before its employment or exercise.

That is why the soul is the first grade of actuality of a natural body having life potentially in it. The body so described is a body which is organized. The parts of plants in spite of their extreme simplicity are 'organs'; e.g. the leaf serves to shelter the pericarp, the pericarp to shelter the fruit, while the roots of plants are analogous to the mouth of animals, both serving for the absorption of food. If, then, we have to give a general formula applicable to all kinds of soul, we must describe it as the first grade of actuality of a natural organized body. That is why we can wholly dismiss as unnecessary the question whether the soul and the body are one: it is as meaningless as to ask whether the wax and the shape given to it by the stamp are one, or generally the matter of a thing and that of which it is the matter. Unity has many senses (as many as 'is' has), but the most proper and fundamental sense of both is the relation of an actuality to that of which it is the actuality. We have now given an answer to the question, 'what is soul?'—an answer which applies to it in its full extent. It is substance in the sense which corresponds to the definitive formula of a thing's essence. That means that it is 'the essential whatness' of a body of the character just assigned. Suppose that what is literally an 'organ', like an axe, were a natural body, its 'essential whatness', would have been its essence, and so its soul; if this disappeared from it, it would have ceased to be an axe, except in name. As it is, it is just an axe; it wants the character which is required to make its whatness or formulable essence a soul; for that, it would have had to be a natural body of a particular kind, viz. one having in itself the power of setting itself in movement and arresting itself. Next, apply this doctrine in the case of the 'parts' of the living body. Suppose that the eye were an animal—sight would have been its soul, for sight is the substance or essence of the eye which corresponds to the formula, the eye being merely the matter of seeing; when seeing is removed the eye is no longer an eye, except in name—it is no more a real eye than the eye of a statue or of a painted figure. We must now extend our consideration from the 'parts' to the whole living body; for what the departmental sense is to the bodily part which is its organ, that the whole faculty of sense is to the whole sensitive body as such.

We must not understand by that which is 'potentially capable of living' what has lost the soul it had, but only what still retains it; but seeds and fruits are bodies which possess the qualification. Consequently, while waking is actuality in a sense corresponding to the cutting and the seeing, the soul is actuality in the sense corresponding to the power of sight and the power in the tool; the body corresponds to what exists in potentiality; as the pupil plus the power of sight constitutes the eye, so the soul plus the body constitutes the animal.

From this it indubitably follows that the soul is inseparable from its body, or at any rate that certain parts of it are (if it has parts) for the actuality of some of them is nothing but the actualities of their bodily parts. Yet some may be separable because they are not the actualities of any body at all. Further, we have no light on the problem whether the soul may not be the actuality of its body in the sense in which the sailor is the actuality of the ship.

This must suffice as our sketch or outline determination of the nature of soul.

Part 2

Since what is clear or logically more evident emerges from what in itself is confused but more observable by us, we must reconsider our results from this point of view. For it is not enough for a definitive formula to express as most now do the mere fact; it must include and exhibit the ground also. At present definitions are given in a form analogous to the conclusion of a syllogism; e.g. 'what is squaring?' The construction of an equilateral rectangle equal to a given oblong rectangle. Such a definition is in form equivalent to a conclusion. One that tells us that squaring is the discovery of a line which is a mean proportional between the two unequal sides of the given rectangle discloses the ground of what is defined.

We resume our inquiry from a fresh starting-point by calling attention to the fact that what has soul in it differs from what has not, in that the former displays life. Now this word has more than one sense, and provided any one alone of these is found in a thing we say that thing is living. Living, that is, may mean thinking or perception or local movement and rest, or movement in the sense of nutrition, decay and growth. Hence we think of plants also as living, for they are observed to possess in themselves an originative power through which they increase or decrease in all spatial directions; they grow up and down, and everything that grows increases its bulk alike in both directions or indeed in all, and continues to live so long as it can absorb nutriment.

This power of self-nutrition can be isolated from the other powers mentioned, but not they from it—in mortal beings at least. The fact is obvious in plants; for it is the only psychic power they possess.

This is the originative power the possession of which leads us to speak of things as living at all, but it is the possession of sensation that leads us for the first time to speak of living things as animals; for even those beings which possess no power of local movement but do possess the power of sensation we call animals and not merely living things.

The primary form of sense is touch, which belongs to all animals, just as the power of self-nutrition can be isolated from touch and sensation generally, so touch can be isolated from all other forms of sense. (By the power of self-nutrition we mean that departmental power of the soul which is common to plants and animals: all animals whatsoever are observed to have the sense of touch.) What the explanation of these two facts is, we must discuss later. At present we must confine ourselves to saying that soul is the source of these phenomena and is characterized by them, viz. by the powers of self-nutrition, sensation, thinking, and motivity.

Is each of these a soul or a part of a soul? And if a part, a part in what sense? A part merely distinguishable by definition or a part distinct in local situation as well? In the case of certain of these powers, the answers to these questions are easy, in the case of others we are puzzled what to say just as in the case of plants which when divided are observed to continue to live though removed to a distance from one another (thus showing that in their case the soul of each individual plant before division was actually one, potentially many), so we notice a similar result in other varieties of soul, i.e. in insects which have been cut in two; each of the segments possesses both sensation and local movement; and if sensation, necessarily also imagination and appetition; for, where there is sensation, there is also pleasure and pain, and, where these, necessarily also desire.

We have no evidence as yet about mind or the power to think; it seems to be a widely different kind of soul, differing as what is eternal from what is perishable; it alone is capable of existence in isolation from all other psychic powers. All the other parts of soul, it is evident from what we have said, are, in spite of certain statements to the contrary, incapable of separate existence though, of course, distinguishable by definition. If opining is distinct from perceiving, to be capable of opining and to be capable of perceiving must be distinct, and so with all the other forms of living above enumerated. Further, some animals possess all these parts of soul, some certain of them only, others one only (this is what enables us to classify animals); the cause must be considered later.' A similar arrangement is found also within the field of the senses; some classes of animals have all the senses, some only certain of them, others only one, the most indispensable, touch.

Since the expression 'that whereby we live and perceive' has two meanings, just like the expression 'that whereby we know'—that may mean either (a) knowledge or (b) the soul, for we can speak of knowing by or with either, and similarly that whereby we are in health may be either (a) health or (b) the body or some part of the body; and since of the two terms thus contrasted knowledge or health is the name of a form, essence, or ratio, or if we so express it an actuality of a recipient matter—knowledge of what is capable of knowing, health of what is capable of being made healthy (for the operation of that which is capable of originating change terminates and has its seat in what is changed or altered); further, since it is the soul by or with which primarily we live, perceive, and think—it follows that the soul must be a ratio or formulable essence, not a matter or subject. For, as we said, word substance has three meanings form, matter, and the complex of both and of these three what is called matter is potentiality, what is called form actuality. Since then the complex here is the living thing, the body cannot be the actuality of the soul; it is the soul which is the actuality of a certain kind of body. Hence the rightness of the view that the soul cannot be without a body, while it cannot be a body; it is not a body but something relative to a body. That is why it is in a body, and a body of a definite kind. It was a mistake, therefore, to do as former thinkers did, merely to fit it into a body without adding a definite specification of the kind or character of that body. Reflection confirms the observed fact; the actuality of any given thing can only be realized in what is already potentially that thing, i.e. in a matter of its own appropriate to it. From all this it follows that soul is an actuality or formulable essence of something that possesses a potentiality of being besouled.

Part 3

Of the psychic powers above enumerated some kinds of living things, as we have said, possess all, some less than all, others one only. Those we have mentioned are the nutritive, the appetitive, the sensory, the locomotive, and the power of thinking. Plants have none but the first, the nutritive, while another order of living things has this plus the sensory. If any order of living things has the sensory, it must also have the appetitive; for appetite is the genus of which desire, passion, and wish are the species; now all animals have one sense at least, viz. touch, and whatever has a sense has the capacity for pleasure and pain and therefore has pleasant and painful objects present to it, and wherever these are present, there is desire, for desire is just appetition of what is pleasant. Further, all animals have the sense for food (for touch is the sense for food); the food of all living things consists of what is dry, moist, hot, cold, and these are the qualities apprehended by touch; all other sensible qualities are apprehended by touch only indirectly. Sounds, colours, and odours contribute nothing to nutriment; flavours fall within the field of tangible qualities. Hunger and thirst are forms of desire, hunger a desire for what is dry and hot, thirst a desire for what is cold and moist; flavour is a sort of seasoning added to both. We must later clear up these points, but at present it may be enough to say that all animals that possess the sense of touch have also appetition.

The case of imagination is obscure; we must examine it later. Certain kinds of animals possess in addition the power of locomotion, and still another order of animate beings, i.e. man and possibly another order like man or superior to him, the power of thinking, i.e. mind. It is now evident that a single definition can be given of soul only in the same sense as one can be given of figure. For, as in that case there is no figure distinguishable and apart from triangle, &c., so here there is no soul apart from the forms of soul just enumerated. It is true that a highly general definition can be given for figure which will fit all figures without expressing the peculiar nature of any figure. So here in the case of soul and its specific forms.

Hence it is absurd in this and similar cases to demand an absolutely general definition which will fail to express the peculiar nature of anything that is, or again, omitting this, to look for separate definitions corresponding to each infima species. The cases of figure and soul are exactly parallel; for the particulars subsumed under the common name in both cases—figures and living beings—constitute a series, each successive term of which potentially contains its predecessor, e.g. the square the triangle, the sensory power the self-nutritive. Hence we must ask in the case of each order of living things, what is its soul, i.e. 'what is the soul of plant, animal, man?' Why the terms are related in this serial way must form the subject of later examination. But the facts are that the power of perception is never found apart from the power of self-nutrition, while—in plants—the latter is found isolated from the former. Again, no sense is found apart from that of touch, while touch is found by itself; many animals have neither sight, hearing, nor smell. Again, among living things that possess sense some have the power of locomotion, some not. Lastly, certain living beings—a small minority—possess calculation and thought, for (among mortal beings) those which possess calculation have all the other powers above mentioned, while the converse does not hold—indeed some live by imagination alone, while others have not even imagination. The mind that knows with immediate intuition presents a different problem.

It is evident that the way to give the most adequate definition of soul is to seek in the case of each of its forms for the most appropriate definition.

Part 4

It is necessary for the student of these forms of soul first to find a definition of each, expressive of what it is, and then to investigate its derivative properties, &c. But if we are to express what each is, viz. what the thinking power is, or the perceptive, or the nutritive, we must go farther back and first give an account of thinking or perceiving, for in the order of investigation the question of what an agent does precedes the question, what enables it to do what it does. If this is correct, we must on the same ground go yet another step farther back and have some clear view of the objects of each; thus we must start with these objects, e.g. with food, with what is perceptible, or with what is intelligible.

It follows that first of all we must treat of nutrition and reproduction, for the nutritive soul is found along with all the others and is the most primitive and widely distributed power of soul, being indeed that one in virtue of which all are said to have life. The acts in which it manifests itself are reproduction and the use of food—reproduction, I say, because for any living thing that has reached its normal development and which is unmutilated, and whose mode of generation is not spontaneous, the most natural act is the production of another like itself, an animal producing an animal, a plant a plant, in order that, as far as its nature allows, it may partake in the eternal and divine. That is the goal towards which all things strive, that for the sake of which they do whatsoever their nature renders possible. The phrase 'for the sake of which' is ambiguous; it may mean either (a) the end to achieve which, or (b) the being in whose interest, the act is done. Since then no living thing is able to partake in what is eternal and divine by uninterrupted continuance (for nothing perishable can forever remain one and the same), it tries to achieve that end in the only way possible to it, and success is possible in varying degrees; so it remains not indeed as the self-same individual but continues its existence in something like itself—not numerically but specifically one.

The soul is the cause or source of the living body. The terms cause and source have many senses. But the soul is the cause of its body alike in all three senses which we explicitly recognize. It is (a) the source or origin of movement, it is (b) the end, it is (c) the essence of the whole living body.

That it is the last, is clear; for in everything the essence is identical with the ground of its being, and here, in the case of living things, their being is to live, and of their being and their living the soul in them is the cause or source. Further, the actuality of whatever is potential is identical with its formulable essence.

It is manifest that the soul is also the final cause of its body. For Nature, like mind, always does whatever it does for the sake of something, which something is its end. To that something corresponds in the case of animals the soul and in this it follows the order of nature; all natural bodies are organs of the soul. This is true of those that enter into the constitution of plants as well as of those which enter into that of animals. This shows that that the sake of which they are is soul. We must here recall the two senses of 'that for the sake of which', viz. (a) the end to achieve which, and (b) the being in whose interest, anything is or is done.

We must maintain, further, that the soul is also the cause of the living body as the original source of local movement. The power of locomotion is not found, however, in all living things. But change of quality and change of quantity are also due to the soul. Sensation is held to be a qualitative alteration, and nothing except what has soul in it is capable of sensation. The same holds of the quantitative changes which constitute growth and decay; nothing grows or decays naturally except what feeds itself, and nothing feeds itself except what has a share of soul in it.

Empedocles is wrong in adding that growth in plants is to be explained, the downward rooting by the natural tendency of earth to travel downwards, and the upward branching by the similar natural tendency of fire to travel upwards. For he misinterprets up and down; up and down are not for all things what they are for the whole Cosmos: if we are to distinguish and identify organs according to their functions, the roots of plants are analogous to the head in animals. Further, we must ask what is the force that holds together the earth and the fire which tend to travel in contrary directions; if there is no counteracting force, they will be torn asunder; if there is, this must be the soul and the cause of nutrition and growth. By some the element of fire is held to be the cause of nutrition and growth, for it alone of the primary bodies or elements is observed to feed and increase itself. Hence the suggestion that in both plants and animals it is it which is the operative force. A concurrent cause in a sense it certainly is, but not the principal cause, that is rather the soul; for while the growth of fire goes on without limit so long as there is a supply of fuel, in the case of all complex wholes formed in the course of nature there is a limit or ratio which determines their size and increase, and limit and ratio are marks of soul but not of fire, and belong to the side of formulable essence rather than that of matter.

Nutrition and reproduction are due to one and the same psychic power. It is necessary first to give precision to our account of food, for it is by this function of absorbing food that this psychic power is distinguished from all the others. The current view is that what serves as food to a living thing is what is contrary to it—not that in every pair of contraries each is food to the other: to be food a contrary must not only be transformable into the other and vice versa, it must also in so doing increase the bulk of the other. Many a contrary is transformed into its other and vice versa, where neither is even a quantum and so cannot increase in bulk, e.g. an invalid into a healthy subject. It is clear that not even those contraries which satisfy both the conditions mentioned above are food to one another in precisely the same sense; water may be said to feed fire, but not fire water. Where the members of the pair are elementary bodies only one of the contraries, it would appear, can be said to feed the other. But there is a difficulty here. One set of thinkers assert that like fed, as well as increased in amount, by like. Another set, as we have said, maintain the very reverse, viz. that what feeds and what is fed are contrary to one another; like, they argue, is incapable of being affected by like; but food is changed in the process of digestion, and change is always to what is opposite or to what is intermediate. Further, food is acted upon by what is nourished by it, not the other way round, as timber is worked by a carpenter and not conversely; there is a change in the carpenter but it is merely a change from not-working to working. In answering this problem it makes all the difference whether we mean by 'the food' the 'finished' or the 'raw' product. If we use the word food of both, viz. of the completely undigested and the completely digested matter, we can justify both the rival accounts of it; taking food in the sense of undigested matter, it is the contrary of what is fed by it, taking it as digested it is like what is fed by it. Consequently it is clear that in a certain sense we may say that both parties are right, both wrong.

Since nothing except what is alive can be fed, what is fed is the besouled body and just because it has soul in it. Hence food is essentially related to what has soul in it. Food has a power which is other than the power to increase the bulk of what is fed by it; so far forth as what has soul in it is a quantum, food may increase its quantity, but it is only so far as what has soul in it is a 'this-somewhat' or substance that food acts as food; in that case it maintains the being of what is fed, and that continues to be what it is so long as the process of nutrition continues. Further, it is the agent in generation, i.e. not the generation of the individual fed but the reproduction of another like it; the substance of the individual fed is already in existence; the existence of no substance is a self-generation but only a self-maintenance.

Hence the psychic power which we are now studying may be described as that which tends to maintain whatever has this power in it of continuing such as it was, and food helps it to do its work. That is why, if deprived of food, it must cease to be.

The process of nutrition involves three factors, (a) what is fed, (b) that wherewith it is fed, (c) what does the feeding; of these (c) is the first soul, (a) the body which has that soul in it, (b) the food. But since it is right to call things after the ends they realize, and the end of this soul is to generate another being like that in which it is, the first soul ought to be named the reproductive soul. The expression (b) 'wherewith it is fed' is ambiguous just as is the expression 'wherewith the ship is steered'; that may mean either (i) the hand or (ii) the rudder, i.e. either (i) what is moved and sets in movement, or (ii) what is merely moved. We can apply this analogy here if we recall that all food must be capable of being digested, and that what produces digestion is warmth; that is why everything that has soul in it possesses warmth.

We have now given an outline account of the nature of food; further details must be given in the appropriate place.

Translated by J. A. Smith

Reading and Discussion Questions

1. How does Aristotle define substances in this reading? How does this add to our understanding of the term as seen in the *Categories*?
2. What, in Aristotle's view, distinguishes animate beings from inanimate objects?
3. What does Aristotle mean by "soul"? How, in Aristotle's view, do the natures that animate plants, animals, and humans differ? Compare and contrast Aristotle's way of thinking about the soul with Plato's understanding and with other uses of this term, including the Judeo-Christian tradition.

The History of Animals (c. 350 BCE)

Aristotle

12

Book I

Part 1

Of the parts of animals some are simple: to wit, all such as divide into parts uniform with themselves, as flesh into flesh; others are composite, such as divide into parts not uniform with themselves, as, for instance, the hand does not divide into hands nor the face into faces.

And of such as these, some are called not parts merely, but limbs or members. Such are those parts that, while entire in themselves, have within themselves other diverse parts: as for instance, the head, foot, hand, the arm as a whole, the chest; for these are all in themselves entire parts, and there are other diverse parts belonging to them.

All those parts that do not subdivide into parts uniform with themselves are composed of parts that do so subdivide, for instance, hand is composed of flesh, sinews, and bones. Of animals, some resemble one another in all their parts, while others have parts wherein they differ. Sometimes the parts are identical in form or species, as, for instance, one man's nose or eye resembles another man's nose or eye, flesh flesh, and bone bone; and in like manner with a horse, and with all other animals which we reckon to be of one and the same species: for as the whole is to the whole, so each to each are the parts severally. In other cases the parts are identical, save only for a difference in the way of excess or defect, as is the case in such animals as are of one and the same genus. By 'genus' I mean, for instance, Bird or Fish, for each of these is subject to difference in respect of its genus, and there are many species of fishes and of birds.

Within the limits of genera, most of the parts as a rule exhibit differences through contrast of the property or accident, such as colour and shape, to which they are subject: in that some are more and some in a less degree the subject of the same property or accident; and also in the way of multitude or fewness, magnitude or parvitude, in short in the way of excess or defect. Thus in some the texture of the flesh is soft, in others firm; some have a long bill, others a short one; some have abundance of feathers, others have only a small quantity. It happens further that some have parts that others have not: for instance, some have spurs and others not, some have crests and others not; but as a general rule, most parts and those that go to make up the bulk of the body are either identical with one another, or differ from one another in the way of contrast and of excess and defect. For 'the more' and 'the less' may be represented as 'excess' or 'defect'.

Once again, we may have to do with animals whose parts are neither identical in form nor yet identical save for differences in the way of excess or defect: but they are the same only in the way of analogy, as, for instance, bone is only analogous to fish-bone, nail to hoof, hand to claw, and scale to feather; for what the feather is in a bird, the scale is in a fish.

The parts, then, which animals severally possess are diverse from, or identical with, one another in the fashion above described. And they are so furthermore in the way of local disposition: for many animals have identical organs that differ in position; for instance, some have teats in the breast, others close to the thighs.

Of the substances that are composed of parts uniform (or homogeneous) with themselves, some are soft and moist, others are dry and solid. The soft and moist are such either absolutely or so long as they are in their natural conditions, as, for instance, blood, serum, lard, suet, marrow, sperm, gall, milk in such as have it flesh and the like; and also, in a different way, the superfluities, as phlegm and the excretions of the belly and the bladder. The dry and solid are such as sinew, skin, vein, hair, bone, gristle, nail, horn (a term which as applied to the part involves an ambiguity, since the whole also by virtue of its form is designated horn), and such parts as present an analogy to these.

Animals differ from one another in their modes of subsistence, in their actions, in their habits, and in their parts. Concerning these differences we shall first speak in broad and general terms, and subsequently we shall treat of the same with close reference to each particular genus.

Differences are manifested in modes of subsistence, in habits, in actions performed. For instance, some animals live in water and others on land. And of those that live in water some do so in one way, and some in another: that is to say, some live and feed in the water, take in and emit water, and cannot live if deprived of water, as is the case with the great majority of fishes; others get their food and spend their days in the water, but do not take in water but air, nor do they bring forth in the water. Many of these creatures are furnished with feet, as the otter, the beaver, and the crocodile; some are furnished with wings, as the diver and the grebe; some are destitute of feet, as the water-snake. Some creatures get their living in the water and cannot exist outside it: but for all that do not take in either air or water, as, for instance, the sea-nettle and the oyster. And of creatures that live in the water some live in the sea, some in rivers, some in lakes, and some in marshes, as the frog and the newt.

Of animals that live on dry land some take in air and emit it, which phenomena are termed 'inhalation' and 'exhalation'; as, for instance, man and all such land animals as are furnished with lungs. Others, again, do not inhale air, yet live and find their sustenance on dry land; as, for instance, the wasp, the bee, and all other insects. And by 'insects' I mean such creatures as have nicks or notches on their bodies, either on their bellies or on both backs and bellies.

And of land animals many, as has been said, derive their subsistence from the water; but of creatures that live in and inhale water not a single one derives its subsistence from dry land.

Some animals at first live in water, and by and by change their shape and live out of water, as is the case with river worms, for out of these the gadfly develops.

Furthermore, some animals are stationary, and some are erratic. Stationary animals are found in water, but no such creature is found on dry land. In the water are many creatures that live in close adhesion to an external object, as is the case with several kinds of oyster. And, by the way, the sponge appears to be endowed with a certain sensibility: as a proof of which it is alleged that the difficulty in detaching it from its moorings is increased if the movement to detach it be not covertly applied.

Other creatures adhere at one time to an object and detach themselves from it at other times, as is the case with a species of the so-called sea-nettle; for some of these creatures seek their food in the night-time loose and unattached.

Many creatures are unattached but motionless, as is the case with oysters and the so-called holothuria. Some can swim, as, for instance, fishes, molluscs, and crustaceans, such as the crawfish. But some of these last move by walking, as the crab, for it is the nature of the creature, though it lives in water, to move by walking.

Of land animals some are furnished with wings, such as birds and bees, and these are so furnished in different ways one from another; others are furnished with feet. Of the animals that are furnished with feet some walk, some creep, and some wriggle. But no creature is able only to move by flying, as the fish is able only to swim, for the animals with leathern wings can walk; the bat has feet and the seal has imperfect feet.

Some birds have feet of little power, and are therefore called Apodes. This little bird is powerful on the wing; and, as a rule, birds that resemble it are weak-footed and strong winged, such as the swallow and the drepanis or Alpine swift; for all these birds resemble one another in their habits and in their plumage, and may easily be mistaken one for another. (The apus is to be seen at all seasons, but the drepanis only after rainy weather in summer; for this is the time when it is seen and captured, though, as a general rule, it is a rare bird.)

Again, some animals move by walking on the ground as well as by swimming in water.

Furthermore, the following differences are manifest in their modes of living and in their actions. Some are gregarious, some are solitary, whether they be furnished with feet or wings or be fitted for a life in the water; and some partake of both characters, the solitary and the gregarious. And of the gregarious, some are disposed to combine for social purposes, others to live each for its own self.

Gregarious creatures are, among birds, such as the pigeon, the crane, and the swan; and, by the way, no bird furnished with crooked talons is gregarious. Of creatures that live in water many kinds of fishes are gregarious, such as the so-called migrants, the tunny, the pelamys, and the bonito.

Man, by the way, presents a mixture of the two characters, the gregarious and the solitary.

Social creatures are such as have some one common object in view; and this property is not common to all creatures that are gregarious. Such social creatures are man, the bee, the wasp, the ant, and the crane.

Again, of these social creatures some submit to a ruler, others are subject to no governance: as, for instance, the crane and the several sorts of bee submit to a ruler, whereas ants and numerous other creatures are everyone his own master.

And again, both of gregarious and of solitary animals, some are attached to a fixed home and others are erratic or nomad.

Also, some are carnivorous, some graminivorous, some omnivorous: whilst some feed on a peculiar diet, as for instance the bees and the spiders, for the bee lives on honey and certain other sweets, and the spider lives by catching flies; and some creatures live on fish. Again, some creatures catch their food, others treasure it up; whereas others do not so.

Some creatures provide themselves with a dwelling, others go without one: of the former kind are the mole, the mouse, the ant, the bee; of the latter kind are many insects and quadrupeds. Further, in respect to locality of dwelling place, some creatures dwell underground, as the lizard and the snake; others live on the surface of the ground, as the horse and the dog make to themselves holes, others do not.

Some are nocturnal, as the owl and the bat; others live in the daylight.

Moreover, some creatures are tame and some are wild: some are at all times tame, as man and the mule; others are at all times savage, as the leopard and the wolf; and some creatures can be rapidly tamed, as the elephant.

Again, we may regard animals in another light. For, whenever a race of animals is found domesticated, the same is always to be found in a wild condition; as we find to be the case with horses, kine, swine, (men), sheep, goats, and dogs.

Further, some animals emit sound while others are mute, and some are endowed with voice: of these latter some have articulate speech, while others are inarticulate; some are given to continual chirping and twittering some are prone to silence; some are musical, and some unmusical; but all animals without exception exercise their power of singing or chattering chiefly in connexion with the intercourse of the sexes.

Again, some creatures live in the fields, as the cushat; some on the mountains, as the hoopoe; some frequent the abodes of men, as the pigeon.

Some, again, are peculiarly salacious, as the partridge, the barn-door cock and their congeners; others are inclined to chastity, as the whole tribe of crows, for birds of this kind indulge but rarely in sexual intercourse.

Of marine animals, again, some live in the open seas, some near the shore, some on rocks.

Furthermore, some are combative under offence; others are provident for defence. Of the former kind are such as act as aggressors upon others or retaliate when subjected to ill usage, and of the latter kind are such as merely have some means of guarding themselves against attack.

Animals also differ from one another in regard to character in the following respects. Some are good-tempered, sluggish, and little prone to ferocity, as the ox; others are quick tempered, ferocious and unteachable, as the wild boar; some are intelligent and timid, as the stag and the hare; others are mean and treacherous, as the snake; others are noble and courageous and high-bred, as the lion; others are thorough-bred and wild and treacherous, as the wolf: for, by the way, an animal is highbred if it come from a noble stock, and an animal is thorough-bred if it does not deflect from its racial characteristics.

Further, some are crafty and mischievous, as the fox; some are spirited and affectionate and fawning, as the dog; others are easy-tempered and easily domesticated, as the elephant; others are cautious and watchful, as the goose; others are jealous and self-conceited, as the peacock. But of all animals man alone is capable of deliberation.

Many animals have memory, and are capable of instruction; but no other creature except man can recall the past at will.

With regard to the several genera of animals, particulars as to their habits of life and modes of existence will be discussed more fully by and by.

. . .

Book II

Part 1

With regard to animals in general, some parts or organs are common to all, as has been said, and some are common only to particular genera; the parts, moreover, are identical with or different from one another on the lines already repeatedly laid down. For as a general rule all animals that are generically distinct have the majority of their parts or organs different in form or species; and some of them they have only analogically similar and diverse in kind or genus, while they have others that are alike in kind but specifically diverse; and many parts or organs exist in some animals, but not in others.

For instance, viviparous quadrupeds have all a head and a neck, and all the parts or organs of the head, but they differ each from other in the shapes of the parts. The lion has its neck composed of one single bone instead of vertebrae; but, when dissected, the animal is found in all internal characters to resemble the dog.

The quadrupedal vivipara instead of arms have forelegs. This is true of all quadrupeds, but such of them as have toes have, practically speaking, organs analogous to hands; at all events, they use these fore-limbs for many purposes as hands. And they have the limbs on the left-hand side less distinct from those on the right than man.

The fore-limbs then serve more or less the purpose of hands in quadrupeds, with the exception of the elephant. This latter animal has its toes somewhat indistinctly defined, and its front legs are much bigger than its hinder ones; it is five-toed, and has short ankles to its hind feet. But it has a nose such in properties and such in size as to allow of its using the same for a hand. For it eats and drinks by lifting up its food with the aid of this organ into its mouth, and with the same organ it lifts up articles to the driver on its back; with this organ it can pluck up trees by the roots, and when walking through water it spouts the water up by means of it; and this organ is capable of being crooked or coiled at the tip, but not of flexing like a joint, for it is composed of gristle.

Of all animals man alone can learn to make equal use of both hands.

All animals have a part analogous to the chest in man, but not similar to his; for the chest in man is broad, but that of all other animals is narrow. Moreover, no other animal but man has breasts in front; the elephant, certainly, has two breasts, not however in the chest, but near it.

Moreover, also, animals have the flexions of their fore and hind limbs in directions opposite to one another, and in directions the reverse of those observed in the arms and legs of man; with the exception of the elephant. In other words, with the viviparous quadrupeds the front legs bend forwards and the hind ones backwards, and the concavities of the two pairs of limbs thus face one another.

The elephant does not sleep standing, as some were wont to assert, but it bends its legs and settles down; only that in consequence of its weight it cannot bend its leg on both sides simultaneously, but falls into a recumbent position on one side or the other, and in this position it goes to sleep. And it bends its hind legs just as a man bends his legs.

In the case of the ovipara, as the crocodile and the lizard and the like, both pairs of legs, fore and hind, bend forwards, with a slight swerve on one side. The flexion is similar in the case of the multipeds; only that the legs in between the extreme ends always move in a manner intermediate between that of those in front and those behind, and accordingly bend sideways rather than backwards or forwards. But man bends his arms and his legs towards the same point, and therefore in opposite ways: that is to say, he bends his arms backwards, with just a slight inclination inwards, and his legs frontwards. No animal bends both its fore-limbs and hind-limbs backwards; but in the case of all animals the flexion of the shoulders is in the opposite direction to that of the elbows or the joints of the forelegs, and the flexure in the hips to that of the knees of the hind-legs: so that since man differs from other animals in flexion, those animals that possess such parts as these move them contrariwise to man.

Birds have the flexions of their limbs like those of the quadrupeds; for, although bipeds, they bend their legs backwards, and instead of arms or front legs have wings which bend frontwards.

The seal is a kind of imperfect or crippled quadruped; for just behind the shoulder-blade its front feet are placed, resembling hands, like the front paws of the bear; for they are furnished with five toes, and each of the toes has three flexions and a nail of inconsiderable size. The hind feet are also furnished with five toes; in their flexions and nails they resemble the front feet, and in shape they resemble a fish's tail.

The movements of animals, quadruped and multiped, are crosswise, or in diagonals, and their equilibrium in standing posture is maintained crosswise; and it is always the limb on the right-hand side that is the first to move. The lion, however, and the two species of camels, both the Bactrian and the Arabian, progress by an amble; and the action so called is when the animal never overpasses the right with the left, but always follows close upon it.

Whatever parts men have in front, these parts quadrupeds have below, in or on the belly; and whatever parts men have behind, these parts quadrupeds have above on their backs. Most quadrupeds have a tail; for even the seal has a tiny one resembling that of the stag. Regarding the tails of the pithecoids we must give their distinctive properties by and by animal.

All viviparous quadrupeds are hair-coated, whereas man has only a few short hairs excepting on the head, but, so far as the head is concerned, he is hairier than any other animal. Further, of hair-coated animals, the back is hairier than the belly, which latter is either comparatively void of hair or smooth and void of hair altogether. With man the reverse is the case.

Man also has upper and lower eyelashes, and hair under the armpits and on the pubes. No other animal has hair in either of these localities, or has an under eyelash; though in the case of some animals a few straggling hairs grow under the eyelid.

Of hair-coated quadrupeds some are hairy all over the body, as the pig, the bear, and the dog; others are especially hairy on the neck and all round about it, as is the case with animals that have a shaggy mane, such as the lion; others again are especially hairy on the upper surface of the neck from the head as far as the withers, namely, such as have a crested mane, as in the case with the horse, the mule, and, among the undomesticated horned animals, the bison.

. . .

Camels have an exceptional organ wherein they differ from all other animals, and that is the so-called 'hump' on their back. The Bactrian camel differs from the Arabian; for the former has two humps and the latter only one, though it has, by the way, a kind of a hump below like the one above, on which, when it kneels, the weight of the whole body rests. The camel has four teats like the cow, a tail like that of an ass, and the privy parts of the male are directed backwards. It has one knee in each leg, and the flexures of the limb are not manifold, as some say, although they appear to be so from the constricted shape of the region of the belly. It has a huckle-bone like that of kine, but meagre and small in proportion to its bulk. It is cloven-footed, and has not got teeth in both jaws; and it is cloven footed in the following way: at the back there is a slight cleft extending as far up as the second joint of the toes; and in front there are small hooves on the tip of the first joint of the toes; and a sort of web passes across the cleft, as in geese. The foot is fleshy underneath, like that of the bear; so that, when the animal goes to war, they protect its feet, when they get sore, with sandals.

The legs of all quadrupeds are bony, sinewy, and fleshless; and in point of fact such is the case with all animals that are furnished with feet, with the exception of man. They are also unfurnished with buttocks; and this last point is plain in an especial degree in birds. It is the reverse with man; for there is scarcely any part of the body in which man is so fleshy as in the buttock, the thigh, and the calf; for the part of the leg called gastroenemia or is fleshy.

Of blooded and viviparous quadrupeds some have the foot cloven into many parts, as is the case with the hands and feet of man (for some animals, by the way, are many-toed, as the lion, the dog, and the pard); others have feet cloven in twain, and instead of nails have hooves, as the sheep, the goat, the deer, and the hippopotamus; others are uncloven of foot, such for instance as the solid-hooved animals, the horse and the mule. Swine are either cloven-footed or uncloven-footed; for there are in Illyria and in Paeonia and elsewhere solid-hooved swine. The cloven-footed animals have two clefts behind; in the solid-hooved this part is continuous and undivided.

Furthermore, of animals some are horned, and some are not so. The great majority of the horned animals are cloven-footed, as the ox, the stag, the goat; and a solid-hooved animal with a pair of horns has never yet been met with. But a few animals are known to be singled-horned and single-hooved, as the Indian ass; and one, to wit the oryx, is single horned and cloven-hooved.

Of all solid-hooved animals the Indian ass alone has an astragalus or huckle-bone; for the pig, as was said above, is either solid-hooved or cloven-footed, and consequently has no well-formed huckle-bone. Of the cloven footed many are provided with a huckle-bone. Of the many-fingered or many-toed, no single one has been observed to have a huckle-bone, none of the others any more than man. The lynx, however, has something like a hemiastragal, and the lion something resembling the sculptor's 'labyrinth'. All the animals that have a huckle-bone have it in the hinder legs. They have also the bone placed straight up in the joint; the upper part, outside; the lower part, inside; the sides called Coa turned towards one another, the sides called Chia outside, and the keraiae or 'horns' on the top. This, then, is the position of the hucklebone in the case of all animals provided with the part.

...

Again, with regard to the breasts and the generative organs, animals differ widely from one another and from man. For instance, the breasts of some animals are situated in front, either in the chest or near to it, and there are in such cases two breasts and two teats, as is the case with man and the elephant, as previously stated. For the elephant has two breasts in the region of the axillae; and the female elephant has two breasts insignificant in size and in no way proportionate to the bulk of the entire frame, in fact, so insignificant as to be invisible in a sideways view; the males also have breasts, like the females, exceedingly small. The she-bear has four breasts. Some animals have two breasts, but situated near the thighs, and teats, likewise two in number, as the sheep; others have four teats, as the cow. Some have breasts neither in the chest nor at the thighs, but in the belly, as the dog and pig; and they have a considerable number of breasts or dugs, but not all of equal size. Thus the shepard has four dugs in the belly, the lioness two, and others more. The she-camel, also, has two dugs and four teats, like the cow. Of solid-hooved animals the males have no dugs, excepting in the case of males that take after the mother, which phenomenon is observable in horses.

Of male animals the genitals of some are external, as is the case with man, the horse, and most other creatures; some are internal, as with the dolphin. With those that have the organ externally placed, the organ in some cases is situated in front, as in the cases already mentioned, and of these some have the organ detached, both penis and testicles, as man; others have penis and testicles closely attached to the belly, some more closely, some less; for this organ is not detached in the wild boar nor in the horse.

The penis of the elephant resembles that of the horse; compared with the size of the animal it is disproportionately small; the testicles are not visible, but are concealed inside in the vicinity of the kidneys; and for this reason the male speedily gives over in the act of intercourse. The genitals of the female are situated where the udder is in sheep; when she is in heat, she draws the organ back and exposes it externally, to facilitate the act of intercourse for the male; and the organ opens out to a considerable extent.

With most animals the genitals have the position above assigned; but some animals discharge their urine backwards, as the lynx, the lion, the camel, and the hare. Male animals differ from one another, as has been said, in this particular, but all female animals are retromingent: even the female elephant like other animals, though she has the privy part below the thighs.

In the male organ itself there is a great diversity. For in some cases the organ is composed of flesh and gristle, as in man; in such cases, the fleshy part does not become inflated, but the gristly part is subject to enlargement. In other cases, the organ is composed of fibrous tissue, as with the camel and the deer; in other cases it is bony, as with the fox, the wolf, the marten, and the weasel; for this organ in the weasel has a bone.

When man has arrived at maturity, his upper part is smaller than the lower one, but with all other blooded animals the reverse holds good. By the 'upper' part we mean all extending from the head down to the parts used for excretion of residuum, and by the 'lower' part else. With animals that have feet the hind legs are to be rated as the lower part in our comparison of magnitudes, and with animals devoid of feet, the tail, and the like.

When animals arrive at maturity, their properties are as above stated; but they differ greatly from one another in their growth towards maturity. For instance, man, when young, has his upper part larger than the lower, but in course of growth he comes to reverse this condition; and it is owing to this circumstance that—an exceptional instance, by the way—he does not progress in early life as he does at maturity, but in infancy creeps on all fours; but some animals, in growth, retain the relative proportion of the parts, as the dog. Some animals at first have the upper part smaller and the lower part larger, and in course of growth the upper part gets to be the larger, as is the case with the bushy-tailed animals such as the horse; for in their case there is never, subsequently to birth, any increase in the part extending from the hoof to the haunch.

Again, in respect to the teeth, animals differ greatly both from one another and from man. All animals that are quadrupedal, blooded and viviparous, are furnished with teeth; but, to begin with, some are double-toothed (or fully furnished with teeth in both jaws), and some are not. For instance, horned quadrupeds are not double-toothed; for they have not got the front teeth in the upper jaw; and some hornless animals, also, are not double toothed, as the camel. Some animals have tusks, like the boar, and some have not. Further, some animals are saw-toothed, such as the lion, the pard, and the dog; and some have teeth that do not interlock but have flat opposing crowns, as the horse and the ox; and by 'saw-toothed' we mean such animals as interlock the sharp-pointed teeth in one jaw between the sharp-pointed ones in the other. No animal is there that possesses both tusks and horns, nor yet do either of these structures exist in any animal possessed of 'saw-teeth'. The front teeth are usually sharp, and the back ones blunt. The seal is saw-toothed throughout, inasmuch as he is a sort of link with the class of fishes; for fishes are almost all saw-toothed.

No animal of these genera is provided with double rows of teeth. There is, however, an animal of the sort, if we are to believe Ctesias. He assures us that the Indian wild beast called the 'martichoras' has a triple row of teeth in both upper and lower jaw; that it is as big as a lion and equally hairy, and that its feet resemble those of the lion; that it resembles man in its face and ears; that its eyes are blue, and its colour vermilion; that its tail is like that of the land-scorpion; that it has a sting in the tail, and has the faculty of shooting off arrow-wise the spines that are attached to the tail; that the sound of its voice is a something between the sound of a pan-pipe and that of a trumpet; that it can run as swiftly as deer, and that it is savage and a man-eater.

Man sheds his teeth, and so do other animals, as the horse, the mule, and the ass. And man sheds his front teeth; but there is no instance of an animal that sheds its molars. The pig sheds none of its teeth at all.

Translated by D'Arcy Wentworth Thompson

Reading and Discussion Questions

1. Reflect on the level of detailed observation in these descriptions. What does this suggest about the level of Aristotle's understanding of anatomy? What characteristics of quadrupeds is Aristotle focusing on here, and how does this relate to his views of which features are accidental or essential?
2. Why does Aristotle call attention to the close analogies between humans and other animals?

On the Parts of Animals (c. 350 BCE)

Aristotle

13

Book I

Part 1

Every systematic science, the humblest and the noblest alike, seems to admit of two distinct kinds of proficiency; one of which may be properly called scientific knowledge of the subject, while the other is a kind of educational acquaintance with it. For an educated man should be able to form a fair off-hand judgement as to the goodness or badness of the method used by a professor in his exposition. To be educated is in fact to be able to do this; and even the man of universal education we deem to be such in virtue of his having this ability. It will, however, of course, be understood that we only ascribe universal education to one who in his own individual person is thus critical in all or nearly all branches of knowledge, and not to one who has a like ability merely in some special subject. For it is possible for a man to have this competence in some one branch of knowledge without having it in all.

It is plain then that, as in other sciences, so in that which inquires into nature, there must be certain canons, by reference to which a hearer shall be able to criticize the method of a professed exposition, quite independently of the question whether the statements made be true or false. Ought we, for instance (to give an illustration of what I mean), to begin by discussing each separate species—man, lion, ox, and the like—taking each kind in hand independently of the rest, or ought we rather to deal first with the attributes which they have in common in virtue of some common element of their nature, and proceed from this as a basis for the consideration of them separately? For genera that are quite distinct yet oftentimes present many identical phenomena, sleep, for instance, respiration, growth, decay, death, and other similar affections and conditions, which may be passed over for the present, as we are not yet prepared to treat of them with clearness and precision. Now it is plain that if we deal with each species independently of the rest, we shall frequently be obliged to repeat the same statements over and over again; for horse and dog and man present, each and all, every one of the phenomena just enumerated. A discussion therefore of the attributes of each such species separately would necessarily involve frequent repetitions as to characters, themselves identical but recurring in animals specifically distinct. (Very possibly also there may be other characters which, though they present specific differences, yet come under one and the same category. For instance, flying, swimming, walking, creeping, are plainly specifically distinct, but yet are all forms of animal progression.) We must, then, have some clear understanding as to the manner in which our investigation is to be conducted; whether, I mean, we are first to deal with the common or generic characters, and afterwards to take into consideration special peculiarities; or whether we are to start straight off with the ultimate species. For as yet no definite rule has been laid down in this matter. So also there is a like uncertainty as to another point now to be mentioned. Ought the writer who deals with the works of nature to follow the plan adopted by the mathematicians in their astronomical demonstrations, and after considering the phenomena presented by animals, and their several parts, proceed subsequently to treat of the causes and the reason why; or ought he to follow some other method? And when these questions are answered, there yet remains another. The causes concerned in the generation of the works of nature are, as we see, more than one. There is the final cause and there is the motor cause. Now we must decide which of these two causes comes first, which second. Plainly, however, that cause is the first which we call the final one. For this is the Reason, and the Reason forms the starting-point, alike in the works of art and in works of nature. For consider how the physician or how the builder sets about his work. He starts by forming for himself a definite picture, in the one case perceptible to mind, in the other to sense, of his end—the physician of health, the builder of a house—and this he holds forward as the reason and explanation of each subsequent step that he takes, and of his acting in this or that way as the case may be. Now in the works of nature the good end and the final cause is still more dominant than in works of art such as these, nor is necessity a factor with the same significance in them all; though almost all writers, while they try to refer their origin to this cause, do so without distinguishing the various senses in which the term necessity is used. For there is absolute necessity, manifested in eternal phenomena; and there is hypothetical necessity, manifested in everything that is generated by nature as in everything that is produced by art, be it a house or what it may. For if a house or other such final object is to be realized, it is necessary that such and such material shall exist; and it is necessary that first this then that shall be produced, and first this and then that set in motion, and so on in continuous succession, until the end and final result is reached, for the sake of which each prior thing is produced and exists. As with these productions of art, so also is it with the productions of nature. The mode of necessity, however, and the mode of ratiocination are different in natural science from what they are in the theoretical sciences; of which we have spoken elsewhere. For in the latter the starting-point is that which is; in the former that which is to be. For it is that which is yet to be—health, let us say, or a man—that, owing to its being of such and such characters, necessitates the pre-existence or previous production of this and that antecedent; and not this or that antecedent which, because it exists or has been generated, makes it necessary that health or a man is in, or shall come into, existence. Nor is it possible to track back the series of necessary antecedents to a starting-point, of which you can say that, existing itself from eternity, it has determined their existence as its consequent. These however again, are matters that have been dealt with in another treatise. There too it was stated in what cases absolute and hypothetical necessity exist; in what cases also the proposition expressing hypothetical necessity is simply convertible, and what cause it is that determines this convertibility.

Another matter which must not be passed over without consideration is, whether the proper subject of our exposition is that with which the ancient writers concerned themselves, namely, what is the process of formation of each animal; or whether it is not rather, what are the characters of a given creature when formed. For there is no small difference between these two views. The best course appears to be that we should follow the method already mentioned, and begin with the phenomena presented by each group of animals, and, when this is done, proceed afterwards to state the causes of those phenomena, and to deal with their evolution. For elsewhere, as for instance in house building, this is the true sequence. The plan of the house, or the house, has this and that form; and because it has this and that form, therefore is its construction carried out in this or that manner. For the process of evolution is for the sake of the thing finally evolved, and not this for the sake of the process. Empedocles, then, was in error when he said that many of the characters presented by animals were merely the results of incidental occurrences during their development; for instance, that the backbone was divided as it is into vertebrae, because it happened to be broken owing to the contorted position of the foetus in the womb. In so saying he overlooked the fact that propagation implies a creative seed endowed with certain formative properties. Secondly, he neglected another fact, namely, that the parent animal pre-exists, not only in idea, but actually in time. For man is generated from man; and thus it is the possession of certain characters by the parent that determines the development of like characters in the child. The same statement holds good also for the operations of art, and even for those which are apparently spontaneous. For the same result as is produced by art may occur spontaneously. Spontaneity, for instance, may bring about the restoration of health. The products of art, however, require the pre-existence of an efficient cause homogeneous with themselves, such as the statuary's art, which must necessarily precede the statue; for this cannot possibly be produced spontaneously. Art indeed consists in the conception of the result to be produced before its realization in the material. As with spontaneity, so with chance; for this also produces the same result as art, and by the same process.

The fittest mode, then, of treatment is to say, a man has such and such parts, because the conception of a man includes their presence, and because they are necessary conditions of his existence, or, if we cannot quite say this, which would be best of all, then the next thing to it, namely, that it is either quite impossible for him to exist without them, or, at any rate, that it is better for him that they should be there; and their existence involves the existence of other antecedents. Thus we should say, because man is an animal with such and such characters, therefore is the process of his development necessarily such as it is; and therefore is it accomplished in such and such an order, this part being formed first, that next, and so on in succession; and after a like fashion should we explain the evolution of all other works of nature.

Now that with which the ancient writers, who first philosophized about Nature, busied themselves, was the material principle and the material cause. They inquired what this is, and what its character; how the universe is generated out of it, and by what motor influence, whether, for instance, by antagonism or friendship, whether by intelligence or spontaneous action, the substratum of matter being assumed to have certain inseparable properties; fire, for instance, to have a hot nature, earth a cold one; the former to be light, the latter heavy. For even the genesis of the universe is thus explained by them. After a like fashion do they deal also with the development of plants and of animals. They say, for instance, that the water contained in the body causes by its currents the formation of the stomach and the other receptacles of food or of excretion; and that the breath by its passage breaks open the outlets of the nostrils; air and water being the materials of which bodies are made; for all represent nature as composed of such or similar substances.

But if men and animals and their several parts are natural phenomena, then the natural philosopher must take into consideration not merely the ultimate substances of which they are made, but also flesh, bone, blood, and all other homogeneous parts; not only these, but also the heterogeneous parts, such as face, hand, foot; and must examine how each of these comes to be what it is, and in virtue of what force. For to say what are the ultimate substances out of which an animal is formed, to state, for instance, that it is made of fire or earth, is no more sufficient than would be a similar account in the case of a couch or the like. For we should not be content with saying that the couch was made of bronze or wood or whatever it might be, but should try to describe its design or mode of composition in preference to the material; or, if we did deal with the material, it would at any rate be with the concretion of material and form. For a couch is such and such a form embodied in this or that matter, or such and such a matter with this or that form; so that its shape and structure must be included in our description. For the formal nature is of greater importance than the material nature.

Does, then, configuration and colour constitute the essence of the various animals and of their several parts? For if so, what Democritus says will be strictly correct. For such appears to have been his notion.

At any rate he says that it is evident to everyone what form it is that makes the man, seeing that he is recognizable by his shape and colour. And yet a dead body has exactly the same configuration as a living one; but for all that is not a man. So also no hand of bronze or wood or constituted in any but the appropriate way can possibly be a hand in more than name. For like a physician in a painting, or like a flute in a sculpture, in spite of its name it will be unable to do the office which that name implies. Precisely in the same way no part of a dead body, such I mean as its eye or its hand, is really an eye or a hand. To say, then, that shape and colour constitute the animal is an inadequate statement, and is much the same as if a woodcarver were to insist that the hand he had cut out was really a hand. Yet the physiologists, when they give an account of the development and causes of the animal form, speak very much like such a craftsman. What, however, I would ask, are the forces by which the hand or the body was fashioned into its shape? The woodcarver will perhaps say, by the axe or the auger; the physiologist, by air and by earth. Of these two answers the artificer's is the better, but it is nevertheless insufficient. For it is not enough for him to say that by the stroke of his tool this part was formed into a concavity, that into a flat surface; but he must state the reasons why he struck his blow in such a way as to effect this, and what his final object was; namely, that the piece of wood should develop eventually into this or that shape. It is plain, then, that the teaching of the old physiologists is inadequate, and that the true method is to state what the definitive characters are that distinguish the animal as a whole; to explain what it is both in substance and in form, and to deal after the same fashion with its several organs; in fact, to proceed in exactly the same way as we should do, were we giving a complete description of a couch.

If now this something that constitutes the form of the living being be the soul, or part of the soul, or something that without the soul cannot exist; as would seem to be the case, seeing at any rate that when the soul departs, what is left is no longer a living animal, and that none of the parts remain what they were before, excepting in mere configuration, like the animals that in the fable are turned into stone; if, I say, this be so, then it will come within the province of the natural philosopher to inform himself concerning the soul, and to treat of it, either in its entirety, or, at any rate, of that part of it which constitutes the essential character of an animal; and it will be his duty to say what this soul or this part of a soul is; and to discuss the attributes that attach to this essential character, especially as nature is spoken of in two senses, and the nature of a thing is either its matter or its essence; nature as essence including both the motor cause and the final cause. Now it is in the latter of these two senses that either the whole soul or some part of it constitutes the nature of an animal; and inasmuch as it is the presence of the soul that enables matter to constitute the animal nature, much more than it is the presence of matter which so enables the soul, the inquirer into nature is bound on every ground to treat of the soul rather than of the matter. For though the wood of which they are made constitutes the couch and the tripod, it only does so because it is capable of receiving such and such a form.

What has been said suggests the question, whether it is the whole soul or only some part of it, the consideration of which comes within the province of natural science. Now if it be of the whole soul that this should treat, then there is no place for any other philosophy beside it. For as it belongs in all cases to one and the same science to deal with correlated subjects—one and the same science, for instance, deals with sensation and with the objects of sense—and as therefore the intelligent soul and the objects of intellect, being correlated, must belong to one and the same science, it follows that natural science will have to include the whole universe in its province. But perhaps it is not the whole soul, nor all its parts collectively, that constitutes the source of motion; but there may be one part, identical with that in plants, which is the source of growth, another, namely the sensory part, which is the source of change of quality, while still another, and this not the intellectual part, is the source of locomotion. I say not the intellectual part; for other animals than man have the power of locomotion, but in none but him is there intellect. Thus then it is plain that it is not of the whole soul that we have to treat. For it is not the whole soul that constitutes the animal nature, but only some part or parts of it. Moreover, it is impossible that any abstraction can form a subject of natural science, seeing that everything that Nature makes is means to an end. For just as human creations are the products of art, so living objects are manifest in the products of an analogous cause or principle, not external but internal, derived like the hot and the cold from the environing universe. And that the heaven, if it had an origin, was evolved and is maintained by such a cause, there is therefore even more reason to believe, than that mortal animals so originated. For order and definiteness are much more plainly manifest in the celestial bodies than in our own frame; while change and chance are characteristic of the perishable things of earth. Yet there are some who, while they allow that every animal exists and was generated by nature, nevertheless hold that the heaven was constructed to be what it is by chance and spontaneity; the heaven, in which not the faintest sign of haphazard or of disorder is discernible! Again, whenever there is plainly some final end, to which a motion tends should nothing stand in the way, we always say that such final end is the aim or purpose of the motion; and from this it is evident that there must be a something or other really existing, corresponding to what we call by the name of Nature. For a given germ does not give rise to any chance living being, nor spring from any chance one; but each germ springs from a definite parent and gives rise to a definite progeny. And thus it is the germ that is the ruling influence and fabricator of the offspring. For these it is by nature, the offspring being at any rate that which in nature will spring from it. At the same time the offspring is anterior to the germ; for germ and perfected progeny are related as the developmental process and the result. Anterior, however, to both germ and product is the organism from which the germ was derived. For every germ implies two organisms, the parent and the progeny. For germ or seed is both the seed of the organism from which it came, of the horse, for instance, from which it was derived, and the seed of the organism that will eventually arise from it, of the mule, for example, which is developed from the seed of the horse. The same seed then is the seed both of the horse and of the mule, though in different ways as here set forth. Moreover, the seed is potentially that which will spring from it, and the relation of potentiality to actuality we know.

There are then two causes, namely, necessity and the final end. For many things are produced, simply as the results of necessity. It may, however, be asked, of what mode of necessity are we speaking when we say this. For it can be of neither of those two modes which are set forth in the philosophical treatises. There is, however, the third mode, in such things at any rate as are generated. For instance, we say that food is necessary; because an animal cannot possibly do without it. This third mode is what may be called hypothetical necessity. Here is another example of it. If a piece of wood is to be split with an axe, the axe must of necessity be hard; and, if hard, must of necessity be made of bronze or iron. Now exactly in the same way the body, which like the axe is an instrument—for both the body as a whole and its several parts individually have definite operations for which they are made—just in the same way, I say, the body, if it is to do its work, must of necessity be of such and such a character, and made of such and such materials.

It is plain then that there are two modes of causation, and that both of these must, so far as possible, be taken into account in explaining the works of nature, or that at any rate an attempt must be made to include them both; and that those who fail in this tell us in reality nothing about nature. For primary cause constitutes the nature of an animal much more than does its matter. There are indeed passages in which even Empedocles hits upon this, and following the guidance of fact, finds himself constrained to speak of the ratio as constituting the essence and real nature of things. Such, for instance, is the case when he explains what is a bone. For he does not merely describe its material, and say it is this one element, or those two or three elements, or a compound of all the elements, but states the ratio of their combination. As with a bone, so manifestly is it with the flesh and all other similar parts.

The reason why our predecessors failed in hitting upon this method of treatment was, that they were not in possession of the notion of essence, nor of any definition of substance. The first who came near it was Democritus, and he was far from adopting it as a necessary method in natural science, but was merely brought to it, spite of himself, by constraint of facts. In the time of Socrates a nearer approach was made to the method. But at this period men gave up inquiring into the works of nature, and philosophers diverted their attention to political science and to the virtues which benefit mankind.

Of the method itself the following is an example. In dealing with respiration we must show that it takes place for such or such a final object; and we must also show that this and that part of the process is necessitated by this and that other stage of it. By necessity we shall sometimes mean hypothetical necessity, the necessity, that is, that the requisite antecedents shall be there, if the final end is to be reached; and sometimes absolute necessity, such necessity as that which connects substances and their inherent properties and characters. For the alternate discharge and re-entrance of heat and the inflow of air are necessary if we are to live. Here we have at once a necessity in the former of the two senses. But the alternation of heat and refrigeration produces of necessity an alternate admission and discharge of the outer air, and this is a necessity of the second kind.

In the foregoing we have an example of the method which we must adopt, and also an example of the kind of phenomena, the causes of which we have to investigate.

Part 2

Some writers propose to reach the definitions of the ultimate forms of animal life by bipartite division. But this method is often difficult, and often impracticable.

Sometimes the final differentia of the subdivision is sufficient by itself, and the antecedent differentiae are mere surplusage. Thus in the series Footed, Two-footed, Cleft-footed, the last term is all-expressive by itself, and to append the higher terms is only an idle iteration. Again it is not permissible to break up a natural group, Birds for instance, by putting its members under different bifurcations, as is done in the published dichotomies, where some birds are ranked with animals of the water, and others placed in a different class. The group Birds and the group Fishes happen to be named, while other natural groups have no popular names; for instance, the groups that we may call Sanguineous and Bloodless are not known popularly by any designations. If such natural groups are not to be broken up, the method of Dichotomy cannot be employed, for it necessarily involves such breaking up and dislocation. The group of the Many-footed, for instance, would, under this method, have to be dismembered, and some of its kinds distributed among land animals, others among water animals.

Part 3

Again, privative terms inevitably form one branch of dichotomous division, as we see in the proposed dichotomies. But privative terms in their character of privatives admit of no subdivision. For there can be no specific forms of a negation, of Featherless for instance or of Footless, as there are of Feathered and of Footed. Yet a generic differentia must be subdivisible; for otherwise what is there that makes it generic rather than specific? There are to be found generic, that is specifically subdivisible, differentiae; Feathered for instance and Footed. For feathers are divisible into Barbed and Unbarbed, and feet into Manycleft, and Twocleft, like those of animals with bifid hoofs, and

Uncleft or Undivided, like those of animals with solid hoofs. Now even with differentiae capable of this specific subdivision it is difficult enough so to make the classification, as that each animal shall be comprehended in some one subdivision and in not more than one; but far more difficult, nay impossible, is it to do this, if we start with a dichotomy into two contradictories. (Suppose for instance we start with the two contradictories, Feathered and Unfeathered; we shall find that the ant, the glow-worm, and some other animals fall under both divisions.) For each differentia must be presented by some species. There must be some species, therefore, under the privative heading. Now specifically distinct animals cannot present in their essence a common undifferentiated element, but any apparently common element must really be differentiated. (Bird and Man for instance are both Two-footed, but their two-footedness is diverse and differentiated. So any two sanguineous groups must have some difference in their blood, if their blood is part of their essence.) From this it follows that a privative term, being insusceptible of differentiation, cannot be a generic differentia; for, if it were, there would be a common undifferentiated element in two different groups.

Again, if the species are ultimate indivisible groups, that is, are groups with indivisible differentiae, and if no differentia be common to several groups, the number of differentiae must be equal to the number of species. If a differentia though not divisible could yet be common to several groups, then it is plain that in virtue of that common differentia specifically distinct animals would fall into the same division. It is necessary then, if the differentiae, under which are ranged all the ultimate and indivisible groups, are specific characters, that none of them shall be common; for otherwise, as already said, specifically distinct animals will come into one and the same division. But this would violate one of the requisite conditions, which are as follows. No ultimate group must be included in more than a single division; different groups must not be included in the same division; and every group must be found in some division. It is plain then that we cannot get at the ultimate specific forms of the animal, or any other, kingdom by bifurcate division. If we could, the number of ultimate differentiae would equal the number of ultimate animal forms. For assume an order of beings whose prime differentiae are White and Black. Each of these branches will bifurcate, and their branches again, and so on till we reach the ultimate differentiae, whose number will be four or some other power of two, and will also be the number of the ultimate species comprehended in the order.

(A species is constituted by the combination differentia and matter. For no part of an animal is purely material or purely immaterial; nor can a body, independently of its condition, constitute an animal or any of its parts, as has repeatedly been observed.)

Further, the differentiae must be elements of the essence, and not merely essential attributes. Thus if Figure is the term to be divided, it must not be divided into figures whose angles are equal to two right angles, and figures whose angles are together greater than two right angles. For it is only an attribute of a triangle and not part of its essence that its angles are equal to two right angles.

Again, the bifurcations must be opposites, like White and Black, Straight and Bent; and if we characterize one branch by either term, we must characterize the other by its opposite, and not, for example, characterize one branch by a colour, the other by a mode of progression, swimming for instance.

Furthermore, living beings cannot be divided by the functions common to body and soul, by Flying, for instance, and Walking, as we see them divided in the dichotomies already referred to. For some groups, Ants for instance, fall under both divisions, some ants flying while others do not. Similarly as regards the division into Wild and Tame; for it also would involve the disruption of a species into different groups. For in almost all species in which some members are tame, there are other members that are wild. Such, for example, is the case with Men, Horses, Oxen, Dogs in India, Pigs, Goats, Sheep; groups which, if double, ought to have what they have not, namely, different appellations; and which, if single, prove that Wildness and Tameness do not amount to specific differences. And whatever single element we take as a basis of division the same difficulty will occur.

The method then that we must adopt is to attempt to recognize the natural groups, following the indications afforded by the instincts of mankind, which led them for instance to form the class of Birds and the class of Fishes, each of which groups combines a multitude of differentiae, and is not defined by a single one as in dichotomy. The method of dichotomy is either impossible (for it would put a single group under different divisions or contrary groups under the same division), or it only furnishes a single ultimate differentia for each species, which either alone or with its series of antecedents has to constitute the ultimate species.

If, again, a new differential character be introduced at any stage into the division, the necessary result is that the continuity of the division becomes merely a unity and continuity of agglomeration, like the unity and continuity of a series of sentences coupled together by conjunctive particles. For instance, suppose we have the bifurcation Feathered and Featherless, and then divide Feathered into Wild and Tame, or into White and Black. Tame and White are not a differentiation of Feathered, but are the commencement of an independent bifurcation, and are foreign to the series at the end of which they are introduced.

As we said then, we must define at the outset by multiplicity of differentiae. If we do so, privative terms will be available, which are unavailable to the dichotomist.

The impossibility of reaching the definition of any of the ultimate forms by dichotomy of the larger group, as some propose, is manifest also from the following considerations. It is impossible that a single differentia, either by itself or with its antecedents, shall express the whole essence of a species. (In saying a single differentia by itself I mean such an isolated differentia as Cleft-footed; in saying a single differentia with antecedent I mean, to give an instance, Manycleft-footed preceded by Cleft-footed. The very continuity of a series of successive differentiae in a division is intended to show that it is their combination that expresses the character of the resulting unit, or ultimate group. But one is misled by the usages of language into imagining that it is merely the final term of the series, Manycleft-footed for instance, that constitutes the whole differentia, and that the antecedent terms, Footed, Cleft-footed, are superfluous. Now it is evident that such a series cannot consist of many terms. For if one divides and subdivides, one soon reaches the final differential term, but for all that will not have got to the ultimate division, that is, to the species.) No single differentia, I repeat, either by itself or with its antecedents, can possibly express the essence of a species. Suppose, for example, Man to be the animal to be defined; the single differentia will be Cleft-footed, either by itself or with its antecedents, Footed and Two-footed. Now if man was nothing more than a Cleft-footed animal, this single differentia would duly represent his essence. But seeing that this is not the case, more differentiae than this one will necessarily be required to define him; and these cannot come under one division; for each single branch of a dichotomy ends in a single differentia, and cannot possibly include several differentiae belonging to one and the same animal.

It is impossible then to reach any of the ultimate animal forms by dichotomous division.

Part 4

It deserves inquiry why a single name denoting a higher group was not invented by mankind, as an appellation to comprehend the two groups of Water animals and Winged animals. For even these have certain attributes in common. However, the present nomenclature is just. Groups that only differ in degree, and in the more or less of an identical element that they possess, are aggregated under a single class; groups whose attributes are not identical but analogous are separated. For instance, bird differs from bird by gradation, or by excess and defect; some birds have long feathers, others short ones, but all are feathered. Bird and Fish are more remote and only agree in having analogous organs; for what in the bird is feather, in the fish is scale. Such analogies can scarcely, however, serve universally as indications for the formation of groups, for almost all animals present analogies in their corresponding parts.

The individuals comprised within a species, such as Socrates and Coriscus, are the real existences; but inasmuch as these individuals possess one common specific form, it will suffice to state the universal attributes of the species, that is, the attributes common to all its individuals, once for all, as otherwise there will be endless reiteration, as has already been pointed out.

But as regards the larger groups—such as Birds—which comprehend many species, there may be a question. For on the one hand it may be urged that as the ultimate species represent the real existences, it will be well, if practicable, to examine these ultimate species separately, just as we examine the species Man separately; to examine, that is, not the whole class Birds collectively, but the Ostrich, the Crane, and the other indivisible groups or species belonging to the class.

On the other hand, however, this course would involve repeated mention of the same attribute, as the same attribute is common to many species, and so far would be somewhat irrational and tedious. Perhaps, then, it will be best to treat generically the universal attributes of the groups that have a common nature and contain closely allied subordinate forms, whether they are groups recognized by a true instinct of mankind, such as Birds and Fishes, or groups not popularly known by a common appellation, but withal composed of closely allied subordinate groups; and only to deal individually with the attributes of a single species, when such species, man, for instance, and any other such, if such there be—stands apart from others, and does not constitute with them a larger natural group.

It is generally similarity in the shape of particular organs, or of the whole body, that has determined the formation of the larger groups. It is in virtue of such a similarity that Birds, Fishes, Cephalopoda, and Testacea have been made to form each a separate class. For within the limits of each such class, the parts do not differ in that they have no nearer resemblance than that of analogy—such as exists between the bone of man and the spine of fish—but differ merely in respect of such corporeal conditions as largeness smallness, softness hardness, smoothness roughness, and other similar oppositions, or, in one word, in respect of degree.

We have now touched upon the canons for criticizing the method of natural science, and have considered what is the most systematic and easy course of investigation; we have also dealt with division, and the mode of conducting it so as best to attain the ends of science, and have shown why dichotomy is either impracticable or inefficacious for its professed purposes.

Having laid this foundation, let us pass on to our next topic.

Translated by William Ogle

Reading and Discussion Questions

1. Does Aristotle think that biologists should proceed as mathematicians and astronomers do or should they follow some other method? For biology might the descriptive project be so large and so difficult that it would be a mistake to "proceed subsequently to treat of the causes and the reason why"?
2. One difficult descriptive question Aristotle wrestles with is the question of how to classify living things. What does he say about the existing debates on the subject of classification, and what distinctions does he think are relevant? What methods of classification does he worry lead us in the wrong direction?

On the Generation of Animals (c. 350 BCE)

Aristotle

14

Book I

1

We have now discussed the other parts of animals, both generally and with reference to the peculiarities of each kind, explaining how each part exists on account of such a cause, and I mean by this the final cause.

There are four causes underlying everything: first, the final cause, that for the sake of which a thing exists; secondly, the formal cause, the definition of its essence (and these two we may regard pretty much as one and the same); thirdly, the material; and fourthly, the moving principle or efficient cause.

We have then already discussed the other three causes, for the definition and the final cause are the same, and the material of animals is their parts of the whole animal the non-homogeneous parts, of these again the homogeneous, and of these last the so-called elements of all matter. It remains to speak of those parts which contribute to the generation of animals and of which nothing definite has yet been said, and to explain what is the moving or efficient cause. To inquire into this last and to inquire into the generation of each animal is in a way the same thing; and, therefore, my plan has united them together, arranging the discussion of these parts last, and the beginning of the question of generation next to them.

Now some animals come into being from the union of male and female, i.e. all those kinds of animal which possess the two sexes. This is not the case with all of them; though in the sanguinea with few exceptions the creature, when its growth is complete, is either male or female, and though some bloodless animals have sexes so that they generate offspring of the same kind, yet other bloodless animals generate indeed, but not offspring of the same kind; such are all that come into being not from a union of the sexes, but from decaying earth and excrements. To speak generally, if we take all animals which change their locality, some by swimming, others by flying, others by walking, we find in these the two sexes, not only in the sanguinea but also in some of the bloodless animals; and this applies in the case of the latter sometimes to the whole class, as the cephalopoda and crustacea, but in the class of insects only to the majority. Of these, all which are produced by union of animals of the same kind generate also after their kind, but all which are not produced by animals, but from decaying matter, generate indeed, but produce another kind, and the offspring is neither male nor female; such are some of the insects. This is what might have been expected, for if those animals which are not produced by parents had themselves united and produced others, then their offspring must have been either like or unlike to themselves. If like, then their parents ought to have come into being in the same way; this is only a reasonable postulate to make, for it is plainly the case with other animals. If unlike, and yet able to copulate, then there would have come into being again from them another kind of creature and again another from these, and this would have gone on to infinity. But Nature flies from the infinite, for the infinite is unending or imperfect, and Nature ever seeks an end.

But all those creatures which do not move, as the testacea and animals that live by clinging to something else, inasmuch as their nature resembles that of plants, have no sex any more than plants have, but as applied to them the word is only used in virtue of a similarity and analogy. For there is a slight distinction of this sort, since even in plants we find in the same kind some trees which bear fruit and others which, while bearing none themselves, yet contribute to the ripening of the fruits of those which do, as in the case of the fig-tree and caprifig.

The same holds good also in plants, some coming into being from seed and others, as it were, by the spontaneous action of Nature, arising either from decomposition of the earth or of some parts in other plants, for some are not formed by themselves separately but are produced upon other trees, as the mistletoe. Plants, however, must be investigated separately.

2

Of the generation of animals we must speak as various questions arise in order in the case of each, and we must connect our account with what has been said. For, as we said above, the male and female principles may be put down first and foremost as origins of generation, the former as containing the efficient cause of generation, the latter the material of it. The most conclusive proof of this is drawn from considering how and whence comes the semen; for there is no doubt that it is out of this that those creatures are formed which are produced in the ordinary course of Nature; but we must observe carefully the way in which this semen actually comes into being from the male and female. For it is just because the semen is secreted from the two sexes, the secretion taking place in them and from them, that they are first principles of generation. For by a male animal we mean that which generates in another, and by a female that which generates in itself; wherefore men apply these terms to the macrocosm also, naming Earth mother as being female, but addressing Heaven and the Sun and other like entities as fathers, as causing generation.

Male and female differ in their essence by each having a separate ability or faculty, and anatomically by certain parts; essentially the male is that which is able to generate in another, as said above; the female is that which is able to generate in itself and out of which comes into being the offspring previously existing in the parent. And since they are differentiated by an ability or faculty and by their function, and since instruments or organs are needed for all functioning, and since the bodily parts are the instruments or organs to serve the faculties, it follows that certain parts must exist for union of parents and production of offspring. And these must differ from each other, so that consequently the male will differ from the female. (For even though we speak of the animal as a whole as male or female, yet really it is not male or female in virtue of the whole of itself, but only in virtue of a certain faculty and a certain part—just as with the part used for sight or locomotion—which part is also plain to sense-perception.)

Now as a matter of fact such parts are in the female the so-called uterus, in the male the testes and the penis, in all the sanguinea; for some of them have testes and others the corresponding passages. There are corresponding differences of male and female in all the bloodless animals also which have this division into opposite sexes. But if in the sanguinea it is the parts concerned in copulation that differ primarily in their forms, we must observe that a small change in a first principle is often attended by changes in other things depending on it. This is plain in the case of castrated animals, for, though only the generative part is disabled, yet pretty well the whole form of the animal changes in consequence so much that it seems to be female or not far short of it, and thus it is clear than an animal is not male or female in virtue of an isolated part or an isolated faculty. Clearly, then, the distinction of sex is a first principle; at any rate, when that which distinguishes male and female suffers change, many other changes accompany it, as would be the case if a first principle is changed.

Book II

1

That the male and the female are the principles of generation has been previously stated, as also what is their power and their essence. But why is it that one thing becomes and is male, another female? It is the business of our discussion as it proceeds to try and point out (1) that the sexes arise from Necessity and the first efficient cause, (2) from what sort of material they are formed. That (3) they exist because it is better and on account of the final cause, takes us back to a principle still further remote.

Now (1) some existing things are eternal and divine whilst others admit of both existence and non-existence. But (2) that which is noble and divine is always, in virtue of its own nature, the cause of the better in such things as admit of being better or worse, and what is not eternal does admit of existence and non-existence, and can partake in the better and the worse. And (3) soul is better than body, and living, having soul, is thereby better than the lifeless which has none, and being is better than not being, living than not living. These, then, are the reasons of the generation of animals. For since it is impossible that such a class of things as animals should be of an eternal nature, therefore that which comes into being is eternal in the only way possible. Now it is impossible for it to be eternal as an individual (though of course the real essence of things is in the individual)—were it such it would be eternal—but it is possible for it as a species. This is why there is always a class of men and animals and plants. But since the male and female essences are the first principles of these, they will exist in the existing individuals for the sake of generation. Again, as the first efficient or moving cause, to which belong the definition and the form, is better and more divine in its nature than the material on which it works, it is better that the superior principle should be separated from the inferior. Therefore, wherever it is possible and so far as it is possible, the male is separated from the female. For the first principle of the movement, or efficient cause, whereby that which comes into being is male, is better and more divine than the material whereby it is female. The male, however, comes together and mingles with the female for the work of generation, because this is common to both.

A thing lives, then, in virtue of participating in the male and female principles, wherefore even plants have some kind of life; but the class of animals exists in virtue of sense-perception. The sexes are divided in nearly all of these that can move about, for the reasons already stated, and some of them, as said before, emit semen in copulation, others not. The reason of this is that the higher animals are more independent in their nature, so that they have greater size, and this cannot exist without vital heat; for the greater body requires more force to move it, and heat is a motive force. Therefore, taking a general view, we may say that sanguinea are of greater size than bloodless animals, and those which move about than those which remain fixed. And these are just the animals which emit semen on account of their heat and size.

So much for the cause of the existence of the two sexes. Some animals bring to perfection and produce into the world a creature like themselves, as all those which bring their young into the world alive; others produce something undeveloped which has not yet acquired its own form; in this latter division the sanguinea lay eggs, the bloodless animals either lay an egg or give birth to a scolex. The difference between egg and scolex is this: an egg is that from a part of which the young comes into being, the rest being nutriment for it; but the whole of a scolex is developed into the whole of the young animal. Of the vivipara, which bring into the world an animal like themselves, some are internally viviparous (as men, horses, cattle, and of marine animals dolphins and the other cetacea); others first lay eggs within themselves, and only after this are externally viviparous (as the cartilaginous fishes). Among the ovipara some produce the egg in a perfect condition (as birds and all oviparous quadrupeds and footless animals, e.g. lizards and tortoises and most snakes; for the eggs of all these do not increase when once laid). The eggs of others are imperfect; such are those of fishes, crustaceans, and cephalopods, for their eggs increase after being produced.

All the vivipara are sanguineous, and the sanguinea are either viviparous or oviparous, except those which are altogether infertile. Among bloodless animals the insects produce a scolex, alike those that are generated by copulation and those that copulate themselves though not so generated. For there are some insects of this sort, which though they come into being by spontaneous generation are yet male and female; from their union something is produced, only it is imperfect; the reason of this has been previously stated.

These classes admit of much cross-division. Not all bipeds are viviparous (for birds are oviparous), nor are they all oviparous (for man is viviparous), nor are all quadrupeds oviparous (for horses, cattle, and countless others are viviparous), nor are they all viviparous (for lizards, crocodiles, and many others lay eggs).

Nor does the presence or absence of feet make the difference between them, for not only are some footless animals viviparous, as vipers and the cartilaginous fishes, while others are oviparous, as the other fishes and serpents, but also among those which have feet many are oviparous and many viviparous, as the quadrupeds above mentioned. And some which have feet, as man, and some which have not, as the whale and dolphin, are internally viviparous. By this character then it is not possible to divide them, nor is any of the locomotive organs the cause of this difference, but it is those animals which are more perfect in their nature and participate in a purer element which are viviparous, for nothing is internally viviparous unless it receive and breathe out air. But the more perfect are those which are hotter in their nature and have more moisture and are not earthy in their composition. And the measure of natural heat is the lung when it has blood in it, for generally those animals which have a lung are hotter than those which have not, and in the former class again those whose lung is not spongy nor solid nor containing only a little blood, but soft and full of blood. And as the animal is perfect but the egg and the scolex are imperfect, so the perfect is naturally produced from the more perfect. If animals are hotter as shown by their possessing a lung but drier in their nature, or are colder but have more moisture, then they either lay a perfect egg or are viviparous after laying an egg within themselves. For birds and scaly reptiles because of their heat produce a perfect egg, but because of their dryness it is only an egg; the cartilaginous fishes have less heat than these but more moisture, so that they are intermediate, for they are both oviparous and viviparous within themselves, the former because they are cold, the latter because of their moisture; for moisture is vivifying, whereas dryness is furthest removed from what has life. Since they have neither feathers nor scales such as either reptiles or other fishes have, all which are signs rather of a dry and earthy nature, the egg they produce is soft; for the earthy matter does not come to the surface in their eggs any more than in themselves. This is why they lay eggs in themselves, for if the egg were laid externally it would be destroyed, having no protection.

Animals that are cold and rather dry than moist also lay eggs, but the egg is imperfect; at the same time, because they are of an earthy nature and the egg they produce is imperfect, therefore it has a hard integument that it may be preserved by the protection of the shell-like covering. Hence fishes, because they are scaly, and crustacea, because they are of an earthy nature, lay eggs with a hard integument.

The cephalopods, having themselves bodies of a sticky nature, preserve in the same way the imperfect eggs they lay, for they deposit a quantity of sticky material about the embryo. All insects produce a scolex. Now all the insects are bloodless, wherefore all creatures that produce a scolex from themselves are so. But we cannot say simply that all bloodless animals produce a scolex, for the classes overlap one another, (1) the insects, (2) the animals that produce a scolex, (3) those that lay their egg imperfect, as the scaly fishes, the crustacea, and the cephalopoda. I say that these form a gradation, for the eggs of these latter resemble a scolex, in that they increase after oviposition, and the scolex of insects again as it develops resembles an egg; how so we shall explain later.

We must observe how rightly Nature orders generation in regular gradation. The more perfect and hotter animals produce their young perfect in respect of quality (in respect of quantity this is so with no animal, for the young always increase in size after birth), and these generate living animals within themselves from the first. The second class do not generate perfect animals within themselves from the first (for they are only viviparous after first laying eggs), but still they are externally viviparous. The third class do not produce a perfect animal, but an egg, and this egg is perfect. Those whose nature is still colder than these produce an egg, but an imperfect one, which is perfected outside the body, as the class of scaly fishes, the crustacea, and the cephalopods. The fifth and coldest class does not even lay an egg from itself; but so far as the young ever attain to this condition at all, it is outside the body of the parent, as has been said already. For insects produce a scolex first; the scolex after developing becomes egg-like (for the so-called chrysalis or pupa is equivalent to an egg); then from this it is that a perfect animal comes into being, reaching the end of its development in the second change.

Some animals then, as said before, do not come into being from semen, but all the sanguinea do so which are generated by copulation, the male emitting semen into the female when this has entered into her the young are formed and assume their peculiar character, some within the animals themselves when they are viviparous, others in eggs.

There is a considerable difficulty in understanding how the plant is formed out of the seed or any animal out of the semen. Everything that comes into being or is made must (1) be made out of something, (2) be made by the agency of something, and (3) must become something. Now that out of which it is made is the material; this some animals have in its first form within themselves, taking it from the female parent, as all those which are not born alive but produced as a scolex or an egg; others receive it from the mother for a long time by sucking, as the young of all those which are not only externally but also internally viviparous. Such, then, is the material out of which things come into being, but we now are inquiring not out of what the parts of an animal are made, but by what agency. Either it is something external which makes them, or else something existing in the seminal fluid and the semen; and this must either be soul or a part of soul, or something containing soul.

Now it would appear irrational to suppose that any of either the internal organs or the other parts is made by something external, since one thing cannot set up a motion in another without touching it, nor can a thing be affected in any way by another if it does not set up a motion in it. Something then of the sort we require exists in the embryo itself, being either a part of it or separate from it. To suppose that it should be something else separate from it is irrational. For after the animal has been produced does this something perish or does it remain in it? But nothing of the kind appears to be in it, nothing which is not a part of the whole plant or animal. Yet, on the other hand, it is absurd to say that it perishes after making either all the parts or only some of them. If it makes some of the parts and then perishes, what is to make the rest of them? Suppose this something makes the heart and then perishes, and the heart makes another organ, by the same argument either all the parts must perish or all must remain. Therefore it is preserved and does not perish. Therefore it is a part of the embryo itself which exists in the semen from the beginning; and if indeed there is no part of the soul which does not exist in some part of the body, it would also be a part containing soul in it from the beginning.

How, then, does it make the other parts? Either all the parts, as heart, lung, liver, eye, and all the rest, come into being together or in succession, as is said in the verse ascribed to Orpheus, for there he says that an animal comes into being in the same way as the knitting of a net. That the former is not the fact is plain even to the senses, for some of the parts are clearly visible as already existing in the embryo while others are not; that it is not because of their being too small that they are not visible is clear, for the lung is of greater size than the heart, and yet appears later than the heart in the original development. Since, then, one is earlier and another later, does the one make the other, and does the later part exist on account of the part which is next to it, or rather does the one come into being only after the other? I mean, for instance, that it is not the fact that the heart, having come into being first, then makes the liver, and the liver again another organ, but that the liver only comes into being after the heart, and not by the agency of the heart, as a man becomes a man after being a boy, not by his agency. An explanation of this is that, in all the productions of Nature or of art, what already exists potentially is brought into being only by what exists actually; therefore if one organ formed another the form and the character of the later organ would have to exist in the earlier, e.g. the form of the liver in the heart. And otherwise also the theory is strange and fictitious.

Yet again, if the whole animal or plant is formed from semen or seed, it is impossible that any part of it should exist ready made in the semen or seed, whether that part be able to make the other parts or no. For it is plain that, if it exists in it from the first, it was made by that which made the semen. But semen must be made first, and that is the function of the generating parent. So, then, it is not possible that any part should exist in it, and therefore it has not within itself that which makes the parts.

But neither can this agent be external, and yet it must needs be one or other of the two. We must try, then, to solve this difficulty, for perhaps some one of the statements made cannot be made without qualification, e.g. the statement that the parts cannot be made by what is external to the semen. For if in a certain sense they cannot, yet in another sense they can. (Now it makes no difference whether we say 'the semen' or 'that from which the semen comes', in so far as the semen has in itself the movement initiated by the other.)

It is possible, then, that A should move B, and B move C; that, in fact, the case should be the same as with the automatic machines shown as curiosities. For the parts of such machines while at rest have a sort of potentiality of motion in them, and when any external force puts the first of them in motion, immediately the next is moved in actuality. As, then, in these automatic machines the external force moves the parts in a certain sense (not by touching any part at the moment, but by having touched one previously), in like manner also that from which the semen comes, or in other words that which made the semen, sets up the movement in the embryo and makes the parts of it by having first touched something though not continuing to touch it. In a way it is the innate motion that does this, as the act of building builds the house. Plainly, then, while there is something which makes the parts, this does not exist as a definite object, nor does it exist in the semen at the first as a complete part.

But how is each part formed? We must answer this by starting in the first instance from the principle that, in all products of Nature or art, a thing is made by something actually existing out of that which is potentially such as the finished product. Now the semen is of such a nature, and has in it such a principle of motion, that when the motion is ceasing each of the parts comes into being, and that as a part having life or soul. For there is no such thing as face or flesh without life or soul in it; it is only equivocally that they will be called face or flesh if the life has gone out of them, just as if they had been made of stone or wood. And the homogeneous parts and the organic come into being together. And just as we should not say that an axe or other instrument or organ was made by the fire alone, so neither shall we say that foot or hand were made by heat alone. The same applies also to flesh, for this too has a function. While, then, we may allow that hardness and softness, stickiness and brittleness, and whatever other qualities are found in the parts that have life and soul, may be caused by mere heat and cold, yet, when we come to the principle in virtue of which flesh is flesh and bone is bone, that is no longer so; what makes them is the movement set up by the male parent, who is in actuality what that out of which the offspring is made is in potentiality. This is what we find in the products of art; heat and cold may make the iron soft and hard, but what makes a sword is the movement of the tools employed, this movement containing the principle of the art. For the art is the starting-point and form of the product; only it exists in something else, whereas the movement of Nature exists in the product itself, issuing from another nature which has the form in actuality.

Has the semen soul, or not? The same argument applies here as in the question concerning the parts. As no part, if it participate not in soul, will be a part except in an equivocal sense (as the eye of a dead man is still called an 'eye'), so no soul will exist in anything except that of which it is soul; it is plain therefore that semen both has soul, and is soul, potentially.

But a thing existing potentially may be nearer or further from its realization in actuality, as e.g. a mathematician when asleep is further from his realization in actuality as engaged in mathematics than when he is awake, and when awake again but not studying mathematics he is further removed than when he is so studying. Accordingly it is not any part that is the cause of the soul's coming into being, but it is the first moving cause from outside. (For nothing generates itself, though when it has come into being it thenceforward increases itself.) Hence it is that only one part comes into being first and not all of them together. But that must first come into being which has a principle of increase (for this nutritive power exists in all alike, whether animals or plants, and this is the same as the power that enables an animal or plant to generate another like itself, that being the function of them all if naturally perfect). And this is necessary for the reason that whenever a living thing is produced it must grow. It is produced, then, by something else of the same name, as e.g. man is produced by man, but it is increased by means of itself. There is, then, something which increases it. If this is a single part, this must come into being first. Therefore if the heart is first made in some animals, and what is analogous to the heart in the others which have no heart, it is from this or its analogue that the first principle of movement would arise.

We have thus discussed the difficulties previously raised on the question what is the efficient cause of generation in each case, as the first moving and formative power.

Translated by Arthur Platt

Reading and Discussion Questions

1. How does Aristotle use the four causes and the distinction between form and matter in discussing the male and female roles in reproduction?
2. How does Aristotle's discussion of the perfection of animals relate to his ideas about final cause, potency, and act?

On the Equilibrium of Planes or the Centres of Gravity of Planes (late third century BCE)

Archimedes

15

Book I

"I POSTULATE the following:

1. Equal weights at equal distances are in equilibrium, and equal weights at unequal distances are not in equilibrium but incline towards the weight which is at the greater distance.
2. If, when weights at certain distances are in equilibrium, something be added to one of the weights, they are not in equilibrium but incline towards that weight to which the addition was made.
3. Similarly, if anything be taken away from one of the weights, they are not in equilibrium but incline towards the weight from which nothing was taken.
4. When equal and similar plane figures coincide if applied to one another, their centres of gravity similarly coincide.
5. In figures which are unequal but similar the centres of gravity will be similarly situated. By points similarly situated in relation to similar figures I mean points such that, if straight lines be drawn from them to the equal angles, they make equal angles with the corresponding sides.
6. If magnitudes at certain distances be in equilibrium, (other) magnitudes equal to them will also be in equilibrium at the same distances.
7. In any figure whose perimeter is concave in (one and) the same direction the centre of gravity must be within the figure."

Proposition 1

Weights which balance at equal distances are equal.

For, if they are unequal, take away from the greater the difference between the two. The remainders will then not balance [*Post.* 3]; which is absurd.

Therefore the weights cannot be unequal.

Proposition 2

Unequal weights at equal distances will not balance but will incline towards the greater weight.

For take away from the greater the difference between the two. The equal remainders will therefore balance [*Post.* 1]. Hence, if we add the difference again, the weights will not balance but incline towards the greater [*Post.* 2].

Proposition 3

Unequal weights will balance at unequal distances, the greater weight being at the lesser distance.

Let A, B be two unequal weights (of which A is the greater) balancing about C at distances AC, BC respectively.

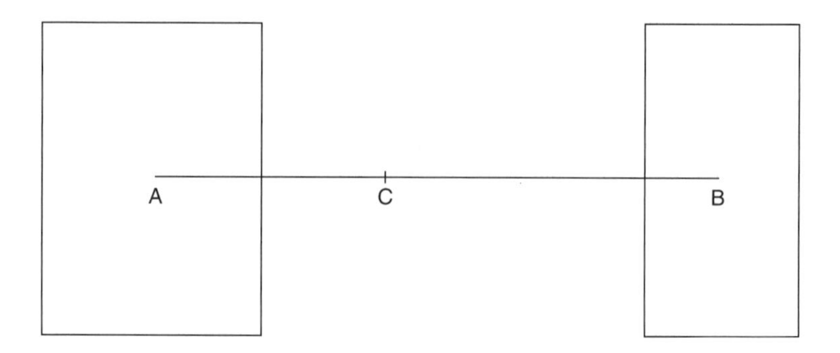

Then shall AC be less than BC. For, if not, take away from A the weight $(A — B)$. The remainders will then incline towards B [*Post.* 3]. But this is impossible, for (1) if $AC = CB$, the equal remainders will balance, or (2) if $AC > CB$; they will incline towards A at the greater distance [*Post.* 1].

Hence $AC < CB$.

Conversely, if the weights balance, and $AC < CB$, then $A > B$.

Proposition 4

If two equal weights have not the same centre of gravity, the centre of gravity of both taken together is at the middle point of the line joining their centres of gravity.

[Proved from Prop. 3 by *reductio ad absurdum.* Archimedes assumes that the centre of gravity of both together is on the straight line joining the centres of gravity of each, saying that this had been proved before. The allusion is no doubt to the lost treatise *On Levers.*]

Proposition 5

If three equal magnitudes have their centres of gravity on a straight line at equal distances, the centre of gravity of the system will coincide with that of the middle magnitude.

[This follows immediately from Prop. 4.]

COR. 1. *The same is true of any odd number of magnitudes if those which are at equal distances from the middle one are equal, while the distances between their centres of gravity are equal.*

COR. 2. *If there be an even number of magnitudes with their centres of gravity situated at equal distances on one straight line, and if the two middle ones be equal, while those which are equidistant from them (on each side) are equal respectively, the centre of gravity of the system is the middle point of the line joining the centres of gravity of the two middle ones.*

Propositions 6, 7

Two magnitudes, whether commensurable [Prop. 6] *or incommensurable* [Prop. 7], *balance at distances reciprocally proportional to the magnitudes.*

I. Suppose the magnitudes A, B to be commensurable, and the points A, B to be their centres of gravity. Let DE be a straight line so divided at C that

$$A:B = DC:CE.$$

We have then to prove that, if A be placed at E and B at D, C is the centre of gravity of the two taken together.

Since A, B are commensurable, so are DC, CE. Let N be a common measure of DC, CE. Make DH, DK each equal to CE, and EL (on CE produced) equal to CD. Then $EH = CD$, since $DH = CE$. Therefore LH is bisected at E, as HK is bisected at D.

Thus LH, HK must each contain N an even number of times.

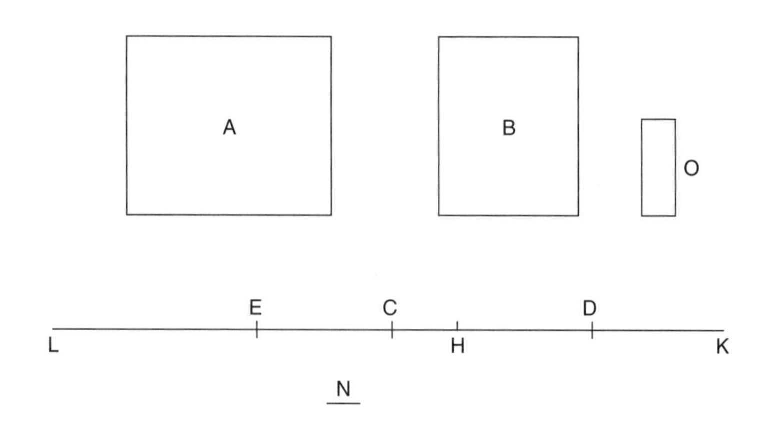

Take a magnitude *O* such that *O* is contained as many times in *A* as *N* is contained in *LH,* whence

$$A{:}O = LH{:}N$$

But $B{:}A = CE{:}DC = HK{:}LH$.

Hence, *ex aequali,* $B{:}O = HK{:}N$, or *O* is contained in *B* as many times as *N* is contained in *HK.*

Thus *O* is a common measure of *A, B.*

Divide *LH, HK* into parts each equal to *N,* and *A, B* into parts each equal to *O.* The parts of *A* will therefore be equal in number to those of *LH,* and the parts of *B* equal in number to those of *HK.* Place one of the parts of *A* at the middle point of each of the parts *N* of *LH,* and one of the parts of *B* at the middle point of each of the parts *N* of *HK.*

Then the centre of gravity of the parts of *A* placed at equal distances on *LH* will be at *E,* the middle point of *LH* [Prop. 5, Cor. 2], and the centre of gravity of the parts of *B* placed at equal distances along *HK* will be at *D,* the middle point of *HK.*

Thus we may suppose *A* itself applied at *E,* and *B* itself applied at D.

But the system formed by the parts *O* of *A* and *B* together is a system of equal magnitudes even in number and placed at equal distances along *LK.* And, since $LE = CD$, and $EC = DK$, $LC = CK$, so that *C* is the middle point of *LK.* Therefore *C* is the centre of gravity of the system ranged along *LK.*

Therefore *A* acting at *E* and *B* acting at *D* balance about the point *C.*

II. Suppose the magnitudes to be incommensurable, and let them be $(A + a)$ and *B* respectively. Let *DE* be a line divided at *C* so that

$$(A+a){:}\ B = DG{:}\ CE.$$

Then, if $(A + a)$ placed at *E* and *B* placed at *D* do not balance about *G,* $(A + a)$ is either too great to balance *B*, or not great enough.

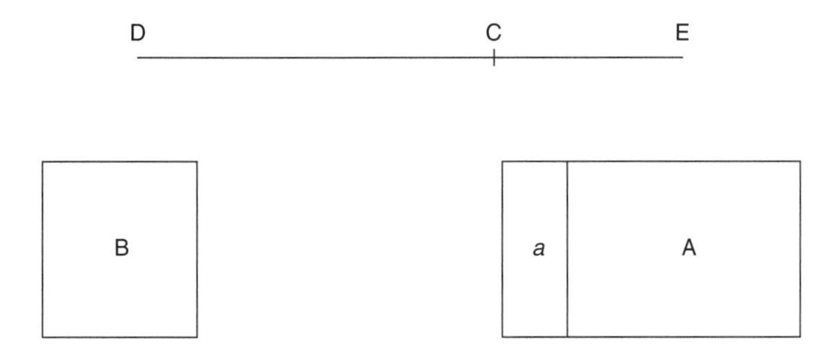

Suppose, if possible, that $(A + a)$ is too great to balance B. Take from $(A + a)$ a magnitude a smaller than the deduction which would make the remainder balance B, but such that the remainder A and the magnitude B are commensurable.

Then, since A, B are commensurable, and

$$A: B < DC: CE,$$

A and B will not balance [Prop. 6], but D will be depressed.

But this is impossible, since the deduction a was an insufficient deduction from $(A + a)$ to produce equilibrium, so that E was still depressed.

Therefore $(A + a)$ is not too great to balance B; and similarly it may be proved that B is not too great to balance $(A + a)$.

Hence $(A + a)$, B taken together have their centre of gravity at C.

Translated by T. L. Heath

Reading and Discussion Question

1. Here Archimedes is using very abstract terms—such as planes, lines, and centers—to describe a physical system with which everyone is familiar. What can we gather about how he thinks about the relationship between mathematics and the physical world?

Elements (c. 350 BCE)

Euclid

16

Book I

Definitions

1. A *point* is that which has no part.
2. A *line* is breadthless length.
3. The extremities of a line are points.
4. A *straight line* is a line which lies evenly with the points on itself.
5. A *surface* is that which has length and breadth only.
6. The extremities of a surface are lines.
7. A *plane surface* is a surface which lies evenly with the straight lines on itself.
8. A *plane angle* is the inclination to one another of two lines in a plane which meet one another and do not lie in a straight line.
9. And when the lines containing the angle are straight, the angle is called *rectilineal*.
10. When a straight line set up on a straight line makes the adjacent angles equal to one another, each of the equal angles is *right*, and the straight line standing on the other is called a *perpendicular* to that on which it stands.
11. An *obtuse angle* is an angle greater than a right angle.

12. An *acute angle* is an angle less than a right angle.
13. A *boundary* is that which is an extremity of anything.
14. A *figure* is that which is contained by any boundary or boundaries.
15. A *circle* is a plane figure contained by one line such that all the straight lines falling upon it from one point among those lying within the figure are equal to one another.
16. And the point is called the *centre* of the circle.
17. A *diameter* of the circle is any straight line drawn through the centre and terminated in both directions by the circumference of the circle, and such a straight line also bisects the circle.
18. A *semicircle* is the figure contained by the diameter and the circumference cut off by it. And the centre of the semicircle is the same as that of the circle.
19. *Rectilineal figures* are those which are contained by straight lines, *trilateral* figures being those contained by three, *quadrilateral* those contained by four, and *multilateral* those contained by more than four straight lines.
20. Of trilateral figures, an *equilateral triangle* is that which has its three sides equal, an *isosceles triangle* that which has two of its sides alone equal, and a *scalene triangle* that which has its three sides unequal.
21. Further, of trilateral figures, a *right-angled triangle* is that which has a right angle, an *obtuse-angled triangle* that which has an obtuse angle, and an *acute-angled triangle* that which has its three angles acute.
22. Of quadrilateral figures, *a square* is that which is both equilateral and right-angled; an *oblong* that which is right-angled but not equilateral; a *rhombus* that which is equilateral but not right-angled; and a *rhomboid* that which has its opposite sides and angles equal to one another but is neither equilateral nor right-angled. And let quadrilaterals other than these be called *trapezia.*
23. *Parallel* straight lines are straight lines which, being in the same plane and being produced indefinitely in both directions, do not meet one another in either direction.

Postulates

Let the following be postulated:

1. To draw a straight line from any point to any point.
2. To produce a finite straight line continuously in a straight line.

3. To describe a circle with any centre and distance.
4. That all right angles are equal to one another.
5. That, if a straight line falling on two straight lines make the interior angles on the same side less than two right angles, the two straight lines, if produced indefinitely, meet on that side on which are the angles less than the two right angles.

Common Notions

1. Things which are equal to the same thing are also equal to one another.
2. If equals be added to equals, the wholes are equal.
3. If equals be subtracted from equals, the remainders are equal.
4. Things which coincide with one another are equal to one another.
5. The whole is greater than the part.

Book I. Propositions

Proposition 1

On a given finite straight line to construct an equilateral triangle.

Let AB be the given finite straight line.

Thus it is required to construct an equilateral triangle on the straight line AB.

With centre A and distance AB let the circle BCD be described; [Post. 3] again, with centre B and distance BA let the circle ACE be described; [Post. 3]

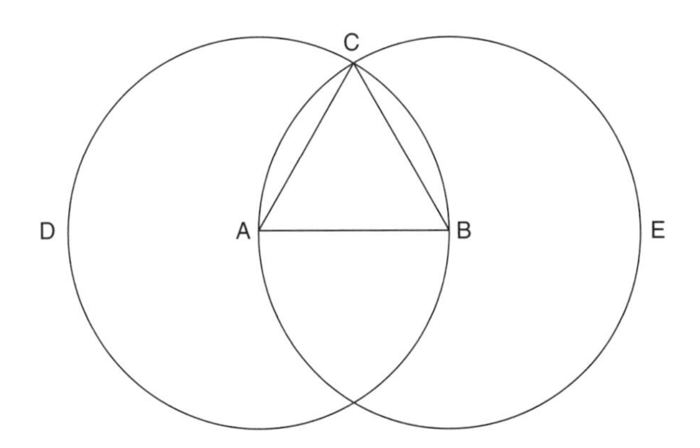

and from the point C, in which the circles cut one another, to the points A, B let the straight lines CA, CB be joined. [Post. 1]

Now, since the point A is the centre of the circle CDB, AC is equal to AB. [Def. 15]

Again, since the point B is the centre of the circle CAE, BC is equal to BA. [Def. 15]

But CA was also proved equal to AB; therefore each of the straight lines CA, CB is equal to AB.

And things which are equal to the same thing are also equal to one another; [C.N. 1] therefore CA is also equal to CB.

Therefore the three straight lines CA, AB, BC are equal to one another.

Therefore the triangle ABC is equilateral; and it has been constructed on the given finite straight line AB.

(Being) what it was required to do.

Proposition 2

To place at a given point (as an extremity) a straight line equal to a given straight line.

Let A be the given point, and BC the given straight line.

Thus it is required to place at the point A (as an extremity) a straight line equal to the given straight line BC.

From the point A to the point B let the straight line AB be joined; [Post. 1] and on it let the equilateral triangle DAB be constructed. [I. 1]

Let the straight lines AE, BF be produced in a straight line with DA, DB; [Post. 2] with centre B and distance BC let the circle CGH be described; [Post. 3] and again, with centre D and distance DG let the circle GKL be described. [Post. 3]

Then, since the point B is the centre of the circle CGH, BC is equal to BG.

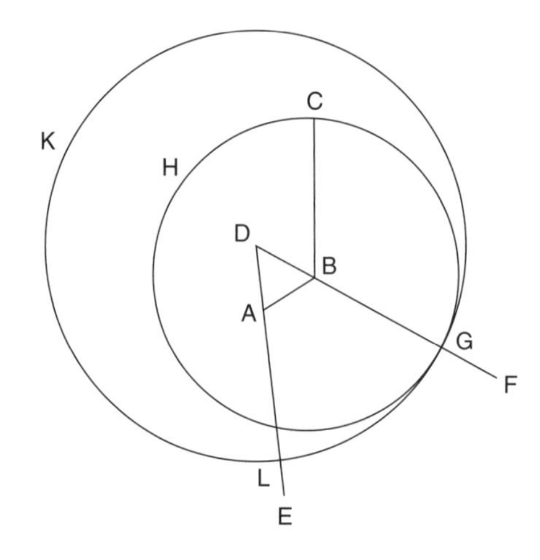

Again, since the point D is the centre of the circle GKL, DL is equal to DG.

And in these DA is equal to DB; therefore the remainder AL is equal to the remainder BG. [C.N. 3]

But BC was also proved equal to BG; therefore each of the straight lines AL, BC is equal to BG.

And things which are equal to the same thing are also equal to one another; [C.N. 1] therefore AL is also equal to BC.

Therefore at the given point A the straight line AL is placed equal to the given straight line BC.

(Being) what it was required to do.

Proposition 3

Given two unequal straight lines, to cut off from the greater a straight line equal to the less.

Let AB, C be the two given unequal straight lines, and let AB be the greater of them.

Thus it is required to cut off from AB the greater a straight line equal to C the less.

At the point A let AD be placed equal to the straight line C; [I. 2] and with centre A and distance AD let the circle DEF be described. [Post. 3]

Now, since the point A is the centre of the circle DEF, AE is equal to AD. [Def. 15] But C is also equal to AD. Therefore each of the straight lines AE, C is equal to AD; so that AE is also equal to C. [C.N. 1]

Therefore, given the two straight lines AB, C, from AB the greater AE has been cut off equal to C the less.

(Being) what it was required to do.

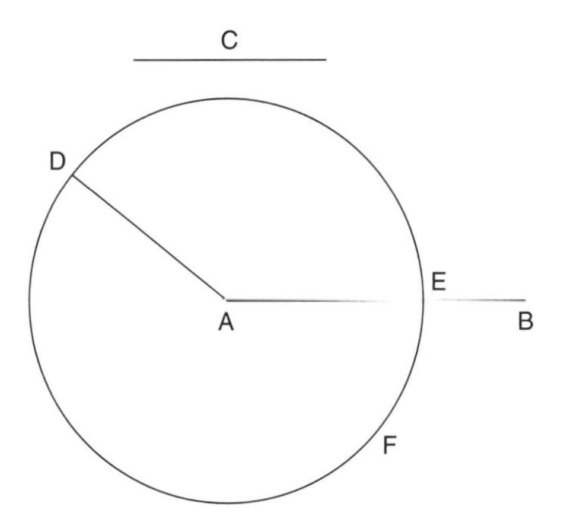

Proposition 4

If two triangles have the two sides equal to two sides respectively, and have the angles contained by the equal straight lines equal, they will also have the base equal to the base, the triangle will be equal to the triangle, and the remaining angles will be equal to the remaining angles respectively, namely those which the equal sides subtend.

Let ABC, DEF be two triangles having the two sides AB, AC equal to the two sides DE, DF respectively, namely AB to DE and AC to DF, and the angle BAC equal to the angle EDF.

I say that the base BC is also equal to the base EF, the triangle ABC will be equal to the triangle DEF, and the remaining angles will be equal to the remaining angles respectively, namely those which the equal sides subtend, that is, the angle ABC to the angle DEF, and the angle ACB to the angle DFE.

For, if the triangle ABC be applied to the triangle DEF, and if the point A be placed on the point D and the straight line AB on DE, then the point B will also coincide with E, because AB is equal to DE.

Again, AB coinciding with DE, the straight line AC will also coincide with DF, because the angle BAC is equal to the angle EDF; hence the point C will also coincide with the point F, because AC is again equal to DF.

But B also coincided with E; hence the base BC will coincide with the base EF.

[For if, when B coincides with E and C with F, the base BC does not coincide with the base EF, two straight lines will enclose a space: which is impossible.

Therefore the base BC will coincide with EF] and will be equal to it. [C.N. 4]

Thus the whole triangle ABC will coincide with the whole triangle DEF, and will be equal to it.

And the remaining angles will also coincide with the remaining angles and will be equal to them, the angle ABC to the angle DEF, and the angle ACB to the angle DFE.

Therefore etc.

(Being) what it was required to prove.

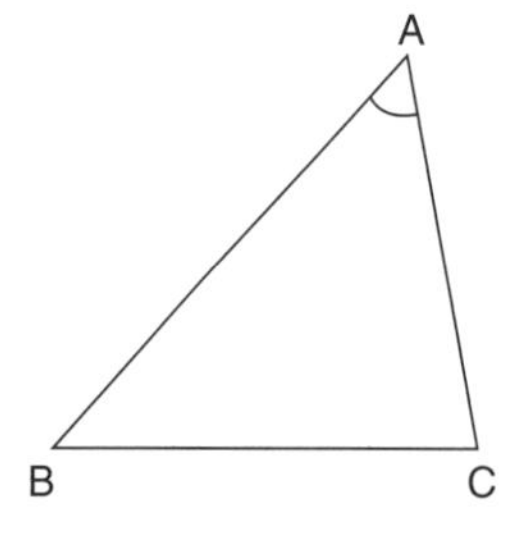

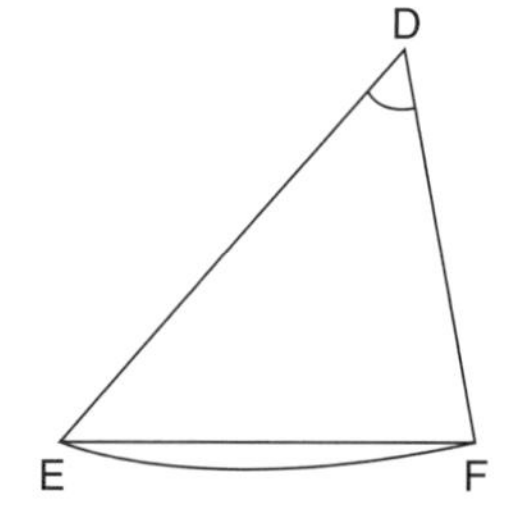

Proposition 5

In isosceles triangles the angles at the base are equal to one another, and, if the equal straight lines be produced further, the angles under the base will be equal to one another.

Let ABC be an isosceles triangle having the side AB equal to the side AC; and let the straight lines BD, CE be produced further in a straight line with AB, AC. [Post. 2]

I say that the angle ABC is equal to the angle ACB, and the angle CBD to the angle BCE.

Let a point F be taken at random on BD; from AE the greater let AG be cut off equal to AF the less; [I. 3] and let the straight lines FC, GB be joined. [Post. 1]

Then, since AF is equal to AG and AB to AC, the two sides FA, AC are equal to the two sides GA, AB, respectively; and they contain a common angle, the angle FAG.

Therefore the base FC is equal to the base GB, and the triangle AFC is equal to the triangle AGB, and the remaining angles will be equal to the remaining angles respectively, namely those which the equal sides subtend, that is, the angle ACF to the angle ABG, and the angle AFC to the angle AGB. [I. 4]

And, since the whole AF is equal to the whole AG, and in these AB is equal to AC, the remainder BF is equal to the remainder CG.

But FC was also proved equal to GB; therefore the two sides BF, FC are equal to the two sides CG, GB respectively; and the angle BFC is equal to the angle CGB, while the base BC is common to them; therefore the triangle BFC is also equal to the triangle CGB, and the remaining angles will be equal to the remaining angles respectively, namely those which the equal sides subtend; therefore the angle FBC is equal to the angle GCB, and the angle BCF to the angle CBG.

Accordingly, since the whole angle ABG was proved equal to the angle ACF, and in these the angle CBG is equal to the angle BCF, the remaining angle ABC is equal to the remaining angle ACB; and they are at the base of the triangle ABC.

But the angle FBC was also proved equal to the angle GCB; and they are under the base.

Therefore etc.

Q. E. D.

...

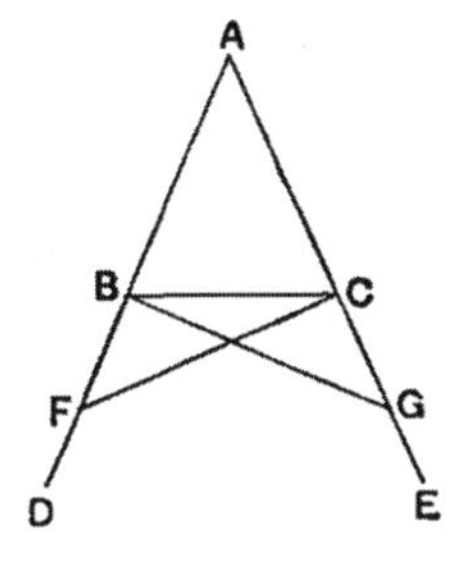

Proposition 47

In right-angled triangles the square on the side subtending the right angle is equal to the squares on the sides containing the right angle.

Let ABC be a right-angled triangle having the angle BAC right; I say that the square on BC is equal to the squares on BA, AC.

For let there be described on BC the square BDEC, and on BA, AC the squares GB, HC; [I. 46] through A let AL be drawn parallel to either BD or CE, and let AD, FC be joined.

Then, since each of the angles BAC, BAG is right, it follows that with a straight line BA, and at the point A on it, the two straight lines AC, AG not lying on the same side make the adjacent angles equal to two right angles; therefore CA is in a straight line with AG. [I. 14]

For the same reason BA is also in a straight line with AH.

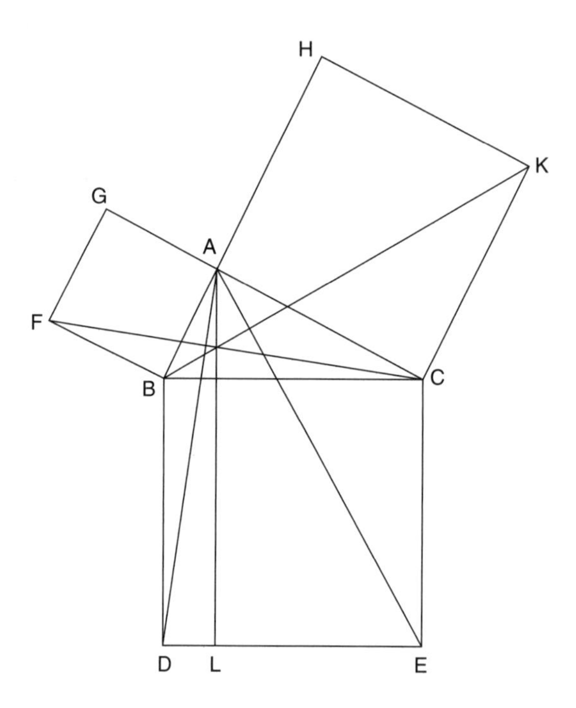

And, since the angle DBC is equal to the angle FBA—for each is right—let the angle ABC be added to each; therefore the whole angle DBA is equal to the whole angle FBC. [C.N. 2]

And, since DB is equal to BC, and FB to BA, the two sides AB, BD are equal to the two sides FB, BC respectively, and the angle ABD is equal to the angle FBC; therefore the base AD is equal to the base FC, and the triangle ABD is equal to the triangle FBC. [I. 4]

Now the parallelogram BL is double of the triangle ABD, for they have the same base BD and are in the same parallels BD, AL. [I. 41]

And the square GB is double of the triangle FBC, for they again have the same base FB and are in the same parallels FB, GC. [I. 41]

[But the doubles of equals are equal to one another.]

Therefore the parallelogram BL is also equal to the square GB.

Similarly, if AE, BK be joined, the parallelogram CL can also be proved equal to the square HC; therefore the whole square BDEC is equal to the two squares GB, HC. [C.N. 2]

And the square BDEC is described on BC, and the squares GB, HC on BA, AC.

Therefore the square on the side BC is equal to the squares on the sides BA, AC.

Therefore etc.

Q. E. D.

Translated by T. L. Heath

Reading and Discussion Questions

1. Describe the structure and methodology of the *Elements* as reflected in this reading. Could any of these items have been presented in a different order?
2. What connections do you see between Euclid's work and Plato's discussion of the Forms, Divided Line, or view of geometry? Do any of the definitions refer to things that exist in the material world?
3. Do each of the postulates and common notions strike you as uncontroversial, certain, or indisputable?
4. To what extent does Euclid's project reflect the ideal of Aristotelian demonstration? What about the structure of this work makes it an exemplar of a deductive axiomatic system?

Treatise on Conic Sections (late third century BCE)

Apollonius of Perga

17

Book I. General Preface

Apollonius to Eudemus, greeting.

If you are in good health and circumstances are in other respects as you wish, it is well; I too am tolerably well. When I was with you in Pergamum, I observed that you were eager to become acquainted with my work in conics; therefore I send you the first book which I have corrected, and the remaining books I will forward when I have finished them to my satisfaction. I daresay you have not forgotten my telling you that I undertook the investigation of this subject at the request of Naucrates the geometer at the time when he came to Alexandria and stayed with me, and that, after working it out in eight books, I communicated thein to him at once, somewhat too hurriedly, without a thorough revision (as he was on the point of sailing), but putting down all that occurred to me, with the intention of returning to them later. Wherefore I now take the opportunity of publishing each portion from time to time, as it is gradually corrected. But, since it has chanced that some other persons also who have been with me have got the first and second books before they were corrected, do not be surprised if you find them in a different shape.

Now of the eight books the first four form an elementary introduction; the first contains the modes of producing the three sections and the opposite branches [of the hyperbola] and their fundamental properties worked out more fully and generally than in the writings of other authors; the second treats of the properties of the diameters and axes of the sections as well as the asymptotes and other things of general importance and necessary for determining limits of possibility, and what I mean by diameters and axes you will learn from this book. The third book contains many remarkable theorems useful for the synthesis of solid loci and determinations of limits; the most and prettiest of these theorems are new, and, when I had discovered them, I observed that Euclid had not worked out the synthesis of the locus with respect to three and four lines, but only a chance portion of it and that not successfully: for it was not possible that the synthesis could have been completed without my additional discoveries. The fourth book shows in how many ways the sections of cones meet one another and the circumference of a circle; it contains other matters in addition, none of which has been discussed by earlier writers, concerning the number of points in which a section of a cone or the circumference of a circle meets (the opposite branches of a hyperbola).

The rest (of the books) are more by way of surplusage: one of them deals somewhat fully with *minima* and *maxima*, one with equal and similar sections of cones, one with theorems involving determination of limits, and the last with determinate conic problems.

When all the books are published it will of course be open to those who read them to judge them as they individually please. Farewell.

...

The Cone

If a straight line indefinite in length, and passing always through a fixed point, be made to move round the circumference of a circle which is not in the same plane with the point, so as to pass successively through every point of that circumference, the moving straight line will trace out the surface of a *double cone,* or two similar cones lying in opposite directions and meeting in the fixed point, which is the *apex* of each cone.

The circle about which the straight line moves is called the *base* of the cone lying between the said circle and the fixed point, and the *axis* is defined as the straight line drawn from the fixed point or the apex to the centre of the circle forming the base.

The cone so described is a *scalene* or *oblique* cone except in the particular case where the axis is perpendicular to the base. In this latter case the cone is a *right* cone.

If a cone be cut by a plane passing through the apex, the resulting section is a triangle, two sides being straight lines lying on the surface of the cone and the third side being the straight line which is the intersection of the cutting plane and the plane of the base.

Let there be a cone whose apex is A and whose base is the circle BC, and let O be the centre of the circle, so that AO is the axis of the cone. Suppose now that the cone is cut by any plane parallel to the plane of the base BC, as DE, and let the axis AO meet the plane DE in o. Let p be any point on the intersection of the plane DE and the surface of the cone. Join Ap and produce it to meet the circumference of the circle BC in P. Join OP, op.

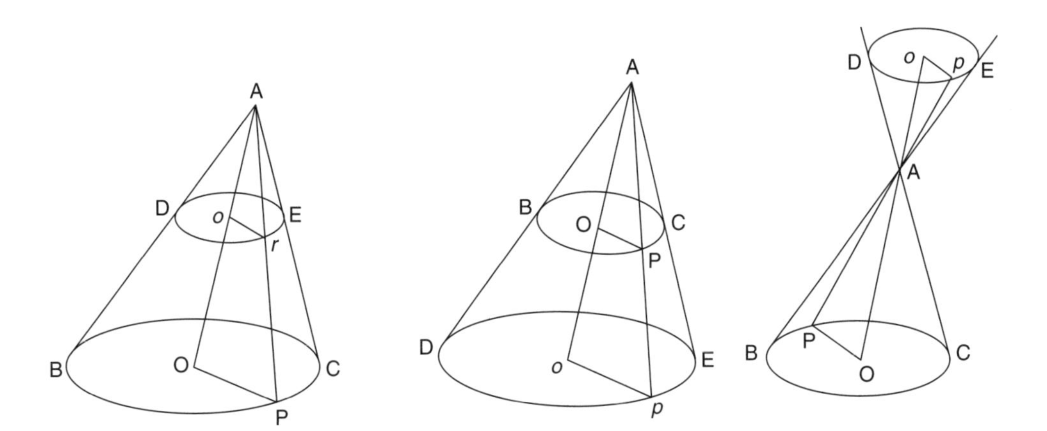

Then, since the plane passing through the straight lines AO, AP cuts the two parallel planes BC, DE in the straight lines OP, op respectively, OP, op are parallel.

$$\therefore \text{Op}: OP = Ao: AO.$$

And, BPC being a circle, OP remains constant for all positions of p on the curve DpE, and the ratio Ao: AO is also constant.

Therefore op is constant for all points on the section of the surface by the plane DE. In other words, that section is a circle.

Hence *all sections of the cone which are parallel to the circular base are circles.* [I. 4.]

. . .

Proposition 1

[I. 11.]

First let the diameter PM of the section be parallel to one of the sides of the axial triangle as AC, and let QV be any ordinate to the diameter PM. Then, if any straight line PL (supposed to be drawn perpendicular to PM in the plane of the section) be taken of such a length that $PL: PA = BC^2: BA. AC$, *it is to be proved that*

$$QV^2 = PL. PV.$$

Let HK be drawn through V parallel to BC. Then, since QV is also parallel to DE, it follows that the plane through H, Q, K is parallel to the base of the cone and therefore produces a circular section whose diameter is *HK*. Also *QV* is at right angles to *HK*.

$$\therefore HV.\ VK = QV^2.$$

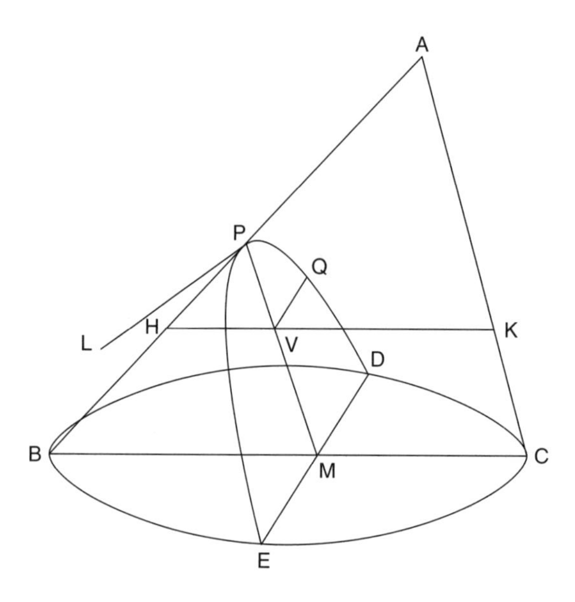

Now, by similar triangles and by parallels,

$$HV: PV = BC: AC$$

$$\text{and } VK: PA = BC: BA.$$

$$\therefore HV.\ YK: PV.\ PA = BC^2: BA.\ AC.$$

$$\text{Hence } QV^2: PV.\ PA = PL:PA$$

$$= PL.\ PV: PV: PA.$$

$$\therefore QV^2 = PL.\ PY.$$

It follows that the square on any ordinate to the fixed diameter *PM* is equal to a rectangle applied to the fixed straight line *PL* drawn at right angles to *PM* with altitude equal to the corresponding abscissa *PV*. Hence the section is called a Parabola.

The fixed straight line *PL* is called the *latus rectum* or *the parameter of the ordinates.*

This parameter, corresponding to the diameter *PM,* will for the future be denoted by the symbol *p*.

$$\text{Thus } QV^2 = p.\ PV,$$

$$\text{or } QV^2 \propto PV.$$

. . .

Proposition 3

[I. 13.]

If PM meets AC in P' and BC in M, draw AF parallel to PM meeting BC produced in F, and draw PL at right angles to PM in the plane of the section and of such a length that PL: PP'= BF. FC: AF². Join P'L and draw VR parallel to PL meeting P'L in R. It will be proved that QV² =PV. VR.

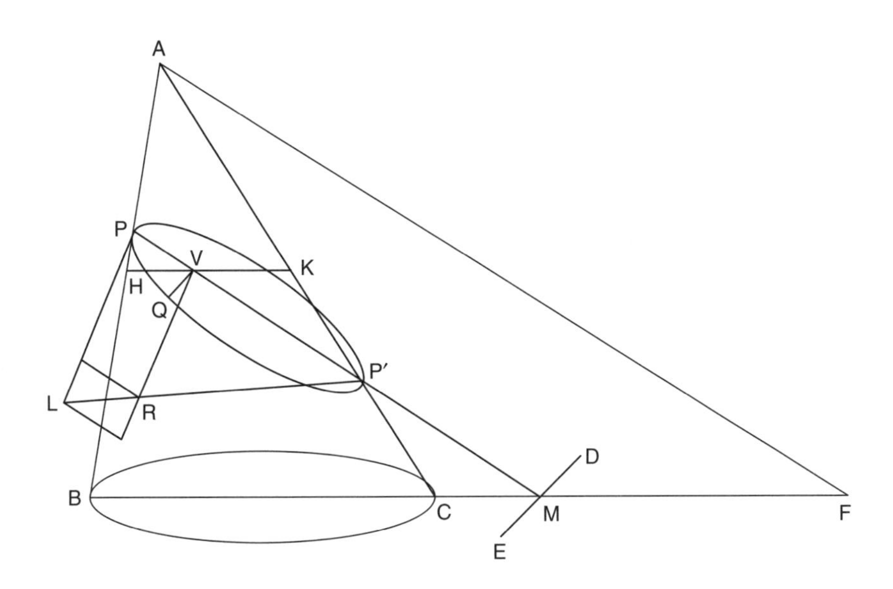

Proposition 4

[I. 14.]

If a plane cuts both parts of a double cone and does not pass through the apex, the sections of the two parts of the cone will both be hyperbolas which will have the same diameter and equal latera recta corresponding thereto. And such sections are called Opposite Branches.

Translated by T. L. Heath

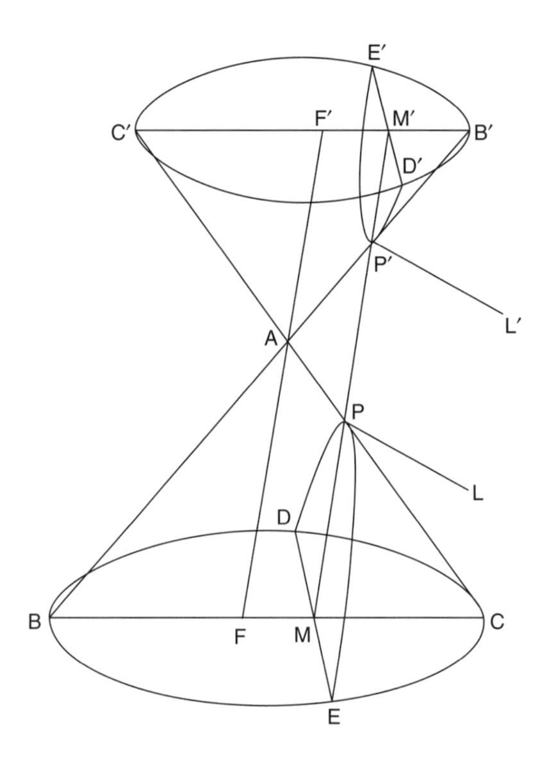

Reading and Discussion Questions

1. Apollonius assumes that we know quite a number of proofs and definitions already. Where do these come from?
2. How does Apollonius define a cone? What sorts of curves could you make by slicing through a cone with a plane?
3. Does Apollonius strike you as someone who is interested in practical applications? Can you think of any applications that his mathematics might have in later physics or astronomy?

On the Sizes and Distances of the Sun and the Moon (third century BCE)

Aristarchus of Samos

18

Lucius thereupon laughed and said: "Do not, my good fellow, bring an action against me for impiety after the manner of Cleanthes, who held that the Greeks ought to indict Aristarchus of Samos on a charge of impiety because he set in motion the hearth of the universe; for he tried to save the phenomena by supposing the heaven to remain at rest, and the earth to revolve in an inclined circle, while rotating at the same time about its own axis."

Plutarch, On the Face in the Moon

Distances of the Sun and the Moon

Hypotheses

1. The moon receives its light from the sun.
2. The earth has the relation of a point and centre to the sphere in which the moon moves.
3. When the moon appears to us halved, the great circle dividing the dark and the bright portions of the moon is in the direction of our eye.
4. When the moon appears to us halved, its distance from the sun is less than a quadrant by one-thirtieth of a quadrant.
5. The breadth of the earth's shadow is that of two moons.
6. The moon subtends one-fifteenth part of a sign of the zodiac.

It may now be proved that *the distance of the sun from the earth is greater than eighteen times, but less than twenty times, the distance of the moon*—this follows from the hypothesis about the halved moon; that *the diameter of the sun has the aforesaid ratio to the diameter of the moon*; and that *the diameter of the sun has to the diameter of the earth a ratio which is greater than* 19:3 *but less than* 43:6—this follows from the ratio discovered about the distances, the hypothesis about the shadow, and the hypothesis that the moon subtends one-fifteenth part of a sign of the zodiac.

. . .

Proposition 7

The distance of the sun from the earth is greater than eighteen times, but less than twenty times, the distance of the moon from the earth.

For let A be the centre of the sun, B that of the earth; let AB be joined and produced; let Γ be the centre of the moon when halved; let a plane be drawn through AB and Γ, and let the section made by it in the sphere on which the centre of the sun moves be the great circle AΔE, let AΓ, ΓB be joined, and let BΓ be produced to Δ.

Then, because the point Γ is the centre of the moon when halved, the angle AΓB will be right. From B let BE be drawn at right angles to BA. Then the arc EΔ will be one-thirtieth of the arc EΔA; for, by hypothesis, when the moon appears to us halved, its distance from the sun is less than a quadrant by one-thirtieth of a quadrant [Hypothesis 4]. Therefore the angle EBΓ is also one-thirtieth of a right angle. Let the parallelogram AE be completed, and let BZ be joined. Then the angle ZBE will be one-half of a right angle. Let the angle ZBE be bisected by the straight line BH; then the angle HBE is one-fourth part of a right angle. But the angle ΔBE is one-thirtieth part of a right angle; therefore angle HBE : angle ΔBE=15 : 2; for, of those parts of which a right angle contains 60, the angle HBE contains 15 and the angle ΔBE contains 2.

Now since HE : EΘ>angle HBE : angle ΔBE, therefore HE : EΘ>15 : 2.

And since BE=EZ, and the angle at E is right, therefore:

	$ZB^2=2BE^2$.
But	$ZB^2 : BE^2 = ZH^2 : HE^2$.
Therefore	$ZH^2=2HE^2$.
Now	$49<2.25$,
so that	$ZH^2 : HE^2>49 : 25$.
Therefore	$ZH : HE>7 : 5$.
Therefore, *componendo*,	$ZE : EH>12 : 5$,
that is,	$ZE : EH>36 : 15$.

But it was also proved that

	HE : EΘ>15 : 2.
Therefore, *ex aequali,*	ZE : EΘ>36 : 2,
that is,	ZE : EΘ>18 : 1.

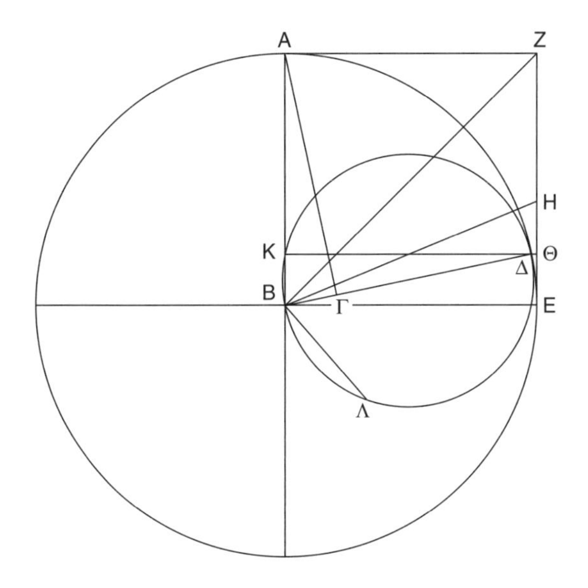

Therefore ZE is greater than eighteen times EΘ. And ZE is equal to BE. Therefore BE is also greater than eighteen times EΘ. Therefore BH is much greater than eighteen times ΘE.

But	BΘ : ΘE = AB : BΓ,

by similarity of triangles. Therefore AB is also greater than eighteen times BΓ. And AB is the distance of the sun from the earth, while ΓB is the distance of the moon from the earth; therefore the distance of the sun from the earth is greater than eighteen times the distance of the moon from the earth.

I say now that it is less than twenty times. For through Δ let ΔK be drawn parallel to EB, and about the triangle ΔKB let the circle ΔKB be drawn; its diameter will be ΔB, by reason of the angle at K being right. Let BΛ, the side of a hexagon, be fitted into the circle. Then, since the angle ΔBE is one-thirtieth of a right angle, therefore the angle BΔK is also one-thirtieth of a right angle. Therefore the arc BK is one-sixtieth of the whole circle. But BΛ is one-sixth part of the whole circle.

Therefore	arc BΛ=10. arc BK.

And the arc BΛ has to the arc BK a ratio greater than that which the straight line BΛ has to the straight line BK.

Therefore	$B\Lambda < 10.\ BK.$
And	$B\Delta = 2\ B\Lambda.$
Therefore	$B\Delta < 20.\ BK.$
But	$B\Delta : BK = AB : B\Gamma.$
Therefore	$AB\ 20. < B\Gamma.$

And AB is the distance of the sun from the earth, while BΓ is the distance of the moon from the earth; therefore the distance of the sun from the earth is less than twenty times the distance of the moon from the earth. And it was proved to be greater than eighteen times.

Translated by T. L. Heath

Reading and Discussion Questions

1. What physical evidence is Aristarchus appealing to in order to calculate the relative distances of the earth, moon, and sun?
2. Aristarchus is remembered for having proposed a heliocentric hypothesis (that the earth revolves around the sun). Given the arguments of Aristotle encountered in previous readings what reasons might ancient natural philosophers have had for rejecting this idea?

Measurement of the Earth (third century BCE) (as described by Cleomedes in *On the Circular Motion of the Heavenly Bodies*, sometime after the mid-first century BCE)

Eratosthenes

19

Eratosthenes' method [of investigating the size of the earth] depends on a geometrical argument, and gives the impression of being more obscure. What he says will, however, become clear if the following assumptions are made. Let us suppose, in this case also, first that Syene and Alexandria lie under the same meridian circle; secondly, that the distance between the two cities is 5000 stades; and thirdly, that the rays sent down from different parts of the sun upon different parts of the earth are parallel; for the geometers proceed on this assumption. Fourthly, let us assume that, as is proved by the geometers, straight lines falling on parallel straight lines make the alternate angles equal, and fifthly, that the arcs subtended by equal angles are similar, that is, have the same proportion and the same ratio to their proper circles—this also being proved by the geometers. For whenever arcs of circles are subtended by equal angles, if any one of these is (say) one-tenth of its proper circle, all the remaining arcs will be tenth parts of their proper circles.

Anyone who has mastered these facts will have no difficulty in understanding the method of Eratosthenes, which is as follows. Syene and Alexandria, he asserts, are under the same meridian. Since meridian circles are great circles in the universe, the circles on the earth which lie under them are necessarily great circles also. Therefore, of whatever size this method shows the circle on the earth through Syene and Alexandria to be, this will be the size of the great circle on the earth. He then asserts, as is indeed the case, that Syene lies under the summer tropic. Therefore, whenever the sun, being in the Crab at the summer solstice, is exactly in the middle of the heavens, the pointers of the sundials necessarily throw no shadows, the sun being in the exact vertical line above them; and this is said to be true over a space 300 stades in diameter. But in Alexandria at the same hour the pointers of the sundials throw shadows, because this city lies farther to the north than Syene. As the two cities lie under the same meridian great circle, if we draw an arc from the extremity of the shadow of the pointer to the base of the pointer of the sundial in Alexandria, the arc will be a segment of a great circle in the bowl of the sundial, since the bowl lies under the great circle. If then we conceive straight lines produced in order from each of the pointers through the earth, they will meet at the centre of the earth. Now since the sundial at Syene is vertically under the sun, if we conceive a straight line drawn from the sun to the top of the pointer of the sundial, the line stretching from the sun to the centre of the earth will be one straight line. If now we conceive another straight line drawn upwards from the extremity of the shadow of the pointer of the sundial in Alexandria, through the top of the pointer to the sun, this straight line and the aforesaid straight line will be parallel, being straight lines drawn through from different parts of the sun to different parts of the earth. Now on these parallel straight lines there falls the straight line drawn from the centre of the earth to the pointer at Alexandria, so that it makes the alternate angles equal; one of these is formed at the centre of the earth by the intersection of the straight lines drawn from the sundials to the centre of the earth; the other is at the intersection of the top of the pointer in Alexandria and the straight line drawn from the extremity of its shadow to the sun through the point where it meets the pointer. Now this latter angle subtends the arc carried round from the extremity of the shadow of the pointer to its base, while the angle at the centre of the earth subtends the arc stretching from Syene to Alexandria. But the arcs are similar since they are subtended by equal angles. Whatever ratio, therefore, the arc in the bowl of the sundial has to its proper circle, the arc reaching from Syene to Alexandria has the same ratio. But the arc in the bowl is found to be the fiftieth part of its proper circle. Therefore the distance from Syene to Alexandria must necessarily be a fiftieth part of the great circle of the earth. And this distance is 5000 stades. Therefore the whole great circle is 250000 stades. Such is the method of Eratosthenes.

Translated by Ivor Thomas

Reading and Discussion Question

1. What observations and mathematical techniques does Eratosthenes draw upon in order to calculate the size of the earth?

Almagest (second century CE)

Ptolemy

20

2. On the Order of the Theorems

A view, therefore, of the general relation of the whole earth to the whole of the heavens will begin this composition of ours. And next, of things in particular, there will first be an account of the ecliptic's position and of the places of that part of the earth inhabited by us, and again of the difference, in order, between each of them according to the inclinations of their horizons. For the theory of these, once understood, facilitates the examination of the rest. And, secondly, there will be an account of the solar and lunar movements and of their incidents. For without a prior understanding of these one could not profitably consider what concerns the stars. The last part, in view of this plan, will be an account of the stars. Those things having to do with the sphere of what are called the fixed stars would reasonably come first, and then those having to do with what are called the five planets. And we shall try and show each of these things using as beginnings and foundations for what we wish to find, the evident and certain appearances from the observations of the ancients and our own, and applying the consequences of these conceptions by means of geometrical demonstrations. And so, in general, we have to state that the heavens are spherical and move spherically; that the earth, in figure, is sensibly spherical also when taken as a whole; in position, lies right in the middle of the heavens, like a geometrical centre; in magnitude and distance, has the ratio of a point with respect to the sphere of the fixed stars, having itself no local motion at all. And we shall go through each of these points briefly to bring them to mind.

3. That the Heavens Move Spherically

It is probable the first notions of these things came to the ancients from some such observation as this. For they kept seeing the sun and moon and other stars always moving from rising to setting in parallel circles, beginning to move upward from below as if out of the earth itself, rising little by little to the top, and then coming around again and going down in the same way until at last they would disappear as if falling into the earth. And then again they would see them, after remaining some time invisible, rising and setting as if from another beginning; and they saw that the times and also the places of rising and setting generally corresponded in an ordered and regular way.

But most of all the observed circular orbit of those stars which are always visible, and their revolution about one and the same centre, led them to this spherical notion. For necessarily this point became the pole of the heavenly sphere; and the stars nearer to it were those that spun around in smaller circles, and those farther away made greater circles in their revolutions in proportion to the distance, until a sufficient distance brought one to the disappearing stars. And then they saw that those near the always-visible stars disappeared for a short time, and those farther away for a longer time proportionately. And for these reasons alone it was sufficient for them to assume this notion as a principle, and forthwith to think through also the other things consequent upon these same appearances, in accordance with the development of the science. For absolutely all the appearances contradict the other opinions.

If, for example, one should assume the movement of the stars to be in a straight line to infinity, as some have opined, how could it be explained that each star will be observed daily moving from the same starting point? For how could the stars turn back while rushing on to infinity? Or how could they turn back without appearing to do so? Or how is it they do not disappear with their size gradually diminishing but on the contrary seem larger when they are about to disappear, being covered little by little as if cut off by the earth's surface? But certainly to suppose that they light up from the earth and then again go out in it would appear most absurd. For if anyone should agree that such an order in their magnitudes and number, and again in the distances, places, and times is accomplished in this way at random and by chance, and that one whole part of the earth has an incandescent nature and another a nature capable of extinguishing, or rather that the same part lights the stars up for some people and puts them out for others, and that the same stars happen to appear to some people either lit up or put out and to others not yet so even if anyone, I say, should accept all such absurdities, what could we say about the always-visible stars which neither rise nor set? Or why don't the stars which light up and go out rise and set for every part of the earth, and why aren't those which are not affected in this way always above the earth for every part of the earth? For in this hypothesis the same stars will not always light up and go out for some people, and never for others. But it is evident to everyone that the same stars rise and set for some parts, and do neither of these things for others.

In a word, whatever figure other than the spherical be assumed for the movement of the heavens, there must be unequal linear distances from the earth to parts of the heavens, wherever or however the earth be situated, so that the magnitudes and angular distances of the stars with respect to each other would appear unequal to the same people within each revolution, now larger now smaller. But this is not observed to happen. For it is not a shorter linear distance which makes them appear larger at the horizon, but the steaming up of the moisture surrounding the earth between them and our eyes, just as things put under water appear larger the farther down they are placed.

The following considerations also lead to the spherical notion: the fact that instruments for measuring time cannot agree with any hypothesis save the spherical one; that, since the movement of the heavenly bodies ought to be the least impeded and most facile, the circle among plane figures offers the easiest path of motion, and the sphere among solids; likewise that, since of different figures having equal perimeters those having the more angles are the greater, the circle is the greatest of plane figures and the sphere of solid figures, and the heavens are greater than any other body.

Moreover, certain physical considerations lead to such a conjecture. For example, the fact that of all bodies the ether has the finest and most homogeneous parts but the surfaces of homogeneous parts must have homogeneous parts, and only the circle is such among plane figures and the sphere among solids. And since the ether is not plane but solid, it can only be spherical. Likewise the fact that nature has built all earthly and corruptible bodies wholly out of rounded figures but with heterogeneous parts, and all divine bodies in the ether out of spherical figures with homogeneous parts, since if they were plane or disc-like they would not appear circular to all those who see them from different parts of the earth at the same time. Therefore it would seem reasonable that the ether surrounding them and of a like nature be also spherical, and that because of the homogeneity of its parts it moves circularly and regularly.

4. That Also the Earth, Taken as a Whole, Is Sensibly Spherical

Now, that also the earth taken as a whole is sensibly spherical, we could most likely think out in this way. For again it is possible to see that the sun and moon and the other stars do not rise and set at the same time for every observer on the earth, but always earlier for those living towards the orient and later for those living towards the occident. For we find that the phenomena of eclipses taking place at the same time, especially those of the moon, are not recorded at the same hours for everyone that is, relatively to equal intervals of time from noon; but we always find later hours recorded for observers towards the orient than for those towards the occident. And since the differences in the hours is found to be proportional to the distances between the places, one would reasonably suppose the surface of the earth spherical, with the result that the general uniformity of curvature would assure every part's covering those following it proportionately. But this would not happen if the figure were any other, as can be seen from the following considerations.

For, if it were concave, the rising stars would appear first to people towards the occident; and if it were flat, the stars would rise and set for all people together and at the same time; and if it were a pyramid, a cube, or any other polygonal figure, they would again appear at the same time for all observers on the same straight line. But none of these things appears to happen. It is further clear that it could not be cylindrical with the curved surface turned to the risings and settings and the plane bases to the poles of the universe, which some think more plausible. For then never would any of the stars be always visible to any of the inhabitants of the curved surface, but either all the stars would both rise and set for observers or the same stars for an equal distance from either of the poles would always be invisible to all observers. Yet the more we advance towards the north pole, the more the southern stars are hidden and the northern stars appear. So it is clear that here the curvature of the earth covering parts uniformly in oblique directions proves its spherical form on every side. Again, whenever we sail towards mountains or any high places from whatever angle and in whatever direction, we see their bulk little by little increasing as if they were arising from the sea, whereas before they seemed submerged because of the curvature of the water's surface.

5. That the Earth Is in the Middle of the Heavens

Now with this done, if one should next take up the question of the earth's position, the observed appearances with respect to it could only be understood if we put it in the middle of the heavens as the centre of the sphere. If this were not so, then the earth would either have to be off the axis but equidistant from the poles, or on the axis but farther advanced towards one of the poles, or neither on the axis nor equidistant from the poles.

The following considerations are opposed to the first of these three positions namely, that if the earth were conceived as placed off the axis either above or below in respect to certain parts of the earth, those parts, in the right sphere, would never have any equinox since the section above the earth and the section below the earth would always be cut unequally by the horizon. Again, if the sphere were inclined with respect to these parts, either they would have no equinox or else the equinox would not take place midway between the summer and winter solstices. The distances would be unequal because the equator which is the greatest of those parallel circles described about the poles would not be cut in half by the horizon; but one of the circles parallel to it, either to the north or to the south, would be so cut in half. It is absolutely agreed by all, however, that these distances are everywhere equal because the increase from the equinox to the longest day at the summer tropic are equal to the decreases to the least days at the winter tropic. And if the deviation for certain parts of the earth were supposed either towards the orient or the occident, it would result that for these parts neither the sizes and angular distances of the stars would appear equal and the same at the eastern and western horizons, nor would the time from rising to the meridian be equal to the time from the meridian to setting. But these things evidently are altogether contrary to the appearances.

As to the second position where the earth would be on the axis but farther advanced towards one of the poles, one could again object that, if this were so, the plane of the horizon in each latitude would always cut into uneven parts the sections of the heavens below the earth and above, different with respect to each other and to themselves for each different deviation. And the horizon could cut into two even parts only in the right sphere. But in the case of the inclined sphere with the nearer pole ever visible, the horizon would always make the part above the earth less and the part below the earth greater with the result that also the great circle through the centre of the signs of the zodiac [ecliptic] would be cut unequally by the plane of the horizon. But this has never been seen, for six of the twelve parts are always and everywhere visible above the earth, and the other six invisible; and again when all these last six are all at once visible, the others are at the same time invisible. And so from the fact that the same semicircles are cut off entirely, now above the earth, now below it is evident that the sections of the zodiac are cut in half by the horizon. And, in general, if the earth did not have its position under the equator but lay either to the north or south nearer one of the poles, the result would be that, during the equinoxes, the shadows of the gnomons at sunrise would never perceptibly be on a straight line with those at sunset in planes parallel to the horizon. But the contrary is everywhere seen to occur. And it is immediately clear that it is not possible to advance the third position since each of the obstacles to the first two would be present here also.

In brief, all the observed order of the increases and decreases of day and night would be thrown into utter confusion if the earth were not in the middle. And there would be added the fact that the eclipses of the moon could not take place for all parts of the heavens by a diametrical opposition to the sun, for the earth would often not be interposed between them in their diametrical oppositions, but at distances less than a semicircle.

6. That the Earth Has the Ratio of a Point to the Heavens

Now, that the earth has sensibly the ratio of a point to its distance from the sphere of the so-called fixed stars gets great support from the fact that in all parts of the earth the sizes and angular distances of the stars at the same times appear everywhere equal and alike, for the observations of the same stars in the different latitudes are not found to differ in the least.

Moreover, this must be added: that sundials placed in any part of the earth and the centres of ancillary spheres can play the role of the earth's true centre for the sightings and the rotations of the shadows, as much in conformity with the hypotheses of the appearances as if they were at the true midpoint of the earth.

And the earth is clearly a point also from this fact: that everywhere the planes drawn through the eye, which we call horizons, always exactly cut in half the whole sphere of the heavens. And this would not happen if the magnitude of the earth with respect to its distance from the heavens were perceptible; but only the plane drawn through the point at the earth's centre would exactly cut the sphere in half, and those drawn through any other part of the earth's surface would make the sections below the earth greater than those above.

7. That the Earth Does Not in Any Way Move Locally

By the same arguments as the preceding it can be shown that the earth can neither move in any one of the aforesaid oblique directions, nor ever change at all from its place at the centre. For the same things would result as if it had another position than at the centre. And so it also seems to me superfluous to look for the causes of the motion to the centre when it is once for all clear from the very appearances that the earth is in the middle of the world and all weights move towards it. And the easiest and only way to understand this is to see that, once the earth has been proved spherical considered as a whole and in the middle of the universe as we have said, then the tendencies and movements of heavy bodies (I mean their proper movements) are everywhere and always at right angles to the tangent plane drawn through the falling body's point of contact with the earth's surface. For because of this it is clear that, if they were not stopped by the earth's surface, they too would go all the way to the centre itself, since the straight line drawn to the centre of a sphere is always perpendicular to the plane tangent to the sphere's surface at the intersection of that line.

All those who think it paradoxical that so great a weight as the earth should not waver or move anywhere seem to me to go astray by making their judgment with an eye to their own affects and not to the property of the whole. For it would not still appear so extraordinary to them, I believe, if they stopped to think that the earth's magnitude compared to the whole body surrounding it is in the ratio of a point to it. For thus it seems possible for that which is relatively least to be supported and pressed against from all sides equally and at the same angle by that which is absolutely greatest and homogeneous. For there is no "above" and "below" in the universe with respect to the earth, just as none would be conceived of in a sphere. And of the compound bodies in the universe, to the extent of their proper and natural motion, the light and subtle ones are scattered in flames to the outside and to the circumference, and they seem to rush in the upward direction relative to each one because we too call "up" from above our heads to the enveloping surface of the universe; but the heavy and coarse bodies move to the middle and centre and they seem to fall downward because again we all call "down" the direction from our feet to the earth's centre. And they properly subside about the middle under the everywhere-equal and like resistance and impact against each other. Therefore the solid body of the earth is reasonably considered as being the largest relative to those moving against it and as remaining unmoved in any direction by the force of the very small weights, and as it were absorbing their fall. And if it had some one common movement, the same as that of the other weights, it would clearly leave them all behind because of its much greater magnitude. And the animals and other weights would be left hanging in the air, and the earth would very quickly fall out of the heavens. Merely to conceive such things makes them appear ridiculous.

Now some people, although they have nothing to oppose to these arguments, agree on something, as they think, more plausible. And it seems to them there is nothing against their supposing, for instance, the heavens immobile and the earth as turning on the same axis from west to east very nearly one revolution a day; or that they both should move to some extent, but only on the same axis as we said, and conformably to the overtaking of the one by the other.

But it has escaped their notice that, indeed, as far as the appearances of the stars are concerned, nothing would perhaps keep things from being in accordance with this simpler conjecture, but that in the light of what happens around us in the air such a notion would seem altogether absurd. For in order for us to grant them what is unnatural in itself, that the lightest and subtlest bodies either do not move at all or no differently from those of contrary nature, while those less light and less subtle bodies in the air are clearly more rapid than all the-more terrestrial ones; and to grant that the heaviest and most compact bodies have their proper swift and regular motion, while again these terrestrial bodies are certainly at times not easily moved by anything else for us to grant these things, they would have to admit that the earth's turning is the swiftest of absolutely all the movements about it because of its making so great a revolution in a short time, so that all those things that were not at rest on the earth would seem to have a movement contrary to it, and never would a cloud be seen to move toward the east nor anything else that flew or was thrown into the air. For the earth would always outstrip them in its eastward motion, so that all other bodies would seem to be left behind and to move towards the west.

For if they should say that the air is also carried around with the earth in the same direction and at the same speed, none the less the bodies contained in it would always seem to be outstripped by the movement of both. Or if they should be carried around as if one with the air, neither the one nor the other would appear as outstripping, or being outstripped by, the other. But these bodies would always remain in the same relative position and there would be no movement or change either in the case of flying bodies or projectiles. And yet we shall clearly see all such things taking place as if their slowness or swiftness did not follow at all from the earth's movement.

Translated by R. Catesby Taliaferro

Reading and Discussion Questions

1. To what extent do Ptolemy's five basic assumptions function like Euclid's axioms? In what way are they different? What arguments does Ptolemy give in support of each assumption? What intellectual traditions or natural philosophers of earlier periods is he indebted to?
2. Of the arguments that he considers for the immobility of the earth, which might you have found most compelling at the time?

Additional Resources

Cohen, I. Bernard. *The Birth of a New Physics.* Revised and Updated ed. New York: W. W. Norton and Company, 1985. A scholarly and accessible presentation of conceptual developments, with valuable comparisons between Ptolemy's astronomical system and that of Copernicus, with excellent diagrams.

Crowe, Michael J. *Mechanics from Aristotle to Einstein.* Santa Fe, NM: Green Lion Press, 2007. A valuable mixture of primary and secondary source material, the first chapter of this text is especially helpful in understanding Oresme.

Cushing, James T. *Philosophical Concepts in Physics: The Historical Relation between Philosophy and Scientific Theories*. Cambridge: Cambridge University Press, 1998. The fourth chapter of this volume provides a clear introduction to Eratosthenes, Aristarchus, and Ptolemy.

Lindberg, David C. *The Beginnings of Western Science: The European Scientific Tradition in Philosophical, Religious, and Institutional Context, Prehistory to A.D. 1450*. 2nd ed. Chicago: Chicago University Press, 2007. Essential background reading for the entire ancient and medieval period, the best secondary source textbook currently available.

Losee, John. *A Historical Introduction to the Philosophy of Science.* 4th ed. Oxford: Oxford University Press, 2001. The first chapter of this text clarifies some of the most difficult concepts of the Aristotelian logic and method.

Robbins, Robert J., ed. *Electronic Scholarly Publishing* (www.esp.org). This website specializes in the history of biology, going all the way back to Aristotle, with a particular focus on genetics.

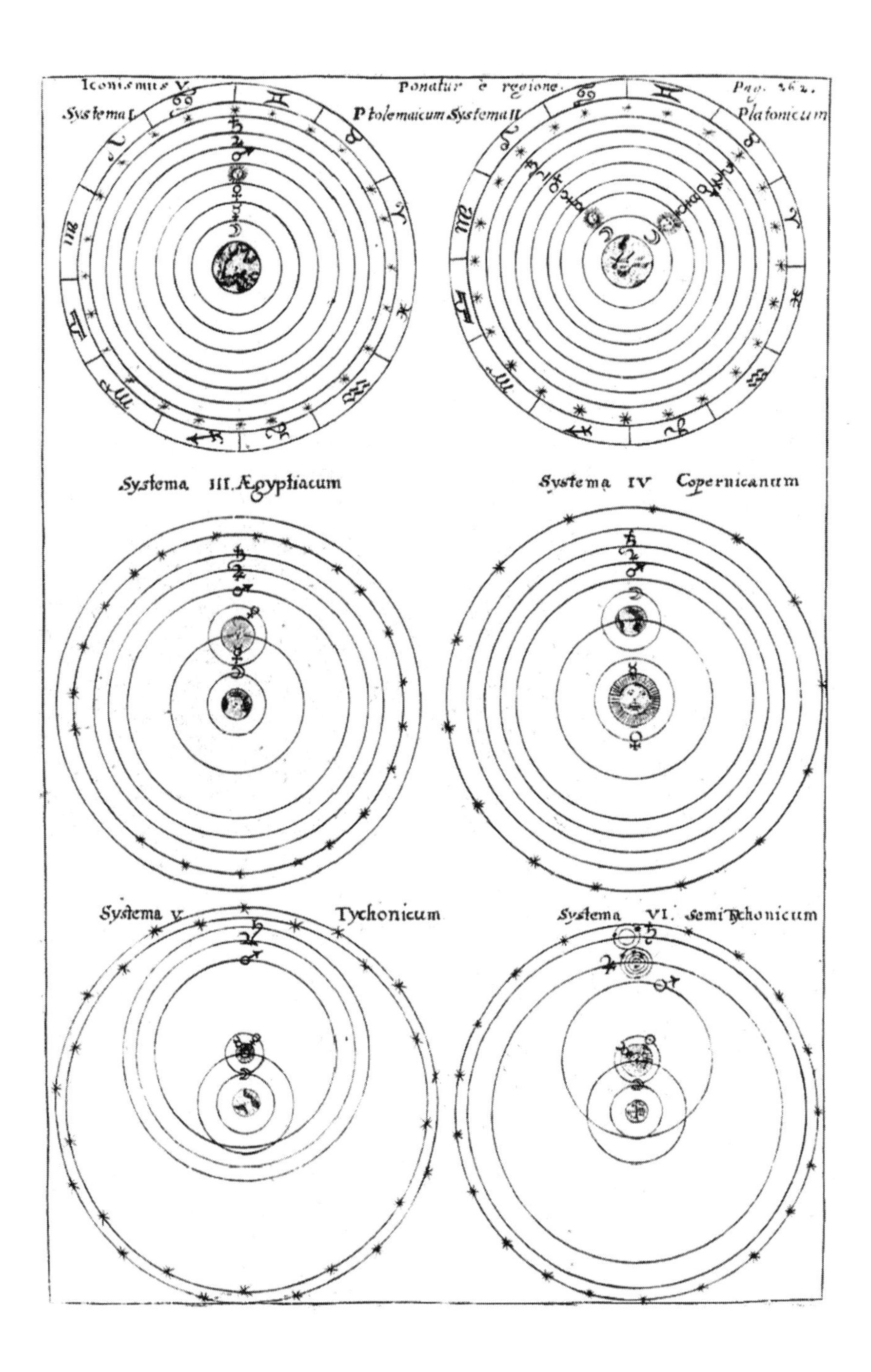

Part II

Translation, Appropriation, and Critical Engagement: Science in the Roman and Medieval Islamic and European Worlds

On the Natural Faculties (second century CE)

Galen

21

Book One

1. Since feeling and voluntary motion are peculiar to animals, whilst growth and nutrition are common to plants as well, we may look on the former as effects of the soul and the latter as effects of the nature. And if there be anyone who allows a share in soul to plants as well, and separates the two kinds of soul, naming the kind in question vegetative, and the other sensory, this person is not saying anything else, although his language is somewhat unusual. We, however, for our part, are convinced that the chief merit of language is clearness, and we know that nothing detracts so much from this as do unfamiliar terms; accordingly we employ those terms which the bulk of people are accustomed to use, and we say that animals are governed at once by their soul and by their nature, and plants by their nature alone, and that growth and nutrition are the effects of nature, not of soul.

2. Thus we shall enquire, in the course of this treatise, from what faculties these effects themselves, as well as any other effects of nature which there may be, take their origin. First, however, we must distinguish and explain clearly the various terms which we are going to use in this treatise, and to what things we apply them; and this will prove to be not merely an explanation of terms but at the same time a demonstration of the effects of nature. When, therefore, such and such a body undergoes no change from its existing state, we say that it is at rest; but, not withstanding, if it departs from this in any respect we then say that in this respect it undergoes motion. Accordingly, when it departs in various ways from its preexisting state, it will be said to undergo various kinds of motion. Thus, if that which is white becomes black, or what is black becomes white, it undergoes motion in respect to colour; or if what was previously sweet now

becomes bitter, or, conversely, from being bitter now becomes sweet, it will be said to undergo motion in respect to flavour; to both of these instances, as well as to those previously mentioned, we shall apply the term qualitative motion. And further, it is not only things which are altered in regard to colour and flavour which, we say, undergo motion; when a warm thing becomes cold, and a cold warm, here too we speak of its undergoing motion; similarly also when anything moist becomes dry, or dry moist. Now, the common term which we apply to all these cases is alteration. This is one kind of motion. But there is another kind which occurs in bodies which change their position, or as we say, pass from one place to another; the name of this is transference. These two kinds of motion, then, are simple and primary, while compounded from them we have growth and decay, as when a small thing becomes bigger, or a big thing smaller, each retaining at the same time its particular form. And two other kinds of motion are genesis and destruction, genesis being a coming into existence, and destruction being the opposite .... The discussion which follows we shall devote entirely, as we originally proposed, to an enquiry into the number and character of the faculties of Nature, and what is the effect which each naturally produces. Now, of course, I mean by an effect that which has already come into existence and has been completed by the activity of these faculties—for example, blood, flesh, or nerve. And activity is the name I give to the active change or motion, and the cause of this I call a faculty. Thus, when food turns into blood, the motion of the food is passive, and that of the vein active. Similarly, when the limbs have their position their position altered, it is the muscle which produces, and the bones which undergo the motion. In these cases I call the motion of the vein and of the muscle an activity, and that of the food and the bones a symptom or affection, since the first group undergoes alteration and the second group is merely transported. One might, therefore, also speak of the activity as an effect of Nature—for example, digestion, absorption, blood-production; one could not, however, in every case call the effect an activity; thus flesh is an effect of Nature, but it is, of course, not an activity. It is, therefore, clear that one of these terms is used in two senses, but not the other.

3. It appears to me, then, that the vein, as well as each of the other parts, functions in such and such a way according to the manner in which the four qualities are mixed. There are, however, a considerable number of not undistinguished men—philosophers and physicians—who refer action to the Warm and the Cold, and who subordinate to these, as passive, the Dry and the Moist; Aristotle, in fact, was the first who attempted to bring back the causes of the various special activities to these principles, and he was followed later by the Stoic school. These latter, of course, could logically make active principles of the Warm and Cold, since they refer the change of the elements themselves into one another to certain diffusions and condensations. This does not hold of Aristotle, however; seeing that he employed the four qualities to explain the genesis of the elements, he ought properly to have also referred the causes of all the special activities to these. How is it that he uses the four qualities in his book "On Genesis

and Destruction," whilst in his "Meteorology," his "Problems," and many other works he uses the uses the two only? Of course, if anyone were to maintain that in the case of animals and plants the Warm and Cold are more active, the Dry and Moist less so, he might perhaps have even Hippocrates on his side; but if he were to say that this happens in all cases, he would, I imagine, lack support, not merely from Hippocrates, but even from Aristotle himself—if, at least, Aristotle chose to remember what he himself taught us in his work "On Genesis and Destruction," not as a matter of simple statement, but with an accompanying demonstration. I have, however, also investigated these questions, in so far as they are of value to a physician, in my work "On Temperaments."

...

5. The effects of Nature, then, while the animal is still being formed in the womb, are all the different parts of its body; and after it has been born, an effect in which all parts share is the progress of each to its full size, and thereafter its maintenance of itself as long as possible. The activities corresponding to the three effects mentioned are necessarily three- one to each- namely, Genesis, Growth, and Nutrition. Genesis, however, is not a simple activity of Nature, but is compounded of alteration and of shaping. That is to say, in order that bone, nerve, veins, and all other [tissues] may come into existence, the underlying substance from which the animal springs must be altered; and in order that the substance so altered may acquire its appropriate shape and position, its cavities, outgrowths, attachments, and so forth, it has to undergo a shaping or formative process. One would be justified in calling this substance which undergoes alteration the material of the animal, just as wood is the material of a ship, and wax of an image. Growth is an increase and expansion in length, breadth, and thickness of the solid parts of the animal (those which have been subjected to the moulding or shaping process). Nutrition is an addition to these, without expansion.

6. Let us speak then, in the first place, of Genesis, which, as we have said, results from alteration together with shaping. The seed having been cast into the womb or into the earth (for there is no difference), then, after a certain definite period, a great number of parts become constituted in the substance which is being generated; these differ as regards moisture, dryness, coldness and warmth, and in all the other qualities which naturally derive therefrom .... Now Nature constructs bone, cartilage, nerve, membrane, ligament, vein, and so forth, at the first stage of the animal's genesis, employing at this task a faculty which is, in general terms, generative and alterative, and, in more detail, warming, chilling, drying, or moistening; or such as spring from the blending of these, for example, the bone-producing, nerve-producing, and cartilage-producing faculties (since for the sake of clearness these names must be used as well). Now the peculiar flesh of the liver is of this kind as well, also that of the spleen, that of the kidneys, that of the lungs, and that of the heart; so also the proper substance of the brain, stomach, gullet, intestines, and uterus is a sensible element, of similar parts all

through, simple, and uncompounded. That is to say, if you remove from each of the organs mentioned its arteries, veins, and nerves, the substance remaining in each organ is, from the point of view of the senses, simple and elementary. As regards those organs consisting of two dissimilar coats, of which each is simple, of these organs the coats are the elements—for example, the coats of the stomach, oesophagus, intestines, and arteries; each of these two coats has an alterative faculty peculiar to it, which has engendered it from the menstrual blood of the mother. Thus the special alterative faculties in each animal are of the same number as the elementary parts; and further, the activities must necessarily correspond each to one of the special parts, just as each part has its special use- for example, those ducts which extend from the kidneys into the bladder, and which are called ureters; for these are not arteries, since they do not pulsate nor do they consist of two coats; and they are not veins, since they neither contain blood, nor do their coats in any way resemble those of veins; from nerves they differ still more than from the structures mentioned. "What, then, are they?" someone asks—as though every part must necessarily be either an artery, a vein, a nerve, or a complex of these, and as though the truth were not what I am now stating, namely, that every one of the various organs has its own particular substance. For in fact the two bladders—that which receives the urine, and that which receives the yellow bile- not only differ from all other organs, but also from one another. Further, the ducts which spring out like kinds of conduits from the gall-bladder and which pass into the liver have no resemblance either to arteries, veins or nerves. But these parts have been treated at a greater length in my work "On the Anatomy of Hippocrates," as well as elsewhere. As for the actual substance of the coats of the stomach, intestine, and uterus, each of these has been rendered what it is by a special alterative faculty of Nature; while the bringing of these together, the therewith of the structures which are inserted into them, the outgrowth into the intestine, the shape of the inner cavities, and the like, have all been determined by a faculty which we call the shaping or formative faculty; this faculty we also state to be artistic—nay, the best and highest art—doing everything for some purpose, so that there is nothing ineffective or superfluous, or capable of being better disposed. This, however, I shall demonstrate in my work "On the Use of Parts."

7. Passing now to the faculty of Growth let us first mention that this, too, is present in the foetus in utero as is also the nutritive faculty, but that at that stage these two faculties are, as it were, handmaids to those already mentioned, and do not possess in themselves supreme authority. When, however, the animal has attained its complete size, then, during the whole period following its birth and until the acme is reached, the faculty of growth is predominant, while the alterative and nutritive faculties are accessory—in fact, act as its handmaids. What, then, is the property of this faculty of growth? To extend in every direction that which has already come into existence—that is to say, the solid parts of the body, the arteries, veins, nerves, bones, cartilages, membranes, ligaments, and the various coats which we have just called elementary, homogeneous, and simple. And

I shall state in what way they gain this extension in every direction, first giving an illustration for the sake of clearness. Children take the bladders of pigs, fill them with air, and then rub them on ashes near the fire, so as to warm, but not to injure them. This is a common game in the district of Ionia, and among not a few other nations. As they rub, they sing songs, to a certain measure, time, and rhythm, and all their words are an exhortation to the bladder to increase in size. When it appears to them fairly well distended, they again blow air into it and expand it further; then they rub it again. This they do several times, until the bladder seems to them to have become large enough. Now, clearly, in these doings of the children, the more the interior cavity of the bladder increases in size, the thinner, necessarily, does its substance become. But, if the children were able to bring nourishment to this thin part, then they would make the bladder big in the same way that Nature does. As it is, however, they cannot do what Nature does, for to imitate this is beyond the power not only of children, but of any one soever; it is a property of Nature alone. It will now, therefore, be clear to you that nutrition is a necessity for growing things. For if such bodies were distended, but not at the same time nourished, they would take on a false appearance of growth, not a true growth. And further, to be distended in all directions belongs only to bodies whose growth is directed by Nature; for those which are distended by us undergo this distension in one direction but grow less in the others; it is impossible to find a body which will remain entire and not be torn through whilst we stretch it in the three dimensions. Thus Nature alone has the power to expand a body in all directions so that it remains unruptured and preserves completely its previous form. Such then is growth, and it cannot occur without the nutriment which flows to the part and is worked up into it.

8. We have, then, it seems, arrived at the subject of Nutrition, which is the third and remaining consideration which we proposed at the outset. For, when the matter which flows to each part of the body in the form of nutriment is being worked up into it, this activity is nutrition, and its cause is the nutritive faculty. Of course, the kind of activity here involved is also an alteration, but not an alteration like that occurring at the stage of genesis. For in the latter case something comes into existence which did not exist previously, while in nutrition the inflowing material becomes assimilated to that which has already come into existence. Therefore, the former kind of alteration has with reason been termed genesis, and the latter, assimilation.

9. Now, since the three faculties of Nature have been exhaustively dealt with, and the animal would appear not to need any others (being possessed of the means for growing, for attaining completion, and for maintaining itself as long a time as possible), this treatise might seem to be already complete, and to constitute an exposition of all the faculties of Nature. If, however, one considers that it has not yet touched upon any of the parts of the animal (I mean the stomach, intestines, liver, and the like), and that it has not dealt with the faculties resident in these, it will seem as though merely a kind of introduction had been given to the practical parts of our teaching. For the whole matter is as follows: Genesis, growth,

and nutrition are the first, and, so to say, the principal effects of Nature; similarly also the faculties which produce these effects—the first faculties—are three in number, and are the most dominating of all. But as has already been shown, these need the service both of each other, and of yet different faculties. Now, these which the faculties of generation and growth require have been stated. I shall now say what ones the nutritive faculty requires.

Translated by Arthur John Brock

Reading and Discussion Questions

1. What point about growth is Galen using the pig's bladder to illustrate?
2. What does this text reflect about the state of anatomy in the second century?

Commentary on Aristotle's Physics (517 CE)

Ioannes Philoponus

22

Weight, then, is the efficient cause of downward motion, as Aristotle himself asserts. This being so, given a distance to be traversed, I mean through a void where there is nothing to impede motion, and given that the efficient cause of the motion differs, the resultant motions will inevitably be at different speeds, even through a void ... Clearly, then, it is the natural weights of bodies, one having a greater and another a lesser downward tendency, that cause differences in motion. For that which has a greater downward tendency divides a medium better. Now air is more effectively divided by a heavier body. To what other cause shall we ascribe this fact than that that which has greater weight has, by its own nature, a greater downward tendency, even if the motion is not through a plenum? ...

And so, if a body cuts through a medium better by reason of its greater downward tendency, then, even if there is nothing to be cut, the body will none the less retain its greater downward tendency .... And if bodies possess a greater or a lesser downward tendency in and of themselves, clearly they will possess this difference in themselves even if they move through a void. The same space will consequently be traversed by the heavier body in shorter time and by the lighter body in longer time, even though the space be void. The result will be due not to greater or lesser interference with the motion but to the greater or lesser downward tendency, in proportion to the natural weight of the bodies in question ....

Sufficient proof has been adduced to show that if motion took place through a void, it would not follow that all bodies would move therein with equal speed. We have also shown that Aristotle's attempt to prove that they would so move does not carry conviction. Now if our reasoning up to this point has been sound it follows that our earlier proposition is also true, namely, that it is possible for motion to take place through a void in finite time ....

Thus, if a certain time is required for each weight, in and of itself, to accomplish a given motion, it will never be possible for one and the same body to traverse a given distance, on one occasion through a plenum and on another through a void, in the same time.

For if a body moves the distance of a stade through air, and the body is not at the beginning and at the end of the stade at one and the same instant, a definite time will be required, dependent on the particular nature of the body in question, for it to travel from the beginning of the course to the end (for, as I have indicated, the body is not at both extremities at the same instant), and this would be true even if the space traversed were a void. But a certain *additional time* is required because of the interference of the medium. For the pressure of the medium and the necessity of cutting through it make motion through it more difficult.

Consequently, the thinner we conceive the air to be through which a motion takes place, the less will be the *additional time* consumed in dividing the air. And if we continue indefinitely to make this medium thinner, the additional time will also be reduced indefinitely, since time is indefinitely divisible. But even if the medium be thinned out indefinitely in this way, the total time consumed will never be reduced to the time which the body consumes in moving the distance of a stade through a void. I shall make my point clearer by examples.

If a stone move the distance of a stade through a void, there will necessarily be a time, let us say an hour, which the body will consume in moving the given distance. But if we suppose this distance of a stade filled with water, no longer will the motion be accomplished in one hour, but a certain additional time will be necessary because of the resistance of the medium. Suppose that for the division of the water another hour is required, so that the same weight covers the distance through a void in one hour and through water in two. Now if you thin out the water, changing it into air, and if air is half as dense as water, the time which the body had consumed in dividing the water will be proportionately reduced. In the case of water the additional time was an hour. Therefore the body will move the same distance through air in an hour and a half. If, again, you make the air half as dense, the motion will be accomplished in an hour and a quarter. And if you continue indefinitely to rarefy the medium, you will decrease indefinitely the time required for the division of the medium, for example, the additional hour required in the case of water. But you will never completely eliminate this additional time, for time is indefinitely divisible.

If, then, by rarefying the medium you will never eliminate this additional time, and if in the case of motion through a plenum there is always some portion of the second hour to be added, in proportion to the density of the medium, clearly the stade will never be traversed by a body through a void in the same time as through a plenum ....

But it is completely false and contrary to the evidence of experience to argue as follows. "If a stade is traversed through a plenum in two hours, and through a void in one hour, then if I take a medium half as dense as the first, the same distance will be traversed through this rarer medium in half the time, that is, in one hour hence the same distance will be traversed through a plenum m the same time as through a void." *For Aristotle wrongly assumes that the ratio of the times required for motion through various media is equal to the ratio of the densities of the media ....*

Now this argument of Aristotle's seems convincing and the fallacy is not easy to detect because it is impossible to find the ratio which air bears to water, in its composition, that is, to find *how* much denser water is than air, or one specimen of air than another. But from a consideration of the moving bodies themselves we are able to refute Aristotle's contention. For if, in the case of one and the same body moving through two different media, the ratio of the times required for the motions were equal to the ratio of the densities of the respective media, then, since differences of velocity are determined not only by the media but also by the moving bodies themselves, the following proposition would be a fair conclusion: "in the case of bodies differing in weight and moving through one and the same medium, the ratio of the times required for the motions is equal to the inverse ratio of the weights." For example, if the weight were doubled, the time would be halved. That is, if a weight of two pounds moved the distance of a stade through air in one-half hour, a weight of one pound would move the same distance in one hour. Conversely, the ratio of the weights of the bodies would have to be equal to the inverse ratio of the times required for the motions.

But this is completely erroneous, and our view may be corroborated by actual observation more effectively than by any sort of verbal argument. *For if you let fall from the same height two weights of which one is many times as heavy as the other, you will see that the ratio of the times required for the motion does not depend on the ratio of the weights, but that the difference in time is a very small one.* And so, if the difference in the weights is not considerable, that is, if one is, let us say, double the other, there will be no difference, or else an imperceptible difference, in time, though the difference in weight is by no means negligible, with one body weighing twice as much as the other.

Now if, in the case of different weights in motion through the same medium, the ratio of the times required for the motions is not equal to the inverse ratio of the weights, and, conversely, the ratio of the weights is not equal to the inverse ratio of the times, the following proposition would surely be reasonable: "If identical bodies move through different media, like air and water, the ratio of the times required for the motions through the air and water, respectively, is not equal to the ratio of the densities of air and water, and conversely."

Now if the ratio of the times is not determined by the ratio of the densities of the media, it follows that a medium half as dense will not be traversed in half the time, but in longer than half. Furthermore, as I have indicated above, in proportion as the medium is rarefied, the shorter is the *additional* time required for the division of the medium. But this additional time is never completely eliminated; it is merely decreased in proportion to the degree of rarefaction of the medium, as has been indicated .... And so, if the *total* time required is not reduced in proportion to the degree of rarefaction of the medium, and if the time added for the division of the medium is diminished in proportion to the rarefaction of the medium, but never entirely eliminated, it follows that a body will never traverse the same distance through a plenum in the same time as through a void.

...

Such, then, is Aristotle's account in which he seeks to show that forced motion and motion contrary to nature could not take place if there were a void. But to me this argument does not seem to carry conviction. For in the first place really nothing has been adduced, sufficiently cogent to satisfy our minds, to the effect that motion contrary to nature or forced motion is caused in one of the ways enumerated by Aristotle.

For in the case of *antiperistasis* there are two possibilities; (1) the air that has been pushed forward by the projected arrow or stone moves back to the rear and takes the place of the arrow or stone, and being thus behind it pushes it on, the process continuing until the impetus of the missile is exhausted, or, (2) it is not the air pushed ahead but the air from the sides that takes the place of the missile ....

Let us suppose that *antiperistasis* takes place according to the first method indicated above, namely, that the air pushed forward by the arrow gets to the rear of the arrow and thus pushes it from behind. On that assumption, one would be hard put to it to say what it is (since there seems to be no counter force) that causes the air, once it has been pushed forward, to move back, that is along the sides of the arrow, and, after it reaches the rear of the arrow, to turn around once more and push the arrow forward. For, on this theory, the air in question must perform three distinct motions: it must be pushed forward by the arrow, then move back, and finally turn and proceed forward once more. Yet air is easily moved, and once set in motion travels a considerable distance. How, then, can the air, pushed by the arrow, fail to move m the direction of the impressed impulse, but instead, turning about, as by some command, retrace its course? Furthermore, how can this air, in so turning about, avoid being scattered into space, but instead impinge precisely on the notched end of the arrow and again push the arrow on and adhere to it? Such a view is quite incredible and borders rather on the fantastic.

Again, the air in front that has been pushed forward by the arrow is, clearly, subjected to some motion, and the arrow, too, moves continuously. How, then, can this air, pushed by the arrow, take the place of the arrow, that is, come into the place which the arrow has left? For before this air moves back, the air from the sides of the arrow and from behind it will come together and, because of the suction caused by the vacuum, will instantaneously fill up the place left by the arrow, particularly so the air moving along with the arrow from behind it. Now one might say that the air pushed forward by the arrow moves back and pushes, in its turn, the air that has taken the place of the arrow, and thus getting behind the arrow pushes it into the place vacated by the very air pushed forward (by the arrow) in the first instance. But in that case the motion of the arrow would have to be discontinuous. For before the air from the sides, which has taken the arrow's place, is itself pushed, the arrow is not moved. For this air does not move it. But if, indeed, it does, what need is there for the air in front to turn about and move back? And in any case, how or by what force could the air that had been pushed forward receive an impetus for motion in the opposite direction? ...

So much, then, for the argument which holds that forced motion is produced when air takes the place of the missile (*antiperistasis*). Now there is a second argument which holds that the air which is pushed in the first instance [i.e., when the arrow is first discharged] receives an impetus to motion, and moves with a more rapid motion than the natural [downward] motion of the missile, thus pushing the missile on while remaining always in contact with it until the motive force originally impressed on this portion of air is dissipated. This explanation, though apparently more plausible, is really no different from the first explanation by *antiperistasis,* and the following refutation will apply also to the explanation by *antiperistasis*.

In the first place we must address the following question to those who hold the views indicated. "When one projects a stone by force, is it by pushing the air behind the stone that one compels the latter to move in a direction contrary to its natural direction? Or does the thrower impart a motive force to the stone, too?" Now if he does not impart any such force to the stone, but moves the stone merely by pushing the air, and if the bowstring moves the arrow in the same way, of what advantage is it for the stone to be in contact with the hand, or for the bowstring to be in contact with the notched end of the arrow?

For it would be possible, without such contact, to place the arrow at the top of a stick, as it were on a thin line, and to place the stone in a similar way, and then, with countless machines, to set a large quantity of air in motion behind these bodies. Now it is evident that the greater the amount of air moved and the greater the force with which it is moved the more should this air push the arrow or stone, and the further should it hurl them. But the fact is that even if you place the arrow or stone upon a line or point quite devoid of thickness and set in motion all the air behind the projectile with all possible force, the projectile will not be moved the distance of a single cubit.

If, then, the air, though moved with a greater force, could not impart motion to the projectile, it is evident that, in the case of the hurling of missiles or the shooting of arrows, it is not the air set in motion by the hand or bowstring that produces the motion of the missile or arrow. For why would such a result be any more likely when the projector is in contact with the projectile than when he is not? And, again, if the arrow is in direct contact with the bowstring and the stone with the hand, and there is nothing between, what air behind the projectile could be moved? If it is the air from the sides that is moved, what has that to do with the projectile? For that air falls outside the [trajectory of the] projectile.

From these considerations and from many others we may see how impossible it is for forced motion to be caused in the way indicated. *Rather is it necessary to assume that some incorporeal motive force is imparted by the projector to the projectile,* and that the air set in motion contributes either nothing at all or else very little to this motion of the projectile. If, then, forced motion is produced as I have suggested, it is quite evident that if one imparts motion "contrary to nature" or forced motion to an arrow or a stone the same degree of motion will be produced much more readily in a void than in a plenum. And there will be no need of any agency external to the projector . . . .

Translated by I. E. Drabkin

Reading and Discussion Questions

1. Reconstruct Philoponus' critique of Aristotle's argument concerning the relationship between an object's rate of fall, its weight, and the density of the medium through which it falls.
2. How does Philoponus criticize Aristotle's theory of why an arrow continues in motion after leaving the bowstring?
3. What aspects of Aristotle's physics does Philoponus seem to accept?

On Medicine (c. 1020)

Ibn Sina (Avicenna)

23

Medicine considers the human body as to the means by which it is cured and by which it is driven away from health. The knowledge of anything, since all things have causes, is not acquired or complete unless it is known by its causes. Therefore in medicine we ought to know the causes of sickness and health. And because health and sickness and their causes are sometimes manifest, and sometimes hidden and not to be comprehended except by the study of symptoms, we must also study the symptoms of health and disease. Now it is established in the sciences that no knowledge is acquired save through the study of its causes and beginnings, if it has had causes and beginnings; nor completed except by knowledge of its accidents and accompanying essentials. Of these causes there are four kinds: material, efficient, formal, and final.

Material causes, on which health and sickness depend, are—the affected member, which is the immediate subject, and the humors; and in these are the elements. And these two are subjects that, according to their mixing together, alter. In the composition and alteration of the substance which is thus composed, a certain unity is attained.

Efficient causes are the causes changing and preserving the conditions of the human body; as airs, and what are united with them; and evacuation and retention; and districts and cities, and habitable places, and what are united with them; and changes in age and diversities in it, and in races and arts and manners, and bodily and animate movings and restings, and sleepings and wakings on account of them; and in things which befall the human body when they touch it, and are either in accordance or at variance with nature.

Formal causes are physical constitutions, and combinations and virtues which result from them. Final causes are operations. And in the science of operations lies the science of virtues, as we have set forth. These are the subjects of the doctrine of medicine; whence one inquires concerning the disease and curing of the human body. One ought to attain perfection in this research; namely, how health may be preserved and sickness cured. And the causes of this kind are rules in eating and drinking, and the choice of air, and the measure of exercise and rest; and doctoring with medicines and doctoring with the hands. All this with physicians is according to three species: the well, the sick, and the medium of whom we have spoken.

Translated by Charles F. Horne et al.

Reading and Discussion Question

1. Describe the way in which Aristotle's four causes and Empedocles' elements have been incorporated into medicine in this period.

The Book of the Remedy: On the Formation of Minerals and Metals (c. 1021)

Ibn Sina (Avicenna)

24

The Fifth Subject of the Physics, consisting of two discourses upon meteorological phenomena. This subject comprises the secondary causes of the inanimate creation such as minerals, meteorological phenomena and the like.

The, First Discourse, upon those things which occur upon the earth.

Section 1. Upon Mountains.

We shall begin by establishing the condition of the formation of mountains and the opinions that must be known upon this subject. The first [topic] is the condition of the formation of stone, the second is the condition of the formation of stones great in bulk or in number, and the third is the condition of the formation of cliffs and heights.

We say that, for the most part, pure earth does not petrify, because the predominance of dryness over [i.e. in] the earth endows it not with coherence but rather with crumbliness. In general, stone is formed in two ways only (a) through the hardening of clay, and (b) by the congelation [of waters]. Many stones in fact, are formed from a substance in which earthiness predominates, and many of them are derived from a substance in which aquosity predominates. Often a clay dries and is changed at first into something intermediate between stone and clay, *viz.* a soft stone, and afterwards is changed into stone [proper]. The clay which most readily lends itself to this is that which is agglutinative, for if it is not agglutinative it usually crumbles before it petrifies. In my childhood I saw, on the bank of the Oxus, deposits of the clay which people use for washing their heads; subsequently I observed that it had become converted into a soft stone, and that was in the space of approximately 23 years.

Stone has also been formed from flowing water in two ways (a) by the congelation of the water as it falls drop by drop or as a whole during its flow, and (b) by the deposition from it, in its course, of something which adheres to the surface of its bed and [then] petrifies. Running waters have been observed, part of which, dripping upon a certain spot, solidifies into stone or pebbles of various colours, and dripping water has been seen which, though not congealing normally, yet immediately petrifies when it falls upon stony ground near its channel. We know therefore that in that ground there must be a congealing petrifying virtue which converts the liquid to the solid. Thus the bases of the formation of stone are [either] a soft clayey substance or a substance in which aquosity predominates. Congelation of the latter variety must he caused by a mineralizing, solidifying virtue, or earthiness must have become predominant in it in the same way in which salt is coagulated, *i.e.* earthiness becomes predominant in it by reason of its [peculiar] virtue and not of its amount. If indeed the earthy quality is not like that in salt, hut is of a different kind, nevertheless the two must be similar in that they are transformed by heat, and in that the advent of heat coagulates them. Or it may be that the virtue is yet another, unknown to us. Alternatively, the converse may be true that its earthiness has prevailed merely by a cold dry virtue.

In short, it is in the nature of water, as you know, to become transformed into earth through a predominating earthy virtue; you know, too, that it is in the nature of earth to become transformed into water through a predominating aqueous virtue. In this connection, there is a substance used by those; folk who have lost their way amid their artful contrivances which, when they are so minded, they call *Virgin's Milk*; it is compounded of two waters which coagulate into a hard solid.[1] This is an indication of the truth of [what I have said above]. They have also many things which they use in liquefaction and coagulation which bear witness to the soundness of these judgments.

Stones are formed, then, either by the hardening of agglutinative clay in the sun, or by the coagulation of aquosity by a desiccative earthy quality, or by reason of a desiccation through heat. If what is said concerning the petrifaction of animals and plants is true, the cause of this [phenomenon] is a powerful mineralizing and petrifying virtue which arises in certain stony spots, or emanates suddenly from the earth during earthquakes and subsidences, and petrifies whatever comes into contact with it. As a matter of fact, the petrifaction of the bodies of animals and plants is not more extraordinary than the transformation of waters.

It is not impossible for compounds to be converted into a single element if the virtue of the latter gets the mastery over them, for each of the elements they contain may be converted into that element. For this reason anything which falls into salt-pans is converted into salt, while objects which fall into the fire are converted into fire. As for the swiftness or slowness of the conversion, that is a matter which necessarily varies according to the variation in the strength of the virtues; if they are very violent they perform the conversion in a short time. In

[1] A thrust at the alchemists.

Arabia there is a tract of volcanic earth which turns to its own colour everyone who lives there and every object which falls upon it. I myself have seen a loaf of bread in the shape of a *raghîf*[2]—baked, thin in the middle, and showing the marks of a bite which had petrified but still retained its original colour, and on one of its sides was the impression of the lines in the oven. I found it thrown away on a mountain near Jâjarm, a town of Khurâsân, and I carried it about with me for a time. These things appear strange only on account of their infrequent occurrence; their natural causes, however, are manifest and well-known.

Certain varieties of stone are formed during the extinction of fire, and frequently ferreous and stony bodies originate during thunderstorms, by reason of the accidental qualities of coldness and dryness which fieriness acquires when it is extinguished. In the country of the Turks there fell, amid thunder and lightning, coppery bodies in the shape of arrowheads with a projection turned back towards the top. A similar one fell in Jil and Dailam,[3] and when it fell it penetrated into the earth. The substance of all these was coppery and dry. I myself undertook, in Khwarazm,[4] the difficult task of fusing a head of that kind, but it would not melt; a greenish fume continued to come off from it until at length an ashy substance remained . . . .

This, then, is one kind of way in which stone is formed. A trustworthy man from among the Shaikhs of the kingdom of Işfahân, Abû Manşûr Hormuz Diyâr ibn Mashakzâr, one in close relation with the illustrious Amir Abû Ja'far Muhammad ibn Dushinanzâr (may God have mercy upon him!), told me that there fell from the sky, in the mountains of Ţabaristân, an object the fall of which resembled the fall of the above-mentioned mass of iron, except that in this case it was a huge stone. This completes the discourse upon the formation of stones.

As for the formation of large stones, this may occur all at once, by intense heat acting suddenly upon a large mass of clay, or little by little with the passage of time.

The formation of heights is brought about by (a) an essential cause and (b) an accidental cause. The essential cause [is concerned] when, as in many violent earthquakes, the wind which produces the earthquake raises a part of the ground and a height is suddenly formed. In the case of the accidental cause, certain parts of the ground become hollowed out while others do not, by the erosive action of winds and Hoods which carry away one part of the earth but not another. That part which suffers the action of the current becomes hollowed out, while that upon which the current does not flow is left as a height. The current continues to penetrate the first-formed hollow until at length it forms a deep valley, while the area from which it has turned aside is left as an eminence. This may be taken as what is definitely known about mountains and the hollows and passes between them.

[2] A round cake of bread.
[3] Two Persian provinces on the south-west shores of the Caspian Sea.
[4] Modern Khiva.

Very often both water and wind would be ineffectual except for the fact that the earth is not uniform, some parts of it being soft and others stony. The soft, earthy parts become hollowed out and the stony parts are left behind as elevations. With the passage of time, the channel is excavated and widened more and more, while the raised portion is left, becoming relatively higher and higher as more earth is hollowed out from [beside] it. These, then, are the principal causes of the three changes [mentioned at the beginning of the *fasl, viz.* the formation of stone, the formation of stones great in bulk or in number, and the formation of cliffs and heights].

Mountains have been formed by one [on other] of the causes of the formation of stone, most probably from agglutinative clay which slowly dried and petrified during ages of which we have no record. It seems likely that this habitable world was in former days uninhabitable and, indeed, submerged beneath the ocean. Then, becoming exposed little by little, it petrified in the course of ages the limits of which history has not preserved; or it may have petrified beneath the waters by reason of the intense heat confined under the sea. The more probable [of these two possibilities] is that petrifaction occurred after the earth had been exposed, and that the condition of the clay, which would then be agglutinative, assisted the petrifaction.

It is for this reason [*i.e.* that the earth was once covered by the sea] that in many stones, when they are broken, are found parts of aquatic animals, such as shells, etc.

It is not impossible that the mineralizing virtue was generated there [*i.e.* in the petrifying clay] and aided the process, while the waters also may have petrified. Most probably, mountains were formed by all these causes.

The abundance of stone in them is due to the abundance, in the sea, of clay which was afterwards exposed. Their elevation is due to the excavating action of floods and winds on the matter which lies between them, for if you examine the majority of mountains you will see that the hollows between them have been caused by floods. This action, however, took place and was completed only in the course of many ages, so that the trace of each individual flood has not been left; only that of the most recent of them can be seen.

At the present time, most mountains are in the stage of decay and disintegration, for they grew and were formed only during their gradual exposure by the waters. Now, however, they are in the grip of disintegration, except those of them which God wills should increase through the petrifaction of waters upon them, or through Hoods which bring them a large quantity of clay that petrifies on them. I have, I believe, heard that this has been observed on certain mountains. As for [the similar phenomenon] which I witnessed upon the banks of the Oxus, that place cannot properly be called a mountain.

Of the land which was exposed by the retreat of the waters, those parts which were of harder clay or more strongly petrified or of greater bulk than the rest remained as elevations and heights when the other parts had been carried away.

As for the veins of clay that are found in mountains, it is possible that these were formed not from the main substance which has undergone petrifaction, but from *debris* of the mountains that turned into dust and filled the valleys and ravines. It then became moistened by streams which flowed upon it, and was covered by the layers of stone forming the mountains, or interlaid with the good clay of the latter. It is possible also that the ancient clay of the sea was not uniform in substance, and that in succession some of it petrified thoroughly, while some did not petrify at all, and some was converted only into a soft stone through a certain quality predominant in it or by reason of some one of innumerable other causes.

It is also possible that the sea may have happened to flow little by little over laud consisting of both plain and mountain and then have ebbed away from it; and so it came to pass that the plain was turned into clay without the same befalling the mountain. Once converted into clay, it was in a fit state to undergo petrifaction when it became exposed, and its petrifaction would be complete and strong. When exposure of the matter which was petrifying took place, it must frequently have happened that the old petrified portions [*i.e.* the mountains] were in a state fit for disintegration, and so would suffer the converse of what was happening to the earth. That is, they became moist and soft and turned into dust again, which itself is in a fit state for petrifaction. For example, when you soak a brick, some earth and some clay in water, and then expose each of them to the fire, the soaking will increase the tendency of the brick to be disintegrated again by the fire, and will also increase the tendency of the earth and the clay to petrify strongly.

It is possible that each time the land was exposed by the ebbing of the sea a layer was left, since we see that, some mountains appear to have been piled up layer by layer, and it, is therefore likely that the clay from which they were formed was itself at one time arranged in layers. One layer was formed first, then, at a different period, a further layer was formed and piled [upon the first, and so on]. Over each layer there spread a substance of different material, which formed a partition between it and the next layer; but when petrifaction took place something occurred to the partition which caused it to break up and disintegrate from between the layers.

As to the bottom of the sea, its clay is either sedimentary or primaeval, the latter not being sedimentary. It is probable that the sedimentary clay was formed by the disintegration of the strata of mountains. Such is the formation of mountains.

...

The Formation of Minerals

The time has now arrived for us to give an account of the properties of mineral substances. We say, therefore, that mineral bodies may be roughly divided into four groups, *viz.* stones, fusible substances, sulphurs and salts. This is for the following reason: some of the mineral bodies are weak in substance and feeble in composition and union, while others are strong in substance. Of the latter, some are malleable and some are not malleable. Of [the former, *i.e.*] those which are feeble in substance, some have the nature of salt and are easily dissolved by moisture, such as alum, vitriol, sal-ammoniac and *qalqand,*[5] while others are oily in nature and are not easily dissolved by moisture alone, such as sulphur and arsenic [sulphides].

Mercury is included in the second group, inasmuch as it is the essential constituent element of malleable bodies or at least is similar to it.

All malleable bodies are fusible, though sometimes only indirectly, whereas most non-malleable substances cannot be fused in the orthodox way or even softened except with difficulty.

The material of malleable bodies is an aqueous substance united so firmly with an earthy substance that the two cannot be separated from one another. This aqueous substance has been congealed by cold after heat has acted upon it and matured it. Included in the group [of malleable bodies], however, are some which are still quick and have not congealed on account of their oily nature; for this reason, too, they are malleable.

As regards the stony kinds of naturally-occurring mineral substances, the material of which they are made is also aqueous, but they have not been congealed by cold alone. Their congelation has, on the contrary, been brought about by dryness which has converted the aquosity into terrestreity. They do not contain a quick, oily humidity and so are non-malleable; and because their solidification has been caused mainly by dryness, the majority of them are infusible unless they are subjected to some physical process which facilitates fusion.

Alum and sal-ammoniac belong to the family of salts, though sal-ammoniac possesses a fieriness in excess of its earthiness, and may therefore be completely sublimed. It consists of water combined with a hot smoke, very tenuous and excessively fiery, and has been coagulated by dryness.

In the case of the sulphurs, their aquosity has suffered a vigorous leavening with earthiness and aeriness under the leavening action of heat, so far as to become oily in nature; subsequently it has been solidified by cold.

The vitriols are composed of a salty principle, a sulphureous principle and stone, and contain the virtue of some of the fusible bodies [metals]. Those of them which resemble *qalqand* and *qalqatâr* are formed from crude vitriols by partial solution, the salty constituent alone dissolving, together with whatever sulphureity there may be. Coagulation follows, after a virtue has been acquired from a metallic ore. Those that acquire the virtue of iron become red or yellow, *e.g. qalqaldr*, while those which acquire the virtue of copper become green. It is for this reason that they are so easily prepared by means of this art.

Mercury seems to be water with which a very tenuous and sulphureous earth has become so intimately mixed that no surface can be separated from it without something of that dryness covering it. Consequently it does not cling to the

[5] Green vitriol, $FeSO_4$.

hand or confine itself closely to the shape of the vessel which contains it, but remains in no particular shape unless it is subdued.[6] Its whiteness is derived from the purity of that aquosity, from the whiteness of the subtle earthiness which it contains, and from the admixture of aeriness with it.

A property of mercury is that it is solidified by the vapours of sulphureous substances; it is therefore quickly solidified by lead or by sulphur vapour. It seems, moreover, that mercury, or something resembling it, is the essential constituent element of all the fusible bodies, for all of them are converted into mercury on fusion. Most of them, however, fuse only at a very high temperature, so that their mercury appears red. In the case of lead, an onlooker does not doubt that this is mercury, since it melts at a lower temperature, but if during the fusion it is heated to the high temperature [mentioned above], its colour becomes the same as that of the other fusible bodies, i.e. fiery-red.

It is for this reason, *viz.* that it is of their substance, that mercury so easily clings to all these bodies. But these bodies differ in their composition from it by reason of variation in the mercury itself—or whatever it is that plays the same part—and also through variation in what is mixed with it and causes its solidification.

If the mercury be pure, and if it be commingled with and solidified by the virtue of a white sulphur which neither induces combustion nor is impure, but on the contrary is more excellent than that prepared by the adepts, then the product is silver. If the sulphur besides being pure is even better than that just described, and whiter, and if in addition it possesses a tinctorial, fiery, subtle and non-combustive virtue—in short, if it is superior to that which the adepts can prepare—it will solidify the mercury into gold.

Then again, if the mercury is of good substance, but the sulphur which solidifies it is impure, possessing on the contrary a property of combustibility, the product will be copper. If the mercury is corrupt, unclean, lacking in cohesion and earthy, and the sulphur is also impure, the product will be iron. As for tin, it is probable that its mercury is good, but that its sulphur is corrupt; and that the commingling [of the two] is not firm, but has taken place, so to speak, layer by layer, for which reason the metal shrieks. Lead, it seems likely, is formed from an impure, heavy, clayey mercury and an impure, fetid and feeble sulphur, for which reason its solidification has not been thorough.

There is little doubt that, by alchemy, the adepts can contrive solidifications in which the qualities of the solidifications of mercury by the sulphurs are perceptible to the senses, though the alchemical qualities are not identical in principle or in perfection with the natural ones, but merely bear a resemblance and relationship to them. Hence the belief arises that their natural formation takes place in this way or in some similar way, though alchemy falls short of nature in this respect and, in spite of great effort, cannot overtake her.

[6] Probably the meaning is: unless it is amalgamated or sublimed or fixed, *i.e.* converted into a compound.

As to the claims of the alchemists, it must be clearly understood that it is not in their power to bring about any true change of species. They can, however, produce excellent imitations, dyeing the red [metal] white so that it closely resembles silver, or dyeing it yellow so that it closely resembles gold. They can, too, dye the white [metal] with any colour they desire, until it bears a close resemblance to gold or copper; and they can free the leads from most of their defects and impurities. Yet in these [dyed metals] the essential nature remains unchanged; they are merely so dominated by induced qualities that errors may be made concerning them, just as it happens that men are deceived by salt, *qalqand*, sal-ammoniac, *etc.*

I do not deny that such a degree of accuracy may be reached as to deceive even the shrewdest, but the possibility of eliminating or imparting the specific difference has never been clear to me. On the contrary, I regard it as impossible, since there is no way of splitting up one combination into another. Those properties which are perceived by the senses are probably not the differences which separate the metals into species, but rather accidents or consequences, the specific differences being unknown. And if a thing is unknown, how is it possible for anyone to endeavour to produce it or to destroy it?

As for the removal or imparting of the dyes or such accidental properties as odours and densities, these are things which one ought not to persist in denying merely because of lack of knowledge concerning them, for there is no proof whatever of their impossibility.

It is likely that the proportion of the elements which enter into the composition of the essential substance of each of the metals enumerated is different from that of any other. If this is so, one metal cannot be converted into another unless the compound is broken up and converted into the composition of that into which its transformation is desired. This, however, cannot be effected by fusion, which maintains the union and merely causes the introduction of some foreign substance or virtue.

There is much I could have said upon this subject if I had so desired, but there is little profit in it nor is there any necessity for it here.

Translated by E. J. Holmyard and D. C. Mandeville

Reading and Discussion Questions

1. What observational evidence does Ibn Sina (Avicenna) provide in describing how geological formations come to be? What does his account of how mountains are formed suggest about his opinion on the age of the earth?
2. In what ways is Ibn Sina building on Aristotle's theory of terrestrial matter as described in *Meteorology*? What other properties of matter is he trying to explain?
3. What reason does Ibn Sina offer for believing that dry parts of the earth, including the mountains, were once covered by sea?

The Book of Instruction in the Elements of the Arts of Astrology (1029)

Al-Bīrūnī

120. The celestial sphere is a body like a ball revolving in its own place; it contains within its interior objects whose movements are different from those of the sphere itself, and we are in the centre of it. It is called *falak* on account of its circular movement like that of the whirl of a spindle, and its name, athir (ether), is current among philosophers.

121. There are eight such spheres enclosed the one within the other, like the skins of an onion; the smallest sphere is that which is nearest to us, within which the moon is always travelling alone, rising and setting, within its limits. To each sphere there is a certain amount of space between the outer and inner boundaries so that the planet to which it belongs has two distances, the one further, the other nearer. The second sphere above that of the moon belongs to Mercury, the third to Venus, the fourth to the Sun, the fifth to Mars, the sixth to Jupiter, the seventh to Saturn. These seven spheres belong to the planets, but above them all is the sphere known as that of the fixed (or desert) stars. The accompanying diagram represents them.

122. A number of people consider that beyond the eighth sphere there is a ninth entirely quiescent; it is this which the Hindus call in their language brahmanda, i.e., the egg of Baraham, because the prime mover must not be moved, and it is on this account that they describe it as motionless. But it is possible that it is not a body like the other spheres, otherwise its existence could be demonstrated, and that to apply this name to it is an error. Many of our ancestors considered that beyond the eight spheres there is an infinite empty space, others, a boundless quiescent substance, while according to Aristotle there is neither substance nor void beyond the revolving bodies . . . .

124. In the centre of the sphere of the moon is the earth, and this centre is in reality the lowest part (and this is a real centre, because all heavy things gravitate towards it). The earth is, as a whole, globular, and in detail is rough-surfaced on

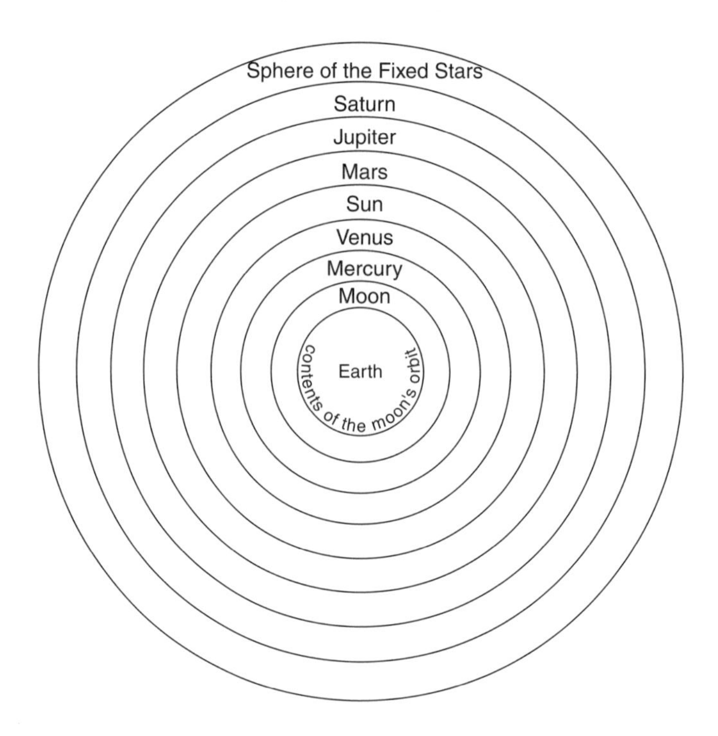

account of the mountains projecting from it and the depressions on its surface, but when considered as a whole it does not depart from the spherical form, for the highest mountains are very small in comparison with the whole globe. Do you not see that a ball of a yard or two in diameter, covered with millet seeds and pitted with depressions of similar size, would still satisfy the definition of a sphere? If the surface of the earth were not so uneven, water coming from all sides would not be retained by it, and would certainly submerge it, so that it would no longer be visible. For water while it shares with earth in having a certain weight, and in falling as low as possible in air, is nevertheless lighter than earth, which therefore settles in water sinking in the form of sediment to the bottom. Moreover water, although it does not penetrate earth itself, sinks into the interstices thereof, and there becomes mixed with air, and as a result of the intimate contact becomes suspended in the air. When the air escapes to the outside, the water regains its natural state in the same way as rain falls from the clouds. On account of the various irregularities projecting from the surface, water tends to collect in the deepest places giving rise to streams.

The earth and the waters together form the one globe, surrounded on all sides by the air; as much of the latter as is in contact with the sphere of the moon becomes heated in consequence of the movement and the friction of the parts in contact. Thus is produced the fire which surrounds the air, less in amount in the proximity of the poles owing to the slackening of the movement there.

125. The fixed stars are those which stud the whole heaven, whose distance from each other is fixed to all eternity, so that they neither approach each other nor separate from each other .... The planets, on the other hand, seven in number, each moving in its own sphere, continually alter their distance from each other and from the fixed stars, sometimes being near and sometimes opposite, in virtue of the difference in the rapidity and character of their movements.

126. Everyone sees that the sun, the moon, and stars are engaged in a first or westward movement; they rise gradually, attain the summit of their course, and then descend little by little till they disappear, thereafter returning to the place where they rose. It is owing to the heavenly bodies that this movement is perceived; it is well-known to animals as they disperse in search of food, more so indeed than to man, for there are animals whose movements correspond with it like the chameleon, which facing the sun turns with it, as do the leaves of many plants, notably: vetch, mash, and liquorice ....

127. The second or eastward movement of all planets is towards the quarter where they arose; but the movement of the fixed stars is very small, and on account of the fact that the distance between them remains the same, they are called fixed, whereas the motion of the planets is much greater, more obvious, and also of varied nature.

It is most obvious in the case of the moon on account of its rapid movement, for, from the time when the moon appears in the west, it moves further away from the sun and any star which is between it and the sun, and approaches any star which is on the other side of the sun from it. When it occults one of these, it does so with its eastern border and clears it with the western. This second movement is common to all the planets; it is an inverted replica of the first, but is not an exact counterpart, for it deviates from it slightly. It is called second because it is different in amount for each planet, while the first is uniform and prevails over all the second movements, although

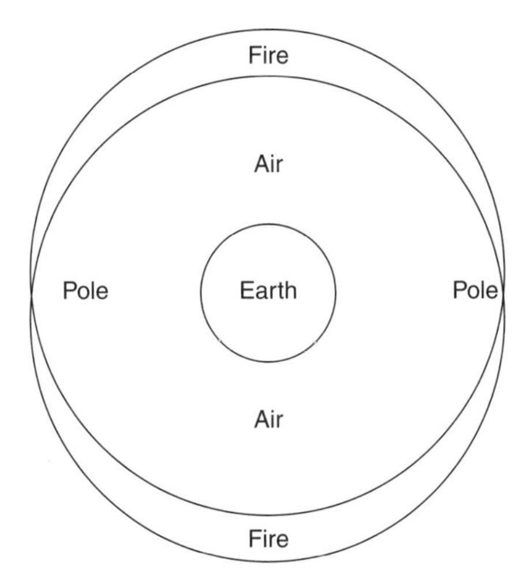

in the opposite direction. It is like the movement of a ship whose passengers may walk in the direction of the current, while they are all being carried up stream by the ship; the uniform movement prevails over the contrary one by reason of greater efficiency.

The second movement is not obvious like the first, but requires some consideration and reasoning based on observation. It is called eastward, on account of its direction towards the rising of the sun.

128. Only about half of the vault of heaven is visible to the observer; it is like a dome placed over the earth, its margin forming a circle round about him. Whatever is above this circle, known as the horizon, is visible to him.

There are two kinds of horizon, the one sensible or visible, the other true or astronomical. The sensible horizon is that already referred to, which we always see when on the surface of the earth, and which divides the celestial sphere into two parts, an upper smaller one, and a lower invisible to us.

The true horizon is parallel to the other, but on a plane passing through the centre of the earth and cutting the celestial sphere into two equal parts. That which is between the two horizons is small in amount so as not to be obvious when the sphere is large [the celestial sphere] but large when the sphere is small [the earth] . . . .

152. Saturn, Jupiter and Mars are superior planets, while Venus and Mercury are inferior, as is the moon; but the moon is not to be reckoned with the other planets. The expressions superior and inferior refer to the position relative to the sun . . . . All the conditions of the planets are certainly referable to the sun, especially the luminosity of the moon and the retrograde movement of the planets.

The difference between the inferior and superior planets is that the distance of the former from the sun is restricted and never exceeds a certain maximum elongation either in the East or in the West. When they precede the sun (are east of it) they leave it so far behind that they become visible after sunset in the evenings. Their visibility increases with the distance from the sun until the greatest eastern elongation is reached. Thereafter their movement becomes slower, and they again begin to approach the sun, when their slow movement comes to a complete stop. This is the stop before retrograding, After this stop, they turn back and their retrograde movement becomes more rapid until, at inferior conjunction, they become invisible in the rays of the sun, their evening occultation. After which, emerging on the other side of the sun, they move more slowly on their retrograde course, and begin to rise before the sun, so as to be visible when they have escaped from its rays; this is called their matutine apparition; then the retrograde movement becomes still slower till the planets reach the second stationary period, before entering on their direct course. Then they soon reach their greatest distance from the sun, their western elongation, and proceed on their direct path till they again approach it, and, at superior conjunction, become invisible in its rays, their matutine occultation. Thereafter, passing through the rays, they again become visible in the west in the evening, thus returning to the sequence of the events described.

But the distance of the superior planets from the sun is not thus restricted; the sun moves quicker and outstrips them so that they escape from its rays and become visible in the east in the morning. Every day their distance from the sun is increased as they proceed on their direct course, until at sunrise they arrive at a point in the heavens, which, if the sun were there, would indicate a time between the early and late afternoon prayers. They then attain the stationary point before retrograding, after which, their distance from the sun increasing every day, they reach the middle point of that course, they are in opposition to the sun, and have thus attained the greatest distance possible within their spheres. They then begin to rise in the east at sunset like the full moon at the fourteenth night of the month. Thereafter, the distance between them and the sun begins to decrease till a point is reached at sunset, which, if the sun were there, would indicate the forenoon. That time corresponds to the stationary period before beginning the direct course; thereafter the sun gradually approaches them till they come within its rays, and they become invisible in the west, a condition described as their occidentality.

The difference, therefore, between the inferior and the superior planets is this, that the former are never further from the sun than the sixth of a circle, and in the middle of their retrograde course are occultated, their apparition and occultation occur both in the east and the west; while the latter attain the greatest possible distance from the sun within their spheres, are not concealed at the middle of their retrograde course but are there in opposition to the sun. Their apparition is only in the east, and their occultation is restricted to the west.

153. A planet is said to be combust, when it comes into conjunction with the sun, the expression being due to the comparison of the sun with fire, and the nonappearance of the planet when it enters the sun's rays, suggesting its combustion or destruction. This phenomenon is common to all the planets, and occurs when they are at the summit or apogee, of the epicycle. The superior planets differ from the inferior ones in that the latter show the same phenomenon at its lowest point or perigee, whereas the former do not, but are then in opposition to the sun.

154. The moon exhibits the same appearance, but this is described as its conjunction. After its first appearance in the west as a slender crescent in the evening at the beginning of the month, the illuminated surface grows with the increasing distance from the sun, till on the seventh evening, halfway between east and west it looks like a half-circle. When the moon has travelled 180° from the sun by the fourteenth evening, it rises at sunset and the whole surface is illuminated. Thereafter as the distance decreases, the bright surface diminishes, so that by the twenty-second evening the dark part is again equal to the bright part; after which the dark part gains on the bright till the crescent shape like that of the new moon is attained, visible in the east in the morning. In all phases the luminosity of the moon comes from that surface which is towards the sun, consequently when it enters the rays of the sun, it is concealed, till after two days it again appears new in the west. During these two days it is in conjunction with the sun, close union,

as Ptolemy describes it in the *Majisti* [Almagest], and so it has come about to speak of this as companionship, rather than as combustion. The opposite position of the moon, full moon, when it confronts the sun, is known as *istiqbal.*

155. The moon is a non-luminous globular body and its brightness is due to the rays of the sun which fall upon it as they do upon the earth, mountains, walls or the like, the other sides of which are not illuminated. When the moon is in conjunction with the sun, it is between us and the sun, because it is lower and the rays fall on that surface which is towards the sun, while we see only the surface facing us, and are unable to distinguish the dark mass of the moon from the blue of the sky on account of the dazzling light of the sun, until the moon moves a little further away from it. Then a small part of the illuminated surface comes into view if the evening twilight is not too bright, and we have the new moon.

Owing to the spherical form of the moon, the margin of the sun's rays which fall upon it is necessarily circular, and so much of the illuminated half as comes into view is also bounded by part of a circle so as a result of the intersection of these two circles on the spherical surface, the interval between them is at first crescentic like the slice of a melon. As the distance from the sun increases, the illuminated surface grows until it equals the dark part, and this is called the first quarter, because the sun and moon are distant from each other by quarter of a circle. This equality of the bright and dark parts occurs also at the second quarter. At full moon, when it is separated from the sun by half a circle, the whole of the surface illuminated by the sun is visible to us, as may be seen from the diagram.

156. Opinions of intelligent people differ as to why this waxing and waning of the moon is not shared by the other planets, and as to whether the planets are

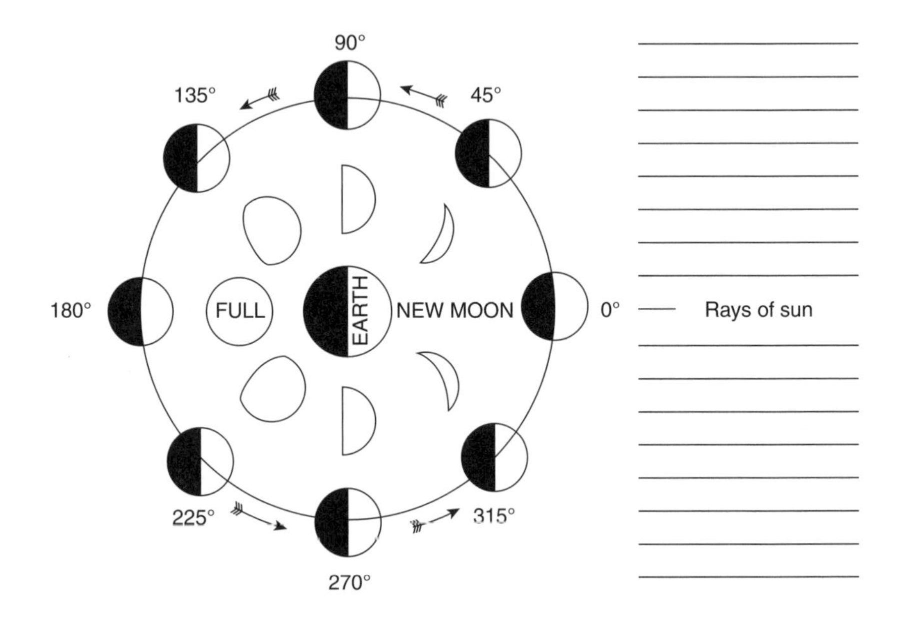

self-luminous like the sun, or merely illuminated by the rays of the sun falling on them.

Many assert that light is exclusively the property of the sun, that all the stars are destitute of it, and that since the movements of the planets are obviously dependent on those of the sun, it may be assumed by analogy that their light is in the same position. But others believe that all the planets are luminous by nature with the exception of the moon, and that its special peculiarities are its paleness and absence of brilliancy. This opinion is more in accord with the truth (as long as there is no evidence to the contrary) and that their concealment under the rays of the sun is just like their non-visibility in diffused daylight, which by its intensity so affects our vision, that we are unable to perceive them. But any one who looks out from the bottom of a deep pit by day may see a planet which happens to pass over the zenith, because his vision is relieved from the intensity of light by the surrounding darkness and strengthened by it, for black concentrates and strengthens vision, while white dissipates and weakens it.

Whether the higher planets are self-luminous or not, they are always to be seen in the same condition. For if the moon were above the sun, it would cease to present the phenomena of waning, and would always appear as full moon. The situation, however, with regard to Venus and Mercury is this, that if they are not luminous, there would be a difference in the amount of their light when at their greatest distance from the sun, and when approaching their disappearance in its rays at conjunction, for indeed they are lower than the sun, and no such difference is observable.

It is therefore preferable to regard the planets as self-luminous, while the special characteristics of the moon and the variety of the phases of its light are due to three things: its captivity (by the sun, conjunction), its pale colour and absence of brilliance, and its position below the sun.

157. The fixed stars in the heavens are so multitudinous that it is impossible to enumerate them, yet those diligent investigators who have endeavoured to recognize them and to determine their positions in longitude in the signs, and their latitude north and south of the ecliptic, observed that they differ in size and have consequently established a scale of magnitudes, to the two first degrees of which astrologers give the name of glory. Of the first magnitude there are fifteen stars, of the second, forty-five, of the third, two hundred and eight, of the fourth, four hundred and seventy-four, and of the fifth, fifty-eight.

Among the stars of the sixth magnitude there are nine stars which Ptolemy described as 'dark', apart from three others not counted with them, which together are called the tresses, (Coma Berenices), the lock-wearer.

Stars which are smaller than the sixth magnitude cannot be separately distinguished by our vision, or if they are can only with difficulty be kept under observation . . .

178. An epicycle, is a small orbit which does not surround the earth, but is entirely outside it. The planet moves on its circumference with the motion peculiar to it.

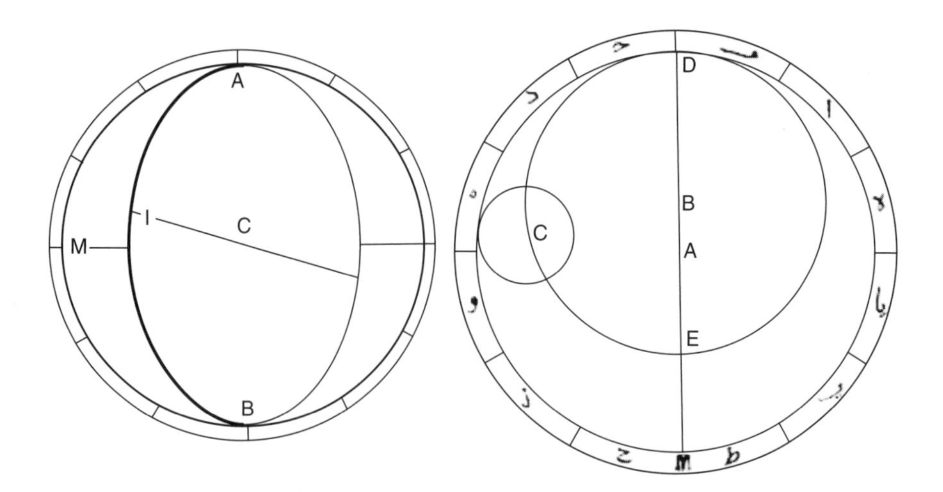

179. The centre of the epicycle travels continuously in the direction of succession of the signs on the circumference of an orbit called the deferent, which is in the plane of the inclined orbit, but like the eccentric, has a different centre from the centre of the world.

180. If the centre of the epicycle traversed equal arcs of the deferent in equal times, then the mean rate of the progress of the planet would be on the deferent, and the angles opposite these arcs would also be equal; the angles of the arcs,

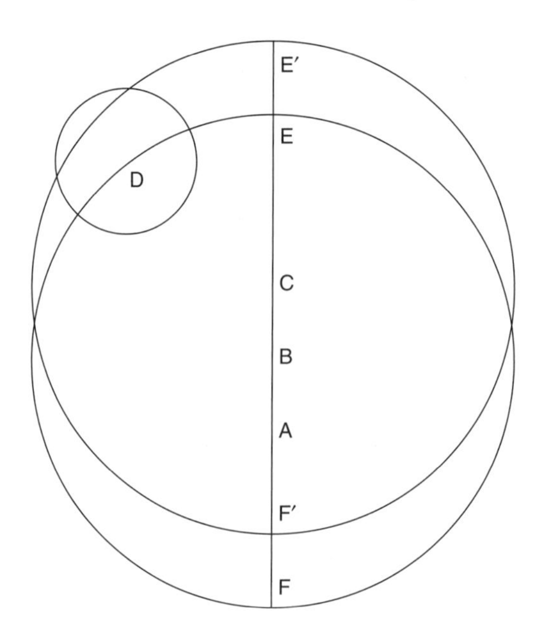

A – Centre of world
B – Centre of deferent
C – Centre of equant
D – Centre of epicycle
E F – Apogee and Perigee of deferent
E'F' – Apogee and Perigee of equant

however, traversed by the centre of the epicycle in equal times are not equal, but are so at a point as far from the centre of the deferent as that is from the centre of the world. This point is the centre of equal progress, the equant, and is the same for Venus and the three superior planets. All three points are in the same straight line. It is necessary to regard this point as the centre of an orbit like the deferent, and to calculate the progress of the planet on its circumference from the position pf the centre of the epicycle, which may be done by lines drawn to this point (without drawing the orbit).

Translated by R. Ramsay Wright

Reading and Discussion Questions

1. How does Al-Bīrūnī, account for the phases of the moon? Does he believe that the planets emit their own light or reflect it? What reasons does he give in support of his position?
2. The epicycles that he is referring to date back to Ptolemy's *Almagest*. What problem are they meant to solve? What developments and additions beyond Aristotelian astronomy are evident in this reading and what remains the same?

Perspectiva (Book of Optics) (c. 1270)

Ibn al-Haytham (Alhazen)

26

It is permissible for us to assert that a transparent body receives something from sight and transmits it to the visible object and that perception occurs through the unbroken succession of that thing to the visible object, between the eye and the visible object. And this is the view of those who suppose that rays issue from the eye. Therefore, [for the sake of argument,] let it be supposed that this is true, that rays issue from the eye and pass through the transparent medium to the object of sight and that perception occurs by means of· those rays. And since, [according to the view now under consideration,] perception occurs in this way, I inquire whether those rays return something to the eye. If perception occurs [only] through [such] rays, and they do not return anything to the eye, then the eye does not perceive. But the eye does perceive the object of sight, and [we have supposed that] it perceives only by the mediation of rays. Therefore, those rays that perceive the visible object [must] transmit something to the eye, by means of which the eye perceives the object. And since the rays transmit something to the eye, by means of which the eye perceives the object, the eye does not perceive the light and color in the visible object unless something comes to the eye from the light and color in the object; and this is delivered by the rays.

Therefore, according to all possibilities, sight does not occur unless something of the visible object comes from the object, whether or not rays issue from the eye. It has already been shown that sight is achieved only if the body intermediate between the eye and the visible object is transparent, and it is not achieved if the medium between them is opaque. And it is evident that a transparent and an opaque body are distinguished only in the aforesaid way. Since, as we have said and as has been demonstrated, the forms of the light and color in the visible object reach the eye (if they were [originally] opposite the eye), that which comes from the visible object to the eye (through which the eye perceives the light and color in the visible object no matter what the situation [with respect to visual rays]) is merely that form, whether or not rays issue [from the eye]. Furthermore, it has been shown that the forms of light and color are always generated in air and in all transparent bodies and are always extended to the opposite regions, whether or not the eye is present. Therefore, the egress of rays [from the eye] is superfluous and useless. Consequently, the eye does not perceive the light and color of the visible object unless form comes from the light and color [to the eye] . . . .

It remains only to consider the view of those who maintain that rays issue from the eye and to indicate what in this view is false and what true. Therefore, let us assert that if sight is due to something issuing from the eye to the visible object, that thing is either corporeal or incorporeal. If it is corporeal, it follows that when we look at the sky and see the stars in it, corporeal substance issues from our eye in that time and fills the whole space between heaven and earth, and [yet] the eye is in no way destroyed; and this is [obviously] false. Therefore, sight does not occur through the passage of corporeal substance from the eye to the visible object. But if that which issues from the eye is incorporeal, it will not perceive the object, since there is no perception except in corporeal things. Therefore, nothing issues from the eye to the visible object to perceive that object. Now, it is evident that sight occurs through the eye; and since this is so, and the eye perceives the visible object only when something issues from the eye to the visible object, and since that which issues forth does not perceive the object, therefore that which issues from the eye to the visible object does not return anything to the eye, by which the eye perceives the object. And [therefore,] that which issues from the eye is not sensible but conjectural, and nothing ought to be believed except through reason or by sight.

However, those who assume that rays issue from the eye suppose this because they ascertain that the eye perceives the visible object and that between the eye and the object is space; and it is known by mankind that there is no perception except by contact. Consequently, they have conjectured that sight does not occur unless something issues from the eye to the visible object, so that the thing issuing forth perceives the object in place of the eye, or rather receives something from the visible object and returns it to the eye; then the eye perceives that thing. And since corporeal substance cannot issue from the eye to perceive the object and nothing perceives a visible object unless it is corporeal, nothing remains except to conjecture that the thing issuing from the eye to the object of vision receives something from the object and returns it to the eye. And since it has been demonstrated that air and transparent bodies receive the form of the visible object and transmit it to the eye and to every facing body, that which they conjecture to return something from the visible object to the eye is nothing but air and the transparent bodies between the eye and the object of vision. Since air and transparent bodies transmit something from the visible object to the eye, regardless of the time and according to all arrangements (when the eye is opposite the visible object), without requiring that something issue from the eye, the reasoning that led from positing rays to maintaining their existence is unnecessary; for that which led them to say that rays exist is their supposition that sight cannot be completed except through something extended from the eye to the visible object, which returns something from the object to the eye. And since air and transparent bodies do this without requiring that something issue from the eye, and moreover, since the air and transparent bodies are extended between the eye and the visible object without defect, it is useless to suppose that something else returns something from the object of vision to the eye. Therefore, it is useless to say that [visual] rays exist.

Moreover, all mathematicians who say there are rays use nothing in their demonstrations except imaginary lines, and they call them radial lines; ... and the belief of those who consider radial lines to be imaginary is true, and the belief of those who suppose that something [really] issues from the eye is false.

Translated by David C. Lindberg

Reading and Discussion Question

1. What theory of perception is Ibn al-Haytham critiquing and what alternative does he espouse?

On Plants (c. 1256)

Albertus Magnus (Albert the Great)

27

Book I, Tract I

Chapter 1 [Objective of Book]

. . . We must begin with the bodies of plants. Concerning this, we intend to present in this book the usual things appropriate to plants according to the whole and its parts, since particulars are infinite and no science can be made of them, as Plato put it very well. Now, since the first common principle appropriate to all plants and their parts is life [itself], which is found in plants, we shall first inquire about the life of plants.

Chapter 6 Concerning the Reasons of Aristotle by Which He Proves That Plants Do Not Sense or Sleep

Now, according to our intent, we must investigate a problem about which we made mention in our preceding discussion, namely whether plants have desire, animal motion, and a soul, and concerning what was said about plants being relaxed by sleep and wakefulness, just as sleep in an animal is caused by what is released by evaporation from the place of digestion. That plants do not have such a spirit, which is drawn by inhaling and exhaling, as Anaxagoras said, we can prove in this way: because we find that there are many animals that do not inhale and exhale; however, such a spirit is more appropriate to animals than plants, because animals are, by their natures, hotter and participate in [that is, contain more of] the higher elements than plants, which are earthy and cold. For this reason animals require animal spirits more than plants.

Similarly, we find that plants do not sleep or wake, since in them evaporation does not occur from the place of digestion to any cool place located ill their bodies. Therefore, the evaporation which is released in them descends and immobilizes the exterior parts of the plant, which by nature always exist immobile and insensible. This is especially so since, as was made clear in *On Sleep and Waking*, to be awake is a certain faculty and force and effect of the senses; sleep, on the contrary, is a certain weakness and loss of the senses. But since plants wholly lack sense they cannot participate in accidents of the senses which involve a force or power [of the senses] and a lack of the senses.

. . .

Chapter 4 On the Diversity of the Essential and Principal Parts of a Plant

... There are parts of all plants ... which grow and increase ... and these are the essential parts of plants: These are roots, branches of roots, stems, and branches. For these parts in plants are like the parts of animals ... .

Barks, wood, and pith, and all such essential parts of this kind arise from the seminal humor [or sap] and from the food of the plant. Some call the pith the mother of the tree because in it the seed seems to be conceived and formed ... . Nodes, veins, and the woody or herbaceous tissues of plants, which fill what is between the veins and nodes, are all constituted of the four elements. Indeed, in these things the powers of the elements are found more than in the parts of animals, because plants are more material and nearer to the elements and in them the elements are altered less than in the bodies of animals.

Chapter 5 In Which the Genus of Plants Is Divided into Species and the Reason for This

If we now consider the plant according to the community and range of its predication, parts of it are designated univocally by name and structure. Of these, some are trees; some lie between trees and herbs and their name shows their intermediate being, for in Greek they are called *ambragyon* and in Latin are called by the common name *arbusta*, bushes. Moreover, some are bushy herbs. Almost every plant falls under these names. I say almost because the arborescent shrub lies between shrub and tree, inclining, however, more toward the tree. And according to this, the highest of the plants is the tree, the lowest the herb. But since nature does not pass instantaneously from extreme to extreme, but passes through all means that have been fitted between them, thus it [nature] has made many means, of which only one is equidistant and middle; and there are others which are closer to the extremes. This will be obvious in the example and definitions of the names introduced.

A tree is that which has a strong stem from its root, and on this stem arise many branches, and on the branches many slender branches, and on the latter arise what are called ultimate branchlets, as in the olive, cypress, and fig trees. Plants that are mean between trees and herbs are called *ambragyon* in Greek and *arbusta* (bush) in Latin. From their roots they send forth many branches in the manner of long branches . . . .

There is another mean plant with much the property of an herb which is called bushy herb by authors. This plant projects many stems from one root and different branches on the stems, but it has little or nothing woody, although, in later age, some of them grow hard like wood. Among this kind are rue, cabbage, and many others. Herbs seem to retain the lowest place among plants; they have one root, from which nothing is produced but leaves, as a clump of scallions, and some others.

In this division of plants there are no fungi, truffles, or mushrooms of any kind, because they seem to possess the least of the life and powers of a plant, so that they lack the force to form leaves but are said to be certain plant-like things that have been exhaled and evaporated from other plants. For this reason they are rarely found except between other plants and they endure for a short time.

Book II, Tract II In Which These Things Are Investigated: What Happens Naturally to Plants Which Contributes to Their Production of Fruit or Generation

Chapter 1 In Which We Discuss What Will Be Said in This Tractate and also about the Nature of Leaves

We shall now examine the common things that happen essentially and naturally to plants with respect to their generation, according as nature intends to save each of their species. We shall not inquire here about the place or mode of generation, but we only desire to know the nature of those parts which remain on them when they are in generation. Now, these parts are the *leaves*, which have the function of covering the fruits, the *flowers*, which are the signs of fruits, and the *fruits* and *seeds.* It is indeed necessary to know the natures which these have in common and their differences, so that after these things have been considered, the causes of the same parts of the body are assigned more carefully and in an orderly manner. For a cause is sought in vain if the reason for which the cause is sought is not known before. Indeed, to philosophize is to investigate the certain, manifest, and true cause of a known effect and to show how it is the cause of it and that it is impossible that it be related otherwise.

Therefore, we say of the leaf in general that the matter of a leaf in all plants is a watery humor mixed somewhat with, and not well separated from, dried earth—not purged from the dregs of earthy matter. The sign of this is that the leaves of large plants having broad and thin leaves suck up moisture quickly and thus almost all leaves are produced full of holes. But where the humor is viscous and more watery, being agitated from the heat that has been retained within its viscosity, the leaves are thicker and cling more. to those plants. But the reason why leaves are frequently thick in herbs and bushes is that the humor is abundantly distributed and multiplied in them because of their nearness to, and continuity with, their roots.

Now, a final purpose [or goal] of leaves is to cover the fruit; [moreover,] nature requires a purging from an excess of watery humidity. Now, since nature is wise and ingenious, it uses the same purging for the protection of the fruits. Thus in many plants nature produces the leaves under the fruit, so that the spread leaf is extended beyond the fruit. However, the production and location of leaves is more general, because the leaf sprouts at the base of the peduncle of the fruit. This is more fitting for the matter and purpose of the leaf. There are two kinds of material vapors in the flesh of the plant and in all other vapors, namely a *humid vapor* and a *dry, windy vapor.*

The humid vapor is the matter of the leaf and the dry, windy vapor is the matter of the fruit. For this reason the fruit is judged to be windy by virtue of its nature. The dry vapor is sharper and distributed in the flesh of the tree; the humid vapor is duller [or weaker]. With its sharpness it cuts the body of the plant so that each of these vapors bursts forth. The vapor of the fruits ascends higher and the vapor of the leaves emerges under it. Nevertheless, since these vapors are mixed together in the flesh of the plant, it is necessary that the humid vapor should have a motion by virtue of the windy vapor mixed with it. And so it happens that because the leaf is generated by a humid vapor it comes forth near the fruit, close to the base of the fruit as in many [plants].

The location of a leaf is varied in three ways. Sometimes it is under the fruit at the base of the peduncle of the fruit, as in pear, apple, bullace, and plum trees, and in many others; sometimes it is opposite the fruit, as in vines; and sometimes it is above the fruit, as in the violet and many other herbs. The cause of the leaf being at the base of [or below] the fruit has been assigned above in terms of the fitness [or suitability] of the matter. This location is also appropriate to its goal [or end], because the leaf that comes forth underneath is extended toward the flower and covers it more usefully and protects the flower.

When the leaf is opposite the fruit, it comes forth in plants, drawing much humor, and especially does the fruit of these plants fill very much with a subtle vaporous and windy humor. From this vapor the watery vapor, which is less distributable and less subtle—just as if it were a contrary—is expelled to the opposite side by a formative power. It is for this reason that the leaf comes forth in an opposite place. This is also appropriate for the purpose of the leaf, since the fruit of such a plant requires a great boiling [action] on the part of the sun because of the abundance of humor. But if the leaf was spread over the fruit it would hinder [the necessary boiling action of the sun on the fruit] . . . .

Leaves that emerge above the fruit especially cover the fruit, so that it is almost always in the shade. And the cause of this is matter, because the fruits have much of earthiness and therefore, when the windiness has been closed off [or hindered], the water humor in them rises higher, and the vapor, from which the fruit is generated, comes forth in the lower parts of it as if drawn with violence. The utility of this effort of nature can be judged by its end [or goal], since such fruits are cold and humid, so that frequently the shade of the leaves helps preserve these two things [namely the coldness and humidity of the fruit]. These, then, are the three general locations of leaves, and from their causes the variation of the location of the leaves is easily known. Nevertheless, perhaps sometimes ways other than those declared here may be found in certain plants.

Chapter 4 On the Nature and Generation of Flowers

The flower, which is the sign of fruit, is found in plants. As in several things, the generation of flowers is of the same substance with fruit. For this reason, also, the flower very frequently adheres to the upper part of the fruit, as in trees; or the fruit is formed in the middle of a flower, as in bushes and herbs. And what we have said about trees to whose fruit the flower adheres when first formed is especially apparent in the flowers of the pomegranate and in pears and apples. But it is not generally suitable, because in almost all fruits having "stones" inside, the flower is formed around the fruit and the fruit is formed in the middle of the flower, as in all plums and acacias. And this is proper for all bushes, as the poppy, and in almost all herbs whose seed vessels where seeds are formed, come forth from the middle of the flower.

From all this one may easily infer [things about flowers], since the flower is produced from the nature of a subtle watery humor well mixed with earth. In substance, then, a flower is of a very solid and smooth substance. Therefore, when it is immersed in water their solidity and compactness prevent the entrance of air and prevent their rise above the water.

That flowers are changed to another color than green happens because of the moist transparency in them and because of the earth that is well distributed and mixed with it. For all the differences of colors are caused by different floatations of earth expanded as vapor in the moistness, or in fiery or clear smoke, or in burned earth, as we said in our discussion of the generation of sensible things.

Generally, the substance of flowers is made from very subtle moisture which appears first from heat and, because of the abundance of water in it, is extended like a leaf. Therefore, since the moisture has a great distribution, the flower is almost universally of good odor. But this would not happen unless the moisture were very well distributed and very subtle [or rare] and the earth in it were very rare and very well mixed with the moisture. For since the creation of fruits is from windy earthy vapor, there is in this vapor something rarer and wetter and of a lesser earthiness which does not easily show, and it is thickened by a distributive heat. And since this is more vaporizable than the rest of what is in the substance of the plant where the bud is located and where the fruit emerges, it comes forth immediately with the first heat and is formed into a flower.

For this reason, the dew, which produces honey and wax clings to the flower. And these are found deep in the interiors of the flowers, for when nature forms a subtle moisture which has been effected by a rare and very well mixed dryness, there flows from it this rare and well-boiled watery moisture—in the manner of a sweet phlegm in the creation of humors in animals—and this [substance], gathered and warmed by the effort of bees, is converted into the nature of honey . . . .

Wax, which is in the lowest parts, is like a purge of yellow bile that trickles from the ears of animals in the purging of the brain of animals. While the flower is formed, what it has of earth is rejected with the easily inflammable fat and like a powder is sprinkled over the insides of the flower, because what is inflammable cannot endure, by natural or alchemical means, the action of natural heat determining and forming the being of things. Indeed, before it can be formed it burns and is converted to the yellowness of yellow bile. It appears to be a formal yellow, especially in the flowers of the poppy, the linden tree, and the white birch. But it is in all flowers more or less, and it clings to the rear legs of bees when they gather honey. Indeed, they build their hives from it for the preservation of the honey . . . .

Book IV, Tract III On the Principles of Generation and the Fecundity of Plants

Chapter 1 On the Five Principles of Generation and Growth of Plants and the Doubts Emerging about Them

Herbs and whatever is planted and grows from a root fixed in the earth require one or more of five things. These are a seed, putridity [rottenness], humor [or sap], water, and planting. Of these five, the first has the formative power of the plant in itself and matter and the power to effect is in it at the same time, as we said in the second book of the *Physics.* The second thing (putridity) receives a formative power from the power of the stars. The humor, which is mixed from the elements, is the food and matter both of generation and of the plant that has been generated; indeed the plant draws this [material] purified from the earth with its first digestion. As in all nutriments, water plays no role except that it is the vehicle of the food. Nor does food flow to the parts of the plant except by means of the motion of water.

Translated by Edward Grant

Reading and Discussion Questions

1. What properties and/or powers of plants does Albertus Magnus take to differentiate them from animals? What are some key characteristics he thinks we should use in the classification of plants?
2. What does his discussion of leaves and fruits suggest to you about the level of detail at which Albertus Magnus is observing? To what extent is he pursuing a causal as well as descriptive project in his account of plant life?

On the Motion of the Heart (De Motu Cordis) (1270)

Thomas Aquinas

28

Since everything that is moved must have a mover, the problem arises: What moves the heart and exactly what kind of movement does it have?

For first of all, it does not seem that any soul moves it. The nutritive soul does not move it, since its activities are generation, nutrition, growth and diminution. But the motion of the heart is none of these. Moreover, the nutritive soul is also in plants, but the motion of the heart belongs to animals only.

Neither do the sensitive and intellectual souls move it, since sense and intellect move only by means of appetite. But the motion of the heart is involuntary.

In fact the heart's motion does not even seem to be natural, since it is made up of opposite types of movements: push and pull. But natural motion is toward one opposite, not both, such as the motion of fire, which is only up, and that of earth, which is down. On the other hand, to say that the motion of the heart is violent is irrational. For obviously if we do away with this motion, we end up doing away with (i.e., killing) the animal, but nothing violent preserves a nature. Indeed, the heart's motion must be most natural, since animal life is inseparably united to it.

Now some who say that it is a natural motion claim that its source is not the particular nature of the animal, but some outside universal nature, or an intelligence.

But this is absurd. For in all natural things, both common and specific properties in them result from an intrinsic principle. Natural things, by definition, have their principle of motion in them. But nothing is more proper to animals than the motion of the heart, for once it stops, the animal dies. Therefore, it follows that the principle of such a motion must be in the animal.

In addition, when the motions in lower bodies are caused by a universal nature, such motions are not always present in them. Take, for example, the ebb and flow of ocean tides, which result from the motion of the moon and change in accord with it. But the motion of the heart is always present in the animal. Therefore, the heart's motion does not result from a separate cause but from an intrinsic principle.

Some others say that the principle of this motion in the animal is heat, which being generated by spirit moves the heart. But this is unreasonable. For the deeper principle is more likely to be the primary cause. But the motion of the heart is a deeper principle in the animal and more contemporaneous with life than even warmth. Therefore warmth is not the cause of the heart's motion, but on the contrary the heart's motion is the cause of warmth. Thus the Philosopher says in *On the Motion of Animals*: "What is about to create motion, not by means of alteration, is of this kind" (c. 10 703a24–25).

There is another way of responding to their opinion: A fully developed animal, one that is capable of moving itself, is more like the whole universe than anything else. This is why man, who is the most fully developed of animals, is called by some a microcosm. Now in the universe the first motion is local motion, which causes alteration and the other motions. So we more clearly see in animals that local motion is the principle of alteration, and not the contrary. As the Philosopher says in the Physics: "For all natural things, to move is to live."

Yet another way: the essential is prior to the accidental. But the first motion of the animal is the motion of the heart. Heat, on the other hand, does not move something else into another place except incidentally. For an essential feature of heat is to warm, and incidentally to move something from one place to another. Therefore, it is ridiculous to say that heat is the principle of the heart's motion. Rather, we need to find a cause that is in its essential makeup a principle of local motion.

Therefore, from this point on we should take as a principle of our investigation what the Philosopher says in *Physics* 8 (254b16–20): "Of those things whose principle of motion is in themselves, we say they are moved by nature. So, even when an animal as a whole moves itself by nature, its body can sometimes be moved both by its own nature and by something outside its nature. For there is a difference between the kind of motion that it happens to undergo and its elemental composition." For when an animal descends it undergoes a motion natural both to it as a whole and to its body, since in the body of an animal the dominant element is heavy, whose nature is to move downward. But when an animal rises it undergoes a motion natural to it as a whole, because its source is an intrinsic principle, namely the soul; nevertheless, this motion is not natural to the heavy body. This is why an animal tires out more in this kind of motion.

Another point to consider is that animals move from place to place because of their desires or intellect, as the Philosopher teaches in the third book of *On the Soul* (433a9-b30).

Therefore, in animals that act only by nature and not by intent, the whole process of motion is natural. For the sparrow naturally makes a nest and the spider a web. But only man acts from intent and not by nature.

Nevertheless, the principle of every human action is natural. For although the conclusions of the theoretical and practical sciences are not naturally known, but rather are discovered through reasoning, nevertheless the first indemonstrable principles are naturally known, and from them we come to know other things. In the same way, the desire for the ultimate goal, happiness, is natural to humans, as is the aversion toward unhappiness. Thus, the desire for things other than what constitute happiness is not natural. The desire for these other things proceeds from the desire of the ultimate goal. For the goal in acts of desire is just like the indemonstrable principles in acts of the intellect, as is said in the second book of *Physics* (200a15–25). And so even though the movements of all the other parts of the body are caused by the heart, as the Philosopher proves in *On the Motion of Animals* (703a14), these movements can still be voluntary, while the first movement, that of the heart, is natural.

Moreover, let us recall that an upward motion is natural to fire as a result of its form, and hence that what generates fire, giving it its form, is essentially a place-to-place mover. In addition, just as a natural motion can result from the form of an element in a natural object, so also nothing prohibits other natural motions resulting from different forms in the same natural object. For example, we see that iron naturally moves toward a magnet, which motion is not natural to it as something heavy, but as something having a particular kind of form. In the same way, therefore, insofar as the animal has a particular kind of form, namely the soul, nothing prohibits it from having a natural motion as a result of that form. And the cause responsible for this motion would be what gives the form.

I myself say that the motion of the heart is a natural motion of the animal. As the Philosopher says in *On the Motion of Animals*, "We should consider the animal as if it was a city under good and legitimate governance. For in a city with this kind of stability of order, there is no need for a separate ruler for each and every event, but instead everyone does everything as planned, and things proceed according to custom. The same thing happens in animals naturally. For every part of the animal is naturally equipped to perform its own special function, so that there is no need for a soul in each and every part as a cause of motion. Rather, with the soul present in the principle of the body, the other parts live and perform their own special work as nature made them."

Thus, the motion of the heart is a natural result of the soul, the form of the living body and principally of the heart.

Perhaps this is why some who have understood this go on to say that the heart's movement is caused by an intelligence, for they think that the soul comes from an intelligence (which is similar to what the Philosopher says in *Physics* Book 8 [255b31–256a3]) about the movement of heavy and light things coming from a generator that gives the form which is the principle of their motion). But it is important to note that every property and movement is a result of a form in a particular condition. So as a result of the form of a subtle element like fire, there is motion to a subtle place, namely upwards motion. Now the most subtle form on earth is the soul, which is most like the principle of the motion of the heavens. Thus, the motion that results from the soul is most like the motion of the heavens. In other words, the heart moves in the animal as the heavenly bodies move in the cosmos.

Nevertheless the heart's motion is not exactly like the heavens', in the same way that what follows from a principle is never exactly like the principle itself. Now as the principle of all the motions in the universe, the motion of the heavens is circular and continuous. For the approach and departure of a heavenly body coordinates with the beginning and end of existence, and by its own continuous movement it preserves the order among moving things that do not exist forever. The motion of the heart, however, is the principle of all movements in the animal. This is why the Philosopher says in the third book of *On the Parts of Animals*, "the movements of pleasure and pain and of all the senses seems to arise there," namely in the heart, and they also end there. Thus, in order for the heart to be the beginning and end of all motions in the animal, it had to have a movement that is like a circle, but not exactly circular, composed namely from a push and pull. And so the Philosopher says in the third book of *On the Soul*, "A natural and organic cause of motion is both the source and termination of the motion. Now since all things are moved by pushes and pulls, it is necessary that something exists in a nearly circular state and that motion arises from it.

We can also say it is a continuous movement as long as the animal lives, unless it is necessary to have a rest in between the push and pull (for it is not a perfectly circular motion).

We are now in a good position to consider objections to the contrary.

For we see that the heart's motion is not natural to it as something having weight, but insofar as it is animated by a particular kind of soul. Moreover, the two motions that make up the complex movement of the heart seem contrary because the heart does not perfectly have the simplicity of circular motion, but it does imitate that motion since where it moves from it also moves toward. Thus, it is not problematic that its motion is in some way to different parts, for even circular motion is like this.

Next, there is no need to say that the heart's motion arises from either sensing or desiring, although it does arise from the sensitive soul. For the heart is not caused to move by the sensitive soul's activities, but insofar as that soul is the form and nature of a particular kind of body.

On the other hand, the progressive motion of an animal is caused by the activities of sensing and desiring. This is why doctors distinguish vital functions from animal functions and say that even when the animal functions cease, the vitals may remain. They call the vitals those functions that are immediately related to the heart's motion, such that when they cease life ceases. This position is reasonable. For to live for living beings is to exist, as is said in the second book of *On the Soul*: the existence of anything is from its own form.

We should note that there is a difference between the principle of the heavenly motion and the soul. The former is not moved in any way at all, neither essentially nor incidentally, but the sensitive soul, although unmoved essentially, is moved incidentally. Thus, different types of sensations and emotions arise in it. So, whereas the heavenly movement is always uniform, the heart's movement varies according to the different emotions and sensations of the soul. For the sensations of the soul are not caused by changes in the heart, but just the opposite is the case. This is why in the passions of the soul, such as anger, there is a formal part that pertains to a feeling, which in this example would be the desire for vengeance. And there is a material part that pertains to the heart's motion, which in the example would be the blood enkindled around the heart.

But in the things of nature, the form is not the result of the matter, but on the contrary, as is evident in the second book of *Physics*, matter has a disposition for form. Therefore, although someone does not desire revenge because his blood is burning around the heart, he is more prone to become angry because of it. But actually being angry is from the desire for vengeance.

Now although some change occurs in the heart's motion because of different sensations and feelings, nevertheless such change is involuntary, for it does not come about through the command of the will. For as the Philosopher says in *On the Cause of the Motion of Animals*, often something will be seen which, without any command of the mind, moves the heart and private parts, the cause of which he says is the natural susceptibility animals have to physical changes. For when its parts undergo change, one part increasing and another decreasing, then naturally the whole animal moves and goes through a sequence of changes.

Now warmth and cold, whether from the outside or occurring naturally within, cause such motions of the heart and private parts in animals, even against reason, by yet another incidental change. For the mind and imagination can cause a feeling of lust or anger or other passions, on account of which the heart is heated or cooled.

And let this be enough said on the motion of the heart.

Translated by Gregory Froelich

Reading and Discussion Questions

1. In this text Aquinas considers a variety of views about what moves the heart and what kind of motion the heart has. Why does he think that the motion of the heart constitutes an especially difficult problem?
2. What Aristotelian resources does he call on in trying to solve this problem?

Opus Majus: On Experimental Science (1268)

Roger Bacon

29

Having laid down the main points of the wisdom of the Latins as regards language, mathematics and optics, I wish now to review the principles of wisdom from the point of view of experimental science, because without experiment it is impossible to know anything thoroughly.

There are two ways of acquiring knowledge, one through reason, the other by experiment. Argument reaches a conclusion and compels us to admit it, but it neither makes us certain nor so annihilates doubt that the mind rests calm in the intuition of truth, unless it finds this certitude by way of experience. Thus many have arguments toward attainable facts, but because they have not experienced them, they overlook them and neither avoid a harmful nor follow a beneficial course. Even if a man that has never seen fire, proves by good reasoning that fire burns, and devours and destroys things, nevertheless the mind of one hearing his arguments would never be convinced, nor would he avoid fire until he puts his hand or some combustible thing into it in order to prove by experiment what the argument taught. But after the fact of combustion is experienced, the mind is satisfied and lies calm in the certainty of truth. Hence argument is not enough, but experience is.

This is evident even in mathematics, where demonstration is the surest. The mind of a man that receives that clearest of demonstrations concerning the equilateral triangle without experiment will never stick to the conclusion nor act upon it till confirmed by experiment by means of the intersection of two circles from either section of which two lines are drawn to the ends of a given line. Then one receives the conclusion without doubt. What Aristotle says of the demonstration by the syllogism being able to give knowledge, can be understood if it is accompanied by experience, but not of the bare demonstration. What he says in the first book of the Metaphysics, that those knowing the reason and cause are wiser than the experienced, he speaks concerning the experienced who know the bare fact only without the cause. But I speak here of the experienced that know the reason and cause through their experience. And such are perfect in their knowledge, as Aristotle wishes to be in the sixth book of the Ethics, whose simple statements are to be believed as if they carried demonstration, as he says in that very place.

Whoever wishes without proof to revel in the truths of things need only know how to neglect experience. This is evident from examples. Authors write many things and the people cling to them through arguments which they make without experiment, that are utterly false. It is commonly believed among all classes that one can break adamant only with the blood of a goat, and philosophers and theologians strengthen this myth. But it is not yet proved by adamant being broken by blood of this kind, as much as it is argued to this conclusion. And yet, even without the blood it can be broken with ease. I have seen this with my eyes; and this must needs be because gems cannot be cut out save by the breaking of the stone. Similarly it is commonly believed that the secretions of the beaver that the doctors use are the testicles of the male, but this is not so, as the beaver has this secretion beneath its breast and even the male as well as the female produces a secretion of this kind. In addition also to this secretion the male has its testicles in the natural place and thus again it is a horrible lie that, since hunters chase the beaver for this secretion, the beaver knowing what they are after, tears out his testicles with his teeth and throws them away. Again it is popularly said that cold water in a vase freezes more quickly than hot; and the argument for this is that contrary is excited by the contrary, like enemies running together. They even impute this to Aristotle in the second book of Meteorology, but he certainly did not say this, but says something like it by which they have been deceived, that if both cold and hot water are poured into a cold place as on ice, the cold freezes quicker (which is true), but if they are placed in two vases, the hot will freeze quicker. It is necessary, then, to prove everything by experience.

Experience is of two kinds. One is through the external senses: such are the experiments that are made upon the heaven through instruments in regard to facts there, and the facts on earth that we prove in various ways to be certain in our own sight. And facts that are not true in places where we are, we know through other wise men that have experienced them. Thus Aristotle with the authority of Alexander, sent 2,000 men throughout various parts of the earth in order to learn at first hand everything on the surface of the world, as Pliny says in his Natural History. And this experience is human and philosophical just as far as a man is able to make use of the beneficent grace given to him, but such experience is not enough for man, because it does not give full certainty as regards corporeal things because of their complexity and touches the spiritual not at all. Hence man's intellect must be aided in another way, and thus the patriarchs and prophets who first gave science to the world secured inner light and did not rest entirely on the senses. So also many of the faithful since Christ. For grace makes many things clear to the faithful, and there is divine inspiration not alone concerning spiritual but even about corporeal things. In accordance with which Ptolemy says in the Centilogium that there is a double way of coming to the knowledge of things, one through the experiments of science, the other through divine inspiration, which latter is far the better as he says.

Of this inner experience there are seven degrees, one through spiritual illumination in regard to scientific things. The second grade consists of virtue, for evil is ignorance as Aristotle says in the second book of the Ethics. And Algazel says in the logic that the mind is disturbed by faults, just as a rusty mirror in which the images of things cannot be clearly seen, but the mind is prepared by virtue like a well polished mirror in which the images of things show clearly. On account of this, true philosophers have accomplished more in ethics in proportion to the soundness of their virtue, denying to one another that they can discover the cause of things unless they have minds free from faults. Augustine relates this fact concerning Socrates in Book VIII, chapter III, of the City of God: to the same purpose Scripture says, to an evil mind, etc., for it is impossible that the mind should lie calm in the sunlight of truth while it is spotted with evil, but like a parrot or magpie it will repeat words foreign to it which it has learned through long practice. And this is our experience, because a known truth draws men into its light for love of it, but the proof of this love is the sight of the result . . . .

The third degree of spiritual experience is the gift of the Holy Spirit, which Isaiah describes. The fourth lies in the beatitudes which our Lord enumerates in the Gospels. The fifth is the spiritual sensibility. The sixth is in such fruits as the peace of God, which passes all understanding. The seventh lies in states of rapture and in the methods of those also, various ones of whom receive it in various ways, that they may see many things which it is not permitted to speak of to man. And whoever is thoroughly practiced in these experiences or in many of them, is able to assure himself and others, not only concerning spiritual things, but all human knowledge . . . .

And because this experimental science is a study entirely unknown by the common people, I cannot convince them of its utility, unless its virtue and characteristics are shown. This alone enables us to find out surely what can be done through nature, what through the application of art, what through fraud, what is the purport and what is mere dream in chance, conjuration, invocations, imprecations, magical sacrifices and what there is in them; so that all falsity may be lifted and the truths we alone of the art retained. This alone teaches us to examine all the insane ideas of the magicians in order not to confirm but to avoid them, just as logic criticizes the art of sophistry. This science has three great purposes in regard to the other sciences: the first is that one may criticize by experiment the noble conclusions of all the other sciences, for the other sciences know that their principles come from experiment, but the conclusions through arguments drawn from the principles discovered, if they care to have the result of their conclusions precise and complete. It is necessary that they have this through the aid of this noble science. It is true that mathematics reaches conclusions in accordance with universal experience about figures and numbers, which indeed apply to all sciences and to this experience, because no science can be known without mathematics. If we would attain to experiments precise, complete and made certain in accordance with the proper method, it is necessary to undertake an examination of the science itself, which is called experimental on our authority. I find an example in the rainbow and in like phenomena, of which nature are the circles about the sun and stars, also the halo beginning from the side of the sun or of a star which seems to be visible in straight lines and is called by Aristotle in the third book of the Meteorology a perpendicular, but by Seneca a halo, and is also called a circular corona, which have many of the colors of the rainbow. Now the natural philosopher discusses these things, and in regard to perspective has many facts to add which are concerned with the operation of seeing which is pertinent in this place. But neither Aristotle or Avicenna have given us knowledge of these things in their books upon Nature, nor Seneca, who wrote a special book concerning them. But experimental science analyzes such things.

The experimenter considers whether among visible things, he can find colors formed and arranged as given in the rainbow. He finds that there are hexagonal crystals from Ireland or India which are called rainbow-hued in Solinus Concerning the Wonders of the World and he holds these in a ray of sunlight falling through the window, and finds all the colors of the rainbow, arranged as in it in the shaded part next the ray. Moreover, the same experimenter places himself in a somewhat shady place and puts the stone up to his eye when it is almost closed, and beholds the colors of the rainbow clearly arranged, as in the bow. And because many persons making use of these stones think that it is on account of some special property of the stones and because of their hexagonal shape the investigator proceeds further and finds this in a crystal, properly shaped, and in other transparent stones. And not only are these Irish crystals in white, but also black, so that the phenomenon occurs in smoky crystal and also in all stones of similar transparency. Moreover, in stones not shaped hexagonally, provided the surfaces are rough, the same as those of the Irish crystals, not entirely smooth and yet not rougher than those—the surfaces have the same quality as nature has given the Irish crystals, for the difference of roughness makes the difference of color. He watches, also, rowers and in the drops falling from the raised oars he finds the same colors, whenever the rays of the sun penetrate the drops.

The case is the same with water falling from the paddles of a water-wheel. And when the investigator looks in a summer morning at the drops of dew clinging to the grass in the field or plane, he sees the same colors. And, likewise, when it rains, if he stands in a shady place and the sun's rays beyond him shine through the falling drops, then in some rather dark place the same colors appear, and they can often be seen at night about a candle. In the summer time, as soon as he rises from sleep while his eyes are not yet fully opened, if he suddenly looks at a window through which the light of the sun is streaming, he will see the colors. Again, sitting outside of the sunlight, if he holds his head covering beyond his eyes, or, likewise, if he closes his eyes, the same thing happens in the shade at the edges, and it also takes place through a glass vase filled with water, sitting in the sunlight. Similarly, if any one holding water in his mouth suddenly sprinkles the water in jets and stands at the side of them; or if through a lamp of oil hanging in the air the rays shine in the proper way, or the light shines upon the surface of the oil, the colors again appear. Thus, in an infinite number of ways, natural as well as artificial, colors of this kind are to be seen, if only the diligent investigator knows how to find them.

. . .

The third value of this science is this—it is on account of the prerogatives through which it looks, not only to the other sciences, but by its own power investigates the secrets of nature, and this takes place in two ways—in the knowledge of future and present events, and in those wonderful works by which it surpasses astronomy commonly so-called in the power of its conclusions. For Ptolemy in the introduction of the Almagest, says that there is another and surer way than the ordinary astronomy; that is, the experimental method which follows after the course of nature, to which many faithful philosophers, such as Aristotle and a vast crowd of the authors of predictions from the stars, are favorable, as he himself says, and we ourselves know through our own experience, which cannot be denied. This wisdom has been found as a natural remedy for human ignorance or imprudence; for it is difficult to have astronomical implements sufficiently exact and more difficult to have tables absolutely verified, especially when the motion of the planets is involved in them. The use of these tables is difficult, but the use of the instruments more so.

Translated by Oliver J. Thatcher et al.

Reading and Discussion Questions

1. What do you think that Roger Bacon means by "experiment"? Give some examples. How is what he has in mind different from or similar to how scientists understand the notion of "experiment" today?
2. What kinds of information does he think are best gained through experiment? What does he take to differentiate external experience from spiritual experience?

A Treatise on the Configuration of Qualities and Motions (c. 1360)

Nicole Oresme

30

III. I. Here begins the third part, on the acquisition and measure of quality and velocity.

Chapter 1. How the acquisition of a quality is to be imagined.

Succession in the acquisition of a quality can take place in two ways, according to extension and according to intension, as was stated above in the fourth chapter of the second part. And so extensive acquisition of linear quality is to be imagined by the motion of a point flowing over that subject line, so that the part [of the line] traversed has the quality (*sit qualificata*) and the part not yet traversed has not the quality. Example: if point *C* were moved over line *AB*, whatever part was traversed by that point would be white and whatever was not yet traversed would not yet be white. Moreover, the extensive acquisition of a surface quality would have to be imagined by the motion of a line dividing the part of the surface altered from the other part which has not yet been altered. In the same manner, the extensive acquisition of corporeal [quality] is to be imagined by the motion of a surface dividing the part altered from the part not yet altered.

Furthermore, the intensive acquisition of a punctual quality is to be imagined by the motion of a point continually ascending above the subject point, while the intensive acquisition of a linear quality is to be imagined by the motion of a line perpendicularly ascending above the subject line; and by its flux or ascent it describes the surface by which the acquired quality is designated. For example: Let *AB* be the subject line (see Fig. 6.5 A). Hence I say that the intension of point *A* is imagined by the motion or perpendicular ascent of point *C*. And the intension of line *AB* or the acquisition of intension is imagined by the ascent of line *CD*. The intensive acquisition of a surface quality in the same way is to be imagined by the ascent of a surface, and by its imagined motion it describes the body by which the quality is designated. And similarly the intensive acquisition of a corporeal quality is imagined by the motion of a surface quality, because

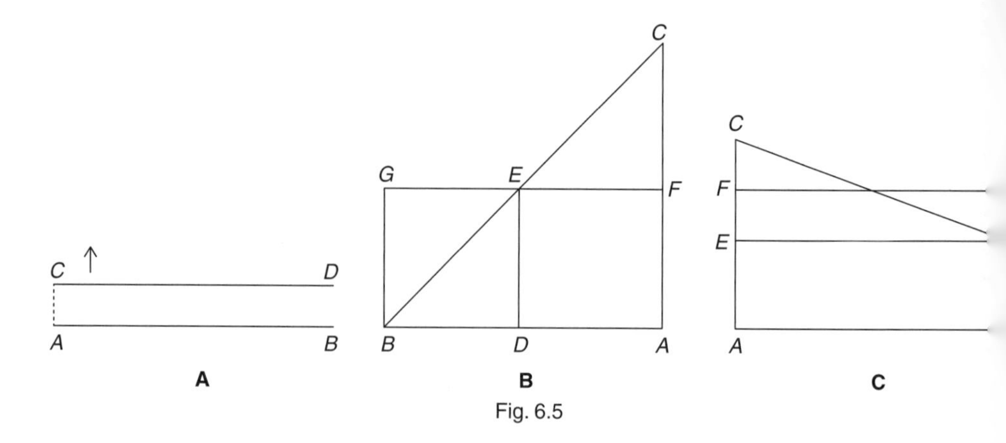

Fig. 6.5

by its imagined flux a surface describes a body. And it is not necessary to give a fourth dimension, as was said in the fourth chapter of the first part.

And what we have just now said about the acquisition of quality ought to be said *and imagined* in the same way about the loss [of quality] ... for such loss is imagined by movements opposite to the aforementioned motions [of acquisition]. What has been just now said about the acquisition or loss of quality is to be imagined as applying in the same way to the acquisition or loss of velocity both in intension and in extension ....

III. 7. On the measure of difform qualities and velocities. Every uniformly difform quality [in a subject] is just as great as would be a quality in the same or equal subject uniform at the degree [of intensity] of the middle point of the same subject; and I understand this [to be so] if the quality is linear. If it is a surface quality, [it would be equal to a quality uniform] at the degree of the mean line; if corporeal, [to one uniform] at the degree of the mean surface, all of them being understood in the same way.

In the first place this is demonstrated for a linear [quality]. Let there be quality imaginable by a triangle *ABC*, which is uniformly difform, and is terminated at zero degree in point B (see Fig. 6.5B); and let D be the middle point of the subject line. The degree of this midpoint, or its intension, is imagined by the line *DE*. Hence the quality which is uniform at degree *DE* throughout the whole subject is imaginable by a quadrangle *AFGB*, as is clear from the tenth chapter of Part I. And it is evident by the twenty-sixth [proposition] of the first [book] of Euclid, that the two small triangles *EFC* and *EGB* are equal. Therefore, the larger triangle *BAC*, which designates the quality uniformly difform, and the quadrangle *AFGD*, which would designate the quality uniform at the degree of the middle point, are equal. Hence the qualities imaginable by a triangle of this kind and a quadrangle are equal; and this was proposed.

In the same way it can be argued with respect to a uniformly difform quality terminated in both extremes at some degree. This quality would be imaginable by a quadrangle *ABCD* (see Fig. 6.5 **C**); for let there be drawn a line *DE* parallel to the subject base [line] and let triangle *CED* he formed. Then there is protracted through the degree of the middle point line *FG*, equal and parallel to the subject base [line]. Also another line *GD* is drawn. Then, as before, it will be proved that triangle *CED* is equal to quandrangle *EFGD*. Therefore, with quadrangle *AEDB* common to both, the two total [areas] are equal, namely, the quadrangle *ACDB*, which designates the uniformly difform quality, and the quadrangle *AFGB*, which designates the quality uniform at the degree of the middle point of the subject *AB*. Therefore, by the tenth chapter of the first part, the qualities representable by these quadrangles are equal.

It can be argued in the same way with respect to surface and even corporeal [quality]. We ought to speak of velocity completely in the same way as linear quality, except that in the place of the middle point [of the subject] the middle instant of time is taken as the measure of this [uniformly difform] velocity ....

III. 8. On the measure and intension to infinity of certain difformities. A finite surface can be made as long and as high as you wish without an [over-all] increase in the area by varying the extension. For such a surface has both longitude and latitude, and so it is possible for one dimension of the surface to be increased at will without increasing the over-all area, so long as the second dimension is decreased proportionally. And it is thus for bodies as well.

Let there be taken, for example (see Fig. 6.6), a rectangular surface whose [altitude] is one foot and whose base line is *AB*, and let there be another surface similar and equal to it whose base is *CD*. This [latter figure] is pictured as being divided into parts continually proportional to infinity, according to a double proportion, on its base *CD*, divided in the same way. Let *E* be the first part,

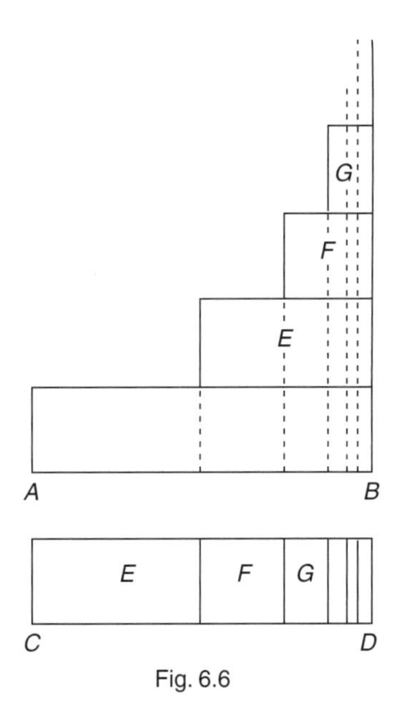

Fig. 6.6

F the second, *G* the third, and so on for the others. Take the first of these parts, namely *E*, which is one-half of the whole [subject], and place it on top of the first surface toward the end *B*. Then on top of both of them put the second part *F*. Then again on top of all of them put the third part *G*, and similarly with the other parts to infinity. When this has been done, the base line *AB* is imagined as being divided into proportional parts continually according to a double proportion and working toward *B*. Then it is immediately clear that above the first proportional part of line *AB* stands a surface with altitude of one foot, upon the second part a surface altitude of two feet, upon the third a surface with altitude of three feet, [upon the fourth part a surface with altitude of four feet], and so forth to infinity. And yet the whole surface as originally conceived with only an altitude of two feet is in no way augmented overall [by this proportional division]. Consequently, the total surface which stands over line *AB* is precisely four times the area of the surface of the part of it which stands over the first proportional part of the same line *AB*. Hence, that quality or velocity which will be proportional in intension as this figure is in altitude would be precisely quadruple to the part of it which would be in the first part of the time or subject according to a dimension of this kind . . . .

Similarly, if any moving body were moved with some velocity in the first proportional part of a period of time so divided [into proportional parts] and were moved with a double velocity in the second [proportional part], with a triple velocity in the third [proportional part], and with a quadruple velocity in the fourth [proportional part], and continually increasing velocity in this way to infinity, the "total velocity" would be precisely quadruple to the velocity of the first part, so that a body in the whole hour would traverse four times as much distance as it traversed in the first half of that hour. And if in the first half or proportional part it traversed one foot, in the whole remaining part it would traverse three feet, and in the whole time it would traverse four feet.

Translated by Marshall Clagett

Reading and Discussion Questions

1. What does Oresme take himself to prove about the relationship between an object moving at a constant velocity as compared to an object that is constantly accelerating over the same period of time? How does he use geometry to establish this conclusion?
2. Might what Oresme says here about rates of change apply to changes in qualities other than velocity, such as the ripening of a fruit or the growth of a flower?

Question on the Eight Books of the Physics of Aristotle (1509)

John Buridan

31

1. BOOK VIII, QUESTION 12. It is sought whether a projectile after leaving the hand of the projector is moved by the air, or by what it is moved.

It is argued that it is not moved by the air, because the air seems rather to resist, since it is necessary that it be divided. Furthermore, if you say that the projector in the beginning moved the projectile and the ambient air along with it, and then that air, having been moved, moves the projectile further to such and such a distance, the doubt will return as to by what the air is moved after the projectile ceases to move. For there is just as much difficulty regarding this (the air) as there is regarding the stone which is thrown.

Aristotle takes the opposite position in the eighth [book] of this work (the *Physics*) thus: "Projectiles are moved further after the projectors are no longer in contact with them, either by antiperistasis, as some say, or by the fact that the air having been pushed, pushes with a movement swifter than the movement of impulsion by which it (the body) is carried towards its own [natural] place." He determines the same thing in the seventh and eighth [books] of this work (the *Physics*) and in the third [book] of the *De caelo*.

2. This question I judge to be very difficult because Aristotle, as it seems to me, has not solved it well. For he touches on two opinions. The first one, which he calls "antiperistasis," holds that the projectile swiftly leaves the place in which it was, and nature, not permitting a vacuum, rapidly sends air in behind to fill up the vacuum. The air moved swiftly in this way and impinging upon the projectile impels it along further. This is repeated continually up to a certain distance . . . . But such a solution notwithstanding, it seems to me that this method of proceeding was without value because of many experiences (*experientie*).

The first experience concerns the top (*trocus*) and the smith's mill (i.e. wheel – *mola fabri*) which are moved for a long time and yet do not leave their places.

Hence, it is not necessary for the air to follow along to fill up the place of departure of a top of this kind and a smith's mill. So it cannot be said [that the top and the smith's mill are moved by the air] in this manner.

The second experience is this: A lance having a conical posterior as sharp as its anterior would be moved after projection just as swiftly as it would be without a sharp conical posterior. But surely the air following could not push a sharp end in this way, because the air would be easily divided by the sharpness.

The third experience is this: a ship drawn swiftly in the river even against the flow of the river, after the drawing has ceased, cannot be stopped quickly, but continues to move for a long time. And yet a sailor on deck does not feel any air from behind pushing him. He feels only the air from the front resisting [him]. Again, suppose that the said ship were loaded with grain or wood and a man were situated to the rear of the cargo. Then if the air were of such an impetus that it could push the ship along so strongly, the man would be pressed very violently between that cargo and the air following it. Experience shows this to be false. Or, at least, if the ship were loaded with grain or straw, the air following and pushing would fold over (*plico*) the stalks which were in the rear. This is all false.

3. Another opinion, which Aristotle seems to approve, is that the projector moves the air adjacent to the projectile [simultaneously] with the projectile and that air moved swiftly has the power of moving the projectile. He does not mean by this that the same air is moved from the place of projection to the place where the projectile stops, but rather that the air joined to the projector is moved by the projector and that air having been moved moves another part of the air next to it, and that [part] moves another (i.e., the next) up to a certain distance. Hence the first air moves the projectile into the second air, and the second [air moves it] into the third air, and so on. Aristotle says, therefore, that there is not one mover but many in turn. Hence he also concludes that the movement is not continuous but consists of succeeding or contiguous entities.

But this opinion and method certainly seems to me equally as impossible as the opinion and method of the preceding view. For this method cannot solve the problem of how the top or smith's mill is turned after the hand [which sets them into motion] has been removed. Because, if you cut off the air on all sides near the smith's mill by a cloth (*linteamine*), the mill does not on this account stop but continues to move for a long time. Therefore it is not moved by the air.

Also a ship drawn swiftly is moved a long time after the haulers have stopped pulling it. The surrounding air does not move it, because if it were covered by a cloth and the cloth with the ambient air were withdrawn, the ship would not stop its motion on this account. And even if the ship were loaded with grain or straw and were moved by the ambient air, then that air ought to blow exterior stalks toward the front. But the contrary is evident, for the stalks are blown rather to the rear because of the resisting ambient air.

Again, the air, regardless of how fast it moves, is easily divisible. Hence it is not evident as to how it would sustain a stone of weight of one thousand pounds projected in a sling or in a machine.

Furthermore, you could, by pushing your hand, move the adjacent air, if there is nothing in your hand, just as fast or faster than if you were holding in your hand a stone which you wish to project. If, therefore, that air by reason of the velocity of its motion is of a great enough impetus to move the stone swiftly, it seems that if I were to impel air toward you equally as fast, the air ought to push you impetuously and with sensible strength. [Yet] we would not perceive this.

Also, it follows that you would throw a feather farther than a stone and something less heavy farther than something heavier, assuming equal magnitudes and shapes. Experience shows this to be false. The consequence is manifest, for the air having been moved ought to sustain or carry or move a feather more easily than something heavier . . . .

4. Thus we can and ought to say that in the stone or other projectile there is impressed something which is the motive force (*virtus motiva*) of that projectile. And this is evidently better than falling back on the statement that the air continues to move that projectile. For the air appears rather to resist. Therefore, it seems to me that it ought to be said that the motor in moving a moving body impresses (*imprimit*) in it a certain impetus (*impetus*) or a certain motive force (*vis motiva*) of the moving body, [which impetus acts] in the direction toward which the mover was moving the moving body, either up or down, or laterally, or circularly. And by the amount the motor moves that moving body more swiftly, by the same amount it will impress in it a stronger impetus. It is by that impetus that the stone is moved after the projector ceases to move. But that impetus is continually decreased (*remittitur*) by the resisting air and by the gravity of the stone, which inclines it in a direction contrary to that in which the impetus was naturally predisposed to move it. Thus the movement of the stone continually becomes slower, and finally that impetus is so diminished or corrupted that the gravity of the stone wins out over it and moves the stone down to its natural place.

This method, it appears to me, ought to be supported because the other methods do not appear to be true and also because all the appearances (*apparentia*) are in harmony with this method.

5. For if anyone seeks why I project a stone farther than a feather, and iron or lead fitted to my hand farther than just as much wood, I answer that the cause of this is that the reception of all forms and natural dispositions is in matter and by reason of matter. Hence by the amount more there is of matter, by that amount can the body receive more of that impetus and more intensely (*intensius*). Now in a dense and heavy body, other things being equal, there is more of prime matter than in a rare and light one. Hence a dense and heavy body receives more of that impetus and more intensely, just as iron can receive more calidity than wood or water of the same quantity. Moreover, a feather receives such an impetus so weakly (*remisse*) that such an impetus is immediately destroyed by the resisting air. And so also if light wood and heavy iron of the same volume and of the same shape are moved equally fast by a projector, the iron will be moved farther because there is impressed in it a more intense impetus, which is not so quickly corrupted as the lesser impetus would be corrupted. This also is the reason why

it is more difficult to bring to rest a large smith's mill which is moving swiftly than a small one, evidently because in the large one, other things being equal, there is more impetus. And for this reason you could throw a stone of one-half or one pound weight farther than you could a thousandth part of it. For the impetus in that thousandth part is so small that it is overcome immediately by the resisting air.

6. From this theory also appears the cause of why the natural motion of a heavy body downward is continually accelerated (*continue velocitatur*). For from the beginning only the gravity was moving it. Therefore, it moved more slowly, but in moving it impressed in the heavy body an impetus. This impetus now [acting] together with its gravity moves it. Therefore, the motion becomes faster; and by the amount it is faster, so the impetus becomes more intense. Therefore, the movement evidently becomes continually faster.

[The impetus then also explains why] one who wishes to jump a long distance drops back a way in order to run faster, so that by running he might acquire an impetus which would carry him a longer distance in the jump. Whence the person so running and jumping does not feel the air moving him, but [rather] feels the air in front strongly resisting him.

Also, since the Bible does not state that appropriate intelligences move the celestial bodies, it could be said that it does not appear necessary to posit intelligences of this kind, because it would be answered that God, when He created the world, moved each of the celestial orbs as He pleased, and in moving them He impressed in them impetuses which moved them without His having to move them any more except by the method of general influence whereby He concurs as a co-agent in all things which take place; "for thus on the seventh day He rested from all work which He had executed by committing to others the actions and the passions in turn." And these impetuses which He impressed in the celestial bodies were not decreased nor corrupted afterwards, because there was no inclination of the celestial bodies for other movements. Nor was there resistance which would be corruptive or repressive of that impetus. But this I do not say assertively, but [rather tentatively] so that I might seek from the theological masters what they might teach me in these matters as to how these things take place . . . .

7. The first [conclusion] is that that impetus is not the very local motion in which the projectile is moved, because that impetus moves the projectile and the mover produces motion. Therefore, the impetus produces that motion, and the same thing cannot produce itself. Therefore, etc.

Also since every motion arises from a motor being present and existing simultaneously with that which is moved, if the impetus were the motion, it would be necessary to assign some other motor from which that motion would arise. And the principal difficulty would return. Hence there would be no gain in positing such an impetus. But others cavil when they say that the prior part of the motion which produces the projection produces another part of the motion which is related successively and that produces another part and so on up to the cessation of the whole movement. But this is not probable, because the "producing something" ought to exist when the something is made, but the prior part of the motion does not exist when the posterior part exists, as was elsewhere stated. Hence, neither does the prior exist when the posterior is made. This consequence is obvious from this reasoning. For it was said elsewhere that motion is nothing else than "the very being produced" (*ipsum fieri*) and the "very being corrupted" (*ipsum corumpi*). Hence motion does not result when it *has been* produced (*factus est*) but when it *is being* produced (*fit*).

8. The second conclusion is that that impetus is not a purely successive thing (*res*), because motion is just such a thing and the definition of motion [as a successive thing] is fitting to it, as was stated elsewhere. And now it has just been affirmed that that impetus is not the local motion.

Also, since a purely successive thing is continually corrupted and produced, it continually demands a producer. But there cannot be assigned a producer of that impetus which would continue to be simultaneous with it.

9. The third conclusion is that that impetus is a thing of permanent nature (*res nature permanentis*), distinct from the local motion in which the projectile is moved. This is evident from the two aforesaid conclusions and from the preceding [statements]. And it is probable (*verisimile*) that that impetus is a quality naturally present and predisposed for moving a body in which it is impressed, just as it is said that a quality impressed in iron by a magnet moves the iron to the magnet. And it also is probable that just as that quality (the impetus) is impressed in the moving body along with the motion by the motor; so with the motion it is remitted, corrupted, or impeded by resistance or a contrary inclination.

10. And in the same way that a luminant generating light generates light reflexively because of an obstacle, so that impetus because of an obstacle acts reflexively. It is true, however, that other causes aptly concur with that impetus for greater or longer reflection. For example, the ball which we bounce with the palm in falling to earth is reflected higher than a stone, although the stone falls more swiftly and more impetuously (*impetuosius*) to the earth. This is because many things are curvable or intracompressible by violence which are innately disposed to return swiftly and by themselves to their correct position or to the disposition natural to them. In thus returning, they can impetuously push or draw something conjunct to them, as is evident in the case of the bow (*arcus*). Hence in this way the ball thrown to the hard ground is compressed into itself by the impetus of its motion; and immediately after striking, it returns swiftly to its sphericity by elevating itself upward. From this elevation it acquires to itself an impetus which moves it upward a long distance.

Also, it is this way with a cither cord which, put under strong tension and percussion, remains a lone time in a certain vibration (*tremulatio*) from which its sound continues a notable time. And this takes place as follows: As a result of striking [the chord] swiftly, it is bent violently in one direction, and so it returns swiftly toward its normal straight position. But on account of the impetus, it crosses beyond the normal straight position in the contrary direction and then again returns. It does this many times. For a similar reason a bell (*campana*), after the ringer ceases to draw [the chord], is moved a long time, first in one direction, now in another. And it cannot be easily and quickly brought to rest.

This, then, is the exposition of the question. I would be delighted if someone would discover a more probable way of answering it. And this is the end.

Translated by Marshall Clagett

Reading and Discussion Questions

1. What observations does Buridan use to critique Aristotle's account of projectile motion?
2. What alternative does Buridan propose?

Additional Resources

Bell, Mark and Sen, Paul, Producers. *Science and Islam* [Documentary]. Britain: BBC, 2009. A sensitive and highly informative three-part introduction to Islamic contributions to natural philosophy in the Medieval period, hosted by Jim Al-Khalili.

Conrad, Lawrence I., et al. *The Western Medical Tradition: 800BC to AD 1800*. Cambridge: Cambridge University Press, 1995.

Crowe, Michael J. *Mechanics from Aristotle to Einstein.* Santa Fe, NM: Green Lion Press, 2007. A valuable mixture of primary and secondary source material, the first chapter of this text is especially helpful in understanding Oresme.

Grant, Edward, ed. *A Source Book in Medieval Science.* Cambridge: Harvard University Press, 1974.

Lindberg, David C. *The Beginnings of Western Science: The European Scientific Tradition in Philosophical, Religious, and Institutional Context, Prehistory to A.D. 1450*. 2nd ed. Chicago: Chicago University Press, 2007. Essential background reading for the entire ancient and medieval period, the best secondary source introduction currently available.

Lindberg, David C. *Theories of Vision from Al Kindi to Kepler*. Chicago: University of Chicago Press, 2006.

Robbins, Robert J. *Electronic Scholarly Publishing*. esp.org This website specializes in the history of biology, going all the way back to Aristotle, with a particular focus on genetics.

Part III

Revolutions in Astronomy and Mechanics: From Copernicus to Newton

The Commentariolus (1515)

Nicolaus Copernicus

32

Our ancestors assumed, I observe, a large number of celestial spheres for this reason especially, to explain the apparent motion of the planets by the principle of regularity. For they thought it altogether absurd that a heavenly body, which is a perfect sphere, should not always move uniformly. They saw that by connecting and combining regular motions in various ways they could make any body appear to move to any position.

Callippus and Eudoxus, who endeavored to solve the problem by the use of concentric spheres, were unable to account for all the planetary movements; they had to explain not merely the apparent revolutions of the planets but also the fact that these bodies appear to us sometimes to mount higher in the heavens, sometimes to descend; and this fact is incompatible with the principle of concentricity.

Therefore it seemed better to employ eccentrics and epicycles, a system which most scholars finally accepted. Yet the planetary theories of Ptolemy and most other astronomers, although consistent with the numerical data, seemed likewise to present no small difficulty. For these theories were not adequate unless certain equants were also conceived; it then appeared that a planet moved with uniform velocity neither on its deferent nor about the center of its epicycle. Hence a system of this sort seemed neither sufficiently absolute nor sufficiently pleasing to the mind.

Having become aware of these defects, I often considered whether there could perhaps be found a more reasonable arrangement of circles, from which every apparent inequality would be derived and in which everything would move uniformly about its proper center, as the rule of absolute motion requires. After I had addressed myself to this very difficult and almost insoluble problem, the suggestion at length came to me how it could be solved with fewer and much simpler constructions than were formerly used, if some assumptions (which are called axioms) were granted me. They follow in this order.

Assumptions

1. There is no one center of all the celestial circles or spheres.
2. The center of the earth is not the center of the universe, but only of gravity and of the lunar sphere.
3. All the spheres revolve about the sun as their mid-point, and therefore the sun is the center of the universe.
4. The ratio of the earth's distance from the sun to the height of the firmament is so much smaller than the ratio of the earth's radius to its distance from the sun that the distance from the earth to the sun is imperceptible in comparison with the height of the firmament.
5. Whatever motion appears in the firmament arises not from any motion of the firmament, but from the earth's motion. The earth together with its circumjacent elements performs a complete rotation on its fixed poles in a daily motion, while the firmament and highest heaven abide unchanged.
6. What appear to us as motions of the sun arise not from its motion but from the motion of the earth and our sphere, with which we revolve about the sun like any other planet. The earth has, then, more than one motion.
7. The apparent retrograde and direct motion of the planets arises not from their motion but from the earth's. The motion of the earth alone, therefore, suffices to explain so many apparent inequalities in the heavens.

Having set forth these assumptions, I shall endeavor briefly to show how uniformity of the motions can be saved in a systematic way. However, I have thought it well, for the sake of brevity, to omit from this sketch mathematical demonstrations, reserving these for my larger work. But in the explanation of the circles I shall set down here the lengths of the radii; and from these the reader who is not unacquainted with mathematics will readily perceive how closely this arrangement of circles agrees with the numerical data and observations.

Accordingly, let no one suppose that I have gratuitously asserted, with the Pythagoreans, the motion of the earth; strong proof will be found in my exposition of the circles. For the principal arguments by which the natural philosophers attempt to establish the immobility of the earth rest for the most part on the appearances; it is particularly such arguments that collapse here, since I treat the earth's immobility as due to an appearance.

The Order of the Spheres

The celestial spheres arc arranged in the following order. The highest is the immovable sphere of the fixed stars, which contains and gives position to all things. Beneath it is Saturn, which Jupiter follows, then Mars. Below Mars is the sphere on which we revolve; then Venus; last is Mercury. The lunar sphere revolves about the center of the earth and moves with the earth like an epicycle. In the same order also, one planet surpasses another in speed of revolution, according as they trace greater or smaller circles. Thus Saturn completes its revolution in thirty years, Jupiter in twelve, Mars in two and one-half, and the earth in one year; Venus in nine months, Mercury in three.

The Apparent Motions of the Sun

The earth has three motions. First, it revolves annually in a great circle about the sun in the order of the signs, always describing equal arcs in equal times; the distance from the center of the circle to the center of the sun is 1/25 of the radius of the circle. The radius is assumed to have a length imperceptible in comparison with the height of the firmament; consequently, the sun appears to revolve with this motion, as if the earth lay in the center of the universe. However, this appearance is caused by the motion not of the sun but of the earth, so that, for example, when the earth is in the sign of Capricornus, the sun is seen diametrically opposite in Cancer, and so on. On account of the previously mentioned distance of the sun from the center of the circle, this apparent motion of the sun is not uniform, the maximum inequality being 2 1/2°. The line drawn from the sun through the center of the circle is invariably directed toward a point of the firmament about 10° west of the more brilliant of the two bright stars in the head of Gemini, therefore when the earth is opposite this point, and the center of the circle lies between them, the sun is seen at is greatest distance from the earth. In this circle, then, the earth revolves together with whatever else is included within the lunar sphere.

The second motion, which is peculiar to the earth, is the daily rotation on the poles in the order of the signs, that is, from west to east. On account of this rotation the entire universe appears to revolve with enormous speed. Thus does the earth rotate together with its circumjacent waters and encircling atmosphere.

The third is the motion in declination. For the axis of the daily rotation is not parallel to the axis of the great circle, but is inclined to it at an angle that intercepts a portion of a circumference, in our time about 23 1/2. Therefore, while the center of the earth always remains in the plane of the ecliptic, that is, in the circumference of the great circle, the poles of the earth rotate, both of them describing small circles about centers equidistant from the axis of the great circle. The period of this motion is not quite a year and is nearly equal to the annual revolution on the great circle. But the axis of the great circle is invariably directed toward the points of the firmament which are called the poles of the ecliptic. In like manner the motion in declination, combined with the annual motion in their joint effect upon the poles of the daily rotation, would keep these poles constantly fixed at the same points of the heavens, if the periods of both motions were exactly equal. Now with the long passage of time is has become clear that this inclination of the earth to the firmament changes. Hence it is the common opinion that the firmament has several motions in conformity with a law not yet sufficiently understood. But the motion of the earth can explain all these changes in a less surprising way. I am not concerned to state what the path of the poles is. I am aware that, in lesser matters, a magnetized iron needle always points in the same direction. It has nevertheless seemed a better view to ascribe the changes to a sphere, whose motion governs the movements of the poles. This sphere must doubtless be sublunar.

Translated by Edward Rosen

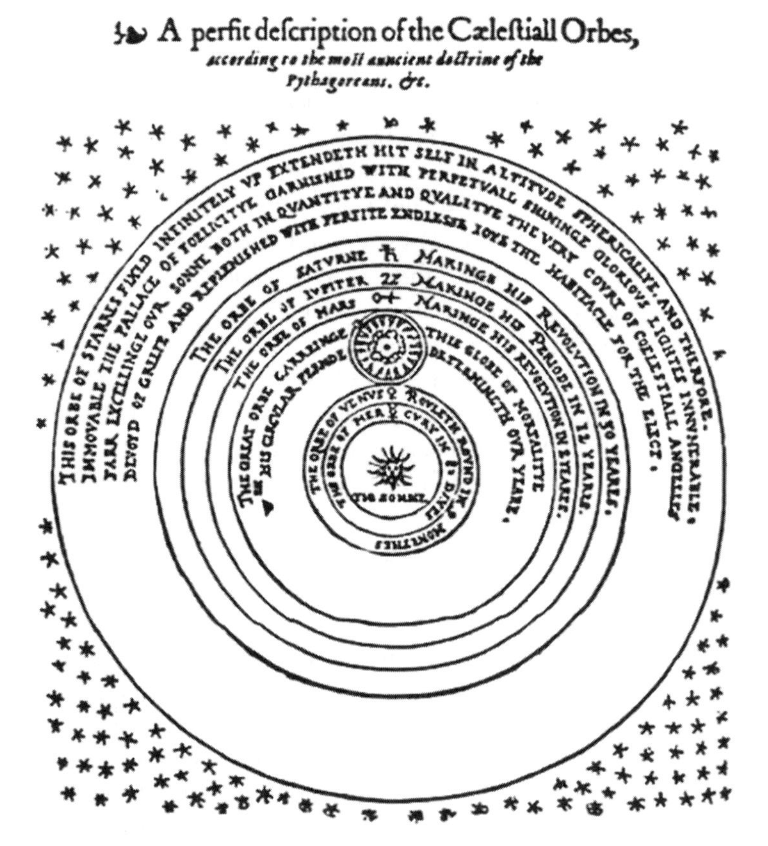

Reading and Discussion Questions

1. What concepts from the ancient world are still present in Copernicus' ideas about the universe?
2. Compare Copernicus' five assumptions to Ptolemy's. What is new or different and how does he think that it improves upon ancient models?

Dedication to On the Revolutions of the Celestial Spheres (1543)

Nicolaus Copernicus

33

To Pope Paul III

I can easily conceive, most Holy Father, that as soon as some people learn that in this book which I have written concerning the revolutions of the heavenly bodies, I ascribe certain motions to the Earth, they will cry out at once that I and my theory should be rejected. For I am not so much in love with my conclusions as not to weigh what others will think about them, and although I know that the meditations of a philosopher are far removed from the judgment of the laity, because his endeavor is to seek out the truth in all things, so far as this is permitted by God to the human reason, I still believe that one must avoid theories altogether foreign to orthodoxy. Accordingly, when I considered in my own mind how absurd a performance it must seem to those who know that the judgment of many centuries has approved the view that the Earth remains fixed as center in the midst of the heavens, if I should, on the contrary, assert that the Earth moves; I was for a long time at a loss to know whether I should publish the commentaries which I have written in proof of its motion, or whether it were not better to follow the example of the Pythagoreans and of some others, who were accustomed to transmit the secrets of Philosophy not in writing but orally, and only to their relatives and friends, as the letter from Lysis to Hipparchus bears witness. They did this, it seems to me, not as some think, because of a certain selfish reluctance to give their views to the world, but in order that the noblest truths, worked out by the careful study of great men, should not be despised by those who are vexed at the idea of taking great pains with any forms of literature except such as would be profitable, or by those who, if they are driven to the study of Philosophy for its own sake by the admonitions and the example of others, nevertheless, on account of their stupidity, hold a place among philosophers similar to that of drones among bees. Therefore, when I considered this carefully, the contempt which I had to fear because of the novelty and apparent absurdity of my view, nearly induced me to abandon utterly the work I had begun.

My friends, however, in spite of long delay and even resistance on my part, withheld me from this decision. First among these was Nicolaus Schonberg, Cardinal of Capua, distinguished in all branches of learning. Next to him comes my very dear friend, Tidemann Giese, Bishop of Culm, a most earnest student, as he is, of sacred and, indeed, of all good learning. The latter has often urged me, at times even spurring me on with reproaches, to publish and at last bring to the light the book which had lain in my study not nine years merely, but already going on four times nine. Not a few other very eminent and scholarly men made the same request, urging that I should no longer through fear refuse to give out my work for the common benefit of students of Mathematics. They said I should find that the more absurd most men now thought this theory of mine concerning the motion of the Earth, the more admiration and gratitude it would command after they saw in the publication of my commentaries the mist of absurdity cleared away by most transparent proofs. So, influenced by these advisors and this hope, I have at length allowed my friends to publish the work, as they had long besought me to do.

But perhaps Your Holiness will not so much wonder that I have ventured to publish these studies of mine, after having taken such pains in elaborating them that I have not hesitated to commit to writing my views of the motion of the Earth, as you will be curious to hear how it occurred to me to venture, contrary to the accepted view of mathematicians, and well-nigh contrary to common sense, to form a conception of any terrestrial motion whatsoever. Therefore I would not have it unknown to Your Holiness, that the only thing which induced me to look for another way of reckoning the movements of the heavenly bodies was that I knew that mathematicians by no means agree in their investigations thereof. For, in the first place, they are so much in doubt concerning the motion of the sun and the moon, that they can not even demonstrate and prove by observation the constant length of a complete year; and in the second place, in determining the motions both of these and of the five other planets, they fail to employ consistently one set of first principles and hypotheses, but use methods of proof based only upon the apparent revolutions and motions. For some employ concentric circles only; others, eccentric circles and epicycles; and even by these means they do not completely attain the desired end. For, although those who have depended upon concentric circles have shown that certain diverse motions can be deduced from these, yet they have not succeeded thereby in laying down any sure principle, corresponding indisputably to the phenomena. These, on the other hand, who have devised systems of eccentric circles, although they seem in great part to have solved the apparent movements by calculations which by these eccentrics are made to fit, have nevertheless introduced many things which seem to contradict the first principles of the uniformity of motion. Nor have they been able to discover or calculate from these the main point, which is the shape of the world and the fixed symmetry of its parts; but their procedure has been as if someone were to collect hands, feet, a head, and other members from various places, all very fine in themselves, but not proportionate to one body, and no single one corresponding in its turn to the others, so that a monster rather than a man would be formed from them. Thus in their process of demonstration which they term a "method," they are found to have omitted something essential, or to have included something foreign and not pertaining to the matter in hand. This certainly would never have happened to them if they had followed fixed principles; for if the hypotheses they assumed were not false, all that resulted therefrom would be verified indubitably. Those things which I am saying now may be obscure, yet they will be made clearer in their proper place.

Therefore, having turned over in my mind for a long time this uncertainty of the traditional mathematical methods of calculating the motions of the celestial bodies, I began to grow disgusted that no more consistent scheme of the movements of the mechanism of the universe, set up for our benefit by that best and most law abiding Architect of all things, was agreed upon by philosophers who otherwise investigate so carefully the most minute details of this world. Wherefore I undertook the task of rereading the books of all the philosophers I could get access to, to see whether any one ever was of the opinion that the motions of the celestial bodies were other than those postulated by the men who taught mathematics in the schools. And I found first, indeed, in Cicero, that Niceta perceived that the Earth moved; and afterward in Plutarch I found that some others were of this opinion, whose words I have seen fit to quote here, that they may be accessible to all:

> Some maintain that the Earth is stationary, but Philolaus the Pythagorean says that it revolves in a circle about the fire of the ecliptic, like the sun and moon. Heraklides of Pontus and Ekphantus the Pythagorean make the Earth move, not changing its position, however, confined in its falling and rising around its own center in the manner of a wheel.

Taking this as a starting point, I began to consider the mobility of the Earth; and although the idea seemed absurd, yet because I knew that the liberty had been granted to others before me to postulate all sorts of little circles for explaining the phenomena of the stars, I thought I also might easily be permitted to try whether by postulating some motion of the Earth, more reliable conclusions could be reached regarding the revolution of the heavenly bodies, than those of my predecessors.

And so, after postulating movements, which, farther on in the book, I ascribe to the Earth, I have found by many and long observations that if the movements of the other planets are assumed for the circular motion of the Earth and are substituted for the revolution of each star, not only do their phenomena follow logically therefrom, but the relative positions and magnitudes both of the stars and all their orbits, and of the heavens themselves, become so closely related that in none of its parts can anything be changed without causing confusion in the other parts and in the whole universe. Therefore, in the course of the work I have followed this plan: I describe in the first book all the positions of the orbits together with the movements which I ascribe to the Earth, in order that this book might contain, as it were, the general scheme of the universe. Thereafter in the remaining books, I set forth the motions of the other stars and of all their orbits together with the movement of the Earth, in order that one may see from this to what extent the movements and appearances of the other stars and their orbits can be saved, if they are transferred to the movement of the Earth. Nor do I doubt that ingenious and learned mathematicians will sustain me, if they are willing to recognize and weigh, not superficially, but with that thoroughness which Philosophy demands above all things, those matters which have been adduced by me in this work to demonstrate these theories. In order, however, that both the learned and the unlearned equally may see that I do not avoid anyone's judgment, I have preferred to dedicate these lucubrations of mine to Your Holiness rather than to any other, because, even in this remote corner of the world where I live, you are considered to be the most eminent man in dignity of rank and in love of all learning and even of mathematics, so that by your authority and judgment you can easily suppress the bites of slanderers, albeit the proverb hath it that there is no remedy for the bite of a sycophant. If perchance there shall be idle talkers, who, though they are ignorant of all mathematical sciences, nevertheless assume the right to pass judgment on these things, and if they should dare to criticise and attack this theory of mine because of some passage of scripture which they have falsely distorted for their own purpose, I care not at all; I will even despise their judgment as foolish. For it is not unknown that Lactantius, otherwise a famous writer but a poor mathematician, speaks most childishly of the shape of the Earth when he makes fun of those who said that the Earth has the form of a sphere. It should not seem strange then to zealous students, if some such people shall ridicule us also. Mathematics are written for mathematicians, to whom, if my opinion does not deceive me, our labors will seem to contribute something to the ecclesiastical state whose chief office Your Holiness now occupies; for when not so very long ago, under Leo X, in the Lateran Council the question of revising the ecclesiastical calendar was discussed, it then remained unsettled, simply because the length of the years and months, and the motions of the sun and moon were held to have been not yet sufficiently determined. Since that time, I have given my attention to observing these more accurately, urged on by a very distinguished man, Paul, Bishop of Fossombrone, who at that time had charge of the matter. But what I may have accomplished herein I leave to the judgment of Your Holiness in particular, and to that of all other learned mathematicians; and lest I seem to Your Holiness to promise more regarding the usefulness of the work than I can perform, I now pass to the work itself.

Translated by Charles William Eliot

Reading and Discussion Questions

1. What are some of the advantages of the Copernican system over the Ptolemaic and/or Aristotelian systems? What are some of the disadvantages?
2. What arguments does Copernicus provide to the Pope about why his work should be published?
3. How does Copernicus seem to think about the relationship between astronomy and scripture?

Preface to Copernicus' *On the Revolution of the Celestial Spheres* (1543)

Andreas Osiander

34

To the Reader Concerning the Hypothesis of This Work

Since the novelty of the hypothesis of this work has already been widely reported, I have no doubt that some learned men have taken serious offence because the book declares that the earth moves, and that the sun is at rest in the center of the universe; these men undoubtedly believe that the liberal arts, established long ago upon a correct basis, should not be thrown into confusion.

But if they are willing to examine the matter closely, they will find that the author of this work has done nothing blameworthy. For it is the duty of an astronomer to compose the history of the celestial motions through careful and skillful observation. Then turning to the causes of these motions or hypotheses about them, he must conceive and devise, since he cannot in any way attain to the true causes, such hypotheses as, being assumed, enable the motions to be calculated correctly from the principles of geometry, for the future as well as for the past . . . .

Now when from time to time there are offered for one and the same motion different hypotheses (as eccentricity and an epicycle for the sun's motion), the astronomer will accept above all others the one which is the easiest to grasp. The philosopher will perhaps rather seek the semblance of the truth. But neither of them will understand or state anything certain, unless it has been divinely revealed to him.

Let us therefore permit these new hypotheses to become known together with the ancient hypotheses, which are no more probable; let us do so especially because the new hypotheses are admirable and also simple, and bring with them a huge treasury of very skillful observations. So far as hypotheses are concerned, let no one expect anything certain from astronomy, which cannot furnish it, lest he accept as the truth ideas conceived for another purpose, and depart from this study a greater fool than when he entered it. Farewell.

Translated by Charles Glenn Wallis

Reading and Discussion Questions

1. How do Osiander's arguments for why the work should be published differ from Copernicus' own?
2. How does this relate to the debate between realism and instrumentalism in the philosophy of science?

Instruments for the Restoration of Astronomy (1598)

Tycho Brahe

35

On That Which We Have Hitherto Accomplished in Astronomy with God's Help, and on That Which with His Gracious Aid Has Yet to Be Completed in the Future

In the year of Our Lord 1563, that is 35 years ago, on the occasion of the great conjunction of the upper planets which took place at the end of Cancer and the beginning of Leo, when I had reached the age of sixteen years, I was occupied with studies of classical literature in Leipzig, where I lived with my governor, supported by my beloved paternal uncle ... By and by I got accustomed to distinguishing the constellations of the sky, and in the course of a month I learnt to know them all, in so far as they were located in that part of the sky which was visible there. For this purpose I made use of a small celestial globe, not greater than a fist, which I used to take with me in the evening without mentioning it to anybody. I learnt this by myself, without any guidance; in fact I never had the benefit of a teacher in Mathematics (Astronomy), otherwise I might have made quicker and better progress in these subjects. Soon my attention was drawn towards the motions of the planets ... .

Later on, in the year 1564, I secretly had a wooden astronomical radius made according to the direction of Gemma Frisius. This instrument was provided with an accurate division utilizing transversal points by Bartholomæus Scultetus, who at the time lived in Leipzig, and with whom I was on intimate terms on account of our mutual interests. Scultetus had been taught the principle of transversal points by his teacher Homelius. When I had got this radius,

I eagerly set about making stellar observations whenever I enjoyed the benefit of a clear sky, and often I stayed awake the whole night through, while my governor slept and knew nothing about it; for I observed the stars through a skylight and entered the observations specially in a small book, which is still in my possession. Soon afterwards I noticed that angular distances, which by the radius had been found to be equal, and which with the help of a mathematical calculation of proportions had been converted into numbers, did not in every respect agree with each other. After I had found the cause of the error, I invented a table by which I could correct the defects of this radius .... But in 1569 and the following year, when I lived in Augsburg, I very often observed the stars, not only with the very large quadrant, which I had made in the garden of the mayor outside the city (about which I have spoken above), but also with another instrument, a wooden sextant that I invented there, and I entered my observations in a special book. I also did this industriously later on, after I had again returned to my fatherland, using another similar, though somewhat larger instrument, particularly when the strange new star, that flared up in 1572, made me give up my chemical investigations which occupied me very much after I had started them in Augsburg and which I continued until that time, and turn towards the study of the celestial phenomena. Having observed it industriously I described it, first in a small book, later conscientiously and thoroughly in a whole volume. In the course of time I had other and yet other astronomical instruments made, some of which I took with me when I travelled again all through Germany and part of Italy. Even on the journey I continued to observe the stars whenever possible.

When at length I had returned to the fatherland about the time of my 28th year ... the noble and mighty Frederick II, King of Denmark and Norway, of illustrious memory, sent one of his young noblemen to me at Knudstrup with a Royal letter bidding me to go to see him immediately wherever he might be dwelling on Sealand. When I had presented myself without delay this excellent King, who cannot be sufficiently praised, of his own accord and according to his most gracious will offered me that island in the far-famed Danish Sound that our countrymen call Hven .... He asked me to erect buildings on this island, and to construct instruments for astronomical investigations as well as for chemical studies, and he graciously promised me that he would abundantly defray the expenses. After I had for some time contemplated the matter and asked some wise men for their advice, I gave up my previous plan and willingly agreed to the King's wish, particularly when I saw that on this island, which is situated all by itself between Scania and Sealand, I could be rid of the disturbances of visitors, and that I could in this way obtain, in my own fatherland to which above other countries I owe so very much, the quiet and the convenient conditions that I had been looking for elsewhere. So, in the year 1576, I began building the castle Uraniborg, suitable for the study of Astronomy, and in the course of time I constructed buildings as well as astronomical instruments of various kinds, fitted for making accurate observations. The most important of these are delineated and explained in this book. Meanwhile I also energetically started observing, and for this work I made use of the assistance of several students who distinguished themselves by talents and a keen vision. I had such students in my house all the time, one class after another, and I taught them this and other sciences. Thus by the grace of God it came about that there was hardly any day or night with clear weather that we did not get a great many, and very accurate, astronomical observations of the fixed stars as well as of all the planets, and also of the comets that appeared during that time, seven of which were carefully observed in the sky from that place. In this way observations were industriously made during 21 years.

These I first collected in some big volumes, but later on I divided them up and distributed them among single books, one for each year, and had fair copies made. The arrangement I followed was such that the fixed stars, in so far as they had been observed during the year in question, had their own place, while the planets all had theirs, first the sun and moon, and next the other five planets in order up to Mercury; for I observed this planet also, although it is very seldom visible. In fact we observed it carefully almost every year, in the morning as well as in the evening .... Being now in possession of the selected and careful observations of 21 years, made in the sky with different ingeniously constructed instruments that I have shown in the preceding pages (not to speak of the observations of the previous 14 years), I hold them as a very rare and costly treasure. Perhaps I shall at some time publish all of them, if God in his grace will permit me to add still more ....

First of all we determined the course of the sun by very careful observations during several years. We not only investigated with great care its entrance into the equinoctial points, but we also considered the positions lying in between these and the solstitial points, particularly in the northern semicircle of the ecliptic since the sun there is not affected by refraction at noon. Observations were made in both cases and repeatedly confirmed, and from these I calculated mathematically both the apogee and the eccentricity corresponding to these times. With regard to the apogee as well as the eccentricity an obvious error has crept into both the Alphonsine tables and Copernicus' work, so that the apogee of the sun is almost three degrees ahead of Copernicus' value. The eccentricity amounts to about 2% when the radius of the eccentric orbit is put equal to 60, while the value of Copernicus is too small by almost a quarter .... This work on the sun was of necessity the first thing that had to be done, since it is on the sun that the motions of the celestial bodies depend, and since it moves in the ecliptic, to which the other motions are referred ....

With regard to the moon we used no less diligence in order to explain its intricate path, which in so many ways is complicated and not so simple and easy to make out as the ancients and Copernicus thought. For it presents another inequality with regard to the longitude, which these astronomers did not notice; nor have they determined the ratios of its revolution with sufficient accuracy. Moreover the limits of its maximum latitude differ from the value determined by Ptolemy, who with regard to this point was too confidently followed by all subsequent astronomers . . . .

After the orbits of both celestial bodies [the sun and the moon] have thus been determined in such a way that they agree with the celestial phenomena, it follows that it will be possible to determine with absolute correctness their eclipses, their relative positions, and their motions and places, the need for which has been long felt. What we have said so far about the course of the sun and the moon, and the question of agreement with celestial phenomena, is clearly presented together with other subjects, in the first chapter of our *Astronomiae instauratae Progymnasmata* . . . .

Further, as far as time and circumstances permitted, we very carefully determined the positions of all fixed stars visible to the naked eye, even those that are denoted as stars of the sixth magnitude, the longitude as well as the latitude. The accuracy was one minute of arc, in some cases even half a minute of arc. In this way we determined the positions of one thousand stars. The ancients were only able to count 22 more in spite of the fact that they lived at a lower geographical latitude where they ought to be able to see as many more as would correspond to the 200 stars that are always hidden from us here. Instead of these we determined a number of others which are very small, and which they did not include on this account. This immense task occupied us for almost 20 years, as we wished to investigate the whole problem carefully with different instruments . . . .

The fact that the latitudes of the fixed stars are also undergoing changes as a consequence of the change in the obliquity of the ecliptic, was first discovered by me. In the chapter mentioned above I have proved it by various examples. Thus we can maintain with ample certainty, and this is confirmed by actual experience, that the positions of the fixed stars have been determined by us with perfect and infallible accuracy. We have even determined a great many of them several times, and with different instruments, too, each leading to the same result . . . .

The only thing which is yet wanting with regard to the stars is to indicate their general motion through all the centuries during which the world has existed. It would not be so difficult to do this carefully, had the observations of the ancients in this field not been accepted, as they actually were. Yet I am convinced that I shall, by suitable corrections, be able to satisfy astronomers in this respect also, as far as that is possible.

One might have wished that the other stars which were catalogued by the ancients, but which are invisible in our latitudes, could have been added to the first thousand that I determined. Further, there are all the others, which were invisible even to the ancients who lived in the regions of Egypt, namely those that are located around the south Pole of the sky. For from the narratives of people who have sailed across the equator we know that there, too, the most beautiful stars are shining. With regard to the first proposition, it would be necessary to go to Egypt or some other similar place in Africa, and there industriously to note all the stars visible from that part of the world. But in order to attain the second goal it would be necessary to sail to South America, or to some other country beyond the equator, whence all the stars around the southern pole are visible, and observe them from there. So, if some mighty noblemen would care to fulfil our own and others' wishes in both these respects, they would do a very good deed that would be ever gloriously remembered. Up to now no one has even tried to do a thing like this in the right way, let alone carried it out, as far as is known. I would be willing to provide the necessary instruments and tools if somebody could organize the work and get the right people for such a deserving enterprise.

With regard finally to the investigations of the intricate course of the five other planets, and attempts at explaining them, I have done all I could. For in this whole field we have assembled, first of all, the apogees as well as the eccentricities, and further the angular motions and the ratios of their orbits and periods, so that they no longer contain all the numerous errors of previous investigations. We have shown that the very apogees of the planets are subject to yet another inequality that had not previously been noted. Further, we have made the discovery that the annual period, which Copernicus explained by a motion of the earth in a large circle, while the ancients explained it by epicycles, is subject to a variation. All this and other matters connected with it we have remedied by means of a special hypothesis [the Tychonic System] that we invented and worked out 14 years ago, basing it on the phenomena ....

With regard to all five planets there remains only one thing to do, namely to construct new and correct tables expressing by numbers all that has been established by more than 25 years of careful celestial observations (without mentioning the observations of the previous 10 years), thereby demonstrating the inaccuracy of the usual tables. We began this work and laid its foundations. It will not be difficult to complete it with the help of a few computers, and the results will then serve as a basis for the calculation of ephemerides for the coming years, as many as desired. The same can be done for the sun and the moon, for which we already have tables. In this way it will be possible with the greatest ease to demonstrate to posterity that the course of the celestial bodies as determined by us agrees with the phenomena, and is correctly given in every respect ....

While we thus with untiring industry through many years observed these eternal celestial bodies which are as old as the world itself, we studied with equal care all new celestial bodies in the ethereal regions that appeared during this time, above all the new and very admirable star that was first seen towards the end of the year 1572 and stayed for 16 months before it became completely invisible. On the subject of this star we wrote a small book describing its appearance, while it was still visible, as I have already indicated. When we resumed this work a few years later, we prepared a whole volume on this same star on account of the wonderful nature of the phenomenon .... We also prepared a special book on the immense comet that appeared five years later. In this we discuss it fully, including in the discussion our own observations and determinations as well as the opinions of others .... Yet I hope that I shall soon, with the help of the gracious God, complete the second part of the second volume also. In this volume I shall clearly demonstrate that all the comets observed by me moved in the ethereal regions of the world and never in the air below the moon as Aristotle and his followers have tried without reason to make us believe for so many centuries; and the demonstration will be clearest for some of the comets, while for others it will be according to the opportunity I had. The reason why I treat the comets in the second volume of the Progymnasmata before I set about the other five planets, which I intend to discuss in the third volume, are given in the same place in the preface ... the comets, the true ethereal nature of which I prove conclusively, show that the entire sky is transparent and clear, and cannot contain any solid and real spheres. For the comets as a rule follow orbits of a kind that no celestial sphere whatever would permit, and consequently it is a settled thing that there is nothing unreasonable in the hypothesis invented by us [the Tychonic System], since we have found that there is no such thing as penetration of spheres and limits of distance, as the solid spheres do not really exist.

In the field of Astrology, too, we carried out work that should not be looked down upon by those who study the influences of the stars. Our purpose was to rid this field of mistakes and superstition, and to obtain the best possible agreement with the experience on which it is based. For I think that it will hardly be possible to find in this field a perfectly accurate theory that can come up to mathematical and astronomical truth. Having in my youth been more interested in this foretelling part of Astronomy that deals with prophesying and builds on conjectures, I later on, feeling that the courses of the stars upon which it builds were insufficiently known, put it aside until I should have remedied this want. After I at length obtained more accurate knowledge of the orbits of the celestial bodies, I took Astrology up again from time to time, and I arrived at the conclusion that this science, although it is considered idle and meaningless not only by laymen but also by most scholars, among which are even several astronomers, is really more reliable than one would think; and this is true not only with regard to meteorological influences and predictions of the weather [natural astrology], but also concerning the predictions by nativities [judicial astrology], provided that the times are determined correctly, and that the courses of the stars and their entrances into definite sections of the sky are utilized in accordance with the actual sky, and that their directions of motion and revolutions are correctly worked ....

I also made with much care alchemical investigations, or chemical experiments. This subject too, I shall occasionally mention here, as the substances treated are somewhat analogous to the celestial bodies and their influences, for which reason I usually call this science terrestrial Astronomy. I have been occupied by this subject as much as by the celestial studies from my 23rd year, trying to gain knowledge and to prepare it, and up to now I have with much labour and at great expense made a great many findings with regard to the metals and minerals as well as the precious stones and plants, and other similar substances. I shall be willing to discuss these questions frankly with princes and noblemen, and other distinguished and learned people, who are interested in this subject and know something about it, and I shall occasionally give them information, as long as I feel sure, of their good intentions and that they will keep it secret. For it serves no useful purpose, and is unreasonable, to make such things generally known. For although many people pretend to understand them, it is not given to everybody to treat these mysteries properly according to the demands of nature, and in an honest and beneficial way.

Translated by H. Ræder, E. Strömgren, and B. Strömgren

Reading and Discussion Questions

1. What improvements in observational method does Brahe make?
2. What observations does Brahe make that no one has made before?
3. What are Brahe's most important criticisms of Copernicus, and how does his own alternative model of the cosmos deal with these problems?

Epitome of Copernican Astronomy (1618–21)

Johannes Kepler

36

Part I

1. On the Principal Parts of the World

What do you judge to be the lay-out of the principal parts of the world?

The Philosophy of Copernicus reckons up the principal parts of the world by dividing the figure of the world into regions. For in the sphere, which is the image of God the Creator and the Archetype of the world—as was proved in Book 1—there are three regions, symbols of the three persons of the Holy Trinity—the centre, a symbol of the Father; the surface, of the Son; and the intermediate space, of the Holy Ghost. So, too, just as many principal parts of the world have been made—the different parts in the different regions of the sphere: the sun in the centre, the sphere of the fixed stars on the surface, and lastly the planetary system in the region intermediate between the sun and the fixed stars.

. . .

Are there solid spheres [orbes] *whereon the planets are carried? And are there empty spaces between the spheres?*

Tycho Brahe disproved the solidity of the spheres by three reasons: the first from the movement of comets; the second from the fact that light is not refracted; the third from the ratio of the spheres.

For if spheres were solid, the comets would not be seen to cross from one sphere into another, for they would be prevented by the solidity; but they cross from one sphere into another, as Brahe shows.

From light thus: since the spheres are eccentric, and since the Earth and its surface—where the eye is—are not situated at the center of each sphere; therefore if. the spheres were solid, that is to say far more dense than that limpid ether, then the rays of the stars would be refracted before they reached our air, as optics teaches; and so the planet would appear irregularly and in places far different from those which could be predicted by the astronomer.

The third reason comes from the principles of Brahe himself; for they bear witness, as do the Copernican, that Mars is sometimes nearer the Earth than the sun is. But Brahe could not believe this interchange to be possible if the spheres were solid, since the sphere of Mars would have to intersect the sphere of the sun.

. . .

3. On the Order of the Movable Spheres

How are the planets divided among themselves?

Into the primary and the secondary. The primary planets are those whose bodies are borne around the sun, as will be shown below; the secondary planets are those whose own circles are arranged not around the sun but around one of the primary planets and who also share in the movement of the primary planet around the sun. Saturn is believed to have two such secondary planets and to draw them around with itself: they come into sight now and then with the help of a telescope. Jupiter has four such planets around itself: *D, E, F, H.* The Earth (*B*) has one (*C*) called the moon. It is not yet clear in the case of Mars, Venus, and Mercury whether they too have such a companion or satellite.

Then how many planets are to be considered in the doctrine on schemata?

No more than seven: the six so-called primary planets: (1) Saturn, (2) Jupiter, (3) Mars, (4) the Earth—the sun to eyesight, (5) Venus, (6) Mercury, and (7) only one of the secondary planets, the moon, because it alone revolves around our home, the Earth; the other secondary planets do not concern us who inhabit the Earth, and we cannot behold them without excellent telescopes.

. . .

What measure does Copernicus use in measuring the intervals of the single planets?

We must use a measure so proportioned that the other spheres can be compared, a measure very closely related to us and thus somehow known to us: such is the amplitude of the sphere whereon the centre of the Earth and the little sphere of the moon

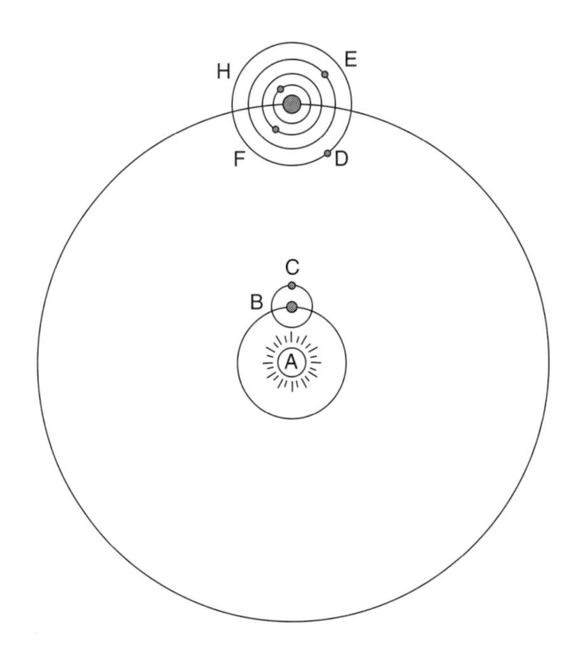

revolve—or its semidiameter, the distance of the Earth from the sun. This distance, like a measuring rod, is suitable for the business. For the Earth is our home; and from it we measure the distances of the heavens; and it occupies the middle position among the planets and for many reasons—on which below—it obtains the proportionality of a beginning among them. But the sun, by the evidence and judgment of our sight, is the principal planet. But by the vote of reason cast above, the sun is the heart of the region of moving planets proposed for measurement. And so our measuring rod has two very signal termini, the Earth and the sun.

How great therefore are the intervals between the single spheres?

The Copernican demonstrations show that the distance of Saturn is a little less than ten times the Earth's from the sun; that of Jupiter, five times; that of Mars, one and one-half times; that of Venus, three-quarters; and that of Mercury, approximately one-third.

And so the diameter of the sphere of Saturn is less than twice the length of its neighbour Jupiter's; the diameter of Jupiter is three times that of the lower planet Mars; the diameter of Mars is one and one-half times that of the terrestrial sphere placed around the sun; the diameter of the Earth's sphere is more than one and one-third that of Venus; and that of Venus is approximately five-thirds or eight-fifths that of Mercury. However, it should be noted that the ratios of the distances are different in other parts of the orbits, especially in the case of Mars and Mercury.

What is *the cause of the planetary intervals upon which the times of the periods follow?*

The archetypal cause of the intervals is the same as that of the number of the primary planets, being six.

I implore you, you do not hope to be able to give the reasons for the number of the planets, do you?

This worry has been resolved, with the help of God, not badly. Geometrical reasons are co-eternal with God—and in them there is first the difference between the curved and the straight line. Above (in Book 1) it was said that the curved somehow bears a likeness to God; the straight line represents creatures. And first in the adornment of the world, the farthest region of the fixed stars has been made spherical, in that geometrical likeness of God because as a corporeal God—worshipped by the gentiles under the name of Jupiter—it had to contain all the remaining things in itself. Accordingly rectilinear magnitudes pertained to the inmost contents of the farthest sphere; and the first and most beautiful magnitudes to the primary contents. But among rectilinear magnitudes the first, the most perfect, the most beautiful, and most simple are those which are called the five regular solids. More than 2,000 years ago Pythagoreans said that these five were the figures of the world, as they believed that the four elements and the heavens—the fifth essence—were conformed to the archetype for these five figures.

But the truer reason for these figures including one another mutually is in order that these five figures may conform to the intervals of the spheres. Therefore, if there are five spherical intervals, it is necessary that there be six spheres: just as with four linear intervals, there must necessarily be five digits.

What are these five regular figures?

The cube, tetrahedron, dodecahedron, icosahedron, and octahedron.

How are these figures divided, and into what classes?

The cube, tetrahedron, and dodecahedron are primary; the octahedron and the icosahedron are secondary.

Why do you make the farmer primary and the latter secondary?

The three former figures have a prior origin, and the most simple angle (*i.e.*, trilinear), and their own proper planes. The two latter have their origin in the primary figures, and a more composite angle made from many lines, and borrowed planes.

. . .

Show now what the place of the sphere of the Earth is among these figures.

The five bodies were distributed into two classes above: into those generated first, and those generated second. The former had a trilinear angle, and the latter a plurilinear. For as Adam was the first-born, and Eve was not his daughter but a part of him—and they are both called the first-made, but Cain and Abel and their sisters are their offspring; so the cube is in the first place, wherefrom have arisen, differently and more simply, the tetrahedronas it were a rib of the cube—and the dodecahedron,

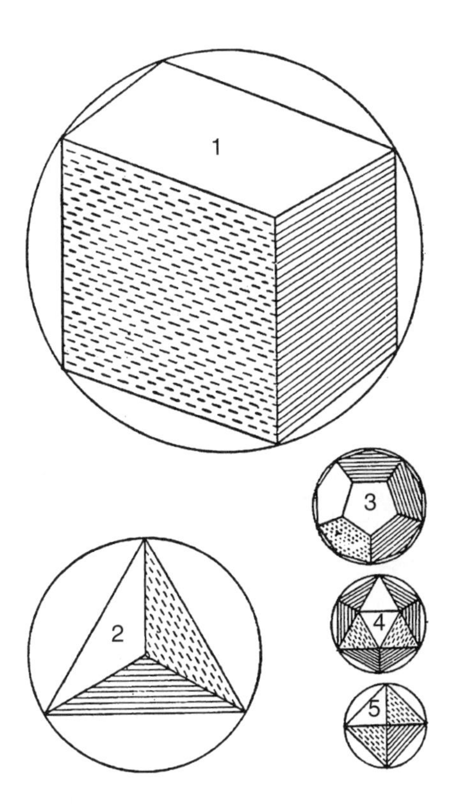

but in such a way that all three remain among the primary figures. The octahedron and the icosahedron, with their triangular planes, are as it were the offspring born of the cube and dodecahedron as fathers and from the tetrahedron as mother; and each of them bears a likeness to its parent.

So the three first figures of the same class had to enclose the circuit of the centre of the Earth and the two figures generated second, as the other class, should be enclosed by the sphere in which the Earth revolves, and so this sphere had to be made a boundary common to both orders, because the Earth, the home of the image of God, was going to be chief among the moving globes. For in this way the nature of being inscribed is kept in the second class and that of circumscribing in the first class. For it is more natural and more fitting that the octahedron should be inscribed in the cube, and the icosahedron in the dodecahedron, than the cube in the octahedron, and the dodecahedron in the icosahedron.

And so in this way the circuit of the centre of the Earth was placed in the middle between the planets; for three planets had to be placed outside, on account of the three primary figures; and two had to be placed inside its circuit on account of the two figures of the second class—to which the sun is added as a third in the inmost embrace of the centre of the mobile spheres. And so Saturn, Jupiter, and Mars were made the higher planets, and Venus, Mercury, and the sun, the lower. But the moon, which has a private movement around the Earth during the same common circuit of the Earth, is among the secondary planets, as was said above.

. . .

Part II

On the Movement of the Bodies of the World

1. How Many and of What Sort Are the Movements?

What was the opinion of Copernicus concerning the movement of bodies? For him, what was in motion and what was at rest?

There are two species of local movement: for either the whole thing turns, while remaining in its place, but with its parts succeeding one another. This movement can be called *δινητις*—lathe-movement, or cone-movement—from the resemblance; or rotation from a rotating pole. Or else the whole thing is borne from place to place circularly. The Greeks call this movement *Φορα,* the Latins *circuitus,* or *circumlatio,* or *ambitus.* But they call both movements generally revolution.

Accordingly Copernicus lays down that the sun is situated at the centre of the world and is motionless as a whole, *viz.,* with respect to its centre and axis. Only a few years ago, however, we grasped by sense that the sun turns with respect to the parts of its body, *i.e.,* around its centre and axis—as reasons had led me to assert for a long time—and with such great speed that one rotation is completed in the space of 25 or 26 days.

Now according as each of the primary bodies is nearer the sun, so it is borne around the sun in a shorter period, under the same common circle of the zodiac, and all in the same direction in which the parts of the solar body precede them—Mercury in the space of three months, Venus in seven and one-half months, the Earth with the lunar heaven in twelve months, Mars in twenty-two and one-half months or less than two years, Jupiter in twelve years, Saturn in thirty years. But for Copernicus the sphere of the fixed stars is utterly immobile.

The Earth meanwhile revolves around its own axis too, and the moon around the Earth—still in the same direction (if you look towards the outer parts of the world) as all the primary bodies.

Now for Copernicus all these movements are direct and continuous, and there are absolutely no stations or retrogradations in the truth of the matter.

. . .

How is the ratio of the periodic times, which you have assigned to the mobile bodies, related to the aforesaid ratio of the spheres wherein those bodies are borne?

The ratio of the times is not equal to the ratio of the spheres, but greater than it, and in the primary planets exactly the ratio of the 3/2th powers. That is to say, if you take the cube roots of the 30 years of Saturn and the 12 years of Jupiter and square them, the true ratio of the spheres of Saturn and Jupiter will exist in these squares. This is the case even if you compare spheres which are not next to one another. For example, Saturn takes 30 years; the Earth takes one year. The cube root of 30 is approximately 3.11. But the cube root of 1 is 1. The squares of these roots are 9.672 and 1. Therefore the sphere of Saturn is to the sphere of the Earth as 9,672 is to 1,000. And a more accurate number will be produced, if you take the times more accurately.

What is gathered from this?

Not all the planets are borne with the same speed, as Aristotle wished, otherwise their times would be as their spheres, and as their diameters; but, according as each planet is higher and farther away from the sun, so it traverses less space in one hour by its mean movement: Saturn—according to the magnitude of the solar sphere believed in by the ancients—traverses 240 German miles (in one hour), Jupiter 320 German miles, Mars 600, the centre of the Earth 740, Venus 800, and Mercury 1,200. And if this is to be according to the solar interval proved by me in the above, the number of miles must everywhere be tripled.

. . .

If there are no solid spheres, then there will seem to be all the more need of intelligences in order to regulate the movements of the heavens, although the intelligences are not gods. For they can be angels or some other rational creature, can they not?

There is no need of these intelligences, as will be proved; and it is not possible for the planetary globe to be carried around by an intelligence alone. For in the first place, mind is destitute of the animal power sufficient to cause movement, and it does not possess any motor force in its assent alone, and it cannot be heard or perceived by the irrational globe; and even if mind were perceived, the material globe would have no faculty of obeying or of moving itself. But before this, it has already been said that no animal force is sufficient for transporting the body from place to place, unless there are organs and some body which is at rest and on which the movement can take place. Therefore the question falls back to the above.

But on the contrary the natural powers which are implanted in the planetary bodies can enable the planet to be transported from place to place. But let it be posited as sufficient for movement that the intelligence should will movement into this or that region: then the discovery of the figure whereon the line of movement is ordered will be irrational. For we are convinced by the astronomical observations which have been taken correctly that the route of a planet is approximately circular and as a matter of fact eccentric—that is, the centre [of the circle] is not at the centre of the world or of some body; and furthermore that during the succession of ages the planet crosses from place to place. Now as many arguments can be drawn up against the discovery of such an orbit as there are parts of it already described.

For firstly, the orbit of the planet is not a perfect circle. But if mind caused the orbit, it would lay out the orbit in a perfect circle, which has beauty and perfection to the mind. On the contrary, the elliptic figure of the route of the planet and the laws of the movements whereby such a figure is caused smell of the nature of the balance or of material necessity rather than of the conception and determination of the mind, as will be shown below.

Finally, in order that we may grant that a different idea from that of a circle shines in the mind of the mover: it is asked by what means the mind can apply this or that [idea] to the regions of the world. Now the circle is described around some one fixed centre, but the ellipse, which is the figure of the planetary orbits, is described around two centres.

. . .

3. On the Revolution of the Solar Body Around Its Axis and Its Effect in the Movement of the Planets

By what reasons are you led to make the sun the moving cause or the source of movement for the planets?

1. Because it is apparent that in so far as any planet is more distant from the sun than the rest, it moves the more slowly—so that the ratio of the periodic times is the ratio of the 3/2th powers of the distances from the sun. Therefore we reason from this that the sun is the source of movement.
2. Below we shall hear the same thing come into use in the case of the single planets—so that the closer any one planet approaches the sun during any time, it is borne with an increase of velocity in exactly the ratio of the square.
3. Nor is the dignity or the fitness of the solar body opposed to this, because it is very beautiful and of a perfect roundness and is very great and is the source of light and heat, whence all life flows out into the vegetables: to such an extent that heat and

light can be judged to be as it were certain instruments fitted to the sun for causing movement in the planets.

4. But in especial, all the estimates of probability are fulfilled by the sun's rotation in its own space around its immobile axis, in the same direction in which all the planets proceed: and in a shorter period than Mercury, the nearest to the sun and fastest of all the planets. For as regards the fact that it is disclosed by the telescope in our time and can be seen every day that the solar body is covered with spots, which cross the disk of the sun or its lower hemisphere within 12 or 13 or 14 days, slowly at the beginning and at the end, but rapidly in the middle, which argues that they are stuck to the surface of the sun and turn with it; I proved in my *Commentaries on Mars*, Chapter 34, by reasons drawn from the very movement of the planets long before it was established by the sun-spots, that this movement necessarily; had to take place.

. . .

Then does the sun by the rotation of its body make the planets revolve? And how can this be since the sun is without hands with which it may lay hold of the planet, which is such a great distance away, and by rotating may make the planet revolve with itself?

Instead of hands there is the virtue of its body, which is emitted in straight lines throughout the whole amplitude of the world, and which—because it is a form of the body—rotates along with the solar body like a very rapid vortex; moving through the total amplitude of the circuit—whatever magnitude it reaches to—with equal speed; and the sun revolves m the narrowest space at the centre.

Could you make the thing clearer by some example?

Indeed there comes to our assistance the attraction between the loadstone and the iron pointer, which has been magnetized by the loadstone and which gets magnetic force by rubbing. Turn the loadstone in the neighbourhood of the pointer; the pointer will turn at the same time. Although the laying hold is of a different kind, nevertheless you see that not even here is there any bodily contact.

. . .

Finally by what arguments do you prove (4) *that* the *centre of the sun, which is at the midpoint of the planetary spheres and bears their whole system—does not revolve in some annual movement, as Brahe wishes, but in accordance with Copernicus sticks immobile in one place, while the centre of the Earth revolves in an annual movement?*

Even though the other necessarily follows from the demonstration of the one, nevertheless certain arguments pertain more closely to the sun and certain to the Earth; and certain others equally to both.

First on this side was the same argument whereby we just now claimed for the sun the midpoint of the spheres: namely, that the superfluous multitude of spheres and movements has been removed. For as it is much more probable that there should be some one system of spheres of the sun and that it should be common to the centre of the sun and to that node of the five spheres, according to Tycho Brahe than that we should believe according to Ptolemy that in any one of the five planets, over and above the spheres which have to do with their proper movements, there is present one whole system of spheres exactly like the sixth system of the sun; so also it is now much more probable that the centre of one Earth should revolve in an annual movement and the sun be at rest, according to Copernicus, than that, according to Brahe, this node of the five systems together with the spheres and planets themselves and the sun as a sixth should have the same annual movement besides the other movements which are proper to each. For even though Brahe removed from the true systems of the planets those five superfluous schemata of Ptolemy, which are like those of the sun, and reduced them to that common node of the systems, hid them, and melted them down into one; nevertheless he left in the world the very thing which was effected by those schemata: that any planet, over and above that movement which must really be granted to it, should be moved by the movement of the sun and should mix both into that one movement. And since there are no solid spheres, from this mixing there are caused in the expanse of the world very involved spirals. See the diagram of this involution in my *Commentaries on Mars,* folium 3.

Copernicus on the contrary by means of this one simple movement of the centre of the Earth stripped the five planets completely of this extrinsic movement of the sun, and made the centres of the six primary planets—that is, the Earth and the remaining five—each describe singly a simple and always similar orbit, or line very close to a circle, in the expanse of the world.

The second argument is from the movement in latitude. If epicycles revolve around an Earth at rest, either according to Ptolemy or according to Brahe; it will be necessary for those epicycles, especially those of the lower planets, in different ways to seek the sides as well as the head and feet, that is, to have a twofold libration. But with the Earth in motion, all the orbital circles have a constant inclination to the ecliptic. See Book v1, Part III where the latitudes of the lower planets supply us with a very clear argument for the movement of the Earth.

Thirdly, just as above, in the doctrine on the sphere, the diurnal revolution of the Earth being granted, the immense sphere of the fixed stars was freed from a diurnal movement of incalculable speed; so now, an annual movement being granted to this same Earth after the model of the other planets, we have ended that very slow movement of the fixed stars, which is called by Copernicus the precession of the equinoxes. See Book VII as regards these things. For it is much more believable to attribute them to the axis of the Earth, a very small body, than to such a great bulk.

Fourthly, the consideration of the ratios of the spheres wars on this side. For it is by no means probable that the centre of a great sphere should revolve in a small sphere. For the proper spheres of the three upper planets are much greater than the sphere of the sun—Saturn's approximately ten times greater; Jupiter's five times; Mars' one and one half times. Therefore these five spheres are not carried around or dislocated from their position; but their centres remain approximately fixed, and, as a consequence, instead of this movement common to them and to the sun, the Earth revolves.

The fifth argument, which is related to the preceding one, is the same as that whereby Brahe tried to disprove the solidity of the spheres. For if Brahe's reasoning holds, as the orbit of Mars is one and one half times the orbit of the sun, so the body of Mars at fixed times returns to that point in the world's expanse where the sun was at other times. And it is quite unbelievable that the regions which the primary planets pass through should be so jumbled together; since in Copernicus they are not only distinct, but are kept separate by very large intervals of emptiness.

I make the sixth argument similar to the fourth: from the magnitude of the movable bodies. For it is more believable that the body around which the smaller bodies revolve should be great. For just as Saturn, Jupiter, Mars, Venus, and Mercury are all smaller bodies than the solar body around which they revolve; so the moon is smaller than the Earth around which the moon revolves; so the four satellites of Jupiter are smaller than the body of Jupiter itself, around which they revolve. But if the sun moves, the sun which is the greatest, and the three higher planets which are all greater than the Earth, will revolve around the Earth which is smaller. Therefore it is more believable that the Earth, a small body, should revolve around the great body of the sun.

The seventh reason is drawn from the reasons for the intervals, which were unfolded above in the first part of this book. These reasons are disturbed and maimed, unless we grant to the Earth too its own sphere, which Copernicus gives to it between the spheres of Mars and of Venus. For even if the interval between Saturn and Jupiter could be deduced from the cube, that of Jupiter and Mars from the tetrahedron, and that of Venus and Mercury from the octahedron, even in Brahe's ordering: yet there would still remain between Mars and Venus a single interval. But there remain two figures in the number of figures of the world. And the interval between Mars and Venus, which is in a greater ratio than double would not square with one of these figures, the dodecahedron or the icosahedron; nor could it be deduced from two figures, not even by the interposition of some sphere between them.

Eighthly, the same things are to be said concerning the harmony of the celestial movements, which are made up of the same numbers and proportions as our musical scale. And if you consider the excellence of the work or the pleasantness of contemplation, or finally the unavoidable force of the persuasion, this harmony can truly be called the soul and life of all astronomy. But this harmony is at last complete only if the Earth in its own place and rank among the planets strikes its own string and as it were sings its own note through a variation of a semitone: otherwise there would be no manifesting of its semitone, and that again is the soul of the song. As a matter of fact, if the semitone of the Earth is gone, there is destroyed from among the celestial movements the manifesting of the genera of song, *i.e.*, the major and the minor modes, the most pleasant, most subtle, and most wonderful thing in this whole discussion. But concerning this in the *Harmonies.*

. . .

Part III

On the Real and True Irregularity of the Planets and Its Causes

. . .

2. On the Causes of Irregularity in Longitude

Then what causes do you bring forward as to why, although all the routes of the primary planets are arranged around the sun, nevertheless the angles—in which as if from the centre of the sun, the different parts of the route of one planet are viewed are not completed by the planet in proportional times?

Two causes concur, the one optical, the other physical, and each of almost equal effect. The first cause is that the route of the planet is not described around the sun at an equal distance everywhere; but one part of it is near the sun, and the opposite part is so much the farther away from the sun. But of equal things, the near are viewed at a greater angle, and the far away, at a smaller; and of those which are viewed at an equal angle, the near are smaller, and the far away are greater.

The other cause is that the planet is really slower at its greater distance from the sun, and faster at its lesser.

Therefore if the two causes are made into one, it is quite clear that of two arcs which are equal to sight, the greater time belongs to the arc which is greater in itself, and a much greater time on account of the real slowness of the planet in that farther arc.

But could not one cause suffice, so that, because generally the orbit of the planet draws as far away from the sun on one side as it draws near on the other, we might make such a great distance that all this apparent irregularity might be explained merely by this unequal distance of the parts of the orbit?

Observations do not allow us to make the inequality of the distances as great as the inequality of the time wherein the planet makes equal angles at the sun; but they bear witness that the inequality of the distances is sufficient to explain merely half of this irregularity: therefore the remainder comes from the real acceleration and slowing up of the planet.

. . .

What is the reason why the sun does not lay hold of the planet with equal strength from far away and from near-by?

The weakening of the form from the solar body is greater in a longer outflow than in a shorter; and although this weakening occurs in the ratio of the squares of the intervals, *i.e.*, both in longitude and in latitude, nevertheless it works only in the simple ratio: the reasons have been stated above.

3. The Causes of the Irregularity in Altitude

But what pushes the planet out into more distant spaces and leads it back towards the sun?

The same which lays hold of the planet, the sun, namely, by means of the virtue of the form which has flowed out from its body throughout all the spaces of the world. For repulsion and attraction are as it were certain elements of this laying hold. For repulsion and attraction take place according to the lines of virtue going out from the centre of the sun; and since these lines revolve along with the sun, it is necessary for the planet too which is repelled and attracted to follow these lines in proportion to their strength in relation to the resistance of the planetary body. So the contrary movements of repulsion and attraction somehow compose this laying hold.

Do you attribute to the simple body of the sun and to its immaterial form the operations of attraction and repulsion which are contrary and so not simple?

The natural action or *ενεργεια* of moving the planetary body for the sake of assimilation or of bringing it back to its primal posture is one [in number]; but it seems to be diverse on account of the diversity of the object. For only in one region is the planetary body in concord with the solar body; in the other region it is discordant. But it belongs to the same simple work to embrace like things and to spit out unlike things. This opinion is strengthened by the case of magnets; for though they are not celestial bodies, nevertheless they do not have that biform virtue from the composition of elements but from a simple bodily form.

Therefore the planetary body itself will be composed of contrary parts?

No, indeed. For it follows only that the planetary globe has an inward configuration of straight lines or threads, like magnetic threads, which happen to be terminated in contrary regions; and in one of these regions, not on account of the body itself but on account of its posture in relation to the sun, there reigns friendship *[familiaritas]* with the sun; and in the other region, discord.

But isn't it unbelievable that the celestial bodies should be certain huge magnets?

Then read the philosophy of magnetism of the Englishman William Gilbert; for in that book, although the author did not believe that the Earth moved among the stars, nevertheless he attributes a magnetic nature to it, by very many arguments, and he teaches that its magnetic threads or filaments extend in straight lines from south to north. Therefore it is by no means absurd or incredible that any one of the primary planets should be what one of the primary planets, namely the Earth, is.

Translated by Charles Glenn Wallis

Reading and Discussion Questions

1. What does Kepler seem to regard as his main goals and accomplishments? From this reading do you think that Kepler was expecting to be remembered primarily for what we call his "three laws of planetary motion" today?
2. What role do the five Platonic solids play in this reading? What importance does Kepler take magnetism to have?
3. What arguments does he provide for the view that the earth moves and that the sun stands still?
4. What are the three systems that he considers as models for the structure of the cosmos? Which system does he favor and why?

Starry Messenger, with Message to Cosimo de 'Medici (1610)

Galileo Galilei

37

THE SIDEREAL MESSENGER

UNFOLDING GREAT AND MARVELLOUS SIGHTS,

AND PROPOSING THEM TO THE ATTENTION OF EVERY ONE,

BUT ESPECIALLY PHILOSOPHERS AND ASTRONOMERS,

BEING SUCH AS HAVE BEEN OBSERVED BY

GALILEO GALILEI

A GENTLEMAN OF FLORENCE,

PROFESSOR OF MATHEMATICS IN THE UNIVERSITY OF PADUA,

WITH THE AID OF A

TELESCOPE

lately invented by him,

Respecting the Moon's Surface, an innumerable number of Fixed Stars,

the Milky Way, and Nebulous Stars, but especially respecting

Four Planets which revolve round the Planet Jupiter at

different distances and in different periodic times, with

amazing velocity, and which, after remaining

unknown to every one up to this day, the

Author recently discovered, and

determined to name the

MEDICEAN STARS.

Venice 1610.

TO THE MOST SERENE

COSMO DE' MEDICI, THE SECOND,

FOURTH GRAND-DUKE OF TUSCANY.

THERE is certainly something very noble and large-minded in the intention of those who have endeavoured to protect from envy the noble achievements of distinguished men, and to rescue their names, worthy of immortality, from oblivion and decay. This desire has given us the lineaments of famous men, sculptured in marble, or fashioned in bronze, as a memorial of them to future ages; to the same feeling we owe the erection of statues, both ordinary and equestrian; hence, as the poet says, has originated expenditure, mounting to the stars, upon columns and pyramids; with this desire, lastly, cities have been built, and distinguished by the names of those men, whom the gratitude of posterity thought worthy of being handed down to all ages. For the state of the human mind is such, that unless it be continually stirred by the counterparts of matters, obtruding themselves upon it from without, all recollection of the matters easily passes away from it.

But others, having regard for more stable and more lasting monuments, secured the eternity of the fame of great men by placing it under the protection, not of marble or bronze, but of the Muses' guardianship and the imperishable monuments of literature. But why do I mention these things, as if human wit, content with these regions, did not dare to advance further; whereas, since she well understood that all human monuments do perish at last by violence, by weather, or by age, she took a wider view, and invented more imperishable signs, over which destroying Time and envious Age could claim no rights; so, betaking herself to the sky, she inscribed on the well-known orbs of the brightest stars—those everlasting orbs—the names of those who, for eminent and god-like deeds, were accounted worthy to enjoy an eternity in company with the stars. Wherefore the fame of Jupiter, Mars, Mercury, Hercules, and the rest of the heroes by whose names the stars are called, will not fade until the extinction of the splendour of the constellations themselves.

But this invention of human shrewdness, so particularly noble and admirable, has gone out of date ages ago, inasmuch as primeval heroes are in possession of those bright abodes, and keep them by a sort of right; into whose company the affection of Augustus in vain attempted to introduce Julius Cæsar; for when he wished that the name of the Julian constellation should be given to a star, which appeared in his time, one of those which the Greeks and the Latins alike name, from their hair-like tails, comets, it vanished in a short time and mocked his too eager hope. But we are able to read the heavens for your highness, most Serene Prince, far more truly and more happily, for scarcely have the immortal graces of your mind begun to shine on earth, when bright stars present themselves in the heavens, like tongues to tell and celebrate your most surpassing virtues to all time. Behold therefore, four stars reserved for your famous name, and those not belonging to the common and less conspicuous multitude of fixed stars, but in the bright ranks of the planets—four stars which, moving differently from each other, round the planet Jupiter, the most glorious of all the planets, as if they were his own children, accomplish the courses of their orbits with marvellous velocity, while all the while with one accord they complete all together mighty revolutions every ten years round the centre of the universe, that is, round the Sun.

But the Maker of the Stars himself seemed to direct me by clear reasons to assign these new planets to the famous name of your highness in preference to all others. For just as these stars, like children worthy of their sire, never leave the side of Jupiter by any appreciable distance, so who does not know that clemency, kindness of heart, gentleness of manners, splendour of royal blood, nobleness in public functions, wide extent of influence and power over others, all of which have fixed their common abode and seat in your highness,—who, I say, does not know that all these qualities, according to the providence of God, from whom all good things do come, emanate from the benign star of Jupiter? Jupiter, Jupiter, I maintain, at the instant of the birth of your highness having at length emerged from the turbid mists of the horizon, and being in possession of the middle quarter of the heavens, and illuminating the eastern angle, from his own royal house, from that exalted throne, looked out upon your most happy birth, and poured forth into a most pure atmosphere all the brightness of his majesty, in order that your tender body and your mind—though that was already adorned by God with still more splendid graces—might imbibe with your first breath the whole of that influence and power. But why should I use only plausible arguments when I can almost absolutely demonstrate my conclusion? It was the will of Almighty God that I should be judged by your most serene parents not unworthy to be employed in teaching your highness mathematics, which duty I discharged, during the four years just passed, at that time of the year when it is customary to take a relaxation from severer studies. Wherefore, since it evidently fell to my lot by God's will, to serve your highness, and so to receive the rays of your surpassing clemency and beneficence in a position near your person, what wonder is it if you have so warmed my heart that it thinks about scarcely anything else day and night, but how I, who am indeed your subject not only by inclination, but also by my very birth and lineage, may be known to be most anxious for your glory, and most grateful to you? And so, inasmuch as under your patronage, most serene Cosmo, I have discovered these stars, which were unknown to all astronomers before me, I have, with very good right, determined to designate them with the most august name of your family. And as I was the first to investigate them, who can rightly blame me if I give them a name, and call them *the Medicean Stars*, hoping that as much consideration may accrue to these stars from this title, as other stars have brought to other heroes? For not to speak of your most serene ancestors, to whose everlasting glory the monuments of all history bear witness, your virtue alone, most mighty sire, can confer on those stars an immortal name;

for who can doubt that you will not only maintain and preserve the expectations, high though they be, about yourself, which you have aroused by the very happy beginning of your government, but that you will also far surpass them, so that when you have conquered others like yourself, you may still vie with yourself, and become day by day greater than yourself and your greatness?

Accept, then, most clement Prince, this addition to the glory of your family, reserved by the stars for you; and may you enjoy for many years those good blessings, which are sent to you not so much from the stars as from God, the Maker and Governor of the stars.

Your Highness's most devoted servant,

Galileo Galilei.
Padua, *March 12, 1610.*

THE ASTRONOMICAL MESSENGER

Containing and setting forth Observations lately made with the aid of a newly invented Telescope *respecting the Moon's Surface, the Milky Way, Nebulous Stars, an innumerable multitude of Fixed Stars, and also respecting Four Planets never before seen, which have been named*

THE COSMIAN STARS.

IN the present small treatise I set forth some matters of great interest for all observers of natural phenomena to look at and consider. They are of great interest, I think, first, from their intrinsic excellence; secondly, from their absolute novelty; and lastly, also on account of the instrument by the aid of which they have been presented to my apprehension.

The number of the Fixed Stars which observers have been able to see without artificial powers of sight up to this day can be counted. It is therefore decidedly a great feat to add to their number, and to set distinctly before the eyes other stars in myriads, which have never been seen before, and which surpass the old, previously known, stars in number more than ten times.

Again, it is a most beautiful and delightful sight to behold the body of the Moon, which is distant from us nearly sixty *semi*-diameters of the Earth, as near as if it was at a distance of only two of the same measures; so that the diameter of this same Moon appears about thirty times larger, its surface about nine hundred times, and its solid mass nearly 27,000 times larger than when it is viewed only with the naked eye; and consequently any one may know with the certainty that is due to the use of our senses, that the Moon certainly does not possess a smooth and polished surface, but one rough and uneven, and, just like the face of the Earth itself, is everywhere full of vast protuberances, deep chasms, and sinuosities.

Then to have got rid of disputes about the Galaxy or Milky Way, and to have made its nature clear to the very senses, not to say to the understanding, seems by no means a matter which ought to be considered of slight importance. In addition to this, to point out, as with one's finger, the nature of those stars which every one of the astronomers up to this time has called *nebulous*, and to demonstrate that it is very different from what has hitherto been believed, will be pleasant, and very fine. But that which will excite the greatest astonishment by far, and which indeed especially moved me to call the attention of all astronomers and philosophers, is this, namely, that I have discovered four planets, neither known nor observed by any one of the astronomers before my time, which have their orbits round a certain bright star, one of those previously known, like Venus and Mercury round the Sun, and are sometimes in front of it, sometimes behind it, though they never depart from it beyond certain limits. All which facts were discovered and observed a few days ago by the help of a telescope devised by me, through God's grace first enlightening my mind.

Perchance other discoveries still more excellent will be made from time to time by me or by other observers, with the assistance of a similar instrument, so I will first briefly record its shape and preparation, as well as the occasion of its being devised, and then I will give an account of the observations made by me.

About ten months ago a report reached my ears that a Dutchman had constructed a telescope, by the aid of which visible objects, although at a great distance from the eye of the observer, were seen distinctly as if near; and some proofs of its most wonderful performances were reported, which some gave credence to, but others contradicted. A few days after, I received confirmation of the report in a letter written from Paris by a noble Frenchman, Jaques Badovere, which finally determined me to give myself up first to inquire into the principle of the telescope, and then to consider the means by which I might compass the invention of a similar instrument, which a little while after I succeeded in doing, through deep study of the theory of Refraction; and I prepared a tube, at first of lead, in the ends of which I fitted two glass lenses, both plane on one side, but on the other side one spherically convex, and the other concave. Then bringing my eye to the concave lens I saw objects satisfactorily large and near, for they appeared one-third of the distance off and nine times larger than when they are seen with the natural eye alone. I shortly afterwards constructed another telescope with more nicety, which magnified objects more than sixty times. At length, by sparing neither labour nor expense, I succeeded in constructing for myself an instrument so superior that objects seen through it appear magnified nearly a thousand times, and more than thirty times nearer than if viewed by the natural powers of sight alone. . . .

Now let me review the observations made by me during the two months just past, again inviting the attention of all who are eager for true philosophy to the beginnings which led to the sight of most important phenomena.

The Moon: Ruggedness of its surface. Existence of lunar mountains and valleys. Let me speak first of the surface of the Moon, which is turned towards us. For the sake of being understood more easily, I distinguish two parts in it, which I call respectively the brighter and the darker. The brighter part seems to surround and pervade the whole hemisphere; but the darker part, like a

sort of cloud, discolours the Moon's surface and makes it appear covered with spots. Now these spots, as they are somewhat dark and of considerable size, are plain to every one, and every age has seen them, wherefore I shall call them *great* or *ancient* spots, to distinguish them from other spots, smaller in size, but so thickly scattered that they sprinkle the whole surface of the Moon, but especially the brighter portion of it. These spots have never been observed by any one before me; and from my observations of them, often repeated, I have been led to that opinion which I have expressed, namely, that I feel sure that the surface of the Moon is not perfectly smooth, free from inequalities and exactly spherical, as a large school of philosophers considers with regard to the Moon and the other heavenly bodies, but that, on the contrary, it is full of inequalities, uneven, full of hollows and protuberances, just like the surface of the Earth itself, which is varied everywhere by lofty mountains and deep valleys. . . .

And here I cannot refrain from mentioning what a remarkable spectacle I observed while the Moon was rapidly approaching her first quarter, a representation of which is given in the same illustration above. A protuberance of the shadow, of great size, indented the illuminated part in the neighbourhood of the lower cusp; and when I had observed this indentation longer, and had seen that it was dark throughout, at length, after about two hours, a bright peak began to arise a little below the middle of the depression; this by degrees increased, and presented a triangular shape, but was as yet quite detached and separated from the illuminated surface. Soon around it three other small points began to shine, until, when the Moon was just about to set, that triangular figure, having now extended and widened, began to be connected with the rest of the illuminated part, and, still girt with the three bright peaks already mentioned, suddenly burst into the indentation of shadow like a vast promontory of light. . . .

If one wishes to revive the old opinion of the Pythagoreans, that the Moon is another Earth, so to say, the brighter portion may very fitly represent the surface of the land, and the darker the expanse of water. Indeed, I have never doubted that if the sphere of the Earth were seen from a distance, when flooded with the Sun's rays, that part of the surface which is land would present itself to view as brighter, and that which is water as darker in comparison. Moreover, the great spots in the Moon are seen to be more depressed than the brighter tracts; for in the Moon, both when crescent and when waning, on the boundary between the light and shadow, which projects in some places round the great spots, the adjacent regions are always brighter, as I have noticed in drawing my illustrations, and the edges of the spots referred to are not only more depressed than the brighter parts, but are more even, and are not broken by ridges or ruggednesses. But the brighter part stands out most near the spots, so that both before the first quarter and about the third quarter also, around a certain spot in the upper part of the figure, that is, occupying the northern region of the Moon, some vast prominences on the upper and lower sides of it rise to an enormous elevation, as the illustrations show. This same spot before the third quarter is seen to be walled round with boundaries of a deeper shade, which just like very lofty mountain summits appear darker on the side away from the Sun, and brighter on the side where they face the Sun; but in the case of the cavities the opposite happens, for the part of them away from the Sun appears brilliant, and that part which lies nearer to the Sun dark and in shadow. After a time, when the enlightened portion of the Moon's surface has diminished in size, as soon as the whole or nearly so of the spot already mentioned is covered with shadow, the brighter ridges of the mountains mount high above the shade. . . .

Calculation to show that the height of some lunar mountains exceeds four Italian miles (22,000 British feet). I think that it has been sufficiently made clear, from the explanation of phenomena which have been given, that the brighter part of the Moon's surface is dotted everywhere with protuberances and cavities; it only remains for me to speak about their size, and to show that the ruggednesses of the Earth's surface are far smaller than those of the Moon's; smaller, I mean, absolutely, so to say, and not only smaller in proportion to the size of the orbs on which they are. And this is plainly shown thus:—As I often observed in various positions of the Moon with reference to the Sun, that some summits within the portion of the Moon in shadow appeared illumined, although at some distance from the boundary of the light (the terminator), by comparing their distance with the complete diameter of the Moon, I learnt that it sometimes exceeded the one-twentieth (1/20th) part of the diameter. Suppose the distance to be exactly 1/20th part of the diameter, and let the diagram represent the Moon's orb, of which C A F is a great circle, E its centre, and C F a diameter,

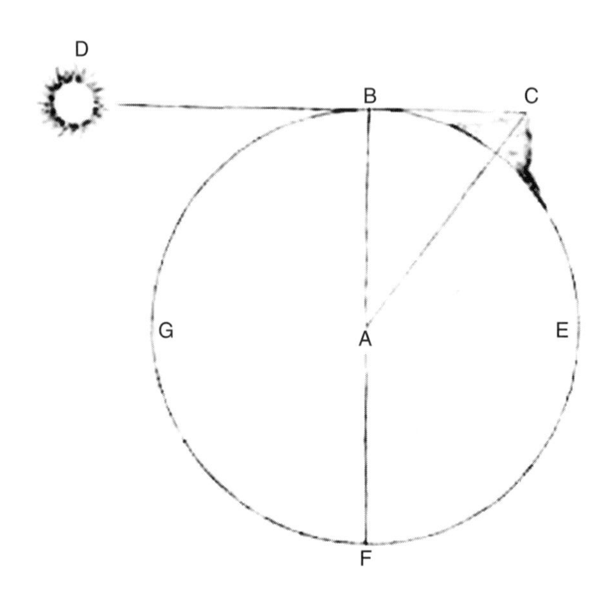

which consequently bears to the diameter of the Earth the ratio 2:7; and since the diameter of the Earth, according to the most exact observations, contains 7000 Italian miles, C F will be 2000, and C E 1000, and the 1/20th part of the whole, C F, 100 miles. Also let C F be a diameter of the great circle which divides the bright part of the Moon from the dark part (for, owing to the very great distance of the Sun from the Moon this circle does not differ sensibly from a great one), and let the distance of A from the point C be 1/20th part of that diameter; let the radius E A be drawn, and let it be produced to cut the tangent line G C D, which represents the ray that illumines the summit, in the point D. Then the arc C A or the straight line C D will be 100 of such units, as C E contains 1000. The sum of the squares of D C, C E is therefore 1,010,000, and the square of D E is equal to this; therefore the whole E D will be more than 1004; and A D will be more than 4 of such units, as C E contained 1000. Therefore the height of A D in the Moon, which represents a summit reaching up to the Sun's ray, G C D, and separated from the extremity C by the distance C D, is more than 4 Italian miles; but in the Earth there are no mountains which reach to the perpendicular height even of one mile. We are therefore left to conclude that it is clear that the prominences of the Moon are loftier than those of the Earth. . . .

When the Moon, both before and after conjunction, is found not far from the Sun, not only does its orb show itself to our sight on the side where it is furnished with shining horns, but a slight and faint circumference is also seen to mark out the circle of the dark part, that part, namely, which is turned away from the Sun, and to separate it from the darker background of the sky. But if we examine the matter more closely, we shall see that not only is the extreme edge of the part in shadow shining with a faint brightness, but that the entire face of the Moon, that side, that is, which does not feel the Sun's glare, is illuminated with a pale light of considerable brightness . . . this tract of the Moon also, although deprived of sunlight, gleams with considerable light, and particularly so if the gloom of the night has already deepened through the absence of the Sun; for with a darker background the same light appears brighter . . . .

This strange brightness has afforded no small perplexity to philosophical minds; and some have published one thing, some another, as the cause to be alleged for it . . . Since, therefore, this kind of secondary brightness is not inherent and the Moon's own, nor borrowed from any of the stars, nor from the Sun, and since there now remains in the whole universe no other body whatever except the Earth, what, pray, must we conclude? What must we assert? Shall we assert that the body of the Moon, or some other dark and sunless orb, receives light from the Earth? Why should it not be the Moon? And most certainly it is. The Earth, with fair and grateful exchange, pays back to the Moon an illumination like that which it receives from the Moon nearly the whole time during the darkest gloom of night.

Stars: Hitherto I have spoken of the observations which I have made concerning the Moon's body; now I will briefly announce the phenomena which have been, as yet, seen by me with reference to the Fixed Stars. And first of all the following fact is worthy of consideration:—The stars, fixed as well as erratic, when seen with a telescope, by no means appear to be increased in magnitude in the same proportion as other objects, and the Moon herself, gain increase of size; but in the case of the stars such increase appears much less, so that you may consider that a telescope, which (for the sake of illustration) is powerful enough to magnify other objects a hundred times, will scarcely render the stars magnified four or five times. But the reason of this is as follows:—When stars are viewed with our natural eyesight they do not present themselves to us of their bare, real size, but beaming with a certain vividness, and fringed with sparkling rays, especially when the night is far advanced; and from this circumstance they appear much larger than they would if they were stripped of those adventitious fringes, for the angle which they subtend at the eye is determined not by the primary disc of the star, but by the brightness which so widely surrounds it. Perhaps you will understand this most clearly from the well-known circumstance that when stars rise just at sunset, in the beginning of twilight, they appear very small, although they may be stars of the first magnitude; and even the planet Venus itself, on any occasion when it may present itself to view in broad daylight, is so small to see that it scarcely seems to equal a star of the last magnitude. It is different in the case of other objects, and even of the Moon, which, whether viewed in the light of midday or in the depth of night, always appears of the same size. We conclude therefore that the stars are seen at midnight in uncurtailed glory, but their fringes are of such a nature that the daylight can cut them off, and not only daylight, but any slight cloud which may be interposed between a star and the eye of the observer. A dark veil or coloured glass has the same effect, for, upon placing them before the eye between it and the stars, all the blaze that surrounds them leaves them at once. A telescope also accomplishes the same result, for it removes from the stars their adventitious and accidental splendours before it enlarges their true discs (if indeed they are of that shape), and so they seem less magnified than other objects, for a star of the fifth or sixth magnitude seen through a telescope is shown as of the first magnitude only.

The difference between the appearance of the planets and the fixed stars seems also deserving of notice. The planets present their discs perfectly round, just as if described with a pair of compasses, and appear as so many little moons, completely illuminated and of a globular shape; but the fixed stars do not look to the naked eye bounded by a circular circumference, but rather like blazes of light, shooting out beams on all sides and very sparkling, and with a telescope they appear of the same shape as when they are viewed by simply looking at them, but so much larger that a star of the fifth or sixth magnitude seems to equal Sirius, the largest of all the fixed stars.

Orion's Belt and Sword; 83 Stars

Pleiades; 36 Stars

Telescopic Stars: As my first example I had determined to depict the entire constellation of Orion, but I was overwhelmed by the vast quantity of stars and by want of time, and so I have deferred attempting this to another occasion, for there are adjacent to, or scattered among, the old stars more than five hundred

new stars within the limits of one or two degrees. For this reason I have selected the three stars in Orion's Belt and the six in his Sword, which have been long well-known groups, and I have added eighty other stars recently discovered in their vicinity, and I have preserved as exactly as possible the intervals between them. The well-known or old stars, for the sake of distinction, I have depicted of larger size, and I have outlined them with a double line; the others, invisible to the naked eye, I have marked smaller and with one line only. I have also preserved the differences of magnitude as much as I could. . . .

Nebulæ resolved into clusters of stars: And whereas that milky brightness, like the brightness of a white cloud, is not only to be seen in the Milky Way, but several spots of a similar colour shine faintly here and there in the heavens, if you turn the telescope upon any of them you will find a cluster of stars packed close together. Further—and you will be more surprised at this,—the stars which have been called by every one of the astronomers up to this day *nebulous*, are groups of small stars set thick together in a wonderful way, and although each one of them on account of its smallness, or its immense distance from us, escapes our sight, from the commingling of their rays there arises that brightness which has hitherto been believed to be the denser part of the heavens, able to reflect the rays of the stars or the Sun.

Discovery of Jupiter's satellites, Jan. 7, 1610: I have now finished my brief account of the observations which I have thus far made with regard to the Moon, the Fixed Stars, and the Galaxy. There remains the matter, which seems to me to deserve to be considered the most important in this work, namely, that I should disclose and publish to the world the occasion of discovering and observing four PLANETS, never seen from the very beginning of the world up to our own times, their positions, and the observations made during the last two months about their movements and their changes of magnitude; and I summon all astronomers to apply themselves to examine and determine their periodic times, which it has not been permitted me to achieve up to this day, owing to the restriction of my time. I give them warning however again, so that they may not approach such an inquiry to no purpose, that they will want a very accurate telescope, and such as I have described in the beginning of this account.

On the 7th day of January in the present year, 1610, in the first hour of the following night, when I was viewing the constellations of the heavens through a telescope, the planet Jupiter presented itself to my view, and as I had prepared for myself a very excellent instrument, I noticed a circumstance which I had never been able to notice before, owing to want of power in my other telescope, namely, that three little stars, small but very bright, were near the planet; and although I believed them to belong to the number of the fixed stars, yet they made me somewhat wonder, because they seemed to be arranged exactly in a straight line, parallel to the ecliptic, and to be brighter than the rest of the stars, equal to them in magnitude. The position of them with reference to one another and to Jupiter was as follows:

On the east side there were two stars, and a single one towards the west. The star which was furthest towards the east, and the western star, appeared rather larger than the third.

I scarcely troubled at all about the distance between them and Jupiter, for, as I have already said, at first I believed them to be fixed stars; but when on January 8th, led by some fatality, I turned again to look at the same part of the heavens, I found a very different state of things, for there were three little stars all west of Jupiter, and nearer together than on the previous night, and they were separated from one another by equal intervals, as the accompanying illustration shows.

East West

Jan.7 * * O *

Jan.8 O * * *

Jan.9 Clouds

Jan.10 * * O

Jan.11 * * O

Jan.12 * * O *

Jan.13 * O * * *

At this point, although I had not turned my thoughts at all upon the approximation of the stars to one another, yet my surprise began to be excited, how Jupiter could one day be found to the east of all the aforesaid fixed stars when the day before it had been west of two of them; and forthwith I became afraid lest the planet might have moved differently from the calculation of astronomers, and so had passed those stars by its own proper motion. I therefore waited for the next night with the most intense longing, but I was disappointed of my hope, for the sky was covered with clouds in every direction.

But on January 10th the stars appeared in the following position with regard to Jupiter; there were two only, and both on the east side of Jupiter, the third, as I thought, being hidden by the planet. They were situated just as before, exactly in the same straight line with Jupiter, and along the Zodiac.

When I had seen these phenomena, as I knew that corresponding changes of position could not by any means belong to Jupiter, and as, moreover, I perceived that the stars which I saw had been always the same, for there were no others either in front or behind, within a great distance, along the Zodiac,—at length, changing from doubt into surprise, I discovered that the interchange of position which I saw belonged not to Jupiter, but to the stars to which my attention had been drawn, and I thought therefore that they ought to be observed henceforward with more attention and precision.

Accordingly, on January 11th I saw an arrangement of the following kind, namely, only two stars to the east of Jupiter, the nearer of which was distant from Jupiter three times as far as from the star further to the east; and the star furthest to the east was nearly twice as large as the other one; whereas on the previous night they had appeared nearly of equal magnitude. I therefore concluded, and decided unhesitatingly, that there are three stars in the heavens moving about Jupiter, as Venus and Mercury round the Sun; which at length was established as clear as daylight by numerous other subsequent observations. These observations also established that there are not only three, but four, erratic sidereal bodies performing their revolutions round Jupiter, observations of whose changes of position made with more exactness on succeeding nights the following account will supply. I have measured also the intervals between them with the telescope in the manner already explained. Besides this, I have given the times of observation, especially when several were made in the same night, for the revolutions of these planets are so swift that an observer may generally get differences of position every hour.

Jan. 12.—At the first hour of the next night I saw these heavenly bodies arranged in this manner: The satellite furthest to the east was greater than the satellite furthest to the west; but both were very conspicuous and bright; the distance of each one from Jupiter was two minutes. A third satellite, certainly not in view before, began to appear at the third hour; it nearly touched Jupiter on the east side, and was exceedingly small. They were all arranged in the same straight line, along the ecliptic.

Jan. 13.—For the first time four satellites were in view in the following position with regard to Jupiter . . . .[1]

These are my observations upon the four Medicean planets, recently discovered for the first time by me; and although it is not yet permitted me to deduce by calculation from these observations the orbits of these bodies, yet I may be allowed to make some statements, based upon them, well worthy of attention. . . . Besides, we have a notable and splendid argument to remove the scruples of those who can tolerate the revolution of the planets round the Sun in the Copernican system, yet are so disturbed by the motion of one Moon about the Earth, while both accomplish an orbit of a year's length about the Sun, that they consider that this theory of the constitution of the universe must be upset as impossible; for now we have not one planet only revolving about another, while both traverse a vast orbit about the Sun, but our sense of sight presents to us four satellites circling about Jupiter, like the Moon about the Earth, while the whole system travels over a mighty orbit about the Sun in the space of twelve years.

Translated by Edward Stafford Carlos Rivingtons

Reading and Discussion Questions

1. What are some of the most important observations Galileo makes through his telescope? Do the telescopic observations provide any evidence for his heliocentric views?
2. What has Galileo discovered that he is calling the "Medician stars"? What reasons does he give for his choice of name, and what does this suggest about the relationship between astronomy and astrology in this period?

[1] There are two months of detailed observations like this, with accompanying images that follow – ed.

Dialogue Concerning the Two Chief World Systems (1632)

Galileo Galilei

38

[Salviati represents the Copernican point of view; the name is that of a friend of Galileo
Simplicio speaks for the Aristotelians; the name is taken from a 6th century follower of Aristotle
Sagredo is supposed to be a neutral voice in the debate—Ed.]

THE SECOND DAY

. . .

SIMPLICIO: I must confess that I have been ruminating all night of what passed yesterday, and, to say the truth, I have met with many acute, new, and plausible notions; yet I remain convinced by the authority of so many great writers and in particular . . . . I see you shaking your head, Sagredus, and grinning to yourself, as if I had uttered some great absurdity.

SAGREDO: I not only grin but actually am ready to burst with holding myself from laughing outright, for you have put me in mind of a very pretty episode that I witnessed not many years since, together with some others of my worthy friends which I could name to you.

SALVIATI: It would be well that you told us what it was, so Simplicius may not still think that he is the point of your laughter.

SAGREDO: Very well. One day at his home in Venice, I found a famous physician to whom some flocked for their studies, while others sometimes came thither out of curiosity to see certain bodies dissected by the hand of a no less learned than careful and experienced anatomist. It chanced upon that day, when I was there, that he was in search of the origin and stem of the nerves, about which there is a famous controversy between the Galenists and Peripatetics. The anatomist shewed how the great trunk of nerves, departing from the brain, their root, passed by the nape of the neck, extended themselves afterwards along the backbone, and branched out through all the body, while only a very small filament, as fine as a thread, went to the heart. Then he turned to a gentleman whom he knew to be a Peripatetic philosopher, and for whose sake he had uncovered and proved everything, and asked if he was satisfied and persuaded that the origin of the nerves was in the brain and not in the heart. The philosopher, after he had stood musing a while, answered: "You have made me see this business so plainly and sensibly that did not the text of Aristotle assert the contrary, which positively affirms the nerves to proceed from the heart, I should be constrained to confess your opinion to be true."

SIMPLICIO: I would have you know, my Masters, that this controversy about the origin of the nerves is not yet so proved and decided, as some may perhaps persuade themselves.

SAGREDO: Nor doubtless shall it ever be if it finds such contradictors· but what you say does not at all lessen the extravagance of the answer of, that Peripatetic, who against such sensible experience did not produce other experiments or reasons of Aristotle but his bare authority and pure *ipse dixit* . . . .

SIMPLICIO: But in case we should give up Aristotle, who is to be our guide in philosophy? Name you some author.

SALVIATI: We need a guide in unknown and uncouth parts, but in clear thoroughfares, and in open plains, only the blind stand in need of a leader; and, for such, it is better that they stay at home. But he who has eyes in his head and in his mind has to use these for his guide. Yet mistake me not, thinking that I speak this because I am against hearing Aristotle; for, on the contrary, I commend the reading and diligent study of him and only blame the servilely giving one's self up a slave to him, so as blindly to subscribe to whatever he delivers, and receive it for an inviolable decree without search of any further reason. This is an abuse that carries with it the other extreme disorder that people will no longer take pains to understand the validity of his demonstrations. And what is more shameful in public disputes than, while someone is treating of demonstrable conclusions, to have someone else come up with a passage of Aristotle, quite often irrelevant, and with that stop the mouth of his opponent? But, if you will continue to study in this manner, I would have you lay aside the name of philosophers and call yourselves either historians or doctors of memory, for it is not fit that those who never philosophize should usurp the honourable title of philosophers. But it is best for us to return to shore and not launch out further into a boundless gulf, out of which we should not be able to get before night. Therefore, Simplicius, come with arguments and demonstrations of your own, or of Aristotle, but bring us no more texts and naked authorities, for our disputes are about the sensible world and not a paper one.

Now in our discourses of yesterday we retrieved the Earth from darkness and exposed it to the open sky, showing that the attempt to number it among those which we call celestial bodies was not a position so compromised and vanquished that it had no life left in it. It follows next that we proceed to examine what probability there is for holding its entire globe fixed and wholly immovable; what likelihood there is for making it movable with some motion; and of what kind that may be. And since I am doubtful in this question, while Simplicius is resolute, as Aristotle is, for its immobility, he shall one by one produce the arguments in favour of their opinion, and I will allege the answers and reasons on the contrary part; next, Sagredus shall tell us his thoughts and to which side he finds himself inclined.

SAGREDO: Content, provided always that I may reserve myself the liberty of alleging what straight natural reason shall sometimes dictate to me . . . .

SALVIATI: Let our contemplation begin therefore with this view: that, whatever motion may be ascribed to the Earth as a whole, it is necessary that it be to us, as partakers of it, altogether imperceptible, so long as we have regard only to terrestrial things. On the other hand, it is equally necessary that the same motion should appear common to all other bodies and visible objects separated from the Earth and that therefore do not participate in it. So that the true method of finding it, and, once found, to know what it is, is to consider and observe if in bodies separated from the Earth one may discover any appearance of motion which equally suits all the rest of them. For a motion that is only seen, for example, in the Moon, and has nothing to do with Venus or Jupiter or any other stars, cannot in any way belong to the Earth or to any other save the Moon alone. Now there is a most general and grand motion above all others, and it is that by which the Sun, the Moon, the other planets, and the fixed stars, and, in a word the whole Universe, except only the Earth, appear to move from the east towards the west in the space of twenty four hours; and this, at first sight, might as well belong to the Earth alone, as, on the other hand, to all the rest of the world, except the Earth; for the same phenomena will appear in the one case as in the other. Hence it is that Aristotle and Ptolemy, having hit upon this consideration when they go about to prove the Earth to be immovable, argue only against this diurnal motion; except that Aristotle hints something in obscure terms against another motion ascribed to it by an Ancient, of whom we shall speak in its place.

SAGREDO: I very well perceive the necessity of your conclusion; but I meet with a doubt from which I do not know how to free myself. It is that, since Copernicus assigns another motion to the Earth beside the diurnal, which, according to the rule even now laid down, ought to be imperceptible to us but visible in all the rest of the world, I might necessarily infer either that he has manifestly erred in assigning the Earth a motion to which there does not appear a general correspondence in heaven or else that, if there be such a congruity therein, Ptolemy has failed in not confuting it, as he has done in the other.

SALVIATI: You have good cause for your doubts, and, when we come to treat of the other motion, you shall see how far Copernicus excelled Ptolemy in clearness and elevation of intellect in that he saw what the other did not; I mean the admirable correspondence whereby this motion reflected itself in all the other celestial bodies. But for the present we will suspend this particular and return to our first consideration. Beginning, then, with more general things, I will propose those reasons which seem to favour the mobility of the Earth and then await the answers which Simplicius shall make. *First,* if we consider only the immense magnitude of the starry sphere compared to the smallness of the terrestrial globe, and weigh the velocity of the motions which must in a day and night make an entire revolution, I cannot persuade myself that there is any man who believes it more reasonable and credible that it is the celestial sphere that turns round, while the terrestrial globe stands still.

SAGREDO: If in the totality of effects which may in Nature depend upon such like motions, there should follow in one hypothesis exactly all the same consequences as in the other. I would esteem at first inspection, that he who should hold it more rational to make the whole Universe move, in order to keep the Earth from moving, is less reasonable than he who being at the top of the dome of your Cathedral in Florence, in order to behold the city and the fields about it, should desire that the whole country might turn round, so that he might not be put to the trouble to turn his head. And surely the conveniences that could be drawn from this position would have to be many and great in order to equate in my mind, and to overcome, this absurdity in such manner as to make it more credible than the former. But perhaps Aristotle, Ptolemy, and Simplicius must find certain advantage therein, which they would do well to communicate to us also, if any such there be; or else they had better declare that there neither is nor can be any . . . .

SALVIATI: . . . A *third* reason which makes the Ptolemaic hypothesis less probable is that it most unreasonably confounds the order that we assuredly see among those celestial bodies of which the circumgyration is not questionable but most certain. And that order is that, the greater a sphere, the longer is its revolution. Thus Saturn, describing a greater circle than all the other planets, completes it in thirty years; Jupiter finishes its, that is less, in twelve years; Mars in two; the Moon runs through its, so much less than the rest, in only a month. Nor do we see less sensibly that that one of the Medicean Stars which is nearest to Jupiter makes its revolution in a very short time, that is, in forty-two hours, or thereabouts, the next to that in three days and a half, the third in seven days, and the most remote in sixteen. And this rate holds well enough, nor will it alter at all when we assign the rotation of twenty-four hours to the terrestrial globe. But, if you would have the Earth immovable, it is necessary that, when you have passed from the short period of the Moon to the others successively bigger, until you come to that of Saturn, which is of thirty years, it is necessary, I say, that you pass to another sphere incomparably greater than that and make this accomplish an entire revolution in twenty-four hours. And this yet is the least disorder that can follow. For if anyone should pass from the sphere of Saturn to the starry orb, and make it so much bigger than that of Saturn, as proportion would require in respect of its own very slow motion of many thousands of years, then it must certainly be a jump much more absurd to skip from this to another still bigger one, the *Primum Mobile,* and to make it revolve in twenty-four hours.

But if the motion of the Earth is granted, the order of the periods will be exactly observed, and from the very slow sphere of Saturn we come to the fixed stars which are wholly immovable and so avoid a *fourth* difficulty we would have if the starry sphere be supposed movable. That is the immense disparity between the motions of those stars themselves, of which some would come to move most swiftly in most vast circles, others most slowly in circles very small, according as the former or the latter should be found nearer or more remote from the Poles. This is accompanied with still more inconveniences because we see that those of whose motion there is no question have been made to move all on a great circle; also, because it seems to be an act done with no good consideration to constitute bodies that are designed to move circularly at immense distances from the centre and afterwards to make them move in very small circles.

And not only the magnitudes of the circles and consequently the velocity of the motions of these stars shall be most different from the circles and motions of those others, but (which shall be the *fifth* inconvenience) the self-same star shall successively vary its circles and velocities. For those which two thousand years since were in the Equator, and consequently described very vast circles, being in our days many degrees distant, must of necessity become slower and be reduced to move in lesser circles. And it is not altogether impossible but that a time may come in which some of them which in aforetime had continually moved shall be reduced by uniting with the pole to a state of rest and then, after some time of cessation, shall return to their motion again; whereas the other stars whose motion none doubt all describe the great circle of their sphere and in that maintain themselves without any variation.

The absurdity is further increased (which is the *sixth* inconvenience) in that no thought can comprehend what ought to be the solidity of that immense sphere, whose depth so steadfastly holds fixed such a multitude of stars, which are with so much concord carried about without ever changing site among themselves with so great disparity of motions. Or else, supposing the heavens to be fluid, as we are with more reason to believe, so that every star wanders to and fro in it by ways of its own, what rules shall regulate their motions, and to what purpose, so that, seen from the Earth, they appear as though they were made by one single sphere? It is my opinion that they might so much more easily and more conveniently do that, by being constituted immovable, than by being made errant, by as much as it is easier to number the blocks in the pavement of a piazza than a rout of boys which run up and down upon them.

Lastly, which is the *seventh* instance, if we attribute the diurnal motion to the highest heaven, it must be constituted of such a force and efficacy as to carry along with it the innumerable multitude of fixed stars, bodies of cast magnitude far bigger than the Earth, and, moreover, all the spheres of the planets, notwithstanding that both the first and the last by their own nature move the contrary way. And, beside all this, it must be granted that also the element of fire and the greater part of the air are likewise forcibly hurried along with the rest and that only the little globe of the Earth contumaciously and pertinaciously stands unmoved against such an impulse. This in my thinking is very difficult; nor can I see how the Earth, a suspended body equilibrated upon its centre, exposed indifferently to either motion or rest, and environed with a liquid ambient, should not yield also like the rest and be carried about. But we find none of these obstacles in making the Earth move, a small body and insensible, compared to the Universe, and therefore unable to offer it any violence . . . .

SIMPLICIO: For this purpose astronomers and philosophers have found another sphere, above all the rest, without stars, to which naturally belongs the diurnal motion; and this they call the *Primum Mobile*; this carries along with it all the inferior spheres, contributing and imparting its motion to them.

SAGREDO: But if, without introducing other spheres unknown and hugely vast, without other motions or communicated raptures, with each sphere having its sole and simple motion, without intermixing contrary motions, but making all turn one way, as they should, depending as they do all upon one principle, if we can find all things proceeding orderly and correspond with most perfect harmony, why do we reject this alternative and give our assent to those prodigious and laborious conditions?

SIMPLICIO: The difficulty lies in finding out this so natural and expeditious a way.

SAGREDO: In my judgment it is found. Make the Earth the *Primum Mobile,* that is, make it turn round its own axis in twenty-four hours and in the same direction as all the other spheres; and, without need of imparting this same motion to any other planet or star, all shall have their risings, settings, and all other appearances.

SIMPLICIO: The business is, to be able to make the Earth move without a thousand inconveniences.

SALVIATI: All the inconveniences shall be removed as fast as you propound them: The things said hitherto are only the primary and more general inducements which give us to believe that the diurnal revolution may not altogether without probability be applied to the Earth rather than to all the rest of the Universe. And these inducements I put forward, not as inviolable axioms, but as hints, which carry with them some likelihood. But I know very well that one sole experiment or conclusive demonstration produced to the contrary suffices to overthrow these and a thousand other probable arguments; hence it is fit not to stay here but to proceed forward and hear what Simplicius answers and what greater probabilities or stronger arguments he alleges on the contrary.

SIMPLICIO: I will first say something in general upon all these considerations together, and then I will descend to some particulars. It seems that you base all you say upon the greater simplicity and facility of producing the same effects. To this I reply that I am also of the same opinion if I think in terms of my own not only finite but feeble power; but, with respect to the strength of the Mover, which is infinite, it is no less easy to move the Universe than the Earth, yea, than a straw. And if his power be infinite, why should he not rather exercise a great part of it than a smaller? Therefore, I hold that your discourse in general is not convincing.

SALVIATI: If I had at any time said that the Universe did not move for want of power in the Mover, I would have erred, and your reproof would have been seasonable; and I grant you that to an infinite power it is as easy to move a hundred thousand as one. But what I did say does not concern the Mover; it refers only to the bodies, and, in them, not only to their resistance, which doubtless is less in the Earth than in the Universe, but to the many other particulars just considered. As to what you say in the next place, that of an infinite power it is better to exercise a great part than a small, I answer that one part of the infinite is not greater than another, if both are finite. Nor can it be said that, of the infinite number, a hundred thousand is a greater part than two, though the first be fifty thousand times greater than the second. If a finite power is necessary to move the Universe, though it is very great in comparison to that which suffices to move the Earth alone, yet it does not involve a greater part of the infinite power, nor is that part less infinite which remains unemployed. So that to apply to a particular effect a little more or a little less power means nothing; besides that, the operation of such virtue has not for its bound or end the diurnal motion only. But there are several other motions in the World that we know of, and there may be many others that are unknown to us. Therefore, with respect to the bodies, and granting it as out of question that it is a shorter and easier way to move the Earth than the Universe, and, moreover, having an eye to the many other abbreviations and aptnesses that are to be obtained only this way, a most true maxim of Aristotle teaches us that *frustra fit per plura, quod potest fieri per pauciora,* "that is done in vain by many means which may be done with fewer." This renders it more probable that the diurnal motion belongs to the Earth alone than to the whole Universe, with the sole exception of the Earth.

SIMPLICIO: In quoting that axiom, you have omitted a small clause, which is as important as all the rest; that is, the words *aeque bene,* "equally well." We ought therefore to examine whether this hypothesis satisfies *equally well* all particulars as the other .... That disposition is not new but very old; and that you may see it is so, Aristotle refutes it; and his refutations are these: "First, if the Earth moves either in itself about its own centre or in a circle around the centre, it is necessary that that motion be violent, for it is not its natural motion; for, if it were, each of its parts would partake thereof; hut each of them moves in a straight line towards the centre. It being therefore violent and preternatural, it could never he perpetual. But the order of the world is perpetual. Therefore, etc. Secondly, all the other movables that move circularly seem to lag behind and to move with more than one motion, the *Primum Mobile* excepted; hence it would he necessary that the Earth also move with two motions. And, if that should he so, it would inevitably follow that mutations should take place in the fixed stars, which, however, are not perceived; nay, without any variation, the same stars always rise from the same places and set in the same places. Thirdly, the motion of the parts is the same with that of the whole and naturally tends towards the centre of the Universe; and for the same cause rests, being arrived thither." He thereupon moves the question whether the motion of the parts has a tendency towards the centre of the Universe or to the centre of the Earth, and he concludes that it goes by proper instinct to the centre of the Universe and *per accidens* to that of the Earth; of this point, we largely discoursed yesterday. He lastly confirms the same with a fourth argument taken from the experiment of heavy bodies which, falling from on high, descend perpendicularly to the Earth's surface; and in the same manner projectiles shot perpendicularly upwards return perpendicularly down again.

All these arguments prove their motion to he towards the centre of the Earth, which without moving at all waits for them and receives them. He intimates in the last place that the astronomers allege other reasons in confirmation of the Earth's being immovable in the centre of the Universe, and instances only in one of them, to wit, that all the phenomena or appearances that are seen in the motions of the stars perfectly agree with the position of the Earth in the centre; this would not he so were the Earth located otherwise. I can give you now, if you please, the rest produced by Ptolemy and the other astronomers, or I can do so after you have spoken what you have to say in answer to Aristotle.

SALVIATI: The arguments brought up are of two kinds: some have respect to the terrestrial accidents without any relation to the stars, and others are taken from the phenomena and observations of things celestial. The arguments of Aristotle are for the most part taken from things near at hand, and he leaves the rest to astronomers; and therefore it is best to examine these taken from experiments touching the Earth and then proceed to those of the other kind. And as Ptolemy, Tycho, and the other philosophers produced certain arguments besides those of Aristotle which they adopted and confirmed, we will combine them, so we will not answer twice to the same objections. Therefore, Simplicius, choose whether you will recite them yourself or have me ease you of this task, for I am ready to serve you.

SIMPLICIO: It is better that you quote them, because, as having taken more pains in the study of them, you can produce them quicker and in greater number.

SALVIATI: For the strongest reason, all allege that heavy bodies falling downwards move by a straight line perpendicular to the surface of the Earth, an argument which is held to prove undeniably that the Earth is immovable. For otherwise a tower from the top of which a stone is let fall, being carried along by the rotation of the Earth in the time that the stone spends in falling, would be transported many hundred yards eastwards, and so far distant from the tower's foot would the stone come to the ground. This effect they back with another experiment: letting a bullet of lead fall from the round top of a ship that lies at anchor and observing the mark it makes where it lights. This they find to be near the foot of the mast, but, if the same bullet be let fall from the same place when the ship is under sail, it will light as far from the former place as the ship has sailed in the time of the lead's descent, because the natural motion of the ball being at liberty is by a straight line towards the centre of the Earth. They fortify this argument with the experiment of a projectile shot on high at a very great distance. For example, a ball shot out of a cannon erected perpendicular to the horizon spends so much time in ascending and falling that, in our parallel, both we and the cannon should be carried by the Earth many miles towards the East, so that that the ball in its return could never come near the piece but would fall as far west as the Earth had run east . . . .

SALVIATI: Before I proceed any further, I must tell Sagredus that in these disputations I personate the Copernican and imitate him as if I were his mask. But what has been effected in *my* private thoughts by these arguments which I seem to allege in his favour I would not have you judge by what I say whilst I am in the heat of acting my part in the fable; after I have laid by my disguise, you may chance to find me different from what you see me upon the stage . . . .

Therefore we may proceed to the *fourth,* upon which it is requisite that we stay some time, for the reason that it is founded upon that experiment from which the greater part of the remaining arguments derive all their strength. Aristotle says that it is a most convincing argument of the Earth's immobility to see that projectiles thrown or shot upright return perpendicularly by the same line unto the same place from whence they were shot or thrown. And this holds true, although the motion be of a very great height. So that hither may be referred the argument taken from a shot fired directly upwards from a cannon, as also that other used by Aristotle and Ptolemy, of the heavy bodies that, falling from on high, are observed to descend by a direct and perpendicular line to the surface of the Earth. Now, that I may begin to untie these knots, I demand this of Simplicius: in case one should deny to Ptolemy and Aristotle that weights in falling freely from on high descend by a right and perpendicular line, that is, directly to the centre, what means would he use to prove it?

SIMPLICIO: The means of the senses, which assure us that that tower or other altitude is upright and perpendicular, and shew us that that stone slides along the wall, without inclining a hair's breadth to one side or another, and lights on the ground just under the place from where it was let fall.

SALVIATI: But if it should happen that the terrestrial globe did move round, and consequently carry the tower also along with it, and that the stone did then also graze and slide along the side of the tower, what must its motion be then?

SIMPLICIO: In this case we may rather say its motions, for it would have one wherewith to descend from the top to the bottom and should then have another to follow the course of the said tower.

SALVIATI: So that its motion should be compounded of two; from this it would follow that the stone would no longer describe that simple straight and perpendicular line but one transverse and perhaps not straight.

SIMPLICIO: I can say nothing of its nonrectitude, but this I know very well: that it would of necessity be transverse.

SALVIATI: You see then that, merely observing the falling stone to glide along the tower, you cannot certainly affirm that it describes a line which is straight and perpendicular unless you first suppose that the Earth stands still.

SIMPLICIO: True; for, if the Earth should move, the stone's motion would be transverse and not perpendicular.

...

SAGREDO: I would defend Aristotle in favour of Simplicius. Should the tower move, it would be impossible that the stone should fall gliding along the side of it; and therefore from its falling in that manner the stability of the Earth is inferred.

SIMPLICIO: It is so; for if you would have the stone in descending to graze along the tower, while being carried around by the Earth, you must allow the stone two natural motions, to wit, the straight motion towards the centre and the circular motion about the centre, which is impossible.

SALVIATI: Aristotle's defense then consists in the impossibility, or at least in his esteeming it an impossibility, that the stone should move with a motion mixed of right and circular. For, if he did not hold it impossible that the stone could move at once to the centre and about the centre, he would have understood that it might come to pass that the falling stone might in its descent graze the tower as well when it moved as when it stood still. Consequently, he ought to have perceived that from this grazing nothing could be inferred touching the mobility or immobility of the Earth. But this does not any way excuse Aristotle; because he ought to have expressed it, if he had had such a notion, it being so material a part of his argument. Also because it cannot be said that such an effect is impossible or that Aristotle did esteem it so. The first cannot be affirmed, for by and by I shall shew that it is not only possible but necessary; nor can the second be averred, for Aristotle himself grants that fire moves naturally upwards in a right line, and moves about with the diurnal motion, imparted by the heavens to the whole element of fire and the greater part of the upper air. If therefore he held it possible to mix the straight motion upwards with the circular communicated to the fire and air from the concave of the sphere of the Moon, much less ought he to account impossible the mixture of the straight motion of the stone downwards with the circular which we presuppose natural to the whole terrestrial globe, of which the stone is a part.

SIMPLICIO: I see no such thing; for, if the element of fire revolves round together with the air, it is a very easy, even a necessary, thing that a spark of fire which mounts upwards from the Earth, in passing through the moving air, should receive the same motion, being a body so thin, light, and easy to be moved. But that a very heavy stone, or a cannon ball, that descends from on high, and that is at liberty to move whither it will, should suffer itself to be transported either by the air of any other thing is altogether incredible. Besides that, we have the experiment which is so proper to our purpose, of the stone let fall from the round top of the mast of a ship, which, when the ship lies still, falls at the foot of the mast, but, when the ship moves, falls as far distant from that place, as how far the ship in the time of the stone falling had run forward.

SALVIATI: There is a great disparity between the case of the ship and that of the Earth, if the terrestrial globe be supposed to have a diurnal motion. For it is manifest that, as the motion of the ship is not natural to it, the motion of all those things that are in it is accidental, whence it is no wonder that the stone which was retained in the round top, being left at liberty, descends downwards without any obligation to follow the motion of the ship. But the diurnal conversion is ascribed to the terrestrial globe for its proper and natural motion, and, consequently, it is so to all parts of the said globe; and, being impressed by Nature, is indelible in them. Therefore that stone that is on the top of the tower has an intrinsic inclination to revolve about the centre of its whole in twenty-four hours, and it exercises this same natural instinct eternally, be it placed in any state whatsoever. To be assured of the truth of this, you have to do no more than alter an antiquated impression made in your mind, and to say: as I, hitherto, held it to be the property of the terrestrial globe to rest immovable about its centre, and never doubted or questioned but that all particles do also naturally remain in the same state of rest; it is reason, in case the terrestrial globe did move round by natural instinct in twenty-four hours, that the intrinsic and natural inclination of all its parts should also he to follow the same revolution. And thus, without running into any inconvenience, one may conclude that, since the motion conferred by the force of oars on the ship, and by it on all the things that are contained within her, is not natural but foreign, it is very reasonable that that stone, being separated from the ship, should bring itself back to its natural disposure and return to exercise its pure simple instinct give it by Nature. To this I add that it is necessary that at least that part of the air which is beneath the greater heights of mountains should be transported and carried round by the roughness of the Earth's surface; or that, as being mixed with many vapours and terrene exhalations, it should naturally follow the diurnal motion, which does not occur in the air about the ship rowed by oars. Therefore your arguing from the ship to the tower has not the force of a conclusion, because the stone which falls from the round top of the mast enters into a medium which is unconcerned in the motion of the ship; but that which departs from the top of the tower finds a medium that has a motion in common with the whole terrestrial globe, so that, rather than being hindered, it is assisted by the motion of the air and may follow the universal course of the Earth.

SIMPLICIO: I cannot conceive that the air can imprint in a very great stone, or in a heavy globe of wood or ball of lead, as suppose of two hundred-weight, the motion wherewith itself is moved, and which it perhaps communicates to feathers, snow, and other very light things. On the contrary, I see that a weight of that nature, being exposed to the most impetuous wind, is not thereby removed an inch from its place; now consider with yourself whether the air will carry it along therewith.

SALVIATI: There is a great difference between your experiment and our case. You introduce the wind blowing against that stone, supposedly in a state of rest, and we expose the stone to the air which already moves, with the same velocity as the stone, so that the air is not to confer a new motion upon it but only to maintain or, to say better, not to hinder the motion already acquired. You would drive the stone with a strange and preternatural motion, and we desire to conserve it in its natural one. If you would produce a more pertinent experiment, you should say that it is observed, if not with the eye of the forehead, yet with that of the mind, what would happen if an eagle that is carried by the course of the wind should let a stone fall from its talons. Now as the stone went along with the wind when it was let go, and after it started on its fall entered into a medium that moved with equal velocity, I am very confident that it would not be seen to descend in its fall perpendicularly but would move with a transverse motion.

SIMPLICIO: But it should first be known how such an experiment may be made; and then one might judge according to the event. In the meantime the effect of the ship hitherto inclines to favour our opinion.

SALVIATI: Well you said "hitherto," for perhaps it may change countenance anon. And that I may no longer hold you in suspense, tell me, Simplicius, do you really believe that the experiment of the ship squares so very well with our purpose that it ought to be believed that that which we see happen in it ought also to take place in the terrestrial globe?

SIMPLICIO: As yet I am of that opinion; and, though you have alleged some small disparities, I do not think them of so great a moment that they should make me change my judgment.

SALVIATI: I rather desire that you would continue in that belief and hold for certain that the effect of the Earth would exactly answer that of the ship, provided that when it shall appear prejudicial to your cause, you will not be of humour to alter your thoughts. You say that, when the ship stands still, the stone falls at the foot of the mast and that, when she is under sail, it lights far from thence; therefore, by conversion, from the stone's falling at the foot is inferred the ship's standing still, and from its falling far from thence is inferred her moving. And because that which occurs to the ship ought likewise to befall the Earth, therefore from the falling of the stone at the foot of the tower is necessarily inferred the immobility of the terrestrial globe. Is this not your argumentation?

SIMPLICIO: It is; and reduced into such conciseness that it is become most easy to he apprehended.

SALVIATI: Now tell me; if the stone let fall from the round top when the ship is in swift course, should fall exactly in the same place of the ship in which it falls when the ship is at anchor, what service would these experiments do you, to the end of ascertaining whether the vessel does stand still or move?

SIMPLICIO: Exactly none. As, for example, from the beating of the pulse one cannot know whether a person is asleep or awake, seeing that the pulse heats in the same manner in sleeping as in waking.

SALVIATI: Very well. Have you ever tried the experiment of the ship?

SIMPLICIO: I have not; but yet I believe that those authors who allege the same have accurately observed it; besides that, the cause of the disparity is so manifestly known that it admits of no questions.

SALVIATI: It may be that those authors believed in it, without having made trial of it, and you yourself are a good witness to the point. For you, without having examined it, allege it as certain and in good faith rely on their authority; as it is now not only possible but obvious that they also relied on their predecessors, without ever arriving at one that had made the experiment. For, mark you, whosoever shall perform it shall find the event succeed quite contrary to what has been written of it. That is, he shall see the stone fall at all times in the same place of the ship, whether it stand still or move with any velocity whatsoever. So that, the same holding true in the Earth as in the ship, one cannot, from the stone's falling perpendicularly at the foot of the tower, conclude anything touching the motion or rest of the Earth.

SIMPLICIO: If you were referring me to any other means than to direct experience, I verily believe our disputations would not come to an end in haste; for this seems to me a thing so remote from all human reason that it leaves not the least place for credulity or probability.

SALVIATI: And yet it has left place in me for both.

SIMPLICIO: How is this? You have not made a hundred tests, no, not even one test, and you so confidently affirm it for true? I for my part will return to my incredulity and to the confidence I had that the experiment has been tried by the principal authors who made use of it and that the event turned out as they affirm.

SALVIATI: I am assured that the effect will ensue as I tell you, without experiment, for so it is necessary that it should, and I further add that you yourself know that it cannot fall out otherwise, however you feign or seem to feign that you know it not. Yet I am so good a broker of minds that I will make you confess the same whether you will or no. But Sagredus stands very quiet, and yet, if I mistake not, I saw him make some move as if to speak.

SAGREDO: I had intended to speak a fleeting something; but my curiosity aroused by your promising that you would force Simplicius to uncover the knowledge which he conceals from us has made me depose all other thoughts. Therefore I pray you to make good your vaunt.

SALVIATI: Provided that Simplicius consents to reply to what I shall ask him, I will not fail to do it.

SIMPLICIO: I will answer what I know, assured that I shall not be much put to it, for, of those things which I hold to be false, I think nothing can he known, since Science concerns truths, not falsehoods.

SALVIATI: I do not desire that you should say that you know anything, save that which you most assuredly know. Therefore, tell me; if you had here a flat surface as polished as a mirror and of a substance as hard as steel that was not horizontal but somewhat inclining, and you put upon it a perfectly spherical ball, say, of bronze, what do you think it would do when released? Do you not believe (as for my part I do) that it would lie still?

SIMPLICIO: If the surface were inclining?

SALVIATI: Yes, as I have already stated.

SIMPLICIO: I cannot conceive how it should lie still. I am confident that it would move towards the declivity with much propenseness.

SALVIATI: Take good heed what you say, Simplicius, for I am confident that it would lie still in whatever place you should lay it.

SIMPLICIO.: So long as you make use of such suppositions, Salviatus, I shall cease to wonder if you conclude most absurd conclusions.

SALVIATI.: Are you assured, then, that it would really move towards the declivity?

SIMPLICIO: Who doubts it?

SALVIATI: And this you verily believe, not because I told you so (for I endeavoured to persuade you to think the contrary), but of yourself, and upon your natural judgment.

SIMPLICIO: Now I see your game; you did not say this really believing it, but to try me, and to wrest words out of my mouth with which to condemn me.

SALVIATI: You are right. And how long and with what velocity would that ball move? But take notice that I gave as the example a ball exactly round, and a plane exquisitely polished, so that all external and accidental impediments might be taken away. Also I would have you remove all obstructions caused by the air's resistance and any other causal obstacles, if any other there can be.

SIMPLICIO: I understand your meaning very well and answer that the ball would continue to move *in infinitum* if the inclination of the plane should last so long, accelerating continually. Such is the nature of ponderous bodies that they acquire strength in going, and, the greater the declivity, the greater the velocity will be.

SALVIATI: But if one should require that that ball should move upwards on that same surface, do you believe that it would do so?

SIMPLICIO: Not spontaneously, but being drawn, or violently thrown, it may.

SALVIATI: And in case it were thrust forward by the impression of some violent impetus from without, what and how great would its motion be?

SIMPLICIO: The motion would be continually decreasing and retarding as being contrary to Nature and would be longer or shorter, according to the strength of the impulse and according to the degree of acclivity.

SALVIATI: It seems, then, that hitherto you have well explained to me the accidents of a body on two different planes. Now tell me, what would befall the same body upon a surface that had neither acclivity nor declivity?

SIMPLICIO: Here you must give me a little time to consider my answer. There being no declivity, there can be no natural inclination to motion; and there being no acclivity, there can be no resistance to being moved. There would then arise an indifference between propulsion and resistance; therefore, I think it ought naturally stand still. But I had forgot myself: it was not long ago that Sagredus gave me to understand that it would do so.

SAGREDO: So I think, provided one did lay it down gently; but if it had an impetus directing it towards any part, what would follow?

SIMPLICIO: That it should move towards that part.

SALVIATI: But with what kind of motion? Continually accelerated, as in declining planes; or successively retarded, as in those ascending?

SIMPLICIO: I cannot tell how to discover any cause of acceleration or retardation, there being no declivity or acclivity.

SALVIATI: Well, if there be no cause of retardation, even less should there be any cause of rest. How long therefore would you have the body move?

SIMPLICIO: As long as that surface, neither inclined nor declined, shall last.

SALVIATI: Therefore if such a space were interminate, the motion upon it would likewise have no termination, that is, would be perpetual.

SIMPLICIO: I think so, if the body is of a durable matter.

SALVIATI: That has been already supposed when it was said that all external and accidental impediments were removed, and the brittleness of the body in this case is one of those accidental impediments. Tell me now, what do you think is the cause that that same ball moves spontaneously upon the inclining plane, and does not, except with violence, upon the plane sloping upwards?

SIMPLICIO: Because the inclination of heavy bodies is to move towards the centre of the Earth and only by violence upwards towards the circumference.

SALVIATI: Therefore a surface which should he neither declining nor nor ascending ought in all its parts to be equally distant from the centre. But is there any such surface in the world?

SIMPLICIO: There is no want of it, such is our terrestrial globe, for example, if it were not rough and mountainous. But you have that of the water, at such time as it is calm and still.

SALVIATI: Then a ship which moves in a calm at sea is one of those bodies that run along one of those surfaces that are neither declining nor ascending and is therefore disposed, in case all obstacles external and accidental were removed, to move incessantly and uniformly with the impulses once imparted.

SIMPLICIO: It should seem to be so.

SALVIATI: And that stone which is on the round top, does it not move as being, together with the ship, carried about by the circumference of a circle about the centre; and therefore consequently by a motion indelible in it, if all external obstacles be removed? And is not this motion as swift as that of the ship?

SIMPLICIO: Up to now all is well. But what follows?

SALVIATI: Then I pray you recant with good grace your last conclusion, if you are satisfied with the truth of all the premises.

...

SALVIATI: For a final proof of the nullity of all the experiments before alleged, I conceive it now a convenient time and place to demonstrate a way how to make an exact trial of them all. Shut yourself up with some friend in the largest room below decks of some large ship and there procure gnats, flies, and such other small winged creatures. Also get a great tub full of water and within it put certain fishes; let also a certain bottle be hung up which, drop by drop, lets forth its water into another narrow necked bottle placed underneath. Then, the ship lying still, observe how those small winged animals fly with like velocity towards all parts of the room; how the fishes swim indifferently towards all sides; and how the distilling drops all fall into the bottle placed underneath. And casting anything towards your friend, you need not throw it with more force one way than another, provided the distances be equal; and jumping broad, you will reach as far one way as another. Having observed all these particulars, though no man doubts that, so long as the vessel stands still, they ought to take place in this manner, make the ship move with what velocity you please, so long as the motion is uniform and not fluctuating this way and that.

You shall not be able to discern the least alteration in all the forenamed effects, nor can you gather by any of them whether the ship moves or stands still.

Translated by Thomas Salusbury

Reading and Discussion Questions

1. What advantages might Galileo see in presenting his work in a dialogue format? What role does each of these characters play in the dialogue?
2. What arguments are offered against Aristotle and/or Aristotelianism?
3. How does Galileo answer the ancient objection that the Earth could not be moving because birds would be left behind?
4. What is the point of the thought experiment about dropping a ball from the mast of a ship at rest or in motion?
5. Explain why Galileo asks readers to imagine taking butterflies into the cabin of a ship.

Dialogues Concerning Two New Sciences (1638)

Galileo Galilei

39

First Day

...

SAGREDO: I quite agree with the peripatetic philosophers in denying the penetrability of matter. As to the vacua I should like to hear a thorough discussion of Aristotle's demonstration in which he opposes them, and what you, Salviati, have to say in reply. I beg of you, Simplicio, that you give us the precise proof of the Philosopher and that you, Salviati, give us the reply.

SIMPLICIO: So far as I remember, Aristotle inveighs against the ancient view that a vacuum is a necessary prerequisite for motion and that the latter could not occur without the former. In opposition to this view Aristotle shows that it is precisely the phenomenon of motion, as we shall see, which renders untenable the idea of a vacuum. His method is to divide the argument into two parts. He first supposes bodies of different weights to move in the same medium; then supposes, one and the same body to move in different media. In the first case, he supposes bodies of different weight to move in one and the same medium with different speeds which stand to one another in the same ratio as the weights; so that, for example, a body which is ten times as heavy as another will move ten times as rapidly as the other. In the second case he assumes that the speeds of one and the same body moving in different media are in inverse ratio to the densities of these media; thus, for instance, if the density of water were ten times that of air, the speed in air would be ten times greater than in water. From this second supposition, he shows that, since the tenuity of a vacuum differs infinitely from that of any medium filled with matter however rare, any body which moves in a plenum through a certain space in a certain time ought to move through a vacuum instantaneously; but instantaneous motion is an impossibility; it is therefore impossible that a vacuum should be produced by motion.

SALVIATI: The argument is, as you see, *ad hominem,* that is, it is directed against those who thought the vacuum a prerequisite for motion. Now if I admit the argument to be conclusive and concede also that motion cannot take place in a vacuum, the assumption of a vacuum considered absolutely and not with reference to motion, is not thereby invalidated. But to tell you what the ancients might possibly have replied and in order to better understand just how conclusive Aristotle's demonstration is, we may, in my opinion, deny both of his assumptions. And as to the first, I greatly doubt that Aristotle ever tested by experiment whether it be true that two stones, one weighing ten times as much as the other, if allowed to fall, at the same instant, from a height of, say, 100 cubits, would so differ in speed that when the heavier had reached the ground, the other would not have fallen more than 10 cubits.

SIMPLICIO: His language would seem to indicate that he had tried the experiment, because he says: *We see the heavier;* now the word *see* shows that he had made the experiment.

SAGREDO: But I, Simplicio, who have made the test can assure you that a cannon ball weighing one or two hundred pounds, or even more, will not reach the ground by as much as a span ahead of a musket ball weighing only half a pound, provided both are dropped from a height of 200 cubits.

SALVIATI: But, even without further experiment, it is possible to prove clearly, by means of a short and conclusive argument, that a heavier body does not move more rapidly than a lighter one provided both bodies are of the same material and in short such as those mentioned by Aristotle. But tell me, Simplicio, whether you admit that each falling body acquires a definite speed fixed by nature, a velocity which cannot be increased or diminished except by the use of force [*violenza*] or resistance.

SIMPLICIO: There can be no doubt but that one and the same body moving in a single medium has a fixed velocity which is determined by nature and which cannot be increased except by the addition of momentum [*impeto*] or diminished except by some resistance which retards it.

SALVIATI: If then we take two bodies whose natural speeds are different, it is clear that on uniting the two, the more rapid one will be partly retarded by the slower, and the slower will be somewhat hastened by the swifter. Do you not agree with me in this opinion?

SIMPLICIO: You are unquestionably right.

SALVIATI: But if this is true, and if a large stone moves with a speed of, say, eight while a smaller moves with a speed of four, then when they are united, the system will move with a speed less than eight; but the two stones when tied together make a stone larger than that which before moved with a speed of eight. Hence the heavier body moves with less speed than the lighter; an effect which is contrary to your supposition. Thus you see how, from your assumption that the heavier body moves more rapidly than the lighter one, I infer that the heavier body moves more slowly.

SIMPLICIO: I am all at sea because it appears to me that the smaller stone when added to the larger increases its weight and by adding weight I do not see how it can fail to increase its speed or, at least, not to diminish it.

SALVIATI: Here again you are in error, Simplicio, because it is not true that the smaller stone adds weight to the larger.

SIMPLICIO: This is, indeed, quite beyond my comprehension.

SALVIATI: It will not be beyond you when I have once shown you the mistake under which you are laboring. Note that it is necessary to distinguish between heavy bodies in motion and the same bodies at rest. A large stone placed in a balance not only acquires additional weight by having another stone placed upon it, but even by the addition of a handful of hemp its weight is augmented six to ten ounces according to the quantity of hemp. But if you tie the hemp to the stone and allow them to fall freely from some height, do you believe that the hemp will press down upon the stone and thus accelerate its motion or do you think the motion will be retarded by a partial upward pressure? One always feels the pressure upon his shoulders when he prevents the motion of a load resting upon him; but if one descends just as rapidly as the load would fall how can it gravitate or press upon him? Do you not see that this would be the same as trying to strike a man with a lance when he is running away from you with a speed which is equal to, or even greater, than that with which you are following him? You must therefore conclude that, during free and natural fall, the small stone does not press upon the larger and consequently does not increase its weight as it does when at rest.

SIMPLICIO: But what if we should place the larger stone upon the smaller? Its weight would be increased if the larger stone moved more rapidly; but we have already concluded that when the small stone moves more slowly it retards to some extent the speed of the larger, so that the combination of the two, which is a heavier body than the larger of the two stones, would move less rapidly, a conclusion which is contrary to your hypothesis. We infer therefore that large and small bodies move with the same speed provided they are of the same specific gravity.

SIMPLICIO: Your discussion is really admirable; yet I do not find it easy to believe that a bird-shot falls as swiftly as a cannon ball.

SALVIATI: Why not say a grain of sand as rapidly as a grindstone? But, Simplicio, I trust you will not follow the example of many others who divert the discussion from its main intent and fasten upon some statement of mine which lacks a hair's-breadth of the truth and, under this hair, hide the fault of another which is as big as a ship's cable. Aristotle says that "an iron ball of one hundred pounds falling from a height of one hundred cubits reaches the ground before a one-pound ball has fallen a single cubit." I say that they arrive at the same time. You find, on making the experiment, that the larger outstrips the smaller by two finger-breadths, that is, when the larger has reached the ground, the other is short of it by two finger-breadths; now you would not hide behind these two fingers the ninety-nine cubits of Aristotle, nor would you mention my small error and at the same time pass over in silence his very large one. Aristotle declares that bodies of different weights, in the same medium, travel (in so far as their motion depends upon gravity) with speeds which are proportional to their weights; this he illustrates by use of bodies in which it is possible to perceive the pure and unadulterated effect of gravity, eliminating other considerations, for example, figure as being of small importance [*minimi momenti*], influences which are greatly dependent upon the medium which modifies the single effect of gravity alone. Thus we observe that gold, the densest of all substances, when beaten out into a very thin leaf, goes floating through the air; the same thing happens with stone when ground into a very fine powder. But if you wish to maintain the general proposition you will have to show that the same ratio of speeds is preserved in the case of all heavy bodies, and that a stone of twenty pounds moves ten times as rapidly as one of two; but I claim that this is false and that, if they fall from a height of fifty or a hundred cubits, they will reach the earth at the same moment.

SIMPLICIO: Perhaps the result would be different if the fall took place not from a few cubits but from some thousands of cubits.

SALVIATI: If this were what Aristotle meant you would burden him with another error which would amount to a falsehood; because, since there is no such sheer height available on earth, it is clear that Aristotle could not have made the experiment; yet he wishes to give us the impression of his having performed it when he speaks of such an effect as one which we see.

...

SALVIATI: The facts set forth by me up to this point and, in particular, the one which shows that difference of weight, even when very great, is without effect in changing the speed of falling bodies, so that as far as weight is concerned they all fall with equal speed: this idea is, I say, so new, and at first glance so remote from fact, that if we do not have the means of making it just as clear as sunlight, it had better not be mentioned; but having once allowed it to pass my lips I must neglect no experiment or argument to establish it.

SAGREDO: Not only this but also many other of your views are so far removed from the commonly accepted opinions and doctrines that if you were to publish them you would stir up a large number of antagonists; for human nature is such that men do not look with favor upon discoveries—either of truth or fallacy—in their own field, when made by others than themselves. They call him an innovator of doctrine, an unpleasant title, by which they hope to cut those knots which they cannot untie, and by subterranean mines they seek to destroy structures which patient artisans have built with customary tools. But as for ourselves who have no such thoughts, the experiments and arguments which you have thus far adduced are fully satisfactory; however if you have any experiments which are more direct or any arguments which are more convincing we will hear them with pleasure.

SALVIATI: The experiment made to ascertain whether two bodies, differing greatly in weight will fall from a given height with the same speed offers some difficulty; because, if the height is considerable, the retarding effect of the medium, which must be penetrated and thrust aside by the falling body, will be greater in the case of the small momentum of the very light body than in the case of the great force [*violenza*] of the heavy body; so that, in a long distance, the light body will be left behind; if the height be small, one may well doubt whether there is any difference; and if there be a difference it will be inappreciable.

It occurred to me therefore to repeat many times the fall through a small height in such a way that I might accumulate all those small intervals of time that elapse between the arrival of the heavy and light bodies respectively at their common terminus, so that this sum makes an interval of time which is not only observable, but easily observable. In order to employ the slowest speeds possible and thus reduce the change which the resisting medium produces upon the simple effect of gravity it occurred to me to allow the bodies to fall along a plane slightly inclined to the horizontal. For in such a plane, just as well as in a vertical plane, one may discover how bodies of different weight behave: and besides this, I also wished to rid myself of the resistance which might arise from contact of the moving body with the aforesaid inclined plane. Accordingly I took two balls, one of lead and one of cork, the former more than a hundred times heavier than the latter, and suspended them by means of two equal fine threads, each four or five cubits long. Pulling each ball aside from the perpendicular, I let them go at the same instant, and they, falling along the circumferences of circles having these equal strings for semi-diameters, passed beyond the perpendicular and returned along the same path. This free vibration [*per lor medesime le andate e le tornate*] repeated a hundred times showed clearly that the heavy body maintains so nearly the period of the light body that neither in a hundred swings nor even in a thousand will the former anticipate the latter by as much as a single moment [*minimo momento*], so perfectly do they keep step. We can also observe the effect of the medium which, by the resistance which it offers to motion, diminishes the vibration of the cork more than that of the lead, but without altering the frequency of either; even when the arc traversed by the cork did not exceed five or six degrees while that of the lead was fifty or sixty, the swings were performed in equal times.

SIMPLICIO: If this be so, why is not the speed of the lead greater than that of the cork, seeing that the former traverses sixty degrees in the same interval in which the latter covers scarcely six?

SALVIATI: But what would you say, Simplicio, if both covered their paths in the same time when the cork, drawn aside through thirty degrees, traverses an arc of sixty, while the lead pulled aside only two degrees traverses an arc of four? Would not then the cork be proportionately swifter? And yet such is the experimental fact. But observe this: having pulled aside the pendulum of lead, say through an arc of fifty degrees, and set it free, it swings beyond the perpendicular almost fifty degrees, thus describing an arc of nearly one hundred degrees; on the return swing it describes a little smaller arc; and after a large number of such vibrations it finally comes to rest. Each vibration, whether of ninety, fifty, twenty, ten, or four degrees occupies the same time: accordingly the speed of the moving body keeps on diminishing since in equal intervals of time, it traverses arcs which grow smaller and smaller.

Precisely the same things happen with the pendulum of cork, suspended by a string of equal length, except that a smaller number of vibrations is required to bring it to rest, since on account of its lightness it is less able to overcome the resistance of the air; nevertheless the vibrations, whether large or small, are all performed in time-intervals which are not only equal among themselves, but also equal to the period of the lead pendulum. Hence it is true that, if while the lead is traversing an arc of fifty degrees the cork covers one of only ten, the cork moves more slowly than the lead; but on the other hand it is also true that the cork may cover an arc of fifty while the lead passes over one of only ten or six; thus, at different times, we have now the cork, now the lead, moving more rapidly. But if these same bodies traverse equal arcs in equal times we may rest assured that their speeds are equal.

SIMPLICIO: I hesitate to admit the conclusiveness of this argument because of the confusion which arises from your making both bodies move now rapidly, now slowly and now very slowly, which leaves me in doubt as to whether their velocities are always equal.

SAGREDO: Allow me, if you please, Salviati, to say just a few words. Now tell me, Simplicio, whether you admit that one can say with certainty that the speeds of the cork and the lead are equal whenever both, starting from rest at the same moment and descending the same slopes, always traverse equal spaces in equal times?

SIMPLICIO: This can neither be doubted nor gainsaid.

SAGREDO: Now it happens, in the case of the pendulums, that each of them traverses now an arc of sixty degrees, now one of fifty, or thirty or ten or eight or four or two, etc.; and when they both swing through an arc of sixty degrees they do so in equal intervals of time; the same thing happens when the arc is fifty degrees or thirty or ten or any other number; and therefore we conclude that the speed of the lead in an arc of sixty degrees is equal to the speed of the cork when the latter also swings through an arc of sixty degrees; in the case of a fifty-degree arc these speeds are also equal to each other; so also in the case of other arcs. But this is not saying that the speed which occurs in an arc of sixty is the same as that which occurs in an arc of fifty; nor is the speed in an arc of fifty equal to that in one of thirty, etc.; but the smaller the arcs, the smaller the speeds; the fact observed is that one and the same moving body requires the same time for traversing a large arc of sixty degrees as for a small arc of fifty or even a very small arc of ten; all these arcs, indeed, are covered in the same interval of time. It is true therefore that the lead and the cork each diminish their speed [*moto*] in proportion as their arcs diminish; but this does not contradict the fact that they maintain equal speeds in equal arcs.

My reason for saying these things has been rather because I wanted to learn whether I had correctly understood Salviati, than because I thought Simplicio had any need of a clearer explanation than that given by Salviati which like everything else of his is extremely lucid, so lucid, indeed, that when he solves questions which are difficult not merely in appearance, but in reality and in fact, he does so with reasons, observations and experiments which are common and familiar to everyone.

In this manner he has, as I have learned from various sources, given occasion to a highly esteemed professor for undervaluing his discoveries on the ground that they are commonplace, and established upon a mean and vulgar basis; as if it were not a most admirable and praiseworthy feature of demonstrative science that it springs from and grows out of principles well-known, understood and conceded by all.

But let us continue with this light diet; and if Simplicio is satisfied to understand and admit that the gravity inherent [*interna gravità*] in various falling bodies has nothing to do with the difference of speed observed among them, and that all bodies, in so far as their speeds depend upon it, would move with the same velocity, pray tell us, Salviati, how you explain the appreciable and evident inequality of motion; please reply also to the objection urged by Simplicio—an objection in which I concur—namely, that a cannon ball falls more rapidly than a bird-shot. From my point of view, one might expect the difference of speed to be small in the case of bodies of the same substance moving through any single medium, whereas the larger ones will descend, during a single pulse-beat, a distance which the smaller ones will not traverse in an hour, or in four, or even in twenty hours; as for instance in the case of stones and fine sand and especially that very fine sand which produces muddy water and which in many hours will not fall through as much as two cubits, a distance which stones not much larger will traverse in a single pulse-beat.

SALVIATI: The action of the medium in producing a greater retardation upon those bodies which have a less specific gravity has already been explained by showing that they experience a diminution of weight. But to explain how one and the same medium produces such different retardations in bodies which are made of the same material and have the same shape, but differ only in size, requires a discussion more clever than that by which one explains how a more expanded shape or an opposing motion of the medium retards the speed of the moving body. The solution of the present problem lies, I think, in the roughness and porosity which are generally and almost necessarily found in the surfaces of solid bodies. When the body is in motion these rough places strike the air or other ambient medium. The evidence for this is found in the humming which accompanies the rapid motion of a body through air, even when that body is as round as possible. One hears not only humming, but also hissing and whistling, whenever there is any appreciable cavity or elevation upon the body. We observe also that a round solid body rotating in a lathe produces a current of air. But what more do we need? When a top spins on the ground at its greatest speed do we not hear a distinct buzzing of high pitch? This sibilant note diminishes in pitch as the speed of rotation slackens, which is evidence that these small rugosities on the surface meet resistance in the air. There can be no doubt, therefore, that in the motion of falling bodies these rugosities strike the surrounding fluid and retard the speed; and this they do so much the more in proportion as the surface is larger, which is the case of small bodies as compared with greater . . . .

Third Day

. . .

SALVIATI: The present does not seem to be the proper time to investigate the cause of the acceleration of natural motion concerning which various opinions have been expressed by various philosophers, some explaining it by attraction to the center, others to repulsion between the very small parts of the body, while still others attribute it to a certain stress in the surrounding medium which closes in behind the falling body and drives it from one of its positions to another. Now, all these fantasies, and others too, ought to be examined; but it is not really worthwhile. At present it is the purpose of our Author merely to investigate and to demonstrate some of the properties of accelerated motion (whatever the cause of this acceleration may be)—meaning thereby a motion, such that the momentum of its velocity [*i momenti della sua velocità*] goes on increasing after departure from rest, in simple proportionality to the time, which is the same as saying that in equal time-intervals the body receives equal increments of velocity; and if we find the properties [of accelerated motion] which will be demonstrated later are realized in freely falling and accelerated bodies, we may conclude that the assumed definition includes such a motion of falling bodies and that their speed [*accelerazione*] goes on increasing as the time and the duration of the motion.

So far as I see at present, the definition might have been put a little more clearly perhaps without changing the fundamental idea, namely, uniformly accelerated motion is such that its speed increases in proportion to the space traversed; so that, for example, the speed acquired by a body in falling four cubits would be double that acquired in falling two cubits and this latter speed would be double that acquired in the first cubit. Because there is no doubt but that a heavy body falling from the height of six cubits has, and strikes with, a momentum [*impeto*] double that it had at the end of three cubits, triple that which it had at the end of one.

SALVIATI: It is very comforting to me to have had such a companion in error; and moreover let me tell you that your proposition seems so highly probable that our Author himself admitted, when I advanced this opinion to him, that he had for some time shared the same fallacy.

But what most surprised me was to see two propositions so inherently probable that they commanded the assent of everyone to whom they were presented, proven in a few simple words to be not only false, but impossible.

SIMPLICIO: I am one of those who accept the proposition, and believe that a falling body acquires force [*vires*] in its descent, its velocity increasing in proportion to the space, and that the momentum [*momento*] of the falling body is doubled when it falls from a doubled height; these propositions, it appears to me, ought to be conceded without hesitation or controversy.

SALVIATI: And yet they are as false and impossible as that motion should be completed instantaneously; and here is a very clear demonstration of it. If the velocities are in proportion to the spaces traversed, or to be traversed, then these spaces are traversed in equal intervals of time; if, therefore, the velocity with which the falling body traverses a space of eight feet were double that with which it covered the first four feet (just as the one distance is double the other) then the time-intervals required for these passages would be equal. But for one and the same body to fall eight feet and four feet in the same time is possible only in the case of instantaneous [discontinuous] motion; but observation shows us that the motion of a falling body occupies time, and less of it in covering a distance of four feet than of eight feet; therefore it is not true that its velocity increases in proportion to the space.

The falsity of the other proposition may be shown with equal clearness. For if we consider a single striking body the difference of momentum in its blows can depend only upon difference of velocity; for if the striking body falling from a double height were to deliver a blow of double momentum, it would be necessary for this body to strike with a doubled velocity; but with this doubled speed it would traverse a doubled space in the same time-interval; observation however shows that the time required for fall from the greater height is longer.

SAGREDO: You present these recondite matters with too much evidence and ease; this great facility makes them less appreciated than they would be had they been presented in a more abstruse manner. For, in my opinion, people esteem more lightly that knowledge which they acquire with so little labor than that acquired through long and obscure discussion.

SALVIATI: If those who demonstrate with brevity and clearness the fallacy of many popular beliefs were treated with contempt instead of gratitude the injury would be quite bearable; but on the other hand it is very unpleasant and annoying to see men, who claim to be peers of anyone in a certain field of study, take for granted certain conclusions which later are quickly and easily shown by another to be false. I do not describe such a feeling as one of envy, which usually degenerates into hatred and anger against those who discover such fallacies; I would call it a strong desire to maintain old errors, rather than accept newly discovered truths. This desire at times induces them to unite against these truths, although at heart believing in them, merely for the purpose of lowering the esteem in which certain others are held by the unthinking crowd. Indeed, I have heard from our Academician many such fallacies held as true but easily refutable; some of these I have in mind.

SAGREDO: You must not withhold them from us, but, at the proper time, tell us about them even though an extra session be necessary. But now, continuing the thread of our talk, it would seem that up to the present we have established the definition of uniformly accelerated motion which is expressed as follows: A motion is said to be equally or uniformly accelerated when, starting from rest, its momentum (*celeritatis momenta*) receives equal increments in equal times.

SALVIATI: This definition established, the Author makes a single assumption, namely, The speeds acquired by one and the same body moving down planes of different inclinations are equal when the heights of these planes are equal. By the height of an inclined plane we mean the perpendicular let fall from the upper end of the plane upon the horizontal line drawn through the lower end of the same plane. Thus, to illustrate, let the line AB be horizontal, and let the planes CA and CD be inclined to it; then the Author calls the perpendicular CB the "height" of the planes CA and CD; he supposes that the speeds acquired by one and the same body, descending along the planes CA and CD to the terminal points A and D are equal since the heights of these planes are the same, CB; and also it must be understood that this speed is that which would be acquired by the same body falling from C to B.

SAGREDO: Your assumption appears to me so reasonable that it ought to be conceded without question, provided of course there are no chance or outside resistances, and that the planes are hard and smooth, and that the figure of the moving body is perfectly round, so that neither plane nor moving body is rough. All resistance and opposition having been removed, my reason tells me at once that a heavy and perfectly round ball descending along the lines CA, CD, CB would reach the terminal points A, D, B, with equal momenta [*impeti eguali*].

SALVIATI: Your words are very plausible; but I hope by experiment to increase the probability to an extent which shall be little short of a rigid demonstration.

This experiment leaves no room for doubt as to the truth of our supposition; for since the two arcs CB and DB are equal and similarly placed, the momentum [*momento*] acquired by the fall through the

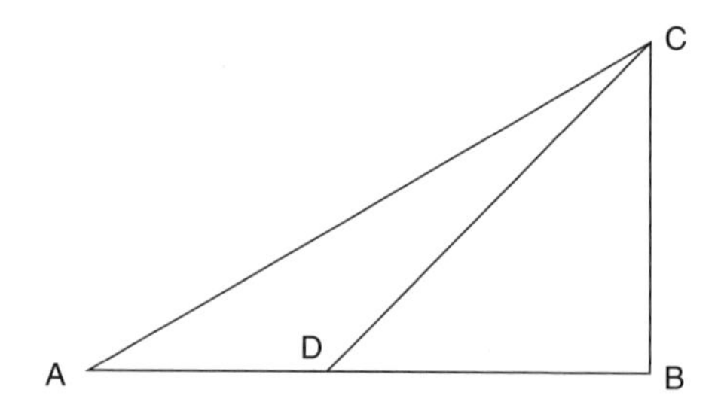

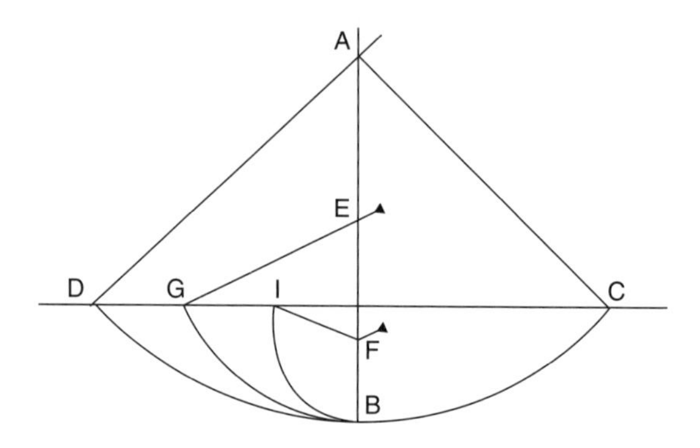

arc CB is the same as that gained by fall through the arc DB; but the momentum [*momento*] acquired at B, owing to fall through CB, is able to lift the same body [*mobile*] through the arc BD; therefore, the momentum acquired in the fall BD is equal to that which lifts the same body through the same arc from B to D; so, in general, every momentum acquired by fall through an arc is equal to that which can lift the same body through the same arc. But all these momenta [*momenti*] which cause a rise through the arcs BD, BG, and BI are equal, since they are produced by the same momentum, gained by fall through CB, as experiment shows. Therefore all the momenta gained by fall through the arcs DB, GB, IB are equal.

SAGREDO: The argument seems to me so conclusive and the experiment so well adapted to establish the hypothesis that we may, indeed, consider it as demonstrated.

SALVIATI: I do not wish, Sagredo, that we trouble ourselves too much about this matter, since we are going to apply this principle mainly in motions which occur on plane surfaces, and not upon curved, along which acceleration varies in a manner greatly different from that which we have assumed for planes.

So that, although the above experiment shows us that the descent of the moving body through the arc CB confers upon it momentum [*momento*] just sufficient to carry it to the same height through any of the arcs BD, BG, BI, we are not able, by similar means, to show that the event would be identical in the case of a perfectly round ball descending along planes whose inclinations are respectively the same as the chords of these arcs. It seems likely, on the other hand, that, since these planes form angles at the point B, they will present an obstacle to the ball which has descended along the chord CB, and starts to rise along the chord BD, BG, BI.

In striking these planes some of its momentum [*impeto*] will be lost and it will not be able to rise to the height of the line CD; but this obstacle, which interferes with the experiment, once removed, it is clear that the momentum [*impeto*] (which gains in strength with descent) will be able to carry the body to the same height. Let us then, for the present, take this as a postulate, the absolute truth of which will be established when we find that the inferences from it correspond to and agree perfectly with experiment. The author having assumed this single principle passes next to the propositions which he clearly demonstrates; the first of these is as follows.

Theorem I, Proposition I

The time in which any space is traversed by a body starting from rest and uniformly accelerated is equal to the time in which that same space would be traversed by the same body moving at a uniform speed whose value is the mean of the highest speed and the speed just before acceleration began.

Let us represent by the line AB the time in which the space CD is traversed by a body which starts from rest at C and is uniformly accelerated; let the final and highest value of the speed gained during the interval AB be represented by the line EB drawn at right angles to AB; draw the line AE, then all lines drawn from equidistant points on AB and parallel to BE will represent the increasing values of the speed, beginning with the instant A. Let the point F bisect the line EB; draw FG parallel to BA, and GA parallel to FB, thus forming a parallel-ogram AGFB which will be equal in area to the triangle AEB, since the side GF bisects the side AE at the point I; for if the parallel lines in the triangle AEB are extended to GI, then the sum of all the parallels contained in the quadrilateral is equal to the sum of those contained in the triangle AEB; for those in the triangle IEF are equal to those contained in the triangle GIA, while those included in the trapezium AIFB are common. Since each and every instant of time in the time-interval AB has its corresponding point on the line AB, from which points parallels drawn in and limited by the triangle AEB represent the increasing values of the growing velocity, and since parallels contained within the rectangle represent the values of a speed which is not increasing, but constant,

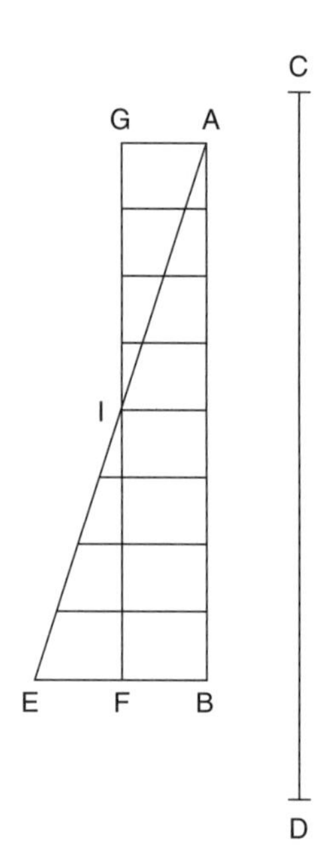

it appears, in like manner, that the momenta [***momenta***] assumed by the moving body may also be represented, in the case of the accelerated motion, by the increasing parallels of the triangle AEB, and, in the case of the uniform motion, by the parallels of the rectangle GB. For, what the momenta may lack in the first part of the accelerated motion (the deficiency of the momenta being represented by the parallels of the triangle AGI) is made up by the momenta represented by the parallels of the triangle IEF.

Hence it is clear that equal spaces will be traversed in equal times by two bodies, one of which, starting from rest, moves with a uniform acceleration, while the momentum of the other, moving with uniform speed, is one-half its maximum momentum under accelerated motion.

q. e. d.

Theorem II, Proposition II

The spaces described by a body falling from rest with a uniformly accelerated motion are to each other as the squares of the time-intervals employed in traversing these distances.

Let the time beginning with any instant A be represented by the straight line AB in which are taken any two time-intervals AD and AE. Let HI represent the distance through which the body, starting from rest at H, falls with uniform acceleration. If HL represents the space traversed during the time-interval AD, and HM that covered during the interval AE, then the space MH stands to the space LH in a ratio

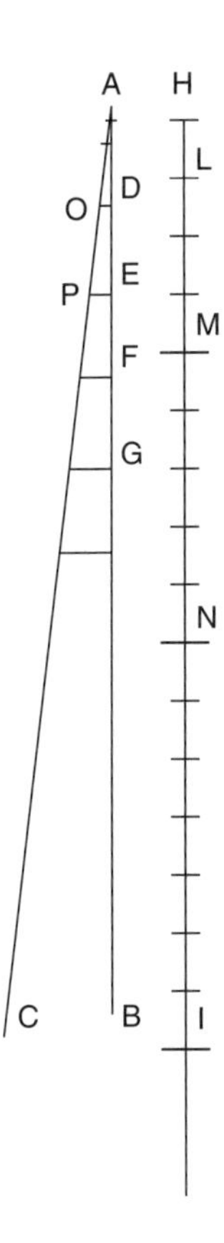

which is the square of the ratio of the time AE to the time AD; or we may say simply that the distances HM and HL are related as the squares of AE and AD.

Draw the line AC making any angle whatever with the line AB; and from the points D and E, draw the parallel lines DO and EP; of these two lines, DO represents the greatest velocity attained during the interval AD, while EP represents the maximum velocity acquired during the interval AE. But it has just been proved that so far as distances traversed are concerned it is precisely the same whether a body falls from rest with a uniform acceleration or whether it falls during an equal time-interval with a constant speed which is one-half the maximum speed attained during the accelerated motion. It follows therefore that the distances HM and HL are the same as would be traversed, during the time-intervals AE and AD, by uniform velocities equal to one-half those represented by DO and EP respectively. If, therefore, one can show that the distances HM and HL are in the same ratio as the squares of the time-intervals AE and AD, our proposition will be proven.

q. e. d.

Evidently then the ratio of the distances is the square of the ratio of the final velocities, that is, of the lines EP and DO, since these are to each other as AE to AD.

Corollary I

Hence it is clear that if we take any equal intervals of time whatever, counting from the beginning of the motion, such as AD, DE, EF, FG, in which the spaces HL, LM, MN, NI are traversed, these spaces will bear to one another the same ratio as the series of odd numbers, 1, 3, 5, 7; for this is the ratio of the differences of the squares of the lines [which represent time], differences which exceed one another by equal amounts, this excess being equal to the smallest line [viz. the one representing a single time-interval]: or we may say [that this is the ratio] of the differences of the squares of the natural numbers beginning with unity.

While, therefore, during equal intervals of time the velocities increase as the natural numbers, the increments in the distances traversed during these equal time-intervals are to one another as the odd numbers beginning with unity.

SAGREDO: Please suspend the discussion for a moment since there just occurs to me an idea which I want to illustrate by means of a diagram in order that it may be clearer both to you and to me.

Let the line AI represent the lapse of time measured from the initial instant A; through A draw the straight line AF making any angle whatever; join the terminal points I and F; divide the time AI in half at C; draw CB parallel to IF. Let us consider CB as the maximum value of the velocity which increases from zero at the beginning, in simple proportionality to the intercepts on the triangle ABC of lines drawn parallel to BC; or what is the same thing, let us suppose the velocity to increase in proportion to the time; then I admit without question, in view of the preceding argument, that the space described by a body falling in the aforesaid manner will be equal to the space traversed by the same body during the same length of time travelling with a uniform speed equal to EC, the half of BC. Further let us imagine that the body has fallen with accelerated motion so that, at the instant C, it has the velocity BC. It is clear that if the body continued to descend with the same speed BC, without acceleration, it would in the next time-interval CI traverse double the distance covered during the interval AC, with the uniform speed EC which is half of BC; but since the falling body acquires equal increments of speed during equal increments of time, it follows that the velocity BC, during the next time-interval CI will be increased by an amount represented by the parallels of the triangle BFG which is equal to the triangle ABC. If, then, one adds to the velocity GI half of the velocity FG, the highest speed acquired by the accelerated motion and determined by the parallels of the triangle BFG, he will have the uniform velocity with which the same space would have been described in the time CI; and since this speed IN is three times as great as EC it follows that the space described during the interval CI is three times as great as that described during the interval AC. Let us imagine the motion extended over another equal time-interval IO, and the triangle extended to APO; it is then evident that if the motion continues during the interval IO, at the constant rate IF acquired by acceleration during the time AI, the space traversed during the interval IO will be four times that traversed during the first interval AC, because the speed IF is four times the speed EC. But if we enlarge our triangle so as to include FPQ which is equal to ABC, still assuming the acceleration to be constant, we shall add to the uniform speed an increment RQ, equal to EC; then the value of the equivalent uniform speed during the time-interval IO will be five times that during the first time-interval AC; therefore the space traversed will be quintuple that during the first interval AC. It is thus evident by simple computation that a moving body starting from rest and acquiring velocity at a rate proportional to the time, will, during equal intervals of time, traverse distances which are related to each other as the odd numbers beginning with unity, 1, 3, 5; or considering the total space traversed, that covered in double time will be quadruple that covered during unit time; in triple time, the space is nine times as great as in unit time. And in general the spaces traversed are in the duplicate ratio of the times, i. e., in the ratio of the squares of the times.

SIMPLICIO: In truth, I find more pleasure in this simple and clear argument of Sagredo than in the Author's demonstration which to me appears rather obscure; so that I am convinced that matters are as described, once having accepted the definition of uniformly accelerated motion. But as to whether this acceleration is that which one meets in nature in the case of falling bodies, I am still doubtful; and it seems to me, not only for my own sake but also for all those who think as I do, that this would be the proper moment to introduce one of those experiments—and there are many of them, I understand—which illustrate in several ways the conclusions reached.

SALVIATI: The request which you, as a man of science, make, is a very reasonable one; for this is the custom—and properly so—in those sciences where mathematical demonstrations are applied to natural phenomena, as is seen in the case of perspective, astronomy, mechanics, music, and others where the principles, once established by well-chosen experiments, become the foundations of the entire superstructure. I hope therefore it will not appear to be a waste of time if we discuss at considerable length this first and most fundamental question upon which hinge numerous consequences of which we have in this book only a small number, placed there by the Author, who has done so much to open a pathway hitherto closed to minds of speculative turn. So far as experiments go they have not been neglected by the Author; and often, in his company, I have attempted in the following manner to assure myself that the acceleration actually experienced by falling bodies is that above described.

A piece of wooden moulding or scantling, about 12 cubits long, half a cubit wide, and three finger-breadths thick, was taken; on its edge was cut a channel a little more than one finger in breadth; having made this groove very straight, smooth, and polished, and having lined it with parchment, also as smooth and polished as possible, we rolled along it a hard, smooth, and very round bronze ball. Having placed this board in a sloping position, by lifting one end some one or two cubits above the other, we rolled the ball, as I was just saying, along the channel, noting, in a manner presently to be described, the time required to make the descent. We repeated this experiment more than once in order to measure the time with an accuracy such that the deviation between two observations never exceeded one-tenth of a pulse-beat. Having performed this operation and having assured ourselves of its reliability, we now rolled the ball only one-quarter the length of the channel; and having measured the time of its descent, we found it precisely one-half of the former. Next we tried other distances, comparing the time for the whole length with that for the half, or with that for two-thirds, or three-fourths, or indeed for any fraction; in such experiments, repeated a full hundred times, we always found that the spaces traversed were to each other as the squares of the times, and this was true for all inclinations of the plane, i.e., of the channel, along which we rolled the ball. We also observed

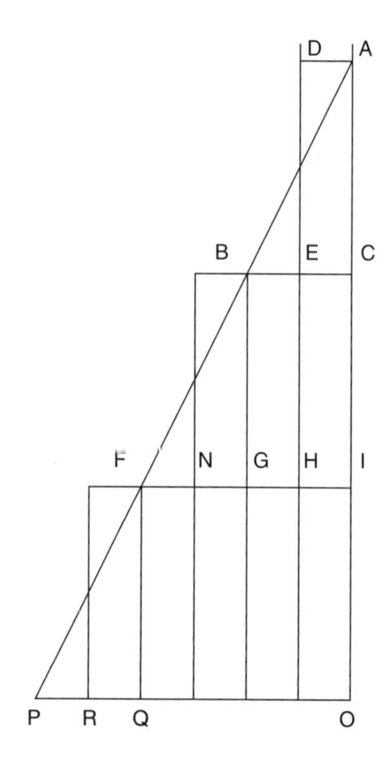

that the times of descent, for various inclinations of the plane, bore to one another precisely that ratio which, as we shall see later, the Author had predicted and demonstrated for them.

For the measurement of time, we employed a large vessel of water placed in an elevated position; to the bottom of this vessel was soldered a pipe of small diameter giving a thin jet of water, which we collected in a small glass during the time of each descent, whether for the whole length of the channel or for a part of its length; the water thus collected was weighed, after each descent, on a very accurate balance; the differences and ratios of these weights gave us the differences and ratios of the times, and this with such accuracy that although the operation was repeated many, many times, there was no appreciable discrepancy in the results.

SIMPLICIO: I would like to have been present at these experiments; but feeling confidence in the care with which you performed them, and in the fidelity with which you relate them, I am satisfied and accept them as true and valid.

SALVIATI: Then we can proceed without discussion.

. . .

Fourth Day

SALVIATI: Once more, Simplicio is here on time; so let us without delay take up the question of motion. The text of our Author is as follows:

The Motion of Projectiles

In the preceding pages we have discussed the properties of uniform motion and of motion naturally accelerated along planes of all inclinations. I now propose to set forth those properties which belong to a body whose motion is compounded of two other motions, namely, one uniform and one naturally accelerated; these properties, well worth knowing, I propose to demonstrate in a rigid manner. This is the kind of motion seen in a moving projectile; its origin I conceive to be as follows:

Imagine any particle projected along a horizontal plane without friction; then we know, from what has been more fully explained in the preceding pages, that this particle will move along this same plane with a motion which is uniform and perpetual, provided the plane has no limits. But if the plane is limited and elevated, then the moving particle, which we imagine to be a heavy one, will on passing over the edge of the plane acquire, in addition to its previous uniform and perpetual motion, a downward propensity due to its own weight; so that the resulting motion which I call projection [*projectio*], is compounded of one which is uniform and horizontal and of another which is vertical and naturally accelerated. We now proceed to demonstrate some of its properties, the first of which is as follows:

Theorem I, Proposition I

A projectile which is carried by a uniform horizontal motion compounded with a naturally accelerated vertical motion describes a path which is a semi-parabola.

. . .

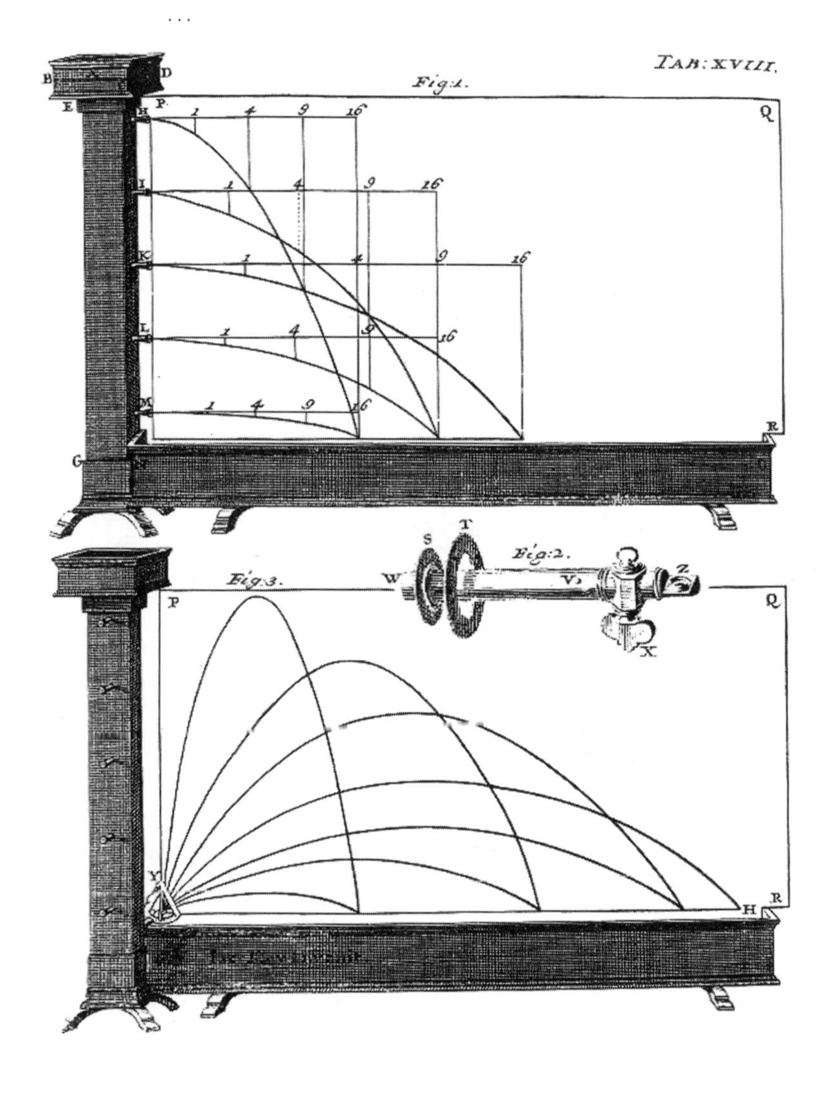

The force of rigid demonstrations such as occur only in mathematics fills me with wonder and delight. From accounts given by gunners, I was already aware of the fact that in the use of cannon and mortars, the maximum range, that is the one in which the shot goes farthest, is obtained when the elevation is 45° or, as they say, at the sixth point of the quadrant; but to understand why this happens far outweighs the mere information obtained by the testimony of others or even by repeated experiment.

. . .

Translated by Henry Crew and Alfonso de Salvio

Reading and Discussion Questions

1. How does Galileo argue against the Aristotelian belief that a body's rate of fall is related to its weight? How does he deal with the problem of air resistance?
2. What does Galileo take himself to have demonstrated about the relationship between acceleration, distance, and time?
3. Because free fall happens so quickly, what are some of the alternative experiments that Galileo employs to examine the phenomenon of acceleration due to gravity?
4. What point does Salviati make about the motion of objects on a smooth flat surface, as part of his discussion of balls accelerating and decelerating on inclined planes?
5. What does Galileo mean in describing the horizontal and vertical motions of a projectile as compounded? How is this account of projectile motion different from the Aristotelian picture?

The World (written c. 1630, published 1664)

René Descartes

40

Chapter Six: Description of a New World, and on the Qualities of the Matter of Which It Is Composed

For a short time, then, allow your thought to wander beyond this world to view another, wholly new one, which I shall cause to unfold before it in imaginary spaces. The philosophers tell us that these spaces are infinite, and they should very well be believed, since it is they themselves who have made the spaces so. Yet, in order that this infinity not impede us and not embarrass us, let us not try to go all the way to the end; let us enter in only so far that we can lose from view all the creatures that God made five or six thousand years ago and, after having stopped there in some fixed place, let us suppose that God creates from anew so much matter all about us that, in whatever direction our imagination can extend itself, it no longer perceives any place that is empty.

Although the sea is not infinite, those who are on some vessel in the middle of it can extend their view seemingly to infinity, and nevertheless there is still water beyond what they see. Thus, even though our imagination seems to be able to extend itself to infinity, and this new matter is not assumed to be infinite, we can nonetheless well suppose that it fills spaces much greater than all those we shall have imagined. Indeed, in order that there be nothing in all this that you could find to blame, let us not permit our imagination to extend itself as far as it could, but let us purposely restrict it to a determinate space that is no greater, say, than the distance between the earth and the principal stars of the firmament, and let us suppose that the matter that God shall have created extends quite far beyond in all directions, out to an indefinite distance. For there is more reason, and we have much better the power, to prescribe limits to the action of our thought than to the works of God.

Now, since we are taking the liberty of imagining this matter to our fancy, let us attribute to it, if you will, a nature in which there is absolutely nothing that anyone cannot know as perfectly as possible. To that end, let us expressly assume that it does not have the form of earth, nor of fire, nor of air, nor any more particular form (such as wood, or a stone, or of a metal); nor does it have the qualities of being hot or cold, dry or moist, light or heavy, or of having some taste, or smell, or sound or color, or light, or suchlike, in the nature of which one could say that there is something that is not clearly known by everyone.

Let us not also think, on the other hand, that our matter is that prime matter of the philosophers that has been so well stripped of all its forms and qualities that nothing more remains that can be clearly understood. Let us rather conceive of it as a real, perfectly solid body, which uniformly fills the entire length, breadth, and depth of the great space at the center of which we have halted our thought. Thus, each of its parts always occupies a part of that space and is so proportioned to its size that it could not fill a larger one nor squeeze itself into a smaller one, nor (while it remains there) suffer another to find a place there.

Let us add further that this matter can be divided into any parts and according to any shapes that we can imagine, and that each of its parts is capable of receiving in itself any motions that we can also conceive. Let us suppose in addition that God truly divides it into many such parts, some larger and some smaller, some of one shape and some of another, as it pleases us to imagine them. It is not that He thereby separates them from one another, so that there is some void in between them; rather, let us think that the entire distinction that He makes there consists in the diversity of the motions He gives to them. From the first instant that they are created, He makes some begin to move in one direction and others in another, some faster and others slower (or indeed, if you wish, not at all); thereafter, He makes them continue their motion according to the ordinary laws of nature. For God has so wondrously established these laws that, even if we suppose that He creates nothing more than what I have said, and even if He does not impose any order or proportion on it but makes of it the most confused and most disordered chaos that the poets could describe, the laws are sufficient to make the parts of that chaos untangle themselves and arrange themselves in such right order that they will have the form of a most perfect world, in which one will be able to see not only light, but also all the other things, both general and particular, that appear in this true world.

But, before I explain this at greater length, stop again for a bit to consider that chaos, and note that it contains nothing that is not so perfectly known to you that you could not even pretend not to know it. For, as regards the qualities that I have posited there, I have, if you have noticed, supposed them to be only such as you can imagine them. And, as regards the matter from which I have composed the chaos, there is nothing simpler nor easier to know among inanimate creatures. The idea of that matter is so included in all those that our imagination can form that you must necessarily conceive of it or you can never imagine anything.

Nonetheless, because the philosophers are so subtle that they can find difficulties in things that appear extremely clear to other men, and because the memory of their prime matter (which they know to be rather difficult to conceive of) could divert them from knowledge of the matter of which I speak, I should say to them at this point that, unless I am mistaken, the whole problem they face with their matter derives only from their wanting to distinguish it from its own proper quantity and from its outward extension, i.e. from the property it has of occupying space. In this, however, I am willing that they think themselves correct, for I have no intention of stopping to contradict them. But they should also not find it strange if I suppose that the quantity of the matter I have described does not differ from its substance any more than number differs from the things numbered. Nor should they find it strange if I conceive of its extension, or the property it has of occupying space, not as an accident, but as its true form and its essence. For they cannot deny that it is quite easy to conceive of it in that way. And my plan is not to set out (as they do) the things that are in fact in the true world, but only to make up as I please from [this matter] a [world] in which there is nothing that the densest minds are not capable of conceiving, and which nevertheless could be created exactly the way I have made it up.

Were I to posit in this new world the least thing that is obscure, it could happen that, within that obscurity, there might be some hidden contradiction I had not perceived, and thus that, without thinking, I might suppose something impossible. Instead, being able to imagine distinctly everything I am positing there, it is certain that, even if there be no such thing in the old world, God can nevertheless create it in a new one; for it is certain that He can create everything we can imagine.

Chapter Seven: On the Laws of Nature of This New World

But I do not want to defer any longer from telling you by what means nature alone could untangle the confusion of the chaos of which I have been speaking, and what the laws of nature are that God has imposed on her.

Know, then, first that by "nature" I do not here mean some deity or other sort of imaginary power. Rather, I use that word to signify matter itself, insofar as I consider it taken together with all the qualities that I have attributed to it, and under the condition that God continues to preserve it in the same way that He created it. For from that alone (i.e. that He continues thus to preserve it) it follows of necessity that there may be many changes in its parts that cannot, it seems to me, be properly attributed to the action of God (because that action does not change) and hence are to be attributed to nature. The rules according to which these changes take place I call the "laws of nature."

To understand this better, recall that, among the qualities of matter, we have supposed that its parts have had diverse motions since the beginning when they were created, and furthermore that they all touch one another on all sides, without there being any void in between. Whence it follows of necessity that from then on, in beginning to move, they also began to change and diversify their motions by colliding with one another. Thus, if God preserves them thereafter in the same way that He created them, He does not preserve them in the same state. That is to say, with God always acting in the same way and consequently always producing the same effect in substance, there occur, as by accident, many diversities in that effect. And it is easy to believe that God, who, as everyone must know, is immutable, always acts in the same way. Without, however, involving myself any further in these metaphysical considerations, I will set out here two or three of the principal rules according to which one must think God to cause the nature of this new world to act and which will suffice, I believe, for you to know all the others.

The first is that each individual part of matter always continues to remain in the same state unless collision with others constrains it to change that state. That is to say, if the part has some size, it will never become smaller unless others divide it; if it is round or square, it will never change that shape without others forcing it to do so; if it is stopped in some place, it will never depart from that place unless others chase it away; and if it has once begun to move, it will always continue with an equal force until others stop or retard it.

There is no one who does not believe that this same rule is observed in the old world with respect to size, shape, rest, and a thousand other like things. But from it the philosophers have exempted motion, which is, however, the thing I most expressly desire to include in it. Do not think thereby that I intend to contradict them. The motion of which they speak is so very different from that which I conceive that it can easily happen that what is true of the one is not true of the other.

They themselves avow that the nature of their motion is very little known. To render it in some way intelligible, they have still not been able to explain it more clearly than in these terms: *motus est actus entis in potentia, prout in potentia est*, which terms are for me so obscure that I am constrained to leave them here in their language, because I cannot interpret them. (And, in fact, the words, "motion is the act of a being in potency, insofar as it is in potency," are no clearer for being in [English].) On the contrary, the nature of the motion of which I mean to speak here is so easy to know that mathematicians themselves, who among all men studied most to conceive very distinctly the things they were considering, judged it simpler and more intelligible than their surfaces and their lines. So it appears from the fact that they explained the line by the motion of a point, and the surface by that of a line.

The philosophers also suppose several motions that they think can be accomplished without any body's changing place, such as those they call *motus ad formam, motus ad calorem, motus ad quantitatem* ("motion to form," "motion to heat," "motion to quantity"), and myriad others. As for me, I conceive of none except that which is easier to conceive of than the lines of mathematicians: the motion by which bodies pass from one place to another and successively occupy all the spaces in between.

Beyond that, the philosophers attribute to the least of these motions a being much more solid and real than they do to rest, which they say is nothing but the privation of motion. As for me, I conceive of rest as being a quality also, which should be attributed to matter while it remains in one place, just as motion is a quality attributed to it while it is changing place.

Finally, the motion of which they speak is of such a strange nature that, whereas all other things have as a goal their perfection and strive only to preserve themselves, it has no other end and no other goal than rest. Contrary to all the laws of nature, it strives on its own to destroy itself. By contrast, the motion I suppose follows the same laws of nature as do generally all the dispositions and all the qualities found in matter, as well those which the scholars call *modos et entia rationis cum fundamento in re* (modes and beings of thought with foundation in the thing) as *qualitates reales* (their real qualities), in which I frankly confess I can find no more reality than in the others.

I suppose as a second rule that, when one of these bodies pushes another, it cannot give the other any motion except by losing as much of its own at the same time; nor can it take away from the other body's motion unless its own is increased by as much. This rule, joined to the preceding, agrees quite well with all experiences in which we see one body begin or cease to move because it is pushed or stopped by some other. For, having supposed the preceding rule, we are free from the difficulty in which the scholars find themselves when they want to explain why a stone continues to move for some time after being out of the hand of him who threw it. For one should ask instead, why does it not continue to move always? Yet the reason is easy to give. For who is there who can deny that the air in which it is moving offers it some resistance? One hears it whistle when it divides the air; and, if one moves in the air a fan or some other very light and very extended body, one will even be able to feel by the weight of one's hand that the air is impeding its motion, far from continuing it, as some have wanted to say. If, however, one fails to explain the effect of the air's resistance according to our second rule, and if one thinks that the more a body can resist the more it is capable of stopping the motion of others (as one can perhaps be persuaded at first), one will in turn have a great deal of trouble explaining why the motion of this stone is weakened more in colliding with a soft body of middling resistance than it is when it collides with a harder one that resists it more. Or also why, as soon as it has made a little effort against the latter, it spontaneously turns on its heels rather than stopping or interrupting the motion it has. Whereas, supposing this rule, there is no difficulty at all in this. For it teaches us that the motion of a body is not retarded by collision with another in proportion to how much the latter resists it, but only in proportion to how much the latter's resistance is surmounted, and to the extent that, in obeying the law, it receives into itself the force of motion that the former surrenders.

Now, even though in most of the motions we see in the true world we cannot perceive that the bodies that begin or cease to move are pushed or stopped by some others, we do not thereby have reason to judge that these two rules are not being observed exactly. For it is certain that those bodies can often receive their agitation from the two elements of air and fire, which are always found among them without being perceptible (as has just been said), or even from the grosser air, which also cannot be perceived. And they can transfer the agitation, sometimes to that grosser air and sometimes to the whole mass of the earth; dispersed therein, it also cannot be perceived.

But, even if all that our senses have ever experienced in the true world seemed manifestly contrary to what is contained in these two rules, the reasoning that has taught them to me seems to me so strong that I would not cease to believe myself obliged to suppose them in the new world I am describing to you. For what more firm and solid foundation could one find to establish a truth (even if one wanted to choose it at will) than to take the very firmity and immutability that is in God?

Now it is the case that those two rules manifestly follow from this alone: that God is immutable and that, acting always in the same way, He always produces the same effect. For, supposing that He placed a certain quantity of motions in all matter in general at the first instant He created it, one must either avow that He always conserves as many of them there or not believe that He always acts in the same way. Supposing in addition that, from that first instant, the diverse parts of matter, in which these motions are found unequally dispersed began to retain them or to transfer them from one to another according as they had the force to do, one must of necessity think that He causes them always to continue the same thing. And that is what those two rules contain.

I will add as a third rule that, when a body is moving, even if its motion most often takes place along a curved line and (as has been said above) can never take place along any line that is not in some way circular, nevertheless each of its individual parts tends always to continue its motion along a straight line. And thus their action, i.e. the inclination they have to move, is different from their motion.

For example, if a wheel is made to turn on its axle, even though its parts go around (because, being linked to one another, they cannot do otherwise), nevertheless their inclination is to go straight ahead, as appears clearly if perchance one of them is detached from the others. For, as soon as it is free, its motion ceases to be circular and continues in a straight line.

By the same token, when one whirls a stone in a sling, not only does it go straight out as soon as it leaves the sling, but in addition, throughout the time it is in the sling, it presses against the middle of the sling and causes the cord to stretch. It clearly shows thereby that it always has an inclination to go in a straight line and that it goes around only under constraint.

This rule rests on the same foundation as the two others and depends only on God's conserving everything by a continuous action and, consequently, on His conserving it not as it may have been some time earlier but precisely as it is at the same instant that He conserves it. Now it is the case that, of all motions, only the straight is entirely simple; its whole nature is understood in an instant. For, to conceive of it, it suffices to think that a body is in the act of moving in a certain direction, and that is the case in each instant that might be determined during the time that it is moving. By contrast, to conceive of circular motion, or of any other possible motion, one must consider at least two of its instants, or rather two of its parts, and the relation between them.

But, so that the philosophers (or rather the sophists) do not find occasion here to exercise their superfluous subtleties, note that I do not thereby say that rectilinear motion can take place in an instant; but only that all that is required to produce it is found in bodies in each instant that might be determined while they are moving, and not all that is required to produce circular motion.

For example, suppose a stone is moving in a sling along the circle marked AB and you consider it precisely as it is at the instant it arrives at point A: you will readily find that it is in the act of moving (for it does not stop there) and of moving in a certain direction (that is, toward C), for it is in that direction that its action is directed in that instant. But you can find nothing there that makes its motion circular. Thus, supposing that the stone then begins to leave tile sling and that God continues to preserve it as it is at that moment, it is certain that He will not preserve it with the inclination to go circularly along the line AB, but with the inclination to go straight ahead toward point C.

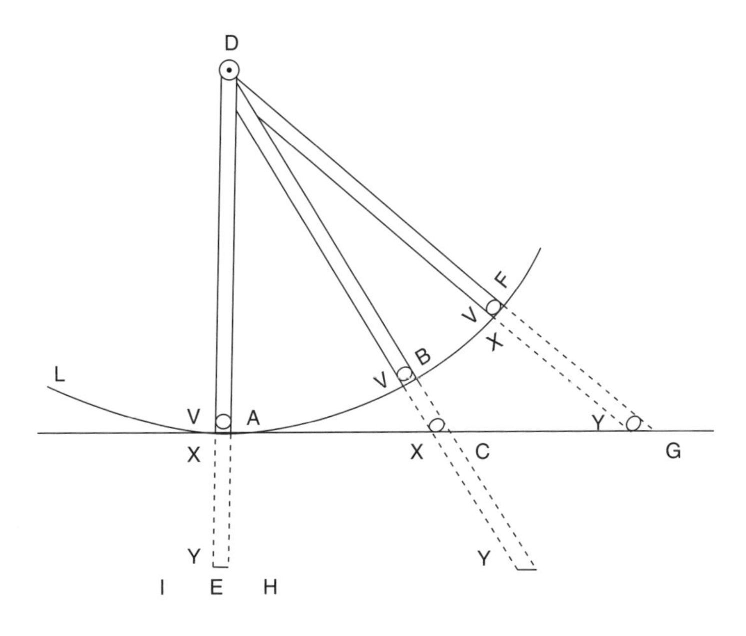

According to this rule, then, one must say that God alone is the author of all the motions in the world, insofar as they exist and insofar as they are straight, but that it is the diverse dispositions of matter that render the motions irregular and curved. So the theologians teach us that God is also the author of all our actions, insofar as they exist and insofar as they have some goodness, but that it is the diverse dispositions of our wills that can render those actions evil.

I could set out here many additional rules for determining in detail when and how and by how much the motion of each body can be diverted and increased or decreased by colliding with others, something that comprises summarily all the effects of nature. But I shall be content with showing you that, besides the three laws that I have explained, I wish to suppose no others but those that most certainly follow from the eternal truths on which the mathematicians are wont to support their most certain and most evident demonstrations; the truths, I say, according to which God Himself has taught us He disposed all things in number, weight, and measure. The knowledge of those laws is so natural to our souls that we cannot but judge them infallible when we conceive them distinctly, nor doubt that, if God had created many worlds, the laws would be as true in all of them as in this one. Thus, those who can examine sufficiently the consequences of these truths and of our rules will be able to know effects by their causes and (to explain myself in the language of the School) will be able to have demonstrations a priori of everything that can be produced in that new world.

And so there will be no exception that impedes this, we will add, if you wish, to our suppositions that God will never mark any miracle in the new world and that the intelligences, or the rational souls, which we might hereafter suppose to be there, will in no way disturb the ordinary course of nature.

Nonetheless, in consequence of this, I do not promise you to set out here exact demonstrations of all the things I will say. It will be enough for me to open to you the path by which you will be able to find them yourselves, whenever you take the trouble to look for them. Most minds lose interest when one makes things too easy for them. And to compose here a setting that pleases you, I must employ shadow as well as bright colors. Thus I will be content to pursue the description I have begun, as if having no other design than to tell you a fable.

Translated by Michael S. Mahoney

Reading and Discussion Questions

1. Descartes is imagining a world created by God that operates according to laws. What laws does he envision? Does he think that God needs to intervene in this world after the laws have been established?
2. How does his second law or rule solve the Aristotelian problem of the continuance of projectile motion?
3. What is Descartes' stance on the value of information derived from the senses? How does that judgement compare with the confidence that he places in reason?

Discourse on the Method of Rightly Conducting the Reason and Seeking Truth in the Sciences (1637)

René Descartes

41

Part I

Good Sense is, of all things among men, the most equally distributed; for every one thinks himself so abundantly provided with it, that those even who are the most difficult to satisfy in everything else, do not usually desire a larger measure of this quality than they already possess. And in this it is not likely that all are mistaken: the conviction is rather to be held as testifying that the power of judging aright and of distinguishing Truth from Error, which is properly what is called Good Sense or Reason, is by nature equal in all men; and that the diversity of our opinions, consequently, does not arise from some being endowed with a larger share of Reason than others, but solely from this, that we conduct our thoughts along different ways, and do not fix our attention on the same objects. For to be possessed of a vigorous mind is not enough; the prime requisite is rightly to apply it. The greatest minds, as they are capable of the highest excellencies, are open likewise to the greatest aberrations; and those who travel very slowly may yet make far greater progress, provided they keep always to the straight road, than those who, while they run, forsake it.

For myself, I have never fancied my mind to be in any respect more perfect than those of the generality; on the contrary, I have often wished that I were equal to some others in promptitude of thought, or in clearness and distinctness of imagination, or in fullness and readiness of memory. And besides these, I know of no other qualities that contribute to the perfection of the mind; for as to the Reason or Sense, inasmuch as it is that alone which constitutes us men, and distinguishes us from the brutes, I am disposed to believe that it is to be found complete in each individual; and on this point to adopt the common opinion of philosophers, who say that the difference of greater and less holds only among the accidents, and not among the forms or natures of individuals of the same species.

I will not hesitate, however, to avow my belief that it "has been my singular good fortune to have very early in life fallen in with certain tracks which have conducted me to considerations and maxims, of which I have formed a Method that gives me the means, as I think, of gradually augmenting my knowledge, and of raising it by little and little to the highest point which the mediocrity of my talents and the brief duration of my life will permit me to reach. For I have already reaped from it such fruits, that, although I have been accustomed to think lowly enough of myself, and although when I look with the eye of a philosopher at the varied courses and pursuits of mankind at large, I find scarcely one which does not appear vain and useless, I nevertheless derive the highest satisfaction from the progress I conceive myself to have already made in the search after truth, and cannot help entertaining such expectations of the future as to believe that if, among the occupations of men as men, there is any one really excellent and important, it is that which I have chosen.

...

Part II

...

But like one walking alone and in the dark, I resolved to proceed so slowly and with such circumspection, that if I did not advance far, I would at least guard against falling. I did not even choose to dismiss summarily any of the opinions that had crept into my belief without having been introduced by Reason, but first of all took sufficient time carefully to satisfy myself of the general nature of the task I was setting myself, and ascertain the true Method by which to arrive at the knowledge of whatever lay within the compass of my powers.

Among the branches of Philosophy, I had, at an earlier period, given some attention to Logic, and among those of the Mathematics to Geometrical Analysis and Algebra,—three Arts or Sciences which ought, as I conceived, to contribute something to my design. But, on examination, I found that, as for Logic, its syllogisms and the majority of its other precepts are of avail rather in the communication of what we already know, or even as the Art of Lully, in speaking without judgment of things of which we are ignorant, than in the investigation of the unknown; and although this Science contains indeed a number of correct and very excellent precepts, there are, nevertheless, so many others, and these either injurious or superfluous, mingled with the former, that it is almost quite as difficult to effect a severance of the true from the false as it is to extract a Diana or a Minerva from a rough block of marble. Then as to the Analysis of the ancients and the Algebra of the moderns, besides that they embrace only matters highly abstract, and, to appearance, of no use, the former is so exclusively restricted to the consideration of figures, that it can exercise the Understanding only on condition of greatly fatiguing the Imagination; and, in the latter, there is so complete a subjection to certain rules and formulas, that there results an art full of confusion and obscurity calculated to embarrass, instead of a science fitted to cultivate the mind. By these considerations I was induced to seek some other Method which would comprise the advantages of the three and be exempt from their defects. And as a multitude of laws often only hampers justice, so that a state is best governed when, with few laws, these are rigidly administered; in like manner, instead of the great number of precepts of which Logic is composed, I believed that the four following would prove perfectly sufficient for me, provided I took the firm and unwavering resolution never in a single instance to fail in observing them.

The first was never to accept anything for true which I did not clearly know to be such; that is to say, carefully to avoid precipitancy and prejudice, and to comprise nothing more in my judgment than what was presented to my mind so clearly and distinctly as to exclude all ground of doubt.

The second, to divide each of the difficulties under examination into as many parts as possible, and as might be necessary for its adequate solution.

The third, to conduct my thoughts in such order that, by commencing with objects the simplest and easiest to know, I might ascend by little and little, and, as it were, step by step, to the knowledge of the more complex; assigning in thought a certain order even to those objects which in their own nature do not stand in a relation of antecedence and sequence.

At the last, in every case to make enumerations so complete, and reviews so general, that I might be assured that nothing was omitted.

The long chains of simple and easy reasonings by means of which geometers are accustomed to reach the conclusions of their most difficult demonstrations, had led me to imagine that all things, to the knowledge of which man is competent, are mutually connected in the same way, and that there is nothing so far removed from us as to be beyond our reach, or so hidden that we cannot discover it, provided only we abstain from accepting the false for the true, and always preserve in our thoughts the order necessary for the deduction of one truth from another. And I had little difficulty in determining the objects with which it was necessary to commence, for I was already persuaded that it must be with the simplest and easiest to know, and considering that of all those who have hitherto sought truth in the Sciences, the mathematicians alone have been able to find any demonstrations, that is, any certain and evident reasons, I did not doubt but that such must have been the rule of their investigations. I resolved to commence, therefore, with the examination of the simplest objects, not anticipating, however, from this any other advantage than that to be found in accustoming my mind to the love and nourishment of truth, and to a distaste for all such reasonings as were unsound. But I had no intention on that account of attempting to master all the particular Sciences commonly denominated Mathematics: but observing that, however different their objects, they all agree in considering only the various relations or proportions subsisting among those objects, I thought it best for my purpose to consider these proportions in the most general form possible, without referring them to any objects in particular, except such as would most facilitate the knowledge of them, and without by any means restricting them to these, that afterward I might thus be the better able to apply them to every other class of objects to which they are legitimately applicable....

Part IV

I am in doubt as to the propriety of making my first meditations, in the place above mentioned, matter of discourse; for these are so metaphysical, and so uncommon, as not, perhaps, to be acceptable to everyone. And yet, that it may be determined whether the foundations that I have laid are sufficiently secure, I find myself in a measure constrained to advert to them. I had long before remarked that, in relation to practice, it is sometimes necessary to adopt, as if above doubt, opinions which we discern to be highly uncertain, as has been already said; but as I then desired to give my attention solely to the search after truth, I thought that a procedure exactly the opposite was called for, and that I ought to reject as absolutely false all opinions in regard to which I could suppose the least ground for doubt, in order to ascertain whether after that there remained aught in my belief that was wholly indubitable. Accordingly, seeing that our senses sometimes deceive us, I was willing to suppose that there existed nothing really such as they presented to us; and because some men err in reasoning, and fall into paralogisms, even on the simplest matters of Geometry, I, convinced that I was as open to error as any other, rejected as false all the reasonings I had hitherto taken for demonstrations; and finally, when I considered that the very same thoughts (presentations) which we experience when awake may also be experienced when we are asleep, while there is at that time not one of them true, I supposed that all the objects (presentations) that had ever entered into my mind when awake, had in them no more truth than the illusions of my dreams. But immediately upon this I observed that, whilst I thus wished to think that all was false, it was absolutely necessary that I, who thus thought, should be somewhat; and as I observed that this truth, I think, hence I am, was so certain and of such evidence, that no ground of doubt, however extravagant, could be alleged by the Sceptics capable of shaking it, I concluded that I might, without scruple, accept it as the first principle of the Philosophy of which I was in search.

In the next place, I attentively examined what I was, and as I observed that I could suppose that I had no body, and that there was no world nor any place in which I might be; but that I could not therefore suppose that I was not; and that, on the contrary, from the very circumstance that I thought to doubt of the truth of all things, it most clearly and certainly followed that I was; while, on the other hand, if I had only ceased to think, although all the other objects which I had ever imagined had been in reality existent, I would have had no reason to believe that I existed; I thence concluded that I was a substance whose whole essence or nature consists only in thinking, and which, that it may exist, has need of no place, nor is dependent on any material thing; so that "I" that is to say, the mind by which I am what I am, is wholly distinct from the body, and is even more easily known than the, latter, and is such, that although the latter were not, it would still continue to be all that it is.

After this I inquired in general into what is essential to the truth and certainty of a proposition; for since I had discovered one which I knew to be true, I thought that I must likewise be able to discover the ground of this certitude. And as I observed that in the words I think, hence I am, there is nothing at all which gives me assurance of their truth beyond this, that I see very clearly that in order to think it is necessary to exist, I concluded that I might take, as a general rule, the principle, that all the things which we very clearly and distinctly conceive are true, only observing, however, that there is some difficulty in rightly determining the objects which we distinctly conceive.

...

Part VI

...

But since I designed to employ my whole life in the search after so necessary a Science, and since I had fallen in with a path which seems to me such, that if any one follow it he must inevitably reach the end desired, unless he be hindered either by the shortness of life or the want of experiments, I judged that there could be no more effectual provision against these two impediments than if I were faithfully to communicate to the public all the little I might myself have found, and incite men of superior genius to strive to proceed farther, by contributing, each according to his inclination and ability, to the experiments which it would be necessary to make, and also by informing the public of all they might discover, so that, by the last beginning where those before them had left off, and thus connecting the lives and labors of many, we might collectively proceed much farther than each by himself could do.

I remarked, moreover, with respect to experiments, that they become always more necessary the more one is advanced in knowledge; for, at the commencement, it is better to make use only of what is spontaneously presented to our senses, and of which we cannot remain ignorant, provided we bestow on it any reflection, however slight, than to concern ourselves about more uncommon and recondite phenomena: the reason of which is, that the more uncommon often only mislead us so long as the causes of the more ordinary are still unknown; and the circumstances upon which they depend are almost always so special and minute as to be highly difficult to detect. But in this I have adopted the following order: first, I have essayed to find in general the principles, or first causes of all that is or can be in the world, without taking into consideration for this end anything but God himself who has created it, and without educing them from any other source than from certain germs of truths naturally existing in our minds. In the second place, I examined what were the first and most ordinary effects that could be deduced from these causes; and it appears to me that, in this way, I have found heavens, stars, and earth, and even on the earth, water, air, fire, minerals, and some other things of this kind, which of all others are the most common and simple, and hence the easiest to know. Afterward, when I wished to descend to the more particular, so many diverse objects presented themselves to me, that I believed it to be impossible for the human mind to distinguish the forms or species of bodies that are upon the earth, from an infinity of others which might have been, if it had pleased God to place them there, or consequently to apply them to our use, unless we rise to causes through their effects, and avail ourselves of many particular experiments. Thereupon, turning over in my mind all the objects that had ever been presented to my senses, I freely venture to state that I have never observed any which I could not satisfactorily explain by the principles I had discovered. But it is necessary also to confess that the power of nature is so ample and vast, and these principles so simple and general, that I have hardly observed a single particular effect which I cannot at once recognize as capable of being deduced in many different modes from the principles, and that my greatest difficulty usually is to discover in which of these modes the effect is dependent upon them; for out of this difficulty I cannot otherwise extricate myself than by again seeking certain experiments, which may be such that their result is not the same, if it is in the one of these modes that we must explain it, as it would be if it were to be explained in the other. As to what remains, I am now in a position to discern, as I think, with sufficient clearness what course must be taken to make the majority of those experiments which may conduce to this end; but I perceive likewise that they are such and so numerous, that neither my hands nor my income, though it were a thousand times larger than it is, would be sufficient for them all; so that, according as henceforward I shall have the means of making more or fewer experiments, I shall in the same proportion make greater or less progress in the knowledge of nature. This was what I had hoped to make known by the Treatise I had written, and so clearly to exhibit the advantage that would thence accrue to the public, as to induce all who have the common good of man at heart, that is, all who are virtuous in truth, and not merely in appearance, or according to opinion, as well to communicate to me the experiments they had already made, as to assist me in those that remain to be made.

Translated by John Veitch

Reading and Discussion Questions

1. Describe and explain the four methodological rules of reasoning that Descartes advances.
2. What attitudes does Descartes express toward logic, geometry, and algebra, and how do these relate to his rules?
3. Does Descartes think that the first principles will be difficult to arrive at?

Principles of Philosophy (1644)

René Descartes

42

Letter of the Author to the French Translator of the *Principles of Philosophy* Serving for a Preface

Sir:—

The version of my Principles which you have been at pains to make, is so elegant and finished as to lead me to expect that the work will be more generally read in French than in Latin, and better understood. The only apprehension I entertain is lest the title should deter some who have not been brought up to letters, or with whom philosophy is in bad repute, because the kind they were taught has proved unsatisfactory; and this makes me think that it will be useful to add a preface to it for the purpose of showing what the matter of the work is, what end I had in view in writing it, and what utility may be derived from it....

After this, that I might lead men to understand the real design I had in publishing them, I should have wished here to explain the order which it seems to me one ought to follow with the view of instructing himself. In the first place, a man who has merely the vulgar and imperfect knowledge which can be acquired by the four means, above explained, ought, before all else, to endeavor to form for himself a code of morals sufficient to regulate the actions of his life, as well for the reason that this does not admit of delay as because it ought to be our first care to live well. In the next place, he ought to study Logic, not that of the schools, for it is only, properly speaking, a dialectic which teaches the mode of expounding to others what we already know, or even of speaking much, without judgment, of what we do not know, by which means it corrupts rather than increases good sense—but the logic which teaches the right conduct of the reason with the view of discovering the truths of which we are ignorant; and, because it greatly depends on usage, it is desirable he should exercise himself for a length of time in practicing its rules on easy and simple questions, as those of the mathematics. Then, when he has acquired some skill in discovering the truth in these questions, he should commence to apply himself in earnest to true philosophy, of which the first part is Metaphysics, containing the principles of knowledge, among which is the explication of the principal attributes of God, of the immateriality of the soul, and of all the clear and simple notions that are in us; the second is Physics, in which, after finding the true principles of material things, we examine, in general, how the whole universe has been framed; in the next place, we consider, in particular, the nature of the earth, and of all the bodies that are most generally found upon it, as air, water, fire, the loadstone and other minerals. In the next place, it is necessary also to examine singly the nature of plants, of animals, and above all of man, in order that we may thereafter be able to discover the other sciences that are useful to us. Thus, all Philosophy is like a tree, of which Metaphysics is the root, Physics the trunk, and all the other sciences the branches that grow out of this trunk, which are reduced to three principal, namely, Medicine, Mechanics, and Ethics. By the science of Morals, I understand the highest and most perfect which, presupposing an entire knowledge of the other sciences, is the last degree of wisdom.

But as it is not from the roots or the trunks of trees that we gather the fruit, but only from the extremities of their branches, so the principal utility of philosophy depends on the separate uses of its parts, which we can only learn last of all. But, though I am ignorant of almost all these, the zeal I have always felt in endeavoring to be of service to the public, was the reason why I published, some ten or twelve years ago, certain Essays on the doctrines I thought I had acquired. The first part of these Essays was a "Discourse on the Method of rightly conducting the Reason, and seeking Truth in the Sciences," in which I gave a summary of the principal rules of logic, and also of an imperfect ethic, which a person may follow provisionally so long as he does not know any better. The other parts were three treatises: the first of Dioptrics, the second of Meteors, and the third of Geometry. In the Dioptrics, I designed to show that we might proceed far enough in philosophy as to arrive, by its means, at the knowledge of the arts that are useful to life, because the invention of the telescope, of which I there gave an explanation, is one of the most difficult that has ever been made. In the treatise of Meteors, I desired to exhibit the difference that subsists between the philosophy I cultivate and that taught in the schools, in which the same matters are usually discussed. In fine, in the Geometry, I professed to demonstrate that I had discovered many things that were before unknown, and thus afford ground for believing that we may still discover many others, with the view of thus stimulating all to the investigation of truth. Since that period, anticipating the difficulty which many would experience in apprehending the foundations of the Metaphysics, I endeavored to explain the chief points of them in a book of Meditations, which is not in itself large, but the size of which has been increased, and the matter greatly illustrated, by the Objections which several very learned persons sent to me on occasion of it, and by the Replies which I made to them. At length, after it appeared to me that those preceding treatises had sufficiently prepared the minds of my readers for the *Principles of Philosophy*, I also published it; and I have divided this work into four parts, the first of which contains the principles of human knowledge, and which may be called the First Philosophy, or Metaphysics. That this part, accordingly, may be properly understood, it will be necessary to read beforehand the book of Meditations I wrote on the same subject. The other three parts contain all that is most general in Physics, namely, the explication of the first laws or principles of nature, and the way in which the heavens, the fixed stars, the planets, comets, and generally the whole universe, were composed; in the next place, the explication, in particular, of the nature of this earth, the air, water, fire, the magnet, which are the bodies we most commonly find everywhere around it, and of all the qualities we observe in these bodies, as light, heat, gravity, and the like. In this way, it seems to me, I have commenced the orderly explanation of the whole of philosophy, without omitting any of the matters that ought to precede the last which I discussed.

But to bring this undertaking to its conclusion, I ought hereafter to explain, in the same manner, the nature of the other more particular bodies that are on the earth, namely, minerals, plants, animals, and especially man; finally to treat thereafter with accuracy of Medicine, Ethics, and Mechanics. I should require to do this in order to give to the world a complete body of philosophy; and I do not yet feel myself so old, I do not so much distrust my strength, nor do I find myself so far removed from the knowledge of what remains, as that I should not dare to undertake to complete this design, provided I were in a position to make all the experiments which I should require for the basis and verification of my reasonings. But seeing that would demand a great expenditure, to which the resources of a private individual like myself would not be adequate, unless aided by the public, and as I have no ground to expect this aid, I believe that I ought for the future to content myself with studying for my own instruction, and posterity will excuse me if I fail hereafter to labor for them.

. . .

I well know, likewise, that many ages may elapse ere all the truths deducible from these principles are evolved out of them, as well because the greater number of such as remain to be discovered depend on certain particular experiments that never occur by chance, but which require to be investigated with care and expense by men of the highest intelligence, as because it will hardly happen that the same persons who have the sagacity to make a right use of them, will possess also the means of making them, and also because the majority of the best minds have formed so low an estimate of philosophy in general, from the imperfections they have remarked in the kind in vogue up to the present time, that they cannot apply themselves to the search after truth.

But in conclusion, if the difference discernible between the principles in question and those of every other system, and the great array of truths deducible from them, lead them to discern the importance of continuing the search after these truths, and to observe the degree of wisdom, the perfection and felicity of life, to which they are fitted to conduct us, I venture to believe that there will not be found one who is not ready to labor hard in so profitable a study, or at least to favor and aid with all his might those who shall devote themselves to it with success.

The height of my wishes is, that posterity may sometime behold the happy issue of it, etc.

The Principles of Philosophy

Part I: Of the Principles of Human Knowledge

I. That in order to seek truth, it is necessary once in the course of our life, to doubt, as far as possible, of all things.

As we were at one time children, and as we formed various judgments regarding the objects presented to our senses, when as yet we had not the entire use of our reason, numerous prejudices stand in the way of our arriving at the knowledge of truth; and of these it seems impossible for us to rid ourselves, unless we undertake, once in our lifetime, to doubt all of those things in which we may discover even the smallest suspicion of uncertainty.

II. That we ought also to consider as false all that is doubtful.

Moreover, it will be useful likewise to esteem as false the things of which we shall be able to doubt, that we may with greater clearness discover what possesses most certainty and is the easiest to know.

III. That we ought not meanwhile to make use of doubt in the conduct of life.

In the meantime, it is to be observed that we are to avail ourselves of this general doubt only while engaged in the contemplation of truth. For, as far as concerns the conduct of life, we are very frequently obliged to follow opinions merely probable, or even sometimes, though of two courses of action we may not perceive more probability in the one than in the other, to choose one or other, seeing the opportunity of acting would not infrequently pass away before we could free ourselves from our doubts.

IV. Why we may doubt of sensible things.

Accordingly, since we now only design to apply ourselves to the investigation of truth, we will doubt, first, whether of all the things that have ever fallen under our senses, or which we have ever imagined, any one really exist; in the first place, because we know by experience that the senses sometimes err, and it would be imprudent to trust too much to what has even once deceived us; secondly, because in dreams we perpetually seem to perceive or imagine innumerable objects which have no existence. And to one who has thus resolved upon a general doubt, there appear no marks by which he can with certainty distinguish sleep from the waking state.

V. Why we may also doubt of mathematical demonstrations.

We will also doubt of the other things we have before held as most certain, even of the demonstrations of mathematics, and of their principles which we have hitherto deemed self-evident; in the first place, because we have sometimes seen men fall into error in such matters, and admit as absolutely certain and self-evident what to us appeared false, but chiefly because we have learned that God who created us is all-powerful; for we do not yet know whether perhaps it was his will to create us so that we are always deceived, even in the things we think we know best; since this does not appear more impossible than our being occasionally deceived, which, however, as observation teaches us, is the case. And if we suppose that an all-powerful God is not the author of our being, and that we exist of ourselves or by some other means, still, the less powerful we suppose our author to be, the greater reason will we have for believing that we are not so perfect as that we may not be continually deceived.

VI. That we possess a free will, by which we can withhold our assent from what is doubtful, and thus avoid error.

But meanwhile, whoever in the end may be the author of our being, and however powerful and deceitful he may be, we are nevertheless conscious of a freedom, by which we can refrain from admitting to a place in our belief aught that is not manifestly certain and undoubted, and thus guard against ever being deceived.

VII. That we cannot doubt of our existence while we doubt, and that this is the first knowledge we acquire when we philosophize in order.

While we thus reject all of which we can entertain the smallest doubt, and even imagine that it is false, we easily indeed suppose that there is neither God, nor sky, nor bodies, and that we ourselves even have neither hands nor feet, nor, finally, a body; but we cannot in the same way suppose that we are not while we doubt of the truth of these things; for there is a repugnance in conceiving that what thinks does not exist at the very time when it thinks. Accordingly, the knowledge, I think, therefore I am, is the first and most certain that occurs to one who philosophizes orderly.

VIII. That we hence discover the distinction between the mind and the body, or between a thinking and corporeal thing.

And this is the best mode of discovering the nature of the mind, and its distinctness from the body: for examining what we are, while supposing, as we now do, that there is nothing really existing apart from our thought, we clearly perceive that neither extension, nor figure, nor local motion, nor anything similar that can be attributed to body, pertains to our nature, and nothing save thought alone; and, consequently, that the notion we have of our mind precedes that of any corporeal thing, and is more certain, seeing we still doubt whether there is any body in existence, while we already perceive that we think.

. . .

LI. What substance is, and that the term is not applicable to God and the creatures in the same sense.

But with regard to what we consider as things or the modes of things, it is worth while to examine each of them by itself. By substance we can conceive nothing else than a thing which exists in such a way as to stand in need of nothing beyond itself in order to its existence. And in truth, there can be conceived but one substance which is absolutely independent, and that is God. We perceive that all other things can exist only by help of the concourse of God. And, accordingly, the term substance does not apply to God and the creatures univocally, to adopt a term familiar in the schools; that is, no signification of this word can be distinctly understood which is common to God and them.

LII. That the term is applicable univocally to the mind and the body, and how substance itself is known.

Created substances, however, whether corporeal or thinking, may be conceived under this common concept; for these are things which, in order to their existence, stand in need of nothing but the concourse of God. But yet substance cannot be first discovered merely from its being a thing which exists independently, for existence by itself is not observed by us. We easily, however, discover substance itself from any attribute of it, by this common notion, that of nothing there are no attributes, properties, or qualities; for, from perceiving that some attribute is present, we infer that some existing thing or substance to which it may be attributed is also of necessity present.

LIII. That of every substance there is one principal attribute, as thinking of the mind, extension of the body. But, although any attribute is sufficient to lead us to the knowledge of substance, there is, however, one principal property of every substance, which constitutes its nature or essence, and upon which all the others depend. Thus, extension in length, breadth, and depth, constitutes the nature of corporeal substance; and thought the nature of thinking substance. For every other thing that can be attributed to body, presupposes extension, and is only some mode of an extended thing; as all the properties we discover in the mind are only diverse modes of thinking. Thus, for example, we cannot conceive figure unless in something extended, nor motion unless in extended space, nor imagination, sensation, or will, unless in a thinking thing. But, on the other hand, we can conceive extension without figure or motion, and thought without imagination or sensation, and so of the others; as is clear to any one who attends to these matters.

...

Part II: Of the Principles of Material Things

I. The grounds on which the existence of material things may be known with certainty.

Although we are all sufficiently persuaded of the existence of material things, yet, since this was before called in question by us, and since we reckoned the persuasion of their existence as among the prejudices of our childhood, it is now necessary for us to investigate the grounds on which this truth may be known with certainty. In the first place, then, it cannot be doubted that every perception we have comes to us from some object different from our mind; for it is not in our power to cause ourselves to experience one perception rather than another, the perception being entirely dependent on the object which affects our senses. It may, indeed, be matter of inquiry whether that object be God, or something different from God; but because we perceive, or rather, stimulated by sense, clearly and distinctly apprehend, certain matter extended in length, breadth, and thickness, the various parts of which have different figures and motions, and give rise to the sensations we have of colors, smells, pain, etc., God would, without question, deserve to be regarded as a deceiver, if he directly and of himself presented to our mind the idea of this extended matter, or merely caused it to be presented to us by some object which possessed neither extension, figure, nor motion. For we clearly conceive this matter as entirely distinct from God, and from ourselves, or our mind; and appear even clearly to discern that the idea of it is formed in us on occasion of objects existing out of our minds, to which it is in every respect similar. But since God cannot deceive us, for this is repugnant to his nature, as has been already remarked, we must unhesitatingly conclude that there exists a certain object extended in length, breadth, and thickness, and possessing all those properties which we clearly apprehend to belong to what is extended. And this extended substance is what we call body or matter.

. . .

XXIV. What motion is, taking the term in its common use.

But motion (viz, local, for I can conceive no other kind of motion, and therefore I do not think we ought to suppose there is any other in nature), in the ordinary sense of the term, is nothing more than the action by which a body passes from one place to another. And just as we have remarked above that the same thing may be said to change and not to change place at the same time, so also we may say that the same thing is at the same time moved and not moved. Thus, for example, a person seated in a vessel which is setting sail, thinks he is in motion if he looks to the shore that he has left, and consider it as fixed; but not if he regard the ship itself, among the parts of which he preserves always the same situation. Moreover, because we are accustomed to suppose that there is no motion without action, and that in rest there is the cessation of action, the person thus seated is more properly said to be at rest than in motion, seeing he is not conscious of being in action.

XXV. What motion is properly *so* called.

But if, instead of occupying ourselves with that which has no foundation, unless in ordinary usage, we desire to know what ought to be understood by motion according to the truth of the thing, we may say, in order to give it a determinate nature, that it is the transporting of one part of matter or of one body from the vicinity of those bodies that are in immediate contact with it, or which we regard as at rest, to the vicinity of other bodies. By a body as a part of matter, I understand all that which is transferred together, although it be perhaps composed of several parts, which in themselves have other motions; and I say that it is the transporting and not the force or action which transports, with the view of showing that motion is always in the movable thing, not in that which moves; for it seems to me that we are not accustomed to distinguish these two things with sufficient accuracy. Further, I understand that it is a mode of the movable thing, and not a substance, just as figure is a property of the thing figured, and repose of that which is at rest.

...

Part IV: Of the Earth

...

CXCVIII. That by our senses we know nothing of external objects beyond their figure [or situation], magnitude, and motion. Besides, we observe no such difference between the nerves as to lead us to judge that one set of them convey to the brain from the organs of the external senses anything different from another, or that anything at all reaches the brain besides the local motion of the nerves themselves. And we see that local motion alone causes in us not only the sensation of titillation and of pain, but also of light and sounds. For if we receive a blow on the eye of sufficient force to cause the vibration of the stroke to reach the retina, we see numerous sparks of fire, which, nevertheless, are not out of our eye; and when we stop our ear with our finger, we hear a humming sound, the cause of which can only proceed from the agitation of the air that is shut up within it. Finally, we frequently observe that heat [hardness, weight], and the other sensible qualities, as far as they are in objects, and also the forms of those bodies that are purely material, as, for example, the forms of fire, are produced in them by the motion of certain other bodies, and that these in their turn likewise produce other motions in other bodies. And we can easily conceive how the motion of one body may be caused by that of another, and diversified by the size, figure, and situation of its parts, but we are wholly unable to conceive how these same things (viz, size, figure, and motion), can produce something else of a nature entirely different from themselves, as, for example, those substantial forms and real qualities which many philosophers suppose to be in bodies; nor likewise can we conceive how these qualities or forms possess force to cause motions in other bodies. But since we know, from the nature of our soul, that the diverse motions of body are sufficient to produce in it all the sensations which it has, and since we learn from experience that several of its sensations are in reality caused by such motions, while we do not discover that anything besides these motions ever passes from the organs of the external senses to the brain, we have reason to conclude that we in no way likewise apprehend that in external objects, which we call light, color, smell, taste, sound, heat or cold, and the other tactile qualities, or that which we call their substantial forms, unless as the various dispositions of these objects which have the power of moving our nerves in various ways.

CXCIX. That there is no phenomenon of nature whose explanation has been omitted in this treatise.

And thus it may be gathered, from an enumeration that is easily made, that there is no phenomenon of nature whose explanation has been omitted in this treatise; for beyond what is perceived by the senses, there is nothing that can be considered a phenomenon of nature. But leaving out of account, motion, magnitude, figure [and the situation of the parts of each body], which I have explained as they exist in body, we perceive nothing out of us by our senses except light, colors, smells, tastes, sounds, and the tactile qualities; and these I have recently shown to be nothing more, at least so far as they are known to us, than certain dispositions of the objects, consisting in magnitude, figure, and motion.

CC. That this treatise contains no principles which are not universally received; and that this philosophy is not new, but of all others the most ancient and common. But I am desirous also that it should be observed that, though I have here endeavored to give an explanation of the whole nature of material things, I have nevertheless made use of no principle which was not received and approved by Aristotle, and by the other philosophers of all ages; so that this philosophy, so far from being new, is of all others the most ancient and common: for I have in truth merely considered the figure, motion, and magnitude of bodies, and examined what must follow from their mutual concourse on the principles of mechanics, which are confirmed by certain and daily experience. But no one ever doubted that bodies are moved, and that they are of various sizes and figures, according to the diversity of which their motions also vary, and that from mutual collision those somewhat greater than others are divided into many smaller, and thus change figure. We have experience of the truth of this, not merely by a single sense, but by several, as touch, sight, and hearing: we also distinctly imagine and understand it. This cannot be said of any of the other things that fall under our senses, as colors, sounds, and the like; for each of these affects but one of our senses, and merely impresses upon our imagination a confused image of itself, affording our understanding no distinct knowledge of what it is.

CCI. That sensible bodies are composed of insensible particles. But I allow many particles in each body that are perceived by none of our senses, and this will not perhaps be approved of by those who take the senses for the measure of the knowable. [We greatly wrong human reason, however, as appears to me, if we suppose that it does not go beyond the eyesight]; for no one can doubt that there are bodies so small as not to be perceptible by any of our senses, provided he only consider what is each moment added to those bodies that are being increased little by little, and what is taken from those that are diminished in the same way. A tree increases daily, and it is impossible to conceive how it becomes greater than it was before, unless we at the same time conceive that some body is added to it. But who ever observed by the senses those small bodies that are in one day added to a tree while growing? Among the philosophers at least, those who hold that quantity is indefinitely divisible, ought to admit that in the division the parts may become so small as to be wholly imperceptible. And indeed it ought not to be a matter of surprise, that we are unable to perceive very minute bodies; for the nerves that must be moved by objects to cause perception are not themselves very minute, but are like small cords, being composed of a quantity of smaller fibers, and thus the most minute bodies are not capable of moving them. Nor do I think that any one who makes use of his reason will deny that we philosophize with much greater truth when we judge of what takes place in those small bodies which are imperceptible from their minuteness only, after the analogy of what we see occurring in those we do perceive [and in this way explain all that is in nature, as I have essayed to do in this treatise], than when we give an explanation of the same things by inventing I know not what novelties, that have no relation to the things we actually perceive [as first matter, substantial forms, and all that grand array of qualities which many are in the habit of supposing, each of which it is more difficult to comprehend than all that is professed to be explained by means of them].

CCII. That the philosophy of Democritus is not less different from ours than from the common.

But it may be said that Democritus also supposed certain corpuscles that were of various figures, sizes, and motions, from the heaping together and mutual concourse of which all sensible bodies arose; and, nevertheless, his mode of philosophizing is commonly rejected by all. To this I reply that the philosophy of Democritus was never rejected by any one, because he allowed the existence of bodies smaller than those we perceive, and attributed to them diverse sizes, figures, and motions, for no one can doubt that there are in reality such, as we have already shown; but it was rejected in the first place, because he supposed that these corpuscles were indivisible, on which ground I also reject it; in the second place, because he imagined there was a vacuum about them, which I show to be impossible; thirdly, because he attributed gravity to these bodies, of which I deny the existence in any body, in so far as a body is considered by itself, because it is a quality that depends on the relations of situation and motion which several bodies bear to each other; and, finally, because he has not explained in particular how all things arose from the concourse of corpuscles alone, or, if he gave this explanation with regard to a few of them, his whole reasoning was far from being coherent [or such as would warrant us in extending the same explanation to the whole of nature]. This, at least, is the verdict we must give regarding his philosophy, if we may judge of his opinions from what has been handed down to us in writing. I leave it to others to determine whether the philosophy I profess possesses a valid coherency [and whether on its principles we can make the requisite number of deductions; and, inasmuch as the consideration of figure, magnitude, and motion has been admitted by Aristotle and by all the others, as well as by Democritus, and since I reject all that the latter has supposed, with this single exception, while I reject generally all that has been supposed by the others, it is plain that this mode of philosophizing has no more affinity with that of Democritus than of any other particular sect].

CCIII. How we may arrive at the knowledge of the figures [magnitude], and motions of the insensible particles of bodies. But, since I assign determinate figures, magnitudes, and motions to the insensible particles of bodies, as if I had seen them, whereas I admit that they do not fall under the senses, some one will perhaps demand how I have come by my knowledge of them. [To this I reply, that I first considered in general all the clear and distinct notions of material things that are to be found in our understanding, and that, finding no others except those of figures, magnitudes, and motions, and of the rules according to which these three things can be diversified by each other, which rules are the principles of geometry and mechanics, I judged that all the knowledge man can have of nature must of necessity be drawn from this source; because all the other notions we have of sensible things, as confused and obscure, can be of no avail in affording us the knowledge of anything out of ourselves, but must serve rather to impede it] Thereupon, taking as my ground of inference the simplest and best known of the principles that have been implanted in our minds by nature, I considered the chief differences that could possibly subsist between the magnitudes, and figures, and situations of bodies insensible on account of their smallness alone, and what sensible effects could be produced by their various modes of coming into contact; and afterward, when I found like effects in the bodies that we perceive by our senses, I judged that they could have been thus produced, especially since no other mode of explaining them could be devised. And in this matter the example of several bodies made by art was of great service to me: for I recognize no difference between these and natural bodies beyond this, that the effects of machines depend for the most part on the agency of certain instruments, which, as they must bear some proportion to the hands of those who make them, are always so large that their figures and motions can be seen: in place of which, the effects of natural bodies almost always depend upon certain organs so minute as to escape our senses. And it is certain that all the rules of mechanics belong also to physics, of which it is a part or species [so that all that is artificial is withal natural]: for it is not less natural for a clock, made of the requisite number of wheels, to mark the hours, than for a tree, which has sprung from this or that seed, to produce the fruit peculiar to it. Accordingly, just as those who are familiar with automata, when they are informed of the use of a machine, and see some of its parts, easily infer from these the way in which the others, that are not seen by them, are made; so from considering the sensible effects and parts of natural bodies, I have essayed to determine the character of their causes and insensible parts.

CCIV. That, touching the things which our senses do not perceive, it is sufficient to explain how they can be [and that this is all that Aristotle has essayed].

But here some one will perhaps reply, that although I have supposed causes which could produce all natural objects, we ought not on this account to conclude that they were produced by these causes; for, just as the same artisan can make two clocks, which, though they both equally well indicate the time, and are not different in outward appearance, have nevertheless nothing resembling in the composition of their wheels; so doubtless the Supreme Maker of things has an infinity of diverse means at his disposal, by each of which he could have made all the things of this world to appear as we see them, without it being possible for the human mind to know which of all these means he chose to employ. I most freely concede this; and I believe that I have done all that was required, if the causes I have assigned are such that their effects accurately correspond to all the phenomena of nature, without determining whether it is by these or by others that they are actually produced. And it will be sufficient for the use of life to know the causes thus imagined, for medicine, mechanics, and in general all

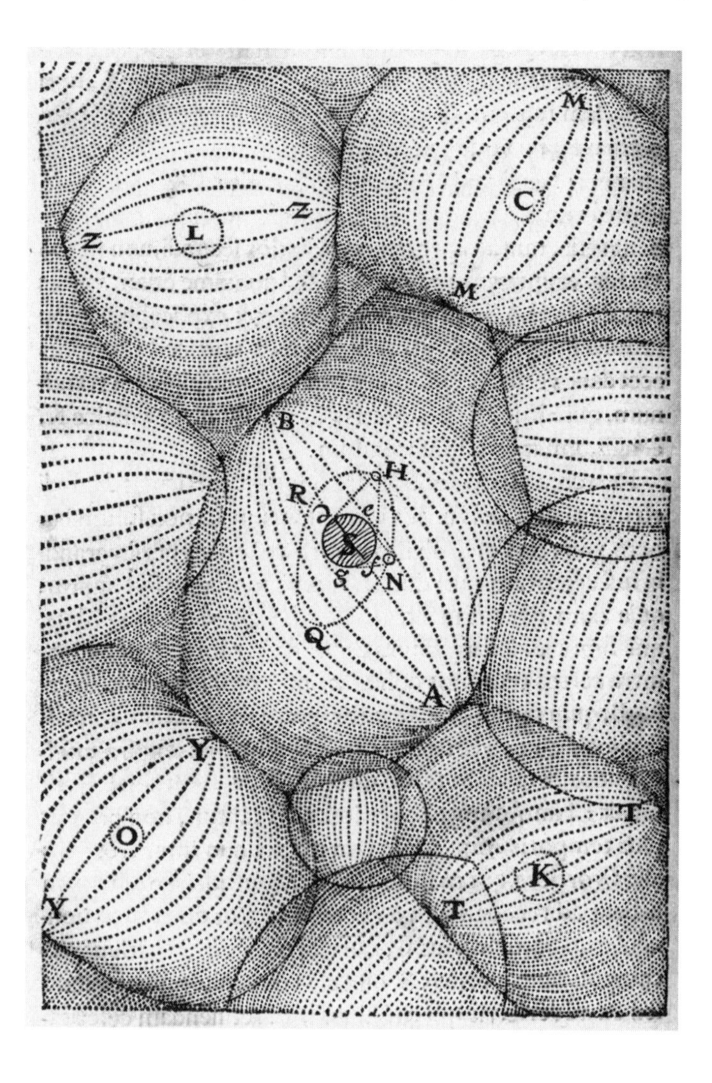

the arts to which the knowledge of physics is of service, have for their end only those effects that are sensible, and that are accordingly to be reckoned among the phenomena of nature. And lest it should be supposed that Aristotle did, or professed to do, anything more than this, it ought to be remembered that he himself expressly says, at the commencement of the seventh chapter of the first book of the Meteorologics, that, with regard to things which are not manifest to the senses, he thinks to adduce sufficient reasons and demonstrations of them, if he only shows that they may be such as he explains them.

Translated by John Veitch

Reading and Discussion Questions

1. How does Descartes use the metaphor of a tree to explain his view of the relationship between various disciplines?
2. What does Descartes mean by "substance"? How does Descartes' use compare to Aristotle's?
3. How many kinds of created substances does Descartes think that there are? What are their essential properties?
4. What arguments does Descartes provide to break down the traditional distinction between motion and rest?

The New Organon, or True Directions Concerning the Interpretation of Nature (1620)

Francis Bacon

43

Author's Preface

Those who have taken upon them to lay down the law of nature as a thing already searched out and understood, whether they have spoken in simple assurance or professional affectation, have therein done philosophy and the sciences great injury. For as they have been successful in inducing belief, so they have been effective in quenching and stopping inquiry; and have done more harm by spoiling and putting an end to other men's efforts than good by their own. Those on the other hand who have taken a contrary course, and asserted that absolutely nothing can be known whether it were from hatred of the ancient sophists, or from uncertainty and fluctuation of mind, or even from a kind of fullness of learning, that they fell upon this opinion have certainly advanced reasons for it that are not to be despised; but yet they have neither started from true principles nor rested in the just conclusion, zeal and affectation having carried them much too far. The more ancient of the Greeks (whose writings are lost) took up with better judgment a position between these two extremes between the presumption of pronouncing on everything, and the despair of comprehending anything; and though frequently and bitterly complaining of the difficulty of inquiry and the obscurity of things, and like impatient horses champing at the bit, they did not the less follow up their object and engage with nature, thinking (it seems) that this very question viz., whether or not anything can be known was to be settled not by arguing, but by trying. And yet they too, trusting entirely to the force of their understanding, applied no rule, but made everything turn upon hard thinking and perpetual working and exercise of the mind.

Now my method, though hard to practice, is easy to explain; and it is this. I propose to establish progressive stages of certainty. The evidence of the sense, helped and guarded by a certain process of correction, I retain. But the mental operation which follows the act of sense I for the most part reject; and instead of it I open and lay out a new and certain path for the mind to proceed in, starting directly from the simple sensuous perception.

. . .

Aphorisms

Book One

I Man, being the servant and interpreter of Nature, can do and understand so much and so much only as he has observed in fact or in thought of the course of nature. Beyond this he neither knows anything nor can do anything.

II Neither the naked hand nor the understanding left to itself can effect much. It is by instruments and helps that the work is done, which are as much wanted for the understanding as for the hand. And as the instruments of the hand either give motion or guide it, so the instruments of the mind supply either suggestions for the understanding or cautions.

III Human knowledge and human power meet in one; for where the cause is not known the effect cannot be produced. Nature to be commanded must be obeyed; and that which in contemplation is as the cause is in operation as the rule.

IV Toward the effecting of works, all that man can do is to put together or put asunder natural bodies. The rest is done by nature working within.

. . .

XXXVIII The idols and false notions which are now in possession of the human understanding, and have taken deep root therein, not only so beset men's minds that truth can hardly find entrance, but even after entrance is obtained, they will again in the very instauration of the sciences meet and trouble us, unless men being forewarned of the danger fortify themselves as far as may be against their assaults.

XXXIX There are four classes of Idols which beset men's minds. To these for distinction's sake I have assigned names, calling the first class *Idols of the Tribe;* the second, *Idols of the Cave;* the third, *Idols of the Market Place*; the fourth, *Idols of the Theater*.

XL The formation of ideas and axioms by true induction is no doubt the proper remedy to be applied for the keeping off and clearing away of idols. To point them out, however, is of great use; for the doctrine of Idols is to the interpretation of nature what the doctrine of the refutation of sophisms is to common logic.

XLI The Idols of the Tribe have their foundation in human nature itself, and in the tribe or race of men. For it is a false assertion that the sense of man is the measure of things. On the contrary, all perceptions as well of the sense as of the mind are according to the measure of the individual and not according to the measure of the universe. And the human understanding is like a false mirror, which, receiving rays irregularly, distorts and discolors the nature of things by mingling its own nature with it.

XLII The Idols of the Cave are the idols of the individual man. For everyone (besides the errors common to human nature in general) has a cave or den of his own, which refracts and discolors the light of nature, owing either to his own proper and peculiar nature; or to his education and conversation with others; or to the reading of books, and the authority of those whom he esteems and admires; or to the differences of impressions, accordingly as they take place in a mind preoccupied and predisposed or in a mind indifferent and settled; or the like. So that the spirit of man (according as it is meted out to different individuals) is in fact a thing variable and full of perturbation, and governed as it were by chance. Whence it was well observed by Heraclitus that men look for sciences in their own lesser worlds, and not in the greater or common world.

XLIII There are also Idols formed by the intercourse and association of men with each other, which I call Idols of the Market Place, on account of the commerce and consort of men there. For it is by discourse that men associate, and words are imposed according to the apprehension of the vulgar. And therefore the ill and unfit choice of words wonderfully obstructs the understanding. Nor do the definitions or explanations wherewith in some things learned men are wont to guard and defend themselves, by any means set the matter right. But words plainly force and overrule the understanding, and throw all into confusion, and lead men away into numberless empty controversies and idle fancies.

XLIV Lastly, there are Idols which have immigrated into men's minds from the various dogmas of philosophies, and also from wrong laws of demonstration. These I call Idols of the Theater, because in my judgment all the received systems are but so many stage plays, representing worlds of their own creation after an unreal and scenic fashion. Nor is it only of the systems now in vogue, or only of the ancient sects and philosophies, that I speak; for many more plays of the same kind may yet be composed and in like artificial manner set forth; seeing that errors the most widely different have nevertheless causes for the most part alike. Neither again do I mean this only of entire systems, but also of many principles and axioms in science, which by tradition, credulity, and negligence have come to be received.

But of these several kinds of Idols I must speak more largely and exactly, that the understanding may be duly cautioned.

XLV The human understanding is of its own nature prone to suppose the existence of more order and regularity in the world than it finds. And though there be many things in nature which are singular and unmatched, yet it devises for them parallels and conjugates and relatives which do not exist. Hence the fiction that all celestial bodies move in perfect circles, spirals and dragons being (except in name) utterly rejected. Hence too the element of fire with its orb is brought in, to make up the square with the other three which the sense perceives. Hence also the ratio of density of the so-called elements is arbitrarily fixed at ten to one. And so on of other dreams. And these fancies affect not dogmas only, but simple notions also.

. . .

LII Such then are the idols which I call *Idols of the Tribe*, and which take their rise either from the homogeneity of the substance of the human spirit, or from its preoccupation, or from its narrowness, or from its restless motion, or from an infusion of the affections, or from the incompetency of the senses, or from the mode of impression.

LIII The *Idols of the Cave* take their rise in the peculiar constitution, mental or bodily, of each individual; and also in education, habit, and accident. Of this kind there is a great number and variety. But I will instance those the pointing out of which contains the most important caution, and which have most effect in disturbing the clearness of the understanding.

LIV Men become attached to certain particular sciences and speculations, either because they fancy themselves the authors and inventors thereof, or because they have bestowed the greatest pains upon them and become most habituated to them. But men of this kind, if they betake themselves to philosophy and contemplation of a general character, distort and color them in obedience to their former fancies; a thing especially to be noticed in Aristotle, who made his natural philosophy a mere bond servant to his logic, thereby rendering it contentious and well-nigh useless. The race of chemists, again out of a few experiments of the furnace, have built up a fantastic philosophy, framed with reference to a few things; and Gilbert also, after he had employed himself most laboriously in the study and observation of the loadstone, proceeded at once to construct an entire system in accordance with his favorite subject.

. . .

LVIII Let such then be our provision and contemplative prudence for keeping off and dislodging the *Idols of the Cave*, which grow for the most part either out of the predominance of a favorite subject, or out of an excessive tendency to compare or to distinguish, or out of partiality for particular ages, or out of the largeness or minuteness of the objects contemplated. And generally let every student of nature take this as a rule: that whatever his mind seizes and dwells upon with peculiar satisfaction is to be held in suspicion, and that so much the more care is to be taken in dealing with such questions to keep the understanding even and clear.

LIX But the *Idols of the Market Place* are the most troublesome of all idols which have crept into the understanding through the alliances of words and names. For men believe that their reason governs words; but it is also true that words react on the understanding; and this it is that has rendered philosophy and the sciences sophistical and inactive. Now words, being commonly framed and applied according to the capacity of the vulgar, follow those lines of division which are most obvious to the vulgar understanding. And whenever an understanding of greater acuteness or a more diligent observation would alter those lines to suit the true divisions of nature, words stand in the way and resist the change. Whence it comes to pass that the high and formal discussions of learned men end oftentimes in disputes about words and names; with which (according to the use and wisdom of the mathematicians) it would be more prudent to begin, and so by means of definitions reduce them to order. Yet even definitions cannot cure this evil in dealing with natural and material things, since the definitions themselves consist of words, and those words beget others. So that it is necessary to recur to individual instances, and those in due series and order, as I shall say presently when I come to the method and scheme for the formation of notions and axioms.

LX The idols imposed by words on the understanding are of two kinds. They are either names of things which do not exist (for as there are things left unnamed through lack of observation, so likewise are there names which result from fantastic suppositions and to which nothing in reality corresponds), or they are names of things which exist, but yet confused and ill-defined, and hastily and irregularly derived from realities. Of the former kind are Fortune, the Prime Mover, Planetary Orbits, Element of Fire, and like fictions which owe their origin to false and idle theories. And this class of idols is more easily expelled, because to get rid of them it is only necessary that all theories should be steadily rejected and dismissed as obsolete.

But the other class, which springs out of a faulty and unskillful abstraction, is intricate and deeply rooted. Let us take for example such a word as *humid* and see how far the several things which the word is used to signify agree with each other, and we shall find the word *humid* to be nothing else than a mark loosely and confusedly applied to denote a variety of actions which will not bear to be reduced to any constant meaning. For it both signifies that which easily spreads itself round any other body; and that which in itself is indeterminate and cannot solidize; and that which readily yields in every direction; and that which easily divides and scatters itself; and that which easily unites and collects itself; and that which readily flows and is put in motion; and that which readily clings to another body and wets it; and that which is easily reduced to a liquid, or being solid easily melts. Accordingly, when you come to apply the word, if you take it in one sense, flame is humid; if in another, air is not humid; if in another, fine dust is humid; if in another, glass is humid. So that it is easy to see that the notion is taken by abstraction only from water and common and ordinary liquids, without any due verification.

There are, however, in words certain degrees of distortion and error. One of the least faulty kinds is that of names of substances, especially of lowest species and well-deduced (for the notion of *chalk* and of *mud* is good, of *earth* bad); a more faulty kind is that of actions, as *to generate, to corrupt, to alter*; the most faulty is of qualities (except such as are the immediate objects of the sense) as *heavy, light, rare, dense*, and the like. Yet in all these cases some notions are of necessity a little better than others, in proportion to the greater variety of subjects that fall within the range of the human sense.

LXI But the *Idols of the Theater* are not innate, nor do they steal into the understanding secretly, but are plainly impressed and received into the mind from the playbooks of philosophical systems and the perverted rules of demonstration. To attempt refutations in this case would be merely inconsistent with what I have already said, for since we agree neither upon principles nor upon demonstrations there is no place for argument. And this is so far well, inasmuch as it leaves the honor of the ancients untouched. For they are no wise disparaged—the question between them and me being only as to the way. For as the saying is, the lame man who keeps the right road outstrips the runner who takes a wrong one. Nay, it is obvious that when a man runs the wrong way, the more active and swift he is, the further he will go astray.

But the course I propose for the discovery of sciences is such as leaves but little to the acuteness and strength of wits, but places all wits and understandings nearly on a level. For as in the drawing of a straight line or a perfect circle, much depends on the steadiness and practice of the hand, if it be done by aim of hand only, but if with the aid of rule or compass, little or nothing; so is it exactly with my plan. But though particular confutations would be of no avail, yet touching the sects and general divisions of such systems I must say something; something also touching the external signs which show that they are unsound; and finally something touching the causes of such great infelicity and of such lasting and general agreement in error; that so the access to truth may be made less difficult, and the human understanding may the more willingly submit to its purgation and dismiss its idols.

LXII Idols of the Theater, or of Systems, are many, and there can be and perhaps will be yet many more. For were it not that now for many ages men's minds have been busied with religion and theology; and were it not that civil governments, especially monarchies, have been averse to such novelties, even in matters speculative; so that men labor therein to the peril and harming of their fortunes—not only unrewarded, but exposed also to contempt and envy doubtless there would have arisen many other philosophical sects like those which in great variety flourished once among the Greeks. For as on the phenomena of the heavens many hypotheses may be constructed, so likewise (and more also) many various dogmas may be set up and established on the phenomena of philosophy. And in the plays of this philosophical theater you may observe the same thing which is found in the theater of the poets, that stories invented for the stage are more compact and elegant, and more as one would wish them to be, than true stories out of history.

In general, however, there is taken for the material of philosophy either a great deal out of a few things, or a very little out of many things; so that on both sides philosophy is based on too narrow a foundation of experiment and natural history, and decides on the authority of too few cases. For the Rational School of philosophers snatches from experience a variety of common instances, neither duly ascertained nor diligently examined and weighed, and leaves all the rest to meditation and agitation of wit.

There is also another class of philosophers who, having bestowed much diligent and careful labor on a few experiments, have thence made bold to educe and construct systems, wresting all other facts in a strange fashion to conformity therewith.

And there is yet a third class, consisting of those who out of faith and veneration mix their philosophy with theology and traditions; among whom the vanity of some has gone so far aside as to seek the origin of sciences among spirits and genii. So that this parent stock of errors this false philosophy is of three kinds: the Sophistical, the Empirical, and the Superstitious.

. . .

XCII But by far the greatest obstacle to the progress of science and to the undertaking of new tasks and provinces therein is found in this that men despair and think things impossible. For wise and serious men are wont in these matters to be altogether distrustful, considering with themselves the obscurity of nature, the shortness of life, the deceitfulness of the senses, the weakness of the judgment, the difficulty of experiment, and the like; and so supposing that in the revolution of time and of the ages of the world the sciences have their ebbs and flows; that at one season they grow and flourish, at another wither and decay, yet in such sort that when they have reached a certain point and condition they can advance no further. If therefore anyone believes or promises more, they think this comes of an ungoverned and unripened mind, and that such attempts have prosperous beginnings, become difficult as they go on, and end in confusion. Now since these are thoughts which naturally present themselves to men grave and of great judgment, we must take good heed that we be not led away by our love for a most fair and excellent object to relax or diminish the severity of our judgment. We must observe diligently what encouragement dawns upon us and from what quarter, and, putting aside the lighter breezes of hope, we must thoroughly sift and examine those which promise greater steadiness and constancy. Nay, and we must take state prudence too into our counsels, whose rule is to distrust, and to take the less favorable view of human affairs. I am now therefore to speak touching hope, especially as I am not a dealer in promises, and wish neither to force nor to ensnare men's judgments, but to lead them by the hand with their good will. And though the strongest means of inspiring hope will be to bring men to particulars, especially to particulars digested and arranged in my Tables of Discovery (the subject partly of the second, but much more of the fourth part of my Instauration), since this is not merely the promise of the thing but the thing itself; nevertheless, that everything may be done with gentleness, I will proceed with my plan of preparing men's minds, of which preparation to give hope is no unimportant part. For without it the rest tends rather to make men sad (by giving them a worse and meaner opinion of things as they are than they now have, and making them more fully to feel and know the unhappiness of their own condition) than to induce any alacrity or to whet their industry in making trial. And therefore it is fit that I publish and set forth those conjectures of mine which make hope in this matter reasonable, just as Columbus did, before that wonderful voyage of his across the Atlantic, when he gave the reasons for his conviction that new lands and continents might be discovered besides those which were known before; which reasons, though rejected at first, were afterwards made good by experience, and were the causes and beginnings of great events.

. . .

XCV Those who have handled sciences have been either men of experiment or men of dogmas. The men of experiment are like the ant, they only collect and use; the reasoners resemble spiders, who make cobwebs out of their own substance. But the bee takes a middle course: it gathers its material from the flowers of the garden and of the field, but transforms and digests it by a power of its own. Not unlike this is the true business of philosophy; for it neither relies solely or chiefly on the powers of the mind, nor does it take the matter which it gathers from natural history and mechanical experiments and lay it up in the memory whole, as it finds it, but lays it up in the understanding altered and digested. Therefore from a closer and purer league between these two faculties, the experimental and the rational (such as has never yet been made), much may be hoped.

. . .

CVIII So much then for the removing of despair and the raising of hope through the dismissal or rectification of the errors of past time. We must now see what else there is to ground hope upon. And this consideration occurs at once that if many useful discoveries have been made by accident or upon occasion, when men were not seeking for them but were busy about other things, no one can doubt but that when they apply themselves to seek and make this their business, and that too by method and in order and not by desultory impulses, they will discover far more. For although it may happen once or twice that a man shall stumble on a thing by accident which, when taking great pains to search for it, he could not find, yet upon the whole it unquestionably falls out the other way. And therefore far better things, and more of them, and at shorter intervals, are to be expected from man's reason and industry and direction and fixed application than from accident and animal instinct and the like, in which inventions have hitherto had their origin.

Translated by James Spedding, Robert Leslie Ellis, and Douglas Denon Heath

Reading and Discussion Questions

1. Describe the Idols of the Tribe, Cave, Marketplace, and Theatre. Try to think of an example of each. How do these serve as a caution to use as we observe the natural world? How does Bacon's view of the significance of the imperfection of our senses differ from Descartes'?
2. Bacon uses the metaphor of the ant, spider, and bee. Which does he think is best? Can you think of figures we have read who might fit these characterizations?
3. How would you characterize Bacon's view of scientific method?

The New Atlantis (1627)

Francis Bacon

44

We sailed from Peru, (where we had continued for the space of one whole year) for China and Japan, by the South Sea; taking with us victuals for twelve months; and had good winds from the east, though soft and weak, for five months space, and more. But the wind came about, and settled in the west for many days, so as we could make little or no way, and were sometime in purpose to turn back. But then again there arose strong and great winds from the south, with a point east, which carried us up (for all that we could do) towards the north; by which time our victuals failed us, though we had made good spare of them. So that finding ourselves, in the midst of the greatest wilderness of waters in the world, without victuals, we gave ourselves for lost men and prepared for death. Yet we did lift up our hearts and voices to God above, who showeth his wonders in the deep, beseeching him of his mercy, that as in the beginning he discovered the face of the deep, and brought forth dry land, so he would now discover land to us, that we might not perish.

And it came to pass that the next day about evening we saw within a kenning before us, towards the north, as it were thick clouds, which did put us in some hope of land; knowing how that part of the South Sea was utterly unknown; and might have islands, or continents, that hitherto were not come to light. Wherefore we bent our course thither, where we saw the appearance of land, all that night; and in the dawning of the next day, we might plainly discern that it was a land; flat to our sight, and full of boscage; which made it show the more dark. And after an hour and a half's sailing, we entered into a good haven, being the port of a fair city; not great indeed, but well built, and that gave a pleasant view from the sea: and we thinking every minute long, till we were on land, came close to the shore, and offered to land. But straightways we saw divers of the people, with bastons in their hands (as it were) forbidding us to land; yet without any cries of fierceness, but only as warning us off, by signs that they made. Whereupon being not a little discomforted, we were advising with ourselves, what we should do.

During which time, there made forth to us a small boat . . . .

When we were come within six yards of their boat, they called to us to stay, and not to approach farther; which we did. And thereupon the man, whom I before described, stood up, and with a loud voice, in Spanish, asked, "Are ye Christians?" We answered, "We were"; fearing the less, because of the cross we had seen in the subscription. At which answer the said person lifted up his right hand towards Heaven, and drew it softly to his mouth (which is the gesture they use, when they thank God;) and then said: "If ye will swear (all of you) by the merits of the Saviour, that ye are no pirates, nor have shed blood, lawfully, nor unlawfully within forty days past, you may have licence to come on land." We said, "We were all ready to take that oath."

. . .

The next morning early, there came to us the same officer that came to us at first with his cane, and told us, He came to conduct us to the Strangers' House . . . .

The Strangers' House is a fair and spacious house, built of brick, of somewhat a bluer colour than our brick; and with handsome windows, some of glass, some of a kind of cambric oiled. He brought us first into a fair parlour above stairs, and then asked us, "What number of persons we were? And how many sick?" We answered, "We were in all, (sick and whole,) one and fifty persons, whereof our sick were seventeen." He desired us to have patience a little, and to stay till he came back to us; . . . . "Ye are to know, that the custom of the land requireth, that after this day and to-morrow, (which we give you for removing of your people from your ship,) you are to keep within doors for three days. But let it not trouble you, nor do not think yourselves restrained, but rather left to your rest and ease. You shall want nothing, and there are six of our people appointed to attend you, for any business you may have abroad." We gave him thanks, with all affection and respect, and said, "God surely is manifested in this land." We offered him also twenty pistolets; but he smiled, and only said; "What? twice paid!" And so he left us.

Soon after our dinner was served in; which was right good viands, both for bread and treat: better than any collegiate diet, that I have known in Europe. We had also drink of three sorts, all wholesome and good; wine of the grape; a drink of grain, such as is with us our ale, but more clear: And a kind of cider made of a fruit of that country; a wonderful pleasing and refreshing drink. Besides, there were brought in to us, great store of those scarlet oranges, for our sick; which (they said) were an assured remedy for sickness taken at sea. There was given us also, a box of small gray, or whitish pills, which they wished our sick should take, one of the pills, every night before sleep; which (they said) would hasten their recovery.

. . .

The next day about ten of the clock, the Governor came to us again, and after salutations, said familiarly; "That he was come to visit us"; and called for a chair, and sat him down: and we, being some ten of us, (the rest were of the meaner sort, or else gone abroad,) sat down with him, And when we were set, he began thus: "We of this island of Bensalem," (for so they call it in their language,) "have this; that by means of our solitary situation; and of the laws of secrecy, which we have for our travellers, and our rare admission of strangers; we know well most part of the habitable world, and are ourselves unknown.

. . .

"There reigned in this land, about nineteen hundred years ago, a king, whose memory of all others we most adore; not superstitiously, but as a divine instrument, though a mortal man; his name was Solamona: and we esteem him as the lawgiver of our nation.

. . .

"Ye shall understand (my dear friends) that amongst the excellent acts of that king, one above all hath the pre-eminence. It was the erection and institution of an Order or Society, which we call Salomon's House; the noblest foundation (as we think) that ever was upon the earth; and the lanthorn of this kingdom. It is dedicated to the study of the works and creatures of God. Some think it beareth the founder's name a little corrupted, as if it should be Solamona's House. But the records write it as it is spoken. So as I take it to be denominate of the king of the Hebrews, which is famous with you, and no stranger to us. For we have some parts of his works, which with you are lost; namely, that natural history, which he wrote, of all plants, from the cedar of Libanus to the moss that groweth out of the wall, and of all things that have life and motion. This maketh me think that our king, finding himself to symbolize in many things with that king of the Hebrews (which lived many years before him), honored him with the title of this foundation. And I am rather induced to be of this opinion, for that I find in ancient records this Order or Society is sometimes called Salomon's House, and sometimes the College of the Six Days Works; whereby I am satisfied that our excellent king had learned from the Hebrews that God had created the world and all that therein is within six days: and therefore he instituting that House for the finding out of the true nature of all things, (whereby God might have the more glory in the workmanship of them, and insert the more fruit in the use of them), did give it also that second name.

"But now to come to our present purpose. When the king had forbidden to all his people navigation into any part that was not under his crown, he made nevertheless this ordinance; that every twelve years there should be set forth, out of this kingdom two ships, appointed to several voyages; That in either of these ships there should be a mission of three of the Fellows or Brethren of Salomon's House; whose errand was only to give us knowledge of the affairs and state of those countries to which they were designed, and especially of the sciences, arts, manufactures, and inventions of all the world; and withal to bring unto us books, instruments, and patterns in every kind: That the ships, after they had landed the brethren, should return; and that the brethren should stay abroad till the new mission. These ships are not otherwise fraught, than with store of victuals, and good quantity of treasure to remain with the brethren, for the buying of such things and rewarding of such persons as they should think fit. Now for me to tell you how the vulgar sort of mariners are contained from being discovered at land; and how they that must be put on shore for any time, color themselves under the names of other nations; and to what places these voyages have been designed; and what places of rendezvous are appointed for the new missions; and the like circumstances of the practique; I may not do it: neither is it much to your desire. But thus you see we maintain a trade not for gold, silver, or jewels; nor for silks; nor for spices; nor any other commodity of matter; but only for God's first creature, which was Light: to have light (I say) of the growth of all parts of the world."

. . .

"Ye are happy men; for the Father of Salomon's House taketh knowledge of your being here, and commanded me to tell you that he will admit all your company to his presence, and have private conference with one of you, that ye shall choose: and for this hath appointed the next day after to-morrow. And because he meaneth to give you his blessing, he hath appointed it in the forenoon."

We came at our day and hour, and I was chosen by my fellows for the private access. We found him in a fair chamber, richly hanged, and carpeted under foot without any degrees to the state. He was set upon a low Throne richly adorned, and a rich cloth of state over his head, of blue satin embroidered. He was alone, save that he had two pages of honour, on either hand one, finely attired in white. His under garments were the like that we saw him wear in the chariot; but instead of his gown, he had on him a mantle with a cape, of the same fine black, fastened about him. When we came in, as we were taught, we bowed low at our first entrance; and when we were come near his chair, he stood up, holding forth his hand ungloved, and in posture of blessing; and we every one of us stooped down, and kissed the hem of his tippet. That done, the rest departed, and I remained. Then he warned the pages forth of the room, and caused me to sit down beside him, and spake to me thus in the Spanish tongue.

"God bless thee, my son; I will give thee the greatest jewel I have. For I will impart unto thee, for the love of God and men, a relation of the true state of Salomon's House. Son, to make you know the true state of Salomon's House, I will keep this order. First, I will set forth unto you the end of our foundation. Secondly, the preparations and instruments we have for our works. Thirdly, the several employments and functions whereto our fellows are assigned. And fourthly, the ordinances and rites which we observe.

"The end of our foundation is the knowledge of causes, and secret motions of things; and the enlarging of the bounds of human empire, to the effecting of all things possible.

"The Preparations and Instruments are these. We have large and deep caves of several depths: the deepest are sunk six hundred fathom: and some of them are digged and made under great hills and mountains: so that if you reckon together the depth of the hill and the depth of the cave, they are (some of them) above three miles deep. For we find, that the depth of a hill, and the depth of a cave from the flat, is the same thing; both remote alike, from the sun and heaven's beams, and from the open air. These caves we call the Lower Region; and we use them for all coagulations, indurations, refrigerations, and conservations of bodies. We use them likewise for the imitation of natural mines; and the producing also of new artificial metals, by compositions and materials which we use, and lay there for many years. We use them also sometimes, (which may seem strange,) for curing of some diseases, and for prolongation of life in some hermits that choose to live there, well accommodated of all things necessary, and indeed live very long; by whom also we learn many things.

"We have burials in several earths, where we put diverse cements, as the Chineses do their porcellain. But we have them in greater variety, and some of them more fine. We have also great variety of composts and soils, for the making of the earth fruitful.

"We have high towers; the highest about half a mile in height; and some of them likewise set upon high mountains; so that the vantage of the hill with the tower is in the highest of them three miles at least. And these places we call the Upper Region; accounting the air between the high places and the low, as a Middle Region. We use these towers, according to their several heights, and situations, for insolation, refrigeration, conservation; and for the view of divers meteors; as winds, rain, snow, hail; and some of the fiery meteors also. And upon them, in some places, are dwellings of hermits, whom we visit sometimes, and instruct what to observe.

"We have great lakes, both salt, and fresh; whereof we have use for the fish and fowl. We use them also for burials of some natural bodies: for we find a difference in things buried in earth or in air below the earth, and things buried in water. We have also pools, of which some do strain fresh water out of salt; and others by art do turn fresh water into salt. We have also some rocks in the midst of the sea, and some bays upon the shore for some works, wherein is required the air and vapor of the sea. We have likewise violent streams and cataracts, which serve us for many motions: and likewise engines for multiplying and enforcing of winds, to set also on going diverse motions.

"We have also a number of artificial wells and fountains, made in imitation of the natural sources and baths; as tincted upon vitriol, sulphur, steel, brass, lead, nitre, and other minerals. And again we have little wells for infusions of many things, where the waters take the virtue quicker and better, than in vessels or basins. And amongst them we have a water which we call Water of Paradise, being, by that we do to it made very sovereign for health, and prolongation of life.

"We have also great and spacious houses where we imitate and demonstrate meteors; as snow, hail, rain, some artificial rains of bodies and not of water, thunders, lightnings; also generations of bodies in air; as frogs, flies, and divers others.

"We have also certain chambers, which we call Chambers of Health, where we qualify the air as we think good and proper for the cure of divers diseases, and preservation of health.

"We have also fair and large baths, of several mixtures, for the cure of diseases, and the restoring of man's body from arefaction: and others for the confirming of it in strength of sinewes, vital parts, and the very juice and substance of the body.

"We have also large and various orchards and gardens; wherein we do not so much respect beauty, as variety of ground and soil, proper for divers trees and herbs: and some very spacious, where trees and berries are set whereof we make divers kinds of drinks, besides the vineyards. In these we practise likewise all conclusions of grafting, and inoculating as well of wild-trees as fruit-trees, which produceth many effects. And we make (by art) in the same orchards and gardens, trees and flowers to come earlier or later than their seasons; and to come up and bear more speedily than by their natural course they do. We make them also by art greater much than their nature; and their fruit greater and sweeter and of differing taste, smell, colour, and figure, from their nature. And many of them we so order, as they become of medicinal use.

"We have also means to make divers plants rise by mixtures of earths without seeds; and likewise to make divers new plants, differing from the vulgar; and to make one tree or plant turn into another.

"We have also parks and enclosures of all sorts of beasts and birds which we use not only for view or rareness, but likewise for dissections and trials; that thereby we may take light what may be wrought upon the body of man. Wherein we find many strange effects; as continuing life in them, though divers parts, which you account vital, be perished and taken forth; resuscitating of some that seem dead in appearance; and the like. We try also all poisons and other medicines upon them, as well of chirurgery, as physic. By art likewise, we make them greater or taller than their kind is; and contrariwise dwarf them, and stay their growth: we make them more fruitful and bearing than their kind is; and contrariwise barren and not generative. Also we make them differ in colour, shape, activity, many ways. We find means to make commixtures and copulations of different kinds; which have produced many new kinds, and them not barren, as the general opinion is. We make a number of kinds of serpents, worms, flies, fishes, of putrefaction; whereof some are advanced (in effect) to be perfect creatures, like bests or birds; and have sexes, and do propagate. Neither do we this by chance, but we know beforehand, of what matter and commixture what kind of those creatures will arise.

"We have also particular pools, where we make trials upon fishes, as we have said before of beasts and birds.

"We have also places for breed and generation of those kinds of worms and flies which are of special use; such as are with you your silk-worms and bees.

"I will not hold you long with recounting of our brewhouses, bake-houses, and kitchens, where are made divers drinks, breads, and meats, rare and of special effects. Wines we have of grapes; and drinks of other juice of fruits, of grains, and of roots; and of mixtures with honey, sugar, manna, and fruits dried, and decocted; Also of the tears or woundings of trees; and of the pulp of canes. And these drinks are of several ages, some to the age or last of forty years. We have drinks also brewed with several herbs, and roots, and spices; yea with several fleshes, and white-meats; whereof some of the drinks are such, as they are in effect meat and drink both: so that divers, especially in age, do desire to live with them, with little or no meat or bread. And above all, we strive to have drink of extreme thin parts, to insinuate into the body, and yet without all biting, sharpness, or fretting; insomuch as some of them put upon the back of your hand will, with a little stay, pass through to the palm, and yet taste mild to the mouth. We have also waters which we ripen in that fashion, as they become nourishing; so that they are indeed excellent drink; and many will use no other. Breads we have of several grains, roots, and kernels; yea and some of flesh and fish dried; with divers kinds of leavenings and seasonings: so that some do extremely move appetites; some do nourish so, as divers do live of them, without any other meat; who live very long. So for meats, we have some of them so beaten and made tender and mortified, yet without all corrupting, as a weak heat of the stomach will turn them into good chylus; as well as a strong heat would meat otherwise prepared. We have some meats also and breads and drinks, which taken by men enable them to fast long after; and some other, that used make the very flesh of men's bodies sensibly more hard and tough and their strength far greater than otherwise it would be.

"We have dispensatories, or shops of medicines. Wherein you may easily think, if we have such variety of plants and living creatures more than you have in Europe, (for we know what you have,) the simples, drugs, and ingredients of medicines, must likewise be in so much the greater variety. We have them likewise of divers ages, and long fermentations. And for their preparations, we have not only all manner of exquisite distillations and separations, and especially by gentle heats and percolations through divers strainers, yea and substances; but also exact forms of composition, whereby they incorporate almost, as they were natural simples.

"We have also divers mechanical arts, which you have not; and stuffs made by them; as papers, linen, silks, tissues; dainty works of feathers of wonderful lustre; excellent dies, and, many others; and shops likewise, as well for such as are not brought into vulgar use amongst us as for those that are. For you must know that of the things before recited, many of them are grown into use throughout the kingdom; but yet, if they did flow from our invention, we have of them also for patterns and principals.

"We have also furnaces of great diversities, and that keep great diversity of heats; fierce and quick; strong and constant; soft and mild; blown, quiet; dry, moist; and the like. But above all, we have heats, in imitation of the Sun's and heavenly bodies' heats, that pass divers inequalities, and (as it were) orbs, progresses, and returns, whereby we produce admirable effects. Besides, we have heats of dungs; and of bellies and maws of living creatures, and of their bloods and bodies; and of hays and herbs laid up moist; of lime unquenched; and such like. Instruments also which generate heat only by motion. And farther, places for strong insulations; and again, places under the earth, which by nature, or art, yield heat. These divers heats we use, as the nature of the operation, which we intend, requireth.

"We have also perspective-houses, where we make demonstrations of all lights and radiations; and of all colours: and out of things uncoloured and transparent, we can represent unto you all several colours; not in rain-bows, (as it is in gems, and prisms,) but of themselves single. We represent also all multiplications of light, which we carry to great distance, and make so sharp as to discern small points and lines. Also all colourations of light; all delusions and deceits of the sight, in figures, magnitudes, motions, colours all demonstrations of shadows. We find also divers means, yet unknown to you, of producing of light originally from divers bodies. We procure means of seeing objects afar off; as in the heaven and remote places; and represent things near as afar off; and things afar off as near; making feigned distances. We have also helps for the sight, far above spectacles and glasses in use. We have also glasses and means to see small and minute bodies perfectly and distinctly; as the shapes and colours of small flies and worms, grains and flaws in gems, which cannot otherwise be seen, observations in urine and blood not otherwise to be seen. We make artificial rain-bows, halo's, and circles about light. We represent also all manner of reflexions, refractions, and multiplications of visual beams of objects.

"We have also precious stones of all kinds, many of them of great beauty, and to you unknown; crystals likewise; and glasses of divers kinds; and amongst them some of metals vitrificated, and other materials besides those of which you make glass. Also a number of fossils, and imperfect minerals, which you have not. Likewise loadstones of prodigious virtue; and other rare stones, both natural and artificial.

"We have also sound-houses, where we practise and demonstrate all sounds, and their generation. We have harmonies which you have not, of quarter-sounds, and lesser slides of sounds. Divers instruments of music likewise to you unknown, some sweeter than any you have, together with bells and rings that are dainty and sweet. We represent small sounds as great and deep; likewise great sounds extenuate and sharp; we make divers tremblings and warblings of sounds, which in their original are entire. We represent and imitate all articulate sounds and letters, and the voices and notes of beasts and birds. We have certain helps which set to the ear do further the hearing greatly. We have also divers strange and artificial echoes, reflecting the voice many times, and as it were tossing it: and some that give back the voice louder than it came, some shriller, and some deeper; yea, some rendering the voice differing in the letters or articulate sound from that they receive. We have also means to convey sounds in trunks and pipes, in strange lines and distances.

"We have also perfume-houses; wherewith we join also practices of taste. We multiply smells, which may seem strange. We imitate smells, making all smells to breathe outs of other mixtures than those that give them. We make divers imitations of taste likewise, so that they will deceive any man's taste. And in this house we contain also a confiture-house; where we make all sweet-meats, dry and moist; and divers pleasant wines, milks, broths, and sallets; in far greater variety than you have.

"We have also engine-houses, where are prepared engines and instruments for all sorts of motions. There we imitate and practise to make swifter motions than any you have, either out of your muskets or any engine that you have: and to make them and multiply them more easily, and with small force, by wheels and other means: and to make them stronger and more violent than yours are; exceeding your greatest cannons and basilisks. We represent also ordnance and instruments of war, and engines of all kinds: and likewise new mixtures and compositions of gun-powder, wild-fires burning in water, and unquenchable. Also fireworks of all variety both for pleasure and use. We imitate also flights of birds; we have some degrees of flying in the air. We have ships and boats for going under water, and brooking of seas; also swimming-girdles and supporters. We have divers curious clocks, and other like motions of return: and some perpetual motions. We imitate also motions of living creatures, by images, of men, beasts, birds, fishes, and serpents. We have also a great number of other various motions, strange for equality, fineness, and subtilty.

"We have also a mathematical house, where are represented all instruments, as well of geometry as astronomy, exquisitely made.

"We have also houses of deceits of the senses; where we represent all manner of feats of juggling, false apparitions, impostures, and illusions; and their fallacies. And surely you will easily believe that we that have so many things truly natural which induce admiration, could in a world of particulars deceive the senses, if we would disguise those things and labour to make them seem more miraculous. But we do hate all impostures, and lies; insomuch as we have severely forbidden it to all our fellows, under pain of ignominy and fines, that they do not show any natural work or thing, adorned or swelling; but only pure as it is, and without all affectation of strangeness.

"These are (my son) the riches of Salomon's House.

"For the several employments and offices of our fellows; we have twelve that sail into foreign countries, under the names of other nations, (for our own we conceal); who bring us the books, and abstracts, and patterns of experiments of all other parts. These we call Merchants of Light.

"We have three that collect the experiments which are in all books. These we call Depredators.

"We have three that collect the experiments of all mechanical arts; and also of liberal sciences; and also of practices which are not brought into arts. These we call Mystery-men.

"We have three that try new experiments, such as themselves think good. These we call Pioneers or Miners.

"We have three that draw the experiments of the former four into titles and tables, to give the better light for the drawing of observations and axioms out of them. These we call Compilers.

"We have three that bend themselves, looking into the experiments of their fellows, and cast about how to draw out of them things of use and practise for man's life, and knowledge, as well for works as for plain demonstration of causes, means of natural divinations, and the easy and clear discovery of the virtues and parts of bodies. These we call Dowry-men or Benefactors.

"Then after divers meetings and consults of our whole number, to consider of the former labours and collections, we have three that take care, out of them, to direct new experiments, of a higher light, more penetrating into nature than the former. These we call Lamps.

"We have three others that do execute the experiments so directed, and report them. These we call Inoculators.

"Lastly, we have three that raise the former discoveries by experiments into greater observations, axioms, and aphorisms. These we call Interpreters of Nature.

"We have also, as you must think, novices and apprentices, that the succession of the former employed men do not fail; besides, a great number of servants and attendants, men and women. And this we do also: we have consultations, which of the inventions and experiences which we have discovered shall be published, and which not: and take all an oath of secrecy, for the concealing of those which we think fit to keep secret: though some of those we do reveal sometimes to the state and some not.

"For our ordinances and rites: we have two very long and fair galleries: in one of these we place patterns and samples of all manner of the more rare and excellent inventions in the other we place the statues of all principal inventors. There we have the statue of your Columbus, that discovered the West Indies: also the inventor of ships: your monk that was the inventor of ordnance and of gunpowder: the inventor of music: the inventor of letters: the inventor of printing: the inventor of observations of astronomy: the inventor of works in metal: the inventor of glass: the inventor of silk of the worm: the inventor of wine: the inventor of corn and bread: the inventor of sugars: and all these, by more certain tradition than you have. Then have we divers inventors of our own, of excellent works; which since you have not seen, it were too long to make descriptions of them; and besides, in the right understanding of those descriptions you might easily err. For upon every invention of value, we erect a statue to the inventor, and give him a liberal and honourable reward. These statues are some of brass; some of marble and touch-stone; some of cedar and other special woods gilt and adorned; some of iron; some of silver; some of gold.

"We have certain hymns and services, which we say daily, of Lord and thanks to God for his marvellous works: and forms of prayers, imploring his aid and blessing for the illumination of our labours, and the turning of them into good and holy uses.

"Lastly, we have circuits or visits of divers principal cities of the kingdom; where, as it cometh to pass, we do publish such new profitable inventions as we think good. And we do also declare natural divinations of diseases, plagues, swarms-of hurtful creatures, scarcity, tempests, earthquakes, great inundations, comets, temperature of the year, and divers other things; and we give counsel thereupon, what the people shall do for the prevention and remedy of them."

And when he had said this, he stood up; and I, as I had been taught, kneeled down, and he laid his right hand upon my head, and said; "God bless thee, my son; and God bless this relation, which I have made. I give thee leave to publish it for the good of other nations; for we here are in God's bosom, a land unknown." And so he left me; having assigned a value of about two thousand ducats, for a bounty to me and my fellows. For they give great largesses where they come upon all occasions.

Reading and Discussion Questions

1. What are the features of the utopia that Bacon depicts here? How is society organized, what kinds of things do they have, what is its leadership, what kinds of subjects are studied (and which are ignored) in Salomon's House? What makes this island such a desirable place to live?
2. What is the purpose of scientific research in Bacon's utopia? What are some of the more remarkable things that he envisions science giving us the ability to do?
3. What does this text reveal about Bacon's view of the relationship between science and religion? What is the relationship between science and political power?

An Attempt to Prove the Motion of the Earth from Observations (1674)

Robert Hooke

45

... And shall only for the present hint that I have in some of my foregoing observations discovered some new Motions even in the Earth it self, which perhaps were not dreamt of before, which I shall hereafter more at large describe, when further tryals have more fully confirmed and compleated these beginings. At which time also I shall explain a System of the World differing in many particulars from any yet known, answering in all things to the common Rules of Mechanical Motions: This depends upon three Suppositions:

First, That all Coelestial Bodies whatsoever, have an attraction or gravitating power towards their own Centers, whereby they attract not only their own parts, and keep them from flying from them, as we may observe the Earth to do, but that they do also attract all the other Coelestial Bodies that are within the sphere of their activity; and consequently that not only the Sun and Moon have an influence upon the body and motion of the Earth, and the Earth upon them, but that also Mercury, Venus, Mars, Jupiter and Saturn by their attractive powers, have a considerable influence upon its motion as in the same manner the corresponding attractive power of the Earth hath a considerable influence upon every one of their motions also.

The second supposition is this, That all bodies whatsoever that are put into a direct and simple motion, will so continue to move forward in a streight line, till they are by some other effectual powers deflected and bent into a Motion, describing a Circle, Ellipsis, or some other more compounded Curve Line.

The third supposition is, That these attractive powers are so much the more powerful in operating, by how much the nearer the body wrought upon is to their own Centers. Now what these several degrees are I have not yet experimentally verified; but it is a notion, which if fully prosecuted as it ought to be, will mightily assist the Astronomer to reduce all the Coelestial Motions to a certain rule, which I doubt will never be done true without it. He that understands the nature of the Circular Pendulum and Circular Motion, will easily understand the whole ground of this Principle, and will know where to find direction in Nature for the true stating thereof.

This I only hint at present to such as have ability and opportunity of prosecuting this Inquiry, and are not wanting of Industry for observing and calculating, wishing heartily such may be found, having my self many other things in hand which I would first compleat, and therefore cannot so well attent it. But this I durst promise the Undertaker, that he will find all the great Motions of the World to be influenced by this Principle, and that the true understanding thereof will be the true perfection of Astronomy.

Reading and Discussion Question

1. Hooke would later claim that Newton should have given him credit for these ideas in the *Principia*. Do you think that Hooke has a good case?

Principia or The Mathematical Principles of Natural Philosophy (1697)

Isaac Newton

46

The Author's Preface

Since the ancients (as we are told by *Pappus*), made great account of the science of mechanics in the investigation of natural things; and the moderns, laying aside substantial forms and occult qualities, have endeavoured to subject the phænomena of nature to the laws of mathematics, I have in this treatise cultivated mathematics so far as it regards philosophy. The ancients considered mechanics in a twofold respect; as rational, which proceeds accurately by demonstration: and practical. To practical mechanics all the manual arts belong, from which mechanics took its name. But as artificers do not work with perfect accuracy, it comes to pass that mechanics is so distinguished from geometry, that what is perfectly accurate is called geometrical, what is less so, is called mechanical. But the errors are not in the art, but in the artificers. He that works with less accuracy is an imperfect mechanic; and if any could work with perfect accuracy, he would be the most perfect mechanic of all; for the description if right lines and circles, upon which geometry is founded, belongs to mechanics. Geometry does not teach us to draw these lines, but requires them to be drawn; for it requires that the learner should first be taught to describe these accurately, before he enters upon geometry; then it shows how by these operations problems may be solved. To describe right lines and circles are problems, but not geometrical problems. The solution of these problems is required from mechanics; and by geometry the use of them, when so solved, is shown; and it is the glory of geometry that from those few principles, brought from without, it is able to produce so many things. Therefore geometry is founded in mechanical practice, and is nothing but that part of universal mechanics which accurately proposes and demonstrates the art of measuring. But since the manual arts are chiefly conversant in the moving of bodies, it comes to pass that geometry is commonly referred to their magnitudes, and mechanics to their motion. In this sense rational mechanics will be the science of motions resulting from any forces whatsoever, and of the forces required to produce any motions, accurately proposed and demonstrated. This part of mechanics was cultivated by the ancients in the five powers which relate to manual arts, who considered gravity (it not being a manual power), no otherwise than as it moved weights by those powers. Our design not respecting arts, but philosophy, and our subject not manual but natural powers, we consider chiefly those things which relate to gravity, levity, elastic force, the resistance of fluids, and the like forces, whether attractive or impulsive; and therefore we offer this work as the mathematical principles if philosophy; for all the difficulty of philosophy seems to consist in this—from the phænomena of motions to investigate the forces of nature, and then from these forces to demonstrate the other phænomena; and to this end the general propositions in the first and second book are directed. In the third book we give an example of this in the explication of the System of the World; for by the propositions mathematically demonstrated in the former books, we in the third derive from the celestial phenomena the forces of gravity with which bodies tend to the sun and the several planets. Then from these forces, by other propositions which are also mathematical, we deduce the motions of the planets, the comets, the moon, and the sea. I wish we could derive the rest of the phænomena of nature by the same kind of reasoning from mechanical principles; for I am induced by many reasons to suspect that they may all depend upon certain forces by which the particles of bodies, by some causes hitherto unknown, are either mutually impelled towards each other, and cohere in regular figures, or are repelled and recede from each other; which forces being unknown, philosophers have hitherto attempted the search of nature in vain; but I hope the principles here laid down will afford some light either to this or some truer method of philosophy.

. . .

ISAAC NEWTON.
Cambridge. Trinity College May 8, 1686

Book I

The Mathematical Principles of Natural Philosophy

Definitions

DEFINITION I.

The quantity of matter is the measure of the same, arising from its density and bulk conjunctly.

Thus air of a double density, in a double space, is quadruple in quantity; in a triple space, sextuple in quantity. The same thing is to be understood of snow, and fine dust or powders, that are condensed by compression or liquefaction; and of all bodies that are by any causes whatever differently condensed. I have no regard in this place to a medium, if any such there is, that freely pervades the interstices between the parts of bodies. It is this quantity that I mean hereafter everywhere under the name of body or mass. And the same is known by the weight of each body; for it is proportional to the weight, as I have found by experiments on pendulums, very accurately made, which shall be shewn hereafter.

DEFINITION II.

The quantity of motion is the measure of the same, arising from the velocity and quantity of matter conjunctly.

The motion of the whole is the sum of the motions of all the parts; and therefore in a body double in quantity, with equal velocity, the motion is double; with twice the velocity, it is quadruple.

DEFINITION III.

The vis insita, *or innate force of matter, is a power of resisting, by which every body, as much as in it lies, endeavours to persevere in its present state, whether it be of rest, or of moving uniformly forward in a right line.*

This force is ever proportional to the body whose force it is; and differs nothing from the inactivity of the mass, but in our manner of conceiving it. A body, from the inactivity of matter, is not without difficulty put out of its state of rest or motion. Upon which account, this *vis insita*, may, by a most significant name, be called *vis inertiæ*, or force of inactivity. But a body exerts this force only, when another force, impressed upon it, endeavours to change its condition; and the exercise of this force may be considered both as resistance and impulse; it is resistance, in so far as the body, for maintaining its present state, withstands the force impressed; it is impulse, in so far as the body, by not easily giving way to the impressed force of another, endeavours to change the state of that other. Resistance is usually ascribed to bodies at rest, and impulse to those in motion; but motion and rest, as commonly conceived, are only relatively distinguished; nor are those bodies always truly at rest, which commonly are taken to be so.

DEFINITION IV.

An impressed force is an action exerted upon a body, in order to change its state, either of rest, or of moving uniformly forward in a right line.

This force consists in the action only; and remains no longer in the body, when the action is over. For a body maintains every new state it acquires, by its *vis inertiæ* only. Impressed forces are of different origins as from percussion, from pressure, from centripetal force.

DEFINITION V.

A centripetal force is that by which bodies are drawn or impelled, or any way tend, towards a point as to a centre.

Of this sort is gravity, by which bodies tend to the centre of the earth magnetism, by which iron tends to the loadstone; and that force, what ever it is, by which the planets are perpetually drawn aside from the rectilinear motions, which otherwise they would pursue, and made to revolve in curvilinear orbits. A stone, whirled about in a sling, endeavours to recede from the hand that turns it; and by that endeavour, distends the sling, and that with so much the greater force, as it is revolved with the greater velocity, and as soon as ever it is let go, flies away. That force which opposes itself to this endeavour, and by which the sling perpetually draws back the stone towards the hand, and retains it in its orbit, because it is directed to the hand as the centre of the orbit, I call the centripetal force. And the same thing is to be understood of all bodies, revolved in any orbits. They all endeavour to recede from the centres of their orbits; and were it not for the opposition of a contrary force which restrains them to, and detains them in their orbits, which I therefore call centripetal, would fly off in right lines, with an uniform motion. A projectile, if it was not for the force of gravity, would not deviate towards the earth, but would go off from it in a right line, and that with an uniform motion, if the resistance of the air was taken away. It is by its gravity that it is drawn aside perpetually from its rectilinear course, and made to deviate towards the earth, more or less, according to the force of its gravity, and the velocity of its motion. The less its gravity is, for the quantity of its matter, or the greater the velocity with which it is projected, the less will it deviate from a rectilinear course, and the farther it will go. If a leaden ball, projected from the top of a mountain by the force of gunpowder with a given velocity, and in a direction parallel to the horizon, is carried in a curve line to the distance of two miles before it falls to the ground; the same, if the resistance of the air were taken away, with a double or decuple velocity, would fly twice or ten times as far. And by increasing the velocity, we may at pleasure increase the distance to which it might be projected, and diminish the curvature of the line, which it might describe, till at last it should fall at the distance of 10, 30, or 90 degrees, or even might go quite round the whole earth before it falls; or lastly, so that it might never fall to the earth, but go forward into the celestial spaces, and proceed in its motion *in infinitum.* And after the same manner that a projectile, by the force of gravity, may be made to revolve in an orbit, and go round the whole earth, the moon also, either by the force of gravity, if it is endued with gravity, or by any other force, that impels it towards the earth, may be perpetually drawn aside towards the earth, out of the rectilinear way, which by its innate force it would pursue; and would be made to revolve in the orbit which it now describes; nor could the moon with out some such force, be retained in its orbit. If this force was too small, it would not sufficiently turn the moon out of a rectilinear course: if it was too great, it would turn it too much, and draw down the moon from its orbit towards the earth. It is necessary, that the force be of a just quantity, and it belongs to the mathematicians to find the force, that may serve exactly to retain a body in a given orbit, with a given velocity; and vice versa, to determine the curvilinear way, into which a body projected from a given place, with a given velocity, may be made to deviate from its natural rectilinear way, by means of a given force.

The quantity of any centripetal force may be considered as of three kinds; absolute, accelerative, and motive.

DEFINITION VI.

The absolute quantity of a centripetal force is the measure of the same proportional to the efficacy of the cause that propagates it from the centre, through the spaces round about.

Thus the magnetic force is greater in one load-stone and less in another according to their sizes and strength of intensity.

DEFINITION VII.

The accelerative quantity of a centripetal force is the measure of the same, proportional to the velocity which it generates in a given time.

Thus the force of the same load-stone is greater at a less distance, and less at a greater: also the force of gravity is greater in valleys, less on tops of exceeding high mountains; and yet less (as shall hereafter be shown), at greater distances from the body of the earth; but at equal distances, it is the same everywhere; because (taking away, or allowing for, the resistance of the air), it equally accelerates all falling bodies, whether heavy or light, great or small.

DEFINITION VIII.

The motive quantity of a centripetal force, is the measure of the same, proportional to the motion which it generates in a given time.

Thus the weight is greater in a greater body, less in a less body; and, in the same body, it is greater near to the earth, and less at remoter distances. This sort of quantity is the centripetency, or propension of the whole body towards the centre, or, as I may say, its weight; and it is always known by the quantity of an equal and contrary force just sufficient to hinder the descent of the body.

...

Scholium

Hitherto I have laid down the definitions of such words as are less known, and explained the sense in which I would have them to be understood in the following discourse. I do not define time, space, place and motion, as being well known to all. Only I must observe, that the vulgar conceive those quantities under no other notions but from the relation they bear to sensible objects. And thence arise certain prejudices, for the removing of which, it will be convenient to distinguish them into absolute and relative, true and apparent, mathematical and common.

I. Absolute, true, and mathematical time, of itself, and from its own nature flows equably without regard to anything external, and by another name is called duration: relative, apparent, and common time, is some sensible and external (whether accurate or unequable) measure of duration by the means of motion, which is commonly used instead of true time; such as an hour, a day, a month, a year.

II. Absolute space, in its own nature, without regard to anything external, remains always similar and immovable. Relative space is some movable dimension or measure of the absolute spaces; which our senses determine by its position to bodies; and which is vulgarly taken for immovable space; such is the dimension of a subterraneous, an æreal, or celestial space, determined by its position in respect of the earth. Absolute and relative space, are the same in figure and magnitude; but they do not remain always numerically the same. For if the earth, for instance, moves, a space of our air, which relatively and in respect of the earth remains always the same, will at one time be one part of the absolute space into which the air passes; at another time it will be another part of the same, and so, absolutely understood, it will be perpetually mutable.

III. Place is a part of space which a body takes up, and is according to the space, either absolute or relative. I say, a part of space; not the situation, nor the external surface of the body. For the places of equal solids are always equal; but their superfices, by reason of their dissimilar figures, are often unequal. Positions properly have no quantity, nor are they so much the places themselves, as the properties of places. The motion of the whole is the same thing with the sum of the motions of the parts; that is, the translation of the whole, out of its place, is the same thing with the sum of the translations of the parts out of their places; and therefore the place of the whole is the same thing with the sum of the places of the parts, and for that reason, it is internal, and in the whole body.

IV. Absolute motion is the translation of a body from one absolute place into another; and relative motion, the translation from one relative place into another. Thus in a ship under sail, the relative place of a body is that part of the ship which the body possesses; or that part of its cavity which the body fills, and which therefore moves together with the ship: and relative rest is the continuance of the body in the same part of the ship, or of its cavity. But real, absolute rest, is the continuance of the body in the same part of that immovable space, in which the ship itself, its cavity, and all that it contains, is moved. Wherefore, if the earth is really at rest, the body, which relatively rests in the ship, will really and absolutely move with the same velocity which the ship has on the earth. But if the earth also moves, the true and absolute motion of the body will arise, partly from the true motion of the earth, in immovable space; partly from the relative motion of the ship on the earth; and if the body moves also relatively in the ship; its true motion will arise, partly from the true motion of the earth, in immovable space, and partly from the relative motions as well of the ship on the earth, as of the body in the ship; and from these relative motions will arise the relative motion of the body on the earth. As if that part of the earth, where the ship is, was truly moved toward the east, with a velocity of 10010 parts; while the ship itself, with a fresh gale, and full sails, is carried towards the west, with a velocity expressed by 10 of those parts; but a sailor walks in the ship towards the east, with 1 part of the said velocity; then the sailor will be moved truly in immovable space towards the east, with a velocity of 10001 parts, and relatively on the earth towards the west, with a velocity of 9 of those parts.

Absolute time, in astronomy, is distinguished from relative, by the equation or correction of the vulgar time. For the natural days are truly unequal, though they are commonly considered as equal, and used for a measure of time; astronomers correct this inequality for their more accurate deducing of the celestial motions. It may be, that there is no such thing as an equable motion, whereby time may be accurately measured. All motions may be accelerated and retarded, but the true, or equable, progress of absolute time is liable to no change. The duration or perseverance of the existence of things remains the same, whether the motions are swift or slow, or none at all: and therefore it ought to be distinguished from what are only sensible measures thereof; and out of which we collect it, by means of the astronomical equation. The necessity of which equation, for determining the times of a phenomenon, is evinced as well from the experiments of the pendulum clock, as by eclipses of the satellites of *Jupiter*.

As the order of the parts of time is immutable, so also is the order of the parts of space. Suppose those parts to be moved out of their places, and they will be moved (if the expression may be allowed) out of themselves. For times and spaces are, as it were, the places as well of themselves as of all other things. All things are placed in time as to order of succession; and in space as to order of situation. It is from their essence or nature that they are places; and that the primary places of things should be moveable, is absurd. These are therefore the absolute places; and translations out of those places, are the only absolute motions.

But because the parts of space cannot be seen, or distinguished from one another by our senses, therefore in their stead we use sensible measures of them. For from the positions and distances of things from any body considered as immovable, we define all places; and then with respect to such places, we estimate all motions, considering bodies as transferred from some of those places into others. And so, instead of absolute places and motions, we use relative ones; and that without any inconvenience in common affairs; but in philosophical disquisitions, we ought to abstract from our senses, and consider things themselves, distinct from what are only sensible measures of them. For it may be that there is no body really at rest, to which the places and motions of others may be referred.

But we may distinguish rest and motion, absolute and relative, one from the other by their properties, causes and effects. It is a property of rest, that bodies really at rest do rest in respect to one another. And therefore as it is possible, that in the remote regions of the fixed stars, or perhaps far beyond them, there may be some body absolutely at rest; but impossible to know, from the position of bodies to one another in our regions whether any of these do keep the same position to that remote body; it follows that absolute rest cannot be determined from the position of bodies in our regions.

It is a property of motion, that the parts, which retain given positions to their wholes, do partake of the motions of those wholes. For all the parts of revolving bodies endeavour to recede from the axis of motion; and the impetus of bodies moving forward, arises from the joint impetus of all the parts. Therefore, if surrounding bodies are moved, those that are relatively at rest within them, will partake of their motion. Upon which account, the true and absolute motion of a body cannot be determined by the translation of it from those which only seem to rest; for the external bodies ought not only to appear at rest, but to be really at rest. For otherwise, all included bodies, beside their translation from near the surrounding ones, partake likewise of their true motions; and though that translation were not made they would not be really at rest, but only seem to be so. For the surrounding bodies stand in the like relation to the surrounded as the exterior part of a whole does to the interior, or as the shell does to the kernel; but, if the shell moves, the kernel will also move, as being part of the whole, without any removal from near the shell.

A property, near akin to the preceding, is this, that if a place is moved, whatever is placed therein moves along with it; and therefore a body, which is moved from a place in motion, partakes also of the motion of its place. Upon which account, all motions, from places in motion, are no other than parts of entire and absolute motions; and every entire motion is composed of the motion of the body out of its first place, and the motion of this place out of its place; and so on, until we come to some immovable place, as in the before-mentioned example of the sailor. Wherefore, entire and absolute motions can be no otherwise determined than by immovable places; and for that reason I did before refer those absolute motions to immovable places, but relative ones to movable places. Now no other places are immovable but those that, from infinity to infinity, do all retain the same given position one to another; and upon this account must ever remain unmoved; and do thereby constitute immovable space.

The causes by which true and relative motions are distinguished, one from the other, are the forces impressed upon bodies to generate motion. True motion is neither generated nor altered, but by some force impressed upon the body moved; but relative motion may be generated or altered without any force impressed upon the body. For it is sufficient only to impress some force on other bodies with which the former is compared, that by their giving way, that relation may be changed, in which the relative rest or motion of this other body did consist. Again, true motion suffers always some change from any force impressed upon the moving body; but relative motion does not necessarily undergo any change by such forces. For if the same forces are likewise impressed on those other bodies, with which the comparison is made, that the relative position may be preserved, then that condition will be preserved in which the relative motion consists. And therefore any relative motion may be changed when the true motion remains unaltered, and the relative may be preserved when the true suffers some change. Upon which accounts, true motion does by no means consist in such relations.

The effects which distinguish absolute from relative motion are, the forces of receding from the axis of circular motion. For there are no such forces in a circular motion purely relative, but in a true and absolute circular motion, they are greater or less, according to the quantity of the motion. If a vessel, hung by a long cord, is so often turned about that the cord is strongly twisted, then filled with water, and held at rest together with the water; after, by the sudden action of another force, it is whirled about the contrary way, and while the cord is untwisting itself, the vessel continues for some time in this motion; the surface of the water will at first be plain, as before the vessel began to move: but the vessel, by gradually communicating its motion to the water, will make it begin sensibly to revolve, and recede by little and little from the middle, and ascend to the sides of the vessel, forming itself into a concave figure (as I have experienced), and the swifter the motion becomes, the higher will the water rise, till at last, performing its revolutions in the same times with the vessel, it becomes relatively at rest in it. This ascent of the water shows its endeavour to recede from the axis of its motion; and the true and absolute circular motion of the water, which is here directly contrary to the relative, discovers itself, and may be measured by this endeavour. At first, when the relative motion of the water in the vessel was greatest, it produced no endeavour to recede from the axis; the water showed no tendency to the circumference, nor any ascent towards the sides of the vessel, but remained of a plain surface, and therefore its true circular motion had not yet begun. But afterwards, when the relative motion of the water had decreased, the ascent thereof towards the sides of the vessel proved its endeavour to recede from the axis; and this endeavour showed the real circular motion of the water perpetually increasing, till it had acquired its greatest quantity, when the water rested relatively in the vessel. And therefore this endeavour does not depend upon any translation of the water in respect of the ambient bodies, nor can true circular motion be defined by such translation. There is only one real circular motion of any one revolving body, corresponding to only one power of endeavouring to recede from its axis of motion, as its proper and adequate effect; but relative motions, in one and the same body, are innumerable, according to the various relations it bears to external bodies, and like other relations, are altogether destitute of any real effect, any otherwise than they may perhaps partake of that one only true motion. And therefore in their system who suppose that our heavens, revolving below the sphere of the fixed stars, carry the planets along with them; the several parts of those heavens, and the planets, which are indeed relatively at rest in their heavens, do yet really move. For they change their position one to another (which never happens to bodies truly at rest), and being carried together with their heavens, partake of their motions, and as parts of revolving wholes, endeavour to recede from the axis of their motions.

Wherefore relative quantities are not the quantities themselves, whose names they bear, but those sensible measures of them (either accurate or inaccurate), which are commonly used instead of the measured quantities themselves. And if the meaning of words is to be determined by their use, then by the names time, space, place and motion, their measures are properly to be understood; and the expression will be unusual, and purely mathematical, if the measured quantities themselves are meant. Upon which account, they do strain the sacred writings, who there interpret those words for the measured quantities. Nor do those less defile the purity of mathematical and philosophical truths, who confound real quantities themselves with their relations and vulgar measures.

It is indeed a matter of great difficulty to discover, and effectually to distinguish, the true motions of particular bodies from the apparent; because the parts of that immovable space, in which those motions are performed, do by no means come under the observation of our senses. Yet the thing is not altogether desperate: for we have some arguments to guide us, partly from the apparent motions, which are the differences of the true motions; partly from the forces, which are the causes and effects of the true motions. For instance, if two globes, kept at a given distance one from the other by means of a cord that connects them, were revolved about their common centre of gravity, we might, from the tension of the cord, discover the endeavour of the globes to recede from the axis of their motion, and from thence we might compute the quantity of their circular motions. And then if any equal forces should be impressed at once on the alternate faces of the globes to augment or diminish their circular motions, from the increase or decrease of the tension of the cord, we might infer the increment or decrement of their motions; and thence would be found on what faces those forces ought to be impressed, that the motions of the globes might be most augmented; that is, we might discover their hindermost faces, or those which, in the circular motion, do follow. But the faces which follow being known, and consequently the opposite ones that precede, we should likewise know the determination of their motions. And thus we might find both the quantity and the determination of this circular motion, even in an immense vacuum, where there was nothing external or sensible with which the globes could be compared. But now, if in that space some remote bodies were placed that kept always a given position one to another, as the fixed stars do in our regions, we could not indeed determine from the relative translation of the globes among those bodies, whether the motion did belong to the globes or to the bodies. But if we observed the cord, and found that its tension was that very tension which the motions of the globes required, we might conclude the motion to be in the globes, and the bodies to be at rest; and then, lastly, from the translation of the globes among the bodies, we should find the determination of their motions. But how we are to collect the true motions from their causes, effects, and apparent differences; and, *vice versa*, how from the motions, either true or apparent, we may come to the knowledge of their causes and effects, shall be explained more at large in the following tract. For to this end it was that I composed it.

Axioms, or Laws of Motion

LAW I.

Every body perseveres in its state of rest, or of uniform motion in a right line, unless it is compelled to change that state by forces impressed thereon.

Projectiles persevere in their motions, so far as they are not retarded by the resistance of the air, or impelled downwards by the force of gravity. A top, whose parts by their cohesion are perpetually drawn aside from rectilinear motions, does not cease its rotation, otherwise than as it is retarded by the air. The greater bodies of the planets and comets, meeting with less resistance in more free spaces, preserve their motions both progressive and circular for a much longer time.

LAW II.

The alteration of motion is ever proportional to the motive force impressed; and is made in the direction of the right line in which that force is impressed.

If any force generates a motion, a double force will generate double the motion, a triple force triple the motion, whether that force be impressed altogether and at once, or gradually and successively. And this motion (being always directed the same way with the generating force), if the body moved before, is added to or subducted from the former motion, according as they directly conspire with or are directly contrary to each other; or obliquely joined, when they are oblique, so as to produce a new motion compounded from the determination of both.

LAW III.

To every action there is always opposed an equal reaction: or the mutual actions of two bodies upon each other are always equal, and directed to contrary parts.

Whatever draws or presses another is as much drawn or pressed by that other. If you press a stone with your finger, the finger is also pressed by the stone. If a horse draws a stone tied to a rope, the horse (if I may so say) will be equally drawn back towards the stone: for the distended rope, by the same endeavour to relax or unbend itself, will draw the horse as much towards the stone, as it does the stone towards the horse, and will obstruct the progress of the one as much as it advances that of the other. If a body impinge upon another, and by its force change the motion of the other, that body also (because of the equality of the mutual pressure) will undergo an equal change, in its own motion, towards the contrary part. The changes made by these actions are equal, not in the velocities but in the motions of bodies; that is to say, if the bodies are not hindered by any other impediments. For, because the motions are equally changed, the changes of the velocities made towards contrary parts are reciprocally proportional to the bodies. This law takes place also in attractions, as will be proved in the next scholium.

COROLLARY I.

A body by two forces conjoined will describe the diagonal of a parallelogram, in the same time that it would describe the sides, by those forces apart.

If a body in a given time, by the force M impressed apart in the place A, should with an uniform motion be carried from A to B; and by the force N impressed apart in the same place, should be carried from A to C; complete the parallelogram ABCD, and, by both forces acting together, it will in the same time be carried in the diagonal from A to D. For since the force N acts in the direction of the line AC, parallel to BD, this force (by the second law) will not at all alter the velocity generated by the other force M, by which the body is carried towards the line BD. The body therefore will arrive at the line BD in the same time, whether the force N be impressed or not; and therefore at the end of that time it will be found somewhere in the line BD. By the same argument, at the end of the same time it will be found somewhere in the line CD. Therefore it will be found in the point D, where both lines meet. But it will move in a right line from A to D, by Law I.

COROLLARY II.

And hence is explained the composition of any one direct force AD, out of any two oblique forces AC and CD; and, on the contrary, the resolution of any one direct force AD into two oblique forces AC and CD: which composition and resolution are abundantly confirmed from mechanics.

. . .

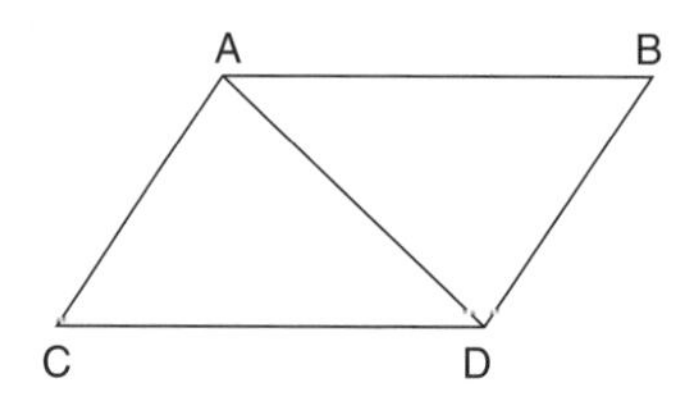

And thus the force of the screw may be deduced from a like resolution of forces; it being no other than a wedge impelled with the force of a lever. Therefore the use of this Corollary spreads far and wide, and by that diffusive extent the truth thereof is farther confirmed. For on what has been said depends the whole doctrine of mechanics variously demonstrated by different authors. For from hence are easily deduced the forces of machines, which are compounded of wheels, pullies, levers, cords, and weights, ascending directly or obliquely, and other mechanical powers; as also the force of the tendons to move the bones of animals.

COROLLARY III.

The quantity of motion, which is collected by taking the sum of the motions directed towards the same parts, and the difference of those that are directed to contrary parts, suffers no change from the action of bodies among themselves.

...

COROLLARY IV.

The common centre of gravity of two or more bodies does not alter its state of motion or rest by the actions of the bodies among themselves; and therefore the common centre of gravity of all bodies acting upon each other (excluding outward actions and impediments) is either at rest, or moves uniformly in a right line.

...

COROLLARY V.

The motions of bodies included in a given space are the same among themselves, whether that space is at rest, or moves uniformly forwards in a right line without any circular motion.

For the differences of the motions tending towards the same parts, and the sums of those that tend towards contrary parts, are, at first (by supposition), in both cases the same; and it is from those sums and differences that the collisions and impulses do arise with which the bodies mutually impinge one upon another. Wherefore (by Law II), the effects of those collisions will be equal in both cases; and therefore the mutual motions of the bodies among themselves in the one case will remain equal to the mutual motions of the bodies among themselves in the other. A clear proof of which we have from the experiment of a ship; where all motions happen after the same manner, whether the ship is at rest, or is carried uniformly forwards in a right line.

COROLLARY VI.

If bodies, any how moved among themselves, are urged in the direction of parallel lines by equal accelerative forces, they will all continue to move among themselves, after the same, manner as if they had been urged by no such forces.

. . .

Scholium

Hitherto I have laid down such principles as have been received by mathematicians, and are confirmed by abundance of experiments. By the first two Laws and the first two Corollaries, Galileo discovered that the descent of bodies observed the duplicate ratio of the time, and that the motion of projectiles was in the curve of a parabola; experience agreeing with both, unless so far as these motions are a little retarded by the resistance of the air. When a body is falling, the uniform force of its gravity acting equally, impresses, in equal particles of time, equal forces upon that body, and therefore generates equal velocities; and in the whole time impresses a whole force, and generates a whole velocity proportional to the time. And the spaces described in proportional times are as the velocities and the times conjunctly; that is, in a duplicate ratio of the times. And when a body is thrown upwards, its uniform gravity impresses forces and takes off velocities proportional to the times; and the times of ascending to the greatest heights are as the velocities to be taken off, and those heights are as the velocities and the times conjunctly, or in the duplicate ratio of the velocities. And if a body be projected in any direction, the motion arising from its projection is compounded with the motion arising from its gravity. As if the body A by its motion of projection alone could describe in a given time the right line AB, and with its motion of falling alone could describe in the same time the altitude AC; complete the parallelogram ABDC, and the body by that compounded motion will at the end of the time be found in the place D; and the curve line AED, which that body describes, will be a parabola, to which the

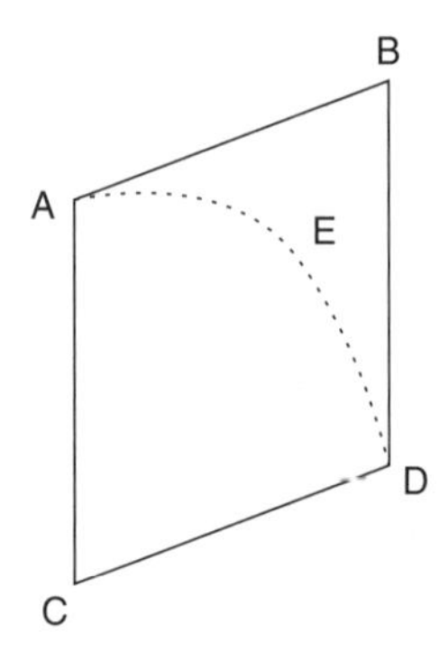

right line AB will be a tangent in A; and whose ordinate BD will be as the square of the line AB. On the same Laws and Corollaries depend those things which have been demonstrated concerning the times of the vibration of pendulums, and are confirmed by the daily experiments of pendulum clocks.

. . .

So those weights are of equal force to move the arms of a balance; which during the play of the balance are reciprocally as their velocities upwards and downwards; that is, if the ascent or descent is direct, those weights are of equal force, which are reciprocally as the distances of the points at which they are suspended from the axis of the balance; but if they are turned aside by the interposition of oblique planes, or other obstacles, and made to ascend or descend obliquely, those bodies will be equipollent, which are reciprocally as the heights of their ascent and descent taken according to the perpendicular; and that on account of the determination of gravity downwards.

And in like manner in the pully, or in a combination of pullies, the force of a hand drawing the rope directly, which is to the weight, whether ascending directly or obliquely, as the velocity of the perpendicular ascent of the weight to the velocity of the hand that draws the rope, will sustain the weight.

In clocks and such like instruments, made up from a combination of wheels, the contrary forces that promote and impede the motion of the wheels, if they are reciprocally as the velocities of the parts of the wheel on which they are impressed, will mutually sustain the one the other.

The force of the screw to press a body is to the force of the hand that turns the handles by which it is moved as the circular velocity of the handle in that part where it is impelled by the hand is to the progressive velocity of the screw towards the pressed body.

The forces by which the wedge presses or drives the two parts of the wood it cleaves are to the force of the mallet upon the wedge as the progress of the wedge in the direction of the force impressed upon it by the mallet is to the velocity with which the parts of the wood yield to the wedge, in the direction of lines perpendicular to the sides of the wedge. And the like account is to be given of all machines.

The power and use of machines consist only in this, that by diminishing the velocity we may augment the force, and the contrary: from whence in all sorts of proper machines, we have the solution of this problem; To move a given weight with a given power, or with a given force to overcome any other given resistance. For if machines are so contrived that the velocities of the agent and resistant are reciprocally as their forces, the agent will just sustain the resistant, but with a greater disparity of velocity will overcome it. So that if the disparity of velocities is so great as to overcome all that resistance which commonly arises either from the attrition of contiguous bodies as they slide by one another, or from the cohesion of continuous bodies that are to be separated, or from the weights of bodies to be raised, the excess of the force remaining, after all those resistances are overcome, will produce an acceleration of motion proportional thereto, as well in the parts of the machine as in the resisting body. But to treat of mechanics is not my present business. I was only willing to show by those examples the great extent and certainty of the third Law of motion. For if we estimate the action of the agent from its force and velocity conjunctly, and likewise the reaction of the impediment conjunctly from the velocities of its several parts, and from the forces of resistance arising from the attrition, cohesion, weight, and acceleration of those parts, the action and reaction in the use of all sorts of machines will be found always equal to one another. And so far as the action is propagated by the intervening instruments, and at last impressed upon the resisting body, the ultimate determination of the action will be always contrary to the determination of the reaction.

Book I

Of the Motion of Bodies

SECTION I.

Of the method of first and last ratios of quantities, by the help whereof we demonstrate the propositions that follow.

LEMMA I.

Quantities, and the ratios of quantities, which in any finite time converge continually to equality, and before the end of that time approach nearer the one to the other than by any given difference, become ultimately equal.

If you deny it, suppose them to be ultimately unequal, and let D be their ultimate difference. Therefore they cannot approach nearer to equality than by that given difference D; which is against the supposition.

LEMMA II.

If in any figure AacE, terminated by the right lines Aa, AE, and the curve acE, there be inscribed any number of parallelograms Ab, Bc, Cd, &c., comprehended under equal bases AB, BC, CD, &c., and the sides, Bb, Cc, Dd, &c., parallel to one side Aa of the figure; and the parallelograms aKbl, bLcm, cMdn, &c., are completed. Then if the breadth of those parallelograms be supposed to be diminished, and their number to be augmented in infinitum; I say, that the ultimate ratios which the inscribed figure AKbLcMdD, the circumscribed figure AalbmcndoE, and curvilinear figure AabcdE, will have to one another, are ratios of equality.

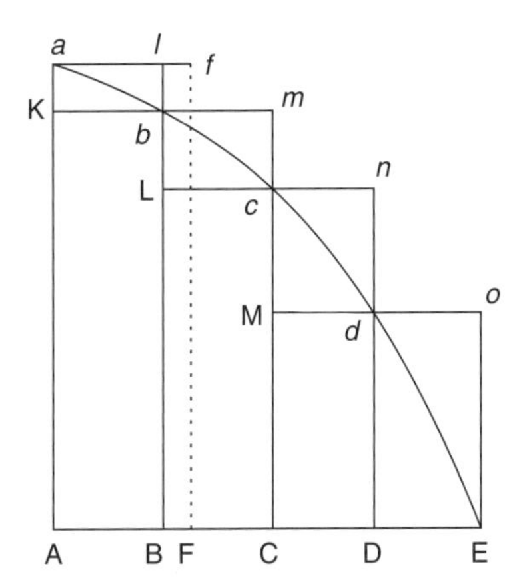

For the difference of the inscribed and circumscribed figures is the sum of the parallelograms Kl, Lm, Mu, Do, that is (from the equality of all their bases), the rectangle under one of their bases Kb and the sum of their altitudes Aa, that is, the rectangle ABla. But this rectangle, because its breadth AB is supposed diminished in infinitum, becomes less than any given space. And therefore (by Lem. I) the figures inscribed and circumscribed become ultimately equal one to the other; and much more will the intermediate curvilinear figure be ultimately equal to either. Q.E.D.

LEMMA III.

The same ultimate ratios are also ratios of equality, when the, breadths, AB, BC, DC, &c., of the parallelograms are unequal, and are all diminished in infinitum.

For suppose AF equal to the greatest breadth, and complete the parallelogram FAaf. This parallelogram will be greater than the difference of the inscribed and circumscribed figures; but, because its breadth AF is diminished in infinitum, it will be come less than any given rectangle. Q.E.D.

. . .

Section II

Of the Invention of Centripetal Forces

PROPOSITION I. THEOREM I.

The areas, which revolving bodies describe by radii drawn to an immovable centre of force do lie in the same immovable planes, and are proportional to the times in which they are described.

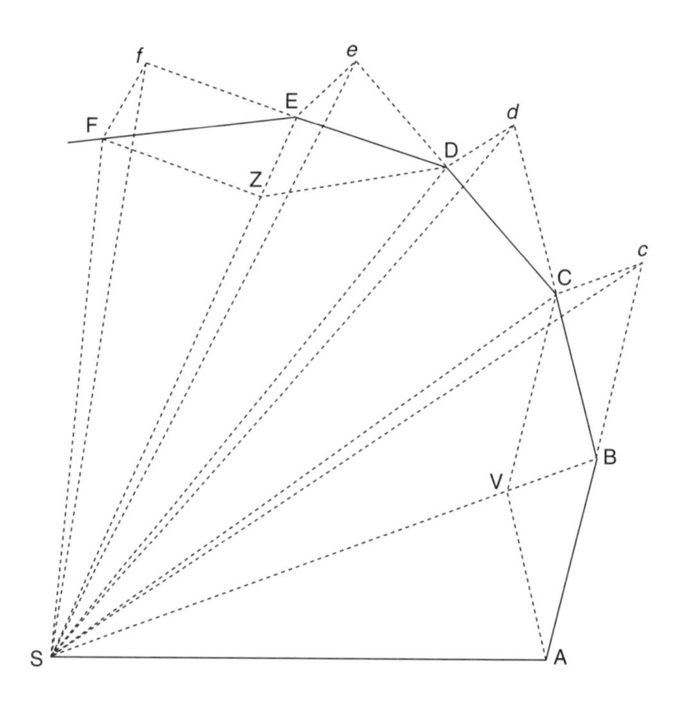

For suppose the time to be divided into equal parts, and in the first part of that time let the body by its innate force describe the right line AB. In the second part of that time, the same would (by Law I.), if not hindered, proceed directly to c, along the line Bc equal to AB; so that by the radii AS, BS, cS, drawn to the centre, the equal areas ASB, BSc, would be described. But when the body is arrived at B, suppose that a centripetal force acts at once with a great impulse; and, turning aside the body from the right line Bc, compels it afterwards to continue its motion along the right line BC. Draw cC parallel to BS meeting BC in C; and at the end of the second part of the time, the body (by Cor. I. of the Laws) will be found in C, in the same plane with the triangle ASB. Join SC, and, because SB and Cc are parallel, the triangle SBC will be equal to the triangle SBc, and therefore also to the triangle SAB. By the like argument, if the centripetal force acts successively in C, D, E. &c.; and makes the body, in each single particle of time, to describe the right lines CD, DE, EF, &c., they will all lie in the same plane; and the triangle SCD will be equal to the triangle SBC, and SDE to SCD, and SEF to SDE. And therefore, in equal times, equal areas are described in one immovable plane: and, by composition, any sums SADS, SAFS, of those areas, are one to the other as the times in which they are described. Now let the number of those triangles be augmented, and their breadth diminished *in infinitum;* and (by Cor. 4, Lem. III.) their ultimate perimeter ADF will be a curve line: and therefore the centripetal force, by which the body is perpetually drawn back from the tangent of this curve, will act continually; and any described areas SADS,

SAFS, which are always proportional to the times of description, will, in this case also, be proportional to those times. Q.E.D.

. . .

Book III

In the preceding Books I have laid down the principles of philosophy, principles not philosophical, but mathematical: such, to wit, as we may build our reasonings upon in philosophical inquiries. These principles are the laws and conditions of certain motions, and powers or forces, which chiefly have respect to philosophy: but, lest they should have appeared of themselves dry and barren, I have illustrated them here and there with some philosophical scholiums, giving an account of such things as are of more general nature, and which philosophy seems chiefly to be founded on; such as the density and the resistance of bodies, spaces void of all bodies, and the motion of light and sounds. It remains that, from the same principles, I now demonstrate the frame of the System of the World. Upon this subject I had, indeed, composed the third Book in a popular method, that it might be read by many; but afterward, considering that such as had not sufficiently entered into the principles could not easily discern the strength of the consequences, nor lay aside the prejudices to which they had been many years accustomed, therefore, to prevent the disputes which might be raised upon such accounts, I chose to reduce the substance of this Book into the form of Propositions (in the mathematical way), which should be read by those only who had first made themselves masters of the principles established in the preceding Books: not that I would advise any one to the previous study of every Proposition of those Books; for they abound with such as might cost too much time, even to readers of good mathematical learning. It is enough if one carefully reads the Definitions, the Laws of Motion, and the first three Sections of the first Book. He may then pass on to this Book, and consult such of the remaining Propositions of the first two Books, as the references in this, and his occasions, shall require.

Rules of Reasoning in Philosophy

RULE I.

We are to admit no more causes of natural things than such as are both time and sufficient to explain their appearances.

To this purpose the philosophers say that Nature does nothing in vain, and more is in vain when less will serve; for Nature is pleased with simplicity, and affects not the pomp of superfluous causes.

RULE II.

> *Therefore to the same natural effects we must, as far as possible, assign the same causes.*

As to respiration in a man and in a beast; the descent of stones in *Europe* and in *America*; the light of our culinary fire and of the sun; the reflection of light in the earth, and in the planets.

RULE III.

> *The qualities of bodies, which admit neither intension nor remission of degrees, and which are found to belong to all bodies within the reach of our experiments, are to be esteemed the universal qualities of all bodies whatsoever.*

For since the qualities of bodies are only known to us by experiments, we are to hold for universal all such as universally agree with experiments; and such as are not liable to diminution can never be quite taken away. We are certainly not to relinquish the evidence of experiments for the sake of dreams and vain fictions of our own devising; nor are we to recede from the analogy of Nature, which uses to be simple, and always consonant to itself. We no other way know the extension of bodies than by our senses, nor do these reach it in all bodies; but because we perceive extension in all that are sensible, therefore we ascribe it universally to all others also. That abundance of bodies are hard, we learn by experience; and because the hardness of the whole arises from the hardness of the parts, we therefore justly infer the hardness of the undivided particles not only of the bodies we feel but of all others. That all bodies are impenetrable, we gather not from reason, but from sensation. The bodies which we handle we find impenetrable, and thence conclude impenetrability to be an universal property of all bodies whatsoever. That all bodies are moveable, and endowed with certain powers (which we call *the vires inertiae*) of persevering in their motion, or in their rest, we only infer from the like properties observed in the bodies which we have seen. The extension, hardness, impenetrability, mobility, and *vis inertiae* of the whole, result from the extension, hardness, impenetrability, mobility, and *vires inertiae* of the parts; and thence we conclude the least particles of all bodies to be also all extended, and hard and impenetrable, and moveable, and endowed with their proper *vires inertia.* And this is the foundation of all philosophy. Moreover, that the divided but contiguous particles of bodies may be separated from one another, is matter of observation; and, in the particles that remain undivided, our minds are able to distinguish yet lesser parts, as is mathematically demonstrated. But whether the parts so distinguished, and not yet divided, may, by the powers of Nature, be actually divided and separated from one another, we cannot certainly determine. Yet, had we the proof of but one experiment that any undivided particle, in breaking a hard and solid body, suffered a division, we might by virtue of this rule conclude that the undivided as well as the divided particles may be divided and actually separated to infinity.

Lastly, if it universally appears, by experiments and astronomical observations, that all bodies about the earth gravitate towards the earth, and that in proportion to the quantity of matter which they severally contain; that the moon likewise, according to the quantity of its matter, gravitates towards the earth; that, on the other hand, our sea gravitates towards the moon; and all the planets mutually one towards another; and the comets in like manner towards the sun; we must, in consequence of this rule, universally allow that all bodies whatsoever are endowed with a principle of mutual gravitation. For the argument from the appearances concludes with more force for the universal gravitation of all bodies than for their impenetrability; of which, among those in the celestial regions, we have no experiments, nor any manner of observation. Not that I affirm gravity to be essential to bodies: by their *vis insita* I mean nothing but their *vis inertiae.* This is immutable. Their gravity is diminished as they recede from the earth.

RULE IV.

In experimental philosophy we are to look upon propositions collected by general induction from phaenomena as accurately or very nearly true, notwithstanding any contrary hypotheses that may be imagined, till such time as other phaenomena occur, by which they may either be made more accurate, or liable to exceptions.

This rule we must follow, that the argument of induction may not be evaded by hypotheses.

Phaenomena, or Appearances

PHAENOMENON I.

That the circumjovial planets, by radii drawn to Jupiter's centre, describe areas proportional to the times of description; and that their periodic times, the fixed stars being at rest, are in the sesquiplicate proportion of their distances from, its centre.

This we know from astronomical observations. For the orbits of these planets differ but insensibly from circles concentric to Jupiter; and their motions in those circles are found to be uniform. And all astronomers agree that their periodic times are in the sesquiplicate proportion of the semi-diameters of their orbits; and so it manifestly appears from the following table.

The periodic times of the satellites of Jupiter.

$1^d.18^h.27'.34''$. $3^d.13^h.13'42''$. $7^d.3^h.42'36''$. $16^d.16^h.32'9''$.

The distances of the satellites from Jupiter's centre.					
From the observations of	1	2	3	4	
Borelli	$5\frac{2}{3}$	$8\frac{2}{3}$	14	$24\frac{2}{3}$	
Townly *by the Micrometer.*	5,52	8,78	13,47	24,72	
Cassini *by the Telescope*	5	8	13	23	semi-diameter of Jupiter.
Cassini *by the eclipse of the satellites.*	$5\frac{2}{3}$	9	$14\frac{23}{60}$	$25\frac{3}{10}$	
From the periodic times	5,667	9,017	14,384	25,299	

Mr. *Pound* has determined, by the help of excellent micrometers, the diameters of Jupiter and the elongation of its satellites after the following manner. The greatest heliocentric elongation of the fourth satellite from Jupiter's centre was taken with a micrometer in a 15 feet telescope, and at the mean distance of Jupiter from the earth was found about 8′ 16″. The elongation of the third satellite was taken with a micrometer in a telescope of 123 feet, and at the same distance of Jupiter from the earth was found 4′ 42″. The greatest elongations of the other satellites, at the same distance of Jupiter from the earth, are found from the periodic times to be 2′ 56″ 47‴, and 1′ 51″ 6‴.

The diameter of Jupiter taken with the micrometer in a 123 feet telescope several times, and reduced to Jupiter's mean distance from the earth, proved always less than 40″, never less than 38″, generally 39″. This diameter in shorter telescopes is 40″, or 41″; for Jupiter's light is a little dilated by the unequal refrangibility of the rays, and this dilatation bears less ratio to the diameter of Jupiter in the longer and more perfect telescopes than in those which are shorter and less perfect. The times in which two satellites, the first and the third, passed over Jupiter's body, were observed, from the beginning of the ingress to the beginning of the egress, and from the complete ingress to the complete egress, with the long telescope. And from the transit of the first satellite, the diameter of Jupiter at its mean distance from the earth came forth $37\frac{1}{8}''$ and from the transit of the third $37\frac{3}{8}''$. There was observed also the time in which the shadow of the first satellite passed over Jupiter's body, and thence the diameter of Jupiter at its mean distance from the earth came out about 37″. Let us suppose its diameter to be $37\frac{1}{4}''$ very nearly, and then the greatest elongations of the first, second, third, and fourth satellite will be respectively equal to 5,965, 9,494, 15,141, and 26,63 semi-diameters of Jupiter.

PHAENOMENON II.

That the circumsaturnal planets, by radii drawn to Saturn's centre, describe areas proportional to the times of description; and that their periodic times, the fixed stars being at rest, are in the sesquiplicate proportion of their distances from its centre.

For, as *Cassini* from his own observations has determined, their distances from Saturn's centre and their periodic times are as follow.

The periodic times of the satellites of Saturn.

$1^d.21^h.18'27''$. $2^d.17^h.41'22''$. 4^d.12h.$25'12''$. $15^d.22^h.41'14''$. $79^d.7^h.48'00''$.

The distances of the satellites from Saturn's centre, in semi-diameters of its ring.

From observations	$1\frac{19}{20}$.	$2\frac{1}{2}$.	$3\frac{1}{2}$.	8.	24.
From the periodic times	1,93.	2,47.	3,45.	8.	23,35.

The greatest elongation of the fourth satellite from Saturn's centre is commonly determined from the observations to be eight of those semi-diameters very nearly. But the greatest elongation of this satellite from Saturn's centre, when taken with an excellent micrometer in Mr. Huygens' telescope of 123 feet, appeared to be eight semi-diameters and $\frac{7}{10}$ of a semi-diameter. And from this observation and the periodic times the distances of the satellites from Saturn's centre in semi-diameters of the ring are 2.1. 2,69. 3,75. 8,7. and 25,35. The diameter of Saturn observed in the same telescope was found to be to the diameter of the ring as 3 to 7; and the diameter of the ring, May 28–29, 1719, was found to be 43″; and thence the diameter of the ring when Saturn is at its mean distance from the earth is 42″, and the diameter of Saturn 18″. These things appear so in very long and excellent telescopes, because in such telescopes the apparent magnitudes of the heavenly bodies bear a greater proportion to the dilatation of light in the extremities of those bodies than in shorter telescopes. If we, then, reject all the spurious light, the diameter of Saturn will not amount to more than 16″.

PHAENOMENON III.

That the five primary planets, Mercury, Venus, Mars, Jupiter, and Saturn, with their several orbits, encompass the sun.

That Mercury and Venus revolve about the sun, is evident from their moonlike appearances. When they shine out with a full face, they are, in respect of us, beyond or above the sun; when they appear half full, they are about the same height on one side or other of the sun; when horned, they are below or between us and the sun; and they are sometimes, *when directly under*, seen like spots traversing the sun's disk. That Mars surrounds the sun, is as plain from its full face when near its conjunction with the sun, and from the gibbous figure which it shews in its quadratures. And the same thing is demonstrable of Jupiter and Saturn, from their appearing full in all situations; for the shadows of their satellites that appear sometimes upon their disks make it plain that the light they shine with is not their own, but borrowed from the sun.

PHAENOMENON IV.

That the fixed stars being at rest, the periodic times of the five primary planets, and (whether of the sun, about the earth, or) of the earth about the sun, are in the sesquiplicate proportion of their mean distances from the sun.

This proportion, first observed by *Kepler,* is now received by all astronomers; for the periodic times are the same, and the dimensions of the orbits are the same, whether the sun revolves about the earth, or the earth about the sun. And as to the measures of the periodic times, all astronomers are agreed about them. But for the dimensions of the orbits, *Kepler* and *Bullialdus*, above all others, have determined them from observations with the greatest accuracy; and the mean distances corresponding to the periodic times differ but insensibly from those

The periodic times with respect to the fixed stars, of the planets and earth revolving about the sun, in days and decimal parts of a day.

♄	♃	♂	♁	♀	☿
10759,275.	4332,514.	686,9785.	365,2565.	224,6176.	87,9692.

The mean distances of the planets and of the earth from the sun.

	♄	♃	♂
According to *Kepler*	951000.	519650.	152350.
According to *Bullialdus*	954198.	522520.	152350.
According to the periodic times	954006.	520096.	152369
	♁	♀	☿
According to *Kepler*	100000.	72400.	38806.
According to *Bullialdus*	100000.	72398.	38585.
According to the periodic times	100000.	72333.	38710

which they have assigned, and for the most part fall in between them; as we may see from the following table.
As to Mercury and Venus, there can be no doubt about their distances from the sun; for they are determined by the elongations of those planets from the sun; and for the distances of the superior planets, all dispute is cut off by the eclipses of the satellites of Jupiter. For by those eclipses the position of the shadow which Jupiter projects is determined; whence we have the heliocentric longitude of Jupiter. And from its heliocentric and geocentric longitudes compared together, we determine its distance.

PHAENOMENON V.

Then the primary planets, by radii drawn to the earth, describe areas no wise proportional to the times; but that the areas which they describe by radii drawn to the sun are proportional to the times of description.

For to the earth they appear sometimes direct, sometimes stationary, nay, and sometimes retrograde. But from the sun they are always seen direct, and to proceed with a motion nearly uniform, that is to say, a little swifter in the perihelion and a little slower in the aphelion distances, so as to maintain an equality in the description of the areas. This a noted proposition among astronomers, and particularly demonstrable in Jupiter, from the eclipses of his satellites; by the help of which eclipses, as we have said, the heliocentric longitudes of that planet, and its distances from the sun, are determined.

PHAENOMENON VI.

That the moon, by a radius drawn to the earth's centre, describes an area proportional to the time of description.

This we gather from the apparent motion of the moon, compared with its apparent diameter. It is true that the motion of the moon is a little disturbed by the action of the sun: but in laying down these Phenomena I neglect those small and inconsiderable errors.

Propositions

PROPOSITION I. THEOREM I.

That the forces by which the circumjovial planets are continually drawn off from rectilinear motions, and retained in their proper orbits, tend to Jupiter's centre; and are reciprocally as the squares of the distances of the places of those planets from that centre.

The former part of this Proposition appears from Phaen. I, and Prop. II or III, Book I; the latter from Phaen. I, and Cor. 6, Prop. IV, of the same Book.

The same thing we are to understand of the planets which encompass Saturn, by Phaen. II.

PROPOSITION II. THEOREM II.

That the forces by which the primary planets are continually drawn off from rectilinear motions, and retained in their proper orbits, tend to the sun; and are reciprocally as the squares of the distances of the places of those planets from the suits centre.

The former part of the Proposition is manifest from Phaen. V, and Prop. II, Book I; the latter from Phaen. IV, and Cor. 6, Prop. IV, of the same Book. But this part of the Proposition is, with great accuracy, demonstrable from the quiescence of the aphelion points; for a very small aberration from the reciprocal duplicate proportion would (by Cor. 1, Prop. XLV, Book I) produce a motion of the apsides sensible enough in every single revolution, and in many of them enormously great.

PROPOSITION III. THEOREM III.

That the force by which the moon is retained in its orbit tends to the earth; and is reciprocally as the square of the distance of its place from the earth's centre.

The former part of the Proposition is evident from Phaen. VI, and Prop. II or III, Book I; the latter from the very slow motion of the moon's apogee; which in every single revolution amounting but to 3° 3′ *in consequentia,* may be neglected. For (by Cor. 1. Prop. XLV, Book I) it appears, that, if the distance of the moon from the earth's centre is to the semi-diameter of the earth as D to 1, the force, from which such a motion will result, is reciprocally as $D^{2\frac{4}{243}}$, i.e., reciprocally as the power of D, whose exponent is $2\frac{4}{243}$; that is to say, in the proportion of the distance something greater than reciprocally duplicate, but which comes $59\frac{3}{4}$ times nearer to the duplicate than to the triplicate proportion. But in regard that this motion is owing to the action of the sun (as we shall afterwards shew), it is here to be neglected. The action of the sun, attracting the moon from the earth, is nearly as the moon's distance from the earth; and therefore (by what we have shewed in Cor. 2, Prop. XLV, Book I) is to the centripetal force of the moon as 2 to 357,45, or nearly so; that is, as 1 to $178\frac{29}{40}$. And if we neglect so inconsiderable a force of the sun, the remaining force, by which the moon is retained in its orb, will be reciprocally as D^2. This will yet more fully appear from comparing this force with the force of gravity, as is done in the next Proposition.

Cor. If we augment the mean centripetal force by which the moon is retained in its orb, first in the proportion of $177\frac{29}{40}$ to $178\frac{29}{40}$, and then in the duplicate proportion of the semi-diameter of the earth to the mean distance of the centres of the moon and earth, we shall have the centripetal force of the moon at the surface of the earth; supposing this force, in descending to the earth's surface, continually to increase in the reciprocal duplicate proportion of the height.

PROPOSITION IV. THEOREM IV.

That the moon gravitates towards the earth, and by the force of gravity is continually drawn off from a rectilinear motion, and retained in its orbit.

The mean distance of the moon from the earth in the syzygies in semi-diameters of the earth, is, according to *Ptolemy* and most astronomers, 59; according to *Vendelin* and *Huygens*, 60; to *Copernicus*, 60⅓; to *Street*, 60⅖; and to *Tycho*, 56½. But *Tycho*, and all that follow his tables of refraction, making the refractions of the sun and moon (altogether against the nature of light) to exceed the refractions of the fixed stars, and that by four or five minutes *near the horizon*, did thereby increase the moon's *horizontal* parallax by a like number of minutes, that is, by a twelfth or fifteenth part of the whole parallax. Correct this error, and the distance will become about 60½ semi-diameters of the earth, near to what others have assigned. Let us assume the mean distance of 60 diameters in the syzygies; and suppose one revolution of the moon, in respect of the fixed stars, to be completed in $27^d.7^h.43'$, as astronomers have determined; and the circumference of the earth to amount to 123249600 *Paris* feet, as the French have found by mensuration. And now if we imagine the moon, deprived of all motion, to be let go, so as to descend towards the earth with the impulse of all that force by which (by Cor. Prop. III) it is retained in its orb, it will in the space of one minute of time, describe in its fall 15$\frac{1}{12}$ *Paris* feet. This we gather by a calculus, founded either upon Prop. XXXVI, Book I, or (which comes to the same thing) upon Cor. 9, Prop. IV, of the same Book. For the versed sine of that arc, which the moon, in the space of one minute of time, would by its mean motion describe at the distance of 60 semi-diameters of the earth, is nearly 15$\frac{1}{12}$ *Paris* feet, or more accurately 15 feet, 1 inch, and 1 line $\frac{4}{9}$. Where fore, since that force, in approaching to the earth, increases in the reciprocal duplicate proportion of the distance, and, upon that account, at the surface of the earth, is 60×60 times greater than at the moon, a body in our regions, falling with that force, ought in the space of one minute of time, to describe $60 \times 60 \times 15\frac{1}{12}$ *Paris* feet; and, in the space of one second of time, to describe 15$\frac{1}{12}$ of those feet; or more accurately 15 feet, 1 inch, and 1 line $\frac{4}{9}$. And with this very force we actually find that bodies here upon earth do really descend; for a pendulum oscillating seconds in the latitude of *Paris* will be 3 *Paris* feet, and 8 lines ½ in length, as Mr. *Huygens* has observed. And the space which a heavy body describes by falling in one second of time is to half the length of this pendulum in the duplicate ratio of the circumference of a circle to its diameter (as Mr. *Huygens* has also shewn), and is therefore 15 *Paris* feet, 1 inch, 1 line $\frac{7}{9}$. And therefore the force by which the moon is retained in its orbit becomes, at the very surface of the earth, equal to the force of gravity which we observe in heavy bodies there. And therefore (by Rule I and II) the force by which the moon is retained in its orbit is that very same force which we commonly call gravity; for, were gravity another force different from that, then bodies descending to the earth with the joint impulse of both forces would fall with a double velocity, and in the space of one second of time would describe 30⅙ *Paris* feet; altogether against experience.

This calculus is founded on the hypothesis of the earth's standing still; for if both earth and moon move about the sun, and at the same time about their common centre of gravity, the distance of the centres of the moon and earth from one another will be 60 ½ semi-diameters of the earth; as may be found by a computation from Prop. LX, Book I.

Scholium

The demonstration of this Proposition may be more diffusely explained after the following manner. Suppose several moons to revolve about the earth, as in the system of Jupiter or Saturn: the periodic times of these moons (by the argument of induction) would observe the same law which *Kepler* found to obtain among the planets; and therefore their centripetal forces would be reciprocally as the squares of the distances from the centre of the earth, by Prop. I, of this Book. Now if the lowest of these were very small, and were so near the earth as almost to touch the tops of the highest mountains, the centripetal force thereof, retaining it in its orb, would be very nearly equal to the weights

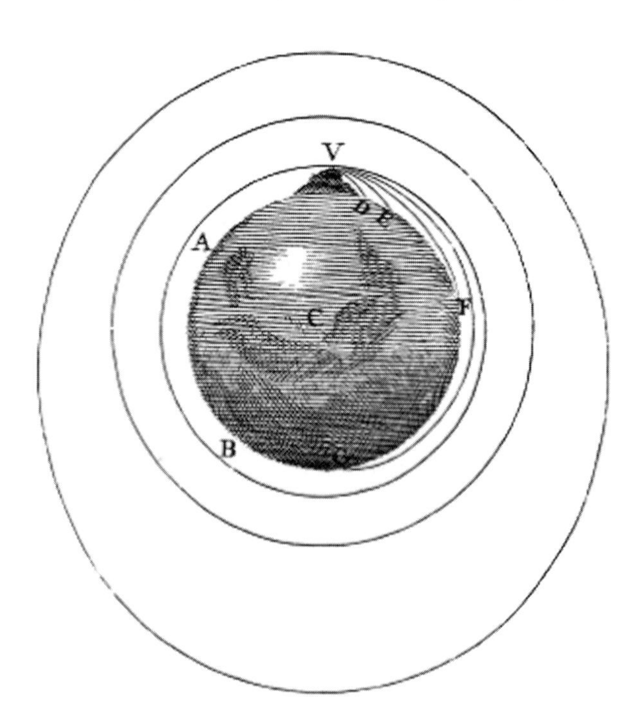

of any *terrestrial* bodies that should be found upon the tops of those mountains, as may be known by the foregoing computation. Therefore if the same little moon should be deserted by its centrifugal force that carries it through its orb; and so be disabled from going onward therein, it would descend to the earth; and that with the same velocity as heavy bodies do actually fall with upon the tops of those very mountains; because of the equality of the forces that oblige them both to descend. And if the force by which that lowest moon would descend were different from gravity, and if that moon were to gravitate towards the earth, as we find terrestrial bodies do upon the tops of mountains, it would then descend with twice the velocity, as being impel led by both these forces conspiring together. Therefore since both these forces, that is, the gravity of heavy bodies, and the centripetal forces of the moons, respect the centre of the earth, and are similar and equal between themselves, they will (by Rule I and II) have one and the same cause. And therefore the force which retains the moon in its orbit is that very force which we commonly call gravity; because otherwise this little moon at the top of a mountain must either be without gravity, or fall twice as swiftly as heavy bodies are wont to do.

PROPOSITION V. THEOREM V.

That the circumjovial planets gravitate towards Jupiter; the circumsaturnal towards Saturn; the circumsolar towards the sun; and by the forces of their gravity are drawn off from rectilinear motions, and retained in curvilinear orbits.

For the revolutions of the circumjovial planets about Jupiter, of the circumsaturnal about Saturn, and of Mercury and Venus, and the other circumsolar planets, about the sun, are appearances of the same sort with the revolution of the moon about the earth; and therefore, by Rule II, must be owing to the same sort of causes; especially since it has been demonstrated, that the forces upon which those revolutions depend tend to the centres of Jupiter, of Saturn, and of the sun; and that those forces, in receding from Jupiter, from Saturn, and from the sun, decrease in the same proportion, and according to the same law, as the force of gravity does in receding from the earth.

Cor. 1. There is, therefore, a power of gravity tending to all the planets; for, doubtless, Venus, Mercury, and the rest, are bodies of the same sort with Jupiter and Saturn. And since all attraction (by Law III) is mutual, Jupiter will therefore gravitate towards all his own satellites, Saturn towards his, the earth towards the moon, and the sun towards all the primary planets.

Cor. 2. The force of gravity which tends to any one planet is reciprocally as the square of the distance of places from that planet's centre.

Cor. 3. All the planets do mutually gravitate towards one another, by Cor. 1 and 2. And hence it is that Jupiter and Saturn, when near their conjunction; by their mutual attractions sensibly disturb each other's motions. So the sun disturbs the motions of the moon; and both sun and moon disturb our sea, as we shall hereafter explain.

Scholium

The force which retains the celestial bodies in their orbits has been hitherto called centripetal force; but it being now made plain that it can be no other than a gravitating force, we shall hereafter call it gravity. For the cause of that centripetal force which retains the moon in its orbit will extend itself to all the planets, by Rule I, II, and IV.

PROPOSITION VI. THEOREM VI.

That all bodies gravitate towards every planet; and that the weights of bodies towards any the same planet, at equal distances from the centre of the planet, are proportional to the quantities of matter which they severally contain.

It has been, now of a long time, observed by others, that all sorts of heavy bodies (allowance being made for the inequality of retardation which they suffer from a small power of resistance in the air) descend to the earth *from equal heights* in equal times; and that equality of times we may distinguish to a great accuracy, by the help of pendulums. I tried the thing in gold, silver, lead, glass, sand, common salt, wood, water, and wheat. I provided two wooden boxes, round and equal: I filled the one with wood, and suspended an equal weight of gold (as exactly as I could) in the centre of oscillation of the other. The boxes hanging by equal threads of 11 feet made a couple of pendulums perfectly equal in weight and figure, and equally receiving the resistance of the air. And, placing the one by the other, I observed them to play together forward and backward, for a long time, with equal vibrations. And therefore the quantity of matter in the gold (by Cor. 1 and 6, Prop. XXIV, Book II) was to the quantity of matter in the wood as the action of the motive force (or *vis motrix*) upon all the gold to the action of the same upon all the wood: that is, as the weight of the one to the weight of the other: and the like happened in the other bodies. By these experiments, in bodies of the same weight, I could manifestly have discovered a difference of matter less than the thousandth part of the whole, had any such been. But, without all doubt, the nature of gravity towards the planets is the same as towards the earth. For, should we imagine our terrestrial bodies removed to the orb of the moon, and there, together with the moon, deprived of all motion, to be let go, so as to fall together towards the earth, it is certain, from what we have demonstrated before, that, in equal times, they would describe equal spaces with the moon, and of consequence are to the moon, in quantity of matter, as their weights to its weight. Moreover, since the satellites of Jupiter perform their revolutions in times which observe the sesquiplicate proportion of their distances from Jupiter's centre, their accelerative gravities towards Jupiter will be reciprocally as the squares of their distances from Jupiter's centre; that is, equal, at equal distances. And, therefore, these satellites, if supposed to fall *towards Jupiter* from equal heights, would describe equal spaces in equal times, in like manner as heavy bodies do on our earth. And, by the same argument, if the circumsolar planets were supposed to be let fall at equal distances from the sun, they would, in their descent *towards the sun*, describe equal spaces in equal times. But forces which equally accelerate unequal bodies must be as those bodies: that is to say, the weights of the planets towards the sun, must be as their quantities of matter. Further, that the weights of Jupiter and of his satellites towards the sun are proportional to the several quantities of their matter, appears from the exceedingly regular motions of the satellites (by Cor. 3, Prop. LXV, Book 1). For if some of those bodies were more strongly attracted to the sun in proportion to their quantity of matter than others, the motions of the satellites would be disturbed by that inequality of attraction (by Cor. 2, Prop. LXV, Book I). If, at equal distances from the sun, any satellite, in proportion to the quantity of its matter, did gravitate towards the sun with a force greater than Jupiter in proportion to his, according to any given proportion, suppose of d to e; then the distance between the centres of the sun and of the satellite's orbit would be always greater than the distance between the centres of the sun and of Jupiter nearly in the subduplicate of that proportion: as by some computations I have found. And if the satellite did gravitate towards the sun with a force, lesser in the proportion of e to d, the distance of the centre of the satellite's orb from the sun would be less than the distance of the centre of Jupiter from the sun in the subduplicate of the same proportion. Therefore if, at equal distances from the sun, the accelerative gravity of any satellite towards the sun were greater or less than the accelerative gravity of Jupiter towards the sun but by one $^1/_{1000}$ part of the whole gravity, the distance of the centre of the satellite's orbit from the sun would be greater or less than the distance of Jupiter from the sun by one $^1/_{2000}$ part of the whole distance; that is, by a fifth part of the distance of the utmost satellite from the centre of Jupiter; an eccentricity of the orbit which would be very sensible. But the orbits of the satellites are concentric to Jupiter, and therefore the accelerative gravities of Jupiter, and of all its satellites towards the sun, are equal among themselves. And by the same argument, the weights of Saturn and of his satellites towards the sun, at equal distances from the sun, are as their several quantities of matter; and the weights of the moon and of the earth towards the sun are either none, or accurately proportional to the masses of matter which they contain. But some they are, by Cor. 1 and 3, Prop. V.

But further; the weights of all the parts of every planet towards any other planet are one to another as the matter in the several parts; for if some parts did gravitate more, others less, than for the quantity of their matter, then the whole planet, according to the sort of parts with which it most abounds, would gravitate more or less than in proportion to the quantity of matter in the whole.

Nor is it of any moment whether these parts are external or internal; for if, for example, we should imagine the terrestrial bodies with us to be raised up to the orb of the moon, to be there compared with its body: if the weights of such bodies were to the weights of the external parts of the moon as the quantities of matter in the one and in the other respectively; but to the weights of the internal parts in a greater or less proportion, then likewise the weights of those bodies would be to the weight of the whole moon in a greater or less proportion; against what we have shewed above.

Cor. 1. Hence the weights of bodies do not depend upon their forms and textures; for if the weights could be altered with the forms, they would be greater or less, according to the variety of forms, in equal matter; altogether against experience.

Cor. 2. Universally, all bodies about the earth gravitate towards the earth; and the weights of all, at equal distances from the earth's centre, are as the quantities of matter which they severally contain. This is the quality of all bodies within the reach of our experiments; and therefore (by Rule III) to be affirmed of all bodies whatsoever. If the *aether*, or any other body, were either altogether void of gravity, or were to gravitate less in proportion to its quantity of matter, then, because (according to *Aristotle, Des Cartes*, and others) there is no difference betwixt that and other bodies but in *mere* form of matter, by a successive change from form to form, it might be changed at last into a body of the same condition with those which gravitate most in proportion to their quantity of matter; and, on the other hand, the heaviest bodies, acquiring the first form of that body, might by degrees quite lose their gravity. And therefore the weights would depend upon the forms of bodies, and with those forms might be changed: contrary to what was proved in the preceding Corollary.

Cor. 3. All spaces are not equally full; for if all spaces were equally full, then the specific gravity of the fluid which fills the region of the air, on account of the extreme density of the matter, would fall nothing short of the specific gravity of quicksilver, or gold, or any other the most dense body; and, therefore, neither gold, nor any other body, could descend in air; for bodies do not descend in fluids, unless they are specifically heavier than the fluids. And if the quantity of matter in a given space can, by any rarefaction, be diminished, what should hinder a diminution to infinity?

Cor. 4. If all the solid particles of all bodies are of the same density, nor can be rarefied without pores, a void, space, or vacuum must be granted. By bodies of the same density, I mean those whose *vires inertiae,* are in the proportion of their bulks.

Cor. 5. The power of gravity is of a different nature from the power of magnetism; for the magnetic attraction is not as the matter attracted. Some bodies are attracted more by the magnet; others less; most bodies not at all. The power of magnetism in one and the same body may be increased and diminished; and is sometimes far stronger, for the quantity of matter, than the power of gravity; and in receding from the magnet decreases not in the duplicate but almost in the triplicate proportion of the distance, as nearly as I could judge from some rude observations.

PROPOSITION VII. THEOREM VII.

That there is a power of gravity tending to all bodies, proportional to the several quantities of matter which they contain.

That all the planets mutually gravitate one towards another, we have proved before; as well as that the force of gravity towards every one of them, considered apart, is reciprocally as the square of the distance of places from the centre of the planet. And thence (by Prop. LXIX, Book I, and its Corollaries) it follows, that the gravity tending towards all the planets is proportional to the matter which they contain.

Moreover, since all the parts of any planet A gravitate towards any other planet B; and the gravity of every part is to the gravity of the whole as the matter of the part to the matter of the whole; and (by Law III) to every action corresponds an equal re-action; therefore the planet B will, on the other hand, gravitate towards all the parts of the planet A; and its gravity towards any one part will be to the gravity towards the whole as the matter of the part to the matter of the whole. Q.E.D.

Cor. 1. Therefore the force of gravity towards any whole planet arises from, and is compounded of, the forces of gravity towards all its parts. Magnetic and electric attractions afford us examples of this; for all attraction towards the whole arises from the attractions towards the several parts. The thing may be easily understood in gravity, if we consider a greater planet, as formed of a number of lesser planets, meeting together in one globe; for *hence it would appear that* the force of the whole must arise from the forces of the component parts. If it is objected, that, according to this law, all bodies with us must mutually gravitate one towards another, whereas no such gravitation any where appears, I answer, that since the gravitation towards these bodies is to the gravitation towards the whole earth as these bodies are to the whole earth, the gravitation towards them must be far less than to fall under the observation of our senses.

Cor. 2. The force of gravity towards the several equal particles of any body is reciprocally as the square of the distance of places from the particles; as appears from Cor. 3, Prop. LXXIV, Book I.

PROPOSITION VIII. THEOREM VIII.

In two spheres mutually gravitating each towards the other, if the matter in places on all sides round about and equi-distant from the centres is similar, the weight of either sphere towards the other will be reciprocally as the square of the distance between their centres.

After I had found that the force of gravity towards a whole planet did arise from and was compounded of the forces of gravity towards all its parts, and towards every one part was in the reciprocal proportion of the squares of the distances from the part, I was yet in doubt whether that reciprocal duplicate proportion did accurately hold, or but nearly so, in the total force compounded of so many partial ones; for it might be that the proportion which accurately enough took place in greater distances should be wide of the truth near the surface of the planet, where the distances of the particles are unequal, and their situation dissimilar. But by the help of Prop. LXXV and LXXVI, Book I, and their Corollaries, I was at last satisfied of the truth of the Proposition, as it now lies before us.

Cor. 1. Hence we may find and compare together the weights of bodies towards different planets; for the weights of bodies revolving in circles about planets are (by Cor. 2, Prop. IV, Book I) as the diameters of the circles directly, and the squares of their periodic times reciprocally; and their weights at the surfaces of the planets, or at any other distances from their centres, are (by this Prop.) greater or less in the reciprocal duplicate proportion of the distances. Thus from the periodic times of Venus, revolving about the sun, in $224^{d}.16\frac{3}{4}^{h}$, of the utmost circumjovial satellite revolving about Jupiter, in $16^{d}.16\frac{8}{15}^{h}$.; of the Huygenian satellite about Saturn in $15^{d}.22\frac{2}{3}^{h}$.; and of the moon about the earth in $27^{d}.7^{h}.43'$; compared with the mean distance of Venus from the sun, and with the greatest heliocentric elongations of the outmost circumjovial satellite from Jupiter's centre, 8′ 16″; of the Huygenian satellite from the centre of Saturn, 3′ 4″; and of the moon from the earth, 10′ 33″: by computation I found that the weight of equal bodies, at equal distances from the centres of the sun, of Jupiter, of Saturn, and of the earth, towards the sun, Jupiter, Saturn, and the earth, were one to another, as 1, $\frac{1}{1067}$, $\frac{1}{3021}$, and $\frac{1}{169282}$ respectively. Then because as the distances are increased or diminished, the weights are diminished or increased in a duplicate ratio, the weights of equal bodies towards the sun, Jupiter, Saturn, and the earth, at the distances 10000, 997, 791, and 109 from their centres, that is, at their very superficies, will be as 10000, 943, 529, and 435 respectively. How much the weights of bodies are at the superficies of the moon, will be shewn hereafter.

Cor. 2. Hence likewise we discover the quantity of matter in the several planets; for their quantities of matter are as the forces of gravity at equal distances from their centres; that is, in the sun, Jupiter, Saturn, and the earth, as 1, $\frac{1}{1067}$, $\frac{1}{3021}$ and $\frac{1}{169282}$ respectively. If the parallax of the sun be taken greater or less than 10″ 30‴, the quantity of matter in the earth must be augmented or diminished in the triplicate of that proportion.

Cor. 3. Hence also we find the densities of the planets; for (by Prop. LXXII, Book I) the weights of equal and similar bodies towards similar spheres are, at the surfaces of those spheres, as the diameters of the spheres and therefore the densities of dissimilar spheres are as those weights applied to the diameters of the spheres. But the true diameters of the Sun, Jupiter, Saturn, and the earth, were one to another as 10000, 997, 791, and 109; and the weights towards the same as 10000, 943, 529, and 435 respectively; and therefore their densities are as 100, 94½, 67, and 400. The density of the earth, which comes out by this computation, does not depend upon the parallax of the sun, but is determined by the parallax of the moon, and therefore is here truly defined. The sun, therefore, is a little denser than Jupiter, and Jupiter than Saturn, and the earth four times denser than the sun; for the sun, by its great heat, is kept in a sort of a rarefied state. The moon is denser than the earth, as shall appear afterward.

Cor. 4. The smaller the planets are, they are, *caeteris paribus*, of so much the greater density; for so the powers of gravity on their several surfaces come nearer to equality. They are likewise, *caeteris paribus*, of the greater density, as they are nearer to the sun. So Jupiter is more dense than Saturn, and the earth than Jupiter; for the planets were to be placed at different distances from the sun, that, according to their degrees of density, they might enjoy a greater or less proportion to the sun's heat. Our water, if it were removed as far as the orb of Saturn, would be converted into ice, and in the orb of Mercury would quickly fly away in vapour; for the light of the sun, to which its heat is proportional, is seven times denser in the orb of Mercury than with us: and by the thermometer I have found that a sevenfold heat of our summer sun will make water boil. Nor are we to doubt that the matter of Mercury is adapted to its heat, and is therefore more dense than the matter of our earth; since, in a denser matter, the operations of Nature require a stronger heat.

PROPOSITION IX. THEOREM IX.

That the force of gravity, considered downward from the surface of the planets, decreases nearly in the proportion of the distances from their centres.

If the matter of the planet were of an uniform density, this Proposition would be accurately true (by Prop. LXXIII. Book I). The error, therefore, can be no greater than what may arise from the inequality of the density.

...

General Scholium

The hypothesis of vortices is pressed with many difficulties. That every planet by a radius drawn to the sun may describe areas proportional to the times of description, the periodic times of the several parts of the vortices should observe the duplicate proportion of their distances from the sun; but that the periodic times of the planets may obtain the sesquiplicate proportion of their distances from the sun, the periodic times of the parts of the vortex ought to be in the sesquiplicate proportion of their distances. That the smaller vortices may maintain their lesser revolutions about *Saturn, Jupiter*, and other planets, and swim quietly and undisturbed in the greater vortex of the sun, the periodic times of the parts of the sun's vortex should be equal; but the rotation of the sun and planets about their axes, which ought to correspond with the motions of their vortices, recede far from all these proportions. The motions of the comets are exceedingly regular, are governed by the same laws with the motions of the planets, and can by no means be accounted for by the hypothesis of vortices; for comets are carried with very eccentric motions through all parts of the heavens indifferently, with a freedom that is incompatible with the notion of a vortex.

Bodies projected in our air suffer no resistance but from the air. Withdraw the air, as is done in Mr. *Boyle's* vacuum, and the resistance ceases; for in this void a bit of line down and a piece of solid gold descend with equal velocity. And the parity of reason must take place in the celestial spaces above the earth's atmosphere; in which spaces, where there is no air to resist their motions, all bodies will move with the greatest freedom; and the planets and comets will constantly pursue their revolutions in orbits given in kind and position, according to the laws above explained; but though these bodies may, indeed, persevere in their orbits by the mere laws of gravity, yet they could by no means have at first derived the regular position of the orbits themselves from those laws.

The six primary planets are revolved about the sun in circles concentric with the sun, and with motions directed towards the same parts, and almost in the same plane. Ten moons are revolved about the earth, Jupiter and Saturn, in circles concentric with them, with the same direction of motion, and nearly in the planes of the orbits of those planets; but it is not to be conceived that mere mechanical causes could give birth to so many regular motions, since the comets range over all parts of the heavens in very eccentric orbits; for by that kind of motion they pass easily through the orbs of the planets, and with great rapidity; and in their aphelions, where they move the slowest, and are detained the longest, they recede to the greatest distances from each other, and thence suffer the least disturbance from their mutual attractions. This most beautiful system of the sun, planets, and comets, could only proceed from the counsel and dominion of an intelligent and powerful Being. And if the fixed stars are the centres of other like systems, these, being formed by the like wise counsel, must be all subject to the dominion of One; especially since the light of the fixed stars is of the same nature with the light of the sun, and from every system light passes into all the other systems: and lest the systems of the fixed stars should, by their gravity, fall on each other mutually, he hath placed those systems at immense distances one from another.

This Being governs all things, not as the soul of the world, but as Lord over all; and on account of his dominion he is wont to be called *Lord God* παντοκράτωρ, or *Universal Ruler*; for *God* is a relative word, and has a respect to servants; and *Deity* is the dominion of God not over his own body, as those imagine who fancy God to be the soul of the world, but over servants. The Supreme God is a Being eternal, infinite, absolutely perfect; but a being, however perfect, without dominion, cannot be said to be Lord God; for we say, my God, your God, the God of *Israel*, the God of Gods, and Lord of Lords; but we do not say, my Eternal, your Eternal, the Eternal of *Israel*, the Eternal of Gods; we do not say, my Infinite, or my Perfect: these are titles which have no respect to servants. The word *God*[1] usually signifies *Lord*; but every lord is not a God. It is the dominion of a spiritual being which constitutes a God: a true, supreme, or imaginary dominion makes a true, supreme, or imaginary God. And from his true dominion it follows that the true God is a living, intelligent, and powerful Being; and, from his other perfections, that he is supreme, or most perfect. He is eternal and infinite, omnipotent and omniscient; that is, his duration reaches from eternity to eternity; his presence from infinity to infinity; he governs all things, and knows all things that are or can be done. He is not eternity or infinity, but eternal and infinite; he is not duration or space, but he endures and is present. He endures for ever, and is every where present; and by existing always and every where, he constitutes duration and space. Since every particle of space is *always*, and every indivisible moment of duration is *every where,* certainly the Maker and Lord of all things cannot be *never* and *no where.* Every soul that has perception is, though in different times and in different organs of sense and motion, still the same indivisible person. There are given successive parts in duration, co-existent parts in space, but neither the one nor the other in the person of a man, or his thinking principle; and much less can they be found in the thinking substance of God. Every man, so far as he is a thing that has perception, is one and the same man during his whole life, in all and each of his organs of sense. God is the same God, always and every where. He is omnipresent not *virtually* only, but also *substantially*; for virtue cannot subsist without substance. In him[2] are all things contained and moved; yet neither affects the other: God suffers nothing from the motion of bodies; bodies find no resistance from the omnipresence of God. It is allowed by all that the Supreme God exists necessarily; and by the same necessity he exists *always* and *every where.* Whence also he is all similar, all eye, all ear, all brain, all arm, all power to perceive, to understand, and to act; but in a manner not at all human, in a manner not at all corporeal, in a manner utterly unknown to us. As a blind mail has no idea of colours, so have we no idea of the manner by which the all-wise God perceives and understands all things. He is utterly void of all body and bodily figure, and can therefore neither be seen, nor heard, nor touched; nor ought he to be worshipped under the representation of any corporeal thing. We have ideas of his attributes, but what the real substance

[1] Dr. *Pocock* derives the Latin word *Deus* from the *Arabic du* (in the oblique case *di*), which signifies *Lord.* And in this sense princes are called *gods, Psal.* lxxxii. ver. 6; and *John* x. ver. 35. And *Moses* is called a *god* to his brother *Aaron*, and a *god* to *Pharaoh* (Exod. iv. ver. 16; and vii. ver. 1). And in the same sense the souls of dead princes were formerly, by the Heathens, called *gods*, but falsely, because of their want of dominion.

[2] This was the opinion of the Ancients. So *Pythagoras*, in *Cicer. de Nat. Deor.* lib. i *Thales, Anaxagoros, Virgil,* Georg. lib. iv. ver. 220; and AEneid, lib. vi. ver. 721. *Philo Allegor,* at the beginning of lib. i. *Aratus,* in his Phaenom. at the beginning. So also the sacred writers; as *St.Paul, Acts,* xvii. ver 27, 28. St. *John's* Gosp. chap. xiv. ver. 2. *Moses.* in *Deut.* iv. ver. 39; and x ver. 14. *David, Psal.* cxxxix. ver. 7, 8, 9. *Solomon, 1 Kings,* viii. ver. 27. *Job,* xxii. ver. 12, 13, 14. *Jeremiah,* xxiii. ver. 23, 24. The Idolaters opposed the sun, moon, and stars, the souls of men, and other parts of the world, to be parts of the Supreme God, and therefore to be worshipped; but erroneously.

of any thing is we know not. In bodies, we see only their figures and colours, we hear only the sounds, we touch only their outward surfaces, we smell only the smells, and taste the savours; but their inward substances are not to be known either by our senses, or by any reflex act of our minds: much less, then, have we any idea of the substance of God. We know him only by his most wise and excellent contrivances of things, and final causes: we admire him for his perfections; but we reverence and adore him on account of his dominion: for we adore him as his servants; and a god without dominion, providence, and final causes, is nothing else but Fate and Nature. Blind metaphysical necessity, which is certainly the same always and every where, could produce no variety of things. All that diversity of natural things which we find suited to different times and places could arise from nothing but the ideas and will of a Being necessarily existing. But, by way of allegory, God is said to see, to speak, to laugh, to love, to hate, to desire, to give, to receive, to rejoice, to be angry, to fight, to frame, to work, to build; for all our notions of God are taken from the ways of mankind by a certain similitude, which, though not perfect, has some likeness, however. And thus much concerning God; to discourse of whom from the appearances of things, does certainly belong to Natural Philosophy.

Hitherto we have explained the phenomena of the heavens and of our sea by the power of gravity, but have not yet assigned the cause of this power. This is certain, that it must proceed from a cause that penetrates to the very centres of the sun and planets, without suffering the least diminution of its force; that operates not according to the quantity of the surfaces of the particles upon which it acts (as mechanical causes use to do), but according to the quantity of the solid matter which they contain, and propagates its virtue on all sides to immense distances, decreasing always in the duplicate proportion of the distances. Gravitation towards the sun is made up out of the gravitations towards the several particles of which the body of the sun is composed; and in receding from the sun decreases accurately in the duplicate proportion of the distances as far as the orb of Saturn, as evidently appears from the quiescence of the aphelions of the planets; nay, and even to the remotest aphelions of the comets, if those aphelions are also quiescent. But hitherto I have not been able to discover the cause of those properties of gravity from phaenomena, and I frame no hypotheses; for whatever is not deduced from the phaenomena is to be called an hypothesis; and hypotheses, whether metaphysical or physical, whether of occult qualities or mechanical, have no place in experimental philosophy. In this philosophy particular propositions are inferred from the phenomena, and afterwards rendered general by induction. Thus it was that the impenetrability, the mobility, and the impulsive force of bodies, and the laws of motion and of gravitation, were discovered. And to us it is enough that gravity does really exist, and act according to the laws which we have explained, and abundantly serves to account for all the motions of the celestial bodies, and of our sea.

And now we might add something concerning a certain most subtle Spirit which pervades and lies hid in all gross bodies; by the force and action of which

Spirit the particles of bodies mutually attract one another at near distances, and cohere, if contiguous; and electric bodies operate to greater distances, as well repelling as attracting the neighbouring corpuscles; and light is emitted, reflected, refracted, inflected, and heats bodies; and all sensation is excited, and the members of animal bodies move at the command of the will, namely, by the vibrations of this Spirit, mutually propagated along the solid filaments of the nerves, from the outward organs of sense to the brain, and from the brain into the muscles. But these are things that cannot be explained in few words, nor are we furnished with that sufficiency of experiments which is required to an accurate determination and demonstration of the laws by which this electric and elastic Spirit operates.

Translated by Andrew Motte

Reading and Discussion Questions

1. Compare and contrast the structure of Newton's *Principia* with Euclid's *Elements*.
2. Read Newton's definitions in Book I. How, for example, does Newton define quantity of matter? Compare and contrast this definition with Descartes' idea that extension is the essence of matter. Can you explain what he means by quantity of motion in your own words?
3. How does the discussion of centripetal force in Definition 5 relate to Galileo's discussion of projectile motion in *On the Two New Sciences*? How does Newton bring terrestrial motion and planetary orbits under a common framework?
4. Explain the argument that Newton gives for the existence of absolute space in the Scholium to the Definitions.
5. Examine the three laws of motion that Newton puts forward in Book I. How closely related are Newton's laws of motion to Descartes' laws? How much is Newton borrowing from his predecessors and what is different?
6. What do the Rules of Philosophizing suggest about how he thinks of scientific method?
7. What does Newton's discussion of God in the General Scholium suggest about how he is thinking about the relationship between science and religion? What does Newton mean when he insists that he does not feign or contrive hypotheses?
8. Give examples of the range of phenomena that Newton is able to account for with his three Laws of Motion and Law of Universal Gravitation. In what ways does Newton make use of Kepler's laws in Book III? To what new phenomena does he apply them?

Natural History: On the Formation of the Planets (1749)

Georges-Louis Leclerc, Comte de Buffon

Article 1: Of the Formation of the Planets

Our subject being Natural History, we would willingly dispense with astronomical observations; but as the nature of the earth is so closely connected with the heavenly bodies, and such observations being calculated to illustrate more fully what has been said, it is necessary to give some general ideas of the formation, motion, and figure of the earth and other planets.

The earth is a globe of about three thousand leagues diameter; it is situated one thousand millions of leagues from the sun, around which it makes its revolution in three hundred and sixty-five days. This revolution is the result of two forces, the one may be considered as an impulse from right to left, or from left to right, and the other an attraction from above downwards, or beneath upwards, to a common center. The direction of these two forces, and their quantities, are so nicely combined and proportioned, that they produce an almost uniform motion in an ellipse, very near to a circle. Like the other planets the earth is opaque, it throws out a shadow; it receives and reflects the light of the sun, round which it revolves in a space of time proportioned to its relative distance and density. It also turns round its own axis once in twenty-four hours, and its axis is inclined 66 [1/2] degrees on the plane of the orbit. Its figure is sphcriodal, the two axes of which differ about [160 1/15th] part from each other, and the smallest axis is that round which the revolution is made.

These are the principal phenomena of the earth, the result of discoveries made by means of geometry, astronomy, and navigation. We shall not here enter into the detail of the proofs and observations by which those facts have been ascertained, but only make a few remarks to clear up what is still doubtful, and at the same time give our ideas respecting the formation of the planets, and the different changes through which it is possible they have passed before they arrived at the state [in which we see them today].

There have been so many systems and hypotheses framed upon the formation of the terrestrial globe, and the changes which it has undergone, that we may presume to add our conjectures to those who have written upon the subject; especially as we mean to support them with a greater degree of probability than has hitherto been done; and we are the more inclined to deliver our opinion upon this subject, from the hope that we shall enable the reader to pronounce on the difference between an hypotheses drawn from possibilities, and a theory founded on facts; between a system such as we are here about to present, on the formation and original state of the earth, and a physical history of its real condition, which has been given in the preceding discourse.

Galileo [discovered] the laws of falling bodies, and Kepler observed, that the area described by the principal planets in moving round the sun, and those of the satellites round the planets to which they belong, are proportional to the time of their revolutions, and that such periods were also in proportion to the square roots of the cubes of their distances from the sun, or principal planets. Newton found that the force which caused heavy bodies to fall on the surface of the earth, extended to the moon, and retained it in its orbit; that this force diminished in the same proportion as the square of the distance increases, and consequently that the moon is attracted by the earth; that the earth and planets are attracted by the sun; and that, in general, all bodies which revolve round a center, and describe areas proportioned to the times of their revolution, are attracted towards that point. This power, known by the name of gravity, is therefore diffused throughout all matter; planets, comets, the sun, the earth, and all nature, is subject to its laws, and it serves as a basis [for] the general harmony which reigns in the universe. Nothing is better proved in physics than the actual [and individualized] existence of this power in every material substance. Observation has confirmed the effects of this power, and geometrical calculations have determined the quantity and relations of it . . . .

[One thing causes us pause, and is in fact independent of this theory. This is the force of impulsion.] We evidently see the force of attraction always draws the planets towards the sun, [and] they would fall in a perpendicular line on that planet, if they were not repelled by some other power that obliges them to move in a straight line, and which impulsive force would compel them to fly off the tangents of their respective orbits, if the force of attraction ceased one moment. The force of impulsion was certainly communicated to the planets by the hand of the Almighty, when he gave motion to the universe; but we ought, as much as possible, to abstain in [natural philosophy] from having recourse to supernatural causes; and it appears that a probable reason may be given for this impulsive force, perfectly accordant with the laws of mechanics, and not by any means more astonishing than the changes and revolutions which may and must happen in the universe.

The sphere of the sun's attraction does not confine itself to the orbs of the planets, but extends to a remote distance, always decreasing in the same ratio as the square of the distance increases; it is demonstrated that the comets which are lost to our sight, in the regions of the sky, obey this power, and by it their motions, like that of the planets, are regulated. All these stars, whose tracks are so different, move round the sun, and describe areas proportioned to the time; the planets in ellipses more or less approaching a circle, and the comets in narrow ellipses of a great extent. Comets and planets move, therefore, by virtue of the force of attraction and impulsion, which continually acting at one time obliges them to describe these courses; but it must be remarked that comets pass over the solar system in all directions, and that the inclinations of their orbits are very different, insomuch, that although subject, like the planets, to the force of attraction, they have nothing in common with respect to their progressive or impulsive motions, but appear, in this respect, independent of each other: the planets, on the contrary, move round the sun in the same direction, and almost in the same plane, never exceeding [7 1/2] degrees of inclination in their planes, the most distant from their orbits. This conformity of position and direction in the motion of the planets, necessarily implies that their impulsive force has been communicated to them by one and the same cause.

May it not be imagined, with some degree of probability, that a comet falling into the body of the sun, will displace and separate some parts from the surface, and communicate to them a motion of impulsion, insomuch, that the planets may formerly have belonged to the body of the sun, and been detached therefrom by an impulsive force, which they still preserve?

This supposition appears to be at least as well founded as the opinion of Leibnitz, who supposes that the earth and planets have formerly been suns; and his system, of which an account will be given in the fifth article, would have been more comprehensive and more agreeable to probability, if he had raised himself to this idea. We agree with him in thinking that this effect was produced at the time when Moses said that God divided light from darkness; for, according to Leibnitz, light was divided from darkness when the planets were [darkened]; but in our supposition there was a physical separation, since the [opaque matter composing the bodies of the planets was actually separated] from the luminous matter which composes the sun.

This idea of the cause of the impulsive force of the planets will be found much less objectionable, when an estimation is made of the analogies and degrees of probability, by which it may be supported. In the first place, the motion of the planets is in the same direction, from West to East, and therefore, according to calculation it is sixty-four to one that such would not have been the case if they had not been indebted to the same cause for their impulsive forces.

This probably will be considerably augmented by the second analogy, viz. that the inclination of the planes of the orbits do not exceed 7 1/2 degrees; for by comparing the spaces, we shall find [it to be] twenty-four to one, that two planets [would be] found in their most distant planes at the same time, and consequently

[24^5], or 7,692,624 to one, that all six would by chance be thus placed; or what amounts to the same, there is a great degree of probability that the planets have been impressed with one common moving force, which has given them this position. But what can have bestowed this common impulsive motion, but the force and direction of the bodies by which it was originally communicated? It may therefore be concluded, with great likelihood, that the planets received their impulsive motion by one single stroke. This probability, which is almost equivalent to a certainty, being established, I seek to know what moving bodies could produce this effect, and I find nothing but comets capable of communicating a motion to such vast bodies. By examining the course of comets, we shall be easily persuaded that it is almost necessary for some of them occasionally to fall into the sun. That of 1680 approached so near, that at its perihelion it was not more distant from the sun than a sixteenth part of [the solar] diameter, and if it returns, as there is every appearance it will, in 2255, it may then possibly fall into the sun. That must depend on the re-encounters it will meet with in its road, and on the retardation it suffers in passing through the atmosphere of the sun.

We may therefore presume, with the great Newton, that comets sometimes fall into the sun; but this fall may be made in different directions. If they fall perpendicular, or in a direction not very oblique, they will remain in the sun, and serve for food to the fire which [consumes that star], and the motion of impulsion which they will have communicated to the sun, will produce no other effect than that of [displacing] it more or less, according as the mass of the comet will be more or less considerable; but if the fall of the comet is in a very oblique direction, which will most frequently happen, then the comet will only graze the surface of the sun, or slightly furrow it; and in this case, it may drive out some parts of matter to which it will communicate a common motion of impulsion, and these parts so forced out of the body of the sun, and even the comet itself, may then become planets, and turn round this luminary in the same direction, and in almost the same plane. We might perhaps calculate, what quantity of matter, velocity, and direction a comet should have to impel from the sun an equal quantity of matter to that which the six planets and their satellites contain; but it will be sufficient to observe here, that all the planets with their satellites, do not make the 650th part of the mass of the sun, because the density of the large planets, Saturn and Jupiter, is less than that of the sun, and although the earth be four times, and the moon near five times more dense than the sun, they are nevertheless but as atoms in comparison with [the mass of this star].

However inconsiderable the 650th part may be, yet it certainly at first appears to require a very powerful comet to separate even that much from the body of the sun; but if we reflect on the prodigious velocity of comets in their perihelion, a velocity so much the greater, as they approach nearer the sun; if besides, we pay attention to the density and solidity of the matter of which they must be composed to suffer, without being destroyed, the inconceivable heat they endure; and consider the bright and solid light which shines through their dark and immense atmospheres, which surround and must obscure it, it cannot be doubted that comets are composed of extremely solid and dense matters, and that they contain a great quantity of matter in a small compass; that consequently a comet of no extraordinary bulk may have sufficient weight and velocity to displace the sun, and give a projectile motion to a quantity of matter, equal to the 650th part of the mass of this luminary. This perfectly agrees with what is known concerning the density of planets, which always decreases as their distance from the sun is increased, they having less heat to support; so that Saturn is less dense than Jupiter, and Jupiter much less than the Earth; therefore, if the density of the planets be as Newton asserts, proportional to the quantity of heat which they have to support, Mercury will be seven times more dense than the earth, and twenty eight times denser than the sun; and the comet of 1680 would be 28,000 times denser than the earth, or 112,000 times denser than the sun, and by supposing it as large as the earth, it would contain nearly an equal quantity of matter to the ninth part of the sun, or by giving it only the 100th part of the size of the earth, its mass would still be equal to the 900th part of the sun. From whence it is easy to conclude, that such a body, though it would be but a small comet, might separate and drive off from the sun a 900th or a 650th part, particularly if we attend to the immense velocity with which comets move when they pass in the vicinity of the sun.

...

The comet, therefore, by its oblique fall upon the surface of the sun, having driven therefrom a quantity of matter equal to the 650th part of its whole mass; this matter, which must be considered in a liquid state, will at first have formed a torrent, the grosser and less dense parts of which will have been driven the farthest, and the smaller and more dense having received only the like impulsion, will remain nearest its source; the force of the sun's attraction would inevitably act upon all the parts detached from [it], and constrain them to circulate around [its] body, and at the same time the mutual attraction of the particles of matter would form themselves into globes at different distances from the sun, the nearest of which necessarily moving with greater rapidity in their orbits than those at a distance.

The earth and planets, at the time of their quitting the sun, were in a state of total liquid fire: in this state they remained only as long as the violence of the heat which had produced it; and which heat necessarily underwent a gradual decay: it was in this state of fluidity that they took their circular forms, and that their regular motions raised the parts of their equators, and lowered their poles. [This figure, which accords so well with the laws of hydrostatics, necessarily supposes that the earth and planets have been in a fluid state, and I am here in agreement with Mr. Leibnitz. This state of fluidity would be caused by violent heating, and] the internal part of the earth must be a vitrifiable matter, of which sand, granite, etc. are the fragments and scoria.

It may, therefore, with some probability be thought, that the planets [originated from] the sun, that they were separated by a single stroke which gave to them a motion of impulsion, and that their position at different distances from the sun, proceeds only from their different densities. It now only remains, to complete this theory, to explain the diurnal motion of the planets, and the formation of the satellites; but this, far from adding difficulties to my hypothesis, seems on the contrary, to confirm it.

For the diurnal motion, or rotation, depends solely on the obliquity of the stroke, an oblique impulse therefore on the surface of a body will necessarily give it a rotative motion; this motion will be equal and always the same, if the body which receives it is homogeneous, and it will be unequal if the body is composed of heterogeneous parts, or of different densities; hence we may conclude, that in all the planets the matter is homogeneous, since their diurnal motions are equal, and regularly performed in the same period of time. [This is another proof of the separation of parts of greater and lesser density when the planets were formed].

But the obliquity of the stroke might be such as to separate from the body of the principal planet a small part of matter, which would of course continue to move in the same direction; these parts would be united, according to their densities, at different distances from the planet, by the force of their mutual attraction, and at the same time follow its course around the sun, by revolving about the body of the planet, nearly in the plane of its orbit. It is plain that those small parts so separated are the satellites: thus the formation, position, and direction of the motions of the satellites perfectly agree with our theory; for they have all the same motion in concentrical circles round their principal planet; their motion is in the same direction, and that nearly in the plane of their orbits. All these effects, which are common to them, and which depend on an impulsive force, can proceed only from one common cause, which is, impulsive motion, communicated to them by one and the same oblique stroke.

Translated by J. S. Barr

Reading and Discussion Questions

1. What Newtonian principles can Buffon assume in making his argument about the formation of the solar system?
2. What scenario does Buffon offer for the origin of the solar system, and what evidence does he give for thinking that this hypothesis might be correct?

Additional Resources

Apsell, Paula, Producer. Oxley, Chris, Director. *Newton's Dark Secrets*. [Documentary] USA: PBS, 2005. An entertaining introduction to Newton's physical, alchemical, and religious writings.

Christianson, John Robert. "Brahe, Tycho." In *The Complete Dictionary of Scientific Biography*. Vol. 19. Detroit: Charles Scribner's Sons, 2008, 380–5.

Cohen, I. Bernard. *The Birth of a New Physics.* 2nd ed. New York: W. W. Norton and Company, 1985. A scholarly and accessible (if occasionally whiggish) presentation of conceptual developments in astronomy. The diagrams are invaluable throughout, and especially helpful in understanding Kepler's three laws of motion.

Crowe, Michael J. *Mechanics from Aristotle to Einstein.* Santa Fe, NM: Green Lion Press, 2007. A valuable mixture of primary and secondary source material, the chapter on Galileo is particularly helpful.

Densmore, Dana, ed. *Selections from Newton's Principia.* Green Lion Press, 2004. The gold standard of readable introductions to the *Principia.*

Donahue, William H., ed. *Selections from Kepler's Astronomia Nova.* Santa Fe, NM: Green Lion Press, 2004. Kepler's argument against the literal interpretation of scripture in the introduction is fascinating.

Ferguson, Kitty. *Tycho and Kepler: The Unlikely Partnership That Forever Changed Our Understanding of the Heavens.* New York: Walker and Company, 2002. A very readable introduction to the lives of these extremely eccentric astronomers.

Gregory, Frederick. *Natural Science in Western History.* Boston, MA: Houghton Mifflin Company, 2008. A solid secondary source textbook, by a past president of the History of Science Society, for the whole period covered by the readings in this volume.

Kuhn, Thomas. *The Structure of Scientific Revolutions.* 3rd ed. Chicago: University of Chicago Press, 1996.

Losee, John. *A Historical Introduction to the Philosophy of Science.* 4th ed. Oxford: Oxford University Press, 2001. Losee's seventh chapter "The Seventeenth Century Attack on Aristotelian Philosophy" helps put this entire section in context.

Machamer, Peter, ed. *The Cambridge Companion to Galileo.* Cambridge: Cambridge University Press, 1998. Contains several important essays, but "Galileo on Science and Scripture" by Ernan McMullin provides vital context for understanding Galileo's trial.

Matthews, Michael R., ed. *The Scientific Background to Modern Philosophy: Selected Readings.* Indianapolis: Hackett Publishing Co., 1989.

Part IV

Investigating the Invisible: Light, Electricity, and Magnetism

Letter to Henry Oldenburg: Draft of "A Theory Concerning Light and Colors" (February 6, 1671/2)

Isaac Newton

48

To perform my late promise to you, I shall without further ceremony acquaint you, that in the beginning of the Year 1666 (at which time I applied myself to the grinding of Optic glasses of other figures than *Spherical*,) I procured me a Triangular glass-Prism, to try therewith the celebrated *Phenomena of Colors.* And in order thereto having darkened my chamber, and made a small hole in my window-shuts, to let in a convenient quantity of the Suns light, I placed my Prism at its entrance, that it might be thereby refracted to the opposite wall. It was at first a very pleasing divertisement, to view the vivid and intense colors produced thereby; but after a while applying myself to consider them more circumspectly, I became surprised to see them in an *oblong* form; which, according to the received laws of Refraction, I expected should have been circular.

They were terminated at the sides with straight lines, but at the ends, the decay of light was so gradual, that it was difficult to determine justly, what was their figure; yet they seemed *semicircular*.

Comparing the length of this colored *Spectrum* with its breadth, I found it about five times greater; a disproportion so extravagant, that it excited me to a more than ordinary curiosity of examining, from whence it might proceed. I could scarce think, that the various *Thickness* of the glass, or the termination with shadow or darkness, could have any Influence on light to Produce such an effect; yet I thought it not amiss to examine first these circumstances, and so tried, what would happen by transmitting light through parts of the glass of diverse thicknesses, or through holes in the window of diverse bignesses, or by setting the Prism without, so that the light might pass through it, and be refracted before it was terminated by the hole: But I found none of those circumstances material. The fashion of the colors was in all these cases the same.

Then I suspected, whether by any *unevenness* in the glass, or other contingent irregularity, these colors might be thus dilated. And to try this, I took another Prism like the former, and so placed it, that the light, passing through them both, might be refracted contrary ways, and so by the latter returned into that course, from which the former had diverted it. For, by this means I thought, the *regular* effects of the first Prism would be destroyed by the second Prism, but the irregular ones more augmented, by the multiplicity of refractions. The event was, that the light, which by the first Prism was diffused into an oblong form, was by the second reduced into an *orbicular* one with as much regularity, as when it did not at all pass through them. So that, whatever was the cause of that length, 'twas not any contingent irregularity.

I then proceeded to examine more critically, what might be effected by the difference of the incidence of Rays coming from diverse parts of the Sun; and to that end, measured the several lines and angles, belonging to the Image. Its distance from the hole or Prism was 22 foot; its utmost length 13¼ inches; its breadth 2⅝ inches; the diameter of the hole a of an inch; the angle, which the Rays, tending towards the middle of the image, made with those lines, in which they would have proceeded without refraction, 44 deg. 56′. And the vertical Angle of the Prism, 63 deg. 12′. Also the Refractions on both sides the Prism, that is, of the Incident, and Emergent Rays, were as near, as I could make them, equal, and consequently about 54 deg. 4′. And the Rays fell perpendicularly upon the wall. Now subducting the diameter of the hole from the length and breadth of the Image, there remains 13 Inches the length, and 2⅜ the breadth, comprehended by those Rays, which passed through the center of the said hole, and consequently the angle at the hole, which that breadth subtended, was about 31′, answerable to the Suns Diameter; but the angle, which its length subtended, was more than five such diameters, namely 2 deg. 49′.

Having made these observations, I first computed from them the refractive power of that glass, and found it measured by the *ratio* of the sines, 20 to 31. And then, by that *ratio*, I computed the Refractions of two Rays flowing from opposite parts of the Sun's *discus*, so as to differ 31′ in their obliquity of Incidence, and found, that the emergent Rays should have comprehended an angle of about 31′, as they did, before they were incident.

But because this computation was founded on the Hypothesis of the proportionality of the *sines* of Incidence, and Refraction, which though by my own & others Experience I could not imagine to be so erroneous, as to make that Angle but 31′, which in reality was 2 deg. 49′; yet my curiosity caused me again to take my Prism. And having placed it at my window, as before, I observed, that by turning it a little about its *axis* to and fro, so as to vary its obliquity to the light, more then by an angle of 4 or 5 degrees, the Colors were not thereby sensibly translated from their place on the wall, and consequently by that variation of Incidence, the quantity of Refraction was not sensibly varied. By this

Experiment therefore, as well as by the former computation, it was evident, that the difference of the Incidence of Rays, flowing from diverse parts of the Sun, could not make them after decussation diverge at a sensibly greater angle, than that at which they before converged; which being, at most, but about 31 or 32 minutes, there still remained some other cause to be found out, from whence it could be 2 deg. 49′.

Then I began to suspect, whether the Rays, after their trajection through the Prism, did not move in curve lines, and according to their more or less curvity tend to diverse parts of the wall. And it increased my suspicion, when I remembered that I had often seen a Tennis-ball, struck with an oblique Racket, describe such a curve line. For, a circular as well as a progressive motion being communicated to it by that streak, its parts on that side, where the motions conspire, must press and beat the contiguous Air more violently than on the other, and there excite a reluctancy and reaction of the Air proportionably greater. And for the same reason, if the Rays of light should possibly be globular bodies, and by their oblique passage out of one medium into another acquire a circulating motion, they ought to feel the greater resistance from the ambient *Aether*, on that side, where the motions conspire, and thence be continually bowed to the other. But notwithstanding this plausible ground of suspicion, when I came to examine it, I could observe no such curvity in them. And besides (which was enough for my purpose) I observed, that the difference betwixt the length of the Image, and diameter of the hole, through which the light was transmitted, was proportionable to their distance.

The gradual removal of these suspicions at length led me to the *Experimentum Crucis*, which was this: I took two boards, and placed one of them close behind the Prism at the window, so that the light might pass through a small hole, made in it for that purpose, and fall on the other board, which I placed at about 12 foot distance, having first made a small hole in it also, for some of that Incident light to pass through. Then I placed another Prism behind this second board, so that the light, trajected through both the boards, might pass through that also, and be again refracted before it arrived at the wall. This done, I took the first Prism in my hand, and turned it to and fro slowly about its Axis, so much as to make the several parts of the Image, cast on the second board, successively pass through the hole in it, that I might observe to what places on the wall the second Prism would refract them. And I saw by the variation of those places, that the light, tending to that end of the Image, towards which the refraction of the first Prism was made, did in the second Prism suffer a Refraction considerably greater than the light tending to the other end. And so the true cause of the length of that Image was detected to be no other, then that *Light* consists of *Rays differently refrangible*, which, without any respect to a difference in their incidence, were, according to their degrees of refrangibility, transmitted towards diverse parts of the wall.

When I understood this, I left off my aforesaid Glass-works; for I saw, that the perfection of Telescopes was hitherto limited, not so much for want of glasses truly figured according to the prescriptions of Optic Authors, which all men have hitherto imagined, as because that Light itself is a *Heterogeneous mixture of differently refrangible Rays*. So that, were a glass so exactly figured, as to collect any one sort of rays into one point, it could not collect those also into the same point, which having the same Incidence upon the same Medium are apt to suffer a different refraction. Nay, I wondered, that seeing the difference of refrangibility was so great, as I found it, Telescopes should arrive to that perfection they are now at. For, measuring the refractions in one of my Prisms, I found, that supposing the common *sine* of Incidence upon one of its planes was 44 parts, the *sine* of refraction of the utmost Rays on the red end of the Colors, made out of the glass into the Air, would be 68 parts, and the *sine* of refraction of the utmost rays on the other end, 89 parts: So that the difference is about a 24th or 25th part of the whole refraction. And consequently, the object-glass of any Telescope cannot collect all the rays, which come from one point of an object so as to make them convene at its *focus* in less room then in a circular space, whose diameter is the 50th part of the Diameter of its Aperture; which is an irregularity, some hundreds of times greater, then a circularly figured *Lens*, of so small a section as the Object glasses of long Telescopes are, would cause by the unfitness of its figure, were Light *uniform*.

This made me take *Reflections* into consideration, and finding them regular, so that the Angle of Reflection of all sorts of Rays was equal to their Angle of Incidence; I understood, that by their mediation Optic instruments might be brought to any degree of perfection imaginable, provided a *Reflecting* substance could be found, which would polish as finely as Glass, and *reflect* as much light, as glass transmits, and the art of communicating to it a *Parabolic* figure be also attained. But these seemed very great difficulties, and I almost thought them insuperable, when I further considered, that every irregularity in a reflecting superficies makes the rays stray 5 or 6 times more out of their due course, than the like irregularities in a refracting one: So that a much greater curiosity would be here requisite, than in figuring glasses for Refraction.

Amidst these thoughts I was forced from *Cambridge* by the Intervening Plague, and it was more than two years, before I proceeded further. But then having thought on a tender way of polishing, proper for metal, whereby, as I imagined, the figure also would be corrected to the last; I began to try, what might be effected in this kind, and by degrees so far perfected an Instrument (in the essential parts of it like that I sent to *London*,) by which I could discern Jupiter's 4 Concomitants, and shewed them diverse times to two others of my acquaintance. I could also discern the Moon-like phase of Venus, but not very distinctly, nor without some niceness in disposing the Instrument.

From that time I was interrupted till this last Autumn, when I made the other. And as that was sensibly better than the first (especially for Day-Objects,) so I doubt not, but they will be still brought to a much greater perfection by their endeavors, who, as you inform me, are taking care about it at *London*.

I have sometimes thought to make a Microscope, which in like manner should have, instead of an Object-glass, a Reflecting piece of metall. And this I hope they will also take into consideration, For those Instruments seem as capable of improvement as *Telescopes*, and perhaps more, because but one reflective piece of metal is requisite in them, and you may perceive by the annexed diagram, where A B represented the object metal, C D the eye glass, F their common Focus, and 0 the other focus of the metal, in which the object is placed.

But to return from this digression, I told you, that Light is not similar, or homogeneal, but consists of *difform* Rays, some of which are more refrangible than others: So that of those, which are alike incident on the same medium, some shall be more refracted than others, and that not by any virtue of the glass, or other external cause, but from a predisposition, which every particular Ray hath to suffer a Particular degree of Refraction.

I shall now proceed to acquaint you with another more notable deformity in its Rays, wherein the *Origin of Colors* is enfolded. A naturalist would scarce expect to see ye science of those become mathematical, & yet I dare affirm that there is as much certainty in it as in any other part of Optics. For what I shall tell concerning them is not a Hypothesis but most rigid consequence, not conjectured by barely inferring 'tis thus because not otherwise or because it satisfies all phenomena (the Philosophers universal Topic), but evinced by ye mediation of experiments concluding directly& without any suspicion of doubt. To continue the historical narration of these experiments would make a discourse too tedious & confused, & therefore I shall rather lay down the *Doctrine* first, and then, for its examination, give you an instance or two of the *Experiments*, as a specimen of the rest.

The Doctrine you will find comprehended and illustrated in the following propositions.

As the Rays of light differ in degrees of Refrangibility, so they also differ in their disposition to exhibit this or that particular color. Colors are not *Qualifications of light*, derived from Refractions, or Reflections of natural Bodies (as 'tis generally believed,) but *Original* and *connate* properties, which in diverse Rays are diverse. Some Rays are disposed to exhibit a red color and no other; some a yellow and no other, some' a green and no other, and so of the rest. Nor are there only Rays proper and particular to the more eminent colors, but even to all their intermediate gradations.

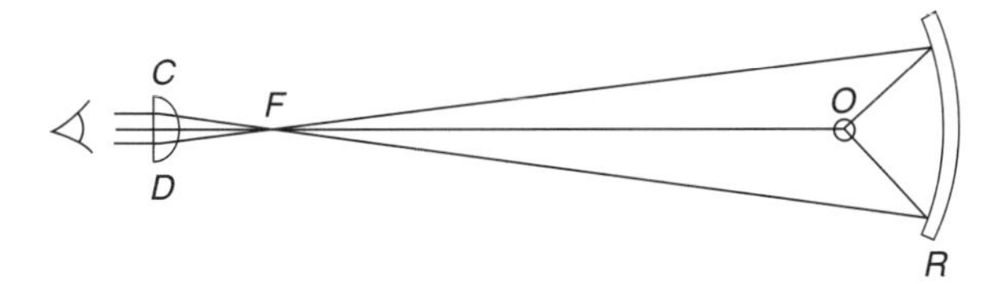

To the same degree of Refrangibility ever belongs the same color, and to the same color ever belongs the same degree of refrangibility. The *least Refrangible* Rays are all disposed to exhibit a *Red* color, and contrarily those Rays, which are disposed to exhibit a Red color, are all the least refrangible: So the most *refrangible* Rays are all disposed to exhibit a deep *Violet* color, and contrarily those which are apt to exhibit such a violet color, are all the most Refrangible. And so to all the intermediate colors in a continued series belong intermediate degrees of refrangibility. And this Analogy 'twixt colors, and refrangibility, is very precise and strict; The Rays always either exactly agreeing in both, or proportionally disagreeing in both.

The species of color, and degree of Refrangibility proper to any particular sort of Rays, is not mutable by Refraction, nor by Reflection from natural bodies, nor by any other cause, that I could yet observe. When any one sort of Rays hath been well parted from those of other kinds, it hath afterwards obstinately retained its color, notwithstanding my utmost endeavors to change it. I have refracted it with Prisms and reflected it with Bodies, which in Day-light were of other colors; I have intercepted it with the colored film of Air interceding two compressed plates of glass; transmitted it through colored Mediums, and through Mediums irradiated with other sort of Rays, and diversely terminated it, and yet could never produce any new color out of it. It would by contracting or dilating become more brisk, or faint, and by the loss of many Rays, in some cases very obscure and dark; but I could never see it changed *in specie*.

Yet seeming transmutations of Colors may be made where there is any mixture of diverse sorts of Rays. For in such mixtures, the component colors appear not, but, by their mutual allaying each other, constitute a middling color. And therefore, if by refraction, or any other of the aforesaid causes, the deform Rays, latent in such a mixture, be separated, there shall emerge colors different from the color of the composition. Which colors are not New generated, but only made Apparent by being parted; for if they be again entirely mixed and blended together, they will again compose that color, which they did before separation. And for the same reason, Transmutations made by the convening of diverse colors are not real; for when the deform Rays are again severed, they will exhibit the very same colors, which they did before they entered the composition; as you see, *Blue* and *Yellow* powders, when finely mixed, appear to the naked eye *Green*, and yet the colors of the Component corpuscles are not thereby really transmuted, but only blended. For, when viewed with a good Microscope, they still appear *Blue* and *Yellow* interspersedly.

There are therefore two sorts of colors. The one original and simple, the other compounded of these. The Original or primary colors are, *Red*, *Yellow*, *Green*, *Blue*, and a *Violet-purple*, together with Orange, Indico, and an indefinite variety of Intermediate gradations.

The same colors in *Specie* with these Primary ones may be also produced by composition: For, a mixture of *Yellow* and *Blue* makes *Green*; of *Red* and *Yellow* makes *Orange*; of *Orange* and *Yellowish green* makes *yellow*. And in general, if any two Colors be mixed, which in the series of those, generated by the Prism, are not too far distant one from another, they by their mutual alloy compound that color, which in the said series appeared in the mid-way between them. But those, which are situated at too great a distance, do not so. *Orange* and *Indico* produce not the intermediate Green, nor Scarlet and Green the intermediate yellow.

But the most surprising and wonderful composition was that of *Whiteness*. There is no one sort of Rays which alone can exhibit this. 'Tis ever compounded, and to its composition are requisite all the aforesaid primary Colors, mixed in a due proportion. I have often with Admiration beheld, that all the Colors of the Prism being made to converge, and thereby to be again mixed as they were in the light before it was Incident upon the Prism, reproduced light, entirely and perfectly white, and not at all sensibly differing from the *direct* Light of the Sun, unless when the glasses, I used, were not sufficiently clear; for then they would a little incline it to *their* color.

Hence therefore it comes to pass, that *Whiteness* is the usual color of *Light*; for, Light is a confused aggregate of Rays indued with all sorts of Colors, as they are promiscuously darted from the various parts of luminous bodies. And of such a confused aggregate, as I said, is generated Whiteness, if there be a due proportion of the Ingredients; but if any one predominate, the Light must incline to that color; as it happens in the blue flame of Brimstone; the yellow flame of a Candle; and the various colors of the Fixed stars.

These things considered, the *manner*, how colors are produced by the Prism, is evident. For, of the Rays, constituting the incident light, since those which differ in Color proportionally differ in refrangibility, *they* by their unequal refractions must be severed and dispersed into an oblong form in an orderly succession from the least refracted Scarlet to the most refracted Violet. And for the same reason it is, that objects, when looked upon through a Prism, appear colored. For, the difform Rays, by their unequal Refractions, are made to diverge towards several parts of the *Retina*, and there express the Images of things colored, as in the former case they did the Suns Image upon a wall. And by this inequality of refractions they become not only colored, but also very confused and indistinct.

Why the Colors of the *Rainbow* appear in falling drops of Rain, is also from hence evident. For, those drops, which refract the Rays, disposed to appear purple, in greatest quantity to the Spectators eye, refract the Rays of other sorts so much less, as to make them pass beside it; and such are the drops in the inside of the *Primary* Bow, and on the outside of the *Second* or Exterior one. So those drops, which refract in greatest plenty the Rays, apt to appear red, toward the Spectators eye, refract those of other sorts so much more, as to make them pass beside it; and such are the drops on the exterior part of the *Primary*, and interior part of the *Secondary* Bow.

The odd Phenomena of an infusion of *Lignum Nephriticum*, *Leaf gold*, *Fragments of colored glass*, and some other transparently colored bodies, appearing in one position of one color, and of another in another, are on these grounds no longer riddles. For, those are substances apt to reflect one sort of light and transmit another; as may be seen in a dark room, by illuminating them with similar or uncompounded light. For then they appear of that color only, with which they are illuminated, but yet in one position more vivid and luminous than in another, accordingly as they are disposed more or less to reflect or transmit the incident color.

From hence also is manifest the reason of an unexpected Experiment, which Mr. *Hook* somewhere in his *Micrographia* relates to have made with two wedge-like transparent vessels, filled the one with a red, the other with a blue liquor: namely, that though they were severally transparent enough, yet both together became opaque; For, if one transmitted only red, and the other only blue, no rays could pass through both.

I might add more instances of this nature, but I shall conclude with this general one, that the Colors of all natural Bodies have no other origin than this, that they are variously qualified to reflect one sort of light in greater plenty then another. And this I have experimented in a dark Room by illuminating those bodies with uncompounded light of diverse colors. For by that means anybody may be made to appear of any color. They have there no appropriate color, but ever appear of the color of the light cast upon them, but yet with this difference, that they are most brisk and vivid in the light of their own day-light color. *Minium* appeared there of any color indifferently, with which 'tis illustrated, but yet most luminous in red, and so *Bise* appeared indifferently of any color with which 'tis illustrated, but yet most luminous in blue. And therefore *Minium* reflected Rays of any color, but most copiously those indued with red; and consequently when illustrated with day-light, that is, with all sorts of Rays promiscuously blended, those qualified with red shall abound most in the reflected light, and by their prevalence cause it to appear of that color. And for the same reason *Bise*, reflecting blue most copiously, shall appear blue by the excess of those Rays in its reflected light; and the like of other bodies. And that this is the entire and adequate cause of their colors, is manifest, because they have no power to change or alter the color of any sort of Rays incident apart, but put on all colors indifferently, with which they are enlightened.

These things being so, it can be no longer disputed, whether there be colors in the dark, nor whether they be the qualities of the objects we see, no nor perhaps, whether Light be a Body. For, since Colors are the *qualities* of Light, having its Rays for their entire and immediate subject, how can we think those Rays *qualities* also, unless one quality may be the subject of and sustain another; which in effect is to call it *substance*. We should not know Bodies for substances, were it not for their sensible qualities, and the Principal of those being now found due to something else, we have as good reason to believe that to be a substance also.

Besides, who ever thought any quality to be a *heterogeneous* aggregate, such as Light is discovered to be. But, to determine more absolutely, what Light is, after what manner refracted, and by what modes or actions it produces in our minds the Phantasms of Colors, is not so easy. And I shall not mingle conjectures with certainties.

Reviewing what I have written, I see the discourse itself will lead to diverse Experiments sufficient for its examination: And therefore I shall not trouble you further, than to describe one of those, which I have already insinuated.

In a darkened Room make a hole in the shut of a window, whose diameter may conveniently be about a third part of an inch, to admit a convenient quantity of the Sun's light: And there place a clear and colorless Prism, to refract the entering light towards the further part of the Room, which, as I said, will thereby be diffused into an oblong colored Image. Then place a *Lens* of about three foot radius (suppose a broad Object-glass of a three foot Telescope,) at the distance of about four or five foot from thence, through which all those colors may at once be transmitted, and made by its Refraction to convene at a further distance of about ten or twelve foot. If at that distance you intercept this light with a sheet of white paper, you will see the colors converted into whiteness again by being mingled. But it is requisite, that the *Prism* and *Lens* be placed steady, and that the paper, on which the colors are cast, be moved to and fro; for, by such motion, you will not only find, at what distance the whiteness is most perfect, but also see, how the colors gradually convene, and vanish into whiteness, and afterwards having crossed one another in that place where they compound Whiteness, are again dissipated, and severed, and in an inverted order retain the same colors, which they had before they entered the composition. You may also see, that, if any of the Colors at the *Lens* be intercepted, the Whiteness will be changed into the other colors. And therefore, that the composition of whiteness be perfect, care must be taken, that none of the colors fall beside the *Lens*.

In the annexed design of this Experiment, A B C represents the Prism set endwise to sight, close by the whole F of the window E G. Its vertical Angle A B C may conveniently be about 60 degrees: M N designs the *Lens*. Its breadth 2½ or 3 inches. S F one of the straight lines, in which difform Rays may be conceived to flow successively from the Sun. F P and F R two of those Rays unequally refracted, which the *Lens* makes to converge towards Q, and after decussation to diverge again. And H I the paper, at diverse distances, on which

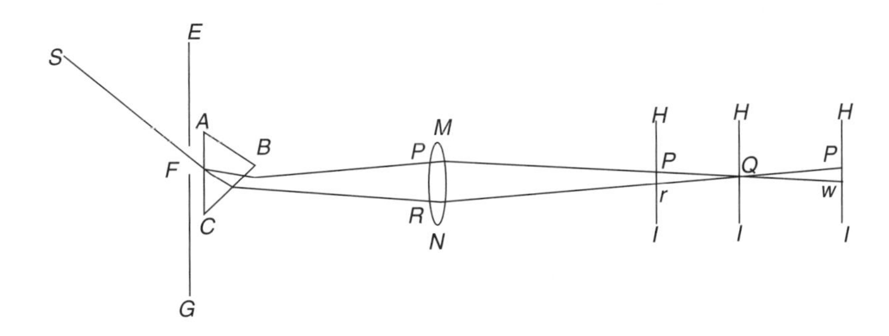

the colors are projected: which in Q constitute *Whiteness*, but are *Red* and *Yellow* in R, r, and p, and *Blue* and *Purple* in P, p, and π.

If you proceed further to try the impossibility of changing any uncompounded color (which I have asserted in the third and thirteenth Propositions,) 'tis requisite that the Room be made very dark, least any scattering light, mixing with the color, disturb and allay it, and render it compound, contrary to the design of the Experiment. 'Tis also requisite, that there be a perfecter separation of the Colors, than, after the manner above described, can be made by the Refraction of one single Prism, and how to make further separations, will scarce be difficult to them, that consider the discovered laws of Refractions. But if trial shall be made with colors not thoroughly separated, there must be allowed changes proportionable to the mixture. Thus if compound Yellow light fall upon blue *Bise*, the Bise will not appear perfectly yellow, but rather green, because there are in the yellow mixture many rays indued with green, and Green being less remote from the usual blue color of Bise than yellow, is the more copiously reflected by it.

In like manner, if any one of the Prismatick colors, suppose Red, be intercepted, on design to try the asserted impossibility of reproducing that Color out of the others, which are pretermitted; 'tis necessary, either that the colors be very well parted before the red be intercepted, or that together with the red the neighboring colors, into which any red is secretly dispersed, (that is, the yellow, and perhaps green too) be intercepted, or else, that allowance be made for the emerging of so much red out of the yellow & green, as may possibly have been diffused, and scatteringly blended in those colors. And if these things be observed, the new Production of Red, or any intercepted color will be found impossible.

This, I conceive, is enough for an Introduction to Experiments of this kind; which if any of the *R. Society* shall be so curious as to prosecute, I should be very glad to be informed with what success: That, if anything seem to be defective, or to thwart this relation, I may have an opportunity of giving further direction about it, or of acknowledging my errors, if I have committed any.

Your humble Servt,
Isaac NEWTON

Reading and Discussion Questions

1. Describe the two prism experiments that Newton performs. What does he take them to show or not show about the nature of light?
2. What does this discussion tell us about how Newton understands the role of hypotheses and experiments in natural philosophy?

Critique of Newton's "Theory of Light and Colors" (delivered to the Royal Society on February 15, 1671/2)

Robert Hooke

49

Mr. Hooke's considerations upon Mr. Newton's discourse on light and colours were read. Mr. Hooke was thanked for the pains taken in bringing in such ingenious reflections; and it was ordered, that this paper should be registered, and a copy of it immediately sent to Mr. Newton: and that in the mean time the printing of Mr. Newton's discourse by itself might go on, and if he did not contradict it; and that Mr. Hooke's paper might be printed afterwards, it not being thought fit to print them together, lest Mr. Newton should look upon it as a disrespect, in printing so sudden a refutation of a discourse of his, which had met with so much applause at the Society but a few days before.

Mr. Hooke's paper was as follows:

"I have perused the discourse of Mr. Newton about colours and refractions, and I was not a little pleased with the niceness and curiosity of his observations. But, tho' I wholly agree with him as to the truth of those he hath alleged, as having, by many hundreds of trials, found them so; yet as to his hypothesis of solving the phenomæna of colours thereby, I confess, I cannot see yet any undeniable argument to convince me of the certainty threof. For all the experiments and observations I have hitherto made, nay, and even those very experiments, which he alledgeth, do seem to me to prove, that *white* is nothing but a pulse or motion, propagated through an homogeneous, uniform and transparent medium: and that colour is nothing but the disturbance of that light, by the communication of that pulse to other transparent mediums, that is, by the refraction thereof: that *whiteness* and *blackness* are nothing but the plenty or scarcity of the undisturbed rays of light: and that the two colours (than the which there are not more uncompounded in nature) are nothing but the effects of a compounded pulse, or disturbed propagation of motion caused by refraction.

But, how certain soever I think myself of my hypothesis (which I did not take up without first trying some hundreds of experiments) yet I should be very glad to meet with one *experimentum crucis* from Mr. Newton, that should divorce me from it. But it is not that, which he so calls, will do the turn; for the same phænomenom will be solved by my hypothesis, as well as by his, without any manner of difficulty or straining: nay, I will undertake to shew another hypothesis, differing from both his and mine, that shall do the same thing.

That the ray of light is as it were split or rarified by refraction, is most certain; and that thereby a differing pulse is propagated, both on those sides, and in all the middle parts of the ray, is easy to be conceived: and also, that differing pulses or compound motions should make differing impressions on the eye, brain, or sense, is also easy to be conceived: and that, whatever refracting medium does again reduce it to its primitive simple motion by destroying the adventitious, does likewise restore it to its primitive whiteness and simplicity.

But why there is a neccessity, that all those motions, or whatever else it be that makes colours, should be originally in the simple rays of light, I do not yet understand the necessity of, no more than that all those sounds must be in the air of the bellows, which are afterwards heard to issue from the organ-pipes; or in the string, which are afterwards, by different stoppings and strikings produced; which string (by the way) is a pretty representation of the shape of a refracted ray to the eye; and the manner of it may be somewhat imagined by the similitude thereof: for the ray is like the string, strained between the luminous object and the eye, and the stop or fingers is like the refracting surface, on the one side of which the string hath no motion, on the other a vibrating one. Now we may say indeed and imagine, that the rest or streightness of the string is caused by the cessation of motions, or coalition of all vibrations; and that all the vibrations are dormant in it: but yet it seems more natural to me to imagine it the other way.

And I am a little troubled, that this supposition should make Mr. Newton wholly lay aside the thoughts of improving telescopes and microscopes by refractions; since it is not improbable, but that he, that hath made so very good an improvement of telescopes by his own trials upon reflection, would, if he had prosecuted it, have done more by refraction. And that reflection is not the only way of improving telescopes, I may possibly hereafter shew some proof of. The truth is, the difficulty of removing that inconvenience of the splitting of the ray, and consequently of the effect of colours, is very great; but yet not insuperable. I have made many trials, both for telescopes and microscopes by reflection, which I have mentioned in my *Micrographia*, but deserted it as to telescopes, when I considered, that the focus of the spherical concave is not a point but a line, and that the rays are less true reflected to a point by a concave, than refracted by a convex; which made me seek that by refraction, which I found could not rationally be expected by reflection: nor indeed could I find any effect of it by one of six foot radius, which, about seven or eight years since, Mr. Reeve made for Mr. Gregory, with which I made several trials; but it now appears it was for want of a good encheiria (from which cause many good experiments have been lost) both which considerations discouraged me from attempting further that way; especially since I found the parabola much more difficult to describe, than the hyperbola or ellipsis. And I was wholly taken from the thoughts of it, by lighting on divers ways, which in theory answered all I could wish for; tho' having much more business, I could not attend to bring them into use for telescopes; tho' for microscopes I have a good while used it. Thus much as to the preamble; I shall now consider the propositions themselves.

First then, Mr. Newton alledgeth, that as rays of light differ in refrangibility, so they differ in their disposition to exibit this or that colour: with which I do in the main agree; that is, that the ray by refraction is, as it were, split or rarified, and that the one side, namely that which is most refracted, gives a *blue*, and that which is least a *red*: the intermediate are the dilutings and intermixtures of those two, which I thus explain. The motion of light in an uniform medium, in which it is generated, is propagated by simple and uniform pulses or waves, which are at right angles with the line of direction; but falling obliquely on the refracting meduim, it receives another impression or motion, which disturbs the former motion, somewhat like the vibration of a string: and that, which was before a line, now becomes a triangular superficies, in which the pulse is not propagated at right angles with its line of direction, but ascew, as I have more at large explained in my *Micrographia*; and that, which makes excursions on the one side, impresses a compound motion on the bottom of the eye, of which we have the imagination of *red*; and that, which makes excursions on the other, causes a sensation, which we imagine a *blue*; and so of all the intermediate dilutings of those colours. Now, that the intermediate are nothing but the dilutings of those two primary, I hope I have sufficiently proved by the experiment of the two wedge-like boxes, described in my Micrographia. Upon this account I cannot assent to the latter part of the proposition, that colours are not qualifications of light, derived from refractions, or reflections of natural bodies, but original and connate properties, &c.

The second proposition I wholly allow, not exactly in the sense there meant, but with my manner of expressing it; that is, that part of the split ray, which is most bent, exhibits a blue, that which is least, a red, and the middle parts middling colours; and that those parts will always exhibit those colours till the compound motions are destroyed, and reduced by other motions to one simple and uniform pulse as it was at first.

And this will easily explain and give a reason of the phænomena of the third proposition, to which I do readily assent in all cases, except where the split ray is made by another refraction, to become intire and uniform, again to diverge and separate, which explains his fourth proposition.

But as to the fifth, that there are an indefinite variety of primary or original colours, amongst which are yellow, green, violet, purple, orange, &c. and an infinite number of intermediate gradations, I cannot assent thereunto, as suppposing it wholly useless to multiply entities without necessity, since I have elsewhere shewn, that all the varieties of colours in the world may be made of two. I agree in the sixth, but cannot approve of his way of explicating the seventh. How the split ray being made doth produce a clear and uniform light, I have before shewed; that is, by being united thereby from a superficial motion, which is susceptible of two, to a lineary, which is susceptible of one only motion; and it is as easy to conceive how all those motions again appear after the rays are again split or rarified. He, that shall but a little consider the undulations on the surface of a small river of water, in a gutter, or the like, will easily see the whole manner curiously exemplified.

The eighth proposition I cannot at all assent to, for the reasons above; and the reasons of the blue flame of brimstone, of the yellow of a candle, the green of copper, and the various colours of the stars, and other luminous bodies, I take to proceed from quite another cause, easily explained by my former hypothesis.

I agree with the observations of the ninth, tenth, and eleventh, though not with his theory, as finding it not absolutely necessary, being as easily and naturally explained and solved by my hypothesis.

The reason of the phænomena of my experiment, which he alledgeth, is as easily solvable by my hypothesis as by his; as are also those, which are mentioned in the thirteenth. I do not therefore see any absolute necessity to believe his theory demonstrated, since I can assure Mr. Newton, I cannot only solve all the phænomena of light and colours by the hypothesis I have formerly printed, and now explicate them by, but by two or three other very differing from it, and from this, which he hath described in his ingenious discourse.

Nor would I be understood to have said all this against his theory, as it is an hypothesis; for I do most readily agree with them in every part thereof, and esteem it very subtil and ingenious, and capable of solving all the phænomena of colours: but I cannot think it to be the only hypothesis, nor so certain as mathematical demonstrations.

But grant his first proposition, that light is a body, and that as many colours as degrees thereof as there may be, so many sorts of bodies there may be, all which compounded together would make white; and grant further, that all luminous bodies are compounded of such substances condensed, and that whilst they shine, they do continually send out an indefinite quantity thereof, every way in orbem, which in a moment of time doth disperse itself to the utmost and most indefinite bounds of the universe; granting these, I say, I do suppose there will be no great difficulty to demonstrate all the rest of his curious theory: though yet, methinks, all the coloured bodies in the world compounded together should not make a white body, and I should be glad to see an experiment of that kind done on the other side. If my supposition be granted, that light is nothing but a simple and uniform motion, or pulse of a homogeneous and adopted (that is a transparent) medium, propagated from the luminous body in orbem, to all imaginable distances in a moment of time, and that that motion is first begun by some other kind of motion in the luminous body; such as by the dissolution of sulphureous bodies by the air, or by the working of the air, or the several component parts one upon another, in rotten wood, or putrifying fish, or by an external stroke, as in diamond, sugar, the sea-water, or two flints or crystal rubbed together; and that this motion is propagated through all bodies susceptible thereof, but is blended or mixt with other adventitious motions, generated by the obliquity of the stroke upon a refracting body; and that, so long as those motions remain distinct in the same part of the medium or propagated ray, so long they produce the same effect, but when blended by other motions, they produce other effects: and supposing, that by a direct contrary motion to the newly impressed, that adventitious one be destroyed and reduced to the first simple motion; I believe Mr. Newton will think it no difficult matter, by my hypothesis, to solve all the phænomena, not only of the prism, tinged liquors, and solid bodies, but of the colours of plated bodies, which seem to have the greatest difficulty. It is true, I can, in my supposition, conceive the white or uniform motion of light to be compounded of the compound motions of all the other colours, as in any one strait and uniform motion may be compounded of thousands of compound motions, in the same manner as Descartes explicates the reason of the refraction; but I see no necessity of it. If Mr. Newton hath any argument, that he supposes as absolute demonstration of his theory, I should be very glad to be convinced by it, the phænomena of light and colours being, in my opinion, as well worthy of contemplation, as any thing else in the world."

Reading and Discussion Questions

1. On what grounds does Hooke critique Newton's theory of light and colors, and what view does he seem to favor?
2. What sort of evidence does Hooke think that Newton would need to provide in order for Hooke to change his mind?

Opticks: Or, a Treatise of the Reflections, Refractions, Inflections, and Colours of Light (originally published in 1704, Query 31 was added to the 1718 edition)

Isaac Newton

50

Query 31

Have not the small Particles of Bodies certain Powers, Virtues, or Forces, by which they act at a distance, not only upon the Rays of Light for reflecting, refracting, and inflecting them, but also upon one another for producing a great Part of the Phænomena of Nature? For it's well known, that Bodies act one upon another by the Attractions of Gravity, Magnetism, and Electricity; and these Instances shew the Tenor and Course of Nature, and make it not improbable but that there may be more attractive Powers than these .... The Attractions of Gravity, Magnetism, and Electricity, reach to very sensible distances, and so have been observed by vulgar Eyes, and there may be others which reach to so small distances as hitherto escape Observation; and perhaps electrical Attraction may reach to such small distances, even without being excited by Friction.

...

All these things being consider'd, it seems probable to me, that God in the Beginning form'd Matter in solid, massy, hard, impenetrable, moveable Particles, of such Sizes and Figures, and with such other Properties, and in such Proportion to Space, as most conduced to the End for which he form'd them; and that these primitive Particles being Solids, are incomparably harder than any porous Bodies compounded of them; even so very hard, as never to wear or break in pieces; no ordinary Power being able to divide what God himself made one in the first Creation. While the Particles continue entire, they may compose Bodies of one and the same Nature and Texture in all Ages: But should they wear away, or break in pieces, the Nature of Things depending on them, would be changed. Water and Earth, composed of old worn Particles and Fragments of Particles, would not be of the same Nature and Texture now, with Water and Earth composed of entire Particles in the Beginning. And therefore, that Nature may be lasting, the Changes of corporeal Things are to be placed only in the various Separations and new Associations and Motions of these permanent Particles; compound Bodies being apt to break, not in the midst of solid Particles, but where those Particles are laid together, and only touch in a few Points.

It seems to me farther, that these Particles have not only a *Vis inertiæ*, accompanied with such passive Laws of Motion as naturally result from that Force, but also that they are moved by certain active Principles, such as is that of Gravity, and that which causes Fermentation, and the Cohesion of Bodies. These Principles I consider, not as occult Qualities, supposed to result from the specifick Forms of Things, but as general Laws of Nature, by which the Things themselves are form'd; their Truth appearing to us by Phænomena, though their Causes be not yet discover'd. For these are manifest Qualities, and their Causes only are occult. And the *Aristotelians* gave the Name of occult Qualities, not to manifest Qualities, but to such Qualities only as they supposed to lie hid in Bodies, and to be the unknown Causes of manifest Effects: Such as would be the Causes of Gravity, and of magnetick and electrick Attractions, and of Fermentations, if we should suppose that these Forces or Actions arose from Qualities unknown to us, and uncapable of being discovered and made manifest. Such occult Qualities put a stop to the Improvement of natural Philosophy, and therefore of late Years have been rejected. To tell us that every Species of Things is endow'd with an occult specifick Quality by which it acts and produces manifest Effects, is to tell us nothing: But to derive two or three general Principles of Motion from Phænomena, and afterwards to tell us how the Properties and Actions of all corporeal Things follow from those manifest Principles, would be a very great step in Philosophy, though the Causes of those Principles were not yet discover'd: And therefore I scruple not to propose the Principles of Motion above-mention'd, they being of very general Extent, and leave their Causes to be found out.

Now by the help of these Principles, all material Things seem to have been composed of the hard and solid Particles above-mention'd, variously associated in the first Creation by the Counsel of an intelligent Agent. For it became him who created them to set them in order. And if he did so, it's unphilosophical to seek for any other Origin of the World, or to pretend that it might arise out of a Chaos by the mere Laws of Nature; though being once form'd, it may continue by those Laws for many Ages. For while Comets move in very excentrick Orbs in all manner of Positions, blind Fate could never make all the Planets move one and the same way in Orbs concentrick, some inconsiderable Irregularities excepted, which may have risen from the mutual Actions of Comets and Planets upon one another, and which will be apt to increase, till this System wants a Reformation. Such a wonderful Uniformity in the Planetary System must be allowed the Effect of Choice. And so must the Uniformity in the Bodies of Animals, they having generally a right and a left side shaped alike, and on either side of their Bodies two Legs behind, and either two Arms, or two Legs, or two Wings before upon their Shoulders, and between their Shoulders a Neck running down into a Back-bone, and a Head upon it; and in the Head two Ears, two Eyes, a Nose, a Mouth, and a Tongue, alike situated. Also the first Contrivance of those very artificial Parts of Animals, the Eyes, Ears, Brain, Muscles, Heart, Lungs, Midriff, Glands, Larynx, Hands, Wings, swimming Bladders, natural Spectacles, and other Organs of Sense and Motion; and the Instinct of Brutes and Insects, can be the effect of nothing else than the Wisdom and Skill of a powerful ever-living Agent, who being in all Places, is more able by his Will to move the Bodies within his boundless uniform Sensorium, and thereby to form and reform the Parts of the Universe, than we are by our Will to move the Parts of our own Bodies. And yet we are not to consider the World as the Body of God, or the several Parts thereof, as the Parts of God. He is an uniform Being, void of Organs, Members or Parts, and they are his Creatures subordinate to him, and subservient to his Will; and he is no more the Soul of them, than the Soul of Man is the Soul of the Species of Things carried through the Organs of Sense into the place of its Sensation, where it perceives them by means of its immediate Presence, without the Intervention of any third thing. The Organs of Sense are not for enabling the Soul to perceive the Species of Things in its Sensorium, but only for conveying them thither; and God has no need of such Organs, he being every where present to the Things themselves. And since Space is divisible *in infinitum*, and Matter is not necessarily in all places, it may be also allow'd that God is able to create Particles of Matter of several Sizes and Figures, and in several Proportions to Space, and perhaps of different Densities and Forces, and thereby to vary the Laws of Nature, and make Worlds of several sorts in several Parts of the Universe. At least, I see nothing of Contradiction in all this.

As in Mathematicks, so in Natural Philosophy, the Investigation of difficult Things by the Method of Analysis, ought ever to precede the Method of Composition. This Analysis consists in making Experiments and Observations, and in drawing general Conclusions from them by Induction, and admitting of no Objections against the Conclusions, but such as are taken from Experiments, or other certain Truths. For Hypotheses are not to be regarded in experimental Philosophy. And although the arguing from Experiments and Observations by Induction be no Demonstration of general Conclusions; yet it is the best way of arguing which the Nature of Things admits of, and may be looked upon as so much the stronger, by how much the Induction is more general. And if no

Exception occur from Phænomena, the Conclusion may be pronounced generally. But if at any time afterwards any Exception shall occur from Experiments, it may then begin to be pronounced with such Exceptions as occur. By this way of Analysis we may proceed from Compounds to Ingredients, and from Motions to the Forces producing them; and in general, from Effects to their Causes, and from particular Causes to more general ones, till the Argument end in the most general. This is the Method of Analysis: And the Synthesis consists in assuming the Causes discover'd, and establish'd as Principles, and by them explaining the Phænomena proceeding from them, and proving the Explanations.

In the two first Books of these Opticks, I proceeded by this Analysis to discover and prove the original Differences of the Rays of Light in respect of Refrangibility, Reflexibility, and Colour, and their alternate Fits of easy Reflexion and easy Transmission, and the Properties of Bodies, both opake and pellucid, on which their Reflexions and Colours depend. And these Discoveries being proved, may be assumed in the Method of Composition for explaining the Phænomena arising from them: An Instance of which Method I gave in the End of the first Book. In this third Book I have only begun the Analysis of what remains to be discover'd about Light and its Effects upon the Frame of Nature, hinting several things about it, and leaving the Hints to be examin'd and improv'd by the farther Experiments and Observations of such as are inquisitive. And if natural Philosophy in all its Parts, by pursuing this Method, shall at length be perfected, the Bounds of Moral Philosophy will be also enlarged. For so far as we can know by natural Philosophy what is the first Cause, what Power he has over us, and what Benefits we receive from him, so far our Duty towards him, as well as that towards one another, will appear to us by the Light of Nature. And no doubt, if the Worship of false Gods had not blinded the Heathen, their moral Philosophy would have gone farther than to the four Cardinal Virtues; and instead of teaching the Transmigration of Souls, and to worship the Sun and Moon, and dead Heroes, they would have taught us to worship our true Author and Benefactor, as their Ancestors did under the Government of *Noah* and his Sons before they corrupted themselves.

Reading and Discussion Questions

1. In the *Opticks* Newton allows a role for "queries." How does this fit with the official position Newton expresses toward hypotheses in the General Scholium in the *Principia*?
2. What kinds of issues do we find Newton speculating about in this text?

Treatise on Light, in Which Are Explained the Causes of That Which Occurs in Reflexion, and in Refraction, and Particularly in the Strange Refraction of Iceland Crystal (1690)

Christiaan Huygens

51

Preface

I wrote this Treatise during my sojourn in France twelve years ago, and I communicated it in the year 1678 to the learned persons who then composed the Royal Academy of Science, to the membership of which the King had done me the honour of calling, me. Several of that body who are still alive will remember having been present when I read it, and above the rest those amongst them who applied themselves particularly to the study of Mathematics; of whom I cannot cite more than the celebrated gentlemen Cassini, Römer, and De la Hire. And, although I have since corrected and changed some parts, the copies which I had made of it at that time may serve for proof that I have yet added nothing to it save some conjectures touching the formation of Iceland Crystal, and a novel observation on the refraction of Rock Crystal. I have desired to relate these particulars to make known how long I have meditated the things which now I publish, and not for the purpose of detracting from the merit of those who, without having seen anything that I have written, may be found to have treated of like matters: as has in fact occurred to two eminent Geometricians, Messieurs Newton and Leibnitz, with respect to the Problem of the figure of glasses for collecting rays when one of the surfaces is given.

One may ask why I have so long delayed to bring this work to the light. The reason is that I wrote it rather carelessly in the Language in which it appears, with the intention of translating it into Latin, so doing in order to obtain greater attention to the thing. After which I proposed to myself to give it out along with another Treatise on Dioptrics, in which I explain the effects of Telescopes and those things which belong more to that Science. But the pleasure of novelty being past, I have put off from time to time the execution of this design, and I know not when I shall ever come to an end if it, being often turned aside either by business or by some new study. Considering which I have finally judged that it was better worthwhile to publish this writing, such as it is, than to let it run the risk, by waiting longer, of remaining lost.

There will be seen in it demonstrations of those kinds which do not produce as great a certitude as those of Geometry, and which even differ much therefrom, since whereas the Geometers prove their Propositions by fixed and incontestable Principles, here the Principles are verified by the conclusions to be drawn from them; the nature of these things not allowing of this being done otherwise.

It is always possible to attain thereby to a degree of probability which very often is scarcely less than complete proof. To wit, when things which have been demonstrated by the Principles that have been assumed correspond perfectly to the phenomena which experiment has brought under observation; especially when there are a great number of them, and further, principally, when one can imagine and foresee new phenomena which ought to follow from the hypotheses which one employs, and when one finds that therein the fact corresponds to our prevision. But if all these proofs of probability are met with in that which I propose to discuss, as it seems to me they are, this ought to be a very strong confirmation of the success of my inquiry; and it must be ill if the facts are not pretty much as I represent them. I would believe then that those who love to know the Causes of things and who are able to admire the marvels of Light, will find some satisfaction in these various speculations regarding it, and in the new explanation of its famous property which is the main foundation of the construction of our eyes and of those great inventions which extend so vastly the use of them.

I hope also that there will be some who by following these beginnings will penetrate much further into this question than I have been able to do, since the subject must be far from being exhausted. This appears from the passages which I have indicated where I leave certain difficulties without having resolved them, and still more from matters which I have not touched at all, such as Luminous Bodies of several sorts, and all that concerns Colours; in which no one until now can boast of having succeeded. Finally, there remains much more to be investigated touching the nature of Light which I do not pretend to have disclosed, and I shall owe much in return to him who shall be able to supplement that which is here lacking to me in knowledge.

The Hague. The 8 January 1690

Chapter I: On Rays Propagated in Straight Lines

As happens in all the sciences in which Geometry is applied to matter, the demonstrations concerning Optics are founded on truths drawn from experience. Such are that the rays of light are propagated in straight lines; that the angles of reflexion and of incidence are equal; and that in refraction the ray is bent according to the law of sines, now so well known, and which is no less certain than the preceding laws.

The majority of those who have written touching the various parts of Optics have contented themselves with presuming these truths. But some, more inquiring, have desired to investigate the origin and the causes, considering these to be in themselves wonderful effects of Nature. In which they advanced some ingenious things, but not however such that the most intelligent folk do not wish for better and more satisfactory explanations. Wherefore I here desire to propound what I have meditated on the subject, so as to contribute as much as I can to the explanation of this department of Natural Science, which, not without reason, is reputed to be one of its most difficult parts. I recognize myself to be much indebted to those who were the first to begin to dissipate the strange obscurity in which these things were enveloped, and to give us hope that they might be explained by intelligible reasoning. But, on the other hand I am astonished also that even here these have often been willing to offer, as assured and demonstrative, reasonings which were far from conclusive. For I do not find that any one has yet given a probable explanation of the first and most notable phenomena of light, namely why it is not propagated except in straight lines, and how visible rays, coming from an infinitude of diverse places, cross one another without hindering one another in any way.

I shall therefore essay in this book, to give, in accordance with the principles accepted in the Philosophy of the present day, some clearer and more probable reasons, firstly of these properties of light propagated rectilinearly; secondly of light which is reflected on meeting other bodies. Then I shall explain the phenomena of those rays which are said to suffer refraction on passing through transparent bodies of different sorts; and in this part I shall also explain the effects of the refraction of the air by the different densities of the Atmosphere.

Thereafter I shall examine the causes of the strange refraction of a certain kind of Crystal which is brought from Iceland. And finally I shall treat of the various shapes of transparent and reflecting bodies by which rays are collected at a point or are turned aside in various ways. From this it will be seen with what facility, following our new Theory, we find not only the Ellipses, Hyperbolas, and other curves which Mr. Descartes has ingeniously invented for this purpose; but also those which the surface of a glass lens ought to possess when its other surface is given as spherical or plane, or of any other figure that may be.

It is inconceivable to doubt that light consists in the motion of some sort of matter. For whether one considers its production, one sees that here upon the Earth it is chiefly engendered by fire and flame which contain without doubt bodies that are in rapid motion, since they dissolve and melt many other bodies, even the most solid; or whether one considers its effects, one sees that when light is collected, as by concave mirrors, it has the property of burning as a fire does, that is to say it disunites the particles of bodies. This is assuredly the mark of motion, at least in the true Philosophy, in which one conceives the causes of all natural effects in terms of mechanical motions. This, in my opinion, we must necessarily do, or else renounce all hopes of ever comprehending anything in Physics.

And as, according to this Philosophy, one holds as certain that the sensation of sight is excited only by the impression of some movement of a kind of matter which acts on the nerves at the back of our eyes, there is here yet one reason more for believing that light consists in a movement of the matter which exists between us and the luminous body.

Further, when one considers the extreme speed with which light spreads on every side, and how, when it comes from different regions, even from those directly opposite, the rays traverse one another without hindrance, one may well understand that when we see a luminous object, it cannot be by any transport of matter coming to us from this object, in the way in which a shot or an arrow traverses the air; for assuredly that would too greatly impugn these two properties of light, especially the second of them. It is then in some other way that light spreads; and that which can lead us to comprehend it is the knowledge which we have of the spreading of Sound in the air.

We know that by means of the air, which is an invisible and impalpable body, Sound spreads around the spot where it has been produced, by a movement which is passed on successively from one part of the air to another; and that the spreading of this movement, taking place equally rapidly on all sides, ought to form spherical surfaces ever enlarging and which strike our ears. Now there is no doubt at all that light also comes from the luminous body to our eyes by some movement impressed on the matter which is between the two; since, as we have already seen, it cannot be by the transport of a body which passes from one to the other. If, in addition, light takes time for its passage—which we are now going to examine—it will follow that this movement, impressed on the intervening matter, is successive; and consequently it spreads, as Sound does, by spherical surfaces and waves: for I call them waves from their resemblance to those which are seen to be formed in water when a stone is thrown into it, and which present a successive spreading as circles, though these arise from another cause, and are only in a flat surface.

To see then whether the spreading of light takes time, let us consider first whether there are any facts of experience which can convince us to the contrary. As to those which can be made here on the Earth, by striking lights at great distances, although they prove that light takes no sensible time to pass over these distances, one may say with good reason that they are too small, and that the only conclusion to be drawn from them is that the passage of light is extremely rapid. Mr. Descartes, who was of opinion that it is instantaneous, founded his views, not without reason, upon a better basis of experience, drawn from the Eclipses of the Moon; which, nevertheless, as I shall show, is not at all convincing. I will set it forth, in a way a little different from his, in order to make the conclusion more comprehensible.

Let A be the place of the sun, BD a part of the orbit or annual path of the Earth: ABC a straight line which I suppose to meet the orbit of the Moon, which is represented by the circle CD, at C.

Now if light requires time, for example one hour, to traverse the space which is between the Earth and the Moon, it will follow that the Earth having arrived at B, the shadow which it casts, or the interruption of the light, will not yet have arrived at the point C, but will only arrive there an hour after. It will then be one hour after, reckoning from the moment when the Earth was at B, that the Moon, arriving at C, will be obscured: but this obscuration or interruption of the light will not reach the Earth till after another hour. Let us suppose that the Earth in these two hours will have arrived at E. The Earth then, being at E, will see the Eclipsed Moon at C, which it left an hour before, and at the same time will see the sun at A. For it being immovable, as I suppose with Copernicus, and the light moving always in straight lines, it must always appear where it is. But one has always observed, we are told, that the eclipsed Moon appears at the point of the Ecliptic opposite to the Sun; and yet here it would appear in arrear of that point by an amount equal to the angle GEC, the supplement of AEC. This, however, is contrary to experience, since the angle GEC would be very sensible, and about 33 degrees. Now according to our computation, which is given in the Treatise on the causes of the phenomena of Saturn, the distance BA between the Earth and the Sun is about twelve thousand diameters of the Earth, and hence four hundred times greater than BC the distance of the Moon, which is 30 diameters. Then the angle ECB will be nearly four hundred times greater than BAE, which is five minutes; namely, the path which the earth travels in two hours along its orbit; and thus the angle BCE will be nearly 33 degrees; and likewise the angle CEG, which is greater by five minutes.

But it must be noted that the speed of light in this argument has been assumed such that it takes a time of one hour to make the passage from here to the Moon. If one supposes that for this it requires only one minute of time, then it is manifest that the angle CEG will only be 33 minutes; and if it requires only ten seconds of time, the angle will be less than six minutes. And then it will not be easy to perceive anything of it in observations of the Eclipse; nor, consequently, will it be permissible to deduce from it that the movement of light is instantaneous.

It is true that we are here supposing a strange velocity that would be a hundred thousand times greater than that of Sound. For Sound, according to what I have observed, travels about 180 Toises in the time of one Second, or in about one beat of the pulse. But this supposition ought not to seem to be an impossibility; since it is not a question of the transport of a body with so great a speed, but of a successive movement which is passed on from some bodies to others. I have then made no difficulty, in meditating on these things, in supposing that the emanation of light is accomplished with time, seeing that in this way all its phenomena can be explained, and that in following the contrary opinion everything is incomprehensible. For it has always seemed to me that even Mr. Descartes, whose aim has been to treat all the subjects of Physics intelligibly, and who assuredly has succeeded in this better than anyone before him, has said nothing that is not full of difficulties, or even inconceivable, in dealing with Light and its properties.

But that which I employed only as a hypothesis, has recently received great seemingness as an established truth by the ingenious proof of Mr. Römer which I am going here to relate, expecting him himself to give all that is needed for its confirmation.

Translated by Silvanus P. Thompson

Reading and Discussion Questions

1. Compare and contrast Huygens' attitude toward conjecture with Newton's attitude toward hypotheses as expressed in the *Principia*.
2. Some scholars have looked to this text as a clear expression of what has come to be known as the hypothetico-deductive method. What is the method that Huygens proposes?
3. How is Huygens thinking about the nature of light?

On the Theory of Light and Colors (1801)

Thomas Young

52

Although the invention of plausible hypotheses, independent of any connection with experimental observations, can be of very little use in the promotion of natural knowledge; yet the discovery of simple and uniform principles, by which a great number of apparently heterogeneous phenomena are reduced to coherent and universal laws, must ever be allowed to be of considerable importance towards the improvement of the human intellect.

The object of the present dissertation is not so much to propose any opinions which are absolutely new, as to refer some theories, which have been already advanced, to their original inventors, to support them by additional evidence, and to apply them to a great number of diversified facts, which have hitherto been buried in obscurity. Nor is it absolutely necessary in this instance to produce a single new experiment; for of experiments there is already an ample store, which are so much the more unexceptionable, as they must have been conducted without the least partiality for the system by which they will be explained; yet some facts, hitherto unobserved, will be brought forwards, in order to show the perfect agreement of that system with the multifarious phenomena of nature ....

Those who are attached, as they may be with the greatest justice, to every doctrine which is stamped with the Newtonian approbation, will probably be disposed to bestow on these considerations *so* much the more of their attention, as they appear to coincide more nearly with Newton's own opinions. For this reason, after having briefly stated each particular position of my theory, I shall collect, from Newton's various writings, such passages as seem to be the most favourable to its admission; and, although I shall quote some papers which may be thought to have been partly retracted at the publication of the optics, yet I shall borrow nothing from them that can be supposed to militate against his maturer judgment.

Hypothesis I

A luminiferous Ether pervades the Universe, rare and elastic in a high degree.

. . .

Hypothesis II

Undulations are excited in this Ether whenever a Body becomes luminous.

Scholium. I use the word undulation, in preference to vibration, because vibration is generally understood as implying a motion which is alternately backwards and forwards, by a combination of the momentum of the body with an accelerating force, and which is naturally more or less permanent; but an undulation is supposed to consist in a vibratory motion, transmitted successively through different parts of a medium, without any tendency in each particle to continue its motion, except in consequence of the transmission of succeeding undulations, from a distinct vibrating body; as, in the air, the vibrations of a chord produce the undulation's constituting sound . . . .

Hypothesis III

The Sensation of different Colours depends on the different frequency of Vibrations, excited by Light in the Retina.

. . . *Scholium.* Since, for the reason here assigned by Newton, it is probable that the motion of the retina is rather of a vibratory than of an undulatory nature, the frequency of the vibrations must be dependent on the constitution of this substance. Now, as it is almost impossible to conceive each sensitive point of the retina to contain an infinite number of particles, each capable of vibrating in perfect unison with every possible undulation, it becomes necessary to suppose the number limited, for instance, to the three principal colours, red, yellow, and blue, of which the undulations are related in magnitude nearly as the numbers 8, 7, and 6; and that each of the particles is capable of being put in motion less or more forcibly, by undulations differing less or more from a perfect unison; for instance, the undulations of green light being nearly in the ratio of 61, will affect equally the particles in unison with yellow and blue, and produce the same effect as a light composed of those two species: and each sensitive filament of the nerve may consist of three portions, one for each principal colour. Allowing this statement, it appears that any attempt to produce a musical effect from colours, must be unsuccessful, or at least that nothing more than a very simple melody could be imitated by them; for the period, which in fact constitutes the harmony of any concord, being a multiple of the periods of the single undulations, would in this case be wholly without the limits of sympathy of the retina, and would lose its effect; in the same manner as the harmony of a third or a fourth is destroyed, by depressing it to the lowest notes of the audible scale. In hearing, there seems to be no permanent vibration of any part of the organ.

Hypothesis IV

All material Bodies have an Attraction for the ethereal Medium, by means of which it is accumulated within their Substance, and for a small Distance around them, in a State of greater Density, but not of greater Elasticity.

It has been shown, that the three former hypotheses, which may be called essential, are literally parts of the more complicated Newtonian system. This fourth hypothesis differs perhaps in some degree from any that have been proposed by former authors, and is diametrically opposite to that of Newton; but, both being in themselves equally probable, the opposition is merely accidental; and it is only to be inquired which is the best capable of explaining the phenomena. Other suppositions might perhaps be substituted for this, and therefore I do not consider it as fundamental, yet it appears to be the simplest and best of any that have occurred to me.

Proposition I

All Impulses are propagated in a homogeneous elastic Medium with an equable Velocity.

Every experiment relative to sound coincides with the observation already quoted from Newton, that all undulations are propagated through the air with equal velocity . . . .

Proposition II

An undulation conceived to originate from the Vibration of a single Particle, must expand through a homogeneous Medium in a spherical Form, but with different quantities of Motion in different Parts.

Proposition III

A Portion of a spherical Undulation, admitted through an Aperture into a quiescent Medium, will proceed to be further propagated rectilinearly in concentric Superficies, terminated laterally, by weak and irregular Portions, of newly diverging Undulations.

At the instant of admission, the circumference of each of the undulations may be supposed to generate a partial undulation, filling up the nascent angle between the radii and the surface, terminating the medium; but no sensible addition will be made to its strength by a divergence of motion from any other parts of the undulation, for want of a coincidence in time, as has already been explained with respect to the various force of a spherical undulation. If indeed the aperature bear but a small proportion to the breadth of an undulation, the newly generated undulation may nearly absorb the whole force of the portion admitted; and this is the case considered by Newton in the Principia. But no experiment can be made under these circumstances with light, on account of the

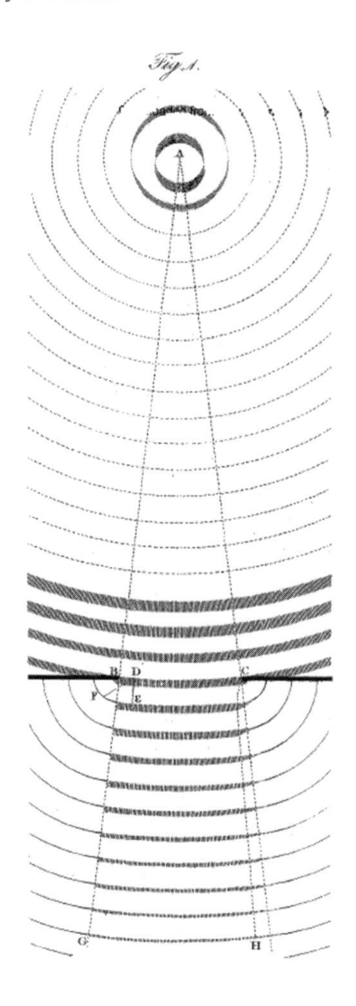

minuteness of its undulations, and the interference of inflection; and yet some faint radiations do actually diverge beyond any probable limits of inflection, rendering the margin of the aperture distinctly visible in all directions; these are attributed by Newton to some un known cause, distinct from inflection; (Optics, Third Book, Obs. 5.) and they fully answer the description of this proposition.

. . .

Proposition V

When an Undulation is transmitted through a Surface terminating different Mediums, it proceeds in such a Direction, that the Sines of the Angles of Incidence and Refraction are in the constant Ratio of the velocity of Propagation in the two Mediums.

Corollary 1. The same demonstrations prove the equality of the angles of reflection and incidence.

Corollary 2. It appears from experiments on the refraction of condensed air that the ratio of the difference of the sines varies simply as the density. Hence it follows, by Schol. I Prop. I. that the excess of the density of the ethereal medium is in the duplicate ratio of the density of the air; each particle cooperating with its neighbours in attracting a greater portion of it.

...

Proposition VIII

From two Undulations, from different Origins, coincide either perfectly or very nearly in Direction, their joint effect is a Combination of the Motions belonging to each.

Since every particle of the medium is affected by each undulation, wherever the directions coincide, the undulations can proceed no otherwise than by uniting their motions, so that the joint motion may be the sum or difference of the separate motions, accordingly as similar or dissimilar parts of the undulations are coincident.

I have, on a former occasion insisted at large on the application of this principle to harmonics (*Phil Trans.* for 1800, p. 130), and it will appear to be of still more extensive utility in explaining the phenomena of colours. The undulations which are now to be compared are those of equal frequency. When the two series coincide exactly in point of time, it is obvious that the united velocity of the particular motions must be greatest, and, in effect at least, double the separate velocities; and also, that it must be smallest, and if the undulations are of equal strength, totally destroyed, when the time of the greatest direct motion belonging to one undulation coincides with that of the greatest retrograde motion of the other. In intermediate states, the joint undulation will be of intermediate strength; but by what laws this intermediate strength must vary, cannot be determined without further data. It *is* well known that a similar cause produces in sound, that effect which *is* called a beat; two series of undulations of nearly equal magnitude cooperating and destroying each other alternately, as they coincide more or less perfectly in the times of performing their respective motions.

...

Proposition IX

Radiant Light consists in Undulations of the luminiferous Ether.

This proposition is the general conclusion form all the preceding; and it is conceived that they conspire to prove it in as satisfactory a manner as can possibly be expected from the nature of the subject. It is clearly granted by Newton, that there are undulations yet he denies that they constitute light; but it is shown in the three first Corollaries of the last Proposition, that all cases of the increase or diminution of light are referable to an increase or diminution of such undulations, and that all the affections to which the undulations would be liable, are distinctly visible in the phenomena of light; it may therefore be very logically inferred, that the undulations are light.

A few detached remarks will serve to obviate some objections which may be raised against this theory.

...

1. Newton has advanced the singular refraction of the Iceland crystal, as an argument that the particles of light must be projected corpuscles; since he thinks it probable that the different sides of these particles must be differently attracted by the crystal, and since Huygens has confessed his inability to account in a satisfactory manner for all the phenomena. But, contrarily to what might have been expected from Newton's usual accuracy and candour, he has laid down a new law for the refraction, without giving a reason for rejecting that of Huygens, which Mr. Hauy has found to be more accurate than Newton's; and, without attempting to deduce from his own system any explanation of the more universal and striking effects of doubling spars, he has omitted to observe that Huygens's most elegant and ingenious theory perfectly accords with these general, effects, in all particulars, and of course derives from them additional pretensions to truth: this he omits, in order to point out a difficulty, for which only a verbal solution can be found in his own theory, and which will probably long remain unexplained by any other.
2. Mr. Mitchell has made some experiments, which appear to show that the rays of light have an actual momentum, by means of which a motion is produced when they fall on a thin plate of copper delicately suspended. (Priestley's Optics) But, taking for granted the exact perpendicularity of the plate, and the absence of any ascending current of air, yet since, in every such experiment, a greater quantity of heat must be communicated to the air at the surface on which the light falls than at the opposite surface, the excess of expansion must necessarily produce an excess of pressure on the first surface, and a very perceptible recession of the plate in the direction of the light. Mr. Bennet has repeated the experiment, with a much more sensible apparatus, and also in the absence of air; and very justly infers from its total failure, an argument in favour of the undulatory system of light. (*Phil. Trans.* for 1792, p. 87) For, granting the utmost imaginable subtlety of the corpuscles of light, their effects might naturally be expected to bear some proportion to the effects of the much less rapid motions of the electrical fluid, which are so very easily perceptible, even in their weakest states.
3. There are some phenomena of the light of solar phosphori, which at first sight might seem to favour the corpuscular system; for instance, its remaining many months as if in a latent state, and its subsequent

> re-emission by the action of heat. But, on further consideration, there is no difficulty in supposing the particles of the phosphori which have been made to vibrate by the action of light, to have this action abruptly suspended by the intervention of cold, whether as contracting the bulk of the substance or otherwise; and again, after the restraint is removed, to proceed in their motion, as a spring would do which had been held fast for a time in an intermediate stage of its vibration; nor is it impossible that heat itself may, in some circumstances, become in a similar manner latent. (*Nicholson's Journal*, vol. ii, p. 399)

But the affections of heat may perhaps hereafter be rendered more intelligible to us; at present, it seems highly probable that light differs from heat only in the frequency of its undulations or vibrations; those undulations which are within certain limits, with respect to frequency, being capable of affecting the optic nerve, and constituting light; and those which are slower, and probably stronger, constituting heat only; that light and heat occur to us, each in two predicaments, the vibratory or permanent, and the undulatory or transient state; vibratory light being the minute motion of ignited bodies, or of solar phosphori, and undulatory or radiant light the motion of the ethereal medium excited by these vibrations; vibratory heat being a motion to which all material substances are liable, and which is more or less permanent; and undulatory heat that motion of the same ethereal medium, which has been shown by Mr. King (Morsels of Criticism, 1786, p. 99), and M. Pictet (*Essais de Physique*, 1790) to be as capable of reflection as light, and by Dr. Herschel to be capable of separate refraction (*Phil. Trans.* for 1800, p. 284).

How much more readily heat is communicated by the free access of colder substances, than either by radiation or by transmission through a quiescent medium, has been shown by the valuable experiments of Count Rumford. It is easy to conceive that some substances, permeable to light, may be unfit for the transmission of heat, in the same manner as particular substances may transmit some kinds of light, while they are opaque with respect to others.

On the whole it appears, that the few optical phenomena which admit of explanation by the corpuscular system, are equally consistent with this theory; that many others, which have long been known, but never understood, become by these means perfectly intelligible; and that several new facts are found to be thus only reducible to a perfect analogy with other facts, and to the simple principles of the undulatory system. It is presumed, that henceforth the second and third books of Newton's Optics will be considered as more fully understood than the first has hitherto been; but, if it should appear to impartial judges, that additional evidence is wanting for the establishment of the theory, it will be easy to enter more minutely into the details of various experiments, and to show the insuperable difficulties attending the Newtonian doctrines, which, without necessity, it would be tedious and invidious to enumerate. The merits of their author in natural philosophy, are great beyond all contest or comparison; his optical discovery of the composition of white light, would alone have immortalised his name; and the very arguments which tend to overthrow his system, give the strongest proofs of the admirable accuracy of his experiments.

Sufficient and decisive as these arguments appear, it cannot be superfluous to seek for further confirmation; which may with considerable confidence be expected, from an experiment very ingeniously suggested by Professor Romson, on the refraction of the light returning to us from the opposite margins of Saturn's ring; for, on the corpuscular theory, the ring must be considerably distorted. When viewed through an achromatic prism; a similar distortion ought also to be observed in the disc of Jupiter; but, if it be found that an equal deviation is produced in the whole light reflected from these planets, there can scarcely be any remaining hope to explain the affections of light, by a comparison with the motions of projectiles.

Reading and Discussion Questions

1. Young seems to be presenting a theory of light which is counter to Newton's. Why, then, is he so careful to suggest how much of his theory aligns with that of his predecessor?
2. What is a luminiferous ether? Why would Young need to postulate such a thing?
3. What evidence does Young provide in favor of this theory that light is an undulation in the luminiferous ether?

Experiments on Electricity, Written to Peter Collinson, Philadelphia (September 1, 1747)

Benjamin Franklin

Sɪʀ:—The necessary trouble of copying long letters, which perhaps, when they come to your hands, may contain nothing new, or worth your reading (so quick is the progress made with you in electricity), half discourages me of writing any more on that subject. Yet I cannot forbear adding a few observations on M. Muschenbroek's wonderful bottle.

1. The non-electric contained in the bottle differs, when electrized, from a non-electric electrized out of the bottle, in this: that the electrical fire of the latter is accumulated *on its surface,* and forms an electrical atmosphere round it of considerable extent; but the electrical fire is crowded *into the substance* of the former, the glass confining it.
2. At the same time that the wire and the top of the bottle, &c., is electrized *positively* or *plus*, the bottom of the bottle is electrized *negatively* or *minus*, in exact proportion; that is, whatever quantity of electrical fire is thrown in at the top, an equal quantity goes out of the bottom. To understand this, suppose the common quantity of electricity in each part of the bottle, before the operation begins, is equal to twenty; and at every stroke of the tube, suppose a quantity equal to one is thrown in; then, after the first stroke, the quantity contained in the wire and upper part of the bottle will be twenty-one, in the bottom nineteen; after the second, the upper part will have twenty-two, the lower eighteen; and so on, till after twenty strokes, the upper part will have a quantity of electrical fire equal to forty, the lower part none; and then the operation ends, for no more can be thrown into the upper part when no more can be driven out of the

lower part. If you attempt to throw more in, it is spewed back through the wire, or flies out in loud cracks through the sides of the bottle.

3. The equilibrium cannot be restored in the bottle by *inward* communication or contact of the parts; but it must be done by a communication formed *without* the bottle, between the top and bottom, by some non-electric, touching or approaching both at the same time; in which case it is restored with a violence and quickness inexpressible; or touching each alternately, in which case the equilibrium is restored by degrees.
4. As no more electrical fire can be thrown into the top of the bottle, when all is driven out of the bottom, so, in a bottle not yet electrized, none can be thrown into the top when none *can* get out at the bottom; which happens either when the bottom is too thick, or when the bottle is placed on an electric *per se.* Again, when the bottle is electrized, but little of the electrical fire can be *drawn out* from the top, by touching the wire, unless an equal quantity can at the same time *get in* at the bottom. Thus, place an electrized bottle on clean glass or dry wax, and you will not, by touching the wire, get out the fire from the top. Place it on a non-electric, and touch the wire, you will get it out in a short time,—but soonest when you form a direct communication as above.

   So wonderfully are these two states of electricity, the *plus* and *minus*, combined and balanced in this miraculous bottle! situated and related to each other in a manner that I can by no means comprehend! If it were possible that a bottle should in one part contain a quantity of air strongly compressed, and in another part a perfect vacuum, we know the equilibrium would be instantly restored *within.* But here we have a bottle containing at the same time a *plenum* of electrical fire and a *vacuum* of the same fire, and yet the equilibrium cannot be restored between them but by a communication *without*, though the *plenum* presses violently to expand, and the hungry vacuum seems to attract as violently in order to be filled.

5. The shock to the nerves (or convulsion rather) is occasioned by the sudden passing of the fire through the body in its way from the top to the bottom of the bottle. The fire takes the shortest course, as Mr. Watson justly observes. But it does not appear from experiment that, in order for a person to be shocked, a communication with the floor is necessary; for he that holds the bottle with one hand and touches the wire with the other, will be shocked as much, though his shoes be dry, or even standing on wax, as otherwise. And on the touch of the wire (or of the gun-barrel, which is the same thing), the fire does not proceed from the touching finger to the wire, as is supposed, but from the wire to the finger, and passes through the body to the other hand, and so into the bottom of the bottle.

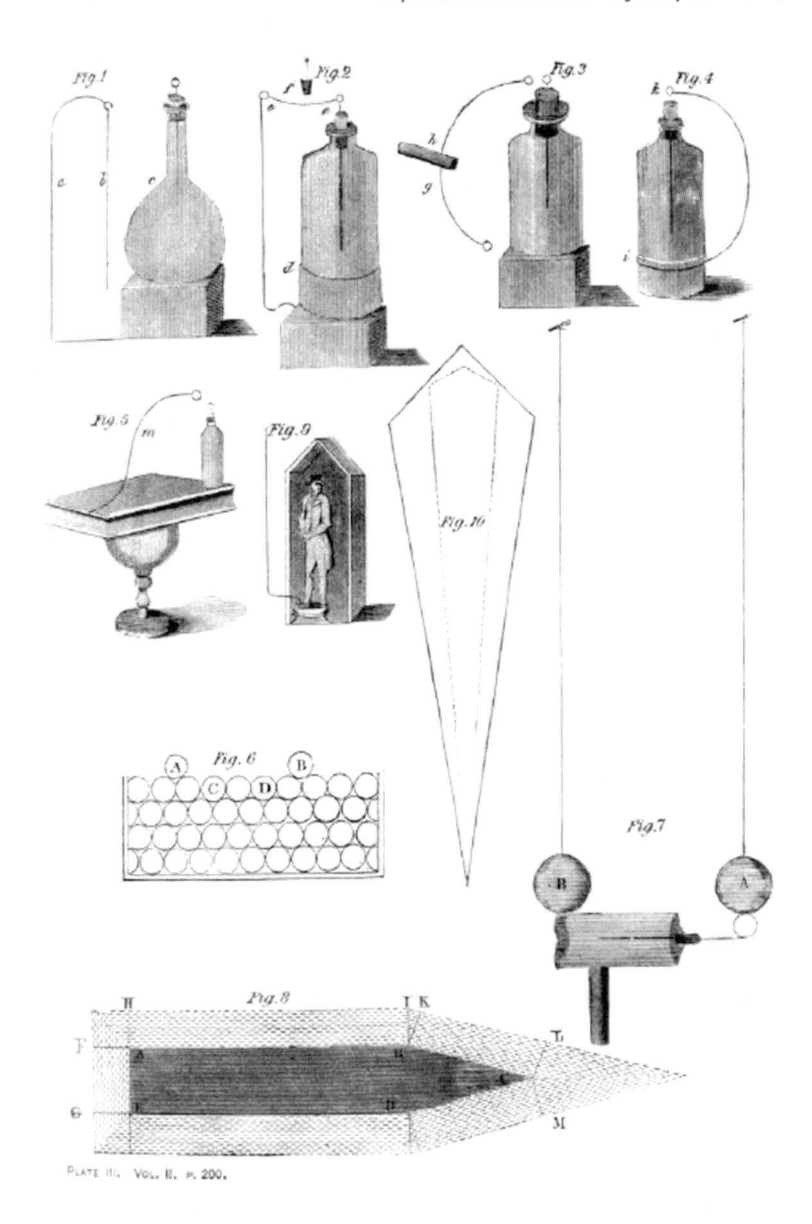

Experiments Confirming the above

EXPERIMENT I Place an electrized phial on wax; a small cork ball, suspended by a dry silk thread, held in your hand and brought near to the wire, will first be attracted and then repelled; when in this state of repellency, sink your hand that the ball may be brought towards the bottom of the bottle. It will be there instantly and strongly attracted till it has parted with its fire.

If the bottle had a *positive* electrical atmosphere, as well as the wire, an electrified cork would be repelled from one as well as from the other.

EXPERIMENT II PLATE III, FIG. 1.—From a bent wire (*a*) sticking in the table, let a small linen thread (*b*) hang down within half an inch of the electrized phial (*c*). Touch the wire or the phial repeatedly with your finger, and at every touch you will see the thread instantly attracted by the bottle. (This is best done by a vinegar-cruet, or some such bellied bottle.) As soon as you draw any fire out from the upper part by touching the wire, the lower part of the bottle draws an equal quantity in by the thread.

EXPERIMENT III FIG. 2.—Fix a wire in the lead, with which the bottom of the bottle is armed (*d*), so as that, bending upwards, its ring-end may be level with the top or ring-end of the wire in the cork (*e*), and at three or four inches distance. Then electrize the bottle and place it on wax. If a cork, suspended by a silk thread (*f*), hang between these two wires, it will play incessantly from one to the other till the bottle is no longer electrized; that is, it fetches and carries fire from the top to the bottom1 of the bottle till the equilibrium is restored.

EXPERIMENT IV FIG. 3.—Place an electrized phial on wax; take a wire (*g*) in form of a *C*, the ends at such a distance, when bent, as that the upper may touch the wire of the bottle when the lower touches the bottom; stick the outer part on a stick of sealing-wax (*h*), which will serve as a handle; then apply the lower end to the bottom of the bottle, and gradually bring the upper end near the wire in the cork. The consequence is, spark follows spark till the equilibrium is restored. Touch the top first, and on approaching the bottom with the other end, you have a constant stream of fire from the wire entering the bottle. Touch the top and bottom together, and the equilibrium will instantly be restored, the crooked wire forming the communication.

EXPERIMENT V FIG. 4.—Let a ring of thin lead or paper surround a bottle (*i*), even at some distance from or above the bottom. From that ring let a wire proceed up till it touch the wire of the cork (*k*). A bottle so fixed cannot by any means be electrized; the equilibrium is never destroyed; for while the communication between the upper and lower parts of the bottle is continued by the outside wire, the fire only circulates; what is driven out at bottom is constantly supplied from the top. Hence a bottle cannot be electrized that is foul or moist on the outside, if such moisture continue up to the cork or wire.

EXPERIMENT VI Place a man on a cake of wax, and present him the wire of the electrified phial to touch, you standing on the floor and holding it in your hand. As often as he touches it he will be electrified *plus;* and any one standing on the floor may draw a spark from him. The fire in this experiment passes out of the wire into him; and at the same time out of your hand into the bottom of the bottle.

EXPERIMENT VII Give him the electrical phial to hold, and do you touch the wire; as often as you touch it he will be electrified *minus,* and may draw a spark from any one standing on the floor. The fire now passes from the wire to you, and from him into the bottom of the bottle.

EXPERIMENT VIII Lay two books on two glasses, back towards back, two or three inches distant. Set the electrified phial on one, and then touch the wire; that book will be electrified *minus,* the electrical fire being drawn out of it by the bottom of the bottle. Take off the bottle, and, holding it in your hand, touch the other with the wire; that book will be electrified *plus;* the fire passing into it from the wire, and the bottle at the same time supplied from your hand. A suspended small cork ball will play between these books till the equilibrium is restored.

EXPERIMENT IX When a body is electrized *plus,* it will repel a positively electrified feather or small cork ball. When *minus* (or when in the common state), it will attract them, but stronger when *minus* than when in the common state, the difference being greater.

EXPERIMENT X Though, as in *Experiment* VI, a man standing on wax may be electrized a number of times by repeatedly touching the wire of an electrized bottle (held in the hand of one standing on the floor), he receiving the fire from the wire each time; yet holding it in his own hand and touching the wire, though he draws a strong spark, and is violently shocked, no electricity remains in him, the fire only passing through him from the upper to the lower part of the bottle. Observe, before the shock, to let some one on the floor touch him to restore the equilibrium of his body; for in taking hold of the bottom of the bottle he sometimes becomes a little electrized *minus,* which will continue after the shock, as would also any *plus* electricity which he might have given him before the shock. For restoring the equilibrium in the bottle does not at all affect the electricity in the man through whom the fire passes; that electricity is neither increased nor diminished.

EXPERIMENT XI The passing of the electrical fire from the upper to the lower part1 of the bottle, to restore the equilibrium, is rendered strongly visible by the following pretty experiment. Take a book whose covering is filleted with gold; bend a wire of eight or ten inches long in the form of (*m*), Fig. 5, slip it on the end of the cover of the book, over the gold line, so as that the shoulder of it may press upon one end of the gold line, the ring up, but leaning towards the other end of the book. Lay the book on a glass or wax, and on the other end of the gold line set the bottle electrized; then bend the springing wire by pressing it with a stick of wax till its ring approaches the ring of the bottle wire; instantly there is a strong spark and stroke, and the whole line of gold, which completes the communication between the top and bottom of the bottle, will appear a vivid flame, like the sharpest lightning. The closer the contact between the shoulder of the wire and the gold at one end of the line, and between the bottom of the bottle and the gold at the other end, the better the experiment succeeds. The room should be darkened. If you would have the whole filleting round the cover appear in fire at once, let the bottle and wire touch the gold in the diagonally opposite corners.

I am, &c.,
B. Franklin.
Benjamin Franklin
1 October, 1747

Reading and Discussion Questions

1. What does the "wonderful bottle" Franklin is experimenting with allow him to do?
2. What are the "two states" of electricity that he describes?

Experiments on the Effect of a Current of Electricity on the Magnetic Needle (1820)

John Christian Oersted

54

The first experiments respecting the subject which I mean at present to explain, were made by me last winter, while lecturing on electricity, galvanism, and magnetism, in the University. It seemed demonstrated by these experiments that the magnetic needle was moved from its position by the galvanic apparatus, but that the galvanic circle must be complete, and not open, which last method was tried in vain some years ago by very celebrated philosophers. But as these experiments were made with a feeble apparatus, and were not, therefore, sufficiently conclusive, considering the importance of the subject, I associated myself with my friend Esmarck to repeat and extend them by means of a very powerful galvanic battery, provided by us in common. Mr. Wleugel, a Knight of the Order of Danneborg, and at the head of the Pilots, was present at, and assisted in, the experiments. There were present likewise Mr. Hauch, a man very well skilled in the Natural Sciences, Mr. Reinhardt, Professor of Natural History, Mr. Jacobsen, Professor of Medicine, and that very skilful chemist, Mr. Zeise, Doctor of Philosophy. I had often made experiments by myself; but every fact which I had observed was repeated in the presence of these gentlemen.

The galvanic apparatus which we employed consists of twenty copper troughs, the length and height of each of which was 12 inches; but the breadth scarcely exceeded 2½ inches. Every trough is supplied with two plates of copper, so bent that they could carry a copper rod, which supports the zinc plate in the water of the next trough. The water of the troughs contained one-sixtieth of its weight of sulphuric acid, and an equal quantity of nitric acid. The portion of each zinc plate sunk in the water is a square whose side is about 10 inches in length. A smaller apparatus will answer provided it be strong enough to heat a metallic wire red hot. The opposite ends of the galvanic battery were joined by a metallic wire, which, for shortness sake, we shall call the *uniting conductor*, or the *uniting wire*. To the effect which takes place in this conductor and in the surrounding space, we shall give the name of the *conflict of electricity*.

Let the straight part of this wire be placed horizontally above the magnetic needle, properly suspended, and parallel to it. If necessary, the uniting wire is bent so as to assume a proper position for the experiment. Things being in this state, the needle will be moved, and the end of it next the negative side of the battery will go westward.

If the distance of the uniting wire does not exceed three-quarters of an inch from the needle, the declination of the needle makes an angle of about 45°. If the distance is increased, the angle diminishes proportionally. The declination likewise varies with the power of the battery.

The uniting wire may change its place, either towards the east or west, provided it continue parallel to the needle, without any other change of the effect than in respect to its quantity. Hence the effect cannot be ascribed to attraction; for the same pole of the magnetic needle, which approaches the uniting wire, while placed on its east side, ought to recede from it when on the west side, if these declinations depended on attractions and repulsions. The uniting conductor may consist of several wires, or metallic ribbons, connected together. The nature of the metal does not alter the effect, but merely the quantity. Wires of platinum, gold, silver, brass, iron, ribbons of lead and tin, a mass of mercury, were employed with equal success. The conductor does not lose its effect, though interrupted by water, unless the interruption amounts to several inches in length.

The effect of the uniting wire passes to the needle through glass, metals, wood, water, resin, stoneware, stones; for it is not taken away by interposing plates of glass, metal or wood. Even glass, metal, and wood, interposed at once, do not destroy, and indeed scarcely diminish the effect. The disc of the electrophorus, plates of porphyry, a stoneware vessel, even filled with water, were interposed with the same result. We found the effects unchanged when the needle was included in a brass box filled with water. It is needless to observe that the transmission of effects through all these matters has never before been observed in electricity and galvanism. The effects, therefore, which take place in the conflict of electricity are very different from the effects of either of the electricities.

If the uniting wire be placed in a horizontal plane under the magnetic needle, all the effects are the same as when it is above the needle, only they are in an opposite direction; for the pole of the magnetic needle next the negative end of the battery declines to the east.

That these facts may be the more easily retained, we may use this formula—the pole *above* which the *negative* electricity enters is turned to the *west; under* which, to the *east*.

If the uniting wire is so turned in a horizontal plane as to form a gradually increasing angle with the magnetic meridian, the declination of the needle *increases*, if the motion of the wire is towards the place of the disturbed needle; but it *diminishes* if the wire moves further from that place.

When the uniting wire is situated in the same horizontal plane in which the needle moves by means of the counterpoise, and parallel to it, no declination is produced either to the east or west; but an *inclination* takes place, so that the pole, next which the negative electricity enters the wire, is *depressed* when the wire is situated on the *west* side, and elevated when situated on the *east* side. If the uniting wire be placed perpendicularly to the plane of the magnetic meridian, whether above or below it, the needle remains at rest, unless it be very near the pole; in that case the pole is *elevated* when the entrance is from the *west* side of the wire, and *depressed*, when from the *east* side.

When the uniting wire is placed perpendicularly opposite to the pole of the magnetic needle, and the upper extremity of the wire receives the negative electricity, the pole is moved towards the east; but when the wire is opposite to a point between the pole and the middle of the needle, the pole is moved towards the west. When the upper end of the wire receives positive electricity, the phenomena are reversed.

If the uniting wire is bent so as to form two legs parallel to each other, it repels or attracts the magnetic poles according to the different conditions of the case. Suppose the wire placed opposite to either pole of the needle, so that the plane of the parallel legs is perpendicular to the magnetic meridian, and let the eastern leg be united with the negative end, the western leg with the positive end of the battery: in that case the nearest pole will be repelled either to the east or west according to the position of the plane of the legs. The eastmost leg being united with the positive, and the westmost with the negative side of the battery, the nearest pole will be attracted. When the plane of the legs is placed perpendicular to the place between the pole and the middle of the needle, the same effects recur but reversed.

A brass needle, suspended like a magnetic needle, is not moved by the effect of the uniting wire. Likewise needles of glass and of gum lac remain unacted on. We may now make a few observations towards explaining these phenomena.

The electric conflict acts only on the magnetic particles of matter. All non-magnetic bodies appear penetrable by the electric conflict, while magnetic bodies, or rather their magnetic particles, resist the passage of this conflict. Hence they can be moved by the impetus of the contending powers.

It is sufficiently evident from the preceding facts that the electric conflict is not confined to the conductor, but dispersed pretty widely in the circumjacent space.

From the preceding facts we may likewise infer that this conflict performs circles; for without this condition it seems impossible that the one part of the uniting wire, when placed below the magnetic pole, should drive it towards the east, and when placed above it towards the west; for it is the nature of a circle that the motions in opposite parts should have an opposite direction. Besides, a motion in circles, joined with a progressive motion, according to the length of the conductor, ought to form a conchoidal or spiral line; but this, unless I am mistaken, contributes nothing to explain the phenomena hitherto observed.

All the effects on the North Pole above-mentioned are easily understood by supposing that negative electricity moves in a spiral line bent towards the right, and propels the North Pole, but does not act on the South Pole. The effects on the South Pole are explained in a similar manner, if we ascribe to positive electricity a contrary motion and power of acting on the South Pole, but not upon the north. The agreement of this law with nature will be better seen by a repetition of the experiments than by a long explanation. The mode of judging of the experiments will be much facilitated if the course of the electricities in the uniting wire be pointed out by marks or figures.

I shall merely add to the above that I have demonstrated in a book published five years ago that heat and light consist of the conflict of the electricities. From the observations now stated, we may conclude that a circular motion likewise occurs in these effects. This I think will contribute very much to illustrate the phenomena to which the appellation of polarization of light has been given.

Reading and Discussion Questions

1. What aspects of the effects Oersted observes on the compass needle does he find surprising? How well do these fit into the Newtonian paradigm? Are there any ways in which the phenomena might be seen as non-Newtonian?
2. What permutations of the experiment does he attempt? What does this suggest about the nature of the force which is moving the compass needle?

Lectures on the Forces of Matter, Lecture V.—Magnetism—Electricity (1859)

Michael Faraday

55

[These lectures are part of Faraday's "Christmas Lectures" series, live demonstrations intended for an audience of children. They present not new research, but extremely valuable background to the state of experimentation and knowledge of phenomena in the field.—Ed.]

I wonder whether we shall be too deep to-day or not. Remember that we spoke of the attraction by gravitation of all bodies to all bodies by their simple approach. Remember that we spoke of the attraction of particles of the same kind to each other—that power which keeps them together in masses—iron attracted to iron, brass to brass, or water to water. Remember that we found, on looking into water, that there were particles of two different kinds attracted to each other; and this was a great step beyond the first simple attraction of gravitation, because here we deal with attraction between different kinds of matter. The hydrogen could attract the oxygen and reduce it to water, but it could not attract any of its own particles, so that there we obtained a first indication of the existence of two attractions.

To-day we come to a kind of attraction even more curious than the last, namely, the attraction which we find to be of a double nature—of a curious and dual nature. And I want, first of all, to make the nature of this doubleness clear to you. Bodies are sometimes endowed with a wonderful attraction, which is not found in them in their ordinary state. For instance, here is a piece of shellac, having the attraction of gravitation, having the attraction of cohesion, and if I set fire to it, it would have the attraction of chemical affinity to the oxygen in the atmosphere. Now all these powers we find in it as if they were parts of its substance; but there is another property which I will try and make evident by means of this ball, this bubble of air [a light India-rubber ball, inflated and suspended by a thread]. There is no attraction between this ball and this shellac at present; there may be a little wind in the rooms slightly moving the ball about, but there is no attraction. But if I rub the shellac with a piece of flannel [rubbing the shellac, and then holding it near the ball], look at the attraction which has arisen out of the shellac simply by this friction, and which I may take away as easily by drawing it gently through my hand. [The lecturer repeated the experiment of exciting the shellac, and then removing the attractive power by drawing it through his hand.] Again, you will see I can repeat this experiment with another substance; for if I take a glass rod, and rub it with a piece of silk covered with what we call amalgam, look at the attraction which it has; how it draws the ball toward it; and then, as before, by quietly rubbing it through the hand, the attraction will be all removed again, to come back by friction with this silk.

But now we come to another fact. I will take this piece of shellac, and make it attraction by friction; and remember that, whenever we get an attraction of gravity, chemical affinity, adhesion, or electricity (as in this case), the body which attracts is attracted also, and just as much as that ball was attracted by the shellac, the shellac was attracted by the ball. Now I will suspend this piece of excited shellac in a little paper stirrup, in this way (Fig. 33), in order to make it move easily, and I will take another piece of shellac, and, after rubbing it with flannel, will bring them near together: you will think that they ought to attract each other; but now what happens? It does not attract; on the contrary, it very strongly repels, and I can thus drive it round to any extent. These, therefore, repel each other, although they are so strongly attractive—repel each other to the extent of driving this heavy piece of shellac round and round in this way. But if I excite this piece of shellac as before, and take this piece of glass and rub it with silk, and then bring them near, what think you will happen? [The lecturer held the excited glass near the excited shellac, when they attracted each other strongly.] You see, therefore, what a difference there is between these two attractions; they are actually two kinds of attraction concerned in this case, quite different to any thing we have met with before, but the force is the same. We have here, then, a double attraction—a dual attraction or force—one attracting and the other repelling.

Again, to show you another experiment which will help to make this clear to you: Suppose I set up this rough indicator again [the excited shellac suspended in the stirrup]: it is rough, but delicate enough for my purpose; and suppose I take this other piece of shellac, and take away the power, which I can do by drawing it gently through the hand; and suppose I take a piece of flannel (Fig. 34), which I have shaped into a cap for it and made dry. I will put this shellac into the flannel, and here comes out a very beautiful result. I will rub this shellac and the flannel together (which I can do by twisting the shellac round), and leave them in contact; and then if I ask, by bringing them near our indicator, what is the attractive force? it is nothing; but if I take them apart, and then ask what will they do when they are separated? why, the shellac is strongly repelled, as it was before, but the cap is strongly attractive; and yet, if I bring them both together again, there is no attraction; it has all disappeared [the experiment was repeated]. Those two bodies, therefore, still contain this attractive power; when they were

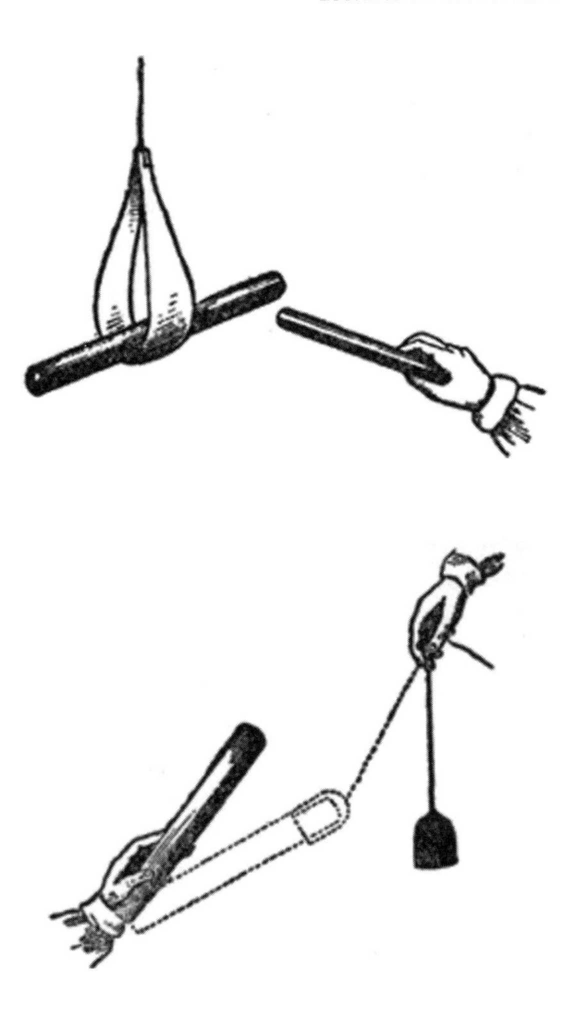

parted, it was evident to your senses that they had it, though they do not attract when they are together.

This, then, is sufficient, in the outset, to give you an idea of the nature of the force which we call ELECTRICITY. There is no end to the things from which you can evolve this power. When you go home, take a stick of sealing-wax—I have rather a large stick, but a smaller one will do—and make an indicator of this sort (Fig. 35).

Take a watch-glass (or your watch itself will do; you only want something which shall have a round face); and now, if you place a piece of flat glass upon that, you have a very easily moved centre; and if I take this lath and put it on the flat glass (you see I am searching for the centre of gravity of this lath; I want to balance it upon the watch-glass), it is very easily moved round; and if I take this piece of sealing-wax and rub it against my coat, and then try whether it is attractive [holding it near the lath], you sec how strong the attraction is; I can even draw it about. Here, then, you have a very beautiful indicator, for I have, with a small piece of sealing-wax and my coat, pulled round a plank of that kind, so you need be in no want of indicators to discover the presence of this attraction. There is scarcely a substance which we may not use. Here are some indicators (Fig. 36). I bend round a strip of paper into a hoop, and we have as good an indicator as can be required. See how it rolls along, traveling after the sealing-wax! If I make them smaller, of course we have them running faster, and sometimes they are actually attracted up into the air. Here, also, is a little collodion balloon. It is so electrical that it will scarcely leave my hand unless to go to the other. See how curiously electrical it is; it is hardly possible for me to touch it without making it electrical; and here is a piece which clings to any thing it is brought near, and which it is not easy to lay down. And here is another substance, gutta-percha, in thin strips: it is astonishing how, by rubbing this in your hands, you make it electrical; but our time forbids us to go farther into this subject at present; you see clearly there are two kinds of electricities which may be obtained by rubbing shellac with flannel or glass with silk.

Now there are some curious bodies in nature (of which I have two specimens on the table) which are called magnets or loadstones; ores of iron, of which there is a great deal sent from Sweden. They have the attraction of gravitation, and attraction of cohesion, and certain chemical attraction; but they also have a great attractive power, for this little key is held up by this stone. Now that is not chemical attraction; it is not the attraction of chemical affinity, or of aggregation of particles, or of cohesion, or of electricity (for it will not attract this ball if I bring it near it), but it is a separate and dual attraction, and, what is more, one which is not readily removed from the substance, for it has existed in it for ages and ages in the bowels of the earth. Now we can make artificial magnets (you

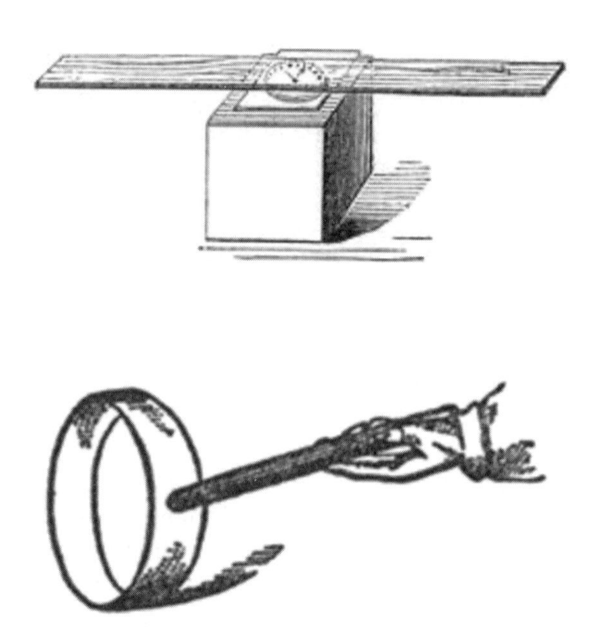

will see me to-morrow make artificial magnets of extraordinary power). And let us take one of these artificial magnets and examine it, and see where the power is in the mass, and whether it is a dual power. You see it attracts these keys, two or three in succession, and it will attract a very large piece of iron. That, then, is a very different thing indeed to what you saw in the case of the shellac, for that only attracted a light ball, but here I have several ounces of iron held up. And if we come to examine this attraction a little more closely, we shall find it presents some other remarkable differences; first of all, one end of this bar (Fig. 37) attracts this key, but the middle does not attract. It is not, then, the whole of the substance which attracts. If I place this little key in the middle it does not adhere; but if I place it there, a little nearer the end, it does, though feebly. Is it not, then, very curious to find that there is an attractive power at the extremities which is not in the middle—to have thus in one bar two places in which this force of attraction resides? If I take this bar and balance it carefully on a point, so that it will be free to move round, I can try what action this piece of iron has on it. Well, it attracts one end, and it also attracts the other end, just as you saw the shellac and the glass did, with the exception of its not attracting in the middle. But if now, instead of a piece of iron, I take a magnet, and examine it in a similar way, you see that one of its ends repels the suspended magnet; the force, then, is no longer attraction, but repulsion; but, if I take the other end of the magnet and bring it near, it shows attraction again.

You will see this better, perhaps, by another kind of experiment. Here (Fig. 38) is a little magnet, and I have colored the ends differently, so that you may distinguish one form the other. Now this end (S) of the magnet (Fig. 37) attracts the uncolored end of the little magnet. You see it pulls toward it with great power; and, as I carry it round, the uncolored end still follows. But now, if I gradually bring the middle of the bar magnet opposite the uncolored end of the needle, it has no effect upon it, either of attraction or repulsion, until, as I come to the opposite extremity (N), you see that it is the colored end of the needle which is pulled toward it. We are now, therefore, dealing with two kinds of power, attracting different ends of the magnet—a double power, already existing in these bodies, which takes up the form of attraction and repulsion. And now, when I put up this label with the word MAGNETISM, you will understand that it is to express this double power.

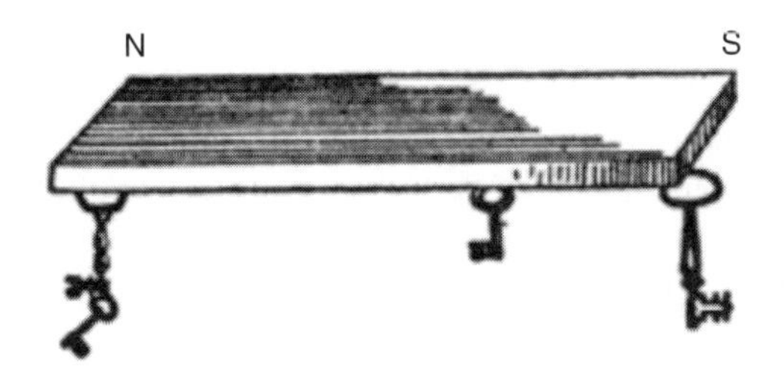

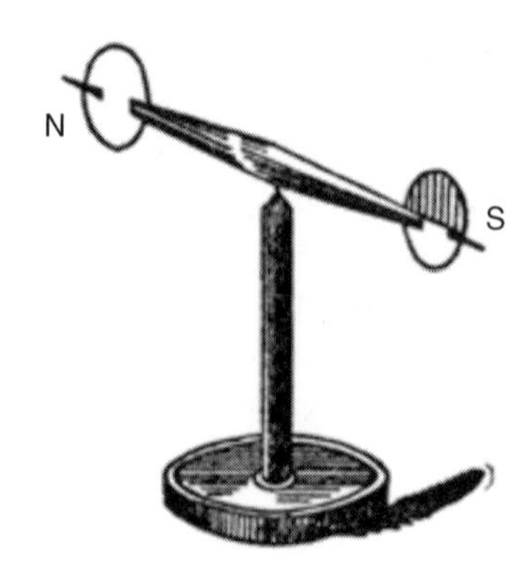

Now with this loadstone you may make magnets artificially. Here is an artificial magnet (Fig. 39) in which both ends have been brought together in order to increase the attraction. This mass will lift that lump of iron, and, what is more, by placing this keeper, as it is called, on the top of the magnet, and taking hold of the handle, it will adhere sufficiently strongly to allow itself to be lifted up, so wonderful is its power of attraction. If you take a needle, and just draw one of its ends along one extremity of the magnet, and then draw the other end along the other extremity, and then gently place it on the surface of some water (the needle will generally float on the surface, owing to the slight greasiness communicated to it by the fingers), you will be able to get all the phenomena of attraction and repulsion by bringing another magnetized needle near to it.

I want you now to observe that, although I have shown you in these magnets that this double power becomes evident principally at the extremities, yet the whole of the magnet is concerned in giving the power. That will at first seem rather strange; and I must therefore show you an experiment to prove that this is not an accidental matter, but that the whole of the mass is really concerned in this force, just as in falling the whole of the mass is really acted upon by the force of gravitation. I have here (Fig. 40) a steel bar, and I am going to make it a magnet by rubbing it on the large magnet (Fig. 39). I have now made the two ends magnetic in opposite ways. I do not at present know one from the other, but we can soon find out. You see, when I bring it near our magnetic needle (Fig. 38), one end repels and the other attracts; and the middle will neither attract nor repel—it can not, because it is half way between the two ends. But now, if I break out that piece (n, s), and then examine it, see how strongly one end (n) pulls at this end (S, Fig. 38), and how it repels the other end (N). And so it can be shown that every part of the magnet contains this power of attraction and repulsion, but that the power is only rendered evident at the end of the mass. You will understand all this in a little while; but what you have now to consider is that every part of this steel is in itself a magnet. Here is a little fragment which I have broken out of the very centre of the bar, and you will still see that one end is attractive and the other is repulsive. Now is not this power a most wonderful thing? And very strange, the means of taking it from one substance and bringing

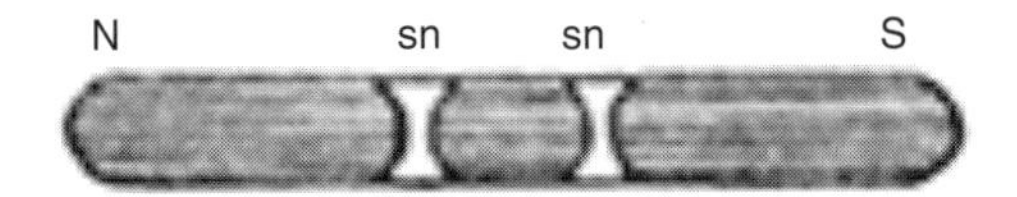

it to other matters. I can not make a piece of iron or any thing else heavier or lighter than it is; its cohesive power it must and does have; but, as you have seen by these experiments, we can add or subtract this power of magnetism, and almost do as we like with it.

And now we will return for a short time to the subject treated of at the commencement of this lecture. You see here (Fig. 41) a large machine arranged for the purpose of rubbing glass with silk, and for obtaining the power called electricity; and the moment the handle of the machine is turned a certain amount of electricity is evolved, as you will see by the rise of the little straw indicator (at A). Now I know, from the appearance of repulsion of the pith ball at the end of the straw, that electricity is present in those brass conductors (BB), and I want you to see the manner in which that electricity can pass away [touching the conductor (B) with his finger, the lecturer drew a spark from it, and the straw electrometer immediately fell]. There, it has all gone; and that I have really taken it away you shall see by an experiment of this sort. If I hold this cylinder of brass by the glass handle, and touch the conductor with it, I take away a little of the electricity. You see the spark in which it passes, and observe that the pith-ball indicator has fallen a little, which seems to imply that so much electricity is lost; but it is not lost; it is here in this brass, and I can take it away and carry it about, not because it has any substance of its own, but by some strange property which we have not before met with as belonging to any other force. Let us see whether we have it here or not. [The lecturer brought the charged cylinder to a jet from which gas was issuing; the spark was seen to pass from the cylinder to the jet, but the gas did not light.] Ah! the gas did not light, but you saw the spark; there is, perhaps, some draught in the room which blew the gas on one side, or else it would light; we will try this experiment afterward. You see from the spark that I can transfer the power from the machine to this cylinder, and then carry it away and give it to some other body.

You know very well, as a matter of experiment, that we can transfer the power of heat from one thing to another; for if I pout my hand near the fire it becomes hot. I can show you this by placing before us this ball, which has just been brought red-hot from the fire. If I press this wire to it some of the heat will be transferred from the ball, and I have only now to touch this piece of gun-cotton with the hot wire, and you see how I can transfer the heat from the ball to the wire, and from the wire to the cotton. So you see that some powers are transferable, and others are not. Observe how long the heat stops in this ball. I might touch it with the wire or with my finger, and if I did so quickly I should merely burn the surface of the skin; whereas, if I touch that cylinder, however rapidly, with my finger, the electricity is gone at once—dispersed on the instant, in a manner wonderful to think of.

I must now take up a little of your time in showing you the manner in which these powers are transferred from one thing to another; for the manner in which force may be conducted or transmitted is extraordinary, and most essential for us to understand. Let us see in what manner these powers travel from place to place. Both heat and electricity can be conducted; and here is an arrangement I have made to show how the former can travel. It consists of a bar of copper (Fig. 42); and if I take a spirit lamp (this is one way of obtaining the power of heat) and place it under that little chimney, the flame will strike against the bar of copper and keep it hot. Now you are aware that power is being transferred from the flame of that lamp to the copper, and you will see by-and-by that it is being conducted along the copper from particle to particle; for inasmuch as I

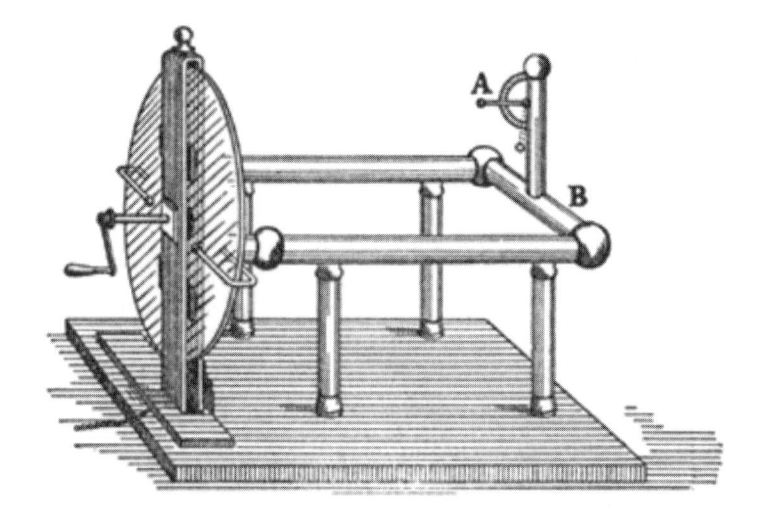

have fastened these wooden balls by a little wax at particular distances from the point where the copper is first heated, first one ball will fall and then the more distant ones, as the heat travels along, and thus you will learn that the heat travels gradually through the copper. You will see that this is a very slow conduction of power as compared with electricity. If I take cylinders of wood and metal, joined together at the ends, and wrap a piece of paper round, and then apply the heat of this lamp to the place where the metal and wood join, you will see how the heat will accumulate where the wood is, and burn the paper with which I have covered it; but where the metal is beneath, the heat is conducted away too fast for the paper to be burned. And so, if I take a piece of wood and a piece of metal joined together, and put it so that the flame shall play equally both upon one and the other, we shall soon find that the metal will become hot before the wood; for if I put a piece of phosphorus on the wood and another piece on the copper, you will find that the phosphorus on the copper will take fire before that on the wood is melted; and this shows you how badly the wood conducts heat. But with regard to the traveling of electricity from place to place, its rapidity is astonishing. I will, first of all, take these pieces of glass and metal, and you will soon understand how it is that the glass does not lose the power which it acquired when it is rubbed by the silk; by one or two experiments I will show you. If I take this piece of brass and bring it near the machine, you see how the electricity leaves the latter and passes to the brass cylinder. And again: if I take a rod of metal and touch the machine with it, I lower the indicator; but when I touch it with a rod of glass, no power is drawn away, showing you that the electricity is conducted by the glass and the metal in a manner entirely different; and, to make you see that more clearly, we will take one of our Leyden jars. Now I must not embarrass your minds with this subject too much, but if I take a piece of metal and bring it against the knob at the top and the metallic coating at the bottom, you will see the electricity passing through the air as a brilliant spark. It takes no sensible time to pass through this; and if I were to take a long metallic wire, no matter what the length, at least as far as we are concerned, and if I make one end of it touch the outside, and the other touch the knob at the top, see how the electricity passes! It has flashed instantaneously through the whole length of this wire. Is not this different from the transmission of heat through this copper bar (Fig. 42) which has taken a quarter of an hour or more to reach the first ball?

Here is another experiment for the purpose of showing the conductibility of this power through some bodies and not through others. Why do I have this arrangement made of brass? [pointing to the brass work of the electrical machine, Fig. 41]. Because it conducts electricity. And why do I have these columns made of glass? Because they obstruct the passage of electricity. And why do I put that paper tassel (Fig. 43) at the top of the pole, upon a glass rod, and connect it with this machine by means of a wire? You see at once that as soon as the handle of the machine is turned, the electricity which is evolved travels along this wire and up the wooden rod, and goes to the tassel at the top, and you see the power of repulsion with which it has endowed these strips of paper, each

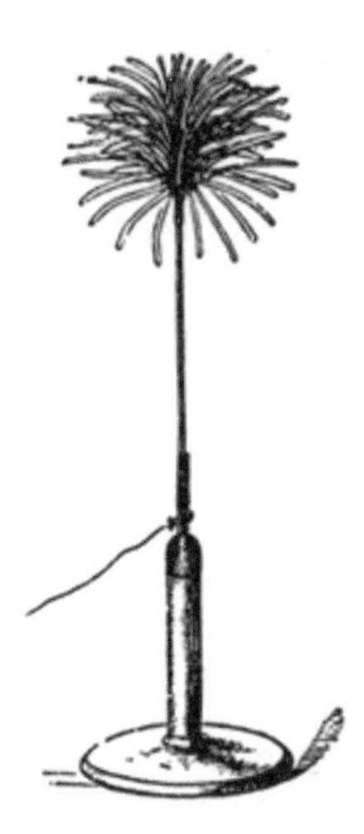

spreading outward to the ceiling and sides of the room. The outside of that wire is covered with gutta-percha; it would not serve to keep the force from you when touching it with your hands, because it would burst through; but it answers our purpose for the present. And so you perceived how easily I can manage to send this power of electricity from place to place by choosing the materials which can conduct the power. Suppose I want to fire a portion of gunpowder, I can readily do it by this transferable power of electricity. I will take a Leyden jar, or any other arrangement which gives us this power, and arrange wires so that they may carry the power to the place I wish; and then placing a little gunpowder on the extremities of the wires, the moment I make the connection by this discharging rod I shall fire the gunpowder [the connection was made and the gunpowder ignited]. And if I were to show you a stool like this, and were to explain to you its construction, you could easily understand that we use glass legs because these are capable of preventing the electricity from going away to the earth. If, therefore, I were to stand on this stool, and receive the electricity through this conductor, I could give it to anything that I touched. [The lecturer stood upon the insulating stool, and placed himself in connection with the conductor of the machine.] Now I am electrified; I can feel my hair rising up, as the paper tassel did just now. Let us see whether I can succeed in lighting gas by touching the jet with my finger. [The lecturer brought his finger near a jet from which gas was issuing, when, after one or two attempts, the spark which came from his finger to the jet set fire to the gas.] You now see how it is that this power of electricity can be transferred from the matter in which it is generated, and conducted along wires and other bodies, and thus be made to serve new purposes, utterly unattainable by the powers we have spoken of on previous days; and you will not now be at a loss to bring this power of electricity into comparison with those which to we have previously examined, and to-morrow we shall be able to go farther into the consideration of these transferable powers.

Reading and Discussion Questions

1. This selection is taken from one of Faraday's "Christmas Lectures," which were given for children. What about the phenomena of electricity and magnetism described here make them interesting and intriguing?
2. What instruments and experimental techniques is Faraday using to demonstrate these phenomena and help children visualize the invisible forces at work?

Experimental Researches in Electricity (1831)

Michael Faraday

56

First Series.

1. The power which electricity of tension possesses of causing an opposite electrical state in its vicinity has been expressed by the general term Induction; which, as it has been received into scientific language, may also, with propriety, be used in the same general sense to express the power which electrical currents may possess of inducing any particular state upon matter in their immediate neighbourhood, otherwise indifferent. It is with this meaning that I purpose using it in the present paper.
2. Certain effects of the induction of electrical currents have already been recognised and described: as those of magnetization; Ampère's experiments of bringing a copper disc near to a flat spiral; his repetition with electro-magnets of Arago's extraordinary experiments, and perhaps a few others. Still it appeared unlikely that these could be all the effects which induction by currents could produce; especially as, upon dispensing with iron, almost the whole of them disappear, whilst yet an infinity of bodies, exhibiting definite phenomena of induction with electricity of tension, still remain to be acted upon by the induction of electricity in motion.
3. Further: Whether Ampère's beautiful theory were adopted, or any other, or whatever reservation were mentally made, still it appeared very extraordinary, that as every electric current was accompanied by a corresponding intensity of magnetic action at right angles to the current, good conductors of electricity, when placed within the sphere of this action, should not have any current induced through them, or some sensible effect produced equivalent in force to such a current.

4. These considerations, with their consequence, the hope of obtaining electricity from ordinary magnetism, have stimulated me at various times to investigate experimentally the inductive effect of electric currents. I lately arrived at positive results; and not only had my hopes fulfilled, but obtained a key which appeared to me to open out a full explanation of Arago's magnetic phenomena, and also to discover a new state, which may probably have great influence in some of the most important effects of electric currents.
5. These results I purpose describing, not as they were obtained, but in such a manner as to give the most concise view of the whole.

§ 1. Induction of Electric Currents.

6. About twenty-six feet of copper wire one twentieth of an inch in diameter were wound round a cylinder of wood as a helix, the different spires of which were prevented from touching by a thin interposed twine. This helix was covered with calico, and then a second wire applied in the same manner. In this way twelve helices were superposed, each containing an average length of wire of twenty-seven feet, and all in the same direction. The first, third, fifth, seventh, ninth, and eleventh of these helices were connected at their extremities end to end, so as to form one helix; the others were connected in a similar manner; and thus two principal helices were produced, closely interposed, having the same direction, not touching anywhere, and each containing one hundred and fifty-five feet in length of wire.
7. One of these helices was connected with a galvanometer, the other with a voltaic battery of ten pairs of plates four inches square, with double coppers and well charged; yet not the slightest sensible reflection of the galvanometer-needle could be observed.
8. A similar compound helix, consisting of six lengths of copper and six of soft iron wire, was constructed. The resulting iron helix contained two hundred and fourteen feet of wire, the resulting copper helix two hundred and eight feet; but whether the current from the trough was passed through the copper or the iron helix, no effect upon the other could be perceived at the galvanometer.
9. In these and many similar experiments no difference in action of any kind appeared between iron and other metals.
10. Two hundred and three feet of copper wire in one length were coiled round a large block of wood; other two hundred and three feet of similar wire were interposed as a spiral between the turns of the first coil, and metallic contact everywhere prevented by twine. One of these helices was connected with a galvanometer, and the other with a

battery of one hundred pairs of plates four inches square, with double coppers, and well charged. When the contact was made, there was a sudden and very slight effect at the galvanometer, and there was also a similar slight effect when the contact with the battery was broken. But whilst the voltaic current was continuing to pass through the one helix, no galvanometrical appearances nor any effect like induction upon the other helix could be perceived, although the active power of the battery was proved to be great, by its heating the whole of its own helix, and by the brilliancy of the discharge when made through charcoal.

11. Repetition of the experiments with a battery of one hundred and twenty pairs of plates produced no other effects; but it was ascertained, both at this and the former time, that the slight deflection of the needle occurring at the moment of completing the connexion, was always in one direction, and that the equally slight deflection produced when the contact was broken, was in the other direction; and also, that these effects occurred when the first helices were used (6. 8.).

12. The results which I had by this time obtained with magnets led me to believe that the battery current through one wire, did, in reality, induce a similar current through the other wire, but that it continued for an instant only, and partook more of the nature of the electrical wave passed through from the shock of a common Leyden jar than of the current from a voltaic battery, and therefore might magnetise a steel needle, although it scarcely affected the galvanometer.

13. This expectation was confirmed; for on substituting a small hollow helix, formed round a glass tube, for the galvanometer, introducing a steel needle, making contact as before between the battery and the inducing wire (7. 10.), and then removing the needle before the battery contact was broken, it was found magnetised.

14. When the battery contact was first made, then an unmagnetised needle introduced into the small indicating helix (13.), and lastly the battery contact broken, the needle was found magnetised to an equal degree apparently as before; but the poles were of the contrary kind.

15. The same effects took place on using the large compound helices first described (6. 8.).

16. When the unmagnetised needle was put into the indicating helix, before contact of the inducing wire with the battery, and remained there until the contact was broken, it exhibited little or no magnetism; the first effect having been nearly neutralised by the second (13. 14.). The force of the induced current upon making contact was found always to exceed that of the induced current at breaking of contact; and if therefore the contact was made and broken many times in succession, whilst the needle remained in the indicating helix, it at last came out

not unmagnetised, but a needle magnetised as if the induced current upon making contact had acted alone on it. This effect may be due to the accumulation (as it is called) at the poles of the unconnected pile, rendering the current upon first making contact more powerful than what it is afterwards, at the moment of breaking contact.

17. If the circuit between the helix or wire under induction and the galvanometer or indicating spiral was not rendered complete *before* the connexion between the battery and the inducing wire was completed or broken, then no effects were perceived at the galvanometer. Thus, if the battery communications were first made, and then the wire under induction connected with the indicating helix, no magnetising power was there exhibited. But still retaining the latter communications, when those with the battery were broken, a magnet was formed in the helix, but of the second kind (14.), i.e. with poles indicating a current in the same direction to that belonging to the battery current, or to that always induced by that current at its cessation.

18. In the preceding experiments the wires were placed near to each other, and the contact of the inducing one with the buttery made when the inductive effect was required; but as the particular action might be supposed to be exerted only at the moments of making and breaking contact, the induction was produced in another way. Several feet of copper wire were stretched in wide zigzag forms, representing the letter W, on one surface of a broad board; a second wire was stretched in precisely similar forms on a second board, so that when brought near the first, the wires should everywhere touch, except that a sheet of thick paper was interposed. One of these wires was connected with the galvanometer, and the other with a voltaic battery. The first wire was then moved towards the second, and as it approached, the needle was deflected. Being then removed, the needle was deflected in the opposite direction. By first making the wires approach and then recede, simultaneously with the vibrations of the needle, the latter soon became very extensive; but when the wires ceased to move from or towards each other, the galvanometer-needle soon came to its usual position.

19. As the wires approximated, the induced current was in the *contrary* direction to the inducing current. As the wires receded, the induced current was in the *same* direction as the inducing current. When the wires remained stationary, there was no induced current (54.).

20. When a small voltaic arrangement was introduced into the circuit between the galvanometer (10.) and its helix or wire, so as to cause a permanent deflection of 30° or 40°, and then the battery of one hundred pairs of plates connected with the inducing wire, there was an instantaneous action as before (11.); but the galvanometer-needle

immediately resumed and retained its place unaltered, notwithstanding the continued contact of the inducing wire with the trough: such was the case in whichever way the contacts were made (33.).

21. Hence it would appear that collateral currents, either in the same or in opposite directions, exert no permanent inducing power on each other, affecting their quantity or tension.

22. I could obtain no evidence by the tongue, by spark, or by heating fine wire or charcoal, of the electricity passing through the wire under induction; neither could I obtain any chemical effects, though the contacts with metallic and other solutions were made and broken alternately with those of the battery, so that the second effect of induction should not oppose or neutralise the first (13. 16.).

23. This deficiency of effect is not because the induced current of electricity cannot pass fluids, but probably because of its brief duration and feeble intensity; for on introducing two large copper plates into the circuit on the induced side (20.), the plates being immersed in brine, but prevented from touching each other by an interposed cloth, the effect at the indicating galvanometer, or helix, occurred as before. The induced electricity could also pass through a voltaic trough (20.). When, however, the quantity of interposed fluid was reduced to a drop, the galvanometer gave no indication.

24. Attempts to obtain similar effects by the use of wires conveying ordinary electricity were doubtful in the results. A compound helix similar to that already described, containing eight elementary helices (6.), was used. Four of the helices had their similar ends bound together by wire, and the two general terminations thus produced connected with the small magnetising helix containing an unmagnetised needle (13.). The other four helices were similarly arranged, but their ends connected with a Leyden jar. On passing the discharge, the needle was found to be a magnet; but it appeared probable that a part of the electricity of the jar had passed off to the small helix, and so magnetised the needle. There was indeed no reason to expect that the electricity of a jar possessing as it does great tension, would not diffuse itself through all the metallic matter interposed between the coatings.

25. Still it does not follow that the discharge of ordinary electricity through a wire does not produce analogous phenomena to those arising from voltaic electricity; but as it appears impossible to separate the effects produced at the moment when the discharge begins to pass, from the equal and contrary effects produced when it ceases to pass (16.), inasmuch as with ordinary electricity these periods are simultaneous, so there can be scarcely any hope that in this form of the experiment they can be perceived.

26. Hence it is evident that currents of voltaic electricity present phenomena of induction somewhat analogous to those produced by electricity of tension, although, as will be seen hereafter, many differences exist between them. The result is the production of other currents, (but which are only momentary,) parallel, or tending to parallelism, with the inducing current. By reference to the poles of the needle formed in the indicating helix (13. 14.) and to the deflections of the galvanometer-needle (11.), it was found in all cases that the induced current, produced by the first action of the inducing current, was in the contrary direction to the latter, but that the current produced by the cessation of the inducing current was in the same direction (19.). For the purpose of avoiding periphrasis, I propose to call this action of the current from the voltaic battery, *volta-electric induction*. The properties of the second wire, after induction has developed the first current, and whilst the electricity from the battery continues to flow through its inducing neighbour (10. 18.), constitute a peculiar electric condition, the consideration of which will be resumed hereafter (60.). All these results have been obtained with a voltaic apparatus consisting of a single pair of plates.

§ 2. Evolution of Electricity from Magnetism.

27. A welded ring was made of soft round bar-iron, the metal being seven-eighths of an inch in thickness, and the ring six inches in external diameter. Three helices were put round one part of this ring, each containing about twenty-four feet of copper wire one twentieth of an inch thick; they were insulated from the iron and each other, and superposed in the manner before described (6.), occupying about nine inches in length upon the ring. They could be used separately or conjointly; the group may be distinguished by the letter A (Pl. I. fig. 1.). On the other part of the ring about sixty feet of similar copper wire in two pieces were applied in the same manner, forming a helix B, which had the same common direction with the helices of A, but being separated from it at each extremity by about half an inch of the uncovered iron.

28. The helix B was connected by copper wires with a galvanometer three feet from the ring. The helices of A were connected end to end so as to form one common helix, the extremities of which were connected with a battery of ten pairs of plates four inches square. The galvanometer was immediately affected, and to a degree far beyond what has been described when with a battery of tenfold power helices *without iron* were used (10.); but though the contact was continued, the effect was not permanent, for the needle soon came to rest in its natural position,

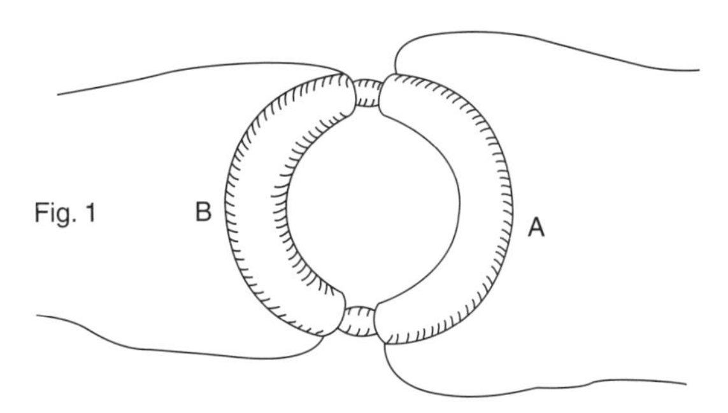

as if quite indifferent to the attached electro-magnetic arrangement. Upon breaking the contact with the battery, the needle was again powerfully deflected, but in the contrary direction to that induced in the first instance.

29. Upon arranging the apparatus so that B should be out of use, the galvanometer be connected with one of the three wires of A (27.), and the other two made into a helix through which the current from the trough (28.) was passed, similar but rather more powerful effects were produced.

30. When the battery contact was made in one direction, the galvanometer-needle was deflected on the one side; if made in the other direction, the deflection was on the other side. The deflection on breaking the battery contact was always the reverse of that produced by completing it. The deflection on making a battery contact always indicated an induced current in the opposite direction to that from the battery; but on breaking the contact the deflection indicated an induced current in the same direction as that of the battery. No making or breaking of the contact at B side, or in any part of the galvanometer circuit, produced any effect at the galvanometer. No continuance of the battery current caused any deflection of the galvanometer-needle. As the above results are common to all these experiments, and to similar ones with ordinary magnets to be hereafter detailed, they need not be again particularly described.

31. Upon using the power of one hundred pairs of plates (10.) with this ring, the impulse at the galvanometer, when contact was completed or broken, was so great as to make the needle spin round rapidly four or five times, before the air and terrestrial magnetism could reduce its motion to mere oscillation.

32. By using charcoal at the ends of the B helix, a minute *spark* could be perceived when the contact of the battery with A was completed. This spark could not be due to any diversion of a part of the current of the battery through the iron to the helix B; for when the battery contact

was continued, the galvanometer still resumed its perfectly indifferent state (28.). The spark was rarely seen on breaking contact. A small platina wire could not be ignited by this induced current; but there seems every reason to believe that the effect would be obtained by using a stronger original current or a more powerful arrangement of helices.

33. A feeble voltaic current was sent through the helix B and the galvanometer, so as to deflect the needle of the latter 30° or 40°, and then the battery of one hundred pairs of plates connected with A; but after the first effect was over, the galvanometer-needle resumed exactly the position due to the feeble current transmitted by its own wire. This took place in whichever way the battery contacts were made, and shows that here again (20.) no permanent influence of the currents upon each other, as to their quantity and tension, exists.

34. Another arrangement was then employed connecting the former experiments on volta-electric induction (6–26.) with the present. A combination of helices like that already described (6.) was constructed upon a hollow cylinder of pasteboard: there were eight lengths of copper wire, containing altogether 220 feet; four of these helices were connected end to end, and then with the galvanometer (7.); the other intervening four were also connected end to end, and the battery of one hundred pairs discharged through them. In this form the effect on the galvanometer was hardly sensible (11.), though magnets could be made by the induced current (13.). But when a soft iron cylinder seven eighths of an inch thick, and twelve inches long, was introduced into the pasteboard tube, surrounded by the helices, then the induced current affected the galvanometer powerfully and with all the phenomena just described (30.). It possessed also the power of making magnets with more energy, apparently, than when no iron cylinder was present.

35. When the iron cylinder was replaced by an equal cylinder of copper, no effect beyond that of the helices alone was produced. The iron cylinder arrangement was not so powerful as the ring arrangement already described (27.).

36. Similar effects were then produced by *ordinary magnets*: thus the hollow helix just described (34.) had all its elementary helices connected with the galvanometer by two copper wires, each five feet in length; the soft iron cylinder was introduced into its axis; a couple of bar magnets, each twenty-four inches long, were arranged with their opposite poles at one end in contact, so as to resemble a horse-shoe magnet, and then contact made between the other poles and the ends of the iron cylinder, so as to convert it for the time into a magnet (fig. 2.): by breaking the magnetic contacts, or reversing them, the magnetism of the iron cylinder could be destroyed or reversed at pleasure.

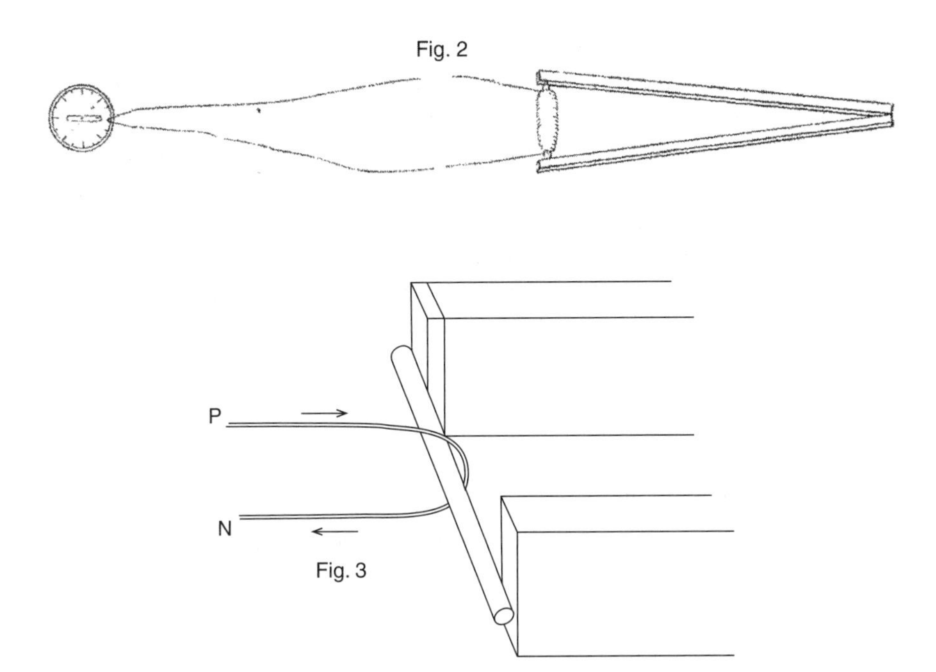

Fig. 2

Fig. 3

37. Upon making magnetic contact, the needle was deflected; continuing the contact, the needle became indifferent, and resumed its first position; on breaking the contact, it was again deflected, but in the opposite direction to the first effect, and then it again became indifferent. When the magnetic contacts were reversed the deflections were reversed.

38. When the magnetic contact was made, the deflection was such as to indicate an induced current of electricity in the opposite direction to that fitted to form a magnet, having the same polarity as that really produced by contact with the bar magnets. Thus when the marked and unmarked poles were placed as in fig. 3, the current in the helix was in the direction represented, P being supposed to be the end of the wire going to the positive pole of the battery, or that end towards which the zinc plates face, and N the negative wire. Such a current would have converted the cylinder into a magnet of the opposite kind to that formed by contact with the poles A and B; and such a current moves in the opposite direction to the currents which in M. Ampère's beautiful theory are considered as constituting a magnet in the position figured.

39. But as it might be supposed that in all the preceding experiments of this section, it was by some peculiar effect taking place during the formation of the magnet, and not by its mere virtual approximation, that the momentary induced current was excited, the following experiment was made. All the similar ends of the compound hollow helix (34.) were

bound together by copper wire, forming two general terminations, and these were connected with the galvanometer. The soft iron cylinder (34.) was removed, and a cylindrical magnet, three quarters of an inch in diameter and eight inches and a half in length, used instead. One end of this magnet was introduced into the axis of the helix (fig. 4.), and then, the galvanometer-needle being stationary, the magnet was suddenly thrust in; immediately the needle was deflected in the same direction as if the magnet had been formed by either of the two preceding processes (34. 36.). Being left in, the needle resumed its first position, and then the magnet being withdrawn the needle was deflected in the opposite direction. These effects were not great; but by introducing and withdrawing the magnet, so that the impulse each time should be added to those previously communicated to the needle, the latter could be made to vibrate through an arc of 180° or more.

40. In this experiment the magnet must not be passed entirely through the helix, for then a second action occurs. When the magnet is introduced, the needle at the galvanometer is deflected in a certain direction; but being in, whether it be pushed quite through or withdrawn, the needle is deflected in a direction the reverse of that previously produced. When the magnet is passed in and through at one continuous motion, the needle moves one way, is then suddenly stopped, and finally moves the other way.

41. If such a hollow helix as that described (34.) be laid east and west (or in any other constant position), and a magnet be retained east and west, its marked pole always being one way; then whichever end of the helix the magnet goes in at, and consequently whichever pole of the magnet enters first, still the needle is deflected the same way: on the other hand, whichever direction is followed in withdrawing the magnet, the deflection is constant, but contrary to that due to its entrance.

42. These effects are simple consequences of the *law* hereafter to be described (114).

43. When the eight elementary helices were made one long helix, the effect was not so great as in the arrangement described. When only one of the

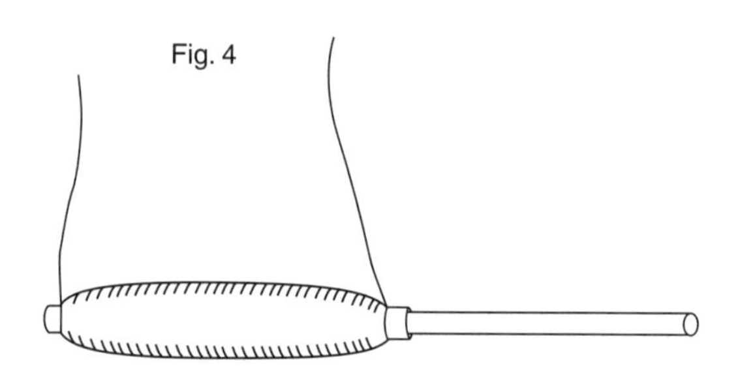

Fig. 5

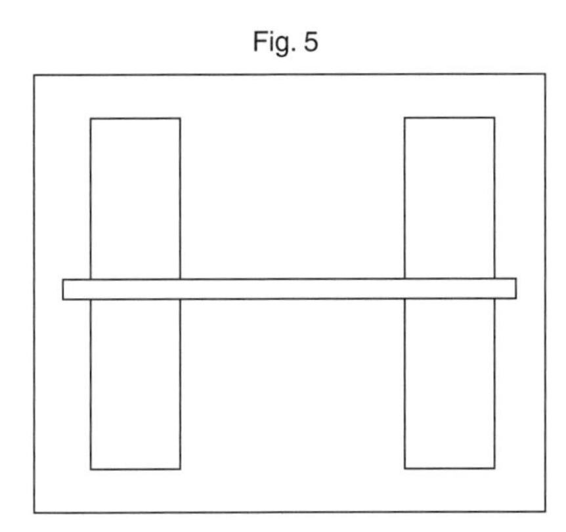

eight helices was used, the effect was also much diminished. All care was taken to guard against tiny direct action of the inducing magnet upon the galvanometer, and it was found that by moving the magnet in the same direction, and to the same degree on the outside of the helix, no effect on the needle was produced.

44. The Royal Society are in possession of a large compound magnet formerly belonging to Dr. Gowin Knight, which, by permission of the President and Council, I was allowed to use in the prosecution of these experiments: it is at present in the charge of Mr. Christie, at his house at Woolwich, where, by Mr. Christie's kindness, I was at liberty to work; and I have to acknowledge my obligations to him for his assistance in all the experiments and observations made with it. This magnet is composed of about 450 bar magnets, each fifteen inches long, one inch wide, and half an inch thick, arranged in a box so as to present at one of its extremities two external poles (fig. 5.). These poles projected horizontally six inches from the box, were each twelve inches high and three inches wide. They were nine inches apart; and when a soft iron cylinder, three quarters of an inch in diameter and twelve inches long, was put across from one to the other, it required a force of nearly one hundred pounds to break the contact. The pole to the left in the figure is the marked pole.[1]

45. The indicating galvanometer, in all experiments made with this magnet, was about eight feet from it, not directly in front of the poles, but about 16° or 17° on one side. It was found that on making or breaking the connexion of the poles by soft iron, the instrument was slightly

[1] To avoid any confusion as to the poles of the magnet, I shall designate the pole pointing to the north as the marked pole; I may occasionally speak of the north and south ends of the needle, but do not mean thereby north and south poles. That is by many considered the true north pole of a needle which points to the south; but in this country it in often called the south pole.

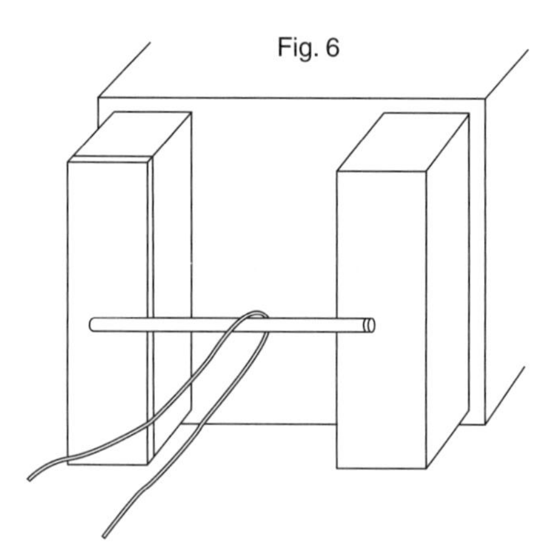
Fig. 6

affected; but all error of observation arising from this cause was easily and carefully avoided.

46. The electrical effects exhibited by this magnet were very striking. When a soft iron cylinder thirteen inches long was put through the compound hollow helix, with its ends arranged as two general terminations (39.), these connected with the galvanometer, and the iron cylinder brought in contact with the two poles of the magnet (fig. 5.), so powerful a rush of electricity took place that the needle whirled round many times in succession.[2]

47. Notwithstanding this great power, if the contact was continued, the needle resumed its natural position, being entirely uninfluenced by the position of the helix (30.). But on breaking the magnetic contact, the needle was whirled round in the opposite direction with a force equal to the former.

48. A piece of copper plate wrapped *once* round the iron cylinder like a socket, but with interposed paper to prevent contact, had its edges connected with the wires of the galvanometer. When the iron was brought in contact with the poles the galvanometer was strongly affected.

49. Dismissing the helices and sockets, the galvanometer wire was passed over, and consequently only half round the iron cylinder (fig. 6.); but even then a strong effect upon the needle was exhibited, when the magnetic contact was made or broken.

50. As the helix with its iron cylinder was brought towards the magnetic poles, but *without making contact*, still powerful effects were produced.

[2] A soft iron bar in the form of a lifter to a horse-shoe magnet, when supplied with a coil of this kind round the middle of it, becomes, by juxta-position with a magnet, a ready source of a brief but determinate current of electricity.

When the helix, without the iron cylinder, and consequently containing no metal but copper, was approached to, or placed between the poles (44.), the needle was thrown 80°, 90°, or more, from its natural position. The inductive force was of course greater, the nearer the helix, either with or without its iron cylinder, was brought to the poles; but otherwise the same effects were produced, whether the helix, &c. was or was not brought into contact with the magnet; i.e. no permanent effect on the galvanometer was produced; and the effects of approximation and removal were the reverse of each other (30.).

51. When a bolt of copper corresponding to the iron cylinder was introduced, no greater effect was produced by the helix than without it. But when a thick iron wire was substituted, the magneto-electric induction was rendered sensibly greater.

52. The direction of the electric current produced in all these experiments with the helix, was the same as that already described (38.) as obtained with the weaker bar magnets.

53. A spiral containing fourteen feet of copper wire, being connected with the galvanometer, and approximated directly towards the marked pole in the line of its axis, affected the instrument strongly; the current induced in it was in the reverse direction to the current theoretically considered by M. Ampère as existing in the magnet (38.), or as the current in an electro-magnet of similar polarity. As the spiral was withdrawn, the induced current was reversed.

54. A similar spiral had the current of eighty pairs of 4-inch plates sent through it so as to form an electro-magnet, and then the other spiral connected with the galvanometer (58.) approximated to it; the needle vibrated, indicating a current in the galvanometer spiral the reverse of that in the battery spiral (18. 26.). On withdrawing the latter spiral, the needle passed in the opposite direction.

55. Single wires, approximated in certain directions towards the magnetic pole, had currents induced in them. On their removal, the currents were inverted. In such experiments the wires should not be removed in directions different to those in which they were approximated; for then occasionally complicated and irregular effects are produced, the causes of which will be very evident in the fourth part of this paper.

56. All attempts to obtain chemical effects by the induced current of electricity failed, though the precautions before described (22.), and all others that could be thought of, were employed. Neither was any sensation on the tongue, or any convulsive effect upon the limbs of a frog, produced. Nor could charcoal or fine wire be ignited (133.). But upon repeating the experiments more at leisure at the Royal Institution, with an armed loadstone belonging to Professor Daniell and capable

of lifting about thirty pounds, a frog was very *powerfully convulsed* each time magnetic contact was made. At first the convulsions could not be obtained on breaking magnetic contact; but conceiving the deficiency of effect was because of the comparative slowness of separation, the latter act was effected by a blow, and then the frog was convulsed strongly. The more instantaneous the union or disunion is effected, the more powerful the convulsion. I thought also I could perceive the *sensation* upon the tongue and the *flash* before the eyes; but I could obtain no evidence of chemical decomposition.

57. The various experiments of this section prove, I think, most completely the production of electricity from ordinary magnetism. That its intensity should be very feeble and quantity small, cannot be considered wonderful, when it is remembered that like thermo-electricity it is evolved entirely within the substance of metals retaining all their conducting power. But an agent which is conducted along metallic wires in the manner described; which whilst so passing possesses the peculiar magnetic actions and force of a current of electricity; which can agitate and convulse the limbs of a frog; and which, finally, can produce a spark[3] by its discharge through charcoal (32.), can only be electricity. As all the effects can be produced by ferruginous electro-magnets (34.), there is no doubt that arrangements like the magnets of Professors Moll, Henry, Ten Eyke, and others, in which as many as two thousand pounds have been lifted, may be used for these experiments; in which case not only a brighter spark may be obtained, but wires also ignited, and, as the current can pass liquids (23.), chemical action be produced. These effects are still more likely to be obtained when the magneto-electric arrangements to be explained in the fourth section are excited by the powers of such apparatus.

58. The similarity of action, almost amounting to identity, between common magnets and either electro-magnets or volta-electric currents, is strikingly in accordance with and confirmatory of M. Ampère's theory, and furnishes powerful reasons for believing that the action is the same in both cases; but, as a distinction in language is still necessary, I propose to call the agency thus exerted by ordinary magnets, *magneto-electric* or *magnelectric* induction (26).

59. The only difference which powerfully strikes the attention as existing between volta-electric and magneto-electric induction, is the

[3] For a mode of obtaining the spark from the common magnet which I have found effectual, see the *Philosophical Magazine* for June 1832, p. 5. In the same Journal for November 1834, vol. v. p. 349, will be found a method of obtaining the magneto-electric spark, still simpler in its principle, the use of soft iron being dispensed with altogether.—*Dec. 1838.*

suddenness of the former, and the sensible time required by the latter; but even in this early state of investigation there are circumstances which seem to indicate, that upon further inquiry this difference will, as a philosophical distinction, disappear (68).[4]

Reading and Discussion Questions

1. Why does Faraday describe in such detail the structure of his experimental apparatus? What about the phenomena he's trying to observe make it necessary for him to use so much wire and such large batteries?
2. Under what circumstances does a current "induce" another current, or cause a magnetic effect? Under what circumstances does a magnet "induce" a current?
3. How is Faraday thinking about the relationship between electricity and magnetism and how does he visualize the "magnetic field" around a magnet or a current-carrying wire?

[4] For important additional phenomena and developments of the induction of electrical currents, see now the ninth series, 1048–1118.—*Dec. 1838.*

A Dynamical Theory of the Electromagnetic Field (1865)

James Clerk Maxwell

57

Part One: Introductory

1. The most obvious mechanical phenomenon in electrical and magnetical experiments is the mutual action by which bodies in certain states set each other in motion while still at a sensible distance from each other. The first step, therefore, in reducing these phenomena into scientific form, is to ascertain the magnitude and direction of the force acting between the bodies, and when it is found that this force depends in a certain way upon the relative position of the bodies and on their electric or magnetic condition, it seems at first sight natural to explain the facts by assuming the existence of some thing either at rest or in motion in each body, constituting its electric or magnetic state, and capable of acting at a distance according to mathematical laws.

In this way mathematical theories of statical electricity, of magnetism, of the mechanical action between conductors carrying currents, and of the induction of currents have been formed. In these theories the force acting between the two bodies is treated with reference only to the condition of the bodies and their relative position, and without any express consideration of the surrounding medium.

These theories assume, more or less explicitly, the existence of substances the particles of which have the property of acting on one another at a distance by attraction or repulsion. The most complete development of a theory of this kind is that of M. W. Weber, who has made the same theory include electrostatic and electromagnetic phenomena.

In doing so, however, he has found it necessary to assume that the force between two electric particles depends on their relative velocity, as well as on their distance.

This theory, as developed by MM. W. Weber and C. Neumann, is exceedingly ingenious, and wonderfully comprehensive in its application to the phenomena of statical electricity, electromagnetic attractions, induction of currents an diamagnetic phenomena; and it comes to us with the more authority, as it has served to guide the speculations of one who has made so great an advance in the practical part of electric science, both by introducing a consistent system of units in electrical measurement, and by actually determining electrical quantities with an accuracy hitherto unknown.

2. The mechanical difficulties, however, which are involved in the assumption of particles acting at a distance with forces which depend on their velocities are such as to prevent me from considering this theory as an ultimate one, though it may have been, and may yet be useful in leading to the coordination of phenomena.

I have therefore preferred to seek an explanation of the fact in another direction, by supposing them to be produced by actions which go on in the surrounding medium as well as in the excited bodies, and endeavouring to explain the action between distant bodies without assuming the existence of forces capable of acting directly at sensible distances.

3. The theory I propose may therefore be called a theory of the *Electromagnetic Field*, because it has to do with the space in the neighbourhood of the electric or magnetic bodies, and it may be called a *Dynamical* Theory, because it assumes that in that space there is matter in motion, by which the observed electromagnetic phenomena are produced.

4. The electromagnetic field is that part of space which contains and surrounds bodies in electric or magnetic conditions.

It may be filled with any kind of matter, or we may endeavour to render it empty of all gross matter, as in the case of Geissleh's tubes and other so-called vacua.

There is always, however, enough of matter left to receive and transmit the undulations of light and heat, and it is because the transmission of these radiations is not greatly altered when transparent bodies of measurable density are substituted for the so-called vacuum, that we are obliged to admit that the undulations are those of an aethereal substance, and not of the gross matter, the presence of which merely modifies in some way the motion of the aether.

We have therefore some reason to believe, from the phenomena of light and heat, that there is an aethereal medium filling space and permeating bodies, capable of being, set in motion and of transmitting that motion from one part to another, and of communicating that motion to gross matter so as to heat it and affect it in various ways.

5. Now the energy communicated to the body in heating it must have formerly existed in the moving medium, for the undulations had left the source of heat some time before they reached the body, and during that time the energy must have been half in the form of motion of the medium and half in the form of elastic resilience. From these considerations Professor W. Thomson has

argued[1] that the medium must have a density capable of comparison with that of gross matter, and has even assigned an inferior limit to that density.

6. We may therefore receive, as a datum derived from a branch of science independent of that with which we have to deal, the existence of a pervading medium, of small but real density, capable of being set in motion, and of transmitting motion from one part to another with great, but not infinite, velocity.

Hence the parts of this medium must be so connected that the motion of one part depends in some way on the motion of the rest; and at the same time these connexions must be capable of a certain kind of elastic yielding, since the communication of motion is not instantaneous, but occupies time.

The medium is therefore capable of receiving and storing up two kinds of energy, namely, the "actual" energy depending on the motions of its parts, and "potential" energy, consisting of the work which the medium will do in recovering from displacement in virtue of its elasticity.

The propagation of undulations consists in the continual transformation of one of these forms of energy into the other alternately, and at any instant the amount of energy in the whole medium is equally divided, so that half is energy of motion, and half is elastic resilience.

7. A medium having such a constitution may be capable of other kinds of motion and displacement than those which produce the phenomena of light and heat, and some of these may be of such a kind that they may be evidenced to our senses by the phenomena they produce.

8. Now we know that the luminiferous medium is in certain cases acted on by magnetism; for Faraday[2] discovered that when a plane polarized ray traverses a transparent diamagnetic medium in the direction of the lines of magnetic force produced by magnets or currents in the neighbourhood, the plane of polarization is caused to rotate.

This rotation is always in the direction in which positive electricity must be carried round the diamagnetic body in order to produce the actual magnetization of the field.

M. Verdet[3] has since discovered that if a paramagnetic body, such as solution of perchloride of iron in ether, be substituted for the diamagnetic body, the rotation is in the opposite direction.

Now Professor W. Thomson[4] has pointed out that no distribution of forces acting between the parts of a medium whose only motion is that of the luminous vibrations, is sufficient to account for the phenomena, but that we must admit the existence of a motion in the medium depending on the magnetization, in addition to the vibratory motion which constitutes light.

[1] "On the Possible Density of the Luminiferous Medium, and on the Mechanical Value of a Cubic Mile of Sunlight," *Transactions of the Royal Society of Edinburgh* (1854), p. 57.

[2] Experimental Researches, Series 19.

[3] *Comptes Rendus* (1856, second half year, p. 529, and 1857, first half year, p. 1209).

[4] *Proceedings of the Royal Society*, June 1856 and June 1861.

It is true that the rotation by magnetism of the plane of polarization has been observed only in media of considerable density; but the properties of the magnetic field are not so much altered by the substitution of one medium for another, or for a vacuum, as to allow us to suppose that the dense medium does anything more than merely modify the motion of the ether. We have therefore warrantable grounds for inquiring whether there may not be a motion of the ethereal medium going on wherever magnetic effects are observed, and we have some reason to suppose that this motion is one of rotation, having the direction of the magnetic force as its axis.

9. We may now consider another phenomenon observed in the electromagnetic field. When a body is moved across the lines of magnetic force it experiences what is called an electromotive force; the two extremities of the body tend to become oppositely electrified, and an electric current tends to flow through the body. When the electromotive force is sufficiently powerful, and is made to act on certain compound bodies, it decomposes them, and causes one of their components to pass towards one extremity of the body, and the other in the opposite direction.

Here we have evidence of a force causing an electric current in spite of resistance; electrifying the extremities of a body in opposite ways, a condition which is sustained only by the action of the electromotive force, and which, as soon as that force is removed, tends, with an equal and opposite force, to produce a counter current through the body and to restore the original electrical state of the body; and finally, if strong enough, tearing to pieces chemical compounds and carrying their components in opposite directions, while their natural tendency is to combine, and to combine with a force which can generate an electromotive force in the reverse direction.

This, then, is a force acting on a body caused by its motion through the electro-magnetic field, or by changes occurring in that field itself; and the effect of the force is either to produce a current and heat the body, or to decompose the body, or, when it can do neither, to put the body in a state of electric polarization,—a state of constraint in which opposite extremities are oppositely electrified, and from which the body tends to relieve itself as soon as the disturbing force is removed.

10. According to the theory which I propose to explain, this "electromotive force" is the force called into play during the communication of motion from one part of the medium to another, and it is by means of this force that the motion of one part causes motion in another part. When electromotive force acts on a conducting circuit, it produces a current, which, as it meets with resistance, occasions a continual transformation of electrical energy into heat, which is incapable of being restored again to the form of electrical energy by any reversal of the process.

11. But when electromotive force acts on a dielectric it produces a state of polarization of its parts similar in distribution to the polarity of the parts of a mass of iron under the influence of a magnet, and like the magnetic polarization,

capable of being described as a state in which every particle has its opposite poles in opposite conditions.[5]

In a dielectric under the action of electromotive force, we may conceive that the electricity in each molecule is so displaced that one side is rendered positively and the other negatively electrical, but that the electricity remains entirely connected with the molecule, and does not pass from one molecule to another. The effect of this action on the whole dielectric mass is to produce a general displacement of electricity in a certain direction. This displacement does not amount to a current, because when it has attained to a certain value it remains constant, but it is the commencement of a current, and its variations constitute currents in the positive or the negative as the displacement is increasing or decreasing. In the interior of the dielectric there is no indication of electrification, because the electrification of the surface of any molecule is neutralized by the opposite electrification of the surface of the molecules in contact with it; but at the bounding surface of the dielectric, where the electrification is not neutralized, we find the phenomena which indicate positive or negative electrification.

The relation between the electromotive force and the amount of electric displacement it produces depends on the nature of the dielectric, the same electromotive force producing generally a greater electric displacement in solid dielectrics, such as glass or sulphur, than in air.

12. Here, then, we perceive another effect of electromotive force, namely, electric displacement, which according to our theory is a kind of elastic yielding to the action of the force, similar to that which takes place in structures and machines owing to the want of perfect rigidity of the connexions.

13. The practical investigation of the inductive capacity of dielectrics is rendered difficult on account of two disturbing phenomena. The first is the conductivity of the dielectric, which, though in many cases exceedingly small, is not altogether insensible. The second is the phenomenon called electric absorption,[6] in virtue of which, when the dielectric is exposed to electromotive force, the electric displacement gradually increases, and when the electromotive force is removed, the dielectric does not instantly return to its primitive state, but only discharges a portion of its electrification, and when left to itself gradually acquires electrification on its surface, as the interior gradually becomes depolarized. Almost all solid dielectrics exhibit this phenomenon, which gives rise to the residual charge in the Leyden jar, and to several phenomena of electric cables described by Mr. F. Jenkin.[7]

14. We have here two other kinds of yielding besides the yielding of the perfect dielectric, which we have compared to a perfectly elastic body. The yielding due to conductivity may be compared to that of a viscous fluid (that is to say, a

[5] Faraday, *Exp. Res.* Series XI.; Mossotti, *Mem. Della Soc. Italiana (Modens)*, vol. xxiv. part 2. p. 49

[6] Faraday, *Exp. Res.* 1233–1250.

[7] *Reports of British Association*, 1859, p. 248; and *Report of Committee of Board of Trade on Submarine Cables*, pp. 136 & 464.

fluid having great internal friction), or a soft solid on which the smallest force produces a permanent alteration of figure increasing with the time during which the force acts. The yielding due to electric absorption may be compared to that of a cellular elastic body containing a thick fluid in its cavities. Such a body, when subjected to pressure, is compressed by degrees on account of the gradual yielding of the thick fluid; and when the pressure is removed it does not at once recover its figure, because the elasticity of the substance of the body has gradually to overcome the tenacity of the fluid before it can regain complete equilibrium.

Several solid bodies in which no such structure as we have supposed can be found, seem to possess a mechanical property of this kind[8]; and it seems probable that the same substances, if dielectrics, may possess the analogous electrical property, and if magnetic, may have corresponding properties relating to the acquisition, retention, and loss of magnetic polarity.

15. It appears therefore that certain phenomena in electricity and magnetism lead to the same conclusion as those of optics, namely, that there is an aethereal medium pervading all bodies, and modified only in degree by their presence; that the parts of this medium are capable of being set in motion by electric currents and magnets; that this motion is communicated from one part of the medium to another by forces arising from the connexions of those parts; that under the action of these forces there is a certain yielding depending on the elasticity of these connexions; and that therefore energy in two different forms may exist in the medium, the one form being the actual energy of motion of its parts, and the other being the potential energy stored up in the connexions, in virtue of their elasticity.

16. Thus, then, we are led to the conception of a complicated mechanism capable of a vast variety of motion, but at the same time so connected that the motion of one part depends, according to definite relations, on the motion of other parts, these motions being communicated by forces arising from the relative displacement of the connected parts, in virtue of their elasticity. Such a mechanism must be subject to the general laws of Dynamics, and we ought to be able to work out all the consequences of its motion, provided we know the form of the relation between the motions of the parts.

17. We know that when an electric current is established in a conducting circuit, the neighbouring part of the field is characterized by certain magnetic properties, and that if two circuits are in the field, the magnetic properties of the field due to the two currents are combined. Thus each part of the field is in connexion with both currents, and the two currents are put in connexion with each other in virtue of their connexion with the magnetization of the field. The first result of this connexion that I propose to examine, is the induction of one current by another, and by the motion of conductors in the field.

[8] As, for instance, the composition of glue, treacle, etc., of which small plastic figures are made, which after being distorted gradually recover their shape.

The second result, which is deduced from this, is the mechanical action between conductors carrying currents. The phenomenon of the induction of currents has been deduced from their mechanical action by Helmholtz[9] and Thomson.[10] I have followed the reverse order, and deduced the mechanical action from the laws of induction. I have then described experimental methods of determining the quantities L, M, N, on which these phenomena depend.

18. I then apply the phenomena of induction and attraction of currents to the exploration of the electromagnetic field, and the laying down systems of lines of magnetic force which indicate its magnetic properties. By exploring the same field with a magnet, I show the distribution of its equipotential magnetic surfaces, cutting the lines of force at right angles.

In order to bring these results within the power of symbolical calculation, I then express them in the form of the General Equations of the Electromagnetic Field. These equations express:

The relation between electric displacement, true conduction, and the total current, compounded of both.

The relation between the lines of magnetic force and the inductive coefficients of a circuit, as already deduced from the laws of induction.

The relation between the strength of a current and its magnetic effects, according to the electromagnetic system of measurement.

The value of the electromotive force in a body, as arising from the motion of body in the field, the alteration of the field itself, and the variation of electric potential from one part of the field to another.

The relation between electric displacement, and the electromotive force which produces it.

The relation between an electric current, and the electromotive force which produces it.

The relation between the amount of free electricity at any point, and the electric displacements in the neighbourhood.

The relation between the increase or diminution of free electricity and the electric currents in the neighbourhood.

There are twenty of these equations in all, involving twenty variable quantities.

19. I then express in terms of these quantities the intrinsic energy of the Electromagnetic Field as depending partly on its magnetic and partly on its electric polarization at every point.

From this I determine the mechanical force acting, 1st, on a moveable conductor carrying an. electric current; 2ndly, on a magnetic pole; 3rdly, on an electrified body.

The last result, namely, the mechanical force acting on an electrified body, gives rise to an independent method of electrical measurement founded on its

[9] "Conservation of Force," *Physical Society of Berlin*, 1847; and Taylor's *Scientific Memoirs*, 1853, p. 114.

[10] *Reports of the British Association*, 1848; *Philosophical Magazine*, Dec. 1851.

electrostatic effects. The relation between the units employed in the two methods is shown to depend on what I have called the "electric elasticity" of the medium, and to be a velocity, which has been experimentally determined by MM. Weber and Kohlrausch.

I then show how to calculate the electrostatic capacity of a condenser, and the specific inductive capacity of a dielectric.

The case of a condenser composed of parallel layers of substances of different electric resistances and inductive capacities is next examined, and it is shown that the phenomenon called electric absorption will generally occur, that is, the condenser, when suddenly discharged, will after a short time show signs of a *residual* charge.

20. The general equations are next applied to the case of a magnetic disturbance propagated through a non-conducting field, and it is shown that the only disturbances which can be so propagated are those which are transverse to the direction of propagation, and that the velocity of propagation is the velocity v, found from experiments such as those of Weber, which expresses the number of electrostatic units of electricity which are contained in one electromagnetic unit.

This velocity is so nearly that of light, that it seems we have strong reason to conclude that light itself (including radiant heat, and other radiations if any) is an electro magnetic disturbance in the form of waves propagated through the electromagnetic field according to electromagnetic laws. If so, the agreement between the elasticity of the medium as calculated from the rapid alternations of luminous vibrations, and as found by the slow processes of electrical experiments, shows how perfect and regular the elastic properties of the medium must be when not encumbered with any matter denser than air. If the same character of the elasticity is retained in dense transparent bodies, it appears that the square of the index of refraction is equal to the product of the specific dielectric capacity and the specific magnetic capacity. Conducting media are shown to absorb such radiations rapidly, and therefore to be generally opaque.

The conception of the propagation of transverse magnetic disturbances to the exclusion of normal ones is distinctly set forth by Professor Faraday in his "Thoughts on Ray Vibrations." The electromagnetic theory of light, as proposed by him, is the same in substance as that which I have begun to develop in this paper, except that in 1846 there were no data to calculate the velocity of propagation.

21. The general equations are then applied to the calculation of the coefficients of mutual induction of two circular currents and the coefficient of self-induction in a coil. The want of uniformity of the current in the different parts of the section of a wire at the commencement of the current is investigated, I believe for the first time, and the consequent correction of the coefficient of self-induction is found.

These results are applied to the calculation of the self-induction of the coil used in the experiments of the Committee of the British Association on

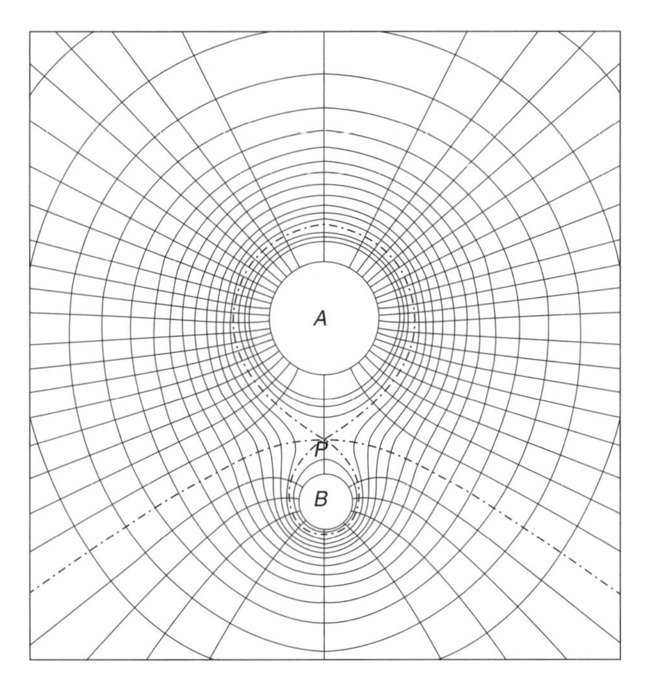

Lines of Force and Equipotential Surfaces.

A - 20. *B - 5.* *P. Point of Equilibrium.* *AP - $\frac{2}{3}$AB.*

Standards of Electric Resistance, and the value compared with that deduced from the experiments.

Reading and Discussion Questions

1. Of what phenomena does Maxwell's theory provide a unified explanation? What about these phenomena does he take his equations to describe? What do Maxwell's equations show about the relationship between electricity, magnetism, and the speed of light?
2. What is an electromagnetic field? Why does Maxwell call his theory of an electromagnetic field a dynamic theory?

Additional Resources

Cantor, G. N. "Physical Optics." In *Companion to the History of Modern Science*. M. J. S. Hodge, R. C. Olby, N. Cantor, and J. R. R. Christie (eds). London: Routledge, 1990.

Cushing, James T. *Philosophical Concepts in Physics: The Historical Relation between Philosophy and Scientific Theories*. Cambridge: Cambridge University Press, 1998. The thirteenth and fourteenth chapters of this volume provide useful background on both aether theories of light and Maxwell's equations.

Fisher, Howard J. *Faraday's Experimental Researches in Electricity: The First Series*. Santa Fe, NM: Green Lion Press, 2004.

Gregory, Frederick. *Natural Science in Western History*. 1st ed. Boston, MA: Houghton Mifflin Company, 2008. A solid secondary source textbook, by a past president of the History of Science Society, for the whole period covered by the readings in this volume. Gregory's introductions to chemistry, electro-magnetism, and thermodynamics are perhaps its greatest strength.

Harman, P. M. *Energy, Force, and Matter: The Conceptual Development of Nineteenth-Century Physics*. Cambridge: University of Cambridge Press, 1982.

Kuhn, Thomas. *The Structure of Scientific Revolutions.* 3rd ed. Chicago: University of Chicago Press, 1996.

Purrington, Robert D. *Physics in the Nineteenth Century*. New Brunswick, NJ: Rutgers University Press 1997.

Shapin, Stephen. "Who Was Robert Hooke?" In *Robert Hooke: New Studies.* Michael Hunter and Simon Schaffer, (eds). Woodbridge: The Boydell Press, 1989.

Smith, Crosby. "Energy." In *Companion to the History of Modern Science*. M. J. S. Hodge, R. C. Olby, N. Cantor, and J. R. R. Christie (eds). London: Routledge, 1990.

Part V

Elements in Transition: Chemistry, Air, Atoms, and Heat

Of the Nature of Things (1537)

Paracelsus (Philippus Aureolus Theophrastus Bombastus von Hohenheim)

58

From Book I: Of the Generations of Naturall Things

The generation of all natural things is twofold: Naturall and without Art; and Artificiall, *viz.* by Alchymie. Although in generall it may bee said that all things are naturally generated of the Earth by means of putrefaction. For Putrefaction is the chiefe degree and first step to Generation. Now Putrefaction is occasioned by a moist heat. For a continuall moist heat causeth putrefaction, and changeth all naturall things from their first form and essence, as also their vertues and efficacy, into another thing. For as putrefaction in the stomach changeth and reduceth all meats into dung; so also putrefaction out of the stomach in a glasse, changeth all things from one form into another, from one essence into another, from one colour into another, from one smell into another, from one vertue into another, from one power into another, from one property into another, and generally from one quality into another. For it is evident and proved by daily experience that many good things which are wholsome and medicinable, become after putrefaction naught, unwholsome, and meer poison. So on the contrary, there are many bad, unwholsome, poisonous, and hurtfull things, which after their putrefaction become good, lose all their unwholsomnesse, and become wonderfull medicinable: because putrefaction produceth great matters, as of this wee have a most famous example in the holy Gospel, where Christ saith: Unless a grain of Wheat bee cast into the Earth, and be putrefied, it cannot bring forth fruit in a hundred fold. Hence also we must know that many things are multiplyed in putrefaction so as to bring forth excellent fruit. For putrefaction is the change and death of all things, and destruction of the first essence of all

Naturall things; whence there ariseth a regeneration, and new generation a thousand times better, &c . . . .

And here wee must take notice of something that is greater and more then this: *viz.* if that living Chicke be in a vessell of glasse like a gourd, and sealed up, burnt to powder, or ashes in the third degree of Fire, and afterward so closed in, be putrefied with the exactest putrefaction of Horse-dung, into a mucilaginous flegm, then that flegm may be brought to maturity and become a renewed and new made Chicke: to wit, if that flegm bee again inclosed in its former shell or receptacle. This is to revive the dead by regeneration and clarification, which indeed is a great and profound miracle of Nature . . . .

Here it is necessary that we speak something of the generation of Metalls; but because we have wrote sufficiently of that in our book of the generation of Metals, wee shall very briefly treat of it here, only briefly adding what was omitted in that book. Know that all the seven Metalls are brought forth after this manner, out of a threefold matter, *viz.* Mercury, Sulphur, & Salt, yet in distinct and peculiar colours. For this reason *Hermes* did not speak amisse when he said, that of three substances are all the seven Metalls produced and compounded, as also the Tinctures and Philosophers Stone. Those 3 substances he calls the Spirit, Soul, and Body: but hee did not shew how this is to be understood, or what hee did mean by this, although haply hee might know the three Principles, but did not make mention of them. Wherefore we do not say that he was here in an error, but only was silent now, that those 3 distinct substances may be rightly understood, *viz.* Spirit, Soul, and Body, we must know, that they signifie nothing else but the three Principles, *i.e.* Mercury, Sulphur, Salt, of which all the seven Metalls are generated. For Mercury is the Spirit, Sulphur the Soule, and Salt the Body, but a Metall is the Soul betwixt the Spirit and the Body (as *Hermes* saith) which Soule indeed is Sulphur; and unites these two contraries, the Body and Spirit, and changeth them into one essence, &c.

Now this is not to bee understood so as that of every Mercury, every Sulphur, or of every Salt, the seven Metalls may be generated, or the Tincture, or the Philosophers Stone by the Art of Alchymie, or industry, with the help of Fire; but all the seven Metalls must be generated in the mountains by the *Archeius*[1] of the Earth. For the Alchymist shall sooner transmute Metalls, then generate or make them.

Yet neverthelesse living Mercury is the Mother of all the seven Metalls, and deservedly it may be called the Mother of the Metalls. For it is an open Metall, and as it contains all colours, which it manifests in the Fire, so also occultly it contains all Metalls in it selfe, but without Fire it cannot shew them, &c.

But generation and renovation of Metalls is made thus: As a man may return into the womb of his Mother, *i.e.* into the Earth, out of which hee was first made a man, and shall again bee raised at the last day: so also all Metalls may

[1] "The immaterial principle supposed by the Paracelsians to produce and preside over the activities of the animal and vegetable economy; vital force" (OED).

returne into living mercury againe, and become and by Fire bee regenerated and purified, if for the space of forty weeks they bee kept in a continuall heat, as an infant is in his Mothers wombe. So that now there are brought forth not common Metalls, but Tinging Metalls. For if Silver bee regenerated (after the manner as wee have spoken) it will afterward tinge all other Metalls into Silver, so win Gold into Gold, and the like is to be understood of all the other Metalls.

Now forasmuch as *Hermes* said that the soule alone is that medium which joines the spirit to the body, it was not without cause hee said so. For seeing Sulphur is that soule, and doth like Fire ripen and digest all things; it can also bind the soule with the body, incorporating and uniting them together, so that from thence may bee produced a most excellent body. Now the common combustible Sulphur is not to bee taken for the soule of metalls, for the soule is another manner of thing then a combustible and corruptible body.

Wherefore it can bee destroyed by no Fire, seeing indeed it is all Fire itself: and indeed it is nothing else but the quintessence of Sulphur, which is extracted out of reverberated Sulphur by the spirit of wine, being of a red colour and as transparent as a Rubie: and which indeed is a great and excellent *Arcanum,* for the transmuting of white metalls, and to coagulate living mercury into fixt and true Gold. Esteeme this as an enriching treasure, and thou maist bee well contented with this onely secret in the Transmutation of Metalls . . . .

Book 2: Of the Growth, and Increase of Naturall Things.

It is sufficiently manifest and knowne to every one, that all naturall things grow and are ripned through heat and moisture, which is sufficiently demonstrated by rain and the heat of the sun. For no man can deny that rain doth make the Earth fruitfull, and it is granted by all that all fruits are ripened by the sun.

Seeing therefore this is by divine ordination naturally possible, who can gain-say or not beleeve that a man is able, through the wise and skilfull Art of Alchymy, to make that which is barren, fruitfull, and that which is crude, to ripen, and all things to grow, and to be increased. . . .

It is possible also that Gold, through industry and skill of an expert Alchymist, may bee so far exalted that it may grow in a glasse like a tree, with many wonderfull boughs and leaves, which indeed is pleasant to behold and most wonderful.

The process is this. Let Gold bee calcined with *Aqua Regis*,[2] till it becomes a kind of chalke, which put into a gourd glasse and poure upon it good new *Aqua Regis*, so that it may cover it foure fingers breadth, then again draw it off with the third degree of fire, until no more ascend. The water that is distilled off, poure

[2] "A mixture of nitric and hydrochloric acids, so called because it can dissolve the 'noble' metals, gold and platinum" (OED).

on againe, then distill it off againe. This doe so long untill thou seest the Gold to rise in the glasse and grow after the manner of a tree, having many boughes and leaves: and so there is made of Gold a wonderful and pleasant shrub, which the Alchymists call their Golden hearb and the Philosophers Tree. In like manner you may proceed with Silver and other Metalls, yet so that their calcination bee made after another manner, by another *Aquafortis*,[3] which I leave to thine experience. If thou art skilled in Alchymie, thou shalt not erre in these things.

Book 4: Of the Life of Naturall Things.

No man can deny that Aire gives life to all things, bodies, and substances that are produced and generated of the Earth. Now you must know what, and what manner of thing the life of every thing in particular is; and it is nothing else then a spirituall essence, a thing that is invisible, impalpable, a spirit, and spirituall. Wherefore there is no corporeall thing which hath not a spirit lying hid in it, as also a life, which, as I said before, is nothing but a spirituall thing. For not only that hath life which moves and stirres, as Men, Animalls, Vermine of the earth. Birds in the Aire, Fish in the sea, but also all corporeall and substantiall things. For here wee must know that God in the beginning of the Creation of all things, created no body at all without its spirit, which it secretly contains in it.

For what is the body without a spirit? Nothing at all. Wherefore the spirit contains in it secretly the vertue and power of the thing, and not the body. For in the body there is death, and the body is the subject of death, neither is any else to be sought for in the body but death.

For that may severall wayes bee destroyed and corrupted, but the spirit cannot. For the living spirit remains for ever, and also is the subject of life: and preserves the body alive; but in the mine of the body it is separated from it, and leaves behind it a dead body, and returnes to its place from whence it came, *viz.* into the Chaos, and the Aire of the upper and lower Firmament. Hence it appears that there are divers spirits, as well as divers bodies.

For there are spirits Celestiall, Infernall, Humane, Metalline, Minerall, of Salts, of Gemmes, of Marcasites, of Arsenides, of Potable things, of Rootes, of Juices, of Flesh, of Blood, of Bones, &c. Wherefore also know that the spirit is most truly the life and balsome[4] of all Corporeall things. But now wee will proceed to the species, and briefly describe to you in this place the life of every naturall thing in particular.

[3] "The early scientific, and still the popular, name of the Nitric Acid of commerce (dilute HNO3), a powerful solvent and corrosive" (OED).

[4] Ruland states that "Balsam is a substance which preserves bodies from putrefaction. It is internal and external … . It is also called a most tempered gluten of the nature of any body to which it belongs. Briefly, it is the liquor of an interior salt most carefully and naturally preserving its body from corruption" (Lexicon 69).

The life therefore of all men is nothing else but an Astrall balsome, a Balsamick impression, and a celestiall invisible Fire, an included Aire, and a tinging spirit of Salt. I cannot name it more plainly, although it bee set out by many names. And seeing wee have declared the best and chiefest, wee shall bee silent in these which are lesse materiall.

The life of Metalls is a secret fatnesse which they have received from Sulphur, which is manifest by their flowing, for every thing that flowes in the fire, flowes by reason of that secret fatnesse that is in it: unlesse that were in it, no Metall could flow, as wee see in Iron and Steel, which have lesse Sulphur and fatnesse then all the other Metalls, wherefore they are of a dryer Nature then all the rest . . . .

Book 7: Of the Transmutation of Naturall Things.

If wee write of the Transmutation of all Naturall things, it is fit and necessary that in the first place wee shew what Transmutation is. Secondly, what bee the degrees to it. Thirdly, by what Mediums, and how it is done.

Transmutation therefore is when a thing loseth its form and is so altered that it is altogether unlike to its former substance and form, but assumes another form, another essence, another colour, another vertue, another nature, or property, as if a Metall bee made glasse or stone: if a stone bee made a coale: if wood be made a coal: clay be made a stone or a brick: a skin bee made glew: cloth be made paper, and many such like things. All these are Transmutations of Naturall things.

After this, it is very necessary also to know the degrees to Transmutation, and how many they be. And they are no more then seven. For although many doe reckon more, yet there are no more but seven which are principall, and the rest may bee reckoned betwixt the degrees, being comprehended under those seven: And they are these. Calcination, Sublimation, Solution, Putrefaction, Distillation, Coagulation, Tincture. If anyone will climbe that Ladder, he shall come into a most wonderfull place, that hee shall see and have experience of many secrets in the Transmutation of Naturall things.

The first degree therefore is Calcination, under which also are comprehended Reverberation, and Cementation. For betwixt these there is but little difference as for Matter of Calcination: wherefore it is here the chiefest degree. For by Reverberation and Cementation, many corporeall things are calcined and brought into Ashes, and especially Metalls. Now what is calcined is not any further reverberated or cemented.

By Calcination therefore all Metalls, Mineralls, Stones, Glasse, &c. and all corporeall things are made a Coal and Ashes, and this is done by a naked strong Fire with blowing, by which all tenacious, soft, and fat earth is hardened into a stone. Also all stones are brought into a Calx, as wee see in a Potters furnace of lime and bricks.

Sublimation is the second degree and one of the most principall for the Transmutation of many Naturall things: under which is contained Exaltation, Elevation, and Fixation; and it is not much unlike Distillation. For as in Distillation the water ascends from all flegmatick and watery things and is separated from its body; so in Sublimation, that which is spirituall is raised from what is corporeall, and is subtilized, volatile from fixed, and that in dry things, as are all Mineralls, and the pure is separated from the impure . . . .

Let that which is sublimed be ground and mixed with its feces, and bee againe sublimed as before, which must bee done so long, till it will no longer sublime, but all will remaine together in the bottom and be fixed.

So there will bee afterward a stone, and oyle when and as oft as thou pleasest, *viz.* if thou puttest it into a cold place, or in the aire in a Glass. For there it will presently bee dissolved into an Oyle. And if thou puttest it againe into the fire, it will againe bee coagulated into a Stone of wonderfull and great vertue. Keep this as a great secret and mystery of Nature, neither discover it to Sophisters . . . .

The third degree is Solution, under which are to bee understood Dissolution and Resolution, and this degree doth most commonly follow Sublimation and Distillation, *viz.* that the matter be resolved which remaines in the bottome. Now Solution is twofold: the one of Cold, the other of Heat; the one without Fire, the other in Fire.

A cold dissolution dissolves all Salts, all Corrosive things, & all calcined things. Whatsoever is of a Salt and Corrosive quality is by it dissolved into Oyle, Liquor, or Water. And this is in a moist, cold cellar or else in the Aire on a marble or in a glasse. For whatsoever is dissolved in the cold contains an Airy spirit of Salt, which oftentimes it gets, and assumes in Sublimation or Distillation. And whatsoever is dissolved in the cold, or in the Aire, may again by the heat of the Fire bee coagulated into powder or a stone . . . .

Putrefaction is the fourth degree, under which is comprehended Digestion and Circulation. Now then Putrefaction is one of the principall degrees, which indeed might deservedly have been the first of all, but that it would be against the true order and mystery, which is here hid and known to few: For those degrees must, as hath been already said, so follow one the other, as links in a chain or steps in a ladder.

For if one of the linkes should bee taken away, the chain is discontinued and broken, and the prisoners would bee at liberty and runne away. So in a ladder, if one step bee taken away in the middle and bee put in the upper or lower part, the ladder would be broken and many would fall down headlong by it with the hazard of their bodies, and lives . . . .

Now putrefaction is of such efficacy, that it abolisheth the old Nature and brings in a new one. All living things are killed in it, all dead things putrefied in it, and all dead things recover life in it.

Putrefaction takes from all Corrosive spirits, the sharpnesse of the Salt and makes them mild and sweet, changeth the colours, and separates the pure from the impure; it places the pure above and the impure beneath.

Distillation is the first degree to the Transmutation of all naturall things. Under it are understood Ascension, Lavation, and Fixation.

By Distillation all Waters, Liquors, and Oyles are subtilized; out of all fat things Oyle is extracted, out of all Liquors, Water, and out of all Flegmaticke things Water and Oyle are separated.

Besides there are many things in Distillation fixed by Cohobation,[5] and especially if the things to bee fixed containe in them Water, as Vitriall doth, which if it bee fixed is called *Colcothar*. . . .

Moreover, in Distillation many bitter, harsh, and sharp things become as sweet as Honey, Sugar, or Manna; and on the contrary, many sweet things, as Sugar, Honey, or Manna may bee made as harsh as Oyle of Vitriall or Vineger, or as bitter as Gall or Gentian, as Eager as a Corrosive . . . .

Coagulation is the sixt degree: now there is a twofold Coagulation, the one by Cold, the other by Heat, i.e. one of the Aire, the other of the Fire: and each of these again is twofold, so that there are foure sorts of Coagulations, two of Cold, and two of Fire ... the Coagulation of Fire, which alone is here to bee taken notice of, is made by an Artificiall and Graduall Fire of the Alchymists, and it is fixed and permanent. For whatsoever such a Fire doth coagulate, the same abides so.

The other Coagulation is done by the Aetnean and Minerall Fire in the Mountains, which indeed the Archeius of the Earth governs and graduates not unlike to the Alchymists, and whatsoever is coagulated by such a Fire is also fixed and constant; as you see in Mineralls and Metalls, which indeed at the beginning are a mucilaginous matter, and are coagulated into Metalls, Stones, Flints, Salts, and other bodies, by the Aetnean fire in the Mountaines, through the Archeius of Earth, and operator of Nature . . . .

Tincture is the seventh and last degree, which concludes the whole worke of our mystery for Transmutation, making all imperfect things perfect and transmuting them into a most excellent essence, and into a most perfect soundnesse, and alters them into another colour.

Tincture therefore is a most excellent matter, wherewith all Minerall and Humane bodies are tinged and are changed into a better and more noble essence and into the highest perfection and purity.

For Tincture colours all things according to its own nature and colour.

Now there are many Tinctures and not only for Metalline but Humane bodies, because every thing which penetrates another matter, or tingeth it with another colour, or essence, so that it bee no more like the former, may bee called a Tincture . . . .

[5] Repeated distillation.

For if a Tincture must tinge, it is necessary that the body or matter which is to bee tinged, bee opened and continue in flux, and unless this should bee so, the Tincture could not operate . . . .

Now these are the Tinctures of Metalls, which it is necessary must bee turned into an Alcool[6] by the first degree of Calcination, then by the second degree of Sublimation, must get an easy and light flux. And lastly, by the degree of Putrefaction and Distillation are made a fixt and incombustible Tincture and of an unchangeable colour.

Now the Tinctures of Mens bodies are that they bee tinged into the highest perfection of health and all Diseases bee expelled from them, that their lost strength and colour bee restored and renewed, and they are these, *viz.* Gold, Pearles, Antimony, Sulphur, Vitriall, and such like, whose preparation wee have diversly taught in other books . . . .

Book 8: Of the Separation of Naturall Things.

In the Creation of the world, the first separation began from the foure Elements, seeing the first matter of the world was one Chaos.

Of this Chaos God made the greater world, being divided into four distinct Elements, *viz.* Fire, Aire, Water, and Earth. Fire is the hot part, Aire the moist. Water the cold, and Earth the dry part of the greater world.

But that you may in brief understand the reason of our purpose in the 8th book, you must know that we doe not purpose to treat here of the Elements of all Naturall things, seeing wee have sufficiently discoursed of those Arcana in the *Archidoxis*[7] of the separation of Naturall things; whereby every one of them is apart and distinctly separated, and divided materially and substantially, *viz.* seeing that two, three, or foure, or more things are mixed into one body, and yet there is seen but one matter. Where it often falls out that the corporeall matter of that thing cannot bee known by any, or signified by any expresse name, untill there bee a separation made. Then sometimes two, three, four, five or more things come forth out of one matter, as is manifest by daily experience in the Art of Alchymie.

As for example, you have an *Electrum*,[8] which of it selfe is no Metall, but yet it hides all Metalls in one Metall. That if it be anatomized by the industry of

[6] An obsolete form of "alcohol." Here the meaning appears to be either a "fine impalpable powder produced by trituration, or especially by sublimation"; or "condensed spirit" (*OED* 2, 3b).

[7] The *Archidoxis* is one of Paracelsus's major alchemical works; first published in Latin translation (from the German) in 1569, it became immediately popular on the Continent, and in England in the seventeenth century.

[8] I.e., an alloy of the seven metals.

Alchymie and separated: all the seven Metalls, *viz. Gold, Silver, Copper, Tinne, Lead, Iron,* and *Quicksilver* come out of it and that pure and perfect.

But that you may understand what Separation is, note that it is nothing else then the severing of one thing from another, whether of two, three, four, or more things mixed together: I say a separation of the three Principles, as of Mercury, Sulphur, and Salt, and the extraction of pure out of the impure: or the pure, excellent spirit and quintessence from a grosse and elementary body; and the preparation of two, three, four, or more out of one: or the dissolution and setting at liberty things that are bound and compact, which are of a contrary nature, acting one against the other untill they destroy one the other . . . .

The first separation of which wee speake must begin from man, because hee is the Microcosme or little world, for whose sake the Macrocosme or greater world was made, *viz.* that hee might be the separator of it.

Now the separation of the Microcosme begins at his death. For in death the two bodies of Man are separated the one from the other, *viz.* his Celestial and Terrestial body, *i.e.* Sacramental and Elementary: one of which ascends on high like an Eagle; the other falls downward to the earth like lead . . . .

After this separation is made, then after the death of the Man three substances, *viz. Body, Soule,* and *Spirit* are divided the one from the other, every one going to its own place, *viz.* its own fountaine, from whence it had its originall, *viz.* the body to the Earth, to the first matter of the Elements: the soul into the first matter of Sacraments, and lastly, the spirit into the first matter of the Airy Chaos . . . .

Of the Separation of Vegetables (Book 8), Concerning Physicians.

All these Separations being made according to the Spagiricall Art, many notable and excellent medicines come from thence, which are to be used as well within as without the body.

But now seeing idlenesse is so much in request amongst Physitians, and all labour and study is turned only to insolency, truly I do not wonder that all such preparations are everywhere neglected, and coales sold at so low a price that if Smiths could be so easily without coales in forging and working their Metals, as Physitians are in preparing their Medicines, certainly Colliers would long since have been brought to extream want.

In the mean time I will give to Spagiricall Physitians[9] their due praise. For they are not given to idlenesse and sloth, nor goe in a proud habit or plush and velvet garments, often shewing their rings upon their fingers, or wearing swords

[9] Physicians who, like Paracelsus, employ the principles and materials of alchemy (e.g., metals and minerals) in effecting cure.s, rather than the herbal preparations of the Galenic school.

with silver hilts by their sides, or fine and gay gloves upon their hands, but diligently follow their labours, sweating whole nights and dayes by their furnaces.

These doe not spend their time abroad for recreation but take delight in their laboratory. They wear Leather garments with a pouch and Apron wherewith they wipe their hands. They put their fingers amongst coales, into clay and dung, not into gold rings. Thy are sooty and black, like Smithes or Colliers, and doe not pride themselves with cleane and beautifull faces. They are not talkative when they come to the sick, neither doe they extoll their Medicines: seeing they well know that the Artificer must not commend his work, but the work the Artificer, and that the sick cannot be cured with fine words.

Therefore laying aside all these kinds of vanities, they delight to bee busied about the fire and to learn the degrees of the science of Alchymie …

[Conclusion of Book 8: On the "Final Separation" / Last Judgment].

And lastly in the end of all things shall bee the last separation, in the third generation, the great day when the Son of God shal come in majesty and glory, before whom shall be carried not swords, garlands, diadems, scepters, &c. and Kingly jewels with which Princes, Kings, Cesars, &c. doe pompously set forth themselves; but his Crosse, his crown of thorns, and nails thrust through his hands and feet, and spear with which his side was pierced, and the reed and spunge in which they gave him vineger to drinke, and the whips wherewith hee was scourged and beaten. He comes not accompanyed with troopes of Horse and beating of Drums, but foure Trumpets shall bee sounded by the Angells towards the foure parts of the world, killing all that are then alive with their horrible noise, in one moment, and then presently raising these again, together with them that are dead and buryed.

For the voice shall bee heard: *Arise yee dead, and come to judgment.* Then shal the twelve Apostles sit down, their seats being prepared in the clouds, and shal judge the twelve Tribes of *Israel.* In that place the holy Angels shall separate the bad from the good, the cursed from the blessed, the goats from the sheep. Then the cursed shall like stones and lead be thrown downward: but the blessed shall like eagles fly on high. Then from the tribunall of God shal go forth this voice to them that stand on his left hand: *Goe yee cursed into everlasting fire prepared for the Devill and his Angells from all eternity: For I was an hungry, and yee fed me not; thirsty, and you gave no drink; sick, in prison, and naked, and you visited me not, freed mee not, cloathed me not, and you shewed no pity towards me, therefore shalt you expect no pity from me.* On the contrary, hee shall speak to them on his right hand: *Come yee blessed, and chosen into my Fathers Kingdome, which hath been prepared for you, and his Angells from the foundation of the world. For I was hungry, and you gave me meat; thirsty, and you gave me drink; I was a stranger, and you took me in; naked, and you covered me; sick, and you visited me; in prison, and you came unto me. Therefore I will receive you into my Fathers Kingdom, where are provided many mansions for the Saints. You took pity on me, therefore will I take pity on you.*[10]

All these being finished and dispatched, all Elementary things will returne to the first matter of the Elements and bee tormented to eternity and never bee consumed, &c. and on the contrary, all holy things shall return to the first matter of Sacraments: *i.e.* shall be purified, and in eternall joy glorifie God their Creator and worship him from age to age, from eternity to eternity. Amen.

Book 9: Of the Signature of Naturall Things (of Minerall Signes).

... But to returne to our purpose concerning Minerall signes, and especially concerning the Coruscation of Metalline veins, we must know that as Metalls which are yet in their first being send forth their Coruscation, i.e. Signes, so also the *Tincture of the Philosophers*, which changeth all imperfect Metalls into Silver and Gold (or White Metalls into Silver, and Red into Gold) puts forth its proper signs like unto Coruscation, if it be Astrally perfected and prepared. For as soon as a small quantity of it is cast upon a fluxil metall, so that they mixe together in the fire, there ariseth a naturall Coruscation and brightnesse, like to that of fine Gold or Silver in a test, which then is a signe that that Gold or Silver is freed and purged without all manner of addition of other Metalls.

But how the Tincture of Philosophers is made Astrall, you must conceive it after this manner: First of all you must know that every Metall, as long as it lies hid in its first being, hath its certaine peculiar stars.

So Gold hath the stars of the Sun, Silver the stars of the Moon, Copper the stars of Venus, Iron the stars of Mars, Tinne the stars of Jupiter, Lead the stars of Saturne, Quicksilver the starres of Mercury.

But as soon as they come to their perfection and are coagulated into a fixt Metalline body, their stars fall off from them, and leave them as a dead body.

Hence it follows that all such bodies are afterwards dead and inefficacious, and that the unconquered star of Metalls doth overcome them all, and converts them into its nature and makes them all Astrall ....

For which cause also our Gold and Silver, which is tinged and prepared with our tincture, is much more excellent and better for the preparation of Medicinall secrets then that which is naturall, which Nature generates in the Mines and afterwards is separated from other Metalls.

Translated by J. F. M. D.

[10] Paracelsus's eschatological scene is drawn from Matt. 25:31–46.

Reading and Discussion Questions

1. Today, alchemy is usually regarded as a pseudoscience. What aspects of this practice look esoteric? What techniques of experimentation have been developed by alchemical researchers?
2. What does Paracelsus regard as the goals of all this experimental effort?
3. Does Paracelsus retain the Aristotelian distinctions between living/non-living and heavenly/earthly? Do you see any evidence of the mercury/sulfur/salt theory developed by Ibn Sina (Avicenna) and other Islamic alchemists?

The Key (c. 1650s–1670s)

Isaac Newton* (George Starkey)[1]

59

First of all know antimony[2] to be a crude and immature mineral having in itself materially what is uniquely metallic, even though otherwise it is a crude and indigested mineral. Moreover, it is truly digested by the sulfur that is found in iron and never elsewhere.

Two parts of antimony [combined] with iron give a regulus[3] which in its fourth fusion exhibits a star; by this sign you may know that the soul of the iron has been made totally volatile by the virtue of the antimony. If this stellate regulus is melted with gold or silver by an ash heat in an earthen pot, the whole regulus is evaporated, which is a mystery. Also, if this regulus is amalgamated with common mercury and is digested in a sealed vessel on a slow fire for a short time—two or three hours—and then ground for 1/8 of an hour in a mortar without

[1] Authorship of *The Key* (*Clavis*) was traditionally attributed to Newton (including by distinguished Newton scholars such as Betty Jo Teeter Dobbs and Richard Westfall), with an estimated composition date of 1675 to 1680. Newton was indeed deeply engaged in the practice of alchemy, however, other science historians have challenged this particular attribution. William Newman, for example, argues that *The Key* was composed by George Starkey probably sometime around 1651. See: Newman, W., & Newton, I. (1987). Newton's *Clavis* as Starkey's *Key*. *Isis*, 78(4), 564–574. Acknowledging the force of this criticism in a more recent book, *The Janus Faces of Genius: The Role of Alchemy in Newton's Thought* (1991), Betty Jo Teeter Dobbs writes "even though one now knows for certain that Newton did not compose the document, the basic alchemical process it describes continues to appear in Newton's later papers, and so the 'Key' continues to provide a useful point of rerefence" (Dobbs 1991, 15).

[2] The *OED's* definition coincides with Newton's statement as to antimony's "crude and immature" original nature and its "digested" potential: "One of the elementary bodies, a brittle metallic substance, of bright bluish white colour and flaky crystalline texture. Its metallic characteristics are less pronounced than those of the metals generally."

[3] A *regulus* is "the metallic form of antimony, so called by early chemists, apparently on account of its ready combination with gold" (*OED*), when brought to a state of liquidity or fusion through the application of heat. But Newton also draws upon its astronomical sense (the bright star in the constellation Leo), in that a star pattern is revealed in the amalgam when it undergoes multiple fusion.

moisture while being warmed moderately, until it spits out its blackness, then it may be washed to deposit the greatest part of its blackness, until the water, which in the beginning becomes quite black, is scarcely more tinged by the blackness. This can be done by flushing it with water many times. Let the amalgam be dried, again placed near the fire, and kept in the above-mentioned heat for three hours. Afterwards let it be ground again as before in a dry and warm mortar. It pushes out new blackness, which must be washed away again; this must be repeated continually until the whole amalgam becomes like shining and cupellated silver,[4] whereas at first it had a dark leaden color.

Then distill this mercury which has been so washed and amalgamate over again seven or nine times, and in each amalgamation see to the heating, grinding, and washing as many times as before. Distill the whole as before. On the seventh time you will have a mercury dissolving all metals, particularly gold."[5] I know whereof I write, for I have in the fire manifold glasses with gold and this mercury. They grow in these glasses in the form of a tree,[6] and by a continued circulation the trees are dissolved again with the work into new mercury. I have such a vessel in the fire with gold thus dissolved, where the gold was visibly not dissolved by a corrosive into atoms, but extrinsically and intrinsically into a mercury as living and mobile as any mercury found in the world. For it makes gold begin to swell, to be swollen, and to putrefy, and to spring forth into sprouts and branches, changing colors daily, the appearances of which fascinate me every day. I reckon this a great secret in Alchemy, and I judge it is not rightly to be sought from artists who have too much wisdom to decide that common mercury ought to be attacked through reiterated cohobation[7] by the regulus of leo [that is, of iron or antimony]. That unique body, that regulus, however, is familial with mercury seeing that it is closest to that mercury you have known and recognized in the whole mineral kingdom, and hence most closely related to gold. And this is the philosophical method of meliorating nature in nature, consanguinity in consanguinity.

With regard to this operation, look at the Letter responding to Thomas of Bologna,[8] and you will find this question fully solved.

[4] I.e., silver that has been purified in a cupel.

[5] Thus the object of the process is production of "philosophical mercury" from common mercury. Dobbs notes that the alchemists' "attempts to dissolve gold seem to have been made with the same notion a modern chemist employs when he analyzes a compound before he attempts to synthesize it: if one knows what a substance is made of, then it is easy enough to make it" (*Foundations of Newton's Alchemy*, 184–5).

[6] The interactive phenomenon Newton describes closely resembles—perhaps is—the alchemists' well known "philosophical tree" or "tree of Hermes," previously noted in Ripley and Philalethes.

[7] Repeated distillation.

[8] A reference to *The Answer of Bernardus Trevisanus, to the Epistle of Thomas of Bononia, Physician to King Charles the Eighth* [sic], which exists in many manuscript copies and had been printed recently in *Aurifontina Chymica: or, A Collection of Fourteen Small Treatises Concerning the First Matter of Philosophers,* trans. John Frederick Houpreght (London, 1680). Thomas of Bologna was a physician, alchemist, and astrologer in the courts of Charles V (1364–80) and Charles VI (1380–1422) of France and father of Christine de Pisan.

Another secret is that you need the mediation of the virgin Diana [quintessence, most pure silver]; otherwise the mercury and the regulus are not united.

The regulus is made from antimony four ounces /nine parts/, iron two ounces / four parts/; this is a good proportion. Do not neglect to have a mass of antimony greater than that of iron, for if an error is made here you will be disappointed. Make the regulus by casting in nitre bit by bit; cast in between three and four ounces of nitre so that the matter may flow.

It is not a good idea to prepare in one crucible a greater quantity than the above measure of antimony. The antimony is ground, then cupelled together with iron, whatever others may say or write.

Little nails may be used and especially the ends of those broken from horn shoes. Let the fire be strong so that the matter may flow [like water], which is easily done. When it flows, cast in a spoonful of nitre; and when that nitre has been destroyed by the fire, cast in another. Continue that process until you have cast in three or four ounces. Then pile up the charcoals about the crucible, taking care that they do not fall into it. Increase the fire as much as the fusion of common silver requires, and keep it in that state for 1/8 of an hour. [The matter ought to be like a subtle water if you have labored correctly] Then pour the matter out into a cone. The regulus wll subside. Separate the ashy scoria from it. Keep the cooled material in a dry vessel.

It is a sign of a good fusion if the iron is completely fused and if the scoriae break up by themselves into powder.

Beat the regulus and add to it two, or at most 2.5, ounces of nitre. Grind the regulus and the nitre together completely and again melt. Throw away the arsenical and useless scoriae.

Grind the regulus a third and fourth time with at most one ounce of nitre and melt in a new crucible, and on the fourth time you will have scoriae tinged with a golden color and a stellate regulus.

NB In the last three times the scoriae must be thrown away because they are arsenical; however, they are useful in surgery.

NB In the last three fusions the regulus must be beaten, and ground and mixed with nitre. Some cast the nitre into the crucible, but this is not recommended, for, firstly, the fusion is as a result prolonged and the regulus is not without some loss of itself by exhalation. Secondly, nitre thrown in in this way stays on the surface and in time it cools the regulus. And since nitre flows easily, it may flow at first and encrust so that it will not flow again without a large fire. If that happens, the best part of the regulus perishes in the conflagration, whence it is that sometimes a star perishes because it is falsely ascribed to a constellation.

You will see that the regulus mixed with nitre in this way flows easily with it; and you will not see it become hard in any manner, except for the difference in the depuration, which is far greater if it is mixed than if the nitre is just tossed in.

Take of this regulus one part, of silver two parts, and melt them together until they are like fused metal. Pour out, and you will have a friable mass of the color of lead.

NB If the regulus is joined with the silver, they flow more easily than either one separately and they remain fused as long as lead even though there are thus two parts of silver, which is then changed into the nature of antimony, friable and leaden.

Beat this friable mass, this lead, and cast it together with the mercury of the vulgar into a marble mortar. The mercury should be washed (say ten times) with nitre and distilled vinegar and likewise dried (twice), and the mortar should be constantly heated just so much as you are able to bear the heat of with your fingers. Grind the mercury 1/4 of an hour with an iron pestle and thus join the mercury, the doves of Diana mediating,[9] with its brother, philosophical gold, from which it will receive spiritual semen. The spiritual semen is a fire which will purge all the superfluities of the mercury, the fermental virtue intervening. Then take a little beaten sal ammoniac and grind with the mercury. When it is fully amalgamated, add just enough humidity to moisten it, and this one philosophical sign will appear to you: that in the very making of the mercury there is a great stink. Finally, wash your mercury by pouring on water, grinding, decanting, and again pouring on fresh water, until few feces appear.

Translated by Betty Jo Teeter Dobbs

Reading and Discussion Questions

1. What instructions does Newton (Starkey) give for repeating this experiment?
2. What barriers would there be to repeating the experiment?

[9] The "doves of Diana" are silver or the female principle that serves to mediate or join mercury with the regulus. Dobbs notes that "the common mercury is receiving 'spiritual semen' from the 'philosophical gold' or star regulus of antimony. Presumably, the 'spiritual semen' has been drawn into the regulus from the 'universal spirit' in the surrounding Neoplatonic 'aire'" (*Foundations of Newton's Alchemy*, 184).

Tracts Written by the Honourable Robert Boyle, Containing New Experiments, Touching the Relation Between Flame and Air, and About Explosions (1672)

Robert Boyle

60

The First Title. Of the Difficulty of Producing Flame without Aire.

Experiment I.

A Way of Kindling Brimstone in vacuo Boyliano Unsuccessfully Tried. We took a small earthen melting Pot, of an almost Cylindrical figure, and well glaz'd (when it was first bak'd) by the heat; and into this we put a small cylinder of Iron of about an inch in thickness, and half as much more in Diameter, made red hot in the fire; and having hastily pump'd out the Air, to prevent the breaking of the Glass; when this vessel seem'd to be well emptied, we let down, by a turning key, a piece of Paper, wherein was put a convenient quantity of flower of Brimstone, under which the iron had been carefully plac'd; so that, being let down, that vehement heat did, as we expected, presently *destroy* the contiguous paper; whence the included Sulphur fell immediately upon the iron, whose upper part was a little concave, that it might contain the flowers when melted. But all the heat of the iron, though it made the Paper and Sulphur smoke, would not actually kindle either of them that we could perceive.

Experiment II. An Ineffectual Attempt to Kindle Sulphur in our Vacuum another Way

Another way I thought of to examine the inflammability of Sulphur without Air; which, though it may prove somewhat hazardous to put it in practice, I resolved to try, and did so after the following manner:

Into a glass-buble of a convenient size, and furnish'd with a neck fit for our purpose, we put a little flower of Brimstone (as likely to be more pure and inflammable than common Sulphur;) and having exhausted the Glass, and secured it against the return of the Air, we laid it upon burning coals, where it did not take fire, but rise all to the opposite part of the glass, in the form of a fine powder; and that part being turned downward and laid on coals, the Brimstone, without kindling, rose again in the form of an expanded substance, which (being removed from the fire) was, for the most part, transparent, not unlike a yellow varnish.

Advertisement.

Though these unsuccessful attempts to kindle Sulphur in our exhausted Receivers, were made more discouraging by some more, that were made another way; yet judging that last way to be rational enough, we persisted somewhat obstinately in our endeavours, and conjecturing that there might be some unperceived difference between Minerals, that do all of them pass, and are sold for common Sulphur, I made trial, according to the way hereafter to be mentioned, with another parcel of brimstone, which differ'd not so much from the former, as to make it worth while to set down a description o it, that probably would not be useful.

But in this place, it may suffice to have given a general intimation of the possibility of the thing. The proof of it you will meet with under the *third Title*, when I come to tell you what use I endeavour'd to make of our sulphureous Flames.

Experiment III. Shewing the Efficacy of Air in the Production of Flame, without Any Actually Flaming or Burning Body.

Having hitherto examin'd by the *presence* of the Air, what interest it has in kindling of Flame; it will not be impertinent to add an Experiment or two, that we tried to shew the same interest of the Air by the effects of its *admission* into our Vacuum. For I thought, it might reasonably be supposed that if such dispositions were introduc'd into a body, as that there should not appear any thing wanting to turn it into Flame but the presence of the Air, an actual ascension of that body might be produced by the admitted Air, without the intervention of any actual Flame, or Fire, or even heated substance; the warrentableness of which supposition may be judged by the two following Experiments.

When we had made the Experiment, ere long to be related in its due place, (*viz. Title II.* Exper. the 2nd) to examine the presumption we had, that even when the Iron was not hot enough to keep the melted Brimstone in such a heat, as was requisite to make it burn without Air, or with very little, it would yet be hot enough to kindle the Sulphur, if the Air had access to it: to examine this (I say) we made two or three several Tryals, and found by them, that if some little while after the flame was extinguished, the Receiver were removed, the sulphur would Presently take fire again, and flame as vigorously as before. But I thought it might without absurdity be doubted, whether or no the agency of the Air in the production of the flame might not be somewhat less than these trials would perswade; because that, by taking off the Receiver, the Sulphur was not only exposed to fresh Air, but also advantaged with a free scope for the avolution of those fumes, which in a close Vessel might be presum'd to have been unfriendly to the Flame.

How far this doubt may, and how far it should, be admitted, we may be assisted to discern by the subjoined experiment, though made in great part for another purpose; which you will perceive by the beginning of the Memorial I made of it, that runs thus;

Experiment IV. A Differing Experiment to the Same Purpose with the Former.

Having a mind to try, at how great a degree of rarefaction of the Air it was possible to make Sulphur lame by the assistance of an adventitious heat, we caused such an experiment as the above mention'd to be reiterated, and the pumping to be continued for some time after the flame of the melted flowers of Brimstone appeared to be quite extinguished, and the Receiver was judged by those that managed the Pump (and that upon probable signs) to be very well exhausted. Then, without stirring the Receiver, we let in at the stop-cock very warily a little

Air, upon which we could perceive, though not a constant flame, yet divers little flashes, as it were, which disclosed themselves by their bleu [sic] colour to be sulphureous flames; and yet the Air, that had suffic'd to re-kindle the Sulphur, was so little, that two exsuctions more drew it out again, and quite depriv'd us of the mentioned flashes. And when a little Air was cautiously let in again at the stop-cock, the like flashes began again to appear, which, upon two executions more did again quite vanish, though, upon the letting in a little fresh Air the third time, they did once more reappear.

Whether and how far such experiments as these may conduce to explicate what is related of Fires suddenly appearing in long undisclosed Vaults or Caves to those that first broke into them, I may perchance elsewhere consider; but shall not here, enquire, especially being not fully satisfied of the truth of the matter of fact. . . .

The Second Title. Of the Difficulty of Preserving Flame without Air

Since it is generally, and in most cases justly, esteemed to be more easy to *preserve* Flame in a body that is already actually kindled, than to *produce* it there at first; we thought fit to try, whether at least bodies already burning might not be kept in that state without the concurrence of Air. And though in some of our formerly published Physico-mechanical experiments it happen'd that actually Flame would scarce last a minute or two in our *large* Pneumatical Receiver; yet because it seem'd not improbable, that mineral bodies once kindled might afford a vigorous and very durable flame; we thought fit to devise and make the following tryals: Whence probably we might receive some new information about the *Diversities*, and some other Phenomena of Flame, and the various *degrees*, wherein the Air is necessary or helpful to them.

Experiment I. Reciting an Attempt to Preserve the Flame of Brimstone without Air

We put upon a thick metalline place a convenient quantity of flowers of Sulphur; and having kindled them in the Air, we nimbly conveyed them into a Receiver, and made haste to pump out some of the included Air, partly for other reasons, and partly that the cavity of the Receiver might be the sooner freed from smok [sic], which would, if plentiful, both injure the flame, and hinder our sight. As soon as the Pump began to be plied, or to be lessen'd at every exsuction of the Air; and in effect, it expir'd before the Air was quite drawn out. Nor did it, upon the early removal of the Receiver, do any more than afford, for a very little while, somewhat more of the smoak in the open Air, than it appear'd to do before.

The reiteration of this experiment presently after, afforded us nothing new, worth mentioning in this place.

Experiment II. Relating a Tryal about the Duration of the Flame of Sulphur in vacuo Boyliano.

To vary a little the foregoing Experiment, and try to save some moments of time, which on these occasions is to be husbanded with the utmost care; having provided a Cylinder of iron larger than the former, that it might be its bulk, being once heated, both contribute to the ascension of the Sulphur, and to the lasting of its flame, we made a tryal, that I find registered to this effect:

We took a pretty big lump of Brimstone, and tied it to the turning-key; and having got what else was necessary in a readiness, we caus'd the iron-plate to be hastily brought red-hot from the fire, and put upon a Pedestal, that the flame might be the more conspicuous; and, having nimbly cemented on the Receiver, we speedily let down the suspended Brimstone, till it rested upon the red-hot iron, by which being kindled, it sent up a Pump, till we had, as we conjectur'd, emptied the Receiver; which we could not do without withdrawing together with the Air much sulphureous smoke, (that was offensive enough both to the eyes and nostrils.) But notwithstanding this pumping out of the Air, though the flame did seem gradually to be somewhat impaired; yet it manifestly continued burning much longer, than by the short duration of other flames in out Receivers (when diligence is us'd to withdraw the air from them) one could have expected. And especially one time, (for the experiment was made more than once) the flame lasted, till the Receiver was judg'd to be well exhausted; and some thought it did so survive the exhaustion, that it went not out so much for want of Air, as Fuel; the Brimstone appearing when we took off the Receiver, either to have been consum'd by the fire that fed on it, or to have casually run off from the Iron, whose heat had kept it constantly melted.

In case you should have a mind to prosecute Experiments of the nature of this and the precedent, it may not prove useless, if I intimate to you the following Advertisements.

1. For the red-hot iron above mentioned, we thought it not amiss to provide, instead of the melting-pot employ'd in the first experiment, a Pedestal (if I may so call it) made of a lump of dryed Tobacco-pipe-clay, that the vehement heat of the iron might neither fill the Receiver with the smoke of what it lean'd on, nor injure the engine, if it should rest immediately upon that; And this Pedestal should be so plac'd, that the iron may be as far, as you can, from the sides of the Receiver, which else the excessive heat would endanger.

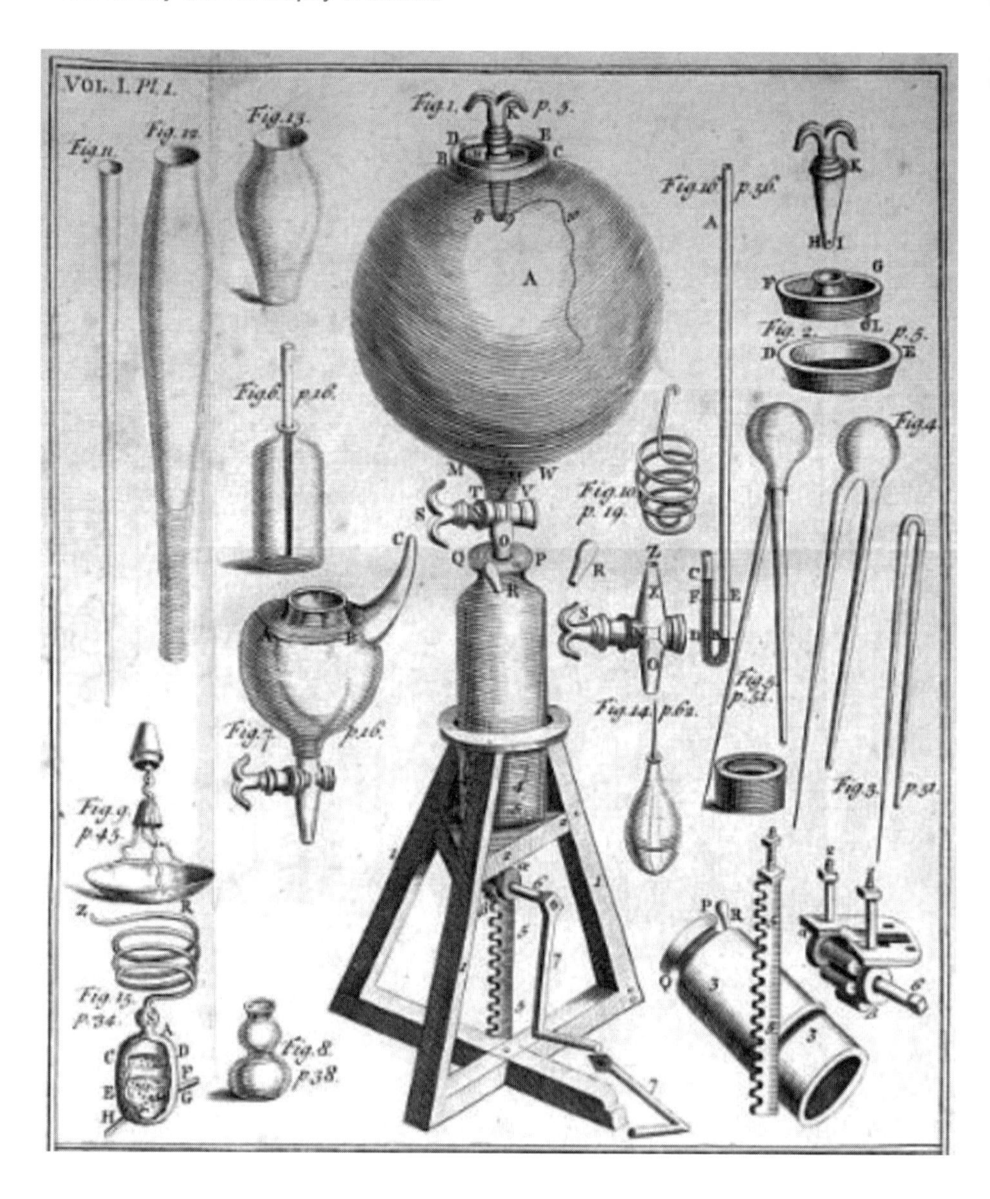

Reading and Discussion Questions

1. What is Boyle attempting to determine about the relationship between fire and air in this series of experiments? What makes this project difficult?
2. In what ways and to what degree does Boyle adopt the Baconian method? What do you think that Boyle's goal is in making so many trials and reporting these results with this degree of detail?

Of the Excellency and Grounds of the Corpuscular or Mechanical Philosophy (1674)

Robert Boyle

61

The importance of the Question, you propose, would oblige me to refer you to the *Dialogue about a good Hypothesis,* and some other Papers of that kind, where you may find my thoughts about the advantages of the *Mechanical* Hypothesis somewhat amply set down, and discoursed of. But, since your desires confine me to deliver in few words, not what I believe resolvedly, but what I think may be probably said for the Preference or the Preeminence of the *Corpuscular* Philosophy above *Aristotle's,* or that of the *Chemists*, you must be content to receive from me, without any Preamble, or exact Method, or ample Discourses, or any other thing that may cost many words, a succinct mention of some of the chief Advantages of the *Hypothesis* we incline to. And I the rather comply, on this occasion, with your Curiosity, because I have often observed you to be alarmed and disquieted, when you hear of any Book that pretends to uphold, or repair the decaying Philosophy of the Schools, or some bold Chemist, that arrogates to those of his Sect the Title of Philosophers, and pretends to build wholly upon Experience, to which he would have all other Naturalists thought strangers. That therefore you may not be so tempted to despond, by the Confidence or Reputation of those Writers, that do some of them applaud, and others censure, what, I fear, they do not understand, (as when the Peripatetics cry up, *Substantial Forms*, and the Chemists, *Mechanical Explications*) of Nature's *Phenomena,* I will propose some Considerations, that, I hope, will not only keep you kind to the Philosophy you have embraced, but perhaps, (by some Considerations which you have not yet met with,) make you think it probable, that the new Attempts you hear of from time to time, will not overthrow the *Corpuscularian Philosophy*, but either be foiled by it, or found reconcilable to it.

But when I speak of the *Corpuscular* or *Mechanical* Philosophy, I am far from meaning with the *Epicureans,* that *Atoms,* meeting together by chance in an infinite *Vacuum,* are able of themselves to produce the World, and all its Phenomena; nor with some Modern Philosophers, that, supposing God to have put into the whole Mass of Matter such an invariable quantity of Motion, he needed do no more to make the World, the material parts being able by their own unguided Motions, to cast themselves into such a System (as we call by that name); But I plead only for such a Philosophy, as reaches but to things purely Corporeal, and distinguishing between the first *original of things,* and the subsequent *course of Nature,* teaches, concerning the *former,* not only that God gave Motion to Matter, but that in the beginning He so guided the various Motions of the parts of it, as to contrive them into the World he designed they should compose, (furnished with the *Seminal* Principles and Structures or Models of Living Creatures,) and established those *Rules of Motion,* and that order amongst things Corporeal, which we are wont to call the *Laws of Nature.* And having told this as to the *former,* it may be allowed as to the *latter* to teach, That the Universe being once framed by God, and the Laws of Motion being settled and all upheld by His incessant concourse and general Providence; the Phenomena of the World thus constituted, are Physically produced by the Mechanical affections of the parts of Matter, and what they operate upon one another according to Mechanical Laws. And now having shewn what kind of *Corpuscular* Philosophy 'tis that I speak of I proceed to the particulars that I thought the most proper to recommend it.

I. The *first* thing that I shall mention to this purpose, is the Intelligibleness or Clearness of Mechanical Principles and Explications. I need not tell you, that among the *Peripatetics,* the Disputes are many and intricate about *Matter, Privation, Substantial Forms,* and their *Eduction,* &c. And the *Chemists* are sufficiently puzzled, (as I have elsewhere shewn,) to give such definitions and accounts of their Hypostatical Principles, as are reconcilable to one another, and even to some obvious *Phenomena.* And much more dark and intricate are their Doctrines about the *Archeus, Astral Beings, Gas, Blass,* and other odd Notions, which perhaps have in part occasioned the darkness and ambiguity of their expressions, that could not be very clear, when their Conceptions were far from being so. And if the Principles of the *Aristotelians* and *Spagyrists* are thus obscure, 'tis not to be expected, the Explications that are made by the help only of such Principles should be clear. And indeed many of them are either so general and slight, or otherwise so unsatisfactory, that granting their Principles, 'tis very hard to understand or admit their applications of them to particular *Phenomena.* And even in some of the more ingenious and subtle of the *Peripatetic* Discourses upon their superficial and narrow Theories, me thinks, the Authors have better plaid the part of *Painters* than *Philosophers*, and have only had the skill, like Drawers of Landships, to make men fancy, they see Castles and Towns, and other Structures that appear solid and magnificent, and to reach to a large extent, when the whole Piece is superficial, and made up of

Colors and Art, and comprised within a Frame perhaps scarce a yard long. But to come now to the *Corpuscular* Philosophy, men do so easily understand one another's meaning, when they talk of *Local Motion, Rest, Bigness, Shape, Order, Situation,* and *Contexture* of Material Substances; and these Principles do afford such clear accounts of those things, that are rightly deduced from them only, that even those *Peripatetics* or *Chemists,* that maintain other Principles, acquiesce in the Explications made by these, when they can be had, and seek not any further, though perhaps the effect be so admirable, as would make it pass for that of a hidden Form, or Occult Quality. Those very *Aristotelians,* that believe the Celestial Bodies to be moved by Intelligences, have no recourse to any peculiar agency of theirs to account for *Eclipses.* And we laugh at those *East-Indians* that, to this day, go out in multitudes, with some Instruments that may relieve the distressed Luminary, whose loss of Light they fancy to proceed from some fainting fit, out of which it must be roused. For no Intelligent man whether *Chemist* or *Peripatetic,* flies to his peculiar Principles, after he is informed, that the Moon is Eclipsed by the interposition of the Earth betwixt her and it, and the Sun by that of the Moon betwixt him and the Earth. And when we see the Image of a Man cast into the Air by a Concave Spherical Looking-glass, though most men are amazed at it, and some suspect it to be no less than an effect of Witchcraft, yet he that is skilled enough in *Catoptricks,* will, without consulting *Aristotle,* or *Paracelsus,* or flying to Hypostatical Principles and Substantial Forms, be satisfied, that the *Phenomenon* is produced by the beams of Light reflected, and thereby made convergent according to Optical, and consequently Mathematical Laws.

But I must not now repeat what I elsewhere say, to shew, that the Corpuscular Principles have been declined by Philosophers of different Sects, not because they think not our Explications clear, if not much more so, than their own; but because they imagine, that the applications of them can be made but to few things, and consequently are insufficient.

II. In the next place I observe, that there cannot be *fewer* Principles than the two grand ones of Mechanical Philosophy, *Matter* and *Motion.* For, Matter alone, unless it be moved, is altogether inactive; and whilst all the parts of a Body continue in one state without any Motion at all, that Body will not exercise any action, nor suffer any alteration itself, though it may perhaps modify the action of other Bodies that move against it.

III. Nor can we conceive any Principles more *primary,* than *Matter* and *Motion.* For, either both of them were immediately created by God, or, (to add that for their sakes that would have Matter to be unproduced,) if *Matter* be eternal, *Motion* must either be produced by some Immaterial Supernatural Agent, or it must immediately flow by way of Emanation from the nature of the matter it appertains to.

IV. Neither can there be any Physical Principles more *simple* than Matter and Motion; neither of them being resoluble into any things, whereof it may be truly, or so much as tolerably, said to be compounded.

V. The next thing I shall name to recommend the Corpuscular Principle, is their great Comprehensiveness. I consider then, that the genuine and necessary effect of the sufficiently strong Motion of one part of Matter against another, is, *either* to drive it on in its entire bulk, *or* else to break or divide it into particles of determinate *Motion, Figure, Size, Posture, Rest, Order,* or *Texture.* The two first of these, for *instance,* are each of them capable of numerous varieties. For the *Figure* of a portion of Matter may *either* be one of the five Regular Figures treated of by Geometricians, or some determinate *Species* of solid Figures, as that of a *Cone, Cylinder,* &c. *or* Irregular, though not perhaps Anonymous, as the Grains of Sand, Hoops, Feathers, Branches, Forks, Files, &c. And as the *Figure,* so the *Motion* of one of these particles may be exceedingly diversified, not only by the determination to this or that part of the world, but by several other things, as particularly by the almost infinitely varying degrees of Celerity, by the manner of its progression with, or without, Rotation, and other modifying Circumstances; and more yet by the Line wherein it moves, as (besides Straight) Circular, Elliptical, Parabolical, Hyperbolical, Spiral, and I know not how many others. For, *as* later Geometricians have shewn, that those crooked Lines may be compounded of several Motions, (that is, traced by a Body whose motion is mixt of, and results from, two or more simpler Motions,) *so* how many more curves may, or rather may not be made by new Compositions and Decompositions of Motion, is no easy task to determine.

Now, since a *single* particle of Matter, by virtue of two only of the Mechanical affections, that belong to it, be diversifiable so many ways; how vast a number of variations may we suppose capable of being produced by the Compositions and Decompositions of *Myriads* of single invisible Corpuscles, that may be contained and contexed in one small Body, and each of them be imbued with more than two or three of the fertile Catholic Principles above mentioned? Especially since the aggregate of those Corpuscles may be farther diversified by the *Texture* resulting from their Convention into a Body, which, as so made up, has its own Bigness, and Shape, and Pores, (perhaps very many, and various) and has also many capacities of acting and suffering upon the score of the place it holds among other Bodies in a World constituted as ours is: So that, when I consider the almost innumerable diversifications, that Compositions and Decompositions may make of a small number, not perhaps exceeding twenty of distinct things, I am apt to look upon those, who think the Mechanical Principles may serve indeed to give an account of the *Phenomena* of this or that particular part of Natural Philosophy, as *Statics, Hydrostatics*, the *Theory of the Planetary Motions,* &c. but can never be applied to all the *Phenomena* of things Corporeal; I am apt, I *say*, to look upon those, otherwise Learned, men, as I would do upon him, that should affirm, that by putting together the Letters of the *Alphabet,* one may indeed make up all the words to be found in one Book, as in *Euclid*, or *Virgil*; or in one Language, as *Latin*, or *English*; but that they can by no means suffice to supply words to all the Books of a great Library, much less to all the Languages in the world.

And whereas there is another sort of Philosophers, that, observing the great efficacy of the bigness, and shape, and situation, and motion, and connection in Engines, are willing to allow, that those Mechanical Principles may have a great stroke in the Operations of Bodies of a sensible bulk, and manifest Mechanism, and therefore may be usefully employed in accounting for the effects and Phenomena of such Bodies, who yet will not admit, that these Principles can be applied to the hidden Transactions that pass among the minute Particles of Bodies; and therefore think it necessary to refer these to what they call *Nature, Substantial Forms, Real Qualities* and the like Un-mechanical Principles and Agents.

But this is not necessary; for, both the Mechanical affections of Matter are to be found, and the Laws of Motion take place, not only in the great Masses, and the middle-sized Lumps, but in the smallest Fragments of Matter; and a lesser portion of it, being as well a Body as a greater, must, as necessarily as it, have its determinate Bulk and Figure: And he that looks upon Sand in a good Microscope, will easily perceive, that each minute Grain of it has as well its own size and shape, as a Rock or Mountain. And when we let fall a great stone and a pebble from the top of a high Building, we find not but that the latter as well as the former moves conformably to the Laws of acceleration in heavy Bodies descending. And the Rules of Motion are observed, not only in Canon Bullets, but in Small Shot; and the one strikes down a Bird according to the same Laws, that the other batters down a Wall. And though *Nature* (or rather its Divine Author) be wont to work with much finer materials, and employ more curious contrivances than *Art*, (whence the Structure even of the rarest Watch is incomparably inferior to that of a Humane Body;) yet an Artist himself, according to the quantity of the matter he employs, the exigency of the design he undertakes, and the bigness and shape of the Instruments he makes use of, is able to make pieces of work of the same nature or kind of extremely differing bulk, where yet the like, though not equal, Art and Contrivance, and oftentimes Motion too, may be observed: As a Smith, who with a Hammer, and other large Instruments, can, out of masses of Iron, forge great Bars or Wedges, and make those strong and heavy Chains that were employed to load Malefactors, and even to secure Streets and Gates, may, with lesser Instruments, make smaller Nails and Filings, almost as minute as Dust; and may yet, with finer Tools, make Links of a strange Slenderness and Lightness, insomuch that good Authors tell us of a Chain of divers Links that was fastened to a Flea, and could be moved by it; and, if I misremember not, I saw something like this, besides other Instances that I beheld with pleasure of the Littleness that Art can give to such pieces of Work, as are usually made of a considerable bigness. And therefore to say, that, though in Natural Bodies, whose bulk is manifest and their structure visible, the Mechanical Principles may be usefully admitted, that are not to be extended to such portions of Matter, whose parts and Texture are invisible; may perhaps look to some, as if a man should allow, that the Laws of Mechanism may take place in a Town-Clock; but cannot in a Pocket-Watch; or (to give you an instance, mixt of Natural and Artificial,) as if, because the Terraqueous Globe is a vast Magnetical Body of seven or eight thousand miles in Diameter, one should affirm, that Magnetical Laws are not to be expected to be of force in a spherical piece of Loadstone that is not perhaps an inch long: And yet Experience shews us, that notwithstanding the inestimable disproportion betwixt these two Globes, the *Terrella*, as well as the *Earth*, hath its Poles, Equator, and Meridians, and in divers other Magnetical Properties, emulates the Terrestrial Globe.

They that, to solve the *Phenomena* of Nature, have recourse to Agents which, though they involve no self-repugnancy in their very Notions, as many of the Judicious think *Substantial Forms* and *Real Qualities* to do; yet are such that we conceive not, how they operate to bring effects to pass: These, I say, when they tell us of such indeterminate Agents, as the *Soul of the World*, the *Universal Spirit*, the *Plastic Power*, and the like; though they may in certain cases tell us some things, yet they tell us nothing that will satisfy the Curiosity of an Inquisitive Person, who seeks not so much to know, what is the *general* Agent, that produces a *Phenomenon*, as, *by what Means*, and after *what Manner*, the *Phenomenon* is produced. The famous *Sennerius*, and some other Learned Physicians, tell us of Diseases which proceed from Incantation; but sure 'tis but a very slight account, that a sober Physician, that comes to visit a Patient reported to be bewitched, receives of the strange Symptoms he meets with, and would have an account of, if he be coldly answered, That 'tis a Witch or the Devil that produces them; and he will never sit down with so short an account, if he can by any means reduce those extravagant Symptoms to any more known and stated Diseases, as *Epilepsies, Convulsions, Hysterical Fits*, &c. and, if he cannot, he will confess his knowledge of this Distemper to come far short of what might be expected and attained in other Diseases, wherein he thinks himself bound to search into the Nature of the Morbific Matter, and will not be satisfied till he can, probably at least, deduce from that, and the structure of an Human Body, and other concurring Physical Causes, the *Phenomena* of the Malady. And it would be but little satisfaction to one, that desires to understand the causes of what occurs to observation in a Watch, and how it comes to point at, and strike, the hours, to be told, That 'twas such a Watch-maker that so contrived it: Or to him that would know the true cause of an *Echo*, to be answered, That 'tis a Man, a Vault, or a Wood that makes it.

And now at length I come to consider that which I observe the most to alienate other Sects from the Mechanical Philosophy; namely, that they think it pretends to have Principles so Universal and so Mathematical, that no other Physical Hypothesis can comport with it, or be tolerated by it.

But this I look upon as an easy indeed, but an important, mistake; because by this very thing, that the Mechanical Principles are so universal, and therefore applicable to so many things, they are rather fitted to *include*, than necessitated to *exclude*, any other Hypothesis that is founded in Nature, as far as it is so. And such *Hypotheses*, if prudently considered by a skillful and moderate person, who is rather disposed to unite Sects than multiply them, will be found, as far as they have Truth in them, to be either Legitimately, (though perhaps not immediately,) deducible from the Mechanical Principles, or fairly reconcilable to them. For, such Hypotheses will probably attempt to account for the *Phenomena* of Nature, *either* by the help of a determinate number of material Ingredients, such as the *Tria Prima* of the Chemists, by participation whereof other Bodies obtain their Qualities; *or* else by introducing some general Agents, as the *Platonic Soul of the World*, or the *Universal Spirit*, asserted by some Spagyrists; *or* by both these ways together.

Now to dispatch *first* those, that I named in the second place; I consider, that the chief thing, that Inquisitive Naturalists should look after in the explicating of difficult *Phenomena*, is not so much what the *Agent* is or does, as, what changes are made in the *Patient*, to bring it to exhibit the *Phenomena* that are proposed; and by what means, and after what manner, those changes are effected. So that the *Mechanical* Philosopher being satisfied, that one part of Matter can act upon another but by virtue of Local Motion, or the effects and consequences of Local Motion, he considers, that *as*, if the proposed Agent be not Intelligible and Physical, it can never Physically *explain* the *Phenomena; so*, if it be Intelligible and Physical, 'twill be reducible to *Matter*, and some or other of those only Catholic affections of Matter, already often mentioned. And, the indefinite divisibility of Matter, the wonderful efficacy of Motion, and the almost infinite variety of Coalitions and Structures, that may be made of minute and insensible Corpuscles, being duly weighed, I see not why a Philosopher should think it impossible, to make out by their help the Mechanical possibility of any corporeal Agent, how subtle, or diffused, or active soever it be, that can be solidly proved to be really existent in Nature, by what name soever it be called or disguised. And though the *Cartesians* be Mechanical Philosophers, yet, according to them, their *Materia Subtilis*, which the very name declares to be a corporeal Substance, is, for ought I know, little (if it be at all) less diffused through the Universe, or less active in it than the Universal Spirit of some Spagyrists, not to say, the *Anima Mundi* of the Platonists. But this upon the by; after which I proceed, and shall venture to add, That whatever be the Physical Agent, whether it be inanimate or living, purely Corporeal, or united to an Intellectual Substance, the above mentioned changes, that are wrought in the Body that is made to exhibit the Phenomena, may be effected by the same or the like means, or after the same or the like manner; as, *for instance*, if Corn be reduced to Meal, the Materials and shape of the Millstones, and their peculiar Motion and Adaptation, will be much of the same kind, and (though they should not, yet) to be sure the grains of Corn will suffer a various contrition and comminution in their passage to the form of Meal; whether the Corn be ground by a Water-mill, or a Wind-mill, or a Horse-mill, or a Hand-mill; that is, by a Mill whose Stones are turned by Inanimate, by Brute, or by Rational, Agents. And, if an Angel himself should work a real change in the nature of a Body, 'tis scarce conceivable to us Men, how he could do it without the assistance of Local Motion; since, if nothing were displaced or otherwise moved than before, (the like happening also to all external Bodies to which it related,) 'tis hardly conceivable, how it should be in itself other, than just what it was before.

But to come now to the other sort of Hypotheses formerly mentioned; if the *Chemists*, or others that would deduce a complete Natural Philosophy from *Salt, Sulphur*, and *Mercury*, or any other set number of Ingredients of things, would well consider what they undertake, they might easily discover, *That* the material parts of Bodies, as such, can reach but to a small part of the *Phenomena* of Nature, whilst these Ingredients are considered but as Quiescent things, and therefore they would find themselves necessitated to suppose them to be active; and *That* things purely Corporeal cannot be but by means of Local Motion, and the effects that may result from that, accompanying variously shaped, sized, and aggregated parts of Matter: So that the Chemists and other Materialists, (if I may so call them,) must (as indeed they are wont to do) leave the greatest part of the *Phenomena* of the Universe unexplicated by the help of the Ingredients, (be they fewer or more than three,) of Bodies, without taking in the Mechanical and more comprehensive affections of Matter, especially Local Motion. I willingly grant, that *Salt, Sulphur*, and *Mercury*, or some Substances analogous to them, are to be obtained by the action of the Fire, from a very great many dissipable Bodies here below; nor would I deny, that, in explicating divers of the *Phenomena* of such Bodies, it may be of use to a skillful Naturalist to know and consider, that this or that Ingredient, as Sulphur, *for instance*, does abound in the Body proposed, whence it may be probably argued, that the Qualities, that usually accompany that Principle when Predominant, may be also, upon its score, found in the Body that so plentifully partakes of it. But not to mention, what I have elsewhere shown, that there are many *Phenomena*, to whose explication this knowledge will contribute very little or nothing at all; I shall only here observe, that, though Chemical Explications be sometimes the most obvious and ready, yet they are not the most fundamental and satisfactory: For, the Chemical Ingredient itself, whether Sulphur or any other, must owe its nature and other qualities to the union of insensible particles in a convenient Size, Shape, Motion or Rest, and Contexture; all which are but Mechanical Affections of convening Corpuscles. And this may be illustrated by what happens in Artificial Fireworks. For, though in most of those many differing sorts that are made either for the use of War, or for Recreation, Gunpowder be a main Ingredient, and divers of the *Phenomena* may be derived from the greater or lesser measure, wherein the Compositions partake of it; yet, besides that there may be Fire-works made without Gun-powder, (as appears by those made of old by the *Greeks* and *Romans*,) Gun-powder itself owes its aptness to be fired and exploded to the Mechanical Contexture of more simple portions of Matter, *Nitre, Charcoal*, and *Sulphur*; and Sulphur itself, though it be by many Chemists mistaken for an Hypostatical Principle, owes its Inflammability to the convention of yet more simple and primary Corpuscles; since Chemists confess, that it has an inflammable Ingredient, and experience shews, that it very much abounds with an acid and uninflammable Salt, and is not quite devoid of Terrestreity. I know, it may be here alleged, that the productions of Chemical Analyses are simple Bodies, and upon that account irresoluble. But, that divers Substances, which Chemists are pleased to call the *Salts*, or *Sulphurs*, or *Mercuries* of the Bodies that afforded them, are not simple and homogeneous, has elsewhere been sufficiently proved; nor is their not being easily dissipable or resoluble a clear proof of their not being made up of more primitive portions of matter. For, compounded and even decompounded Bodies, may be as difficultly resoluble, as most of those that Chemists obtain by what they call their *Analysis* by the Fire; witness common green Glass, which is far more durable and irresoluble than many of those that pass for Hypostatical Substances. And we see, that some *Amels* will be several times even vitrified in the Fire, without losing their Nature, or oftentimes so much as their color; and yet *Amel* is manifestly not only a compounded, but a decompounded Body, consisting of Salt and Powder of Pebbles or Sand, and calcined Tinn, and, if the *Amel* be not white, usually of some tinging Metal or Mineral. But how indestructible soever the Chemical Principles be supposed, divers of the Operations ascribed to them will never be well made out, without the help of Local Motion, (and that diversified too;) without which, we can little better give an account of the *Phenomena* of many Bodies, by knowing what Ingredients compose them, than we can explain the Operations of a Watch, by knowing of how many and of what Metals the Balance, the Wheels, the Chain, and other parts, are made; or than we can derive the Operations of a Wind-mill from the bare knowledge, that 'tis made up of Wood, and Stone; and Canvas, and Iron. And here let me add, that it would not at all overthrow the Corpuscularian Hypothesis, though either by more exquisite Purifications, or by some other Operations than the usual *Analysis* of the Fire, it should be made appear, that the Material Principles or Elements of mixt Bodies should not be the *Tria Prima* of the vulgar Chemists, but either Substances of another nature, or else fewer, or more in number; as would be, if that were true, which some Spagyrists affirm, (but I could never find,) that from all sorts of mixt Bodies, five, and but five, differing similar Substances can be separated: Or, as if it were true, that the *Helmontians* had such a resolving Menstruum as the *Alkahest* of their Master, by which he affirms, that he could reduce Stones into Salt of the same weight with the Mineral, and bring both that Salt and all other kind of mixt and tangible Bodies into insipid Water. For, whatever be the number or qualities of the Chemical Principles, if they be really existent in Nature, it may very possibly be shewn, that they may be made up of insensible Corpuscles of determinate bulks and shapes; and by the various Coalitions and Contextures of such Corpuscles, *not only* three or five, *but* many more material Ingredients, may be composed or made to result: But, though the *Alkahestical* Reductions newly mentioned should be admitted, yet the Mechanical Principles might well be accommodated, even to them. For, the Solidity, Taste, &c. of Salt, may be fairly accounted for, by the Stiffness, Sharpness, and other Mechanical Affections of the minute

Particles, whereof Salts consist; and if, by a farther action of the *Alkahest,* the Salt or any other solid Body, be reduced into insipid Water, this also may be explicated by the same Principles, supposing a further Comminution of the parts, and such an attrition, as wears off the edges and points that enabled them to strike briskly the Organ of Taste: For, as to Fluidity and Firmness, those mainly depend upon two of our grand Principles, *Motion* and *Rest.* And I have else-where shewn, by several proofs, that the Agitation or Rest, and the looser contact, or closer cohesion, of the particles, is able to make the same portion of Matter, at one time a firm, and at another time, a fluid Body. So that, though the further Sagacity and Industry of Chemists (which I would by no means discourage) should be able to obtain from mixt Bodies homogeneous substances differing in number, or nature, or both, from their vulgar Salt, Sulphur, and Mercury; yet the Corpuscular Philosophy is so *general* and *fertile,* as to be fairly reconcilable to such a Discovery; and also so *useful,* that these new material Principles will, as well as the old *Tria Prima,* stand in need of the more Catholic Principles of the *Corpuscularians,* especially Local Motion. And indeed, whatever Elements or Ingredients men have (that I know of) pitched upon, yet if they take not in the Mechanical Affections of Matter, their Principles have been so deficient, that I have usually observed, that the Materialists, without at all excepting the Chemists, do not only, as I was saying, leave many things unexplained, to which their narrow Principles will not extend; but, even in the particulars they presume to give an account of, they either content themselves to assign such common and indefinite Causes, as are too general to signify much towards an inquisitive man's satisfaction; or if they venture to give particular Causes, they assign precarious or false ones, and liable to be easily disproved by Circumstances, or Instances, whereto their Doctrine will not agree, as I have often elsewhere had occasion to shew. And yet the Chemists need not be frighted from acknowledging the Prerogative of the Mechanical Philosophy, since that may be reconcilable with the Truth of their own Principles, as far as these agree with the *Phenomena* they are applied to. For these more confined *Hypotheses* may be subordinated to those more general and fertile Principles, and there can be no Ingredient assigned, that has a real existence in Nature, that may not be derived either immediately, or by a row of Decompositions, from the Universal Matter, modified by its Mechanical Affections. For, if with the same Bricks, diversely put together and ranged, several Walls, Houses, Furnaces, and other Structures, as Vaults, Bridges, Pyramids, &c. may be built, merely by a various contrivement of parts of the same kind; how much more may great variety of Ingredients be produced by, or, according to the institution of Nature, result from, the various coalitions and contextures of Corpuscles, that need not be supposed, like Bricks, all of the same, or near the same, size and shape, but may have amongst them, both of the one and the other, as great a variety as need be wished for, and indeed a greater than can easily be so much as imagined. And the primary and minute Concretions that belong to these Ingredients, may, without Opposition from the Mechanical Philosophy, be supposed to have their particles so minute and strongly coherent, that Nature of herself does scarce ever tear them asunder; as we see, that *Mercury* and *Gold* may be successively made to put on a multitude of disguises, and yet so retain their nature, as to be reducible to their pristine forms. And you know, I lately told you, that common Glass and good Amels, though both of them but factitious Bodies, and not only mixed, but decompounded Concretions, have yet their component parts so strictly united by the skill of illiterate Tradesmen, as to maintain their union in the vitrifying violence of the Fire. Nor do we find, that common Glass will be wrought upon by *Aqua fortis,* or *Aqua Regis,* though the former of them will dissolve *Mercury,* and the later *Gold.*

From the foregoing Discourse it may (probably at least) result, That if, besides Rational Souls, there are any Immaterial Substances (such as the Heavenly Intelligences, and the Substantial Forms of the *Aristotelians*) that regularly are to be numbered among Natural Agents, their way of working being unknown to us, they can but help to constitute and effect things, but will very little help us to conceive *how* things are effected; so that, by whatever Principles Natural things be *constituted,* 'tis by the Mechanical Principles that their *Phenomena* must be clearly *explicated.* As for instance, though we should grant the *Aristotelians,* that the Planets are made of a quintessential matter, and moved by Angels, or Immaterial Intelligences; yet, to explain the Stations, Progressions, and Retrogradations, and other *Phenomena* of the Planets, we must have recourse either to Eccentrics, Epicycles, &c. or to motions made in Elliptical or other peculiar Lines; and, in a word, to Theories, wherein the Motion, and Figure, Situation, and other Mathematical or Mechanical Affections of Bodies are mainly employed. But if the Principles proposed be corporeal things, they will be then fairly Reducible, or Reconcilable, to the Mechanical Principles; these being so general and pregnant, that, among things corporeal, there is nothing *real,* (and I meddle not with *Chimerical* Beings, such as some of *Paracelsus*',) that may not be derived from, or be brought to, a subordination to such comprehensive Principles. And when the Chemists shall shew, that mixed Bodies owe their qualities to the predominance of this or that of their three grand Ingredients, the *Corpuscularians* will shew, that the very Qualities of this or that Ingredient flow from its peculiar Texture, and the Mechanical affections of the Corpuscles 'tis made up of. And to affirm, that, because the Furnaces of Chemists afford a great number of uncommon Productions and Phenomena, there are Bodies or Operations amongst things purely Corporeal, that cannot be derived from, or reconciled to, the comprehensive and pregnant Principles of the Mechanical Philosophy, is, as if, because there are a great number and variety of Anthems, Hymns, Pavins, Threnodies, Courants, Gavots, Branles, Sarabands, Jigs, and other (grave and sprightly) Tunes to be met with in the Books and Practices of Musicians, one should maintain, that there are in them a great many Tunes, or at least Notes, that have no dependence on the Scale of Music; or, as if, because, besides Rhombuses, Rhomboids, Trapeziums, Squares, Pentagons,

Chiliagons, Myriagons, and innumerable other *Polygons,* Regular and Irregular, one should presume to affirm, that there are among them some Rectilinear Figures, that are not reducible to Triangles, or have Affections that will overthrow what *Euclid* has taught of *Triangles* and *Polygons.*

To what has been said, I shall add but one thing more; That, *as,* according to what I formerly intimated, Mechanical Principles and Explications are for their clearness preferred, even by Materialists themselves, to others in the cases where they can be had; *so,* the Sagacity and Industry of modern Naturalists and Mathematicians, having happily applied them to several of those difficult *Phenomena*, (in *Hydrostatics,* the practical part of *Optics, Gunnery,* &c.) that before were, or might be referred as Qualities, 'tis probable, that, when this Philosophy is deeplier searched into, and farther improved, it will be found applicable to the solution of more and more of the *Phenomena* of Nature. And on this occasion let me observe, that 'tis not always necessary, though it be always desirable, that he that propounds an *Hypothesis* in Astronomy, Chemistry, Anatomy, or other part of Physics, be able, *à priori*, to prove his *Hypothesis* to be true, or demonstratively to shew, that the other *Hypotheses* proposed about the same subject must be false. For *as,* if I mistake not, *Plato* said, That the World was God's Epistle written to Mankind, & might have added, consonantly to another saying of his, 'twas written in Mathematical Letters: *So,* in the Physical Explications of the Parts and System of the World, me thinks, there is somewhat like what happens, when men conjecturally frame several Keys to enable us to understand a Letter written in Cyphers. For, though one man by his sagacity have found out the right Key, it will be very difficult for him, *either* to prove otherwise than by trial, that this or that word is not such as 'tis guessed to be by others according to their Keys; *or* to evince, *à priori*, that theirs are to be rejected, and his to be preferred; yet, if due trial being made, the Key he proposes, shall be found so agreeable to the Characters of the Letter, as to enable one to understand them, and make a coherent sense of them, its suitableness to what it should decipher, is, without either confutations, or extraneous positive proofs, sufficient to make it be accepted as the right Key of that Cypher. And so, in Physical *Hypotheses*, there are some, that, without noise, or falling foul upon others, peaceably obtain discerning men's approbation only by their fitness to solve the *Phenomena*, for which they were devised, without crossing any known Observation or Law of Nature. And therefore, if the Mechanical Philosophy go on to explicate things Corporeal at the rate it has of late years proceeded at, 'tis scarce to be doubted, but that in time unprejudiced persons will think it sufficiently recommended by its consistency with itself, and its applicableness to so many Phenomena of Nature.

Reading and Discussion Questions

1. How would you characterize what Boyle calls the corpuscular or mechanical philosophy?

2. What are Boyle's attitudes toward the Aristotelians and chymists? How is Boyle thinking about the relation between chymical explanations and the corpuscular philosophy?
3. What place does Boyle allow for hypothesis and conjecture in natural philosophy?

Concerning the First Principle of Metals and Stones (1669)

Johann Joachim Becher

62

I say that there are three different earths in metals and stones; the first, aside from its own mixtures, is found in stones and alkali salt; the second in niter; the third in common salt. When these three earths are mixed together without any other additions, they constitute true and genuine metals, and also stones, according to the manner of formation. Hence, I conclude that stones and metals naturally belong together, as we will specifically show in the following chapters of this section and in Book 2; for this is our opinion, based on practice, that metals and stones are made up of three simple earths, and that the evidence for these comes not from the resulting bodies, since these are already mixed or can exist unmixed, but from the beginnings; in so far as these are miscible and mixed, they determine the body of which they constitute the principles. Thus, in this kind of philosophizing it is necessary to argue from the beginnings to the results, since the subjects are subterraneous, homogeneous, and insoluble, unless some new decomposition occurs. Now, our opinion having been explained in general, we will hurry on to the explanation of the three principles, namely, the three earths, according to our mind. . . .

Concerning the First Principle of Metals and Stones, Which Is called the Vitreous Stone, or Stony Earth (terra lapidia) and Improperly Salt

Up to now, we have been treating in general the principles of those things that are subterraneous . . . . Now we are going to explain each of the principles, of which there are three (that is three earths), and in this chapter we are going to begin with the first, which is the mother and source of the other two. The kind of stone which melts in fire, and melting, produces glass, is found in all species of stones, of which there are three: for some stones melt, others do not melt, but in a strong fire are reduced to a calx, and still others do not melt nor are they reduced to a calx, but remain intact even in the strongest fire; still further, some of these always remain when glowing hot but when made ice cold break apart and crumble, but others get red hot or ice cold as often as you wish, and always remain unchanged.

Concerning the first species of stone, namely stones that melt .... These are recognized by diverse properties, for mud, sand, flint, and many other stones melt in fire; but by this species we understand that which is the noblest of all, and is often called calx by mineralogists. Without it no other mineral is of any value, or sends forth any fertility, for that stone is so necessary to minerals that, existing in the mountains either raw or without any other metal, it becomes an infallible sign of future metal ... therefore we state and acknowledge this earth or stone ... as the first principle of all metals, minerals, stones, and gems .... It is actually present in all metals and minerals, and also in all stones and gems ....

Concerning the Second Principle of Minerals, Which Is Fatty Earth (terra pinguis), Improperly called Sulfur

We find the three earths in animals, vegetables, and minerals .... These three have a great affinity one for the other, and a great analogy. While it (i.e. terra pinguis) is related to the other earth which we have already treated and whence metals and stones obtain their liquidity and fusibility, it most certainly has a great relationship and analogy with the earth of vegetables, that, namely, which is present in the calcination of vegetables and the lixiviation of ashes .... It can also be prepared from all vegetables, for this earth, although it seems to be useless and of no value, has a very great analogy with the preceding earth of minerals and magnetism. This is apparent in the glassmaker's art, for when the glassmaker makes glass out of stone and flint, that is, from the aforementioned earth of minerals, which is thickened in flowing, either because of a lack of salt, or because of an excess of fire, so that it is again made hard and coagulates, nevertheless as soon as the preceding earth is injected from the ashes, not only is the whole mass of glass made more flowing, but also ... takes on growth and augment. This not only proves the great affinity of vegetable earth with mineral, but even a great likeness, harmony, and analogy, since this earth per se can be made into glass.

Translated by Henry M. Leicester and Herbert S. Klickstein

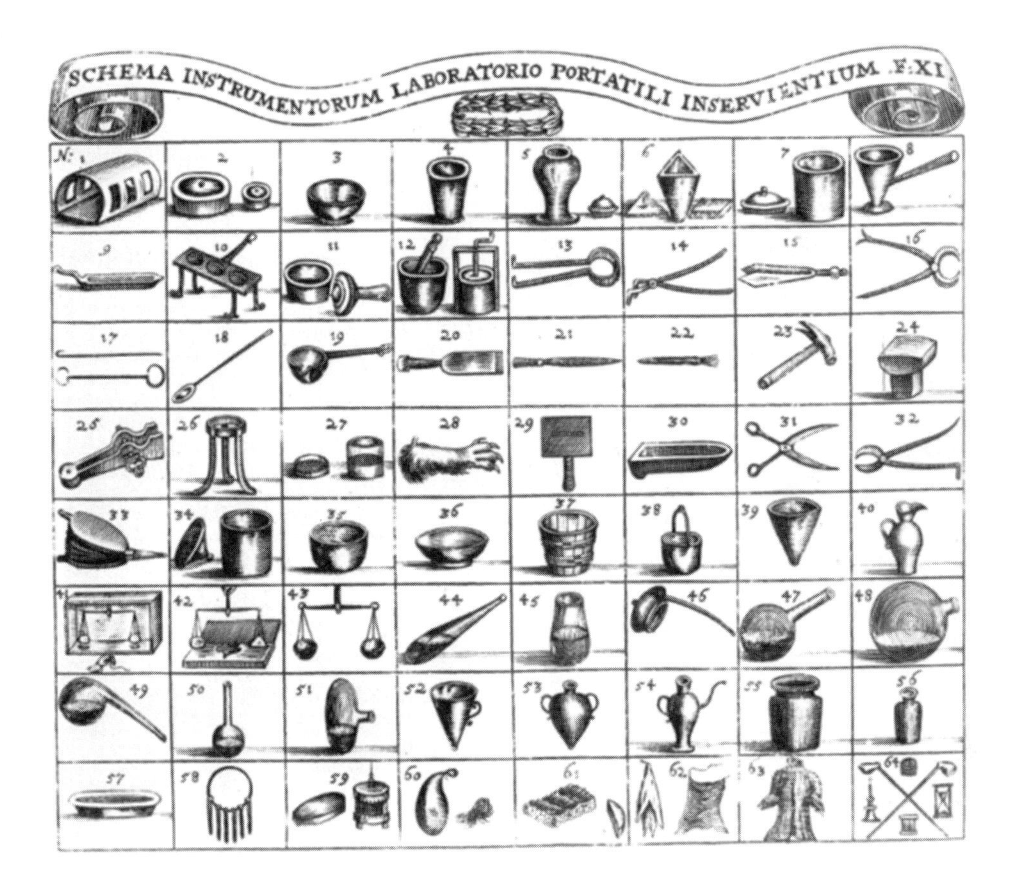

Reading and Discussion Questions

1. What properties is Becher trying to explain by categorizing "earths" into three different kinds?
2. What is it about metals that makes them especially intriguing to the early chemists?

Foundation of the Fermentative Art (1697)

Georg Ernst Stahl

63

The same thing works very well with sulfur, when certainly two parts, or better, three parts of alkali salt and one of pulverized sulfur are successively poured into and fused in a crucible. There is formed liver of sulfur. This, in the space of a quarter of an hour more or less, by fire alone, without any addition, can be converted to such a salt as is obtained from oil of sulfur *per campanum* and salt of tartar, that which is commonly called *vitriolated tartar*. There is no more trace of sulfur or alkali salt, and in place of the red color of the liver, this salt is most white; in place of the very evil taste of the liver, this salt is very bitter; in place of the easy solution, nay, the spontaneous deliquescence of the liver, by reason of its alkali salt, this salt is the most difficult of all salts except tartar of wine to be dissolved; in place of the impossibility of crystallizing the liver, this is very prone to form almost octahedral crystals; in place of the fusibility of the liver, this is devoid of all fusion.

If this new salt, from the acid of sulfur and alkaline salt formed as stated above when the phlogiston has been used up, is treated with charcoal, in the space of a quarter of an hour the original liver of sulfur reappears, and this can be so converted a hundred times . . . .

I can indeed show by various other experiments how phlogiston from fatty substances and charcoal enters very promptly into metals themselves and regenerates them from the burned calx into their own *fusible*, *malleable*, and *amalgamable* state.

Translated by Henry M. Leicester and Herbert S. Klickstein

Reading and Discussion Question

1. In his *The Book of the Remedy* (c. 1021), we saw Ibn Sina (Avicenna) working with Islamic Mercury, Sulphur, and Salt theory—a set of ideas that we have also seen picked up, for example, in Paracelsus' *Of the Nature of Things* (1537) and in Robert Boyle's *Excellency and Grounds of the Corpuscular or Mechanical Philosophy* (1674). What ontology of matter does Stahl seem to be working with in 1697, and what kinds of experimental practices and techniques is he using?

Random Thoughts and Useful Concerns (1718)

Georg Ernst Stahl

64

Now the first thing to consider concerning the principle of sulfur is its properties, as follows:

Behavior toward fire
Display of colors
Subtle and intimate mixing with other metal substances
Behavior toward water and humidity
Its own great and wonderful subtlety
Its own form in the dry or fluid state
Where it can be found or occurs

According to these conditions and intentions, I now have demonstrable grounds to say, first,

Toward fire, this sulfur principle behaves in such a manner that it is not only suitable for the movement of fire but is also one and the same being, yes, even created and designed for it.

But also, according to a reasonable manner of speaking, it is the corporeal fire, the essential fire material, the true basis of fire movement in all inflammable compounds.

However, except in compounds, no fire at all occurs, but it dissipates and volatilizes in invisible particles, or at least, develops and forms a finely divided and invisible fire, namely, heat.

On the other hand, it is very important to note that this fire material, of and by itself and apart from other things, especially air and water, is not found united and active, either as a liquid or in an attenuated state.

But if once by the movement of fire, with the addition of free air, it is attenuated and volatilized, then by this in all such conditions it is lost through unrecognizable subtlety and immeasurable attenuation, so that from this point on no science known to man, no human art, can collect it together or bring it into narrow limits, especially if this occurred rapidly and in quantity.

But how enormously attenuated and subtle material becomes through the movement of fire is shown by experience, which furnishes a field for thought and which also delights us.

From all these various conditions, therefore, I have believed that it should be given a name, as the first, unique, basic, inflammable principle. But since it cannot, until this hour, be found by itself, outside of all compounds and unions with other materials, and so there are no grounds or basis for giving a descriptive name based on properties, I have felt that it is most fitting to name it from its general action, which it customarily shows in all its compounds. And therefore I have chosen the Greek name phlogiston, in German, *Brennlich*. ...

The seventh and last consideration was where it could be found or occurred. The answer to this is now also in part easy to give from the discussion already presented, and from consideration that all corporeal compounded things have more or less of this substance, in all the so-called "kingdoms": vegetable, animal, and mineral. As then in the first two kingdoms there is contained a great amount of this principle, and all their parts are intimately penetrated and combined with it (except the watery parts which occur in them, but which still are not entirely free from it as long as they are in the body), then it is chiefly found in the fatty materials of both kingdoms.

In the mineral kingdom there is nothing but water, common salt, pure vitriolic salts, and light sand and stones in which the substance is little or not at all found. On the other hand, coal and bitumen are full of it; sulfur, not indeed in weight, but in the number of its finest particles, is completely possessed with it. Not less is it found in all inflammable, incomplete, and so-called "unripe" metals.

...

Translated by Henry M. Leicester and Herbert S. Klickstein

Reading and Discussion Question

1. What is phlogiston and how is Stahl thinking about its role in combustion, and its relationship to heat?

Dogmatic and Experiential Foundations of Chemistry (1723)

Georg Ernst Stahl

65

Preliminaries

Universal chemistry is the Art of resolving *mixt*, *compound*, or *aggregate* Bodies into their *Principles*; and of composing *such Bodies* from those *Principles*.

It has for its *Subject* all the *mix'd*, *compound*, and *aggregate Bodies* that are *resolvable* and *combinable*; and *Resolution* and *Combination*, or Destruction and Generation, for its *Object*.

Its *Means* in general are either *remote* or *immediate*; that is, either *Instruments* or the *Operations* themselves.

Its *End* is either *philosophical* and *theoretical*; or *medicinal*, *mechanical*, *economical*, and *practical*.

Its *efficient Cause* is the *Chemist*.

. . .

The Structure of Simple, Mix'd, Compound and Aggregate Matter

As *mix'd*, *compound*, and *aggregate Bodies* are, according to our Definition, the Subject of *Chemistry*, 'tis necessary that we here consider their *chemical Structure*.

All *natural Bodies* are either *simple* or *compounded*: the *simple* do not consist of *physical parts*; but the *compounded* do. The *simple* are *Principles*, or the first material causes of *Mixts*; and the *compounded*, according to the difference of their mixture, are either *mix'd, compound* or *aggregate*: *mix'd*, if composed merely of Principles; *compound*, if formed of Mixts into any determinable single thing; and *aggregate*, when several such things form any other entire parcel of matter, whatsoever it be.

A *Principle* is defined, *à priori*, that in mix'd matter, which *first existed*; and *à posteriori*, that into which it is *at last resolved*.

Both these definitions are exact, if we allow of a *pure, natural resolution*: but as this is not easily obtainable from the Chemistry of these days, and so can hardly be come at by Art, a difference, at present, prevails between the *physical* and *chemical Principles* of mix'd Bodies.

Those are called *physical Principles* whereof a Mixt is really composed, but they are not hitherto settled: for the four *Peripatetical Elements*, according to their vulgar acceptation, do not deserve this title. And those are usually termed *chemical Principles*, into which all Bodies are found reducible by the chemical operations hitherto known.

These *chemical Principles* are called *Salt, Sulphur*, and *Mercury*; the analogy being taken from Minerals: or, *Salt, Oil*, and *Spirit*; to which Dr. *Willis* adds *Phlegm* and *Earth*; but improperly, since *Phlegm* is comprehended under *Spirit*: for inflammable Spirits cannot be here meant; these consisting manifestly of Water, Oil and Salt, as we shall see hereafter.

But as the four *Peripatetic Elements*, howsoever understood, cannot have place, if supposed specifically the same in all Subjects; so neither can the *Chemical Principles*: for no-one has hitherto pretended to shew that these *Principles* are specifically the same in all Bodies. But if consider'd only as to their *generical qualities*, they may be allow'd in *Compounds*.

We say particularly in *Compounds*, because all the darkness and disputes about *Principles* arise from a neglect of that *real distinction* between *original* and *secondary Mixts*, or *Mixts consisting of Principles* and *Bodies compounded of Mixts*. Whilst these two are confounded, and supposed to be resolved by an operation that is contrary to Nature, the common *chemical Principles* of vegetables, animals and minerals are produced, and prove in reality *artificial Mixts*: but when *Compounds* are separated by bare resolution, without the least combination, their *Principles* are *natural Mixts*.

By justly distinguishing between *Mixts* and *Compounds*, without directly undertaking to exhibit the *first Principles* of the latter, we may easily settle this affair. *Helmont* and *Becher* have attempted it; the former taking *Water* for the first and only *material Principle* of all things; and the other, *Water* and *Earth*; but distinguishing the *Earth* into *three kinds*.

Translated by Peter Shaw

Reading and Discussion Question

1. What is Stahl doing to try to differentiate chemistry from alchemy? Why is he so concerned here with proper definitions of terms?

Experiments and Observations on Different Kinds of Air (1775)

Joseph Priestley

66

The contents of this section will furnish a very striking illustration of the truth of a remark, which I have more than once made in my philosophical writings, and which can hardly be too often repeated, as it tends greatly to encourage philosophical investigations; viz. that more is owing to what we call *chance*, that is, philosophically speaking, to the observation of *events arising from unknown causes*, than to any proper *design*, or pre-conceived *theory* in this business. This does not appear in the works of those who write *synthetically* upon these subjects; but would, I doubt not, appear very strikingly in those who are the most celebrated for their philosophical acumen, did they write *analytically* and ingenuously.

For my own part, I will frankly acknowledge, that, at the commencement of the experiments recited in this section, I was so far from having formed any hypothesis that led to the discoveries I made in pursuing them, that they would have appeared very improbable to me had I been told of them; and when the decisive facts did at length obtrude themselves upon my notice, it was very slowly, and with great hesitation, that I yielded to the evidence of my senses. And yet, when I re-consider the matter, and compare my last discoveries relating to the constitution of the atmosphere with the first, I see the closest and the easiest connexion in the world between them, so as to wonder that I should not have been led immediately from the one to the other. That this was not the case, I attribute to the force of prejudice, which, unknown to ourselves, biases not only our *judgments*, properly so called, but even the perceptions of our senses: for we may take a maxim so strongly for granted, that the plainest evidence of sense will not entirely change, and often hardly modify our persuasions; and the more ingenious a man is, the more effectually he is entangled in his errors; his ingenuity only helping him to deceive himself, by evading the force of truth.

There are, I believe, very few maxims in philosophy that have laid firmer hold upon the mind, than that air, meaning atmospherical air (free from various foreign matters, which were always supposed to be dissolved, and intermixed with it) is a *simple elementary substance*, indestructible, and unalterable, at least as much so as water is supposed to be. In the course of my inquiries, I was, however, soon satisfied that atmospherical air is not an unalterable thing; for that the phlogiston with which it becomes loaded from bodies burning in it, and animals breathing it, and various other chemical processes, so far alters and depraves it, as to render it altogether unfit for inflammation, respiration, and other purposes to which it is subservient; and I had discovered that agitation in water, the process of vegetation, and probably other natural processes, by taking out the superfluous phlogiston, restore it to its original purity. But I own I had no idea of the possibility of going any farther in this way, and thereby procuring air purer than the best common air. I might, indeed, have naturally imagined that such would be air that should contain less phlogiston than the air of the atmosphere; but I had no idea that such a composition was possible.

It will be seen in my last publication, that, from the experiments which I made on the marine acid air, I was led to conclude that common air consisted of some acid (and I naturally inclined to the acid that I was then operating upon) and phlogiston; because the union of this acid vapour and phlogiston made inflammable air; and inflammable air, by agitation in water, ceases to be inflammable, and becomes respirable. And though I could never make it quite so good as common air, I thought it very probable that vegetation, in more favourable circumstances than any in which I could apply it, or some other natural process, might render it more pure.

Upon this, which no person can say was an improbable supposition, was founded my conjecture, of volcanos having given birth to the atmosphere of this planet, supplying it with a permanent air, first inflammable, then deprived of its inflammability by agitation in water, and farther purified by vegetation.

Several of the known phenomena of the *nitrous acid* might have led me to think, that this was more proper for the constitution of the atmosphere than the marine acid: but my thoughts had got into a different train, and nothing but a series of observations, which I shall now distinctly relate, compelled me to adopt another hypothesis, and brought me, in a way of which I had then no idea, to the solution of the great problem, which my reader will perceive I have had in view ever since my discovery that the atmospheric air is alterable, and therefore that it is not an elementary substance, but a *composition*, viz. what this composition is, or *what is the thing that we breathe*, and how is it to be made from its constituent principles.

At the time of my former publication, I was not possessed of a *burning lens* of any considerable force; and for want of one, I could not possibly make many of the experiments that I had projected, and which, in theory, appeared very promising. I had, indeed, a *mirror* of force sufficient for my purpose. But the nature of this instrument is such, that it cannot be applied, with effect, except upon substances that are capable of being suspended or resting on a very slender support. It cannot be directed at all upon any substance in the form of a *powder*, nor hardly upon any thing that requires to be put into a vessel of quicksilver; which appears to me to be the most accurate method of extracting air from a great variety of substances, as was explained in the Introduction to this volume. But having afterwards procured a lens of twelve inches diameter, and twenty inches focal distance, I proceeded with great alacrity to examine, by the help of it, what kind of air a great variety of substances, natural and factitious, would yield, putting them into the vessels represented fig. *a*, which I filled with quicksilver, and kept inverted in a bason of the same. Mr. Warltire, a good chemist, and lecturer in natural philosophy, happening to be at that time in Calne, I explained my views to him, and was furnished by him with many substances, which I could not otherwise have procured.

With this apparatus, after a variety of other experiments, an account of which will be found in its proper place, on the 1st of August, 1774, I endeavoured to extract air from *mercurius calcinatus per se*; and I presently found that, by means of this lens, air was expelled from it very readily. Having got about three or four times as much as the bulk of my materials, I admitted water to it, and found that it was not imbibed by it. But what surprised me more than I can well express, was, that a candle burned in this air with a remarkably vigorous flame, very much like that enlarged flame with which a candle burns in nitrous air, exposed to iron or liver of sulphur; but as I had go nothing like this remarkable appearance from any kind of air besides this particular modification of nitrous air, and I knew no nitrous acid was used in the preparation of *mercurius calcinatus*, I was utterly at a loss how to account for it.

In this case, also, though I did not give sufficient attention to the circumstance at that time, the flame of the candle, besides being larger, burned with more splendor and heat than in that species of nitrous air; and a piece of red-hot wood sparkled in it, exactly like paper dipped in a solution of nitre, and it consumed very fast; an experiment which I had never thought of trying with nitrous air.

At the same time that I made the above mentioned experiment, I extracted a quantity of air, with the very same property, from the common *red precipitate*, which being produced by a solution of mercury in spirit of nitre, made me conclude that this peculiar property, being similar to that of the modification of nitrous air above mentioned, depended upon something being communicated to it by the nitrous acid; and since the *mercurius calcinatus* is produced by exposing mercury to a certain degree of heat, where common air has access to it, I likewise concluded that this substance had collected something of *nitre*, in that state of heat, from the atmosphere.

This, however, appearing to me much more extraordinary than it ought to have done, I entertained some suspicion that the *mercurius calcinatus*, on which I had made my experiments, being bought at a common apothecary's, might, in fact, be nothing more than red precipitate; though, had I been any thing of a practical chemist, I could not have entertained any such suspicion. However, mentioning this suspicion to Mr. Warltire, he furnished me with some that he had kept for a specimen of the preparation, and which, he told me, he could warrant to be genuine. This being treated in the same manner as the former, only by a longer continuance of heat, I extracted much more air from it than from the other.

This experiment might have satisfied any moderate sceptic: but, however, being at Paris in the October following, and knowing that there were several very eminent chemists in that place, I did not omit the opportunity, by means of my friend Mr. Magellan, to get an ounce of *mercurius calcinatus* prepared by Mr. Cadet, of the genuineness of which there could not possibly be any suspicion; and at the same time, I frequently mentioned my surprise at the kind of air which I had got from this preparation to Mr. Lavoisier, Mr. le Roy, and several other philosophers, who honoured me with their notice in that city; and who, I dare say, cannot fail to recollect the circumstance.

At the same time, I had no suspicion that the air which I had got from the *mercurius calcinatus* was even wholesome, so far was I from knowing what it was that I had really found; taking it for granted, that it was nothing more than such kind of air as I had brought nitrous air to be by the processes above mentioned; and in this air I have observed that a candle would burn sometime quite naturally, and sometimes with a beautiful enlarged flame, and yet remain perfectly noxious.

At the same time that I had got the air above mentioned from *mercurius calcinatus* and the red precipitate, I had got the same kind from *red lead* or *minium*. In this process, that part of the minium on which the focus of the lens had fallen, turned yellow. One third of the air, in this experiment, was readily absorbed by water, but, in the remainder, a candle burned very strongly, and with a crackling noise.

That fixed air is contained in red lead I had observed before; for I had expelled it by the heat of a candle, and had found it to be very pure. I imagine it requires more heat than I then used to expel any of the other kind of air.

This experiment with *red lead* confirmed me more in my suspicion, that the *mercurius calcinatus* must get the property of yielding this kind of air from the atmosphere, the process by which that preparation, and this of red lead is made, being similar. As I never make the least secret of anything I observe, I mentioned this experiment also, as well as those with the *mercurius calcinatus*, and the red precipitate, to all my philosophical acquaintance at Paris, and elsewhere; having no idea, at that time, to what these remarkable facts would lead.

Presently after my return from abroad, I went to work upon the *mercurius calcinatus*, which I had procured from Mr. Cadet; and, with a very moderate degree of heat, I got from about one fourth of an ounce of it, an ounce-measure of air, which I observed to be not readily imbibed, either by the substance itself from which it had been expelled (for I suffered them to continue a long time together before I transferred the air to any other place) or by water, in which I suffered this air to stand a considerable time before I made any experiment upon it.

In this air, as I had expected, a candle burned with a vivid flame; but what I observed new at this time (Nov. 19), and which surprised me no less than the fact I had discovered before, was, that, whereas a few moments agitation in water will deprive the modified nitrous air of its property of admitting a candle to burn in it; yet, after more than ten times as much agitation as would be sufficient to produce this alteration in the nitrous air, no sensible change was produced in this. A candle still burned in it with a strong flame; and it did not, in the least, diminish common air, which I have observed that nitrous air, in this state, in some measure, does.

But I was much more surprised, when, after two days, in which this air had continued in contact with water (by which it was diminished about one twentieth of its bulk) I agitated it violently in water about five minutes, and found that a candle still burned in it as well as in common air. The same degree of agitation would have made phlogisticated nitrous air fit for respiration indeed, but it would certainly have extinguished a candle.

These facts fully convinced me, that there must be a very material difference between the constitution of the air from *mercurius calcinatus*, and that of phlogisticated nitrous air, notwithstanding their resemblance in some particulars. But though I did not doubt that the air from *mercurius calcinatus* was fit for respiration, after being agitated in water, as every kind of air without exception, on which I had tried the experiment, had been, I still did not suspect that it was respirable in the first instance; so far was I from having any idea of this air being, what it really was, much superior, in this respect, to the air of the atmosphere.

In this ignorance of the real nature of this kind of air, I continued from this time (November) to the 1st of March following; having, in the meantime, been intent upon my experiments on the vitriolic acid air above recited, and the various modifications of air produced by spirit of nitre, an account of which will follow. But in the course of this month, I not only ascertained the nature of this kind of air, though very gradually, but was led by it to the complete discovery of the constitution of the air we breathe.

Till this 1st of March, 1775, I had so little suspicion of the air from *mercurius calcinatus*, etc. being wholesome, that I had not even thought of applying to it the test of nitrous air; but thinking (as my reader must imagine I frequently must have done) on the candle burning in it after long agitation in water, it occurred to me at last to make the experiment; and putting one measure of nitrous air to two measures of this air, I found, not only that it was diminished, but that it was diminished quite as much as common air, and that the redness of the mixture was likewise equal to that of a similar mixture of nitrous and common air.

After this I had no doubt but that the air from *mercurius calcinatus* was fit for respiration, and that it had all the other properties of genuine common air. But I did not take notice of what I might have observed, if I had not been so fully possessed by the notion of there being no air better than common air, that the redness was really deeper, and the diminution something greater than common air would have admitted.

Moreover, this advance in the way of truth, in reality, threw me back into error, making me give up the hypothesis I had first formed, viz. that the *mercurius calcinatus* had extracted spirit of nitre from the air; for I now concluded, that all the constituent parts of the air were equally, and in their proper proportion, imbibed in the preparation of this substance, and also in the process of making red lead. For at the same time that I made the above-mentioned experiment on the air from *mercurius calcinatus*, I likewise observed that the air which I had extracted from red lead, after the fixed air was washed out of it, was of the same nature, being diminished by nitrous air like common air: but, at the same time, I was puzzled to find that air from the red precipitate was diminished in the same manner, though the process for making this substance is quite different from that of making the two others. But to this circumstance I happened not to give much attention.

I wish my reader be not quite tired with the frequent repetition of the word *surprise*, and others of similar import; but I must go on in that style a little longer. For the next day I was more surprised than ever I had been before, with finding that, after the above-mentioned mixture of nitrous air and the air from *mercurius calcinatus*, had stood all night, (in which time the whole diminution must have taken place; and, consequently, had it been common air, it must have been made perfectly noxious, and entirely unfit for respiration or inflammation) a candle burned in it, and even better than in common air.

I cannot, at this distance of time, recollect what it was that I had in view in making this experiment; but I know I had no expectation of the real issue of it. Having acquired a considerable degree of readiness in making experiments of this kind, a very slight and evanescent motive would be sufficient to induce me to do it. If, however, I had not happened for some other purpose, to have had a lighted candle before me, I should probably never have made the trial; and the whole train of my future experiments relating to this kind of air might have been prevented.

Still, however, having no conception of the real cause of this phenomenon, I considered it as something very extraordinary; but as a property that was peculiar to air extracted from these substances, and *adventitious*; and I always spoke of the air to my acquaintance as being substantially the same thing with common air. I particularly remember my telling Dr. Price, that I was myself perfectly satisfied of its being common air, as it appeared to be so by the test of nitrous air; though, for the satisfaction of others, I wanted a mouse to make the proof quite complete.

On the 8th of this month I procured a mouse, and put it into a glass vessel, containing two ounce-measures of the air from *mercurius calcinatus*. Had it been common air, a full-grown mouse, as this was, would have lived in it about a quarter of an hour. In this air, however, my mouse lived a full half hour; and though it was taken out seemingly dead, it appeared to have been only exceedingly chilled; for, upon being held to the fire, it presently revived, and appeared not to have received any harm from the experiment.

By this I was confirmed in my conclusion, that the air extracted from *mercurius calcinatus*, etc. was, *at least, as good* as common air; but I did not certainly conclude that it was any *better*; because, though one mouse would live only a quarter of an hour in a given quantity of air, I knew it was not impossible that another mouse might have lived in it half an hour; so little accuracy is there in this method of ascertaining the goodness of air: and indeed I have never had recourse to it for my own satisfaction, since the discovery of that most ready, accurate, and elegant test that nitrous air furnishes. But in this case I had a view to publishing the most generally-satisfactory account of my experiments that the nature of the thing would admit of.

This experiment with the mouse, when I had reflected upon it some time, gave me so much suspicion that the air into which I had put it was better than common air, that I was induced, the day after, to apply the test of nitrous air to a small part of that very quantity of air which the mouse had breathed so long; so that, had it been common air, I was satisfied it must have been very nearly, if not altogether, as noxious as possible, so as not to be affected by nitrous air; when, to my surprise again, I found that though it had been breathed so long, it was still better than common air. For after mixing it with nitrous air, in the usual proportion of two to one, it was diminished in the proportion of 4½ to 3½; that is, the nitrous air had made it two ninths less than before, and this in a very short space of time; whereas I had never found that, in the longest time, any common air was reduced more than one fifth of its bulk by any proportion of nitrous air, nor more than one fourth by any phlogistic process whatever. Thinking of this extraordinary fact upon my pillow, the next morning I put another measure of nitrous air to the same mixture, and, to my utter astonishment, found that it was farther diminished to almost one half of its original quantity. I then put a third measure to it; but this did not diminish it any farther: but, however, left it one measure less than it was even after the mouse had been taken out of it.

Being now fully satisfied that this air, even after the mouse had breathed it half an hour, was much better than common air; and having a quantity of it still left, sufficient for the experiment, viz. an ounce-measure and a half, I put the mouse into it; when I observed that it seemed to feel no shock upon being put into it, evident signs of which would have been visible, if the air had not been very wholesome; but that it remained perfectly at its ease another full half hour, when I took it out quite lively and vigorous. Measuring the air the next day, I found it to be reduced from 1½ to ⅔ of an ounce-measure. And after this, if I remember well (for in my *register* of the day I only find it noted, that it was *considerably diminished* by nitrous air) it was nearly as good as common air. It was evident, indeed, from the mouse having been taken out quite vigorous, that the air could not have been rendered very noxious.

For my farther satisfaction I procured another mouse, and putting it into less than two ounce-measures of air extracted from *mercurius calcinatus* and air from red precipitate (which, having found them to be of the same quality, I had mixed together) it lived three quarters of an hour. But not having had the precaution to set the vessel in a warm place, I suspect that the mouse died of cold. However, as it had lived three times as long as it could probably have lived in the same quantity of common air, and I did not expect much accuracy from this kind of test, I did not think it necessary to make any more experiments with mice.

Being now fully satisfied of the superior goodness of this kind of air, I proceeded to measure that degree of purity, with as much accuracy as I could, by the test of nitrous air; and I began with putting one measure of nitrous air to two measures of this air, as if I had been examining common air; and now I observed that the diminution was evidently greater than common air would have suffered by the same treatment. A second measure of nitrous air reduced it to two thirds of its original quantity, and a third measure to one half. Suspecting that the diminution could not proceed much farther, I then added only half a measure of nitrous air, by which it was diminished still more; but not much, and another half measure made it more than half of its original quantity; so that, in this case, two measures of this air took more than two measures of nitrous air, and yet remained less than half of what it was. Five measures brought it pretty exactly to its original dimensions.

At the same time, air from *red precipitate* was diminished in the same proportion as that from *mercurius calcinatus*, five measures of nitrous air being received by two measures of this without any increase of dimensions. Now as common air takes about one half of its bulk of nitrous air, before it begins to receive any addition to its dimensions from more nitrous air, and this air took more than four half-measures before it ceased to be diminished by more nitrous air, and even five half-measures made no addition to its original dimensions, I conclude that it was between four and five times as good as common air. It will be seen that I have since procured air better than this, even between five and six times as good as the best common air that I have ever met with.

Section Seven: Of the Purification of Air by Plants and the Influence of Light on That Process

One of my earliest observations on the subject of air, but made casually, when, in fact, I expected a contrary result from the process, was the purification of air injured by respiration or putrefaction, by the vegetation of plants. But at that time I was altogether ignorant of the part that light had to act in the business. At the publication of the experiments recited in the last section, I had fully ascertained the influence of light in the production of dephlogisticated air in water by means of a green substance, which I at first supposed to be a plant, but not being able to discover the form of one, I contented myself with calling it simply *green matter*.

Several of my friends, however, better skilled in botany than myself, never entertained any doubt of its being a plant; and I had afterwards the fullest conviction that it must be one. Mr. Bewly has lately observed the regular form of it by a microscope. My own eyes having always been weak, I have, as much as possible, avoided the use of a microscope.

The principle reason that made me question whether this green matter was a plant, besides my not being able to discover the form of it, was its being produced, as I then thought, in a vial close stopped. But this being only with a common cork, the seeds of this plant, which must float invisibly in the air, might have insinuated themselves through some unperceived fracture in it; or the seeds might have been contained in the water previous to its being put into the phial. Both Mr. Bewly and myself found, in the course of the last summer, that when distilled water was exposed to the sun, in phials filled in part with quicksilver, and in part with distilled water, and inverted into basons of quicksilver, none of this green matter was ever produced; no seed of this plant having been able to penetrate through the mercury, to reach the water incumbent upon it, though, in several cases, it will be seen, that these seeds diffuse and insinuate themselves, in a manner that is truly wonderful.

Without light, it is well known, that no plant can thrive; and if it do grow at all in the dark, it is always white, and is, in all other respects, in a weak and sickly state. Healthy plants are probably in a state similar to that of *sleep* in the absence of light, and do not resume their proper functions, but by the influence of light, and especially the action of the rays of the sun. This was the reason why no green matter was ever produced by means of mere *warmth* in my former experiments, and that in jars standing in the same exposure, but covered so that the light had no access to them, no pure air was collected, none of the green matter being then found in them.

This I verified most completely by covering the greatest part of a glass jar with black scaling-wax, which made it thoroughly opaque; and besides answering that purpose better than brown paper, as I made the experiment before mentioned, did not imbibe any of the water, and therefore did not promote the evaporation of it. To be able to observe whether any air was collected in these jars, or not, the upper part of them was not coated with scaling-wax, but had a thick movable cap of paper, which I could easily take off, and then inspect the surface of the water.

In order to satisfy myself as fully as possible with respect to this remarkable circumstance, I also made the following experiments, the results of which are indeed, very decisive in favour of the influence of *light* in this case.

Having a large trough of water, full of recent green matter, giving air very copiously, so that all the surface of it was covered with froth, and jars filled with it, and inverted, collected great quantities of it, and very fast; I filled a jar with it, and inverting it in a basin of the same, I placed it in a dark room. From that instant no more air was yielded by it, and in a few days it had a very offensive smell, the green vegetable matter with which it abounded being then all dead, and putrid.

Again, having filled a receiver with fresh pump water, and having waited till it was in a state of giving air copiously, I removed it into a dark room; and from that time the production of air from it entirely ceased. When I placed it again in the sun, it gave no air till about ten days after, when it had more green matter, the former plants being probably all dead; and no air could be produced till new ones formed . . . .

It appears from these experiments that air combined with water is liable to be phlogisticated by respiration, and to be dephlogisticated by vegetation, as much as air in an elastic state, out of water. For fishes, as I shall observe, foul the air contained in the water in which they are confined, and water plants now appear to purify it. This is no doubt one of the great uses of weeds, and other aquatic plants, with which fresh water lakes, and even seas abound, as well as their serving for food to a great number of fishes.

Reading and Discussion Questions

1. How would you describe the narrative style of this account of scientific research? Do you find that Priestley's style makes the piece more or less accessible than some of the experimental reports we encounter in other readings?
2. Priestly is "surprised" to find that air is not a simple elementary substance but rather a mixture. What does he think that it is a mixture of and how does he arrive at that conclusion?
3. What does Priestly conclude about the relationship between plants and phlogiston?

Elements of Chemistry, in a New Systematic Order, Containing All Modern Discoveries (1789)

Antoine-Laurent Lavoisier

67

Preface of the Author

When I began the following Work, my only object was to extend and explain more fully the Memoir which I read at the public meeting of the Academy of Sciences in the month of April 1787, on the necessity of reforming and completing the Nomenclature of Chemistry. While engaged in this employment, I perceived, better than I had ever done before, the justice of the following maxims of the Abbé de Condillac, in his System of Logic, and some other of his works.

"We think only through the medium of words. Languages are true analytical methods. Algebra, which is adapted to its purpose in every species of expression, in the most simple, most exact, and best manner possible, is at the same time a language and an analytical method. The art of reasoning is nothing more than a language well arranged."

Thus, while I thought myself employed only in forming a Nomenclature, and while I proposed to myself nothing more than to improve the chemical language, my work transformed itself by degrees, without my being able to prevent it, into a treatise upon the Elements of Chemistry.

The impossibility of separating the nomenclature of a science from the science itself, is owing to this, that every branch of physical science must consist of three things; the series of facts which are the objects of the science, the ideas which represent these facts, and the words by which these ideas are expressed. Like three impressions of the same seal, the word ought to produce the idea, and the idea to be a picture of the fact. And, as ideas are preserved and communicated by means of words, it necessarily follows that we cannot improve the language of any science without at the same time improving the science itself; neither can we, on the other hand, improve a science, without improving the language or nomenclature which belongs to it. However certain the facts of any science may be, and, however just the ideas we may have formed of these facts, we can only communicate false impressions to others, while we want words by which these may be properly expressed.

. . .

The only method of preventing such errors from taking place, and of correcting them when formed, is to restrain and simplify our reasoning as much as possible. This depends entirely upon ourselves, and the neglect of it is the only source of our mistakes. We must trust to nothing but facts: These are presented to us by Nature, and cannot deceive. We ought, in every instance, to submit our reasoning to the test of experiment, and never to search for truth but by the natural road of experiment and observation. Thus mathematicians obtain the solution of a problem by the mere arrangement of data, and by reducing their reasoning to such simple steps, to conclusions so very obvious, as never to lose sight of the evidence which guides them.

Thoroughly convinced of these truths, I have imposed upon myself, as a law, never to advance but from what is known to what is unknown; never to form any conclusion which is not an immediate consequence necessarily flowing from observation and experiment; and always to arrange the facts, and the conclusions which are drawn from them, in such an order as shall render it most easy for beginners in the study of chemistry thoroughly to understand them. Hence I have been obliged to depart from the usual order of courses of lectures and of treatises upon chemistry, which always assume the first principles of the science, as known, when the pupil or the reader should never be supposed to know them till they have been explained in subsequent lessons. In almost every instance, these begin by treating of the elements of matter, and by explaining the table of affinities, without considering, that, in so doing, they must bring the principal phenomena of chemistry into view at the very outset: They make use of terms which have not been defined, and suppose the science to be understood by the very persons they are only beginning to teach. It ought likewise to be considered, that very little of chemistry can be learned in a first course, which is hardly sufficient to make the language of the science familiar to the ears, or the apparatus familiar to the eyes. It is almost impossible to become a chemist in less than three or four years of constant application.

. . .

It will, no doubt, be a matter of surprise, that in a treatise upon the elements of chemistry, there should be no chapter on the constituent and elementary parts of matter; but I shall take occasion, in this place, to remark, that the fondness for reducing all the bodies in nature to three or four elements, proceeds from a prejudice which has descended to us from the Greek Philosophers. The notion of four elements, which, by the variety of their proportions, compose all the known substances in nature, is a mere hypothesis, assumed long before the first principles of experimental philosophy or of chemistry had any existence. In those days, without possessing facts, they framed systems; while we, who have collected facts, seem determined to reject them, when they do not agree with our prejudices. The authority of these fathers of human philosophy still carry great weight, and there is reason to fear that it will even bear hard upon generations yet to come.

It is very remarkable, that, notwithstanding of the number of philosophical chemists who have supported the doctrine of the four elements, there is not one who has not been led by the evidence of facts to admit a greater number of elements into their theory. The first chemists that wrote after the revival of letters, considered sulphur and salt as elementary substances entering into the composition of a great number of substances; hence, instead of four, they admitted the existence of six elements. Becher assumes the existence of three kinds of earth, from the combination of which, in different proportions, he supposed all the varieties of metallic substances to be produced. Stahl gave a new modification to this system; and succeeding chemists have taken the liberty to make or to imagine changes and additions of a similar nature. All these chemists were carried along by the influence of the genius of the age in which they lived, which contented itself with assertions without proofs; or, at least, often admitted as proofs the slighted degrees of probability, unsupported by that strictly rigorous analysis required by modern philosophy.

All that can be said upon the number and nature of elements is, in my opinion, confined to discussions entirely of a metaphysical nature. The subject only furnishes us with indefinite problems, which may be solved in a thousand different ways, not one of which, in all probability, is consistent with nature. I shall therefore only add upon this subject, that if, by the term *elements*, we mean to express those simple and indivisible atoms of which matter is composed, it is extremely probable we know nothing at all about them; but, if we apply the term *elements*, or *principles of bodies*, to express our idea of the last point which analysis is capable of reaching, we must admit, as elements, all the substances into which we are capable, by any means, to reduce bodies by decomposition. Not that we are entitled to affirm, that these substances we consider as simple may not be compounded of two, or even of a greater number of principles; but, since these principles cannot be separated, or rather since we have not hitherto discovered the means of separating them, they act with regard to us as simple substances, and we ought never to suppose them compounded until experiment and observation has proved them to be so.

The foregoing reflections upon the progress of chemical ideas naturally apply to the words by which these ideas are to be expressed. Guided by the work which, in the year 1787, Messrs de Morveau, Berthollet, de Fourcroy, and I composed upon the Nomenclature of Chemistry, I have endeavoured, as much as possible, to denominate simple bodies by simple terms, and I was naturally led to name these first. It will be recollected, that we were obliged to retain that name of any substance by which it had been long known in the world, and that in two cases only we took the liberty of making alterations; first, in the case of those which were but newly discovered, and had not yet obtained names, or at least which had been known but for a short time, and the names of which had not yet received the sanction of the public; and, secondly, when the names which had been adopted, whether by the ancients or the moderns, appeared to us to express evidently false ideas, when they confounded the substances, to which they were applied, with others possessed of different, or perhaps opposite qualities. We made no scruple, in this case, of substituting other names in their room, and the greatest number of these were borrowed from the Greek language. We endeavoured to frame them in such a manner as to express the most general and the most characteristic quality of the substances; and this was attended with the additional advantage both of assisting the memory of beginners, who find it difficult to remember a new word which has no meaning, and of accustoming them early to admit no word without connecting with it some determinate idea.

To those bodies which are formed by the union of several simple substances we gave new names, compounded in such a manner as the nature of the substances directed; but, as the number of double combinations is already very considerable, the only method by which we could avoid confusion, was to divide them into classes. In the natural order of ideas, the name of the class or genus is that which expresses a quality common to a great number of individuals: The name of the species, on the contrary, expresses a quality peculiar to certain individuals only.

These distinctions are not, as some may imagine, merely metaphysical, but are established by Nature. "A child," says the Abbé de Condillac, "is taught to give the name *tree* to the first one which is pointed out to him. The next one he sees presents the same idea, and he gives it the same name. This he does likewise to a third and a fourth, till at last the word *tree*, which he first applied to an individual, comes to be employed by him as the name of a class or a genus, an abstract idea, which comprehends all trees in general. But, when he learns that all trees serve not the same purpose, that they do not all produce the same kind of fruit, he will soon learn to distinguish them by specific and particular names." This is the logic of all the sciences, and is naturally applied to chemistry.

The acids, for example, are compounded of two substances, of the order of those which we consider as simple; the one constitutes acidity, and is common to all acids, and, from this substance, the name of the class or the genus ought to be taken; the other is peculiar to each acid, and distinguishes it from the rest, and from this substance is to be taken the name of the species. But, in the greatest number of acids, the two constituent elements, the acidifying principle, and that which it acidifies, may exist in different proportions, constituting all the possible points of equilibrium or of saturation. This is the case in the sulphuric and the sulphurous acids; and these two states of the same acid we have marked by varying the termination of the specific name.

Metallic substances which have been exposed to the joint action of the air and of fire, lose their metallic lustre, increase in weight, and assume an earthy appearance. In this state, like the acids, they are compounded of a principle which is common to all, and one which is peculiar to each. In the same way, therefore, we have thought proper to class them under a generic name, derived from the common principle; for which purpose, we adopted the term *oxide*; and we distinguish them from each other by the particular name of the metal to which each belongs.

Combustible substances, which in acids and metallic oxides are a specific and particular principle, are capable of becoming, in their turn, common principles of a great number of substances. The sulphurous combinations have been long the only known ones in this kind. Now, however, we know, from the experiments of Messrs Vandermonde, Monge, and Berthollet, that charcoal may be combined with iron, and perhaps with several other metals; and that, from this combination, according to the proportions, may be produced steel, plumbago, etc. We know likewise, from the experiments of M. Pelletier, that phosphorus may be combined with a great number of metallic substances. These different combinations we have classed under generic names taken from the common substance, with a termination which marks this analogy, specifying them by another name taken from that substance which is proper to each.

. . .

In short, we have advanced so far, that from the name alone may be instantly found what the combustible substance is which enters into any combination; whether that combustible substance be combined with the acidifying principle, and in what proportion; what is the state of the acid; with what basis it is united; whether the saturation be exact, or whether the acid or the basis be in excess.

It may be easily supposed that it was not possible to attain all these different objects without departing, in some instances, from established custom, and adopting terms which at first sight will appear uncouth and barbarous. But we considered that the ear is soon habituated to new words, especially when they are connected with a general and rational system. The names, besides, which were formerly employed, such as *powder of algaroth*, *salt of alembroth*, *pompholix*, *phagadenic water*, *turbith mineral*, *colcathar*, and many others, were neither less barbarous nor less uncommon. It required a great deal of practice, and no small degree of memory, to recollect the substances to which they were applied, much more to recollect the genus of combination to which they belonged. The names of *oil of tartar per deliquium*, *oil of vitriol*, *butter of arsenic and of antimony*, *flowers of zinc*, etc. were still more improper, because they suggested false ideas: For, in the whole mineral kingdom, and particularly in the metallic class, there exists no such thing as butters, oils, or flowers; and, in short, the substances to which they give these fallacious names, are nothing less than rank poisons.

. . .

There is an objection to the work which I am going to present to the public, which is perhaps better founded, that I have given no account of the opinion of those who have gone before me; that I have stated only my own opinion, without examining that of others. By this I have been prevented from doing that justice to my associates, and more especially to foreign chemists, which I wished to render them. But I beseech the reader to consider, that, if I had filled an elementary work with a multitude of quotations; if I had allowed myself to enter into long dissertations on the history of the science, and the works of those who have studied it, I must have lost sight of the true object I had in view, and produced a work, the reading of which must have been extremely tiresome to beginners. It is not to the history of the science, or of the human mind, that we are to attend in an elementary treatise: Our only aim ought to be ease and perspicuity, and with the utmost care to keep every thing out of view which might draw aside the attention of the student; it is a road which we should be continually rendering more smooth, and from which we should endeavour to remove every obstacle which can occasion delay ....

The remarks I have made on the order which I thought myself obliged to follow in the arrangement of proofs and ideas, are to be applied only to the first part of this work. It is the only one which contains the general sum of the doctrine I have adopted, and to which I wished to give a form completely elementary.

The second part is composed chiefly of tables of the nomenclature of the neutral salts. To these I have only added general explanations, the object of which was to point out the most simple processes for obtaining the different kinds of known acids. This part contains nothing which I can call my own, and presents only a very short abridgment of the results of these processes, extracted from the works of different authors.

In the third part, I have given a description, in detail, of all the operations connected with modern chemistry. I have long thought that a work of this kind was much wanted, and I am convinced it will not be without use. The method of performing experiments, and particularly those of modern chemistry, is not so generally known as it ought to be; and had I, in the different memoirs which I have presented to the Academy, been more particular in the detail of the manipulations of my experiments, it is probable I should have made myself better understood, and the science might have made a more rapid progress. The order of the different matters contained in this third part appeared to me to be almost arbitrary; and the only one I have observed was to class together, in each of the chapters of which it is composed, those operations which are most connected with one another. I need hardly mention that this part could not be borrowed from any other work, and that, in the principal articles it contains, I could not derive assistance from any thing but the experiments which I have made myself.

Part One

Of the Formation and Decomposition of Aëriform Fluids—of the Combustion of Simple Bodies—and the Formation of Acids.

Chapter One

Of the Combinations of Caloric, and the Formation of Elastic Aëriform Fluids

That every body, whether solid or fluid, is augmented in all its dimensions by any increase of its sensible heat, was long ago fully established as a physical axiom, or universal proposition, by the celebrated Boerhaave. Such facts as have been adduced for controverting the generality of this principle offer only fallacious results, or, at least, such as are so complicated with foreign circumstances as to mislead the judgment: But, when we separately consider the effects, so as to deduce each from the cause to which they separately belong, it is easy to perceive that the separation of particles by heat is a constant and general law of nature.

When we have heated a solid body to a certain degree, and have thereby caused its particles to separate from each other, if we allow the body to cool, its particles again approach each other in the same proportion in which they were separated by the increased temperature; the body returns through the same degrees of expansion which it before extended through; and, if it be brought back to the same temperature from which we set out at the commencement of the experiment, it recovers exactly the same dimensions which it formerly occupied. But, as we are still very far from being able to arrive at the degree of absolute cold, or deprivation of all heat, being unacquainted with any degree of coldness which we cannot suppose capable of still farther augmentation, it follows, that we are still incapable of causing the ultimate particles of bodies to approach each other as near as is possible; and, consequently, that the particles of all bodies do not touch each other in any state hitherto known, which, tho' a very singular conclusion, is yet impossible to be denied.

It is supposed, that, since the particles of bodies are thus continually impelled by heat to separate from each other, they would have no connection between themselves; and, of consequence, that there could be no solidity in nature, unless they were held together by some other power which tends to unite them, and, so to speak, to chain them together; which power, whatever be its cause, or manner of operation, we name Attraction.

Thus the particles of all bodies may be considered as subjected to the action of two opposite powers, the one repulsive, the other attractive, between which they remain *in equilibrio*. So long as the attractive force remains stronger, the body must continue in a state of solidity; but if, on the contrary, heat has so far removed these particles from each other, as to place them beyond the sphere of attraction, they lose the adhesion they before had with each other, and the body ceases to be solid.

Water gives us a regular and constant example of these facts; whilst below zero[1] of the French thermometer, or 32° of Fahrenheit, it remains solid, and is called ice. Above that degree of temperature, its particles being no longer held together by reciprocal attraction, it becomes liquid; and, when we raise its temperature above

[1] Whenever the degree of heat occurs in this work, it is stated by the author according to Reaumur's scale. The degrees within brackets are the correspondent degrees of Fahrenheit's scale, added by the translator.

80°, (212°) its particles, giving way to the repulsion caused by the heat, assume the state of vapour or gas, and the water is changed into an aëriform fluid.

The same may be affirmed of all bodies in nature: They are either solid or liquid, or in the state of elastic aëriform vapour, according to the proportion which takes place between the attractive force inherent in their particles, and the repulsive power of the heat acting upon these; or, what amounts to the same thing, in proportion to the degree of heat to which they are exposed.

It is difficult to comprehend these phenomena, without admitting them as the effects of a real and material substance, or very subtile fluid, which, insinuating itself between the particles of bodies, separates them from each other; and, even allowing the existence of this fluid to be hypothetical, we shall see in the sequel, that it explains the phenomena of nature in a very satisfactory manner.

This substance, whatever it is, being the cause of heat, or, in other words, the sensation which we call *warmth* being caused by the accumulation of this substance, we cannot, in strict language, distinguish it by the term *heat*; because the same name would then very improperly express both cause and effect. For this reason, in the memoir which I published in 1777,[2] I gave it the names of *igneous fluid* and *matter of heat*. And, since that time, in the work[3] published by Mr de Morveau, Mr Berthollet, Mr de Fourcroy, and myself, upon the reformation of chemical nomenclature, we thought it necessary to banish all periphrastic expressions, which both lengthen physical language, and render it more tedious and less distinct, and which even frequently does not convey sufficiently just ideas of the subject intended. Wherefore, we have distinguished the cause of heat, or that exquisitely elastic fluid which produces it, by the term of *caloric*. Besides, that this expression fulfils our object in the system which we have adopted, it possesses this farther advantage, that it accords with every species of opinion, since, strictly speaking, we are not obliged to suppose this to be a real substance; it being sufficient, as will more clearly appear in the sequel of this work, that it be considered as the repulsive cause, whatever that may be, which separates the particles of matter from each other; so that we are still at liberty to investigate its effects in an abstract and mathematical manner.

In the present state of our knowledge, we are unable to determine whether light be a modification of caloric, or if caloric be, on the contrary, a modification of light. This, however, is indisputable, that, in a system where only decided facts are admissible, and where we avoid, as far as possible, to suppose any thing to be that is not really known to exist, we ought provisionally to distinguish, by distinct terms, such things as are known to produce different effects. We therefore distinguish light from caloric; though we do not therefore deny that these have certain qualities in common, and that, in certain circumstances, they combine with other bodies almost in the same manner, and produce, in part, the same effects.

What I have already said may suffice to determine the idea affixed to the word *caloric*; but there remains a more difficult attempt, which is, to give a just

[2] Collections of the French Academy of Sciences for that year, p. 420.

[3] *Chemical Nomenclature*.

conception of the manner in which caloric acts upon other bodies. Since this subtile matter penetrates through the pores of all known substances; since there are no vessels through which it cannot escape, and, consequently, as there are none which are capable of retaining it, we can only come at the knowledge of its properties by effects which are fleeting, and difficultly ascertainable. It is in these things which we neither see nor feel, that it is especially necessary to guard against the extravagancy of our imagination, which forever inclines to step beyond the bounds of truth, and is very difficultly restrained within the narrow line of facts.

...

All these facts, which could be easily multiplied if necessary, give me full right to assume, as a general principle, that almost every body in nature is susceptible of three several states of existence, solid, liquid, and aëriform, and that these three states of existence depend upon the quantity of caloric combined with the body. Henceforwards I shall express these elastic aëriform fluids by the generic term *gas*; and in each species of gas I shall distinguish between the caloric, which in some measure serves the purpose of a solvent, and the substance, which in combination with the caloric, forms the base of the gas.

To these bases of the different gases, which are hitherto but little known, we have been obliged to assign names; these I shall point out in chapter four of this work, when I have previously given an account of the phenomena attendant upon the heating and cooling of bodies, and when I have established precise ideas concerning the composition of our atmosphere.

We have already shown, that the particles of every substance in nature exist in a certain state of equilibrium, between that attraction which tends to unite and keep the particles together, and the effects of the caloric which tends to separate them. Hence the caloric not only surrounds the particles of all bodies on every side, but fills up every interval which the particles of bodies leave between each other. We may form an idea of this, by supposing a vessel filled with small spherical leaden bullets, into which a quantity of fine sand is poured, which, insinuating into the intervals between the bullets, will fill up every void. The balls, in this comparison, are to the sand which surrounds them exactly in the same situation as the particles of bodies are with respect to the caloric; with this difference only, that the balls are supposed to touch each other, whereas the particles of bodies are not in contact, being retained at a small distance from each other, by the caloric.

If, instead of spherical balls, we substitute solid bodies of a hexahedral, octohedral, or any other regular figure, the capacity of the intervals between them will be lessened, and consequently will no longer contain the same quantity of sand. The same thing takes place, with respect to natural bodies; the intervals left between their particles are not of equal capacity, but vary in consequence of the different figures and magnitude of their particles, and of the distance at which these particles are maintained, according to the existing proportion between their inherent attraction, and the repulsive force exerted upon them by the caloric.

In this manner we must understand the following expression, introduced by the English philosophers, who have given us the first precise ideas upon this subject; *the capacity of bodies for containing the matter of heat*. As comparisons with sensible objects are of great use in assisting us to form distinct notions of abstract ideas, we shall endeavour to illustrate this, by instancing the phenomena which take place between water and bodies which are wetted and penetrated by it, with a few reflections.

If we immerge equal pieces of different kinds of wood, suppose cubes of one foot each, into water, the fluid gradually insinuates itself into their pores, and the pieces of wood are augmented both in weight and magnitude: But each species of wood will imbibe a different quantity of water; the lighter and more porous woods will admit a larger, the compact and closer grained will admit of a lesser quantity; for the proportional quantities of water imbibed by the pieces will depend upon the nature of the constituent particles of the wood, and upon the greater or lesser affinity subsisting between them and water. Very resinous wood, for instance, though it may be at the same time very porous, will admit but little water. We may therefore say, that the different kinds of wood possess different capacities for receiving water; we may even determine, by means of the augmentation of their weights, what quantity of water they have actually absorbed; but, as we are ignorant how much water they contained, previous to immersion, we cannot determine the absolute quantity they contain, after being taken out of the water.

The same circumstances undoubtedly take place, with bodies that are immersed in caloric; taking into consideration, however, that water is an incompressible fluid, whereas caloric is, on the contrary, endowed with very great elasticity; or, in other words, the particles of caloric have a great tendency to separate from each other, when forced by any other power to approach; this difference must of necessity occasion very considerable diversities in the results of experiments made upon these two substances.

Having established these clear and simple propositions, it will be very easy to explain the ideas which ought to be affixed to the following expressions, which are by no means synonimous, but possess each a strict and determinate meaning, as in the following definitions:

Free caloric, is that which is not combined in any manner with any other body. But, as we live in a system to which caloric has a very strong adhesion, it follows that we are never able to obtain it in the state of absolute freedom.

Combined caloric, is that which is fixed in bodies by affinity or elective attraction, so as to form part of the substance of the body, even part of its solidity.

By the expression *specific caloric* of bodies, we understand the respective quantities of caloric requisite for raising a number of bodies of the same weight to an equal degree of temperature. This proportional quantity of caloric depends upon the distance between the constituent particles of bodies, and their greater or lesser degrees of cohesion; and this distance, or rather the space or void resulting from it, is, as I have already observed, called the *capacity of bodies for containing caloric*.

Heat, considered as a sensation, or, in other words, sensible heat, is only the effect produced upon our sentient organs, by the motion or passage of caloric, disengaged from the surrounding bodies. In general, we receive impressions only in consequence of motion, and we might establish it as an axiom, *That*, without motion, there is no sensation. This general principle applies very accurately to the sensations of heat and cold: When we touch a cold body, the caloric which always tends to become in equilibrio in all bodies, passes from our hand into the body we touch, which gives us the feeling or sensation of cold. The direct contrary happens, when we touch a warm body, the caloric then passing from the body into our hand, produces the sensation of heat. If the hand and the body touched be of the same temperature, or very nearly so, we receive no impression, either of heat or cold, because there is no motion or passage of caloric; and thus no sensation can take place, without some correspondent motion to occasion it.

When the thermometer rises, it shows, that free caloric is entering into the surrounding bodies: The thermometer, which is one of these, receives its share in proportion to its mass, and to the capacity which it possesses for containing caloric. The change therefore which takes place upon the thermometer, only announces a change of place of the caloric in those bodies, of which the thermometer forms one part; it only indicates the portion of caloric received, without being a measure of the whole quantity disengaged, displaced, or absorbed ....

Chapter Two

Analysis of Atmospheric Air, and its Division into two Elastic Fluids; the one fit for Respiration, the other incapable of being respired

From what has been premised, it follows, that our atmosphere is composed of a mixture of every substance capable of retaining the gaseous or aëriform state in the common temperature, and under the usual pressure which it experiences. These fluids constitute a mass, in some measure homogeneous, extending from the surface of the earth to the greatest height hitherto attained, of which the density continually decreases in the inverse ratio of the superincumbent weight. But, as I have before observed, it is possible that this first stratum is surmounted by several others consisting of very different fluids ....

Chapter Four

Nomenclature of the several Constituent Parts of Atmospheric Air

Hitherto I have been obliged to make use of circumlocution, to express the nature of the several substances which constitute our atmosphere, having provisionally used the terms of *respirable* and *noxious*, or *non-respirable parts of the air*. But the investigations I mean to undertake require a more direct mode of expression; and, having now endeavoured to give simple and distinct ideas of the different substances which enter into the composition of the atmosphere, I shall henceforth express these ideas by words equally simple.

The temperature of our earth being very near to that at which water becomes solid, and reciprocally changes from solid to fluid, and as this phenomenon takes place frequently under our observation, it has very naturally followed, that, in the languages of at least every climate subjected to any degree of winter, a term has been used for signifying water in the state of solidity, when deprived of its caloric. The same, however, has not been found necessary with respect to water reduced to the state of vapour by an additional dose of caloric; since those persons who do not make a particular study of objects of this kind, are still ignorant that water, when in a temperature only a little above the boiling heat, is changed into an elastic aëriform fluid, susceptible, like all other gasses, of being received and contained in vessels, and preserving its gaseous form so long as it remains at the temperature of 80° (212°), and under a pressure not exceeding 28 inches of the mercurial barometer. As this phenomenon has not been generally observed, no language has used a particular term for expressing water in this state[4]; and the same thing occurs with all fluids, and all substances, which do not evaporate in the common temperature, and under the usual pressure of our atmosphere.

For similar reasons, names have not been given to the liquid or concrete states of most of the aëriform fluids: These were not known to arise from the combination of caloric with certain bases; and, as they had not been seen either in the liquid or solid states, their existence, under these forms, was even unknown to natural philosophers.

We have not pretended to make any alteration upon such terms as are sanctified by ancient custom; and, therefore, continue to use the words *water* and *ice* in their common acceptation: We likewise retain the word *air*, to express that collection of elastic fluids which composes our atmosphere; but we have not thought it necessary to preserve the same respect for modern terms, adopted by latter philosophers, having considered ourselves as at liberty to reject such as appeared liable to occasion erroneous ideas of the substances they are meant to express, and either to substitute new terms, or to employ the old ones, after modifying them in such a manner as to convey more determinate ideas. New words have been drawn, chiefly from the Greek language, in such a manner as to make their etymology convey some idea of what was meant to be represented; and these we have always endeavoured to make short, and of such a nature as to be changeable into adjectives and verbs.

Following these principles, we have, after Mr Macquer's example, retained the term *gas*, employed by Vanhelmont, having arranged the numerous class of elastic aëriform fluids under that name, excepting only atmospheric air. *Gas*, therefore, in our nomenclature, becomes a generic term, expressing the fullest degree of saturation in any body with caloric; being, in fact, a term expressive of a mode of existence. To distinguish each species of gas, we employ a second term from the name of the base, which, saturated with caloric, forms each particular gas. Thus, we name water combined to saturation with caloric, so as to form an elastic fluid, *aqueous gas*; ether, combined in the same manner, *etherial gas*; the combination of alkohol with caloric, becomes *alkoholic gas*; and, following the same principles, we have *muriatic acid gas*, *ammoniacal gas*, and so on of every substance susceptible of being combined with caloric, in such a manner as to assume the gaseous or elastic aëriform state.

[4] In English, the word *steam* is exclusively appropriated to water in the state of vapour.

We have already seen, that the atmospheric air is composed of two gasses, or aëriform fluids, one of which is capable, by respiration, of contributing to animal life, and in which metals are calcinable, and combustible bodies may burn; the other, on the contrary, is endowed with directly opposite qualities; it cannot be breathed by animals, neither will it admit of the combustion of inflammable bodies, nor of the calcination of metals. We have given to the base of the former, or respirable portion of the air, the name of *oxygen* . . . .

The chemical properties of the noxious portion of atmospheric air being hitherto but little known, we have been satisfied to derive the name of its base from its known quality of killing such animals as are forced to breathe it, giving it the name of *azote*, from the Greek privative particle α and ξαη, vita; hence the name of the noxious part of atmospheric air is *azotic gas*; the weight of which, in the same temperature, and under the same pressure, is 1 *oz.* 2 *gros.* and 48 *grs.* to the cubical foot, or 0.4444 of a grain to the cubical inch. We cannot deny that this name appears somewhat extraordinary; but this must be the case with all new terms, which cannot be expected to become familiar until they have been some time in use. We long endeavoured to find a more proper designation without success; it was at first proposed to call it *alkaligen gas*, as, from the experiments of Mr Berthollet, it appears to enter into the composition of ammoniac, or volatile alkali; but then, we have as yet no proof of its making one of the constituent elements of the other alkalies; beside, it is proved to compose a part of the nitric acid, which gives as good reason to have called it *nitrogen*. For these reasons, finding it necessary to reject any name upon systematic principles, we have considered that we run no risk of mistake in adopting the terms of *azote*, and *azotic gas*, which only express a matter of fact, or that property which it possesses, of depriving such animals as breathe it of their lives.

I should anticipate subjects more properly reserved for the subsequent chapters, were I in this place to enter upon the nomenclature of the several species of gasses: It is sufficient, in this part of the work, to establish the principles upon which their denominations are founded. The principal merit of the nomenclature we have adopted is, that, when once the simple elementary substance is distinguished by an appropriate term, the names of all its compounds derive readily, and necessarily, from this first denomination.

Chapter Seven

Of the Decomposition of Oxygen Gas by means of Metals, and the Formation of Metallic Oxides

Oxygen has a stronger affinity with metals heated to a certain degree than with caloric; in consequence of which, all metallic bodies, excepting gold, silver, and platina, have the property of decomposing oxygen gas, by attracting its base from the caloric with which it was combined. We have already shown in what manner this decomposition takes place, by means of mercury and iron; having observed, that, in the case of the first, it must be considered as a kind of gradual combustion, whilst, in the latter, the combustion is extremely rapid, and attended with a brilliant flame. The use of the heat employed in these operations is to separate the particles of the metal from each other, and to diminish their attraction of cohesion or aggregation, or, what is the same thing, their mutual attraction for each other.

The absolute weight of metallic substances is augmented in proportion to the quantity of oxygen they absorb; they, at the same time, lose their metallic splendour, and are reduced into an earthy pulverulent matter. In this state metals must not be considered as entirely saturated with oxygen, because their action upon this element is counterbalanced by the power of affinity between it and caloric. During the calcination of metals, the oxygen is therefore acted upon by two separate and opposite powers, that of its attraction for caloric, and that exerted by the metal, and only tends to unite with the latter in consequence of the excess of the latter over the former, which is, in general, very inconsiderable. Wherefore, when metallic substances are oxygenated in atmospheric air, or in oxygen gas, they are not converted into acids like sulphur, phosphorus, and charcoal, but are only changed into intermediate substances, which, though approaching to the nature of salts, have not acquired all the saline properties. The old chemists have affixed the name of *calx* not only to metals in this state, but to every body which has been long exposed to the action of fire without being melted. They have converted this word *calx* into a generical term, under which they confound calcareous earth, which, from a neutral salt, which it really was before calcination, has been changed by fire into an earthy alkali, by *losing* half of its weight, with metals which, by the same means, have joined themselves to a new substance, whose quantity often *exceeds* half their weight, and by which they have been changed almost into the nature of acids. This mode of classifying substances of so very opposite natures, under the same generic name, would have been quite contrary to our principles of nomenclature, especially as, by retaining the above term for this state of metallic substances, we must have conveyed very false ideas of its nature. We have, therefore, laid aside the expression *metallic calx* altogether, and have substituted in its place the term *oxide*, from the Greek word οξυς.

By this may be seen, that the language we have adopted is both copious and expressive. The first or lowest degree of oxygenation in bodies, converts them into *oxyides*; a second degree of additional oxygenation constitutes the class of acids, of which the specific names, drawn from their particular bases, terminate in *ous*, as the *nitrous* and *sulphurous* acids; the third degree of oxygenation changes these into the species of acids distinguished by the termination in *ic*, as the *nitric* and *sulphuric* acids; and, lastly, we can express a fourth, or highest degree of oxygenation, by adding the word *oxygenated* to the name of the acid, as has been already done with the *oxygenated muriatic* acid.

We have not confined the term *oxide* to expressing the combinations of metals with oxygen, but have extended it to signify that first degree of oxygenation in all bodies, which, without converting them into acids, causes them to approach to the nature of salts. Thus, we give the name of *oxide of sulphur* to that soft substance into which sulphur is converted by incipient combustion; and we call the yellow matter left by phosphorus, after combustion, by the name of *oxide of phosphorus*. In the same manner, nitrous gas, which is azote in its first degree of oxygenation, is the *oxide of azote*. We have likewise oxides in great numbers from the vegetable and animal kingdoms; and I shall show, in the sequel, that this new language throws great light upon all the operations of art and nature.

We have already observed, that almost all the metallic oxides have peculiar and permanent colours. These vary not only in the different species of metals, but even according to the various degrees of oxygenation in the same metal. Hence we are under the necessity of adding two epithets to each oxide, one of which indicates the metal *oxidated*, while the other indicates the peculiar colour of the oxide. Thus, we have the black oxide of iron, the red oxide of iron, and the yellow oxide of iron; which expressions respectively answer to the old unmeaning terms of martial ethiops, colcothar, and rust of iron, or ochre. We have likewise the gray, yellow, and red oxides of lead, which answer to the equally false or insignificant terms, ashes of lead, massicot, and minium.

These denominations sometimes become rather long, especially when we mean to indicate whether the metal has been oxidated in the air, by detonation with nitre, or by means of acids; but then they always convey just and accurate ideas of the corresponding object which we wish to express by their use. All this will be rendered perfectly clear and distinct by means of the tables which are added to this work.

TABLE OF SIMPLE SUBSTANCES.

Simple substances belonging to all the kingdoms of nature, which may be considered as the elements of bodies.

New Names.	Correspondent Old Names.
Light	Light.
Caloric	Heat.
	Principle or element of heat.
	Fire. Igneous fluid.
	Matter of fire and of heat.
Oxygen	Depholgisticated air.
	Empyreal air.
	Vital air, or
	Base of vital air.
Azote	Phlogisticated air or gas.
	Mephitis, or its base.
Hydrogen	Inflammable air or gas,
	or the base of inflammable air.

Oxideable and Acidifiable simple Substances not Metallic.

New Names.	Correspondent Old Names.
Sulphur	The same names.
Phosphorus	
Charcoal	
Muriatic radical	Still unknown.
Fluoric radical	
Boracic radical	

Oxideable and Acidifiable simple Metallic Bodies.

New Names.	Correspondent Old Names.	
Antimony	Regulus of	Antimony.
Arsenic	" "	Arsenic
Bismuth	" "	Bismuth
Cobalt	" "	Cobalt
Copper	" "	Copper
Gold	" "	Gold
Iron	" "	Iron
Lead	" "	Lead
Manganese	" "	Manganese
Mercury	" "	Mercury
Molybdena	" "	Molybdena
Nickel	" "	Nickel
Platina	" "	Platina
Silver	" "	Silver
Tin	" "	Tin
Tungstein	" "	Tungstein
Zinc	" "	Zinc

Salifiable simple Earthy Substances

New Names.	Correspondent Old Names.
Lime	Chalk, calcareous earth.
	Quicklime.
Magnesia	Magnesia, base of Epsom salt.
	Calcined or caustic magnesia.
Barytes	Barytes, or heavy earth.
Argill	Clay, earth of alum.
Silex	Siliceous or vitrifiable earth.

Translated by Robert Kerr

Reading and Discussion Questions

1. What does Lavoisier describe as the main goals of this project in the preface? What failures of the past is he trying to avoid? What steps does he take in his attempt to impose order and method in the field of chemistry?
2. Pay close attention to Lavoisier's list of elements. How does his list compare to some of the other theories of matter that we have encountered? Which elements look familiar to the modern reader and which are surprising? To what degree does he succeed in his attempt to avoid nomenclature which expresses false theory?
3. What is caloric? What phenomenon does Lavoisier use it to explain?
4. How does Lavoisier understand the phenomenon of combustion? How does his view differ from Stahl and Priestley's phlogiston theory? What evidence does he give to support this new theory?

Memoir on Heat (1780)

Antoine-Laurent Lavoisier and Pierre-Simon Laplace

68

THIS MEMOIR IS the result of experiments upon heat which M. de Laplace and I made together during the last winter; the mildness of the season did not permit us to perform a greater number .... We have devised the following apparatus [for making measurements of the heat which is developed by combustion, the respiration of animals, combinations of oil of vitriol with water, and the like], all of which have been impossible by the means hitherto known.

The plate represents a vertical section showing the interior of the device. Its volume is divided into three chambers: we shall distinguish them by the terms *inner chamber, middle chamber,* and *outer chamber.*

The inner chamber *f* is constructed of a meshwork of iron wire reinforced by strips of the same metal; the experimental object is placed in this chamber; the top of the chamber is fitted with a cover which is entirely open above; its bottom consists of an iron wire netting.

The middle chamber is intended to contain ice which entirely surrounds the inner chamber and which is melted by the heat of the experimental object; this ice is supported and retained by a grill under which is a sieve. In proportion as the ice is melted by the heat of the object in the innermost vessel, the water runs down through the grill and the sieve; it then runs down the cone and the tube and is collected in the vessel placed under the apparatus; the stopcock permits one to stop the outflow of the water at will. Finally, the outer chamber is designed to be filled with ice, the purpose of which is to prevent the entrance of heat from the external air or surrounding objects; the water produced by the melting of this ice runs down pipe *s-t*, which can be opened or closed by means of stopcock *r.* The entire apparatus is covered by a lid entirely open above and closed beneath; it is constructed of tin painted with oil to prevent rust.

In performing an experiment, one fills the middle chamber and the inner cover with crushed ice, as well as the outer chamber and the outside lid. One must be careful to crush the ice fine and to pack it down well into the apparatus.

The inside ice (as we shall call the ice enclosed in the middle chamber and inner lid) is allowed to drain; when it has drained sufficiently, the device is opened, the desired object is placed inside, and the lid is replaced immediately. One waits until the object is completely cooled and the apparatus sufficiently drained; then one weighs the water which has collected in the vessel. Its weight exactly measures the heat emitted by the object; for clearly all its heat has been absorbed by the inner ice which has been protected from the effects of any other heat by the ice contained in the lid and in the outer container.

It is essential that there be no communication between the middle chamber and the outer chamber, which can be easily tested by filling the outer chamber with water. If there were communication between these two chambers, the ice melted by the atmosphere, the heat of which affects the wall of the outer chamber, might pass into the middle chamber, and then the water flowing out from the latter would no longer be a measure of the heat lost by the experimental object.

When the atmospheric temperature is above zero, heat enters the middle cavity only with difficulty because it is stopped by the ice in the lid and the outer chamber; but if the external temperature is below zero, the atmosphere may cool the inner ice; it is therefore essential to work at temperatures above zero; in cold weather the apparatus must be kept in a heated room; furthermore, the ice used must not be colder than zero degrees; if it is, it must be crushed and spread out in thin layers for a time in a place where the temperature is above zero.

The inner ice always retains a small amount of water which adheres to its surface, and one might think that this water would affect the experimental results; but it must be pointed out that at the beginning of each experiment the ice is already saturated with all the water it can thus retain, so that though a small part of the melted ice remains adhering to the inner mass of ice, a very nearly equal quantity of water, originally adhering to the surface of the ice, must run down into the collecting vessel, since the surface of the inner ice changes very little during the course of the experiment . . . .

We have had two machines constructed as described; one of them is intended for experiments in which it is not necessary to change the air within the chamber; the other apparatus is designed for experiments in which the air must be renewed, such as those involving combustion and respiration; the latter apparatus differs from the former only in that the two covers are pierced by two holes through which pass two thin pipes which serve for the passage of air between the outside and the inside; by this means it is possible to blow atmospheric air upon combustible objects . . . .

Experiments on Heat, Carried out by This Method

We took a small earthen vessel which had been dried; after having placed it on a balance and tared it very exactly, we placed glowing coals in it, blowing upon them to keep them at red heat; at the instant when their weight was one ounce [30.59 grams] we transferred them quickly to one of our machines; their combustion, in the interior of the apparatus, was maintained by means of a bellows; they were consumed in 32 minutes. At the beginning of the experiment the outside thermometer stood at 1.5° and it rose to 2.5° during the experiment; the apparatus when well-drained yielded 6 pounds, 2 ounces [2,998 grams] of melted ice; this was produced by the combustion of one ounce of carbon.

The outside thermometer being at 1.5°, we placed a guinea pig into one of our machines; its internal body temperature was about 32° [40° C.], i.e., not very different from that of the human body. To prevent its suffering during the experiment, we placed the animal in a little basket lined with cotton, the temperature of which was zero; the animal remained for 5 hours and 36 minutes in the apparatus; during this period we gave it four or five changes of air by means of a bellows. After removing the animal, we left the basket in the apparatus and waited until it had cooled off; the well-drained machine yielded about 7 ounces [214 grams] of melted ice. In a second experiment, the outside thermometer was still at 1.5°; the same guinea pig remained in the apparatus for 10 hours and 36 minutes, the air being renewed only three times; the machine yielded 14 ounces, 5 gros [447.38 grams]. The animal did not appear to suffer at all during these experiments.

According to the first experiment, the amount of ice which the animal could melt in 10 hours would be 12 ounces, 4 gros [382.38 grams]; according to the second experiment this quantity, for the same interval, would be 13 ounces, 6 gros, 27 grains [422.05 grams]; the average of these two results is 13 ounces, 1 gros, 13.5 grains [402.21 grams].

Combustion and Respiration

Until recently, only vague and imperfect ideas were current regarding the phenomena of the heat liberated in combustion and respiration. Experience had shown that bodies could not burn, nor could animals respire, in the absence of atmospheric air; but nothing was known of the manner in which it influences these two important natural processes and the resulting changes which the air undergoes. The most widespread opinion attributed to the air only the functions of cooling the blood as it passed through the lungs, and of holding the fire against a combustible object by its pressure. The important discoveries which have been made during the last few years on the nature of aerial fluids have greatly extended our knowledge of this subject; it is established that a single kind of air, known as *dephlogisticated air, pure air,* or *vital air*, is concerned in combustion, respiration, and the calcination of metals; this type of air comprises only about a quarter of the atmospheric air, and it is either absorbed, or altered, or converted into fixed air by the addition of a principle which we shall name the *base of fixed air*, in order to avoid any discussion as to its nature; thus, the air does not act simply as a mechanical force, but as an agency of new combinations. M. Lavoisier, having observed these phenomena, suspected that the heat and light liberated in combustion are due, at least to a great extent, to changes which the pure air undergoes. The facts pertaining to combustion and respiration are explained in such a natural and simple manner on this hypothesis, that he did not hesitate to propose it, if not as a demonstrated truth, at least as a very reasonable conjecture, worthy of the attention of natural philosophers. . . .

We have confined ourselves here to a comparison of the quantities of heat which are liberated in combustion and in respiration with the corresponding alterations of the pure air, without going into the question whether this heat comes from the air, or from the combustible substances and respiring animals. With the object of studying these alterations, we have performed the following experiments:

Upon a large trough filled with mercury we set a bell-jar full of dephlogisticated air [oxygen]; this air was not perfectly pure; it contained 16 parts of pure

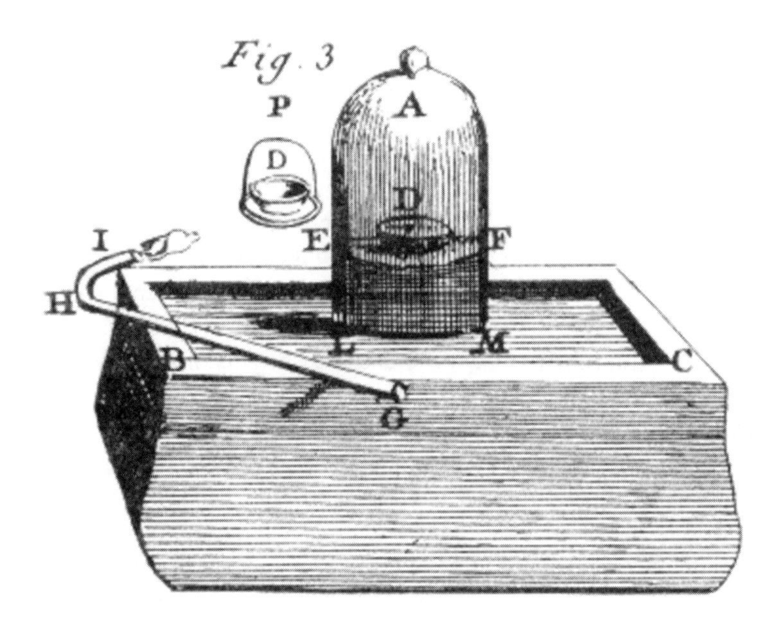

air [oxygen] per 19 parts and it included about 1/57th of its volume of fixed air [carbon dioxide].

[The bell jar image above is from a similar experiment, and is helpful in understanding what follows. Note however, that its letter markings do not correspond to the description below—Ed.]

We introduced under the bell-jar a small earthen jar, filled with coal which had previously been freed of its inflammable air by strong heat; upon the coal we placed a little tinder upon which was a small fragment of phosphorus, weighing at most a tenth of a grain [5 milligrams]. The earthen jar and all its contents had been weighed very exactly; we then raised the mercury within the bell-jar up to a marked level (E) by suction applied to the interior, in order that the expansion of the air produced by the burning of the carbon would not lower the level of the mercury too much below that of the outside mercury, which would have permitted the escape of air from within the bell-jar. Next, by means of a red-hot iron, passed very quickly through the mercury, we ignited the phosphorus, which set fire to the tinder and thus to the coal. Combustion lasted for 20 or 25 minutes, and when the ember was extinguished, and the inside air had cooled to room temperature, we marked a second line at the level (E') where the mercury had risen by diminution of the volume of the enclosed air. We then introduced some caustic alkali under the bell-jar; all the fixed air was absorbed, and having allowed sufficient time for this to occur, when the mercury had ceased to rise in the bell-jar, we marked a third line (E") at the level of the surface of the caustic alkali; we took care to observe, at the three positions E, E', and E", the heights of the mercury in the bell-jar above its level in the trough. Atmospheric air introduced into the bell-jar by means of a glass tube had the effect of lowering the mercury level to that of the outside. We then removed the earthen vessel, which we dried and weighed very exactly; the loss of weight gave us the quantity of carbon consumed. The external temperature varied very little during the course of the experiment, and the barometric pressure was about 28 inches.

In order to determine the volumes of air contained [at levels E, E', and E"], we filled them with plain water, the respective weights of which gave the volumes of these spaces in cubic inches. But, since the enclosed air had been unequally compressed as a result of the different heights of the mercury in the bell-jar, we reduced the volumes, by computation from the observed heights of the mercury, to that which the air would have occupied if it had been compressed by a 28-inch column of mercury. Finally, we reduced all our experimental results to the values which would have been obtained had the external temperature been 10°, utilizing the fact that, at a temperature of 10°, air expands 1/215th for each degree of temperature increase; therefore, the volumes of air which we shall report must be taken as the values for a temperature of 10° and a pressure of 28 inches of mercury.

In the preceding experiment, the bell-jar had contained 202.35 inches [4,026.8 cm^3] of dephlogisticated air; its volume, by the sole combustion of carbon, was reduced to 170.59 inches [3,394.7 cm^3]. After absorption of the fixed air by the caustic alkali, the volume of the remaining air was only 73·93 inches [1471.2 cm^3]; the weight of the carbon consumed, apart from its ash, was 17.2 grains 1 [0.912 grams]; the weights of the tinder and the phosphorus together might have been half a grain [26 milligrams]; moreover, we have found, through many experiments, that the weight of ash formed by the coal is approximately 10 grains per ounce [17.4%]; it might therefore be estimated very nearly that 18 grains [0.954 grams] of carbon were consumed in the experiment, taking into account its ash.

The dephlogisticated air which we used contained about 1/57th of its volume of fixed air which had not been absorbed by the water over which it had been stored for several months; this intimate adhesion of fixed air to the pure air has led us to believe that, even after the fixed air was absorbed by caustic alkali in our experiments, the remaining air still contained a little fixed air, which we may without appreciable error estimate at 1/57th of its total volume. According to this hypothesis, in order to obtain the volume of all the pure air consumed by the carbon, one must take the difference between the volume of the air before combustion and the volume of the air remaining after absorption with caustic alkali, and then subtract 1/57th. Making a similar correction for the volume of air absorbed by the alkali, one may obtain the volume of fixed air formed in combustion; it will thus be found that one ounce of carbon, in burning, consumes 4037.5 inches of pure air and forms 3021.1 inches of fixed air. If one designates the volume of pure air consumed as unity, its volume, after combustion, would be reduced by 0.74828.

In order to estimate the weight of these volumes of pure air and fixed air, the weight of a cubic inch of each of these airs must be known; now, it has been observed that pure air is a little heavier than atmospheric air, approximately by a ratio of 187 to 185. The weight of atmospheric air has been determined very exactly by M. de Luc. Utilizing these determinations, it is found that at a temperature of 10° and a barometric pressure of 28 inches, a cubic inch of dephlogisticated air weighs 0.47317 grains. M. Lavoisier has observed that at the same temperature and pressure, a cubic inch of fixed air weighs very nearly 0.7 of a grain. According to these results, an ounce of carbon, in burning, consumes 3.3167 ounces of pure air and forms 3.6715 ounces of fixed air. Thus, in ten parts by volume of fixed air, there are about nine parts of pure air and one part of a principle supplied by the carbon, which is the base of fixed air; but a determination of such delicacy requires a greater number of experiments.

We have previously seen that an ounce of carbon, in burning, melts 6 pounds, 2 ounces of ice, from which it may be readily concluded that, in the combustion of carbon, the alteration of an ounce of pure air is capable of melting 29.547 ounces of ice, and that the production of one ounce of fixed air is capable of melting 26.692 ounces.

It is with the greatest circumspection that we present these results on the quantities of heat liberated by the alteration of an ounce of pure air by the combustion of carbon. We have performed only one experiment on the heat liberated by this combustion, and although it was carried out under quite favorable conditions, nevertheless we shall not be quite confident of its exactness until we have repeated it a number of times. As we have said before, and we cannot stress this too much, it is not so much the result of our experiments, as the method we have employed that we present to the natural philosophers, inviting them, if the method appears to offer advantages, to confirm these experiments, which we intend to repeat ourselves with the greatest care. . . .

In order to determine the alterations which the respiration of animals brings about in pure air, we filled the bell-jar of the apparatus previously described with this gas, and we introduced into it various guinea-pigs of nearly the same weight as the one used in our experiment on animal heat. In one of these experiments, the bell-jar contained 248.01 inches of pure air before the guinea pig was put in; the animal was kept there for an hour and a quarter. In order to introduce it into the bell-jar, we passed it through the mercury; it was removed in the same manner. After the inside air had been allowed to cool to room temperature, its volume was slightly diminished to 240.25 inches; finally after the fixed air had been absorbed by caustic alkali, 200.56 inches of air remained. In this experiment, 46.62 inches of pure air had been altered, and 37.96 inches of fixed air produced, correcting for the small amount of fixed air contained by the dephlogisticated air in the bell-jar. If the volume of altered pure air be designated as unity, the reduction in volume due to respiration would be 0.814; in the combustion of carbon the volume of air was diminished by a ratio of 1 to 0.74828; this difference may be ascribed in part to errors of measurement, but it also results from a cause which we had not at first suspected and which those who wish to repeat these experiments might well be warned against.

In order to keep the bell-jar stable in the trough, we raised the level of the mercury inside slightly above the outside level; now, in introducing the animal and in removing it from the bell-jar, we observed that a small amount of air was carried in along the body of the animal, although it was partly immersed in the mercury; the mercury does not adhere closely enough to the hair and skin to prevent all communication between the outside air and the air under the bell-jar; thus the air appears to be less reduced by respiration than is actually the case.

The weight of the fixed air produced in the previous experiment is 26.572 grains; from which it follows that in an interval of ten hours the animal would have produced 212.576 grains of fixed air.

At the beginning of the experiment, the animal, breathing an air much purer than atmospheric air, might in a given time produce a larger quantity of fixed air; but at the end it breathes with difficulty, because the fixed air, accumulating by its weight at the bottom of the bell-jar where the animal is located, displaces the pure air which rises to the top of the bell-jar, and probably also because the fixed air is itself noxious to animals. It may therefore be assumed, with no appreciable error, that the amount of fixed air produced is the same as if the animal had been breathing atmospheric air, the quality of which is about the average between that of the air in the bottom of the bell-jar at the beginning and at the end of the experiment.

We then determined directly the amount of fixed air produced by a guinea-pig breathing air the same as the atmosphere. For this purpose, we placed one in a jar through which we had set up a current of atmospheric air; the air, compressed in a suitable apparatus, entered the vessel through a glass tube, and emerged through a second curved tube, the concave part of which was immersed in mercury, and the lower end of which terminated in a second flask filled with caustic alkali. The air was then led by a third tube into a second flask full of caustic alkali, and thence out into the atmosphere. The fixed air formed by the animal in the jar was in large part retained by the caustic alkali of the first flask; whatever escaped was absorbed by the alkali of the second flask; the increase in weight of the flasks gave us the weight of the fixed air there combined. During a three-hour interval, the weight of the first flask increased 63 grains; that of the second increased 8 grains; thus the total weight of the two flasks increased by 71 grains. Assuming that this quantity of fixed air is due solely to the respiration of the animal, it would, in ten hours, have formed 236.667 grains of fixed air, which differs by about one-ninth from the results obtained in the previous experiment. This difference could be attributed to the difference in size and strength of the two animals and to their momentary state during the experiment.

If the vapors [expired water] of respiration, carried by the air current, had been deposited in the flasks, the increase in weight of the caustic alkali would not have given the amount of fixed air produced by the animal; it was to avoid this inconvenience that we used a curved tube with its concave part immersed in mercury; the vapors of respiration condensed against the walls of this part of the tube and collected in its concavity, with the result that the air entering the first flask contained no appreciable amount of moisture, as shown by the fact that the part of the tube entering the flask remained transparent; it may therefore be assumed that though the weight of the flasks might have been augmented by these vapors, this increase would have been compensated for by the evaporation of water from the alkali. It might still be feared that a part of the fixed air combined came from the atmospheric air itself. To reassure ourselves in this regard we repeated the same experiment, but without placing a guinea-pig in the vessel; no increase then occurred in the weight of the flasks; that of the second flask diminished by 4 or 5 grains, no doubt owing to evaporation of water from its alkali.

A third experiment performed on a guinea-pig in dephlogisticated air gave 226 grains as the amount of fixed air produced in ten hours.

Taking an average of these experiments and several similar ones performed with· a number of guinea-pigs, both in dephlogisticated air and in atmospheric air, we have obtained an estimate of 224 grains as the amount of fixed air produced in ten hours by the guinea-pig on which we had experimented in our apparatus to determine its animal heat.

Inasmuch as these experiments were carried out at a temperature of 14°–15°, it is possible that the amount of fixed air produced by respiration is a little less than at a temperature of zero degrees, which is that of the interior of our apparatus; for greater precision it would therefore be necessary to determine the production of fixed air at the latter temperature; we shall take up this question in further experiments which we intend to carry out.

The foregoing experiments are contrary to those which MM. Scheele and Priestley have reported on the alterations of pure air by the respiration of animals. Respiration, according to these two excellent naturalists, produces very little fixed air and a large amount of vitiated air, which the latter has designated as *phlogisticated,* but upon investigating the effect of respiration of birds and guinea-pigs on pure air with the greatest possible care, by a great number of experiments, we have constantly observed that the transformation of this air into fixed air is the main alteration produced by the respiration of animals. By having guinea-pigs breathe a large amount of pure air and observing by means of caustic alkali the amount of fixed air produced by their respiration, and by subsequently causing the residual air to be breathed by birds and absorbing the newly-formed fixed air once more by caustic alkali, we have been able to convert into fixed air a large part of the pure air which we have been using; the remaining air had nearly the same quality which it would have had assuming that the transformation of pure air into fixed air is the only effect of respiration upon the air. It therefore seems certain to us that if respiration produces other alterations in pure air, they are inconsiderable, and we have no doubt that any naturalist performing the same experiments with a large mercury apparatus will be led to the same conclusion. It was previously seen that in the combustion of carbon, the formation of an ounce of fixed air can melt 26.692 ounces of ice; on the basis of this result, it is found that the formation of 224 grains of fixed air must melt 10.38 ounces. This amount of melted ice consequently represents the heat produced by the respiration of a guinea-pig during ten hours.

In the experiment on animal heat of a guinea-pig, this animal emerged from our apparatus with nearly the same heat with which it entered, for it is known that the internal heat of animals is always nearly constant. Without the constant renewal of its heat, all the heat which it had at first would have been gradually dissipated, and we should have found it cold upon taking it out of the apparatus like all the inanimate objects which we have used in our experiments. But the animal's vital functions continually restore to it the heat which it gives off to its environment, and which in our experiment is diffused into the inner ice, of which it melted 13 ounces in ten hours. This amount of melted ice thus represents approximately the amount of heat renewed during this time interval by the vital functions of the guinea-pig. Perhaps an ounce or two should be subtracted, or maybe more, on account of the fact that the extremities of the body of the animal were chilled in the apparatus, although the interior of the body retained nearly the same temperature; furthermore, the moisture which its internal heat had evaporated melted a small amount of ice as it cooled, adding to the water draining out of the apparatus.

On subtracting about 2.5 ounces from this quantity of ice, one obtains the amount melted by the effect of the respiration of the animal upon the air. Now, if one considers the inevitable errors in these experiments and in the factors which were the starting point for our calculations, it will be seen that it is not possible to hope for a more perfect agreement between these results. Thus the heat which is liberated in the transformation of pure to fixed air by respiration may be regarded as the principal cause of the conservation of animal heat and if other causes are involved, they are of lesser significance.

Respiration is therefore a combustion, very slow to be sure, but perfectly similar to that of carbon. It occurs in the interior of the lungs, without the liberation of any perceptible light because the fire, as fast as it is freed, is absorbed by the humidity of these organs. The heat developed by this combustion is transferred to the blood which passes through the lungs, and thence is transmitted throughout the animal system. Thus the air which we breathe serves two purposes equally necessary for our preservation: it removes from the blood the base of fixed air, an excess of which would be most injurious; and the heat which this combination releases in the lungs replaces the constant loss of heat into the atmosphere and surrounding bodies to which we are subject.

Translated by M. L. Gabriel

Reading and Discussion Questions

1. What instruments and methods do Lavoisier and Laplace develop or employ for the purposes of this experiment? How does this setup and approach compare to what you might encounter in a present day chemistry lab?

2. Describe the series of experiments that Lavoisier and Laplace report in this reading. What are they trying to accomplish in each case?
3. Despite the fact that the observed amount of ice melted differs from the predicted value, Lavoisier and Laplace attribute this discrepancy to various sources of experimental error and interpret the result as evidence for their hypothesis. Does this seem plausible, or should they have taken the results to falsify their theory? What insights might this episode give us into the nature of scientific reasoning and methodology?

A New System of Chemical Philosophy (1808)

John Dalton

69

Chapter 2: On the Constitution of Bodies.

There are three distinctions in the kinds of bodies, or three states, which have more especially claimed the attention of philosophical chemists; namely, those which are marked by the terms *elastic fluids, liquids, and solids*. A very famous instance is exhibited to us in water, of a body, which, in certain circumstances, is capable of assuming all the three states. In steam we recognise a perfectly elastic fluid, in water a perfect liquid, and in ice a complete solid. These observations have tacitly led to the conclusion which seems universally adopted, that all bodies of sensible magnitude, whether liquid or solid, are constituted of a vast number of extremely small particles, or atoms of matter bound together by a force of attraction, which is more or less powerful according to circumstances, and which as it endeavours to prevent their separation, is very properly called in that view, *attraction of cohesion*; but as it collects them from a dispersed state (as from steam into water) it is called, *attraction of aggregation*, or more simply *affinity*. Whatever names it may go by, they still signify one and the same power. It is not my design to call in question this conclusion, which appears completely satisfactory; but to shew that we have hitherto made no use of it, and that the consequence of the neglect, has been a very obscure view of chemical agency, which is daily growing more so in proportion to the new lights attempted to be thrown upon it.

The opinions I more particularly allude to, are those of Berthollet on the Laws of chemical affinity; such as that chemical affinity is proportional to the mass, and that in all chemical unions, there exist insensible gradations in the proportions of the constituent principles. The inconsistence of these opinions, both with reason and observation, cannot, I think, fail to strike every one who takes a proper view of the phenomena.

Whether the ultimate particles of a body, such as water, are all alike, that is, of the same figure, weight, etc. is a question of some importance. From what is known, we have no reason to apprehend a diversity in the particulars: if it does exist in water, it must equally exist in the elements constituting water, namely, hydrogen and oxygen. Now it is scarcely possible to conceive how the aggregates of dissimilar particles should be so uniformly the same. If some of the particles of water were heavier than others, if a parcel of the liquid on any occasion were constituted principally of these heavier particles, it must be supposed to affect the specific gravity of the mass, a circumstance not known. Similar observations may be made on other substances. Therefore we may conclude that *the ultimate particles of all homogeneous bodies are perfectly alike in weight, figure, etc.* In other words, every particle of water is like every other particle of water; every particle of hydrogen is like every other particle of hydrogen, etc.

Besides the force of attraction, which, in one character or another, belongs universally to ponderable bodies, we find another force that is likewise universal, or acts upon all matter which comes under our cognisance, namely, a force of repulsion. This is now generally, and I think properly, ascribed to the agency of heat. An atmosphere of this subtle fluid constantly surrounds the atoms of all bodies, and prevents them from being drawn into actual contact. This appears to be satisfactorily proved by the observation, that the bulk of a body may be diminished by abstracting some of its heat: But from what has been stated in the last section, it should seem that enlargement and diminution of bulk depend perhaps more on the arrangement, than on the size of the ultimate particles. Be this as it may, we cannot avoid inferring from the preceding doctrine on heat, and particularly from the section on the natural zero of temperature, that solid bodies, such as ice, contain a large portion, perhaps 4/5 of the heat which the same are found to contain in an elastic state, as steam.

We are now to consider how these two great antagonist powers of attraction and repulsion are adjusted, so as to allow of the three different states of *elastic fluids, liquids, and solids*. We shall divide the subject into four Sections; namely, first, *on the constitution of pure elastic fluids*; second, *on the constitution of mixed elastic fluids*; third, *on the constitution of liquids*, and fourth, *on the constitution of solids*.

Chapter 3: On Chemical Synthesis.

When any body exists in the elastic state, its ultimate particles are separated from each other to a much greater distance than in any other state; each particle occupies the centre of a comparatively large sphere, and supports its dignity by keeping all the rest, which by their gravity, or otherwise are disposed to encroach upon it, at a respectful distance. When we attempt to conceive the *number* of particles in an atmosphere, it is somewhat like attempting to conceive the number of stars in the universe; we are confounded with the thought. But if we limit the subject, by taking a given volume of any gas, we seem persuaded that, let the divisions be ever so minute, the number of particles must be finite; just as in a given space of the universe, the number of stars and planets cannot be infinite.

Chemical analysis and synthesis go no farther than to the separation of particles one from another, and to their reunion. No new creation or destruction of matter is within the reach of chemical agency. We might as well attempt to introduce a new planet into the solar system, or to annihilate one already in existence, as to create or destroy a particle of hydrogen. All the changes we can produce, consist in separating particles that are in a state of cohesion or combination, and joining those that were previously at a distance.

In all chemical investigations, it has justly been considered an important object to ascertain the relative *weights* of the simples which constitute a compound. But unfortunately the enquiry has terminated here; whereas from the relative weights in the mass, the relative weights of the ultimate particles or atoms of the bodies might have been inferred, from which their number and weight in various other compounds would appear, in order to assist and to guide future investigations, and to correct their results. Now it is one great object of this work, to shew the importance and advantage of ascertaining *the relative weights of the ultimate particles, both of simple and compound bodies, the number of simple elementary particles which constitute one compound particle, and the number of less compound particles which enter into the formation of one more compound particle.*

If there are two bodies, A and B, which are disposed to combine, the following is the order in which the combinations may take place, beginning with the most simple: namely,

1 atom of A + 1 atom of B = 1 atom of C, binary.
1 atom of A + 2 atoms of B = 1 atom of D, ternary.
2 atoms of A + 1 atom of B = 1 atom of E, ternary.
1 atom of A + 3 atoms of B = 1 atom of F, quarternary.
3 atoms of A + 1 atom of B = 1 atom of G, quarternary.
etc.

The following general rules may be adopted as guides in all our investigations respecting chemical synthesis:

When only one combination of two bodies can be obtained, it must be presumed to be a binary one, unless some other cause appear to the contrary.

When two combinations are observed, they must be presumed to be a *binary* and a *ternary*.

When three combinations are observed, they must be presumed to be a *binary*, and the other two *ternary*.

When four combinations are observed, we should expect one *binary*, two *ternary*, and one *quarternary*, etc.
A *binary* compound should always be specifically heavier than the mere mixture of its two ingredients.
A *ternary* compound should be specifically heavier than the mixture of a binary and a simple, which would, if combined, constitute it; etc.

The above rules and observations equally apply, when two bodies, such as C and D, D and E, etc., are combined.

From the application of these rules, to the chemical facts already well ascertained, we deduce the following conclusions; 1st. That water is a binary compound of hydrogen and oxygen, and the relative weights of the two elementary atoms are as 1:7, nearly; 2d. That ammonia is a binary compound of hydrogen and azote, and the relative weights of the two atoms are as 1:5, nearly; 3d. That nitrous gas is a binary compound of azote and oxygen, the atoms of which weigh 5 and 7 respectively; that nitric acid is a binary or ternary compound according as it is derived, and consists of one atom of azote and two of oxygen, together weighing 19; that nitrous oxide is a compound similar to nitric acid, and consists of one atom of oxygen and two of azote, weighing 17; that nitrous acid is a binary compound of nitric acid and nitrous gas, weighing 31; that oxynitric acid is a binary compound of nitric acid with oxygen, weighing 26; 4th. That carbonic oxide is a binary compound, consisting of one atom of charcoal, and one of oxygen, together weighing nearly 12; that carbonic acid is a ternary compound, (but sometimes binary) consisting of one atom of charcoal, and two of oxygen, weighing 19; etc. In all these cases the weights are expressed in atoms of hydrogen, each of which is denoted by unity.

In the sequel, the facts and experiments from which these conclusions are derived, will be detailed; as well as a great variety of others from which are inferred the constitution and weight of the ultimate particles of the principal acids, the alkalis, the earths, the metals, the metallic oxides and sulphurets, the long train of neutral salts, and in short, all the chemical compounds which have hitherto obtained a tolerably good analysis. Several of the conclusions will be supported by original experiments.

From the novelty as well as importance of the ideas suggested in this chapter, it is deemed expedient to give plates, exhibiting the mode of combination in some of the more simple cases. A specimen of these accompanies this first part. The elements or atoms of such bodies as are conceived at present to be simple, are denoted by a small circle, with some distinctive mark; and the combinations consist in the juxta-position of two or more of these; when three or more particles of elastic fluids are combined together in one, it is supposed that the particles of the same kind repel each other, and therefore take their stations accordingly.

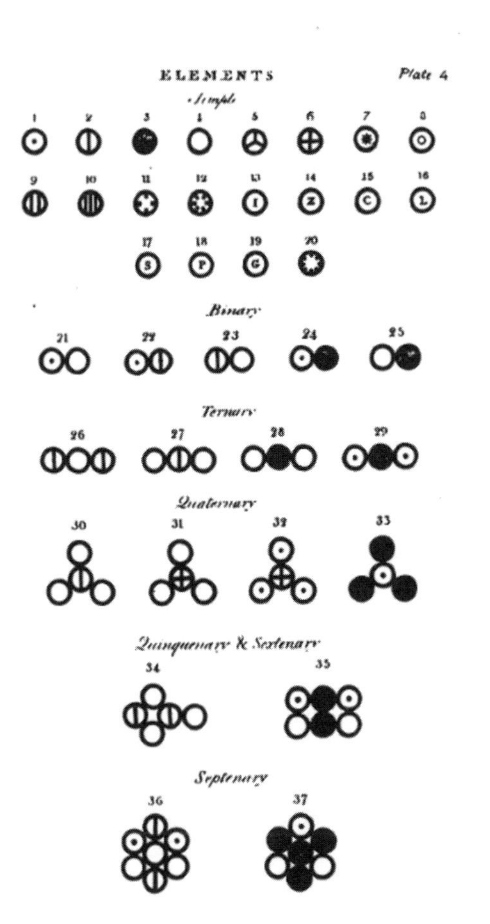

PLATE IV. This plate contains the arbitrary marks or signs chosen to represent the several chemical elements or ultimate particles.

Fig.		Fig.	
1 Hydrog. its rel. weight	1	11 Strontites - - -	46
2 Azote, - - - - -	5	12 Barytes - - - -	68
3 Carbone or charcoal, -	5	13 Iron - - - - -	38
4 Oxygen, - - - -	7	14 Zinc - - - - -	56
5 Phosphorus, - - -	9	15 Copper - - - -	56
6 Sulphur, - - - -	13	16 Lead - - - - -	95
7 Magnesia, - - - -	20	17 Silver - - - - -	100
8 Lime, - - - - -	23	18 Platina - - - -	100
9 Soda, - - - - -	28	19 Gold - - - - -	140
10 Potash, - - - -	42	20 Mercury - - - -	167

21. An atom of water or steam, composed of 1 of oxygen and 1 of hydrogen, retained in physical contact by a strong affinity, and supposed to be surrounded by a common atmosphere of heat; its relative weight = - - - - - 8
22. An atom of ammonia, composed of 1 of azote and 1 of hydrogen - - - - - - - - - - 6
23. An atom of nitrous gas, composed of 1 of azote and 1 of oxygen - - - - - - - - - 12
24. An atom of olefiant gas, composed of 1 of carbone and 1 of hydrogen - - - - - - - - - 6
25 An atom of carbonic oxide composed of 1 of carbone and 1 of oxygen - - - - - - - - 12
26. An atom of nitrous oxide, 2 azote + 1 oxygen - 17
27. An atom of nitric acid, 1 azote + 2 oxygen - - 19
28. An atom of carbonic acid, 1 carbone + 2 oxygen 19
29. An atom of carburetted hydrogen, 1 carbone + 2 hydrogen - - - - - - - - - - - - - 7
30. An atom of oxynitric acid, 1 azote + 3 oxygen 26
31. An atom of sulphuric acid, 1 sulphur + 3 oxygen 34
32. An atom of sulphuretted hydrogen, 1 sulphur + 3 hydrogen - - - - - - - - - - - - 16
33. An atom of alcohol, 3 carbone + 1 hydrogen - 16
34. An atom of nitrous acid, 1 nitric acid + 1 nitrous gas - - - - - - - - - - - - - - - 31
35. An atom of acetous acid, 2 carbone + 2 water - 26
36. An atom of nitrate of ammonia, 1 nitric acid + 1 ammonia + 1 water - - - - - - - - 33
37. An atom of sugar, 1 alcohol + 1 carbonic acid - 35

Reading and Discussion Questions

1. What sets Dalton's atomic theory apart from earlier Greek atomist and seventeenth century corpuscularian ideas?
2. What ideas and methods has Dalton adopted from Lavoisier's *Elements of Chemistry*?
3. What measurements is Dalton paying most close attention to?

Memoir on the Combination of Gaseous Substances with Each Other (1809)

Joseph Louis Gay-Lussac

70

Substances, whether in the solid, liquid, or gaseous state, possess properties which are independent of the force of cohesion; but they also possess others which appear to be modified by this force (so variable in its intensity), and which no longer follow any regular law. The same pressure applied to all solid or liquid substances would produce a diminution of volume differing in each case, while it would be equal for all elastic fluids. Similarly, heat expands all substances; but the dilations of liquids and solids have hitherto presented no regularity, and it is only those of elastic fluids which are equal and independent of the nature of each gas. The attraction of the molecules in solids and liquids is, therefore, the cause which modifies their special properties; and it appears that it is only when the attraction is entirely destroyed, as in gases, that bodies under similar conditions obey simple and regular laws. At least, it is my intention to make known some new properties in gases, the effects of which are regular, by showing that these substances combine amongst themselves in very simple proportions, and that the contraction of volume which they experience on combination also follows a regular law. I hope by this means to give a proof of an idea advanced by several very distinguished chemists—that we are perhaps not far removed from the time when we shall be able to submit the bulk of chemical phenomena to calculation.

It is a very important question in itself, and one much discussed amongst chemists, to ascertain if compounds are formed in all sorts of proportions. M. Proust, who appears first to have fixed his attention on this subject, is of opinion that the metals are susceptible of only two degrees of oxidation, a *minimum* and a *maximum*; but led away by this seductive theory, he has seen himself forced to entertain principles contrary to physics in order to reduce to two oxides all those which the same metal sometimes presents. M. Berthollet thinks, on the other hand—reasoning from general considerations and his own experiments—that compounds are always formed in very variable proportions, unless they are determined by special causes, such as crystallisation, insolubility, or elasticity. Lastly, Dalton has advanced the idea that compounds of two bodies are formed in such a way that one atom of the one unites with one, two, three, or more atoms of the other. It would follow from this mode of looking at compounds that they are formed in constant proportions, the existence of intermediate bodies being excluded, and in this respect Dalton's theory would resemble that of M. Proust; but M. Berthollet has already strongly opposed it in the Introduction he has written to Thomson's *Chemistry*, and we shall see that in reality it is not entirely exact. Such is the state of the question now under discussion; it is still very far from receiving its solution, but I hope that the facts which I now proceed to set forth, facts which have entirely escaped the notice of chemists, will contribute to its elucidation.

Suspecting, from the exact ratio of 100 of oxygen to 200 of hydrogen, which M. Humboldt and I had determined for the proportions of water, that other gases might also combine in simple ratios, I have made the following experiments. I prepared fluoboric, muriatic, and carbonic gases, and made them combine successively with ammonia gas. 100 parts of muriatic gas saturate precisely 100 parts of ammonia gas, and the salt which is formed from them is perfectly neutral, whether one or the other of the gases is in excess. Fluoboric gas, on the contrary, unites in two proportions with ammonia gas. When the acid gas is put first into the graduated tube, and the other gas is then passed in, it is found that equal volumes of the two condense, and that the salt formed is neutral. But if we begin by first putting the ammonia gas into the tube, and then admitting the fluoboric gas in single bubbles, the first gas will then be in excess with regard to the second, and there will result a salt with excess of base, composed of 100 of fluoboric gas and 200 ammonia gas. If carbonic gas is brought into contact with ammonia gas, by passing it sometimes first, sometimes second, into the tube, there is always formed a sub-carbonate composed of 100 parts of carbonic gas and 200 of ammonia gas. It may, however, be proved that neutral carbonate of ammonia would be composed of equal volumes of each of these components. M. Berthollet, who has analysed this salt, obtained by passing carbonic gas into the sub-carbonate, found that it was composed of 73.34 parts by weight of carbonic gas and 26.66 of ammonia gas. Now, if we suppose it to be composed of equal volumes of its components, we find from their known specific gravity, that it contains by weight,

71.81 of carbonic acid
28.19 of ammonia
100.00

a proportion differing only slightly from the preceding.

. . .

It is not less remarkable that, whether we obtain a neutral salt or a *sub-salt*, their elements combine in simple ratios which may be considered as limits to their proportions. Accordingly, if we accept the specific gravity of muriatic acid determined by M. Biot and myself, and those of carbonic gas and ammonia given by M. Biot and Arago, we find that dry muriate of ammonia is composed of

Ammonia,	100.0 or 38.35
Muriatic acid,	160.7 or 61.65
	100.00

a proportion very far from that of M. Berthollet—

100 of ammonia
213 of acid

In the same way, we find that sub-carbonate of ammonia contains

Ammonia,	100.0 or 43.98
Carbonic acid,	127.3 or 56.02
	100.00

and the neutral carbonate

Ammonia,	100.0 or 28.19
Carbonic acid,	254.6 or 71.81
	100.00

It is easy from the preceding results to ascertain the ratios of the capacity of fluoboric, muriatic, and carbonic acids; for since these three gases saturate the same volume of ammonia gas, their relative capacities will be inversely as their densities, allowance having been made for the water contained in muriatic acid.

We might even now conclude that gases combine with each other in very simple ratios; but I shall still give some fresh proofs.

According to the experiments of M. Amédée Berthollet, ammonia is composed of

100 of nitrogen
300 of hydrogen

by volume.

I have found (1st vol. of the Société d'Arceuil) that sulphuric acid is composed of

100 of sulphurous gas,
50 of oxygen gas.

When a mixture of 50 parts of oxygen and 100 of carbonic oxide (formed by the distillation of oxide of zinc with strongly calcined charcoal) is inflamed, these two gases are destroyed and their place taken by 100 parts of carbonic acid gas. Consequently carbonic acid may be considered as being composed of

100 of carbonic oxide gas,
50 of oxygen gas.

...

Thus it appears evident to me that gases always combine in the simplest proportions when they act on one another; and we have seen in reality in all the preceding examples that the ratio of combinations is 1 to 1, 1 to 2, or 1 to 3. It is very important to observe that in considering weights there is no simple and finite relation between the elements of any one compound; it is only when there is a second compound between the same elements that the new proportion of the element that has been added is a multiple of the first quantity. Gases, on the contrary, in whatever proportions they may combine, always give rise to compounds whose elements by volume are multiples of each other.

...

According to Dalton's ingenious idea, that combinations are formed from atom to atom, the various compounds which two substances can form would be produced by the union of one molecule of the one with one molecule of the other, or with two, or with a greater number, but always without intermediate compounds. Thomson and Wollaston have indeed described experiments which appear to confirm this theory. Thomson has found that super-oxalate of potash contains twice as much acid as is necessary to saturate the alkali; and Wollaston, that the sub-carbonate of potash contains, on the other hand, twice as much alkali as is necessary to saturate the acid.

The numerous results I have brought forward in this Memoir are also very favorable to the theory. But M. Berthollet, who thinks that combinations are made continuously, cites in proof of his opinion the acid sulphates, glass alloys, mixtures of various liquids,–all of which are compounds with very variable proportions, and he insists principally on the identity of the force which produces chemical compounds and solutions.

Each of these two opinions has, therefore, a large number of facts in its favour; although they are apparently utterly opposed, it is easy to reconcile them.

We must first of all admit, with M. Berthollet, that chemical action is exercised indefinitely in a continuous manner between the molecules of substances, whatever their number and ratio may be, and that in general we can obtain compounds with very variable proportions. But then we must admit at the same time that—apart from insolubility, cohesion, and elasticity, which tend to produce compounds in fixed proportions—chemical action is exerted more powerfully when the elements are in simple ratios or in multiple proportions among themselves, and that compounds are thus produced which separate out more easily. In this way we reconcile the two opinions, and maintain the great chemical law, that whenever two substances are in presence of each other they act in their sphere of activity according to their masses, and give rise in general to compounds with very variable proportions, unless these proportions are determined by special circumstances.

Conclusion

I have shown in this Memoir that the compounds of gaseous substances with each other are always formed in very simple ratios, so that representing one of the terms by unity, the other is 1, or 2, or at most 3. These ratios by volume are not observed with solid or liquid substances, nor when we consider weights, and they form a new proof that it is only in the gaseous state that substances are in the same circumstances and obey regular laws. It is remarkable to see that ammonia gas neutralizes exactly its own volume of gaseous acids; and it is probable that if all acids and alkalies were in the elastic state, they would all combine in equal volumes to produce neutral salts. The capacity of saturation of acids and alkalies measured by volume would then be the same, and this might perhaps be the true manner of determining it. The apparent contraction of volume suffered by gases on combination is also very simply related to the volume of one of them, and this property likewise is peculiar to gaseous substances.

Translation by The Alembic Club

Reading and Discussion Questions

1. What reasoning does Gay-Lussac provide for focusing on the reactions between gases?
2. What measurements is Gay-Lussac paying most close attention to? How does this differ from Dalton?

Essay on a Manner of Determining the Relative Masses of the Elementary Molecules of Bodies, and the Proportions in Which They Enter into These Compounds (1811)

Amedeo Avogadro

71

I.

M. Gay-Lussac has shown in an interesting Memoir (Mémoires de la Société d'Arcueil, Tome II.) that gases always unite in a very simple proportion by volume, and that when the result of the union is a gas, its volume also is very simply related to those of its components. But the quantitative proportions of substances in compounds seem only to depend on the relative number of molecules which combine, and on the number of composite molecules which result. It must then be admitted that very simple relations also exist between the volumes of gaseous substances and the numbers of simple or compound molecules which form them. The first hypothesis to present itself in this connection, and apparently even the only admissible one, is the supposition that the number of integral molecules in any gases is always the same for equal volumes, or always proportional to the volumes. Indeed, if we were to suppose that the number of molecules contained in a given volume were different for different gases, it would scarcely be possible to conceive that the law regulating the distance of molecules could give in all cases relations as simple as those which the facts just detailed compel us to acknowledge between the volume and the number of molecules. On the other hand, it is very well conceivable that the molecules of gases being at such a distance that their mutual attraction cannot be exercised, their varying attraction for caloric may be limited to condensing the atmosphere formed by this fluid having any greater extent in the one case than in the other, and, consequently, without the distance between the molecules varying; or, in other words, without the number of molecules contained in a given volume being different. Dalton, it is true, has proposed a hypothesis directly opposed to this, namely that the quantity of caloric is always the same for the molecules of all bodies whatsoever in the gaseous state, and that the greater or less attraction for caloric only results in producing a greater or less condensation of this quantity around the molecules, and thus varying the distance between the molecules themselves. But in our present ignorance of the manner in which this attraction of the molecules for caloric is exerted, there is nothing to decide us à priori in favour of the one of these hypotheses rather than the other; and we should rather be inclined to adopt a neutral hypothesis, which would make the distance between the molecules and the quantities of caloric vary according to unknown laws, were it not that the hypothesis we have just proposed is based on that simplicity of relation between the volumes of gases on combination, which would appear to be otherwise inexplicable.

Setting out from this hypothesis, it is apparent that we have the means of determining very easily the relative masses of the molecules of substances obtainable in the gaseous state, and the relative number of these molecules in compounds; for the ratios of the masses of the molecules are then the same as those of the densities of the different gases at equal temperature and pressure, and the relative number of molecules in a compound is given at once by the ratio of the volumes of the gases that form it. For example, since the numbers 1.10359 and 0.07321 express the densities of the two gases oxygen and hydrogen compared to that of atmospheric air as unity, and the ratio of the two numbers consequently represents the ratio between the masses of equal volumes of these two gases, it will also represent on our hypothesis the ratio of the masses of their molecules. Thus the mass of the molecule of oxygen will be about 15 times that of the molecule of hydrogen, or, more exactly as 15.074 to 1. In the same way the mass of the molecule of nitrogen will be to that of hydrogen as 0.96913 to 0.07321, that is, as 13, or more exactly 13.238, to 1. On the other hand, since we know that the ratio of the volumes of hydrogen and oxygen in the formation of water is 2 to 1, it follows that water results from the union of each molecule of oxygen with two molecules of hydrogen. Similarly, according to the proportions by volume established by M. Gay-Lussac for the elements of ammonia, nitrous oxide, nitrous gas, and nitric acid, ammonia will result from the union of one molecule of nitrogen with three of hydrogen, nitrous oxide from one molecule of oxygen with two of nitrogen, nitrous gas from one molecule of nitrogen with one of oxygen, and nitric acid from one of nitrogen with two of oxygen.

II.

There is a consideration which appears at first sight to be opposed to the admission of our hypothesis with respect to compound substances. It seems that a molecule composed of two or more elementary molecules should have its mass equal to the sum of the masses of these molecules; and that in particular, if in a compound one molecule of one substance unites with two or more molecules of another substance, the number of compound molecules should remain the same as the number of molecules of the first substance. Accordingly, on our hypothesis, when a gas combines with two or more times its volume of another gas, the resulting compound, if gaseous, must have a volume equal to that of the first of these gases. Now, in general, this is not actually the case. For instance, the volume of water in the gaseous state is, as M. Gay-Lussac has shown, twice as great as the volume of oxygen which enters into it, or, what comes to the same thing, equal to that of the hydrogen instead of being equal to that of the oxygen. But a means of explaining facts of this type in conformity with our hypothesis presents itself naturally enough; we suppose, namely, that the constituent molecules of any simple gas whatever (i.e., the molecules which are at such a distance from each other that they cannot exercise their mutual action) are not formed of a solitary elementary molecule, but are made up of a certain number of these molecules united by attraction to form a single one; and further, that when molecules of another substance unite with the former to form a compound molecule, the integral molecule which should result splits itself into two or more parts (or integral molecules) composed of half, quarter, &c., the number of elementary molecules going to form the constituent molecule of the first substance, combined with half, quarter, &c., the number of constituent molecules of the second substance that ought to enter into combination with one constituent molecule of the first substance (or, what comes to the same thing, combined with a number equal to this last of half-molecules, quarter-molecules, &c., of the second substance); so that the number of integral molecules of the compound becomes double, quadruple, &c., what it would have been if there had been no splitting up, and exactly what is necessary to satisfy the volume of the resulting gas.[1]

[1] Thus, for example, the integral molecule of water will be composed of a half-molecule of oxygen with one molecule, or what is the same thing, two half-molecules of hydrogen.

On reviewing the various compound gases most generally known, I only find examples of duplication of the volume relatively to the volume of that one of the constituents which combines with one or more volumes in the other. We have already seen this for water. In the same way, we know that the volume of ammonia gas is twice that of the nitrogen which enters into it. M. Gay-Lussac has also shown that the volume of nitrous oxide is equal to that of the nitrogen which forms part of it, and consequently is twice that of the oxygen. Finally, nitrous gas, which contains equal volumes of nitrogen and oxygen, has a volume equal to the sum of the two constituent gases, that is to say, double that of each of them. Thus in all these cases there must be a division of the molecule into two; but it is possible that in other cases the division might be into four, eight, &c. The possibility of this division of compound molecules might have been conjectured à priori; for otherwise the integral molecules of bodies composed of several substances with a relatively large number of molecules, would come to have a mass excessive in comparison with the molecules of simple substances. We might therefore imagine that nature had some means of bringing them back to the order of the latter, and the facts have pointed out to us the existence of such means. Besides, there is another consideration which would seem to make us admit in some cases the division in question; for how could one otherwise conceive a real combination between two gaseous substances uniting in equal volumes without condensation, such as takes place in the formation of nitrous gas? Supposing the molecules to remain at such a distance that the mutual attraction of those of each gas could not be exercised, we cannot imagine that a new attraction could take place between the molecules of one gas and those of the other. But on the hypothesis of division of the molecule, it is easy to see that the combination really reduces two different molecules to one, and that there would be contraction by the whole volume of one of the gases if each compound molecule did not split up into two molecules of the same nature. M. Gay-Lussac clearly saw that, according to the facts, the diminution of volume on the combination of gases cannot represent the approximation of their elementary molecules. The division of molecules on combination explains to us how these two things may be made independent of each other.

III.

Dalton, on arbitrary suppositions as to the most likely relative number of molecules in compounds, has endeavoured to fix ratios between the masses of the molecules of simple substances. Our hypothesis, supposing it well-founded, puts us in a position to confirm or rectify his results from precise data, and, above all, to assign the magnitude of compound molecules according to the volumes of the gaseous compounds, which depend partly on the division of molecules entirely unexpected by this physicist.

Thus Dalton supposes[2] that water is formed by the union of hydrogen and oxygen, molecule to molecule. From this, and from the ratio by weight of the two components, it would follow that the mass of the molecule of oxygen would be to that of hydrogen as 7½ to 1 nearly, or, according to Dalton's evaluation, as 6 to 1. This ratio on our hypothesis is, as we saw, twice as great, namely, as 15 to 1. As for the molecule of water, its mass ought to be roughly expressed by 15+2=17 (taking for unity that of hydrogen), if there were no division of the molecule into two; but on account of this division it is reduced to half, 8½, or more exactly 8.537, as may also be found by dividing the density of aqueous vapour 0.625 (Gay-Lussac) by the density of hydrogen 0.0732. This mass differs from 7, that assigned to it by Dalton, by the difference in the values for the composition of water; so that in this respect Dalton's result is approximately correct from the combination of two compensating errors—the error in the mass of the molecule of oxygen, and his neglect of the division of the molecule.

Dalton supposes that in nitrous gas the combination of nitrogen and oxygen is molecule to molecule: we have seen on our hypothesis that this is actually the same. Thus Dalton would have found the same molecular mass for nitrogen as we have, always supposing that of hydrogen to be unity, if he had not set out from a different value for that of oxygen, and if he had taken precisely the same value for the quantities of the elements in nitrous gas by weight. But supposing the molecule of oxygen to be less than half what we find, he has been obliged to make that of nitrogen also equal to less than half the value we have assigned to it, viz., 5 instead of 13. As regards the molecule of nitrous gas itself, his neglect of the division of the molecule again makes his result approach ours; he has made it 6+5=11, whilst according to us it is about (15+13)/2=14, or more exactly (15.074+13.238)/2=14.156, as we also find by dividing 1.03636, the density of nitrous gas according to Gay-Lussac, by 0.07321. Dalton has likewise fixed in the same manner as the facts has given us, the relative number of molecules in nitrous oxide and in nitric acid, and in the first case the same circumstance has rectified his result for the magnitude of the molecule. He makes it 6+(2x5)=16, whilst according to our method it should be (15.074+2x13.238)/2=20.775, a number which is also obtained by dividing 1.52092, Gay-Lussac's value for the density of nitrous oxide, by the density of hydrogen.

In the case of ammonia, Dalton's supposition as to the relative number of molecules in its composition is on our hypothesis entirely at fault. He supposes nitrogen and hydrogen to be united in it molecule to molecule, whereas we have seen that one molecule of nitrogen unites with three molecules of hydrogen. According to him the molecule of ammonia would be 5+1=6; according to us it should be (13+3)/2=8, or more exactly 8.119, as may also be deduced directly from the density of ammonia gas. The division of the molecule, which does not

[2] In what follows I shall make use of the exposition of Dalton's ideas given in Thomson's System of Chemistry.

enter into Dalton's calculations, partly corrects in this case also the error which would result from his other suppositions.

All the compounds we have just discussed are produced by the union of one molecule of one of the components with one or more molecules of the other. In nitrous acid we have another compound of two of the substances already spoken of, in which the terms of the ratio between the number of molecules both differ from unity. From Gay-Lussac's experiments (Société d'Arcueil, same volume) it appears that this acid is formed from 1 part by volume of oxygen and 3 of nitrous gas, or, what comes to the same thing, of 3 parts of nitrogen and 5 of oxygen; hence it would follow, on our hypothesis, that its molecule should be composed of 3 molecules of nitrogen and 5 of oxygen, leaving the possibility of division out of account. But this mode of combination can be referred to the preceding simpler forms by considering it as the result of the union of 1 molecule of oxygen with 3 of nitrous gas, i.e. with 3 molecules, each composed of a half-molecule of oxygen and a half-molecule of nitrogen, which thus already included the division of some of the molecules of oxygen which enter into that of nitrous acid. Supposing there to be no other division, the mass of this last molecule would be 57.542, that of hydrogen being taken as unity, and the density of nitrous acid gas would be 4.21267, the density of air being taken as unity. But it is probable that there is at least another division into two, and consequently a reduction of the density to half: we must wait until this density has been determined by experiment....

VIII.

It will have been in general remarked on reading this Memoir that there are many points of agreement between our special results and those of Dalton, although we set out from a general principle, and Dalton has only been guided by considerations of detail. This agreement is an argument in favour of our hypothesis, which is at bottom merely Dalton's system furnished with a new means of precision from the connection we have found between it and the general fact established by M. Gay-Lussac. Dalton's system supposes that compounds are made in general in fixed proportions, and this is what experiment shows with regard to the more stable compounds and those most interesting to the chemist. It would appear that it is only combinations of this sort that can take place amongst gases, on account of the enormous size of the molecules which would result from ratios expressed by larger numbers, in spite of the division of the molecules, which is in all probability confined within narrow limits. We perceive that the close packing of the molecules in solids and liquids, which only leaves between the integral molecules distances of the same order as those between the elementary molecules, can give rise to more complicated ratios, and even to combinations in all proportions; but these compounds will be so to speak of a different type from those with which we have been concerned, and this distinction may serve to reconcile M. Berthollet's ideas as to compounds with the theory of fixed proportions.

Translated by The Alembic Club

Reading and Discussion Questions

1. What is Avogadro's hypothesis?
2. In what ways does this hypothesis help to reconcile Dalton's results with those of Gay-Lussac?

Reflections on the Motive Power of Heat and on Engines Suitable for Developing This Power (1824)

Sadi Carnot

72

It is well known that heat may be used as a cause of motion, and that the motive power which may be obtained from it is very great. The steam-engine, now in such general use, is a manifest proof of this fact.

To the agency of heat may be ascribed those vast disturbances which we see occurring everywhere on the earth; the movements of the atmosphere, the rising of mists, the fall of rain and other meteors, the streams of water which channel the surface of the earth, of which man has succeeded in utilizing only a small part. To heat are due also volcanic eruptions and earthquakes.

From this great source we draw the moving force necessary for our use. Nature, by supplying combustible material everywhere, has afforded us the means of generating heat and the motive power which is given by it, at all times and in all places, and the steam-engine has made it possible to develop and use this power.

The study of the steam-engine is of the highest interest, owing to its importance, its constantly increasing use, and the great changes it is destined to make in the civilized world. It has already developed mines, propelled ships, and dredged rivers and harbors. It forges iron, saws wood, grinds grain, spins and weaves stuffs, and transports the heaviest loads. In the future it will most probably be the universal motor, and will furnish the power now obtained from animals, from waterfalls, and from air-currents.

Over the first of these motors it has the advantage of economy, and over the other two the in calculable advantage that it can be used everywhere and always, and that its work need never be interrupted.

If in the future the steam-engine is so perfected as to render it less costly to construct it and to supply it with fuel, it will unite all desirable qualities and will promote the development of the industrial arts to an extent which it is difficult to foresee. It is, indeed, not only a powerful and convenient motor, which can be set up or transported anywhere, and substituted for other motors already in use, but it leads to the rapid extension of those arts in which it is used, and it can even create arts hitherto unknown.

The most signal service which has been rendered to England by the steam-engine is that of having revived the working of her coal-mines, which had languished and was threatened with extinction on account of the increasing difficulty of excavation and extraction of the coal. We may place in the second rank the services rendered in the manufacture of iron, as much by furnishing an abundant supply of coal, which took the place of wood as the wood began to be exhausted, as by the powerful machines of all kinds the use of which it either facilitated or made possible.

Iron and fire, as everyone knows, are the mainstays of the mechanical arts. Perhaps there is not in all England a single industry whose existence is not dependent on these agents, and which does not use them extensively. If England were to-day to lose its steam-engines it would lose also its coal and iron, and this loss would dry up all its sources of wealth and destroy its prosperity; it would annihilate this colossal power. The destruction of its navy, which it considers its strongest support, would be, perhaps, less fatal.

The safe and rapid navigation by means of steamships is an entirely new art due to the steam-engine. This art has already made possible the establishment of prompt and regular communication on the arms of the sea, and on the great rivers of the old and new continents. By means of the steam-engine regions still savage have been traversed which but a short time ago could hardly have been penetrated. The products of civilization have been taken to all parts of the earth, which they would otherwise not have reached for many years. The navigation due to the steam-engine has in a measure drawn together the most distant nations. It tends to unite the peoples of the earth as if they all lived in the same country. In fact, to diminish the duration, the fatigue, the uncertainty and danger of voyages is to lessen their length.

The discovery of the steam-engine, like most human inventions, owes its birth to crude attempts which have been attributed to various persons and of which the real author is not known. The principal discovery consists indeed less in these first trials than in the successive improvements which have brought it to its present perfection. There is almost as great a difference between the first structures where expansive force was developed and the actual steam-engine as there is between the first raft ever constructed and a man-of-war.

If the honor of a discovery belongs to the nation where it acquired all its development and improvement, this honor cannot in this case be withheld from England: Savery, Newcomen, Smeaton, the celebrated Watt, Woolf, Trevithick, and other English engineers, are the real inventors of the steam-engine. At their hands it received each successive improvement. It is natural that an invention should be made, improved, and perfected where the need of it is most strongly felt.

In spite of labor of all sorts expended on the steam-engine, and in spite of the perfection to which it has been brought, its theory is very little advanced, and the attempts to better this state of affairs have thus far been directed almost at random.

The question has often been raised whether the motive power of heat is limited or not; whether there is a limit to the possible improvements of the steam-engine which, in the nature of the case, cannot be passed by any means; or if, on the other hand, these improvements are capable of indefinite extension. Inventors have tried for a long time, and are still trying, to find whether there is not a more efficient agent than water by which to develop the motive power of heat; whether, for example, atmospheric air does not offer great advantages in this respect. We propose to submit these questions to a critical examination.

The phenomenon of the production of motion by heat has not been considered in a sufficiently general way. It has been treated only in connection with machines whose nature and mode of action do not admit of a full investigation of it. In such machines the phenomenon is, in a measure, imperfect and incomplete; it thus becomes difficult to recognize its principles and study its laws.

To examine the principle of the production of motion by heat in all its generality, it must be conceived in dependently of any mechanism or of any particular agent; it is necessary to establish proofs applicable not only to steam-engines but to all other heat-engines, irrespective of the working substance and the manner in which it acts.

The machines which are not worked by heat for instance, those worked by men or animals, by water-falls, or by air currents can be studied to their last details by the principles of mechanics. All possible cases may be anticipated, all imaginable actions are subject to general principles already well established and applicable in all circumstances. The theory of such machines is complete. Such a theory is evidently lacking for heat-engines. We shall never possess it until the laws of physics are so extended and generalized as to make known in advance all the effects of heat acting in a definite way on any body whatsoever.

We shall take for granted in what follows a knowledge, at least a superficial one, of the various parts which compose an ordinary steam-engine. We think it unnecessary to describe the fire-box, the boiler, the steam-chest, the piston, the condenser, etc.

The production of motion in the steam-engine is always accompanied by a circumstance which we should particularly notice. This circumstance is the re-establishment of equilibrium in the caloric—that is, its passage from one body where the temperature is more or less elevated to another where it is lower. What happens, in fact, in a steam-engine at work? The caloric developed in the fire-box as an effect of combustion passes through the wall of the boiler and produces steam, incorporating itself with the steam in some way. This steam, carrying the caloric with it, transports it first into the cylinder, where it fulfils some function, and thence into the condenser, where the steam is precipitated by coming in contact with cold water. As a last result the cold water in the condenser receives the caloric developed by combustion. It is warmed by means of the steam, as if it had been placed directly on the fire-box. The steam is here only a means of transporting caloric; it thus fulfils the same office as in the heating of baths by steam, with the exception that in the case in hand its motion is rendered useful.

We can easily perceive, in the operation which we have just described, the re-establishment of equilibrium in the caloric and its passage from a hotter to a colder body. The first of these bodies is the heated air of the fire-box; the second, the water of condensation. The re-establishment of equilibrium of the caloric is accomplished between them if not completely, at least in part; for, on the one hand, the heated air after having done its work escapes through the smoke-stack at a much lower temperature than that which it had acquired by the combustion; and, on the other hand, the water of the condenser, after having precipitated the steam, leaves the engine with a higher temperature than that which it had when it entered.

The production of motive power in the steam-engine is therefore not due to a real consumption of the caloric, but to its transfer from a hotter to a colder body that is to say, to the re-establishment of its equilibrium, which is assumed to have been destroyed by a chemical action such as combustion, or by some other cause. We shall soon see that this principle is applicable to all engines operated by heat.

Translated by William Francis Magie

Reading and Discussion Questions

1. What political situation inspires Carnot's theoretical reflections on steam power?
2. How does the use of Lavoisier's caloric theory of heat as a fluid influence Carnot's description of the work produced by heat?
3. Carnot is usually included among several scientists who are given credit for articulating a version of what would later be called the second law of thermodynamics. Why?

On the Nature of the Motion Which We Call Heat (1857)

Rudolf Clausius

73

1. Before writing my first memoir on heat, which was published in 1850,[1] and in which heat is assumed to be a motion, I had already formed for myself a distinct conception of the nature of this motion, and had even employed the same in several investigations and calculations. In my former memoirs I intentionally avoided mentioning this conception, because I wished to separate the conclusions which are deducible from certain general principles from those which presuppose a particular kind of motion, and because I hoped to be able at some future time to devote a separate memoir to my notion of this motion and to the special conclusions which flow therefrom. The execution of this project, however, has been retarded longer than I at first expected, inasmuch as the difficulties of the subject, as well as other occupations, have hitherto prevented me from giving to its development that degree of completeness which I deemed necessary for publication.

A memoir has lately been published by Krönig, under the title *Gundzüge einer Theorie der Gase*,[2] in which I have recognized some of my own views. Seeing that Krönig has arrived at these views just as independently as I have, and has published them before me, all claim to priority on my part is of course out of the question; nevertheless, the subject having once been mooted in this memoir, I feel myself induced to publish those parts of my own views which I have not yet found in it. For the present, I shall confine myself to a brief indication of a few principal points, and reserve a more complete analysis for another time.[3]

[1] *Ann. Phys.* 79, 368, 500 (1850).

[2] This was first printed separately by A. W. Hayn in Berlin, and afterwards appeared in Poggendorff's *Annalen*, vol. xcix, p. 315.

[3] I must not omit to mention here, that some time ago Mr William Siemens of London, when on a visit in Berlin, informed me that Joule had also expressed similar ideas in the *Memoirs of the Literary*

2. Krönig assumes that the molecules of gas do not oscillate about definite positions of equilibrium, but that they move with constant velocity in right lines until they strike against other molecules, or against some surface which to them is impermeable. I share this view completely, and I also believe that the expansive force of the gas arises from this motion. On the other hand, I am of opinion that this is not the only motion present.

In the first place, the hypothesis of a rotary as well as a progressive motion of the molecules at once suggests itself; for at every impact of two bodies, unless the same happens to be central and rectilinear, a rotary as well as a translatory motion ensues.

I am also of an opinion that vibrations take place within the several masses in a state of progressive motion. Such vibrations are conceivable in several ways. Even if we limit ourselves to the consideration of the atomic masses solely, and regard these as absolutely rigid, it is still possible that a molecule, which consists of several atoms, may not also constitute an absolutely rigid mass, but that within it the several atoms are to a certain extent moveable, and thus capable of oscillating with respect to each other.

I may also remark, that by thus ascribing a movement to the atomic masses themselves, we do not exclude the hypothesis that each atomic mass may be provided with a quantity of finer matter, which, without separating from the atom, may still be moveable in its vicinity.

By means of a mathematical investigation given at the end of the present memoir, it may be proved that the *vis viva* of the translatory motion alone is too small to represent the whole heat present in the gas; so that without entering into the probability of the same, we are thus compelled to assume one or more motions of another kind. According to this calculation, the excess of the whole *vis viva* over that of the translatory motion alone is particularly important in gases of a complicated chemical constitution, in which each molecule consists of a great number of atoms.

3. In one and the same gas the translatory motion of the whole molecules will always have a constant relation to the several motions which, in addition to the above, the constituents of the molecules likewise possess. For brevity I will call the latter the *motions of the constituents*.

Conceive a number of molecules whose constituents are in active motion, but which have no translatory motion. It is evident the latter will commence as soon as two molecules in contact strike against each other in consequence of the motion of their constituents. The translatory motion thus originated will of course occasion a corresponding loss of *vis viva* in the motion of the

and Philosophical Society of Manchester. My views being consequently no longer completely new, this was an additional reason why I should hasten their publication less than I otherwise should have done. Hitherto I have not been able to procure the memoir of Joule in question, and therefore I am ignorant how far he has pursued the subject, and whether his views coincide with mine in all points. It is to be regretted that Joule did not publish his memoir in a more widely circulated periodical.

constituents. On the other hand, if the constituents of a number of molecules in a state of translatory motion were motionless, they could not long remain so, in consequence of the collisions between the molecules themselves, and between them and fixed sides or walls. It is only when all possible motions have reached a certain relation towards one another, which relation will depend upon the constitution of the molecules, that they will cease mutually to increase or diminish each other.

When two molecules whose constituents are in motion come into collision they will not rebound, like two elastic balls, according to the ordinary laws of elasticity; for their velocities and directions after collision will depend, not only upon the motion which the whole molecules had before impact, but also upon the motion of those constituents which are nearest each other at the moment of collision. After the equalization of the several motions, however, when the translatory motion is, on the whole, neither increased nor diminished by the motions of the constituents, we may, in our investigation of the total action of a great number of molecules, neglect the irregularities occurring at the several collisions, and assume that, in reference to the translatory motion, the molecules follow the common laws of elasticity.

4. The explanation of the expansive force of gases and its dependence upon volume and temperature, as given by Krönig, suffers no essential modification through the introduction of other motions. The pressure of the gas against a fixed surface is caused by the molecules in great number continually striking against and rebounding from the same. The force which must thence arise is, in the first place, by equal velocity of motion inversely proportional to the volume of the given quantity of gas; and secondly, by equal volume proportional to the *vis viva* of the translatory motion: the other motions do not here immediately come into consideration.

On the other hand, from Gay-Lussac's law we know that, under constant volume, the pressure of a perfect gas increases in the same ratio as temperature calculated from -273°C, which we call the absolute temperature. Hence, according to the above, it follows that the absolute temperature is proportional to the *vis viva* of the translatory motion of the molecules. But as, according to a former remark, the several motions in one and the same gas bear a constant relation to each other, it is evident that the *vis viva* of the translatory motion forms an aliquot part of the total *vis viva*, so that the absolute temperature is also proportional to the whole *vis viva* in the gas.

These considerations, together with others connected therewith to be given hereafter, induced me, in my memoir "On the Moving Force of Heat," to express the opinion that the specific heat of gases was constant; which opinion was in opposition to the experiments then known.[4] The quantity of heat which must be imparted to a gas, under constant volume, in order to raise its temperature is to be considered as the increase of the *vis viva* in the gas, inasmuch as in this case no work is done whereby heat could be consumed. The specific heat *under constant*

[4] Poggendorff's *Annalen*, vol. lxxix, p. 393. *Phil. Mag.*, vol. ii, pp. 1, 102.

volume, therefore, is in a perfect gas the magnitude which Rankine calls the *true* specific heat. Now the assertion that the true specific heat of a gas is constant, is simply equivalent to the assertion that *the total* vis viva *in the gas has a constant ratio to the* vis viva *of the translatory motion which serves us as a measure of the temperature*. With respect to the specific heat under constant pressure, I have proved in the memoir before cited, and by means of a hypothesis proceeding from the same considerations, that it differs only by a constant magnitude from the true specific heat.

5. The foregoing is true for permanent gases only, and even for these only approximatively. In general, the small deviations which present themselves can be easily accounted for.

In order that Mariotte's and Gay-Lussac's laws, as well as others in connexion with the same, may be strictly fulfilled, the gas must satisfy the following conditions with respect to its molecular condition:—

(1) The space actually filled by the molecules of the gas must be infinitesimal in comparison to the whole space occupied by the gas itself.

(2) The duration of an impact, that is to say, the time required to produce the actually occurring change in the motion of a molecule when it strikes another molecule or a fixed surface, must be infinitesimal in comparison to the interval of time between two successive collisions.

(3) The influence of the molecular forces must be infinitesimal. Two conditions are herein involved. In the first place, it is requisite that the force with which all the molecules at their mean distances attract each other, vanish when compared to the expansive force due to motion. But the molecules are not always at their mean distances asunder; on the contrary, during their motion a molecule is often brought into close proximity to another, or to a fixed surface consisting of active molecules, and in such moments the molecular forces will of course commence their activity. The second condition requires, therefore, that those parts of the path described by a molecule under the influence of the molecular forces, when the latter are capable of altering appreciably the direction or velocity of the molecule's motion, should vanish when compared with those parts of its path with respect to which the influence of these forces may be regarded as zero.

If these conditions are not fulfilled, deviations in several ways from the simple laws of gases necessarily arise; and these deviations become more important the less the molecular condition of the gas fulfils the conditions in question.

On becoming acquainted with the celebrated investigations of Regnault on the deviations of gases from Mariotte's and Gay-Lussac's laws, I attempted, by means of the principles above intimated, to deduce some conclusions with respect to the molecular condition of several gases from the nature of the deviations which Regnault detected in the same. A description of this method, however, would be too prolix; and even the results, in consequence of the many difficulties encountered in actual calculation, are too uncertain to merit being here adduced.

Whenever, therefore, in the sequel a gas is spoken of, we shall, as before, conceive it to be one which *perfectly* fulfils the above conditions, and which Regnault calls an *ideal gas*, inasmuch as all known gases present but an approximation to this condition.

6. After these considerations on the *gaseous* condition, the question at once arises in what manner the *solid* and *liquid* conditions differ from the gaseous. Although a definition of these states of aggregation, in order to be satisfactory in all its details, would require a more complete knowledge than we at present possess of the condition of the individual molecules, yet it appears to me that several fundamental distinctions may be advanced with tolerable probability.

A motion of the molecules takes place in all three states of aggregation.

In the *solid* state, the motion is such that the molecules move about certain positions of equilibrium without ever forsaking the same, unless acted upon by foreign forces. In solid bodies, therefore, the motion may be characterized as a vibrating one, which may, however, be of a very complicated kind. In the first place, the constituents of a molecule may vibrate among themselves; and secondly, the molecule may vibrate as a whole: again, the latter vibrations may consist in oscillations to and fro of the centre of gravity, as well as in rotatory oscillations around this centre of gravity. In cases where external forces act on the body, as in concussions, the molecules may also be permanently displaced.

In the *liquid* state the molecules have no longer any definite position of equilibrium. They can turn completely around their centres of gravity; and the latter, too, may be moved completely out of its place. The separating action of the motion is not, however, sufficiently strong, in comparison to the mutual attraction between the molecules, to be able to separate the latter entirely. Although a molecule no longer adheres to definite neighbouring molecules, still it does not spontaneously forsake the latter, but only under the united actions of forces proceeding from other molecules, with respect to which it then occupies the same position as it formerly did with respect to its neighboring molecules. In liquids, therefore, an oscillatory, a rotatory, and a translatory motion of the molecules takes place, but in such a manner that these molecules are not thereby separated from each other, but, even in the absence of external forces, remain within a certain volume.

Lastly, in the *gaseous* state the motion of the molecules entirely transports them beyond the spheres of their mutual attraction, causing them to recede in right lines according to the ordinary laws of motion. If two such molecules come into collision during their motion, they will in general fly asunder again with the same vehemence with which they moved towards each other; and this will the more readily occur, since a molecule will be attracted with much less force by another single molecule than by all the molecules which in the liquid or solid state surround it.

7. The phaenomenon of *evaporation* appearing peculiarly interesting to me, I have attempted to account for the same in the following manner.

It has been stated above, that in liquids a molecule, during its motion, either remains within the sphere of attraction of its neighbouring molecules, or only leaves the same in order to take up a corresponding position with respect to other neighbouring molecules. This applies only to the mean value of the motions, however; and as the latter are quite irregular, we must assume that the velocities of the several molecules deviate within wide limits on both sides of this mean value.

Taking next the surface of a liquid into consideration, I assume that, amongst the varied motions to and fro, it happens that under the influence of a favourable cooperation of the translatory, oscillatory, and rotatory motions, a molecule separates itself with such violence from its neighboring molecules that it has already receded from the sphere of their action before losing all its velocity under the influence of their attracting forces, and thus that it continues its flight into the space above the liquid.

Conceive this space to be enclosed, and at the commencement empty; it will gradually become more and more filled with these expelled molecules, which will now deport themselves in the space exactly as a gas, and consequently in their motion strike against the enclosing surfaces. The liquid itself, however, will form one of these surfaces; and when a molecule strikes against the same, it will not in general be driven back, but rather retained, and, as it were, absorbed in consequence of the renewed attraction of the other molecules into whose vicinity it has been driven. A state of equilibrium will ensue when the number of molecules in the superincumbent space is such, that on the average as many molecules strike against, and are retained by the surface of the liquid in a given time, as there are molecules expelled from it in the same time. The resulting state of equilibrium, therefore, is not a state of rest or a cessation of evaporation, but a state in which evaporation and condensation continually take place and compensate each other in consequence of their equal intensity.

The density of the vapour necessary for this compensation depends upon the number of molecules expelled from the surface of the liquid in the unit of time; and this number is again evidently dependent upon the activity of the motion within the liquid, that is to say, upon its temperature. I have not yet succeeded in deducing from these considerations the law according to which the pressure of vapour must increase with the temperature.

The preceding remarks on the deportment of the surface of the liquid towards the superincumbent vapour, apply in a similar manner to the other surfaces which enclose the space filled with vapour. The vapour is in the first place condensed on these surfaces, and the liquid thus produced then suffers evaporation, so that here also a state must be attained in which condensation and evaporation become equal. The requisite quantity of condensed vapour on these surfaces depends upon the density of the vapour in the enclosed space, upon the temperature of the vapour and of the enclosing surfaces, and upon the force with which the molecules of vapour are attracted toward these surfaces. In this respect a maximum will occur when the enclosing surfaces are completely moistened with the condensed liquid; and as soon as this takes place, these surfaces deport themselves exactly like a single surface of the same liquid.

8. The reason why the presence of another gas above the liquid cannot impede the evaporation of the same may now be immediately explained.

The pressure of the gas on the liquid arises solely from the fact, that here and there single molecules of gas strike against the surface of the liquid. In other respects, however, inasmuch as the molecules of gas themselves actually fill but a very small pert of the superincumbent space, the latter must be considered as empty, and as offering a free passage to the molecules of the liquid. In general these molecules will only come into collision with those of the gas at comparatively great distances from the surface, and the former will then deport themselves towards the latter as would the molecules of any other admixed gas. We must conclude, therefore, that the liquid also expels its molecules into the space filled with gas; and that in this case also the quantity of vapour thus mixed with the gas continues to increase until, on the whole, as many molecules of vapour strike against and are absorbed by the surface of the liquid as the latter itself expels; and the number of molecules of vapour to the unit of volume requisite hereto, is the same whether the space does or does not contain additional molecules of gas.

The pressure of the gas, however, exercises a different influence on the interior of the liquid. Here also, or at places where the mass of liquid is bounded by a side of the vessel, it may happen that the molecules separate from each other with such force that for a moment the continuity of the mass is broken. The small vacuum thus produced, however, is surrounded on all sides by masses which do not admit of the passage of the moved molecules; and hence this vacuum will only then become magnified into a bubble of vapour, and be able to continue as such, when the number of molecules expelled from its enclosing liquid walls is sufficient to produce an internal vapour-pressure capable of holding in equilibrium the pressure which acts externally and tends to compress the bubble again. Hence the expansive force of the enclosed vapour must be greater, the greater the pressure to which the liquid is exposed, and thus is explained the relation which exists between the pressure and the temperature of the boiling-point.

The relations will be more complicated when the gas above the liquid is itself condensable, and forms a liquid which mixes with the given one, for then of course the tendency of the two kinds of matter to mix enters as a new force. I shall not here enter into these phaenomena.

As in liquids, so also in solids the possibility of an evaporation may be comprehended; nevertheless it does not follow from this that, on the contrary, an evaporation *must* take place on the surface of all bodies. It is, in fact, readily conceivable that the mutual cohesion of the molecules of a body may be so great, that, so long as the temperature does not exceed a certain limit, even the most favourable combination of the several molecular motions is not able to overcome this cohesion.

...

11. Lastly, I must mention a phaenomenon the explanation of which appears to me to be of great importance, viz. *when two gases combine with each other, or when a gas combines with another body, and the combination is also gaseous, the volume of the compound gas bears a simple ratio to the volumes of the single constituents, at least when the latter are gaseous.*

Krönig has already proved that the pressure exerted by a gas on the unit of its enclosing surface must be proportional to the number of molecules contained in the unit of volume, and to the *vis viva* of the several molecules arising from their translatory motion, the only one which Krönig considers.

If we apply this to simple gases, and assume that, when pressure and temperature are the same, equal volumes contain the same number of atoms—a hypothesis which for other reasons is very probable—it follows that, in reference to their translatory motion, the atoms of different gases must have the same *vis viva*.

We will next examine in what manner this theorem remains true when applied to the molecules of compound gases.

12. In the first place, let us compare compound gases amongst themselves, *e.g.* two gases to form which the constituents have combined in ratios of volume respectively equal to 1:1 and 1:2. Nitric acid and nitrous acid may serve as examples.

With respect to these two gases, we know that quantities containing the same amount of oxygen occupy the same volume. Hence here, too, equal volumes contain the same number of molecules, although in the one gas each molecule consists of two, and in the other of three atoms; and we must further conclude, that even these differently constituted molecules have the same *vis viva* with respect to their translatory motion.

In most other compound gases we are led to the same conclusion; and in cases which do not submit themselves to this rule, it does not appear to me impossible that the discrepancy may be accounted for in one or both of two ways: either the gas was not sufficiently removed from its points of condensation when its volume was determined, or the chemical formula hitherto employed does not properly represent the manner in which the atoms are combined to form molecules.

On comparing compound and simple gases, however, an unmistakable deviation from the foregoing rule shows itself, inasmuch as the space corresponding to an atom of the simple gas does not correspond to a molecule of the compound one. When two simple gases combine in equal volumes, it is well known that no change of volume takes place, whilst according to the above rule the volume ought to be diminished in the ratio of 2:1. Again, when a volume of one gas combines with two or three volumes of another, the combination is found to occupy two volumes, whereas according to rule it ought only to occupy one volume, and so on.

13. On seeking to explain these curious anomalies, and especially to find a common law governing the relations of volume in gases, I was led to adopt the following view as being most plausible. I beg to offer the same to the scientific public as a hypothesis which is at least worthy of further examination.

I assume that the force which determines chemical combination, and which probably consists in a kind of polarity of the atoms, is already active in simple substances, and that *in these likewise two or more atoms are combined to form one molecule*. For instance, let equal volumes of oxygen and nitrogen be given. A mixture of these gases contains a certain number of molecules, which consist either of two atoms of oxygen or of two atoms of nitrogen. Conceive the mixture to pass into a chemical compound, and the latter then contains just as many molecules, which are merely constituted in a different manner, inasmuch as each consists of an atom of oxygen and an atom of nitrogen. Hence there is no reason why a change of volume should take place. If, on the other hand, one volume of oxygen and two of nitrogen are given, then in the mixture each molecule consists of two, and in the compound of three atoms. The chemical combination, therefore, has caused the number of molecules to diminish in the ratio of 3:2, and consequently the volume ought to diminish in the same ratio.

It is well known that some simple substances do not, in the gaseous form, occupy the volume which their atomic weights and the volumes of their combinations would lead us to anticipate, but another, and in most cases a smaller volume, which bears to the former a simple ratio. A special investigation of these substances would here be out of place, more especially as two of them, sulfur and phosphorus, deport themselves in other respects in so remarkable a manner, in consequence of the variety of conditions they are capable of assuming, that we may reasonably expect further discoveries from chemistry with respect to these bodies; and then, perhaps, besides other irregularities, those of the volumes of their vapors will be explained. Nevertheless I may here recall one circumstance which in some cases may possibly facilitate this explanation. I refer to the fact, that the above hypothesis, according to which the molecules of simple substances each consist of *two* atoms, may not be the only possible one.

On comparing with each other all cases of simple and compound gases, we must not expect to find immediately a perfect agreement throughout. I am of opinion, however, that, under the present uncertainty with respect to the inner constitution of several bodies, and particularly of those which possess a complicated chemical composition, too great weight ought not to be laid upon individual anomalies; and I deem it probable, that, by means of the above hypothesis respecting the molecules of simple substances, all relations of volume in gases may be referred back to the theorem, *that the several molecules of all gases possess equal* vis viva *in reference to their translatory motion.*

Translated by John Tyndall

Reading and Discussion Questions

1. If Clausius says heat is a motion, what is it a motion of? How does this differ from Lavoisier's caloric theory?
2. Clausius describes three kinds of motion, all of which contribute to the phenomena we perceive as heat. What observations does having each of these three kinds of movement help explain?

Letter of Professor Stanislao Cannizzaro to Professor S. de Luca: Sketch of a Course of Chemical Philosophy (1858)

Stanislao Cannizzaro

74

I believe that the progress of science made in these last years has confirmed the hypothesis of Avogadro, of Ampère, and of Dumas on the similar constitution of substances in the gaseous state; that is, that equal volumes of these substances, whether simple or compound, contain an equal number of molecules: not however an equal number of atoms, since the molecules of the different substances, or those of the same substance in its different states, may contain a different number of atoms, whether of the same or of diverse nature.

In order to lead my students to the conviction which I have reached myself, I wish to place them on the same path as that by which I have arrived at it—the path, that is, of the historical examination of chemical theories.

I commence, then, in the first lecture by showing how, from the examination of the physical properties of gaseous bodies, and from the law of Gay-Lussac on the volume relations between components and compounds, there arose almost spontaneously the hypothesis alluded to above, which was first of all enunciated by Avogadro, and shortly afterwards by Ampère. Analysing the conception of these two physicists, I show that it contains nothing contradictory to known facts, provided that we distinguish, as they did, molecules from atoms; provided that we do not confuse the criteria by which the number and the weight of the former are compared, with the criteria which serve to deduce the weight of the latter; provided that, finally, we have not fixed in our minds the prejudice that whilst the molecules of compound substances may consist of different numbers of atoms, the molecules of the various simple substances must all contain either one atom, or at least an equal number of atoms.

In the second lecture I set myself the task of investigating the reasons why this hypothesis of Avogadro and Ampère was not immediately accepted by the majority of chemists. I therefore expound rapidly the work and the ideas of those who examined the relationships of the reacting quantities of substances without concerning themselves with the volumes which these substances occupy in the gaseous state; and I pause to explain the ideas of Berzelius, by the influence of which the hypothesis above cited appeared to chemists out of harmony with the facts.

I examine the order of the ideas of Berzelius, and show how on the one hand he developed and completed the dualistic theory of Lavoisier by his own electro-chemical hypothesis, and how on the other hand, influenced by the atomic theory of Dalton (which had been confirmed by the experiments of Wollaston), he applied this theory and took it into agreement with the dualistic electro-chemical theory, whilst at the same time he extended the laws of Richter and tried to harmonise them with the results of Proust. I bring out clearly the reason why he was led to assume that the atoms, whilst separate in simple bodies, should unite to form the atoms of a compound of the first order, and these in turn, uniting in simple proportions, should form composite atoms of the second order, and why (since he could not admit that when two substances give a single compound, a molecule of the one and a molecule of the other, instead of uniting to form a single molecule, should change into two molecules of the same nature) he could not accept the hypothesis of Avogadro and of Ampère, which in many cases leads to the conclusion just indicated.

I then show how Berzelius, being unable to escape from his own dualistic ideas, and yet wishing to explain the simple relations discovered by Gay-Lussac between the volumes of gaseous compounds and their gaseous components, was led to formulate a hypothesis very different from that of Avogadro and of Ampère, namely, that equal volumes of simple substances in the gaseous state contain the same number of atoms, which in combination unite intact; how, later, the vapour densities of many simple substances having been determined, he had to restrict this hypothesis by saying that only simple substances which are permanent gases obey this law; how, not believing that composite atoms even of the same order could be equidistant in the gaseous state under the same conditions, he was led to suppose that in the molecules of hydrochloric, hydriodic, and hydrobromic acids, and in those of water and sulphuretted hydrogen, there was contained the same quantity of hydrogen, although the different behaviour of these compounds confirmed the deductions from the hypothesis of Avogadro and of Ampère.

I conclude this lecture by showing that we have only to distinguish atoms from molecules in order to reconcile all the experimental results known to Berzelius, and have no need to assume any difference in constitution between permanent and coercible, or between simple and compound gases, in contradiction to the physical properties of all elastic fluids.

In the third lecture I pass in review the various researches of physicists on gaseous bodies, and show that all the new researches from Gay-Lussac to Clausius confirm the hypothesis of Avogadro and of Ampère that the distances between the molecules, so long as they remain in the gaseous state, do not depend on their nature, nor on their mass, nor on the number of atoms they contain, but only on their temperature and on the pressure to which they are subjected.

In the fourth lecture I pass under review the chemical theories since Berzelius: I pause to examine how Dumas, inclining to the idea of Ampère, had habituated chemists who busied themselves with organic substances to apply this idea in determining the molecular weights of compounds; and what were the reasons which had stopped him half way in the application of this theory. I then expound, in continuation of this, two different methods—the one due to Berzelius, the other to Ampère and Dumas—which were used to determine formulae in inorganic and in organic chemistry respectively until Laurent and Gerhardt sought to bring both parts of the science into harmony. I explain clearly how the discoveries made by Gerhardt, Williamson, Hofmann, Wurtz, Berthelot, Frankland, and others, on the constitution of organic compounds confirm the hypothesis of Avogadro and Ampère, and how that part of Gerhardt's theory which corresponds best with facts and best explains their connection, is nothing but the extension of Ampère's theory, that is, its complete application, already begun by Dumas.

I draw attention, however, to the fact that Gerhardt did not always consistently follow the theory which had given him such fertile results; since he assumed that equal volumes of gaseous bodies contain the same number of molecules, only in the majority of cases, but not always.

I show how he was constrained by a prejudice, the reverse of that of Berzelius, frequently to distort the facts. Whilst Berzelius, on the one hand, did not admit that the molecules of simple substance could be divided in the act of combination, Gerhardt supposes that all the molecules of simple substances could be divided in the act of combination, Gerhardt supposes that all the molecules of simple substances are divisible in chemical action. This prejudice forces him to suppose that the molecule of mercury and all the metals consists of two atoms, like that of hydrogen, and therefore that the compounds of all the metals are of the same type as those of hydrogen. This error even persists in the minds of chemists, and has prevented them from discovering amongst the metals the existence of biatomic radicals perfectly analogous to those lately discovered by Wurtz in organic chemistry.

From the historical examination of chemical theories as well as from physical researches, I draw the conclusion that to bring harmony all the branches of chemistry we must have recourse to the complete application of the theory of Avogadro and Ampère in order to compare the weights and the numbers of the molecules; and I propose in the sequel to show that the conclusions drawn from it are invariably in accordance with all physical and chemical laws hitherto discovered.

I begin in the fifth lecture by applying the hypothesis of Avogadro and Ampère to determine the weights of molecules even before their composition in known.

On the basis of the hypothesis cited above, the weights of the molecules are proportional to the densities of vapours to express the weights of the molecules, it is expedient to refer them all to the density of a simple gas taken as unity, rather than to the weight of a mixture of two gases such as air.

Hydrogen being the lightest gas, we may take it as the unit to which we refer the densities of other gaseous bodies, which in such a case express the weights of the molecules compared to the weight of the molecule of hydrogen = 1.

Since I prefer to take as common unit for the weights of the molecules and for their fractions, the weight of a half and not of the whole molecule of hydrogen, I therefore refer the densities of the various gaseous bodies to that of hydrogen = 2. If the densities are referred to air = 1, it is sufficient to multiply by 14.438 to change them to those referred to that of hydrogen = 1; and by 28.87 to refer them to the density of hydrogen = 2.

I write the two series of number, expressing these weights in the following manner:

Names of Substances	Densities of weights of one volume, the volume of Hydrogen being made=1, i.e., weights of the molecules referred to the weight of a whole molecule of Hydrogen taken as unity.	Densities referred to that of Hydrogen = 2, i.e., weights of the molecules referred to the weight of half a molecule of Hydrogen taken as unity.
Hydrogen	1	2
Oxygen, ordinary	16	32
Oxygen, electrised	64	128
Sulphur below 1000°	96	192
Sulphur* above 1000°	32	64
Chlorine	35.5	71
Bromine	80	160
Arsenic	150	300
Mercury	100	200
Water	9	18
Hydrochloric Acid	18.25	36.50**
Acetic Acid	30	60

* This determination was made by Bineau, but I believe it requires confirmation.

** The numbers expressing the densities are approximate: we arrive at a closer approximation by comparing them with those derived from chemical data, and bringing the two into harmony.

Whoever wishes to refer the densities to hydrogen = 1 and the weights of the molecules to the weight of half a molecule of hydrogen, can say that the weights of the molecule are all represented by the weight of two volumes.

I myself, however, for simplicity of exposition, prefer to refer the densities to that of hydrogen = 2, and so the weights of the molecules are all represented by the weight of one volume.

From the few examples contained in the table, I show that the same substance in its different allotropic states can have different molecular weights, without concealing the fact that the experimental data on which this conclusion is founded still require confirmation.

I assume that the study of the various compounds has been begun by determining the weights of the molecules, i.e., their densities in the gaseous state, without enquiring if they are simple or compound.

I then come to the examination of the composition of these molecules. If the substance is undecomposable, we are forced to admit that its molecule is entirely made up by the weight of one and the same kind of matter. If the body is composite, its elementary analysis is made, and thus we discover the constant relations between the weights of its components: then the weight of the molecule s divided into parts proportional to the numbers expressing the relative weights of the components, and thus we obtain the quantities of these components contained in the molecule of the compound referred to the same unit as that to which we refer the weights of all the molecules. By this method I have constructed the following table:

Name of Substance	Weight of one volume, i.e., weight of the molecule referred to the weight of half a molecule of Hydrogen = 1	Component weights of one volume, i.e., components weights of the molecule, all referred to the weight of half a molecule of Hydrogen = 1
Hydrogen	2	2 Hydrogen
Oxygen, ordinary	32	32 Oxygen
" electrised	128	128 "
Sulphur below 1000°	192	192 Sulphur
Sulphur above 1000° (?)	64	64 "
Phosphorus	124	124 Phosphorus

Name of Substance	Weight of one volume, i.e., weight of the molecule referred to the weight of half a molecule of Hydrogen = 1	Component weights of one volume, i.e., components weights of the molecule, all referred to the weight of half a molecule of Hydrogen = 1
Chlorine	71	71 Chlorine
Bromine	160	160 Bromine
Iodine	254	254 Iodine
Nitrogen	28	28 Nitrogen
Arsenic	300	300 Arsenic
Mercury	200	200 Mercury
Hydrochloric Acid	36.5	35.5 Chlorine 1 Hydrogen
Hydrobromic Acid	81	80 Bromine 1 Hydrogen
Hydriodic Acid	128	127 Iodine 1 Hydrogen
Water	18	16 Oxygen 2 Hydrogen
Ammonia	17	14 Nitrogen 3 Hydrogen
Arseniuretted Hyd.	78	75 Arsenic 3 Hydrogen
Phosphuretted Hyd.	35	32 Phosphorus 3 Hydrogen
Calomel	235.6	35.5 Chlorine 200 Mercury
Corrosive Sublimate	271	71 " 200 "
Arsenic Trichloride	181.5	106.5 " 75 Arsenic
Protochloride of Phosphorus	138.5	106.5 " 32 Phosphorus
Perchloride of Iron	325	213 " 112 Iron
Protoxide of Nitrogen	44	16 Oxygen 28 Nitrogen
Binoxide of Nitrogen	30	16 " 14 "
Carbonic Acid	28	16 " 12 Carbon
" Acid	44	32 " 12 "

Name of Substance	Weight of one volume, i.e., weight of the molecule referred to the weight of half a molecule of Hydrogen = 1	Component weights of one volume, i.e., components weights of the molecule, all referred to the weight of half a molecule of Hydrogen = 1
Ethylene	28	4 Hydrogen 24 "
Propylene	42	6 " 36 "
Acetic Acid, hydrated	60	4 " 32 Oxygen 24 Carbon
" anhydrous	102	6 Hydrogen 48 Oxygen 48 Carbon
Alcohol	46	6 Hydrogen 16 Oxygen 24 Carbon
Ether	74	10 Hydrogen 16 Oxygen 48 Carbon

All the numbers contained in the preceding table are comparable amongst themselves, being referred to the same units. And to fix this well in the minds of pupils, I have recourse to a very simple artifice: I say to them, namely, "Suppose it to be shown that half molecule of hydrogen weighs a millionth of a milligram, then all the numbers of the preceding table become concrete numbers, expressing in millionth of a milligram the concrete weights of the molecules and their components: the same thing would follow if the common unit had any other concrete value," and so I lead them to gain a clear conception of the comparability of these numbers, whatever be the concrete value of the common unit.

Once this artifice has served its purpose, I hasten to destroy it by explaining how it is not possible in reality to know the concrete value of this unit; but the clear ideas remain in the minds of my pupils whatever may be their degree of mathematical knowledge. I proceed pretty much as engineers do when they destroy the wooden scaffolding which has served them to construct their bridges, as soon as these can support themselves. But I fear that you will say, "Is it worth the trouble and the waste of time and ink to tell me of this very common artifice?" I am, however, constrained to tell you that I have paused to do so because I have become attached to this pedagogic expedient, having had such great success with it amongst my pupils, and thus I recommend it to all those who, like myself, must teach chemistry to youths not well accustomed to the comparison of quantities.

Once my students have become familiar with the importance of the numbers as they are exhibited the preceding table, it is easy to lead them to discover the law which results from their comparison. "Compare," I say to them, "the various quantities of the same element contained in the molecule of the free substance and in those of all its different compounds, and you will not be able to escape the following law: The different quantities of the same element contained in different molecules are all whole multiples of on and the same quantity, which, always being entire, as the right to be called an atom."

Translated by The Alembic Club

Reading and Discussion Questions

1. Why does Avogadro play such an important role in Cannizzaro's chemical philosophy? Why is he so concerned with the determination of accurate atomic and molecular weights?
2. What issues of nomenclature are under debate in this period?

The Periodic Law of the Chemical Elements (1889)

Dmitri Mendeleev

75

The high honour bestowed by the Chemical Society in inviting me to pay a tribute to the world-famous name of Faraday by delivering this lecture has induced me to take for its subject the Periodic Law of the Elements—this being a generalisation in chemistry which has of late attracted much attention.

While science is pursuing a steady onward movement, it is convenient from time to time to cast a glance back on the route already traversed, and especially to consider the new conceptions which aim at discovering the general meaning of the stock of facts accumulated from day to day in our laboratories. Owing to the possession of laboratories, modern science now bears a new character, quite unknown not only to antiquity but even to the preceding century. Bacon's and Descartes' idea of submitting the mechanism of science simultaneously to experiment and reasoning has been fully realised in the case of chemistry, it having become not only possible but always customary to experiment. Under the all-penetrating control of experiment, a new theory, even if crude, is quickly strengthened, provided it be founded on a sufficient basis; the asperities are removed, it is amended by degrees, and soon loses the phantom light of a shadowy form or of one founded on mere prejudice; it is able to lead to logical conclusions and to submit to experimental proof. Willingly or not, in science we all must submit not to what seems to us attractive from one point of view or from another, but to what represents an agreement between theory and experiment; in other words, to demonstrated generalisation and to the approved experiment. Is it long since many refused to accept the generalisations involved in the law of Avogadro and Ampère, so widely extended by Gerhardt? We still may hear the voices of its opponents; they enjoy perfect freedom, but vainly will their voices rise so long as they do not use the language of demonstrated facts. The striking observations with the spectroscope which have permitted us to analyse the chemical constitution of distant worlds, seemed, at first, applicable to the task of determining the nature of the atoms themselves; but the working out of the idea in the laboratory soon demonstrated that the characters of spectra are determined—not directly by the atoms, but by the molecules into which the atoms are packed; and so it became evident that more verified facts must be collected before it will be possible to formulate new generalisations capable of taking their place beside those ordinary ones based upon the conception of simple bodies and atoms. But as the shade of the leaves and roots of living plants, together with the relics of a decayed vegetation, favour the growth of the seedling and serve to promote its luxurious development, in like manner sound generalisations—together with the relics of those which have proved to be untenable—promote scientific productivity, and ensure the luxurious growth of science under the influence of rays emanating from the centres of scientific energy. Such centres are scientific associations and societies. Before one of the oldest and most powerful of these I am about to take the liberty of passing in review the 20 years' life of a generalisation which is known under the name of the Periodic Law. It was in March, 1869, that I ventured to lay before the then youthful Russian Chemical Society the ideas upon the same subject, which I had expressed in my just written "Principles of Chemistry."

Without entering into details, I will give the conclusions I then arrived at, in the very words I used:

1. The elements, if arranged according to their atomic weights, exhibit an evident *periodicity* of properties.
2. Elements which are similar as regards their chemical properties have atomic weights which are either of nearly the same value (*e.g.*, platinum, iridium, osmium) or which increase regularly (*e.g.*, potassium, rubidium, caesium).
3. The arrangement of the elements, or of groups of elements in the order of their atomic weights corresponds to their so-called *valencies* as well as, to some extent, to their distinctive chemical properties—as is apparent among other series in that of lithium, beryllium, barium, carbon, nitrogen, oxygen and iron.[1]
4. The elements which are the most widely diffused have *small* atomic weights.
5. The *magnitude* of the atomic weight determines the character of the element just as the magnitude of the molecule determines the character of a compound body.
6. We must expect the discovery of many yet *unknown* elements, for example, elements analogous to aluminium and silicon, whose atomic weight would be between 65 and 75.

[1] The printed speech in *J. Chem. Soc.* says barium and iron. Obviously boron (B) and fluorine (F) are meant. Mendeleev's 1869 paper lists the symbols B and F rather than the names of the elements.— CJG

7. The atomic weight of an element may sometimes be amended by a knowledge of those of the contiguous elements. Thus, the atomic weight of tellurium must lie between 123 and 126, and cannot be 128.
8. Certain characteristic properties of the elements can be foretold from their atomic weights.

The aim of this communication will be fully attained if I succeed in drawing the attention of investigators to those relations which exist between the atomic weights of dissimilar elements, which, as far as I know, have hitherto been almost completely neglected. I believe that the solution of some of the most important problems of our science lies in researches of this kind.

To-day, 20 years after the above conclusions were formulated, they may still be considered as expressing the essence of the now well-known periodic law.

Reverting to the epoch terminating with the sixties, it is proper to indicate three series of data without the knowledge of which the periodic law could not have been discovered, and which rendered its appearance natural and intelligible.

In the first place, it was at that time that the numerical value of atomic weights became definitely known. Ten years earlier such knowledge did not exist, as may be gathered from the fact that in 1860 chemists from all parts of the world met at Karlsruhe in order to come to some agreement, if not with respect to views relating to atoms, at any rate as regards their definite representation. Many of those present probably remember how vain were the hopes of coming to an understanding, and how much ground was gained at that Congress by the followers of the unitary theory so brilliantly represented by Cannizzaro. I vividly remember the impression produced by his speeches, which admitted of no compromise, and seemed to advocate truth itself, based on the conceptions of Avogadro, Gerhardt and Regnault, which at that time were far from being generally recognised. And though no understanding could be arrived at, yet the objects of the meeting were attained, for the ideas of Cannizzaro proved, after a few years, to be the only ones which could stand criticism, and which represented an atom as—"the smallest portion of an element which enters into a molecule of its compound." Only such real atomic weights—not conventional ones—could afford a basis for generalisation. It is sufficient, by way of example, to indicate the following cases in which the relation is seen at once and is perfectly clear:

$$K = 39 \quad Rb = 85 \quad Cs = 133$$

$$Ca = 40 \quad Sr = 87 \quad Ba = 137$$

whereas with the equivalents then in use—

$$K = 39 \quad Rb = 85 \quad Cs = 133$$

$$Ca = 20 \quad Sr = 43.5 \quad Ba = 68.5$$

the consecutiveness of change in atomic weight, which with the true values is so evident, completely disappears.

Secondly, it had become evident during the period 1860–70, and even during the preceding decade, that the relations between the atomic weights of analogous elements were governed by some general and simple laws ....

In such attempts at arrangement and in such views are to be recognised the real forerunners of the periodic law; the ground was prepared for it between 1860 and 1870, and that it was not expressed in a determinate form before the end of the decade, may, I suppose, be ascribed to the fact that only analogous elements had been compared. The idea of seeking for a relation between the atomic weights of all the elements was foreign to the ideas then current, so that neither the *vis tellurique* of De Chancourtois, nor the *law of octaves* of Newlands, could secure anybody's attention. And yet both De Chancourtois and Newlands, like Dumas and Strecker, more than Lenssen and Pettenkofer, had made an approach to the periodic law and had discovered its germs. The solution of the problem advanced but slowly, because the facts, and not the law, stood foremost in all attempts; and the law could not awaken a general interest so long as elements, having no apparent connection with each other, were included in the same octave, as for example:

1st octave of Newlands	H F Cl Co & Ni Br	Pd I Pt & Ir
7th Ditto	O S Fe Se	Rh & Ru Te Au Os or Th

Analogies of the above order seemed quite accidental, and the more so as the octave contained occasionally 10 elements instead of eight, and when two such elements as Ba and V, Co and Ni, or Rh and Ru, occupied one place in the octave. Nevertheless, the fruit was ripening, and I now see clearly that Strecker, De Chancourtois and Newlands stood foremost in the way toward the discovery of the periodic law, and that they merely wanted the boldness necessary to place the whole question at such a height that its reflection on the facts could be clearly seen.

A third circumstance which revealed the periodicity of chemical elements was the accumulation, by the end of the sixties, of new information respecting the rare elements, disclosing their many-sided relations to the other elements and to each other. The researches of Marignac on niobium, and those of Roscoe on vanadium were of special moment. The striking analogies between vanadium and phosphorus on the one hand, and between vanadium and chromium on the other, which became so apparent in the investigations connected with that element, naturally induced the comparison of V = 51 with Cr = 52, Nb = 94 with Mo = 96, and Ta = 192 with W = 194; while, on the other hand, P = 31 could be compared with S = 32, As = 75 with Se = 79, and Sb = 120 with Te = 125. From such approximations there remained but one step to the discovery of the law of periodicity.

The law of periodicity was thus a direct outcome of the stock of generalisations and established facts which had accumulated by the end of the decade 1860–1870: it is an embodiment of those data in a more or less systematic expression. Where, then, lies the secret of the special importance which has since been attached to the periodic law, and has raised it to the position of a generalisation which has already given to chemistry unexpected aid, and which promises to be far more fruitful in the future and to impress upon several branches of chemical research a peculiar and original stamp? ...

In the remaining part of my communication I shall endeavour to show, and as briefly as possible, in how far the periodic law contributes to enlarge our range of vision. Before the promulgation of this law the chemical elements were mere fragmentary, incidental facts in Nature; there was no special reason to expect the discovery of new elements, and the new ones which were discovered from time to time appeared to be possessed of quite novel properties. The law of periodicity first enabled us to perceive undiscovered elements at a distance which formerly was inaccessible to chemical vision; and long ere they were discovered new elements appeared before our eyes possessed of a number of well-defined properties. We now know three cases of elements whose existence and properties were foreseen by the instrumentality of the periodic law. I need but mention the brilliant discovery of *gallium*, which proved to correspond to eka-aluminium of the periodic law, by Lecoq de Boisbaudran; of *scandium*, corresponding to eka-boron, by Nilson; and of *germanium*, which proved to correspond in all respects to eka-silicium, by Winckler. When, in 1871, I described to the Russian Chemical Society the properties, clearly defined by the periodic law, which such elements ought to possess, I never hoped that I should live to mention their discovery to the Chemical Society of Great Britain as a confirmation of the exactitude and the generality of the periodic law. Now, that I have had the happiness of doing so, I unhesitatingly say that although greatly enlarging our vision, even now the periodic law needs further improvements in order that it may become a trustworthy instrument in further discoveries. ...

The alteration of the atomic weight of uranium from U = 120 into U = 240 attracted more attention, the change having been made on account of the periodic law, and for no other reason. Now that Roscoe, Rammelsberg, Zimmermann, and several others have admitted the various claims of the periodic law in the case of uranium, its high atomic weight is received without objection, and it endows that element with a special interest.

While thus demonstrating the necessity of modifying the atomic weights of several insufficiently known elements, the periodic law enabled us also to detect errors in the determination of the atomic weights of several elements whose valencies and true position among other elements were already well known. Three such cases are especially noteworthy: those of tellurium, titanium and platinum. Berzelius had determined the atomic weight of *tellurium* to be 128, while the periodic law claimed for it an atomic weight below that of iodine, which had been fixed by Stas at 126.5, and which was certainly not higher than

127. Brauner then undertook the investigation, and he has shown that the true atomic weight of tellurium is lower than that of iodine, being near to 125 . . . .

The foregoing account is far from being an exhaustive one of all that has already been discovered by means of the periodic law telescope in the boundless realms of chemical evolution. Still less is it an exhaustive account of all that may yet be seen, but I trust that the little which I have said will account for the philosophical interest attached in chemistry to this law. Although but a recent scientific generalisation, it has already stood the test of laboratory verification and appears as an instrument of thought which has not yet been compelled to undergo modification; but it needs not only new applications, but also improvements, further development, and plenty of fresh energy. All this will surely come, seeing that such an assembly of men of science as the Chemical Society of Great Britain has expressed the desire to have the history of the periodic law described in a lecture dedicated to the glorious name of Faraday.

Reading and Discussion Questions

1. Which predecessors does Mendeleev credit with providing the information necessary to develop his periodic table?
2. Around what properties of matter is the periodic table organized? What about it is periodic? What does the existence of his periodic table allow Mendeleev to predict?

On the Dynamical Theory of Heat, with Numerical Results Deduced from Mr Joule's Equivalent of a Thermal Unit, and M. Regnault's Observations on Steam (1851–2)

Kelvin (William Thomson)

76

Introductory Notice

1. Sir Humphry Davy, by his experiment of melting two pieces of ice by rubbing them together, established the following proposition: "The phenomena of repulsion are not dependent on a peculiar elastic fluid for their existence, or caloric does not exist." And he concludes that heat consists of a motion excited among the particles of bodies. "To distinguish this motion from others, and to signify

the cause of our sensation of heat," and of the expansion or expansive pressure produced in matter by heat, "the name *repulsive* motion has been adopted."[1]

2. The dynamical theory of heat, thus established by Sir Humphry Davy, is extended to radiant heat by the discovery of phenomena, especially those of the polarization of radiant heat, which render it excessively probable that heat propagated through "vacant space," or through diathermanic substances, consists of waves of transverse vibrations in an all-pervading medium.

3. The recent discoveries made by Mayer and Joule,[2] of the generation of heat through the friction of fluids in motion, and by the magneto-electric excitation of galvanic currents, would either of them be sufficient to demonstrate the immateriality of heat; and would so afford, if required, a perfect confirmation of Sir Humphry Davy's views.

4. Considering it as thus established, that heat is not a substance, but a dynamical form of mechanical effect, we perceive that there must be an equivalence between mechanical work and heat, as between cause and effect. The first published statement of this principle appears to be in Mayer's *Bemerkungen über die Kräfte der unbelebten Natur*,[3] which contains some correct views regarding the mutual convertibility of heat and mechanical effect, along with a false analogy between the approach of a weight to the earth and a diminution of the volume of a continuous substance, on which an attempt is founded to find numerically the mechanical equivalent of a given quantity of heat. In a paper published about fourteen months later, "On the Calorific Effects of Magneto-Electricity and the Mechanical Value of Heat,"[4] Mr. Joule, of Manchester, expresses very distinctly the consequences regarding the mutual convertibility of heat and mechanical effect which follow from the fact that heat is not a substance but a state of motion; and investigates on unquestionable principles the "absolute numerical relations," according to which heat is connected with mechanical power; verifying experimentally, that whenever heat is generated from purely mechanical action, and no other effect produced, whether it be by means of the friction of fluids or by the magneto-electric excitation of galvanic currents, the same quantity is generated by the same amount of work spent; and determining the actual amount of work, in foot-pounds, required to generate a unit of heat, which he calls "the mechanical equivalent of heat." Since the publication of that paper,

[1] From Davy's first work, entitled *An Essay on Heat, Light, and the Combinations of Light*, published in 1799, in "Contributions to Physical and Medical Knowledge, principally from the West of England, collected by Thomas Beddoes, M.D.," and republished in Dr Davy's edition of his brother's collected works, Vol. II. Lond. 1836.

[2] In May, 1842, Mayer announced in the *Annalen* of Wöhler and Liebig, that he had raised the temperature of water from 12° to 13° Cent. by agitating it. In August, 1843, Joule announced to the British Association "That heat is evolved by the passage of water through narrow tubes;" and that he had "obtained one degree of heat per lb. of water from a mechanical force capable of raising 770 lbs. to the height of one foot;" and that heat is generated when work is spent in turning a magneto-electric machine, or an electro-magnetic machine. (See his paper "On the Calorific Effects of Magneto-Electricity, and on the Mechanical Value of Heat."—*Phil. Mag.*, Vol. XXIII., 1843.)

[3] *Annalen* of Wöhler and Leibig, May, 1842.

[4] British Association, August, 1843; and *Phil. Mag.*, September, 1843.

Mr. Joule has made numerous series of experiments for determining with as much accuracy as possible the mechanical equivalent of heat so defined, and has given accounts of them in various communications to the British Association, to the *Philosophical Magazine*, to the Royal Society, and to the French Institute.

5. Important contributions to the dynamical theory of heat have recently been made by Rankine and Clausius; who, by mathematical reasoning analogous to Carnot's on the motive power of heat, but founded on an axiom contrary to his fundamental axiom, have arrived at some remarkable conclusions. The researches of these authors have been published in the *Transactions* of this Society, and in Poggendorff's *Annalen*, during the past year; and they are more particularly referred to below in connection with corresponding parts of the investigations at present laid before the Royal Society.

. . .

6. The object of the present paper is threefold:

(1) To show what modifications of the conclusions arrived at by Carnot, and by others who have followed his peculiar mode of reasoning regarding the motive power of heat, must be made when the hypothesis of the dynamical theory, contrary as it is to Carnot's fundamental hypothesis, is adopted.

(2) To point out the significance in the dynamical theory, of the numerical results deduced from Regnault's observations on steam, and communicated about two years ago to the Society, with an account of Carnot's theory, by the author of the present paper; and to show that by taking these numbers (subject to correction when accurate experimental data regarding the density of saturated steam shall have been afforded), in connection with Joule's mechanical equivalent of a thermal unit, a complete theory of the motive power of heat, within the temperature limits of the experimental data, is obtained.

(3) To point out some remarkable relations connecting the physical properties of all substances, established by reasoning analogous to that of Carnot, but founded in part on the contrary principle of the dynamical theory.

Part One

Fundamental Principles in the Theory of the Motive Power of Heat

7. According to an obvious principle, first introduced, however, into the theory of the motive power and heat by Carnot, mechanical effect produced in

any process cannot be said to have been derived from a purely thermal source, unless at the end of the process all the materials used are in precisely the same physical and mechanical circumstances as they were at the beginning. In some conceivable "thermo-dynamic engines," as, for instance, Faraday's floating magnet, or Barlow's "wheel and axle," made to rotate and perform work uniformly by means of a current continuously excited by heat communicated to two metals in contact, or the thermo-electric rotatory apparatus devised by Marsh, which has been actually constructed, this condition is fulfilled at every instant. On the other hand, in all thermo-dynamic engines, founded on electrical agency, in which discontinuous galvanic currents, or pieces of soft iron in a variable state of magnetization, are used, and in all engines founded on the alternate expansions and contractions of media, there are really alterations in the condition of materials; but, in accordance with the principle stated above, these alterations must be strictly periodical. In any such engine the series of motions performed during a period, at the end of which the materials are restored to precisely the same condition as that in which they existed at the beginning, constitutes what will be called a complete cycle of its operations. Whenever in what follows, the work done or the mechanical effect produced by a thermo-dynamic engine is mentioned without qualification, it must be understood that the mechanical effect produced, either in a non-varying engine, or in a complete cycle, or any number of complete cycles of a periodical engine, is meant.

8. The source of heat will always be supposed to be a hot body at a given constant temperature put in contact with some part of the engine; and when any part of the engine is to be kept from rising in temperature (which can only be done by drawing off whatever heat is deposited in it), this will be supposed to be done by putting a cold body, which will be called the refrigerator, at a given constant temperature in contact with it.

9. The whole theory of the motive power of heat is founded on the two following propositions, due respectively to Joule, and to Carnot and Clausius.

Prop. I. (Joule).—When equal quantities of mechanical effect are produced by any means whatever from purely thermal sources, or lost in purely thermal effects, equal quantities of heat are put out of existence or are generated.

Prop. II. (Carnot and Clausius).—If an engine be such that, when it is worked backwards, the physical and mechanical agencies in every part of its motions are all reversed, it produces as much mechanical effect as can be produced by any thermo-dynamic engine, with the same temperatures of source and refrigerator, from a given quantity of heat.

10. The former proposition is shown to be included in the general "principle of mechanical effect," and is so established beyond all doubt by the following demonstration.

11. By whatever direct effect the heat gained or lost by a body in any conceivable circumstances is tested, the measurement of its quantity may always be founded on a determination of the quantity of some standard substance, which it or any equal quantity of heat could raise from one standard temperature to

another; the test of equality between two quantities of heat being their capability of raising equal quantities of any substance from any temperature to the same higher temperatures. Now, according to the dynamical theory of heat, the temperature of a substance can only be raised by working upon it in some way so as to produce increased thermal motions within it, besides effecting any modifications in the mutual distances or arrangements of its particles which may accompany a change of temperature. The work necessary to produce this total mechanical effect is of course proportional to the quantity of the substance raised from one standard temperature to another; and therefore when a body, or a group of bodies, or a machine, parts with or receives heat, there is in reality mechanical effect produced from it, or taken into it, to an extent precisely proportional to the quantity of heat which it emits or absorbs. But the work which any external forces do upon it, the work done by its own molecular forces, and the amount by which the half *vis viva* of the thermal motions of all its parts is diminished, must together be equal to the mechanical effect produced from it; and, consequently, to the mechanical equivalent of the heat which it emits (which will be positive or negative, according as the sum of those terms is positive or negative). Now let there be either no molecular change or alteration of temperature in any part of the body, or, by a cycle of operations, let the temperature and physical condition be restored exactly to what they were at the beginning; the second and third of the three parts of the work which it has to produce vanish; and we conclude that the heat which it emits or absorbs will be the thermal equivalent of the work done upon it by external forces, or done by it against external forces; which is the proposition to be proved.

12. The demonstration of the second proposition is founded on the following axiom:—

> *It is impossible, by means of inanimate material agency, to derive mechanical effect from any portion of matter by cooling it below the temperature of the coldest of the surrounding objects.*[5]

13. To demonstrate the second proposition, let A and B be two thermo-dynamic engines, of which B satisfies the conditions expressed in the enunciation; and let, if possible A derive more work from a given quantity of heat than B, when their sources and refrigerators are at the same temperatures, respectively. Then on account of the condition of complete reversibility in all its operations which it fulfills, B may be worked backwards, and made to restore any quantity of heat to its source, by the expenditure of the amount of work which, by its forward action, it would derive from the same quantity of heat. If, therefore, B be worked backwards, and made to restore to the source of A (which we may suppose to be adjustable to the engine B) as much heat as has been drawn from it during a

[5] If this axiom be denied for all temperature, it would have to be admitted that a self-acting machine might be set to work and produce mechanical effect by cooling the sea or earth, with no limit but the total loss of heat from the earth and sea, or, in reality, from the whole material world.

certain period of the working of *A*, a smaller amount of work will be spent thus than was gained by the working of *A*. Hence, if such a series of operations of *A* forwards and of *B* backwards be continued, either alternately or simultaneously, there will result a continued production of work without any continued abstraction of heat from the source; and, by Prop. I., it follows that there must be more heat abstracted from the refrigerator by the working of *B* backwards than is deposited in it by *A*. Now it is obvious that *A* might be made to spend part of its work in working *B* backwards, and the whole might be made self-acting. Also, there being no heat either taken from or given to the source on the whole, all the surrounding bodies and space except the refrigerator might, without interfering with any of the conditions which have been assumed, be made of the same temperature as the source, whatever that may be. We should thus have a self acting machine, capable of drawing heat constantly from a body surrounded by others of a higher temperature, and converting it into mechanical effect. But this is contrary to the axiom, and therefore we conclude that the hypothesis that *A* derives more mechanical effect from the same quantity of heat drawn from the source than *B* is false. Hence no engine whatever, with source and refrigerator at the same temperatures, can get more work from a given quantity of heat introduced than any engine which satisfies the condition of reversibility, which was to be proved.

14. This proposition was first enunciated by Carnot, being the expression of his criterion of a perfect thermo-dynamic engine.[6] He proved it by demonstrating that a negation of it would require the admission that there might be a self-acting machine constructed which would produce mechanical effect indefinitely, without any source either in heat or the consumption of materials, or any other physical agency; but this demonstration involves, fundamentally, the assumption that, in "a complete cycle of operations," the medium parts with exactly the same quantity of heat as it receives. A very strong expression of doubt regarding the truth of this assumption, as a universal principle, is given by Carnot himself;[7] and that it is false, where mechanical work is, on the whole, either gained or spent in the operations, may (as I have tried to show above) be considered to be perfectly certain. It must then be admitted that Carnot's original demonstration utterly fails, but we cannot infer that the proposition concluded is false. The truth of the conclusion appeared to me, indeed so probable that I took it in connection with Joule's principle, on account of which Carnot's demonstration of it fails, as the foundation of an investigation of the motive power of heat[8] in air-engines or steam-engines through finite ranges of temperature, and obtained about a year ago results, of which the substance is given in the second part of the paper at present communicated to the Royal Society. It was not until the commencement of the present year that I found the demonstration

[6] Account of Carnot's *Theory*, §13.
[7] Account of Carnot's *Theory*, §6.
[8] Poggendorff's *Annalen*, referred to above.

given above, by which the truth of the proposition is established upon an axiom, which I think will be generally admitted. It is with no wish to claim priority that I make these statements, as the merit of first establishing the proposition upon correct principles is entirely due to Clausius, who published his demonstration of it in the month of May last year, in the second part of his paper on the motive power of heat. I may be allowed to add that I have given the demonstration exactly as it occurred to me before I knew that Clausius had either enunciated or demonstrated the proposition. The following is the axiom on which Clausius's demonstration is founded:—

> *It is impossible for a self-acting machine, unaided by any external agency, to convey heat from one body to another at a higher temperature.*

It is easily shown that, although this and the axiom I have used are different in form, either is a consequence of the other. The reasoning in each demonstration is strictly analogous to that which Carnot originally gave.

Reading and Discussion Questions

1. What evidence does Kelvin provide against the caloric theory of heat?
2. Whose work does he think most contributes to his own dynamical theory?
3. Kelvin writes that "*It is impossible for a self-acting machine, unaided by any external agency, to convey heat from one body to another at a higher temperature.*" What connection does Kelvin take this axiom to have with the work of Clausius and Carnot?

On the Interaction of Natural Forces (1854)

Hermann von Helmholtz

77

A new conquest of very general interest has been recently made by natural philosophy. In the following pages I will endeavor to give a notion of the nature of this conquest. It has reference to a new and universal natural law, which rules the action of natural forces in their mutual relations towards each other, and is as influential on our theoretic views of natural processes as it is important in their technical applications.

Among the practical arts which owe their progress to the development of the natural sciences, from the conclusion of the middle ages downwards, practical mechanics, aided by the mathematical science which bears the same name, was one of the most prominent. The character of the art was, at the time referred to, naturally very different from its present one. Surprised and stimulated by its own success, it thought no problem beyond its power, and immediately attacked some of the most difficult and complicated. Thus it was attempted to build automaton figures which should perform the functions of men and animals . . . .

From these efforts to imitate living creatures, another idea, also by a misunderstanding, seems to have developed itself, which, as it were, formed the new philosopher's stone of the seventeenth and eighteenth centuries. It was now the endeavor to construct a perpetual motion. Under this term was understood a machine, which, without being wound up, without consuming in the working of it falling water, wind, or any other natural force, should still continue in motion, the motive power being perpetually supplied by the machine itself. Beasts and human beings seemed to correspond to the idea of such an apparatus, for they moved themselves energetically and incessantly as long as they lived, and were never wound up; nobody set them in motion. A connexion between the taking-in of nourishment and the development of force did not make itself apparent. The nourishment seemed only necessary to grease, as it were, the wheel work of the animal machine, to replace what was used up, and to renew the old. The development of force out of itself seemed to be the essential peculiarity, the real quintessence of organic life. If, therefore, men were to be constructed, a perpetual motion must first be found.

Another hope also seemed to take up incidentally the second place, which in our wiser age would certainly have claimed the first rank in the thoughts of men. The perpetual motion was to produce work inexhaustibly without corresponding consumption, that is to say, out of nothing. Work, however, is money. Here, therefore, the great practical problem which the cunning heads of all centuries have followed in the most diverse ways, namely, to fabricate money out of nothing, invited solution. The similarity with the philosopher's stone sought by the ancient chemists was complete. That also was thought to contain the quintessence of organic life, and to be capable of producing gold . . . .

We have here arrived at the idea of the driving force or power of a machine, and shall have much to do with it in future. I must therefore give an explanation of it. The idea of work is evidently transferred to machines by comparing their arrangements with those of men and animals, to replace which they were applied. We still reckon the work of steam-engines according to horse power. The value of manual labor is determined partly by the force which is expended in it (a strong laborer is valued more highly than a weak one), partly, however, by the skill which is brought into action. A machine, on the contrary, which executes work skillfully, can always be multiplied to any extent; hence its skill has not the high value of human skill in domains where the latter cannot be supplied by machines. Thus the idea of the quantity of work in the case of machines has been limited to the consideration of the expenditure of force; this was the more important,. as indeed most machines are constructed for the express purpose of exceeding, by the magnitude of their effects, the powers of men and animals. Hence, in a mechanical sense, the idea of work is become identical with that of the expenditure of force, and in this way I will apply it in the following pages.

How, then, can *we* measure this expenditure, and compare it in the case of different machines?

I must here conduct you a portion of the way—as short a portion as possible—over the uninviting field of mathematico-mechanical ideas, in order to bring you to a point of view from which a more rewarding prospect will open. And though the example which I shall here choose, namely, that of a water-mill with iron hammer, appears to be tolerably romantic, still, alas, I must leave the dark forest valley, the spark-emitting anvil, and the black Cyclops wholly out of sight, and beg a moment's attention to the less poetic side of the question, namely, the machinery. This is driven by a water-wheel, which in its turn is set in motion by the falling water. The axle of the water-wheel has at certain places small projections, thumbs, which, during the rotation, lift the heavy hammer and permit it to fall again. The falling hammer belabors the mass of metal, which is introduced beneath it. The work therefore done by the machine consists, in this case, in the lifting of the hammer, to do which the gravity of the latter must be overcome. The expenditure of force will in the first place, other circumstances being equal, be proportional to the weight of the hammer; it will, for example, be double when the weight of the hammer is doubled. But the action of the hammer depends not upon its weight alone, but also upon the height from which it falls. If it falls through two feet, it will produce a greater effect than if it falls through only one foot. It is, however, clear that if the machine, with a certain expenditure of force, lifts the hammer a foot in height, the same amount of force must be expended to raise it a second foot in height. The work is therefore not only doubled when the weight of the hammer is increased twofold, but also when the space through which it falls is doubled. From this it is easy to see that the work must be measured by the product of the weight into the space through which it ascends. And in this way, indeed, do we measure in mechanics. The unit of work is a foot-pound, that is, a pound weight raised to the height of one foot .... if we multiply the weight of the falling water by the height through which it falls, and regard, as before, the product as the measure of the work, then the work performed by the machine in raising the hammer, can, in the most favorable case, be only equal to the number of foot-pounds of water which have fallen in the same time. In practice, indeed, this ratio is by no means attained: a great portion of the work of the falling water escapes unused, inasmuch as part of the force is willingly sacrificed for the sake of obtaining greater speed ....

Our machinery, therefore, has in the first place done nothing more than make use of the gravity of the falling water in order to overpower the gravity of the hammer, and to raise the latter. When it has lifted the hammer to the necessary height, it again liberates it, and the hammer falls upon the metal mass which is pushed beneath it. But why docs the falling hammer here exercise a greater force than when it is permitted simply to press with its own weight on the mass of metal? Why is its power greater as the height from which it falls is increased? We find, in fact, that the work performed by the hammer is determined by its velocity. In other cases, also, the velocity of moving masses is a means of producing great effects. I only remind you of the destructive effects of musket-bullets, which in a state of rest arc the most harmless things in the world. I remind you of the windmill, which derives its force from the moving air. It may appear surprising that motion, which we are accustomed to regard as a non-essential and transitory endowment of bodies, can produce such great effects. But the fact is, that motion appears to us under ordinary circumstances transitory, because the movement of all terrestrial bodies is resisted perpetually by other forces, friction, resistance of the air, &c., so that the motion is incessantly weakened and finally neutralized. A body, however, which is opposed by no resisting force, when once set in motion, moves onward eternally with undiminished velocity. Thus we know that the planetary bodies have moved without change through space for thousands of years. Only by resisting forces can motion be diminished or destroyed. A moving body, such as the hammer or the musket-ball, when it strikes against another, presses the latter together, or penetrates it, until the sum of the resisting forces which the body struck presents to its pressure, or to the separation of its particles, is sufficiently great to destroy the motion of the hammer or of the bullet.

The motion of a mass regarded as taking the place of working force is called the living force (*vis viva*) of the mass. The word "living" has of course here no reference whatever to living things, but is intended to represent solely the force of the motion as distinguished from the state of unchanged rest—from the gravity of a motionless body, for example, which produces an incessant pressure against the surface which supports it, but does not produce any motion.

In the case before us, therefore, we had first power in the form of a falling mass of water, then in the form of a lifted hammer, and thirdly in the form of the living force of the falling hammer. We should transform the third form into the second, if we for example, permitted the hammer to fall upon a highly elastic steel beam strong enough to resist the shock. The hammer would rebound, and in the most favorable case would reach a height equal to that from which it fell, but would never rise higher. In this way its mass would ascend; and at the moment when its highest point has been attained it would represent the same number of raised foot-pounds as before it fell, never in greater number; that is to say, living force can generate the same amount of work as that expended in its production. It is therefore equivalent to this quantity of work.

Our clocks are driven by means of sinking weights, and our watches by means of the tension of springs. A weight which lies on the ground, an elastic spring which is without tension, can produce no effects: to obtain such we must first raise the weight or impart tension to the spring, which is accomplished when we wind up our clocks and watches. The man who winds the clock or watch communicates to the weight or to the spring a certain amount of power, and exactly so much as is thus communicated is gradually given out again during the following twenty-four hours, the original force being thus slowly consumed to overcome the friction of the wheels and the resistance which the pendulum encounters from the air. The wheelwork of the clock therefore exhibits no working force which was not previously communicated to it, but simply distributes the force given to it uniformly over a longer time.

Into the chamber of an air-gun we squeeze, by means of a condensing air-pump, a great quantity of air. When we afterwards open the cock of a gun and admit the compressed air into the barrel, the ball is driven out of the latter with a force similar to that exerted by ignited powder. Now we may determine the work consumed in the pumping-in of the air, and the living force which, upon firing, is communicated to the ball, but we shall never find the latter greater than the former. The compressed air has generated no working force, but simply gives to the bullet that which has been previously communicated to it. And while we have pumped for perhaps a quarter of an hour to charge the gun, the force is expended in a few seconds when the bullet is discharged; but because the action is compressed into so short a time, a much greater velocity is imparted to the ball than would be possible to communicate to it by the unaided effort of the arm in throwing it.

From these examples you observe, and the mathematical theory has corroborated this for all purely mechanical, that is to say, for moving forces, that all our machinery and apparatus generate no force, but simply yield up the power communicated to them by natural forces,- falling water, moving wind, or by the muscles of men and animals. After this law had been established by the great mathematicians of the last century, a perpetual motion, which should only make use of pure mechanical forces, such as gravity, elasticity, pressure of liquids and gases, could only be sought after by bewildered and ill-instructed people. But there are still other natural forces which are not reckoned among the purely moving forces,-heat, electricity, magnetism, light, chemical forces, all of which nevertheless stand in manifold relation to mechanical processes. There is hardly a natural process to be found which is not accompanied by mechanical actions, or from which mechanical work may not be derived. Here the question of a perpetual motion remained open; the decision of this question marks the progress of modern physics, regarding which I promised to address you.

In the case of the air-gun, the work to be accomplished in the propulsion of the ball was given by the arm of the man who pumped in the air. In ordinary firearms, the condensed mass of air which propels the bullet is obtained in a totally different manner, namely, by the combustion of the powder. Gunpowder is transformed by combustion for the most part into gaseous products, which endeavor to occupy a much larger space than that previously taken up by the volume of the powder. Thus you see, that, by the use of gunpowder, the work which the human arm must accomplish in the case of the air-gun is spared.

In the mightiest of our machines, the steam-engine, it is a strongly compressed aeriform body, water vapor, which, by its effort to expand, sets the machine in motion. Here also we do not condense the steam by means of an external mechanical force, but by communicating heat to a mass of water in a closed boiler, we change this water into steam, which, in consequence of the limits of the space, is developed under strong pressure. In this case, therefore, it is the heat communicated which generates the mechanical force. The heat thus necessary for the machine we might obtain in many ways: the ordinary method is to procure it from the combustion of coal.

Combustion is a chemical process. A particular constituent of our atmosphere, oxygen, possesses a strong force of attraction, or, as it is named in chemistry, a strong affinity for the constituents of the combustible body, which affinity, however, in most cases can only exert itself at high temperatures. As soon as a portion of the combustible body, for example the coal, is sufficiently heated, the carbon unites itself with great violence to the oxygen of the atmosphere and forms a peculiar gas, carbonic acid, the same which we see foaming from beer and champagne. By this combination, light and heat are generated: heat is generally developed by any combination of two bodies of strong affinity for each other; and when the heat is intense enough, light appears. Hence in the steam-engine it is chemical processes and chemical forces which produce the astonishing work of these machines. In like manner the combustion of gunpowder is a chemical process, which in the barrel of the gun communicates living force to the bullet.

While now the steam-engine develops for us mechanical work out of heat, we can conversely generate heat by mechanical forces. A skillful blacksmith can render an iron wedge red-hot by hammering. The axles of our carriages must be protected by careful greasing from ignition through friction. And lately this property has been applied on a large scale. In some factories, where a surplus of water-power is at hand, this surplus is applied to cause a strong iron plate to rotate swiftly upon another, so that they become strongly heated by the friction. The heat so obtained warms the room, and thus a stove without fuel is provided. Now could not the heat generated by the plates be applied to a small steam-engine, which in its turn should be able to keep the rubbing plates in motion? The perpetual motion would thus be at length found. This question might be asked, and could not be decided by the older mathematico-mechanical investigations. I will remark beforehand, that the general law which I will lay before you answers the question in the negative.

By a similar plan, however, a speculative American set some time ago the industrial world of Europe in excitement. The magneto-electric machines often made use of in the case of rheumatic disorders are well known to the public. By imparting a swift rotation to the magnet of such a machine we obtain powerful currents of electricity. If those be conducted through water, the latter will be reduced into its two components, oxygen and hydrogen. By the combustion of hydrogen, water is again generated. If this combustion takes place, not in atmospheric air, of which oxygen only constitutes a fifth part, but in pure oxygen, and if a bit of chalk be placed in the flame, the chalk will be raised to a white heat, and give us the sun-like Drummond's light. At the same time the flame develops a considerable quantity of heat. Our American proposed to utilize in this way the gases obtained from electrolytic decomposition, and asserted, that by the combustion a sufficient amount of heat was generated to keep a small steam-engine in action, which again drove his magneto-electric machine, decomposed the water, and thus continually prepared its own fuel. This would certainly have been the most splendid of all discoveries; a perpetual motion which, besides the force that kept it going, generated light like the sun, and warmed all around it. The matter was by no means badly cogitated. Each practical step in the affair was known to be possible; but those who at that time were acquainted with the physical investigations which bear upon this subject, could have affirmed, on first hearing the report, that the matter was to be numbered among the numerous stories of the fable-rich America; and indeed a fable it remained.

. . .

Now it is clear that if by any means we could succeed, through mechanical forces, as the above American professed to have done, to excite chemical, electrical, or other natural processes, which, by any circuit whatever, and without altering permanently the active masses in the machine, could produce mechanical force in greater quantity than that at first applied, a portion of the work thus gained might be made use of to keep the machine in motion, while the rest of the work might be applied to any other purpose whatever. The problem was to find, in the complicated net of reciprocal actions, a track through chemical, electrical, magnetical, and thermic processes, back to mechanical actions, which might be followed with a final gain of mechanical work: thus would the perpetual motion be found.

But, warned by the futility of former experiments, the public had become wiser. On the whole, people did not seek much after combinations which promised to furnish a perpetual motion, but the question was inverted. It was no more asked, how can I make use of the known and unknown relations of natural forces so as to construct a perpetual motion? but it was asked, if a perpetual motion be impossible, what are the relations which must subsist between natural forces? Everything was gained by this inversion of the question. The relations of natural forces rendered necessary by the above assumption, might be easily and completely stated. It was found that all known relations of forces harmonize with the consequences of that assumption, and a series of unknown relations were discovered at the same time, the correctness of which remained to be proved. If a single one of them could be proved false, then a perpetual motion would be possible.

The first who endeavored to travel this way was a Frenchman named Carnot, in the year 1824. In spite of a too limited conception of his subject, and an incorrect view as to the nature of heat, which led him to some erroneous conclusions, his experiment was not quite unsuccessful. He discovered a law which now bears his name, and to which I will return further on. His labors remained for a long time without notice, and it was not till eighteen years afterwards, that is, in 1842, that different investigators in different countries, and independent of Carnot, laid hold of the same thought. The first who saw truly the general law here referred to, and expressed it correctly, was a German physician, J. R. Mayer of Heilbronn, in the year 1842. A little later, in 1843, a Dane named Colding, presented a memoir to the Academy of Copenhagen, in which the same law found utterance, and some experiments were described for its further corroboration. In England, Joule began about the same time to make experiments having reference to the same subject. We often find, in the case of questions to the solution of which the development of science points, that several heads, quite independent of each other, generate exactly the same series of reflections.

I myself, without being acquainted with either Mayer or Colding, and having first made the acquaintance of Joule's experiments at the end of my investigation, followed the same path. I endeavored to ascertain all the relations between the different natural processes, which followed from our regarding them from the above point of view. My inquiry was made public in 1847, in a small pamphlet bearing the title, "On the Conservation of Force."

Since that time the interest of the scientific public for this subject has gradually augmented, particularly in England, of which I had an opportunity of convincing myself during a visit last summer. A great number of the essential consequences of the above manner of viewing the subject, the proof of which was wanting when the first theoretic notions were published, have since been confirmed by experiment, particularly by those of Joule; and during the last year the most eminent physicist of France, Regnault, has adopted the new mode of regarding the question, and by fresh investigations on the specific heat or gases has contributed much to its support. For some important consequences the experimental proof is still wanting, but the number of confirmations is so predominant, that I have not deemed it too early to bring the subject before even a non-scientific audience.

How the question has been decided you may already infer from what has been stated. In the series of natural processes there is no circuit to be found, by which mechanical force can be gained without a corresponding consumption. The perpetual motion remains impossible. Our reflections, however, gain thereby a higher interest.

We have thus far regarded the development of force by natural processes, only in its relation to its usefulness to man, as mechanical force. You now see that we have arrived at a general law, which holds good wholly independent of the application which man makes of natural forces; we must therefore make the expression of our law correspond to this more general significance. It is in the first place clear, that the work which, by any natural process whatever, is performed under favorable conditions by a machine, and which may be measured in the way already indicated, may be used as a measure of force common to all. Further, the important question arises, if the quantity of force cannot be augmented except by corresponding consumption, can it be diminished or lost? For the purposes of our machines it certainly can, if we neglect the opportunity to convert natural processes to use, but as investigation has proved, not for nature as a whole.

In the collision and friction of bodies against each other, the mechanics of former years assumed simply that living force was lost. But I have already stated that each collision and each act of friction generates heat; and, moreover, Joule has established by experiment the important law, that for every foot-pound of force which is lost a definite quantity of heat is always generated, and that when work is performed by the consumption of heat, for each foot-pound thus gained a definite quantity of heat disappears. The quantity of heat necessary to raise the temperature of a pound of water a degree of the Centigrade thermometer, corresponds to a mechanical force by which a pound weight would be raised to the height of 1350 feet: we name this quantity the mechanical equivalent of heat. I may mention here that these facts conduct of necessity to the conclusion, that heat is not, as was formerly imagined, a fine imponderable substance, but that, like light, it is a peculiar vibratory motion of the ultimate particles of bodies. In collision and friction, according to this manner of viewing the subject, the motion of the mass of a body which is apparently lost is converted into a motion of the ultimate particles of the body; and conversely, when mechanical force is generated by heat, the motion of the ultimate particles is converted into a motion of the mass.

Chemical combinations generate heat, and the quantity of this heat is totally independent of the time and steps through which the combination has been effected, provided that other actions are not at the same time brought into play. If, however, mechanical work is at the same time accomplished, as in the case of the steam-engine, we obtain as much less heat as is equivalent to this work. The quantity of work produced by chemical force is in general very great. A pound of the purest coal gives, when burnt, sufficient heat to raise the temperature of 8086 lbs. of water one degree of the Centigrade thermometer; from this we can calculate that the magnitude of the chemical force of attraction between the particles of a pound of coal and the quantity of oxygen that corresponds to it, is capable of lifting a weight of 100 pounds to a height of twenty miles. Unfortunately in our steam-engines we have hitherto been able to gain only the smallest portion of this work, the greater part being lost in the shape of heat. The best expansive engines give back as mechanical work only eighteen percent of the heat generated by the fuel.

From a similar investigation of all the other known physical and chemical processes, we arrive at the conclusion, that nature, as a whole, possesses a store of force which cannot in any way be either increased or diminished, and that therefore the quantity of force in nature is just as eternal and unalterable as the quantity of matter. Expressed in this form, I have named the general law "The Principle of the Conservation of Force."

We can not create mechanical force, but we may help ourselves from the general storehouse of nature. The brook and the wind, which drive our mills, the forest and the coal-bed, which supply our steam-engines and warm our rooms, are to us the bearers of a small portion of the great natural supply which we draw upon for our purposes, and the actions of which we can apply as we think fit. The possessor of a mill claims the gravity of the descending rivulet, or the living force of the moving wind, as his possession. These portions of the store of nature are what give his property its chief value.

Further, from the fact that no portion of force can be absolutely lost, it does not follow that a portion may not be inapplicable to human purposes. In this respect the inferences drawn by William Thomson from the law of Carnot are of importance. This law, which was discovered by Carnot during his endeavors to ascertain the relations between heat and mechanical force—which, however, by no means belongs to the necessary consequences of the conservation of force, and which Clausius was the first to modify in such a manner that it no longer contradicted the above general law—expresses a certain relation between the compressibility, the capacity for heat, and the expansion by heat, of all bodies. It is not yet considered as actually proved, but some remarkable deductions having been drawn from it and afterwards proved to be facts by experiment, it has attained thereby a great degree of probability. Besides the mathematical form in which the law was first expressed by Carnot, we can give it the following more general expression: "Only when heat passes from a warmer to a colder body, and even then only partially, can it be converted into mechanical work."

The heat of a body which we cannot cool further, cannot be changed into another form of force; into the electric or chemical force, for example. Thus in our steam-engines we convert a portion of the heat of the glowing coal into work, by permitting it to pass to the less warm water of the boiler. If, however, all the bodies in nature bad the same temperature, it would be impossible to convert any portion of their heat into mechanical work. According to this we can divide the total force store of the universe into two parts, one of which is heat, and must continue to be such; the other, to which a portion of the heat of the warmer bodies, and the total supply of chemical, mechanical, electrical, and magnetical forces belong, is capable of the most varied changes of form, and constitutes the whole wealth of change which takes place in nature.

But the heat of the warmer bodies strives perpetually to pass to bodies less warm by radiation and conduction, and thus to establish an equilibrium of temperature. At each motion of a terrestrial body a portion of mechanical force passes by friction or collision into heat, of which only a part can be converted back again into mechanical force. This is also generally the case in every electrical and chemical process. From this it follows, that the first portion of the store of force, the unchangeable heat, is augmented by every natural process, while the second portion, mechanical, electrical, and chemical force, must be diminished; so that if the universe be delivered over to the undisturbed action of its physical processes, all force will finally pass into the form of heat, and all heat come into a state of equilibrium. Then all possibility of a further change would be at an end, and the complete cessation of all natural processes must set in. The life of men, animals, and plants, could not of course continue if the sun had lost his high temperature, and with it his light, if all the components of the earth's surface had ceased those combinations which their affinities demand. In short, the universe from that time forward would be condemned to a state of eternal rest.

These consequences of the law of Carnot are of course only valid, provided that the law, when sufficiently tested, proves to be universally correct. In the mean time there is little prospect of the law being proved incorrect. At all events we must admire the sagacity of Thomson, who, in the letters of a long-known little mathematical formula, which only speaks of the heat, volume and pressure of the bodies, was able to discern consequences which threatened the universe, though certainly after an infinite period of time, with eternal death.

...

Translated by John Tyndall

Reading and Discussion Questions

1. Helmholtz begins by describing how different mechanical motions can be converted into one another. What other forms of energy can be converted into motion and vice versa?

2. On what grounds does he argue that perpetual motion is impossible? What about these laws of energy "threatened the universe ... with eternal death?"

Additional Resources

Chang, Hasok. "We Have Never Been Whiggish About Phlogiston." *Centaurus*, 51(4) 2009, 239–64. A controversial but historically unimpeachable reassessment of the Priestley/Lavoisier debate.

Dugan, David, producer and director. *Absolute Zero.* USA: PBS 2007. A two-part series, the first of which has a good historical introduction to theories of heat and the development of thermodynamics

Fisher, Howard J. *Faraday's Experimental Researches in Electricity: The First Series.* Santa Fe, NM: Green Lion Press, 2004.

Giunta, Carmen, ed. *Classic Chemistry* (https://web.lemoyne.edu/giunta/). Giunta has done the history of chemistry community a tremendous service by providing hundreds of carefully edited, and in some cases personally annotated, classic papers from the history of chemistry online. Organized both by topic and by author, this website is a necessary first stop for anyone interested in the history of chemistry.

Gregory, Frederick. *Natural Science in Western History*. Boston, MA: Houghton Mifflin Company, 2008. A solid secondary source textbook, by a past president of the History of Science Society, for the whole period covered by the readings in this volume. Gregory's introductions to chemistry, electromagnetism, and thermodynamics are perhaps its greatest strength.

Linden, Stanton J. *The Alchemy Reader*. Cambridge: Cambridge University Press, 2003.

Lyons, Stephen, producer and director. *The Mystery of Matter: Search for the Elements*. [Documentary]. USA: PBS, 2015. The best historical introduction to the development of the periodic table available on video. A great help with this section.

Merchant, Carolyn. "Secrets of Nature: The Bacon Debates Revisited." *Journal of the History of Ideas*, 69(1) 2008, 147–62.

Perrin, Carleton E. "The Chemical Revolution." In *Companion to the History of Modern Science*. M. J. S. Hodge, R. C. Olby, N. Cantor, and J. R. R. Christie (eds). London: Routledge, 1990.

Purrington, Robert D. *Physics in the Nineteenth Century*. New Brunswick, NJ: Rutgers University Press, 1997.

Stanley, Matthew. "The Pointsman: Maxwell's Demon, Victorian Free Will, and the Boundaries of Science." *Journal of the History of Ideas*, 69(3), (July 2008). Maxwell's thermodynamic publications are too mathematically complex for this volume, but Stanley provides an approachable introduction to the serious philosophical problems raised by statistical mechanics.

Uglow, Jenny. *The Lunar Men*. New York: Farrar, Straus and Giroux, 2002.

Part VI

The Earth and All Its Creatures: Developments in Geology and Biology

On the Motion of the Heart and Blood in Animals (1628)

William Harvey

78

Prefatory Remarks

As we are about to discuss the motion, action, and use of the heart and arteries, it is imperative on us first to state what has been thought of these things by others in their writings, and what has been held by the vulgar and by tradition, in order that what is true may be confirmed, and what is false set right by dissection, multiplied experience, and accurate observation . . . .

But I should like to be informed why, if the pulmonary vein were destined for the conveyance of air, it has the structure of a blood-vessel here. Nature had rather need of annular tubes, such as those of the bronchi in order that they might always remain open, and not be liable to collapse; and that they might continue entirely free from blood, lest the liquid should interfere with the passage of the air, as it so obviously does when the lungs labour from being either greatly oppressed or loaded in a less degree with phlegm, as they are when the breathing is performed with a sibilous or rattling noise.

Still less is that opinion to be tolerated which, as a two-fold material, one aerial, one sanguineous, is required for the composition of vital spirits, supposes the blood to ooze through the septum of the heart from the right to the left ventricle by certain hidden porosities, and the air to be attracted from the lungs through the great vessel, the pulmonary vein; and which, consequently, will have it, that there are numerous porosities in the septum of the heart adapted for the transmission of the blood. But by Hercules! no such pores can be demonstrated, nor in fact do any such exist. For the septum of the heart is of a denser and more compact structure than any portion of the body, except the bones and sinews. But even supposing that there were foramina or pores in this situation, how could one of the ventricles extract anything from the other—the left, e.g., obtain blood from the right, when we see that both ventricles contract and dilate simultaneously?

. . .

Chapter I: The Author's Motives for Writing

When I first gave my mind to vivisections, as a means of discovering the motions and uses of the heart, and sought to discover these from actual inspection, and not from the writings of others, I found the task so truly arduous, so full of difficulties, that I was almost tempted to think, with Fracastorius, that the motion of the heart was only to be comprehended by God. For I could neither rightly perceive at first when the systole and when the diastole took place, nor when and where dilatation and contraction occurred, by reason of the rapidity of the motion, which in many animals is accomplished in the twinkling of an eye, coming and going like a flash of lightning; so that the systole presented itself to me now from this point, now from that; the diastole the same; and then everything was reversed, the motions occurring, as it seemed, variously and confusedly together. My mind was therefore greatly unsettled nor did I know what I should myself conclude, nor what believe from others . . . .

At length, by using greater and daily diligence and investigation, making frequent inspection of many and various animals, and collating numerous observations, I thought that I had attained to the truth, that I should extricate myself and escape from this labyrinth, and that I had discovered what I so much desired, both the motion and the use of the heart and arteries. From that time I have not hesitated to expose my views upon these subjects, not only in private to my friends, but also in public, in my anatomical lectures, after the manner of the Academy of old.

. . .

Chapter II: On the Motions of the Heart, as Seen in the Dissection of Living Animals

In the first place, then, when the chest of a living animal is laid open and the capsule that immediately surrounds the heart is slit up or removed, the organ is seen now to move, now to be at rest; there is a time when it moves, and a time when it is motionless.

These things are more obvious in the colder animals, such as toads, frogs, serpents, small fishes, crabs, shrimps, snails, and shellfish. They also become more distinct in warm-blooded animals, such as the dog and hog, if they be attentively noted when the heart begins to flag, to move more slowly, and, as it were, to die: the movements then become slower and rarer, the pauses longer, by which it is made much more easy to perceive and unravel what the motions really are, and how they are performed. In the pause, as in death, the heart is soft, flaccid, exhausted, lying, as it were, at rest.

In the motion, and interval in which this is accomplished, three principal circumstances are to be noted:

1. That the heart is erected, and rises upwards to a point, so that at this time it strikes against the breast and the pulse is felt externally.
2. That it is everywhere contracted, but more especially towards the sides so that it looks narrower, relatively longer, more drawn together. The heart of an eel taken out of the body of the animal and placed upon the table or the hand, shows these particulars; but the same things are manifest in the hearts of all small fishes and of those colder animals where the organ is more conical or elongated.
3. The heart being grasped in the hand, is felt to become harder during its action. Now this hardness proceeds from tension, precisely as when the forearm is grasped, its tendons are perceived to become tense and resilient when the fingers are moved.
4. It may further be observed in fishes, and the colder blooded animals, such as frogs, serpents, etc., that the heart, when it moves, becomes of a paler color, when quiescent of a deeper blood-red color.

From these particulars it appears evident to me that the motion of the heart consists in a certain universal tension—both contraction in the line of its fibres, and constriction in every sense. It becomes erect, hard, and of diminished size during its action; the motion is plainly of the same nature as that of the muscles when they contract in the line of their sinews and fibres; for the muscles, when in action, acquire vigor and tenseness, and from soft become hard, prominent, and thickened: and in the same manner the heart.

We are therefore authorized to conclude that the heart, at the moment of its action, is at once constricted on all sides, rendered thicker in its parietes and smaller in its ventricles, and so made apt to project or expel its charge of blood. This, indeed, is made sufficiently manifest by the preceding fourth observation in which we have seen that the heart, by squeezing out the blood that it contains, becomes paler, and then when it sinks into repose and the ventricle is filled anew with blood, that the deeper crimson colour returns. But no one need remain in doubt of the fact, for if the ventricle be pierced the blood will be seen to be forcibly projected outwards upon each motion or pulsation when the heart is tense.

These things, therefore, happen together or at the same instant: the tension of the heart, the pulse of its apex, which is felt externally by its striking against the chest, the thickening of its parietes, and the forcible expulsion of the blood it contains by the constriction of its ventricles.

Hence the very opposite of the opinions commonly received appears to be true; inasmuch as it is generally believed that when the heart strikes the breast and the pulse is felt without, the heart is dilated in its ventricles and is filled with blood; but the contrary of this is the fact, and the heart, when it contracts (and the impulse of the apex is conveyed through the chest wall), is emptied. Whence the motion which is generally regarded as the diastole of the heart, is in truth its systole. And in like manner the intrinsic motion of the heart is not the diastole but the systole; neither is it in the diastole that the heart grows firm and tense, but in the systole, for then only, when tense, is it moved and made vigorous.

. . .

Neither is it true, as vulgarly believed, that the heart by any dilatation or motion of its own, has the power of drawing the blood into the ventricles; for when it acts and becomes tense, the blood is expelled; when it relaxes and sinks together it receives the blood in the manner and wise which will by-and-by be explained.

. . .

Chapter V. Of the Motion, Action and Office of the Heart

From these and other observations of a similar nature, I am persuaded it will be found that the motion of the heart is as follows:

First of all, the auricle contracts, and in the course of its contraction forces the blood (which it contains in ample quantity as the head of the veins, the storehouse and cistern of the blood) into the ventricle, which, being filled, the heart raises itself straightway, makes all its fibres tense, contracts the ventricles, and performs a beat, by which beat it immediately sends the blood supplied to it by the auricle into the arteries. The right ventricle sends its charge into the lungs by the vessel which is called vena arteriosa, but which in structure and function, and all other respects, is an artery. The left ventricle sends its charge into the aorta, and through this by the arteries to the body at large.

These two motions, one of the ventricles, the other of the auricles, take place consecutively, but in such a manner that there is a kind of harmony or rhythm preserved between them, the two concurring in such wise that but one motion is apparent, especially in the warmer blooded animals, in which the movements in question are rapid. Nor is this for any other reason than it is in a piece of machinery, in which, though one wheel gives motion to another, yet all the wheels seem to move simultaneously; or in that mechanical contrivance which is adapted to firearms, where, the trigger being touched, down comes the flint, strikes against the steel, elicits a spark, which falling among the powder, ignites it, when the flame extends, enters the barrel, causes the explosion, propels the ball, and the mark is attained—all of which incidents, by reason of the celerity with which they happen, seem to take place in the twinkling of an eye . . . .

The motion of the heart, then, is entirely of this description, and the one action of the heart is the transmission of the blood and its distribution, by means of the arteries, to the very extremities of the body; so that the pulse which we feel in the arteries is nothing more than the impulse of the blood derived from the heart.

Whether or not the heart, besides propelling the blood, giving it motion locally, and distributing it to the body, adds anything else to it—heat, spirit, perfection,—must be inquired into by-and-by, and decided upon other grounds. So much may suffice at this time, when it is shown that by the action of the heart the blood is transfused through the ventricles from the veins to the arteries, and distributed by them to all parts of the body.

. . . The grand cause of doubt and error in this subject appears to me to have been the intimate connexion between the heart and the lungs. When men saw both the pulmonary artery and the pulmonary veins losing themselves in the lungs, of course it became a puzzle to them to know how or by what means the right ventricle should distribute the blood to the body, or the left draw it from the venæ cavæ. . . .

Had anatomists only been as conversant with the dissection of the lower animals as they are with that of the human body, the matters that have hitherto kept them in a perplexity of doubt would, in my opinion, have met them freed from every kind of difficulty.

. . .

And first, in fishes, in which the heart consists of but a single ventricle, being devoid of lungs, the thing is sufficiently manifest. Here the sac, which is situated at the base of the heart, and is the part analogous to the auricle in man, plainly forces the blood into the heart, and the heart, in its turn, conspicuously transmits it by a pipe or artery, or vessel analogous to an artery; these are facts which are confirmed by simple ocular inspection, as well as by a division of the vessel, when the blood is seen to be projected by each pulsation of the heart.

The same thing is also not difficult of demonstration in those animals that have, as it were, no more than a single ventricle to the heart, such as toads, frogs, serpents, and lizards, which have lungs in a certain sense, as they have a voice. I have many observations by me on the admirable structure of the lungs of these animals, and matters appertaining, which, however, I cannot introduce in this place. Their anatomy plainly shows us that the blood is transferred in them from the veins to the arteries in the same manner as in higher animals, viz., by the action of the heart; the way, in fact, is patent, open, manifest; there is no difficulty, no room for doubt about it; for in them the matter stands precisely as it would in man were the septum of his heart perforated or removed, or one ventricle made out of two; and this being the case, I imagine that no one will doubt as to the way by which the blood may pass from the veins into the arteries.

. . .

Chapter VIII. Of the Quantity of Blood Passing Through the Heart from the Veins to the Arteries; and of the Circular Motion of the Blood

. . . Doctrine once sown strikes deep its root, and respect for antiquity influences all men. Still the die is cast, and my trust is in my love of truth and the candour of cultivated minds. And sooth to say, when I surveyed my mass of evidence, whether derived from vivisections, and my various reflections on them, or from the study of the ventricles of the heart and the vessels that enter into and issue from them, the symmetry and size of these conduits—for nature doing nothing in vain, would never have given them so large a relative size without a purpose—or from observing the arrangement and intimate structure of the valves in particular, and of the other parts of the heart in general, with many things besides, I frequently and seriously bethought me, and long revolved in my mind, what might be the quantity of blood which was transmitted, in how short a time its passage might be effected, and the like. But not finding it possible that this could be supplied by the juices of the ingested aliment without the veins on the one hand becoming drained, and the arteries on the other getting ruptured through the excessive charge of blood, unless the blood should somehow find its way from the arteries into the veins, and so return to the right side of the heart, I began to think whether there might not be a MOTION, AS IT WERE, IN A CIRCLE. Now, this I afterwards found to be true; and I finally saw that the blood, forced by the action of the left ventricle into the arteries, was distributed to the body at large, and its several parts, in the same manner as it is sent through the lungs, impelled by the right ventricle into the pulmonary artery, and that it then passed through the veins and along the vena cava, and so round to the left ventricle in the manner already indicated. This motion we may be allowed to call circular, in the same way as Aristotle says that the air and the rain emulate the circular motion of the superior bodies; for the moist earth, warmed by the sun, evaporates; the vapours drawn upwards are condensed, and descending in the form of rain, moisten the earth again. By this arrangement are generations of living things produced; and in like manner are tempests and meteors engendered by the circular motion, and by the approach and recession of the sun.

And similarly does it come to pass in the body, through the motion of the blood, that the various parts are nourished, cherished, quickened by the warmer, more perfect, vaporous, spirituous, and, as I may say, alimentive blood; which, on the other hand, owing to its contact with these parts, becomes cooled, coagulated, and so to speak effete. It then returns to its sovereign, the heart, as if to its source, or to the inmost home of the body, there to recover its state of excellence or perfection. Here it renews its fluidity, natural heat, and becomes powerful, fervid, a kind of treasury of life, and impregnated with spirits, it might be said with balsam. Thence it is again dispersed. All this depends on the motion and action of the heart.

The heart, consequently, is the beginning of life; the sun of the microcosm, even as the sun in his turn might well be designated the heart of the world; for it is the heart by whose virtue and pulse the blood is moved, perfected, and made nutrient, and is preserved from corruption and coagulation; it is the household divinity which, discharging its function, nourishes, cherishes, quickens the whole body, and is indeed the foundation of life, the source of all action. But of these things we shall speak more opportunely when we come to speculate upon the final cause of this motion of the heart.

As the blood-vessels, therefore, are the canals and agents that transport the blood, they are of two kinds, the cava and the aorta; and this not by reason of there being two sides of the body, as Aristotle has it, but because of the difference of office, not, as is commonly said, in consequence of any diversity of structure, for in many animals, as I have said, the vein does not differ from the artery in the thickness of its walls, but solely in virtue of their distinct functions and uses. A vein and an artery, both styled veins by the ancients, and that not without reason, as Galen has remarked, for the artery is the vessel which carries the blood from the heart to the body at large, the vein of the present day bringing it back from the general system to the heart; the former is the conduit from, the latter the channel to, the heart; the latter contains the cruder, effete blood, rendered unfit for nutrition; the former transmits the digested, perfect, peculiarly nutritive fluid.

Chapter IX. That There Is a Circulation of the Blood Is Confirmed from the First Proposition

...

Let us assume, either arbitrarily or from experiment, the quantity of blood which the left ventricle of the heart will contain when distended, to be, say, two ounces, three ounces, or one ounce and a half—in the dead body I have found it to hold upwards of two ounces. Let us assume further how much less the heart will hold in the contracted than in the dilated state; and how much blood it will project into the aorta upon each contraction; and all the world allows that with the systole something is always projected, a necessary consequence demonstrated in the third chapter, and obvious from the structure of the valves; and let us suppose as approaching the truth that the fourth, or fifth, or sixth, or even but the eighth part of its charge is thrown into the artery at each contraction; this would give either half an ounce, or three drachms, or one drachm of blood as propelled by the heart at each pulse into the aorta; which quantity, by reason of the valves at the root of the vessel, can by no means return into the ventricle. Now, in the course of half an hour, the heart will have made more than one thousand beats, in some as many as two, three, and even four thousand. Multiplying the number of drachms propelled by the number of pulses, we shall have either one thousand half ounces, or one thousand times three drachms, or a like proportional quantity of blood, according to the amount which we assume as propelled with each stroke of the heart, sent from this organ into the artery—a larger quantity in every case than is contained in the whole body! ...

Upon this supposition, therefore, assumed merely as a ground for reasoning, we see the whole mass of blood passing through the heart, from the veins to the arteries, and in like manner through the lungs.

...

This truth, indeed, presents itself obviously before us when we consider what happens in the dissection of living animals; the great artery need not be divided, but a very small branch only (as Galen even proves in regard to man), to have the whole of the blood in the body, as well that of the veins as of the arteries, drained away in the course of no long time—some half-hour or less. Butchers are well aware of the fact and can bear witness to it; for, cutting the throat of an ox and so dividing the vessels of the neck, in less than a quarter of an hour they have all the vessels bloodless—the whole mass of blood has escaped. The same thing also occasionally occurs with great rapidity in performing amputations and removing tumors in the human subject.

...

Chapter XIII. The Third Position Is Confirmed: And the Circulation of the Blood Is Demonstrated from It

The celebrated Hieronymus Fabricius of Aquapendente, a most skilful anatomist, and venerable old man, or, as the learned Riolan will have it, Jacobus Silvius, first gave representations of the valves in the veins, which consist of raised or loose portions of the inner membranes of these vessels, of extreme delicacy, and a sigmoid or semilunar shape. They are situated at different distances from one another, and diversely in different individuals; they are connate at the sides of the veins; they are directed upwards towards the trunks of the veins; the two—for there are for the most part two together—regard each other, mutually touch, and are so ready to come into contact by their edges, that if anything attempts to pass from the trunks into the branches of the veins, or from the greater vessels into the less, they completely prevent it; they are farther so arranged, that the horns of those that succeed are opposite the middle of the convexity of those that precede, and so on alternately.

The discoverer of these valves did not rightly understand their use, nor have succeeding anatomists added anything to our knowledge: for their office is by no means explained when we are told that it is to hinder the blood, by its weight, from all flowing into inferior parts; for the edges of the valves in the jugular veins hang downwards, and are so contrived that they prevent the blood from rising upwards; the valves, in a word, do not invariably look upwards, but always toward the trunks of the veins, invariably towards the seat of the heart. I, and indeed others, have sometimes found valves in the emulgent veins, and in those of the mesentery, the edges of which were directed towards the vena cava and vena portæ . . . .

But the valves are solely made and instituted lest the blood should pass from the greater into the lesser veins, and either rupture them or cause them to become varicose; lest, instead of advancing from the extreme to the central parts of the body, the blood should rather proceed along the veins from the centre to the extremities; but the delicate valves, while they readily open in the right direction, entirely prevent all such contrary motion, being so situated and arranged, that if anything escapes, or is less perfectly obstructed by the cornua of the one above, the fluid passing, as it were, by the chinks between the cornua, it is immediately received on the convexity of the one beneath, which is placed transversely with reference to the former, and so is effectually hindered from getting any farther.

And this I have frequently experienced in my dissections of the veins: if I attempted to pass a probe from the trunk of the veins into one of the smaller branches, whatever care I took I found it impossible to introduce it far any way, by reason of the valves; whilst, on the contrary, it was most easy to push it along in the opposite direction, from without inwards, or from the branches towards the trunks and roots. In many places two valves are so placed and fitted, that when raised they come exactly together in the middle of the vein, and are there united by the contact of their margins; and so accurate is the adaptation, that neither by the eye nor by any other means of examination, can the slightest chink along the line of contact be perceived. But if the probe be now introduced from the extreme towards the more central parts, the valves, like the floodgates of a river, give way, and are most readily pushed aside. The effect of this arrangement plainly is to prevent all motion of the blood from the heart and vena cava, whether it be upwards towards the head, or downwards towards the feet, or to either side towards the arms, not a drop can pass; all motion of the blood, beginning in the larger and tending towards the smaller veins, is opposed and resisted by them; whilst the motion that proceeds from the lesser to end in the larger branches is favoured, or, at all events, a free and open passage is left for it.

But that this truth may be made the more apparent, let an arm be tied up above the elbow as if for phlebotomy (A, A, fig. 1). At intervals in the course of the veins, especially in labouring people and those whose veins are large, certain knots or elevations (B, C, D, E, F) will be perceived, and this not only at the places where a branch is received (E, F), but also where none enters (C, D): these knots or risings are all formed by valves, which thus show themselves externally. And now if you press the blood from the space above one of the valves, from H to O, (fig. 2,) and keep the point of a finger upon the vein inferiorly, you will see no influx of blood from above; the portion of the vein between the point of the finger and the valve O will be obliterated; yet will the vessel continue sufficiently distended above the valve (O, G). The blood being thus pressed out and the vein emptied, if you now apply a finger of the other hand upon the distended part of the vein above the valve O, (fig. 3,) and press downwards, you will find that you cannot force the blood through or beyond the valve; but the greater effort you use, you will only see the portion of vein that is between the finger and the valve become more distended, that portion of the vein which is below the valve remaining all the while empty (H, O, fig. 3).

It would therefore appear that the function of the valves in the veins is the same as that of the three sigmoid valves which we find at the commencement of the aorta and pulmonary artery, viz., to prevent all reflux of the blood that is passing over them.

Farther, the arm being bound as before, and the veins looking full and distended, if you press at one part in the course of a vein with the point of a finger (L, fig. 4), and then with another finger streak the blood upwards beyond the next valve (N), you will perceive that this portion of the vein continues empty (L. N), and that the blood cannot retrograde, precisely as we have already seen the case to be in fig. 2; but the finger first applied (H, fig. 2, L, fig. 4), being removed, immediately the vein is filled from below, and the arm becomes as it appears at D C, fig. 1. That the blood in the veins therefore proceeds from inferior or more remote parts, and towards the heart, moving in these vessels in this and not in the contrary direction, appears most obviously. And although in some places the valves, by not acting with such perfect accuracy, or where there is but a single valve, do not seem totally to prevent the passage of the blood from the centre, still the greater number of them plainly do so; and then, where things appear contrived more negligently, this is compensated either by the more frequent occurrence or more perfect action of the succeeding valves, or in some other way: the veins in short, as they are the free and open conduits of the blood returning to the heart, so are they effectually prevented from serving as its channels of distribution from the heart.

But this other circumstance has to be noted: The arm being bound, and the veins made turgid, and the valves prominent, as before, apply the thumb or finger over a vein in the situation of one of the valves in such a way as

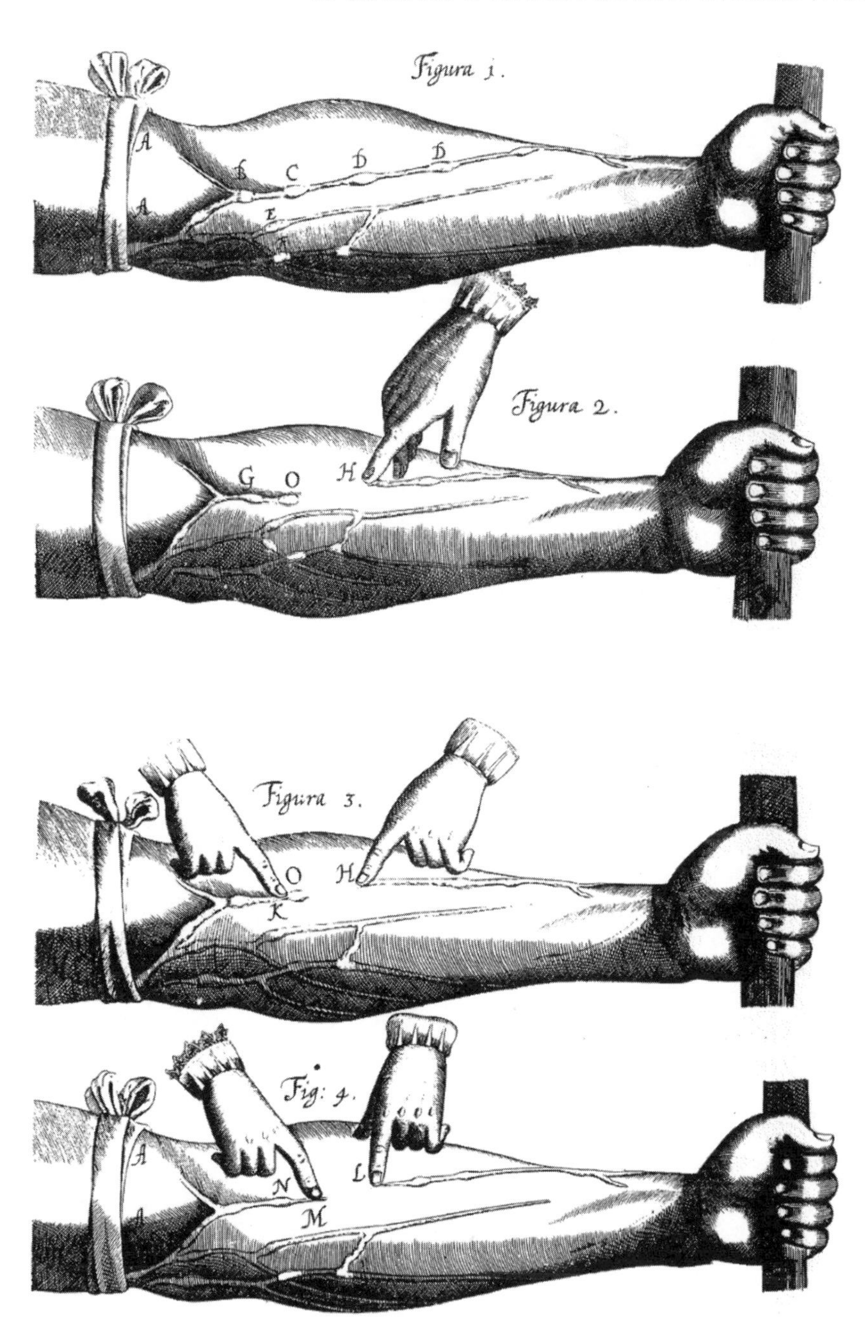

to compress it, and prevent any blood from passing upwards from the hand; then, with a finger of the other hand, streak the blood in the vein upwards till it has passed the next valve above (N, fig. 4), the vessel now remains empty; but the finger at L being removed for an instant, the vein is immediately filled from below; apply the finger again, and having in the same manner streaked the blood upwards, again remove the finger below, and again the vessel becomes distended as before; and this repeat, say a thousand times, in a short space of time. And now compute the quantity of blood which you have thus pressed up beyond the valve, and then multiplying the assumed quantity by one thousand, you will find that so much blood has passed through a certain portion of the vessel; and I do now believe that you will find yourself convinced of the circulation of the blood, and of its rapid motion. But if in this experiment you say that a violence is done to nature, I do not doubt but that, if you proceed in the same way, only taking as great a length of vein as possible, and merely remark with what rapidity the blood flows upwards, and fills the vessel from below, you will come to the same conclusion.

Translated by Robert Willis

Reading and Discussion Questions

1. Harvey acknowledges that the analysis of the motions of the heart is a terribly difficult project. What makes the motion of the heart so hard to observe and analyze? What experimental subjects and methods does Harvey employ to make his investigations possible?
2. On what two grounds does Harvey argue that the blood must be circulating, and not constantly being both produced and consumed in the body? What is the importance of valves in the second argument?
3. Is Harvey a mechanist about biology, like Descartes, or does he evidence a more Aristotelian approach?

Treatise on Man (1664)

René Descartes

79

These men will be comprised like ourselves of a soul and a body. First, it is necessary that I describe for you the body in itself, then afterward the soul in itself, and finally that I show you how these two natures would have to be joined and united in order to comprise some men who resemble us.

I make the supposition that the body is nothing else but a statue or earthen machine, that God has willed to form entire, in order to make it as similar to us as is possible. Thus he not only would have given it the external color and shape of our members, but also he put in the interior all the parts which are required to make it walk, eat, respire, and that it imitate, in the end, all of our functions which can be imagined to proceed from matter alone, and depend only on the disposition of the organs.

We see clocks, artificial fountains, mills, and other similar machines, which, being only made by men, nevertheless do not lack the force to move themselves in several diverse means. And it seems to me that I could not imagine as many kinds of movements in the latter as I suppose to be made by the hand of God, nor attribute to him only so much craftsmanship as we could think of.

Now I will not stop to describe to you the bones, nerves, muscles, veins, arteries, stomach, liver, spleen, heart and brain, nor all the other diverse parts of which this statue must be comprised. I assume these to be entirely similar to the parts of our body which bear the same name, and that could be shown to us by some learned anatomist, at least those large enough to be seen, if you do not already know them sufficiently from yourselves. Those which on account of their minuteness are invisible, I will be more easily and more clearly be able to make these known to you in speaking of the motions which depend on them. Thus there is only the need here to explicate these motions in order, [and] by the same means tell you that they are the same as those functions of ours they represent.

First off, the foodstuffs are digested in the stomach of this machine, by the force of specific fluids, which glide between their particles, separate them, agitate them, and heat them, just like ordinary water does to quicklime, or acid does to metals. Furthermore, as these liquids are brought very quickly to the heart by the arteries, as I will speak later on, they must be very hot. And even the foodstuffs are ordinarily such that they can be broken down and heated by their own power, just like new hay in a barn, if it is shut up before it is dry.

Know as well that the agitation that the little particles of these foodstuffs receive in being heated, conjoined to that of the stomach and bowels which contain them, and by the disposition of the small fibers which compose the intestines, cause them to descend little by little toward the orifice by which the most coarse of these must be exited, in the degree to which they are digested. However, the most subtle and active particles encounter here and there an infinity of small holes, by which they pass into the network of a great vein [renal portal vein] which carries them to the liver, and into other veins which carry them elsewhere. There is, in addition to this, nothing else which separates the smallest from the largest particles except these holes, just as when one shakes meal in a sack the purest part escapes, and only the smallness of the holes prevents the bran from following.

These most subtle particles of the foodstuffs, being unequal and still imperfectly mixed together, comprise a fluid [chyle] which would remain turbulent and whitish, except that a part of it is mixed immediately with the mass of the blood contained in the ramifications of the Portal vein, which receives this liquid from the intestines, and passes into the ramifications of the *Vena Cava*, which conducts it toward the heart, and into the liver, as if in a single vessel.

Likewise, it should be noted here that the pores of the liver are so disposed that when this fluid enters into it, it is refined, elaborated, given its color, and acquires the form of blood, just as the juice of black grapes, which is white, is converted into red wine when it is allowed to ferment on the vine.

Now the blood thereby contained in the veins has only one distinct passage by which it can leave them, namely that which conducts it into the concavity of the right heart [i.e., right ventricle]. And one knows that the flesh of the heart contains in its pores one of those lightless fires of which I have spoken above. This makes the heart so hot and fiery, that in proportion to the entrance of the blood into one of the two chambers or concavities which compose it, it is quickly expanded and dilated there, just as you could show by experience what would happen to the blood or milk of some animal if you pour it drop by drop into a very hot flask. And the fire which is in the heart of the machine I am describing serves only to dilate, heat and subtilize the blood in this way, which falls continually, drop by drop, by a passage from the vena cava, into the right cavity, from which it is exhaled into the lung. And from the pulmonary vein, which the anatomists have named the *venous artery*, it flows into the other cavity, from which it is distributed to all the body.

The flesh of the lung is so rarified and soft, and at all times so refreshed by respired air, that in proportion as the vapors of the blood leave the right cavity of the heart and enter the lung by the artery named by the anatomists the *arterial vein* [i.e., pulmonary vein], they are there thickened and converted into the blood once again. Then from there the blood falls drop by drop into the left cavity of the heart. If they were to enter without being thus thickened once again, the blood would not serve sufficiently to nourish the fire which is there.

Thus you see that respiration, which only functions in this machine to thicken these vapors, is no less necessary for the maintenance of this fire, than it is for the conservation of life in ourselves, at least in independent human beings, since in the fetus still in the womb cannot draw in fresh air by respiration, and there are two conduits which supply this defect. By one, the blood [in the fetus] from the vena cava passes into the venous artery [i.e. the pulmonary artery via the *foramen ovale*], and by the other, the vapors or rarefied blood from the arterial vein are exhaled and move into the Great Artery [*aorta*]. For the animals which have no lung, they have only a single cavity in the heart, or if they have several, they are all in a consecutive series to one another.

The pulse, or beating of the arteries, depends on eleven little membranes [i.e. mitral, tricuspid and semi-lunar valves], which like little doors, close and open the entrances of the four vessels which open upon the two cavities of the heart. At the moment one of these beats ends, and another is almost beginning, the little doors at the entrance of the two arteries are found to be precisely closed, and those at the openings of the two veins are open. Necessarily, two discharges of blood fall immediately from these two veins into each cavity of the heart. Then by their rarefication and expansion suddenly into a greater space than they previously occupied, press and close the little doors at the entrance of the two veins, by this means preventing further descent of blood into the heart, while pressing and opening those of the two arteries. Into these they enter quickly and with force, causing the inflation of the heart and all the arteries at the same moment. But immediately afterward, the rarified blood is condensed once again, or penetrates other parts of the body. Thus the heart and arteries are deflated, the little doors at the entrances of the two arteries are closed, and those at the opening of the two veins are reopened, admitting two other discharges of blood, which once again cause the inflation of the heart and arteries in the same way as the preceding.

Thus knowing the cause of the pulse, it is easy to understand that it isn't so much the blood contained in the veins of this machine, and that freshly coming from the liver, which attaches itself to various parts, and repairs what their continual action, and the diverse activities of the other bodies which surround them detach from them, but that blood which is contained in its arteries, and already distilled in its heart. Because the blood which is in the veins always flows, little by little, from the extremities toward the heart. Furthermore, the arrangement of specific little doors or small valves, that anatomists have noticed in several

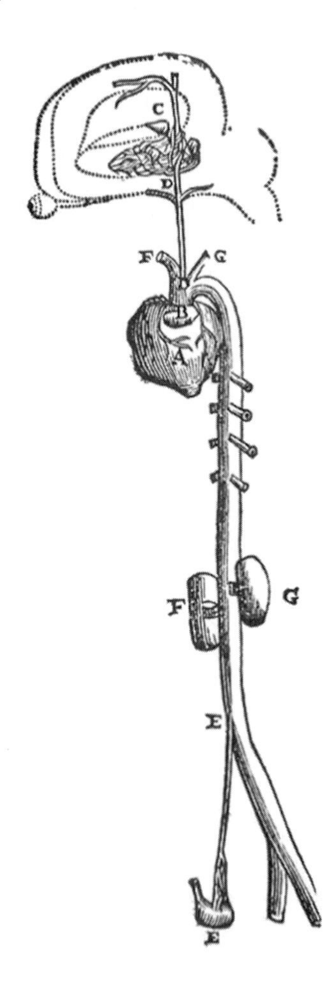

Descartes' figure inserted into the 1664 edition showing relation of heart, brain, kidney and reproductive system.

places along our veins, must sufficiently persuade us that the same thing happens in ourselves. On the other hand, that which is in the arteries is pushed out of the heart with force, in separate little thrusts, toward the extremities of the arteries. Hence, it can be joined and united to all the members, and thereby maintain them, or even cause them to grow, if this [model] is to represent the body of a man arranged in this way.

At the moment that the arteries are inflated, the small particles of the blood that they contain here and there strike the roots of certain small filaments which then move from the extremities of the small branches of these arteries to comprise the bone, flesh, skin, nerves, brain, and all the remainder of the solid parts, according to the diverse means by which they are joined together or interlaced. Thus these particles have the power to push these filaments gradually ahead, and push them into their proper place. Then, at the moment that the arteries deflate, each of these particles stops at the point it is at, and by this means alone is joined and united to those parts it touches, according to what has just been said.

But what must primarily be observed is that all the most rapid, forceful and most subtle particles of the blood are transmitted into the cavity of the brain, inasmuch as the arteries which carry them there [carotids] are those which are most in a right line from the heart, and as you know, all bodies which are moved tend, as much as possible, to continue their motion in a right line.

Observe, for example [in the preceding figure] the heart "A," and consider that when the blood leaves it with force by the opening "B," all of its particles tend toward "C," the cavities of the brain. But the vessel being restricted in size and unable to carry all of them, the most weakly impelled are deflected by the stronger, which by this means are alone moved there.

You should also note in passing that after those particles which enter the brain, there are none stronger or more active than those which go to the reproductive parts. For example, if those which have sufficient force to reach "D" cannot move further to "C," because there is not enough room for them, they return preferably toward "E," rather than toward "F" and "G" [i.e. the kidneys], because the route to "E" is more direct. In consequence of this, I could possibly have you see how one could form a new machine, entirely similar to the first, by the way in which the humor assembles at "E." However, I cannot say more on this matter.

As for those particles of the blood which penetrate up to the brain, they function not only to nourish it and enter its substance, but more primarily to produce there a certain very wind, or rather a very lively and pure flame, that one terms the *animal spirits*. It is necessary to understand that the arteries which bring the blood from the heart, after being divided into an infinity of small branches and composing the fine tissues which are extended like tapestries at the base of the ventricles of the brain, are reassembled around a specific little *gland*, situated near the middle of the substance of the brain [the pineal gland], just at the entrance of its ventricles. At this place are a great number of small openings, by means of which the most subtle particles of the blood contained in the arteries can flow into this gland. But the arteries are so narrow that they allow no passage to the grosser particles.

It is also necessary to know that the arteries do not terminate at that point, but with several combined into one, they extend upward in a right line, then enter the great blood vessel which, like Euripos, bathes the entire superficial exterior of the brain [see "H" on diagram]. Furthermore, it is necessary to notice grossest particles of the blood can lose much of their motion in the windings of the fine tissues through which they pass. In the same proportion, they have the force to push the smaller particles among these, and thus transfer some of their motion to them. But the smaller particles cannot transfer their motion in the same way, inasmuch as their motion is increased by that transferred from the largest particles, and there are no other bodies around them to which they could as easily transfer it.

Whence it is easy to conceive that when the coarsest mount in a right line toward the superficial exterior of the brain, from whence they serve to nourish its substance, they cause the smallest and most agitated particles to be deflected, and all of them enter the [pineal] gland, which must be imagined as an overflowing spring, from which they flow simultaneously on all sides into the ventricles of the brain. Thus, without any further preparation or changes, except the separation of these from the coarser particles, but retaining the extreme velocity that has been given them by the heat of the heart, they cease to have the form of blood particles, and are called the *animal spirits*.

Now, to the degree that these animal spirits thus enter into the ventricles of the brain, they pass from there into the pores in the brain substance, and from these pores into the nerves. And according as they enter, or tend to enter, one or the other of these, they have the power to alter the shape of the muscles into which these nerves are inserted, and by this means make the members move, just as you may have seen in the grottos and fountains of our King, in which the simple force imparted to the water in leaving the fountain is sufficient for the motions of different machines, even making them play musical instruments, or speak words according to the diverse disposition of the tubes conducting the water.

In truth, one can make a strong comparison between the nerves of the machine I am describing to you and the tubes in these water-machines. Also, this holds for the other diverse machinery and springs which serve to move the water machine, and the muscles and tendons of the other, and also for the animal spirits and the driving water, with the heart as the fountain, and the ventricles of the brain the water main. Furthermore, respiration and other such natural and ordinary actions, which depend on the flow of the spirits, are like the movements of a clock or mill that the flow of water can render continuous. External objects, which by their presence alone, act upon the sense organs of this machine, and by this means force it to move in several different ways, are like intruders into one of these fountain grottoes, cause without thought the movements which are made in their presence. Because they cannot enter except by walking upon certain tiles, specially placed, so that, for example, if they approach the Bathing Diana, they will cause her to hide in the reeds. And if they attempt to pursue her, they will cause Neptune to move toward them, menacing them with his Trident. If they move to some other side, a marine monster will arise, who will spit water on their face, or something similar, all according to the skill of the investors who have made the statues.

Finally, when the *rational soul* is put in this machine, it will have its principal location in the brain, and will be there like a fountain director, who must be in the fountain-house from which emerge all the tubes of these machines into which the water is directed when one wishes to excite, inhibit or in some manner change their motions.

. . .

[In conclusion], I would like you to reflect, after the preceding, on how all the functions that I have attributed to this machine, such as the digestion of food, the beating of the heart and arteries, the nutrition and growth of the members, respiration, waking and sleep, reception of light, sound, smell, taste, heat and such qualities by the external sense organs, the impression of these sensory ideas on the organ of common sense and imagination [i.e. the pineal gland], and the retention or imprinting of these ideas in the memory, occur. Similarly, reflect on the internal motions of the appetites and passions, and finally on the external motions of all the members, which follow with reference both to the objects presented to the senses, and to the passions and impressions contained in the memory, which are imitated as closely as possible those of a true man. Thus, I say, when you reflect on how these functions follow completely naturally in this machine solely from the disposition of the organs, no more nor less than those of a clock or other automaton from its counterweights and wheels, then it is not necessary to conceive on this account any other vegetative soul, nor sensitive one, nor any other principle of motion and life, than its blood and animal spirits, agitated by the heat of the continually burning fire in the heart, and which is of the same nature as those fires found in inanimate bodies.

Translated by P. R. Sloan

Reading and Discussion Questions

1. To what kinds of machines does Descartes compare the body and various parts of the body? Can you think of examples of similar kinds of comparisons between the body and machines that you have encountered in contemporary science?
2. What role do "animal spirits" and the pineal gland play in Cartesian physiology?
3. Why does he think we can dispose of the concepts of vegetative and sensitive souls?

Micrographia or Some Physiological Descriptions of Minute Bodies Made by Magnifying Glasses with Observations and Inquiries Thereupon (1665)

Robert Hooke

80

The Preface

It is the great prerogative of Mankind above other Creatures, that we are not only able to behold the works of Nature, or barely to sustain our lives by them, but we have also the power of considering, comparing, altering, assisting, and improving them to various uses. And as this is the peculiar privilege of human Nature in general, so is it capable of being so far advanced by the helps of Art, and Experience, as to make some Men excel others in their Observations, and Deductions, almost as much as they do Beasts. By the addition of such artificial Instruments and methods, there may be, in some manner, a reparation made for the mischiefs, and imperfection, mankind has drawn upon itself, by negligence, and intemperance, and a willful and superstitious deserting the Prescripts and Rules of Nature, whereby every man, both from a derived corruption, innate and born with him, and from his breeding and converse with men, is very subject to slip into all sorts of errors.

The only way which now remains for us to recover some degree of those former perfections, seems to be, by rectifying the operations of the Sense, the Memory, and Reason, since upon the evidence, the strength, the integrity, and the right correspondence of all these, all the light, by which our actions are to be guided is to be renewed, and all our command over things it to be established.

It is therefore most worthy of our consideration, to recollect their several defects, that so we may the better understand how to supply them, and by what assistances we may enlarge their power, and secure them in performing their particular duties.

As for the actions of our Senses, we cannot but observe them to be in many particulars much outdone by those of other Creatures, and when at best, to be far short of the perfection they seem capable of: And these infirmities of the Senses arise from a double cause, either from the disproportion of the Object to the Organ, whereby an infinite number of things can never enter into them, or else from error in the Perception, that many things, which come within their reach, are not received in a right manner.

The like frailties are to be found in the Memory; we often let many things slip away from us, which deserve to be retained, and of those which we treasure up, a great part is either frivolous or false; and if good, and substantial, either in tract of time obliterated, or at best so overwhelmed and buried under more frothy notions, that when there is need of them, they are in vain sought for.

The two main foundations being so deceivable, it is no wonder, that all the succeeding works which we build upon them, of arguing, concluding, defining, judging, and all the other degrees of Reason, are liable to the same imperfection, being, at best, either vain, or uncertain: So that the errors of the understanding are answerable to the two other, being defective both in the quantity and goodness of its knowledge; for the limits, to which our thoughts are confined, are small in respect of the vast extent of Nature itself; some parts of it are too large to be comprehended, and some too little to be perceived. And from thence it must follow, that not having a full sensation of the Object, we must be very lame and imperfect in our conceptions about it, and in all the proportions which we build upon it; hence, we often take the shadow of things for the substance, small appearances for good similitudes, similitudes for definitions; and even many of those, which we think, to be the most solid definitions, are rather expressions of our own misguided apprehensions then of the true nature of the things themselves.

The effects of these imperfections are manifested in different ways, according to the temper and disposition of the several minds of men, some they incline to gross ignorance and stupidity, and others to a presumptuous imposing on other men's Opinions, and a confident dogmatizing on matters, whereof there is no assurance to be given.

Thus all the uncertainty, and mistakes of humane actions, proceed either from the narrowness and wandering of our Senses, from the slipperiness or delusion of our Memory, from the confinement or rashness of our Understanding, so that 'tis no wonder, that our power over natural causes and effects is so slowly improved, seeing we are not only to contend with the obscurity and difficulty of the things whereon we work and think, but even the forces of our own minds conspire to betray us.

These being the dangers in the process of humane Reason, the remedies of them all can only proceed from the real, the mechanical, the experimental Philosophy, which has this advantage over the Philosophy of discourse and disputation, that whereas that chiefly aims at the subtlety of its Deductions and Conclusions, without much regard to the first ground-work, which ought to be well laid on the Sense and Memory; so this intends the right ordering of them all, and the making them serviceable to each other.

The first thing to be undertaken in this weighty work, is a watchfulness over the failings and an enlargement of the dominion, of the Senses.

To which end it is requisite, first, that there should be a scrupulous choice, and a strict examination, of the reality, constancy, and certainty of the Particulars that we admit: This is the first rise whereon truth is to begin, and here the most severe, and most impartial diligence, must be employed; the storing up of all, without any regard to evidence or use, will only tend to darkness and confusion. We must not therefore esteem the riches of our Philosophical treasure by the number only, but chiefly by the weight; the most vulgar Instances are not to be neglected, but above all, the most instructive are to be entertained; the footsteps of Nature are to be traced, not only in her ordinary course, but when she seems to be put to her shifts, to make many doublings and turnings, and to use some kind of art in endeavoring to avoid our discovery.

The next care to be taken, in respect of the Senses, is a supplying of their infirmities with Instruments, and, as it were, the adding of artificial Organs to the natural; this in one of them has been of late years accomplished with prodigious benefit to all sorts of useful knowledge, by the invention of Optical Glasses. By the means of Telescopes, there is nothing so far distant but may be represented to our view; and by the help of Microscopes, there is nothing so small, as to escape our inquiry; hence there is a new visible World discovered to the understanding. By this means the Heavens are opened, and a vast number of new Stars, and new Motions, and new Productions appear in them, to which all the ancient Astronomers were utterly Strangers. By this the Earth itself, which lies so near us, under our feet, shews quite a new thing to us, and in every little particle of its matter; we now behold almost as great a variety of Creatures, as we were able before to reckon up in the whole Universe itself.

It seems not improbable, but that by these helps the subtlety of the composition of Bodies, the structure of their parts, the various texture of their matter, the instruments and manner of their inward motions, and all the other possible appearances of things, may come to be more fully discovered; all which the ancient Peripateticks were content to comprehend in two general and (unless further explained) useless words of Matter and Form. From whence there may arise many admirable advantages, towards the increase of the Operative, and the

Mechanic Knowledge, to which this Age seems so much inclined, because we may perhaps be enabled to discern all the secret workings of Nature, almost in the same manner as we do those that are the productions of Art, and are managed by Wheels, and Engines, and Springs, that were devised by humane Wit.

In this kind I here present to the World my imperfect Endeavors; which though they shall prove no other way considerable, yet, I hope, they may be in some measure useful to the main Design of a reformation in Philosophy, if it be only by shewing, that there is not so much required towards it, any strength of Imagination, or exactness of Method, or depth of Contemplation (though the addition of these, where they can be had, must needs produce a much more perfect composure) as a sincere Hand, and a faithful Eye, to examine, and to record, the things themselves as they appear.

And I beg my Reader, to let me take the boldness to assure him, that in this present condition of knowledge, a man so qualified, as I have endeavored to be, only with resolution, and integrity, and plain intentions of employing his Senses aright, may venture to compare the reality and the usefulness of his services, towards the true Philosophy, with those of other men, that are of much stronger, and more acute speculations, that shall not make use of the same method by the Senses.

The truth is, the Science of Nature has been already too long made only a work of the Brain and the Fancy: It is now high time that it should return to the plainness and soundness of Observations on material and obvious things. It is said of great Empires, that the best way to preserve them from decay, is to bring them back to the first Principles, and Arts, on which they did begin. The same is undoubtedly true in Philosophy, that by wandering far away into invisible Notions, has almost quite destroyed itself, and it can never be recovered, or continued, but by returning into the same sensible paths, in which it did at first proceed.

If therefore the Reader expects from me any infallible Deductions, or certainty of Axioms, I am to say for myself, that those stronger Works of Wit and Imagination are above my weak Abilities; or if they had not been so, I would not have made use of them in this present Subject before me: Whenever he finds that I have ventured at any small Conjectures, at the causes of the things that I have observed, I beseech him to look, upon them only as doubtful Problems, and uncertain guesses, and not as unquestionable Conclusions, or matters of unconfutable Science; I have produced nothing here, with intent to bind his understanding to an implicit consent; I am so far from that, that I desire him, not absolutely to rely upon these Observations of my eyes, if he finds them contradicted by the future Ocular Experiments of other and impartial Discoverers.

As for my part, I have obtained my end, if these my small Labours shall be thought fit to take up some place in the large stock, of natural Observations, which so many hands are busy in providing. If I have contributed the meanest foundations whereon others may raise nobler Superstructures, I am abundantly satisfied; and all my ambition is, that I may serve to the great Philosophers of this Age, as the makers and the grinders of my Glasses did to me; that I may prepare and furnish them with some Materials, which they may afterwards order and manage with better skill, and to far greater advantage ....

...

If once this method were followed with diligence and attention, there is nothing that lies within the power of human Wit (or which is far more effectual) of human Industry, which we might not compass; we might not only hope for Inventions to equalize those of Copernicus, Galileo, Gilbert, Harvy, and of others, whose Names are almost lost, that were the Inventors of Gun-powder, the Seamans Compass, Printing, Etching, Graving, Microscopes, &c. but multitudes that may far exceed them: for even those discoveries seem to have been the products of some such method, though but imperfect; What may not be therefore expected from it if thoroughly prosecuted? Talking and contention of Arguments would soon be turned into labours; all the fine dreams of Opinions, and universal metaphysical natures, which the luxury of subtle Brains has devised, would quickly vanish, and give place to solid Histories, Experiments and Works. And as at first, mankind fell by tasting of the forbidden Tree of Knowledge, so we, their Posterity, may be in part restored by the same way, not only by beholding and contemplating, but by tasting too those fruits of Natural knowledge, that were never yet forbidden.

From hence the World may be assisted with variety of Inventions, new matter for Sciences may be collected, the old improved, and their rust rubbed away; and as it is by the benefit of Senses that we receive all our Skill in the works of Nature, so they also may be wonderfully benefited by it, and may be guided to an easier and more exact performance of their Offices; 'tis not unlikely, but that we may find out wherein our Senses are deficient, and as easily find ways of repairing them....

Micrographia, or Some Physiological Descriptions of Minute Bodies, Made by Magnifying Glasses; with Observations and Inquiries Thereupon

Observ. I. Of the Point of a Sharp Small Needle.

As in Geometry, the most natural way of beginning is from a Mathematical point; so is the same method in Observations and Natural history the most genuine, simple, and instructive. We must first endeavor to make letters, and draw single strokes true, before we venture to write whole Sentences, or to draw large Pictures. And in Physical Enquiries, we must endeavor to follow Nature in the more plain and easy ways she treads in the most simple and uncompounded bodies, to trace her steps, and be acquainted with her manner of walking there, before we venture ourselves into the multitude of meanders she has in bodies of a more complicated nature; lest, being unable to distinguish and judge of our way, we quickly lose both Nature our Guide, and our selves too, and are left to wander in the labyrinth of groundless opinions; wanting both judgment, that light, and experience, that clew, which should direct our proceedings.

We will begin these our Inquiries therefore with the Observations of Bodies of the most simple nature first, and so gradually proceed to those of a more compounded one. In prosecution of which method, we shall begin with a Physical point; of which kind the Point of a Needle is commonly reckoned for one; and is indeed, for the most part, made so sharp, that the naked eye cannot distinguish any parts of it: It very easily pierces, and makes its way through all kind of bodies softer then itself: But if viewed with a very good Microscope, we may find that the top of a Needle (though as to the sense very sharp) appears abroad, blunt, and very irregular end; not resembling a Cone, as is imagined, but only a piece of a tapering body, with a great part of the top removed, or deficient. The Points of Pins are yet more blunt, and the Points of the most curious Mathematical Instruments do very seldom arrive at so great a sharpness; how much therefore can be built upon demonstrations made only by the productions of the Ruler and Compasses, he will be better able to consider that shall but view those points and lines with a Microscope.

Now though this point be commonly accounted the sharpest (whence when we would express the sharpness of a point the most superlatively, we say, As sharp as a Needle) yet the Microscope can afford us hundreds of Instances of Points many thousand times sharper: such as those of the hairs, and bristles, and claws of multitudes of Insects; the thorns, or crooks, or hairs of leaves, and other small vegetables; nay, the ends of the striæ or small parallelipipeds of Amianthus, and alumen plumosum; of many of which, though the Points are so sharp as not to be visible, though viewed with a Microscope (which magnifies the Object, in bulk, above a million of times) yet I doubt not, but were we able practically to make Microscopes according to the theory of them, we might find hills, and dales, and pores, and a sufficient breadth, or expansion, to give all those parts elbow-room, even in the blunt top of the very Point of any of these so very sharp bodies. For certainly the quantity or extension of anybody may be Divisible in infinitum, though perhaps not the matter.

But to proceed: The Image we have here exhibited in the first Figure, was the top of a small and very sharp Needle, whose point aa nevertheless appeared through the Microscope above a quarter of an inch broad, not round nor flat, but irregular and uneven; so that it seemed to have been big enough to have afforded a hundred armed Mites room enough to be ranged by each other without endangering the breaking one another's necks, by being thrust off on either side. The surface of which, though appearing to the naked eye very smooth, could not nevertheless hide a multitude of holes and scratches and ruggednesses from being discovered by the Microscope to invest it, several of which inequalities (as A, B, C, seemed holes made by some small specks of Rust; and D some adventitious body, that stuck very close to it) were casual. All the rest that roughen the surface, were only so many marks of the rudeness and bungling

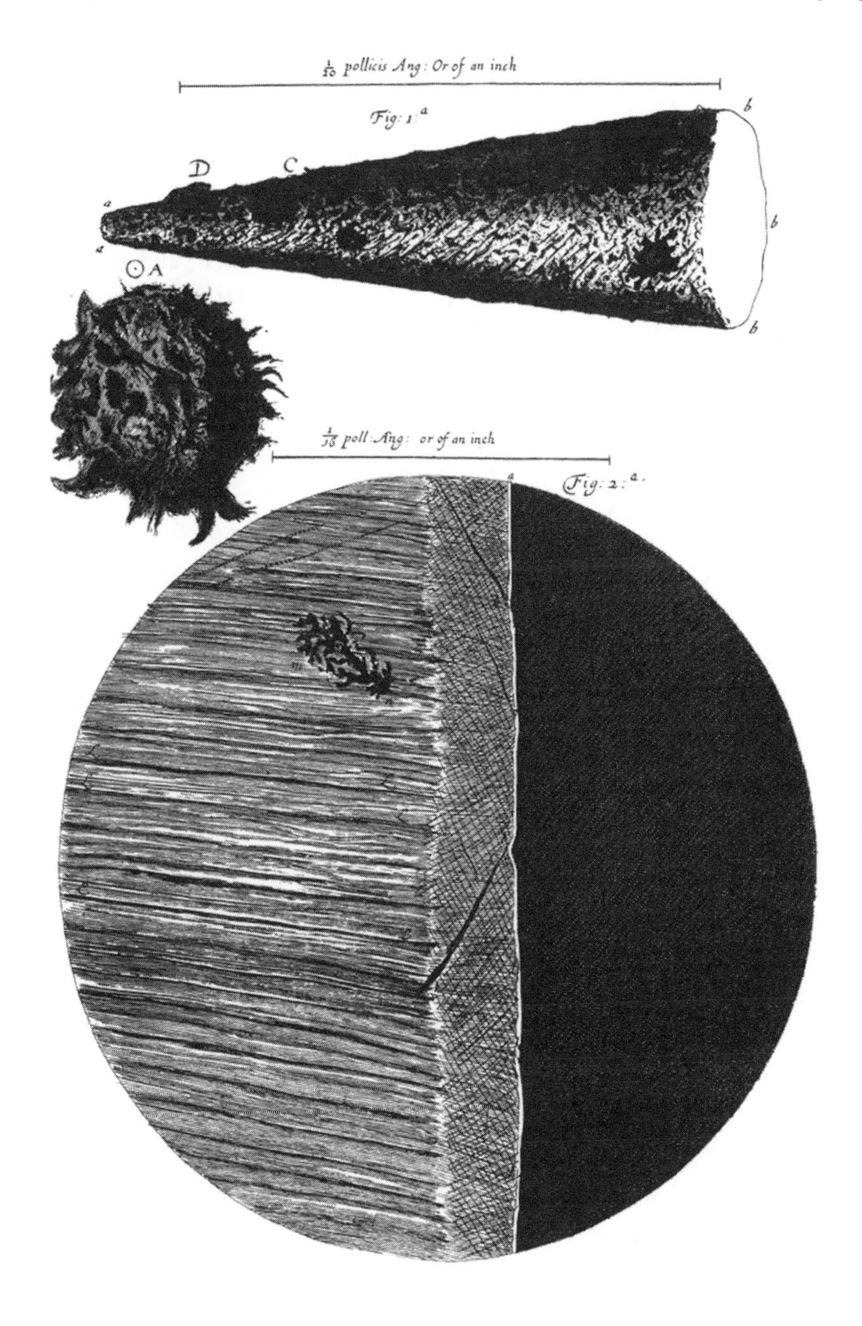

of Art. So unaccurate is it, in all its productions, even in those which seem most neat, that if examined with an organ more acute then that by which they were made, the more we see of their shape, the less appearance will there be of their beauty: whereas in the works of Nature, the deepest Discoveries shew us the greatest Excellencies. An evident Argument, that he that was the Author of all these things, was no other than Omnipotent; being able to include as great a variety of parts and contrivances in the yet smallest Discernable Point, as in those vaster bodies (which comparatively are called also Points) such as the Earth, Sun, or Planets. Nor need it seem strange that the Earth itself may be by Analogy called a Physical Point: For as its body, though now so near us as to fill our eyes and fancies with a sense of the vastness of it, may by a little Distance, and some convenient Diminishing Glasses, be made vanish into a scarce visible Speck, or Point (as I have often tried on the Moon, and (when not too bright) on the Sun it self.) So, could a Mechanical contrivance successfully answer our Theory, we might see the least spot as big as the Earth itself; and Discover, as Descartes also conjectures (Diop. ch. 10. § 9.), as great a variety of bodies in the Moon, or Planets, as in the Earth.

Observ. XVIII. Of the Schematisme or Texture of Cork, and of the Cells and Pores of Some Other such Frothy Bodies.

I took a good clear piece of Cork, and with a Pen-knife sharpened as keen as a Razor, I cut a piece of it off, and thereby left the surface of it exceeding smooth, then examining it very diligently with a Microscope, me thought I could perceive it to appear a little porous; but I could not so plainly distinguish them, as to be sure that they were pores, much less what Figure they were of: But judging from the lightness and yielding quality of the Cork, that certainly the texture could not be so curious, but that possibly, if I could use some further diligence, I might find it to be discernable with a Microscope, I with the same sharp Penknife, cut off from the former smooth surface an exceeding thin piece of it, and placing it on a black object Plate, because it was itself a white body, and casting the light on it with a deep plano-convex Glass, I could exceeding plainly perceive it to be all perforated and porous, much like a Honey-comb, but that the pores of it were not regular; yet it was not unlike a Honey-comb in these particulars.

First, in that it had a very little solid substance, in comparison of the empty cavity that was contained between, as does more manifestly.

Fig. 1. appear by the Figure A and B of the XI. Scheme, for the Interstitia, or walls (as I may so call them) or partitions of those pores were near as thin in proportion to their pores, as those thin films of Wax in a Honey-comb (which enclose and constitute the sexangular celts) are to theirs.

Next, in that these pores, or cells, were not very deep, but consisted of a great many little Boxes, separated out of one continued long pore, by certain

Diaphragms, as is visible by the Figure B, which represents a sight of those pores split the long-ways.

I no sooner discerned these (which were indeed the first microscopical pores I ever saw, and perhaps, that were ever seen, for I had not met with any Writer or Person, that had made any mention of them before this) but me thought I had with the discovery of them, presently hinted to me the true and intelligible reason of all the Phænomena of Cork . . . .

First, if I enquired why it was so exceeding light a body? my Microscope could presently inform me that here was the same reason evident that there is found for the lightness of froth, an empty Honey-comb, Wool, a Spunge, a Pumice-stone, or the like; namely, a very small quantity of a solid body, extended into exceeding large dimensions.

Next, it seemed nothing more difficult to give an intelligible reason, why Cork is a body so very unapt to suck and drink in Water, and consequently preserves itself, floating on the top of Water, though left on it never so long: and why it is able to stop and hold air in a Bottle, though it be there very much condensed and consequently presses very strongly to get a passage out, without suffering the least bubble to pass through its substance. For, as to the first, since our Microscope informs us that the substance of Cork is altogether filled with Air, and that that Air is perfectly enclosed in little Boxes or Cells distinct from one another. It seems very plain, why neither the Water, nor any other Air can easily insinuate itself into them, since there is already within them an *intus existens*, and consequently, why the pieces of Cork become so good floats for Nets, and stopples for Viols, or other close Vessels.

And thirdly, if we enquire why Cork has such a springiness and swelling nature when compressed? and how it comes to suffer so great a compression, or seeming penetration of dimensions, so as to be made a substance as heavy again and more, bulk for bulk, as it was before compression, and yet suffered to return, is found to extend itself again into the same space? Our Microscope will easily inform us, that the whole mass consists of an infinite company of small Boxes or Bladders of Air, which is a substance of a springy nature, and that will suffer a considerable condensation (as I have several times found by divers trials, by which I have most evidently condensed it into less than a twentieth part of its usual dimensions near the Earth, and that with no other strength than that of my hands without any kind of forcing Engine, such as Racks, Leavers, Wheels, Pullies, or the like, but this only by and by) and besides, it seems very probable that those very films or sides of the pores, have in them a springing quality, as almost all other kind of Vegetable substances have, so as to help to restore themselves to their former position.

And could we so easily and certainly discover the Schematisme and Texture even of these films, and of several other bodies, as we can these of Cork; there seems no probable reason to the contrary, but that we might as readily render the true reason of all their Phænomena; as namely, what were the cause of the springingess, and toughness of some, both as to their flexibility and restitution. What, of the friability or brittleness of some others, and the like; but till such time as our Microscope, or some other means, enable us to discover the true Schematism and Texture of all kinds of bodies, we must grope, as it were, in the dark, and only guess at the true reasons of things by similitudes and comparisons.

But, to return to our Observation. I told several lines of these pores, and found that there were usually about threescore of these small Cells placed end-ways in the eighteenth part of an Inch in length, whence I concluded there must be near eleven hundred of them, or somewhat more than a thousand in the length of an Inch, and therefore in a square Inch above a Million, or 1166400, and in a Cubic Inch, above twelve hundred Millions, or 1259712000, a thing almost incredible, did not our Microscope assure us of it by ocular demonstration; nay, did it not discover to us the pores of a body, which were they diaphragmed, like those of Cork, would afford us in one Cubic Inch, more than ten times as many little Cells, as is evident in several charred Vegetables; so prodigiously curious are the works of Nature, that even these conspicuous pores of bodies, which seem to be the channels or pipes through which the *Succus nutritius*, or natural juices of Vegetables are conveyed, and seem to correspond to the veins, arteries and other Vessels in sensible creatures, that these pores I say, which seem to be the Vessels of nutrition to the vastest body in the World, are yet so exceeding small, that the Atoms which Epicurus fancied would go near to prove too big to enter them, much more to constitute a fluid body in them. And how infinitely smaller then must be the Vessels of a Mite, or the pores of one of those little Vegetables I have discovered to grow on the back-side of a Rose-leaf, and shall anon more fully describe, whose bulk is many millions of times less than the bulk of the small shrub it grows on; and even that shrub, many millions of times less in bulk then several trees (that have heretofore grown in England, and are this day flourishing in other hotter Climates, as we are very credibly informed) if at least the pores of this small Vegetable should keep any such proportion to the body of it, as we have found these pores of other Vegetables to do to their bulk. But of these pores I have said more elsewhere.

...

Nor is this kind of Texture peculiar to Cork only; for upon examination with my Microscope, I have found that the pith of an Elder, or almost any other Tree, the inner pulp or pith of the Cany hollow stalks of several other Vegetables: as of Fennel, Carrots, Daucus, Burdocks, Teasels, Fearn, some kinds of Reeds, &c. have much such a kind of Schematisme, as I have lately shewn that of Cork, save only that here the pores are ranged the long-ways, or the same ways with the length of the Cane, whereas in Cork they are transverse.

Observ. LIII. Of a Flea.

The strength and beauty of this small creature, had it no other relation at all to man, would deserve a description.

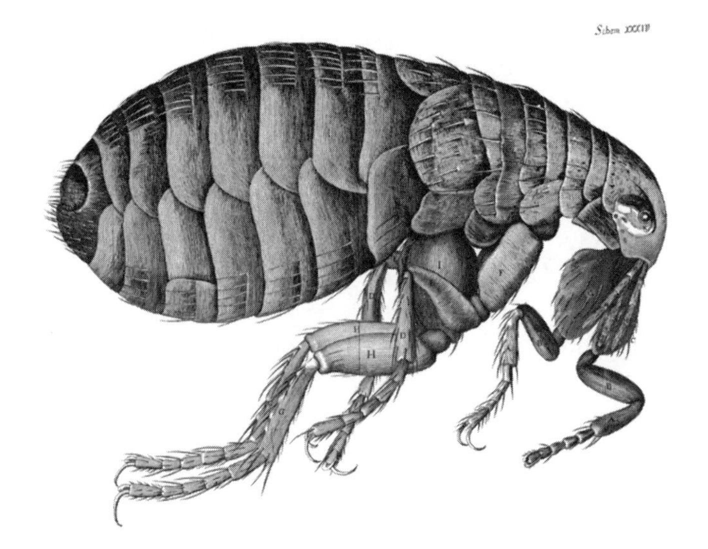

For its strength, the Microscope is able to make no greater discoveries of it then the naked eye, but only the curious contrivance of its legs and joints, for the exerting that strength, is very plainly manifested, such as no other creature, I have yet observed, has anything like it; for the joints of it are so adapted, that he can, as 'twere, fold them short one within another, and suddenly stretch, or spring them out to their whole length, that is, of the fore-legs, the part A, of the 34. Scheme, lies within B, and B within C, parallel to, or side by side each other; but the parts of the two next, lie quite contrary, that is, D without E, and E without F, but parallel also; but the parts of the hinder legs, G, H and I, bend one within another, like the parts of a double jointed Ruler, or like the foot, leg and thigh of a man; these six legs he clitches up altogether, and when he leaps, springs them all out, and thereby exerts his whole strength at once.

But, as for the beauty of it, the Microscope manifests it to be all over adorned with a curiously polished suit of sable Armour, neatly jointed, and beset with multitudes of sharp pins, shaped almost like Porcupine's Quills, or bright conical Steel-bodkins; the head is on either side beautified with a quick and round black eye K, behind each of which also appears a small cavity, L, in which he seems to move to and fro a certain thin film beset with many small transparent hairs, which probably may be his ears; in the forepart of his head, between the two fore-legs, he has two small long jointed feelers, or rather smellers, MM, which have four joints, and are hairy, like those of several other creatures; between these, it has a small proboscis, or probe, NNO, that seems to consist of a tube NN, and a tongue or sucker O, which I have perceived him to slip in and out. Besides these, it has also two chaps or biters PP, which are somewhat like those of an Ant, but I could not perceive them toothed; these were shaped very like the blades of a pair of round-topped scissors, and were opened and shut just after the same manner; with these Instruments does this little busy Creature bite and pierce the skin, and suck out the blood of an Animal, leaving the skin inflamed with a small round red spot. These parts are very difficult to be discovered, because, for the most part, they lye covered between the fore-legs. There are many other particulars, which, being more obvious, and affording no great matter of information, I shall pass by, and refer the Reader to the Figure.

Observ. LIV. Of a Louse.

This is a Creature so officious, that 'twill be known to everyone at one time or other, so busy, and so impudent, that it will be intruding itself in every ones company, and so proud and aspiring withal, that it fears not to trample on the best, and affects nothing so much as a Crown; feeds and lives very high, and that makes it so saucy, as to pull any one by the ears that comes in its way, and will never be quiet till it has drawn blood: it is troubled at nothing so much as at a man that scratches his head, as knowing that man is plotting and contriving some mischief against it, and that makes it oftentime sculk into some meaner and lower place, and run behind a man's back, though it go very much against the hair; which ill conditions of it having made it better known then trusted, would exempt me from making any further description of it, did not my faithful Mercury, my Microscope, bring me other information of it. For this has discovered to me, by means of a very bright light cast on it, that it is a Creature of a very odd shape; it has a head shaped like that expressed in 35. Scheme marked with A, which seems almost Conical, but is a little flatted on the upper and under sides, at the biggest part of which, on either side behind the head (as it were, being the place where other Creatures ears stand) are placed its two black shining goggle eyes BB, looking backwards, and fenced round with several small cilia, or hairs that encompass it, so that it seems this Creature has no very good foresight: It does not seem to have any eye-lids, and therefore perhaps its eyes were so placed, that it might the better cleanse them with its fore-legs; and perhaps this may be the reason, why they so much avoid and run from the light behind them, for being made to live in the shady and dark recesses of the hair, and thence probably their eye having a great aperture, the open and clear light, especially that of the Sun, must needs very much offend them; to secure these eyes from receiving any injury from the hairs through which it passes, it has two horns that grow before it, in the place where one would have thought the eyes should be; each of these CC hath four joints, which are fringed, as 'twere, with small bristles, from which to the tip of its snout D, the head seems very round and tapering, ending in a very sharp nose D, which seems to have a small hole, and to be the passage through which he sucks the blood. Now whereas if it be placed on its back, with its belly upwards, as it is in the 35. Scheme, it seems in several Positions to have a resemblance of chaps, or jaws, as is represented in the Figure by EE, yet in other postures those dark strokes disappear; and having kept several of them in a box for two or three days, so that for all that time they had nothing to feed on,

I found, upon letting one creep on my hand, that it immediately fell to sucking, and did neither seem to thrust its nose very deep into the skin, nor to open any kind of mouth, but I could plainly perceive a small current of blood, which came directly from its snout, and past into its belly; and about A there seemed a contrivance, somewhat resembling a Pump, pair of Bellows, or Heart, for by a very swift systole and diastole the blood seemed drawn from the nose, and forced into the body. It did not seem at all, though I viewed it a good while as it was sucking, to thrust more of its nose into the skin then the very snout D, nor did it cause the least discernable pain, and yet the blood seemed to run through its head very quick and freely, so that it seems there is no part of the skin but the blood is dispersed into, nay, even into the cuticula; for had it thrust its whole nose in from D to CC, it would not have amounted to the supposed thickness of that tegument, the length of the nose being not more than a three hundredth part of an inch. It has six legs, covered with a very transparent shell, and jointed exactly like a Crab's, or Lobster's; each leg is divided into six parts by these joints, and those have here and there several small hairs; and at the end of each leg it has two claws, very properly adapted for its peculiar use, being thereby enabled to walk very securely both on the skin and hair; and indeed this contrivance of the feet is very curious, and could not be made more commodiously and compendiously, for performing both these requisite motions, of walking and climbing up the hair of a man's head, then it is: for, by having the lesser claw (a) set so much short of the bigger (b) when it walks on the skin the shorter touches not, and then the feet are the same with those of a Mite, and several other small Insects, but by means of the small joints of the longer claw it can bend it round, and so with both claws take hold of a hair, in the manner represented in the Figure, the long transparent Cylinder FFF, being a Man's hair held by it.

The Thorax seemed cased with another kind of substance then the belly, namely, with a thin transparent horny substance, which upon the fasting of the Creature did not grow flaccid; through this I could plainly see the blood, sucked from my hand, to be variously distributed, and moved to and fro; and about G there seemed a pretty big white substance, which seemed to be moved within its thorax; besides, there appeared very many small milk-white vessels, which crossed over the breast between the legs, out of which, on either side, were many small branchings, these seemed to be the veins and arteries, for that which is analogous to blood in all Insects is milk-white.

The belly is covered with a transparent substance likewise, but more resembling a skin then a shell, for 'tis grained all over the belly just like the skin in the palms of a man's hand, and when the belly is empty, grows very flaccid and wrinkled; at the upper end of this is placed the stomach HH, and perhaps also the white spot II may be the liver or pancreas, which, by the peristaltick motion of the guts, is a little moved to and fro, not with a systole and diastole, but rather with a thronging or jostling motion. Viewing one of these Creatures, after it had fasted two days, all the hinder part was lank and flaccid, and the white spot II hardly moved, most of the white branchings disappeared, and most also of the

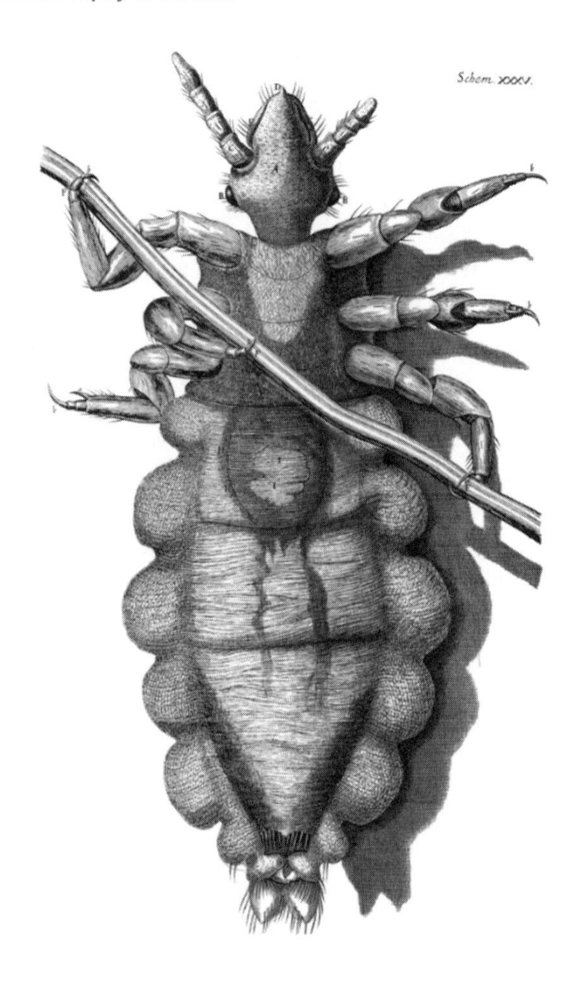

redness or sucked blood in the guts, the peristaltick motion of which was scarce discernable; but upon the suffering it to suck, it presently filled the skin of the belly, and of the six scalloped embossments on either side, as full as it could be stuffed, the stomach and guts were as full as they could hold; the peristaltick motion of the gut grew quick, and the jostling motion of II accordingly; multitudes of milk-white vessels seemed quickly filled, and turgid, which were perhaps the veins and arteries and the Creature was so greedy, that though it could not contain more, yet it continued sucking as fast as ever, and as fast emptying itself behind: the digestion of this Creature must needs be very quick, for though I perceived the blood thicker and blacker when sucked, yet, when in the guts, it was of a very lovely ruby colour, and that part of it, which was digested into the veins, seemed white; whence it appears, that a further digestion of blood may make it milk, at least of a resembling colour: What is else observable in the figure of this Creature, may be seen by the 35.

Reading and Discussion Questions

1. What is Hooke's opinion of data gathered from our senses? What is the importance of observation? What are its pitfalls? Do we have any way of overcoming the limitations of our senses?
2. Why does Hooke begin his observations with an account of examining a needle? What does he conclude from this examination about the products of human artifice? How does this compare to the discoveries about the productions of nature he observes in the flea and the louse?
3. What does Hooke discover in his examination of cork?

Microscopical Researches into the Accordance in the Structure and Growth of Animals and Plants (1839)

Theodor Schwann

81

Theory of the Cells

The various opinions entertained with respect to the fundamental powers of an organized body may be reduced to two, which are essentially different from one another. The first is, that every organism originates with an inherent power, which models it into conformity with a predominant idea, arranging the molecules in the relation necessary for accomplishing certain purposes held forth by this idea. Here, therefore, that which arranges and combines the molecules is a power acting with a definite purpose. A power of this kind would be essentially different from all the powers of inorganic nature, because action goes on in the latter quite blindly. A certain impression is followed of necessity by a certain change of quality and quantity, without regard to any purpose. In this view, however, the fundamental power of the organism (or the soul, in the sense employed by Stahl) would, inasmuch as it works with a definite individual purpose, be much more nearly allied to the immaterial principle, endued with consciousness which we must admit operates in man.

The other view is, that the fundamental powers of organized bodies agree essentially with those of inorganic nature, that they work altogether blindly according to laws of necessity and irrespective of any purpose, that they are powers which are as much established with the existence of matter as the physical powers are. It might be assumed that the powers which form organized bodies do not appear at all in inorganic nature, because this or that particular combination of molecules, by which the powers are elicited, does not occur in inorganic nature, and yet they might not be essentially distinct from physical and chemical powers. It cannot, indeed, be denied that adaptation to a particular purpose, in some individuals even in a high degree, is characteristic of every organism; but, according to this view, the source of this adaptation does not depend upon each organism being developed by the operation of its own power in obedience to that purpose, but it originates as in inorganic nature, in the creation of the matter with its blind powers by a rational Being . . . .

The first view of the fundamental powers of organized bodies may be called the *teleological,* the second the *physical* view. An example will show at once, how important for physiology is the solution of the question as to which is to be followed. If, for instance, we define inflammation and suppuration to be the effort of the organism to remove a foreign body that has been introduced into it; or fever to be the effort of the organism to eliminate diseased matter, and both as the result of the "autocracy of the organism," then these explanations accord with the teleological view. For, since by these processes the obnoxious matter is actually removed, the process which effects them is one adapted to an end; and as the fundamental power of the organism operates in accordance with definite purposes, it may either set these processes in action primarily, or may also summon further powers of matter to its aid, always, however, remaining itself the "*primum movens.*" On the other hand, according to the physical view, this is just as little an explanation as it would be to say, that the motion of the earth around the sun is an effort of the fundamental power of the planetary system to produce a change of seasons on the planets, or to say, that ebb and flood are the reaction of the organism of the earth upon the moon.

In physics, all those explanations which were suggested by a teleological view of nature, as "*horror vacui,*" and the like, have long been discarded. But in animated nature, adaptation—individual adaptation—to a purpose is so prominently marked, that it is difficult to reject all teleological explanations. Meanwhile it must be remembered that those explanations, which explain at once all and nothing, can be but the last resources, when no other view can possibly be adopted; and there is no such necessity for admitting the teleological view in the case of organized bodies . . . .

In any case it conduces much more to the object of science to strive, at least, to adopt the physical explanation. And I would repeat that, when speaking of a physical explanation of organic phenomena, it is not necessary to understand an explanation by known physical powers, such, for instance, as that universal refuge electricity, and the like; but an explanation by means of powers which operate like the physical powers, in accordance with strict laws of blind necessity, whether they be also to be found in inorganic nature or not.

We set out, therefore, with the supposition that an organized body is not produced by a fundamental power which is guided in its operation by a definite idea, but is developed, according to blind laws of necessity, by powers which, like those of inorganic nature, are established by the very existence of matter. As the elementary materials of organic nature are not different from those of the inorganic kingdom, the source of the organic phenomena can only reside in another combination of these materials, whether it be in a peculiar mode of union of the elementary atoms to form atoms of the second order, or in the arrangement of these conglomerate molecules when forming either the separate morphological elementary parts of organisms, or an entire organism. We have here to do with the latter question solely, whether the cause of organic phenomena lies in the whole organism, or in its separate elementary, parts. If this question can be answered, a further inquiry still remains as to whether the organism or its elementary parts possess this power through the peculiar mode of combination of the conglomerate molecules, or through the mode in which the elementary atoms are united into conglomerate molecules.

We may, then, form the two following ideas of the cause of organic phenomena, such as growth, etc. First, that the cause resides in the totality of the organism. By the combination of the molecules into a systematic whole, such as the organism is in every stage of its development, a power is engendered, which enables such an organism to take up fresh material from without, and appropriate it either to the formation of new elementary parts, or to the growth of those already present. Here, therefore, the cause of the growth of the elementary parts resides in the totality of the organism. The other mode of explanation is, that growth does not ensue from a power resident in the entire organism, but that each separate elementary part is possessed of an independent power, an independent life, so to speak; in other words, the molecules in each separate elementary part are so combined as to set free a power by which it is capable of attracting new molecules, and so increasing, and the whole organism subsists only by means of the reciprocal action of the single elementary parts. So that here the single elementary parts only exert an active influence on nutrition, and totality of the organism may indeed be a condition, but is not in this view a cause.

In order to determine which of these two views is the correct one, we must summon to our aid the results of the previous investigation. We have seen that all organized bodies are composed of essentially similar parts, namely, of cells; that these cells are formed and grow in accordance with essentially similar laws; and, therefore, that these processes must, in every instance, be produced by the same powers. Now, if we find that some of these elementary parts, not differing from the others, are capable of separating themselves from the organism, and pursuing an independent growth, we may thence conclude that each of the other elementary parts, each cell, is already possessed of power to take up fresh molecules and grow; and that, therefore, every elementary part possesses a power of its own, an independent life, by means of which it would be enabled to develop itself independently, if the relations which it bore to external parts were but similar to those in which it stands in the organism. The ova of animals afford us examples of such independent cells, growing apart from the organism. It may, indeed, be said of the ova of higher animals, that after impregnation the ovum is essentially different from the other cells of the organism; that by impregnation there is a something conveyed to the ovum, which is more to it than an external condition for vitality, more than nutrient matter; and that it might thereby have first received its peculiar vitality, and therefore that nothing can be inferred from it with respect to the other cells. But this fails in application to those classes which consist only of female individuals, as well as with the spores of the lower plants; and, besides, in the inferior plants any given cell may be separated from the plant, and then grow alone. So that here are whole plants consisting of cells, which can be positively proved to have independent vitality. Now, as all cells grow according to the same laws, and consequently the cause of growth cannot in one case lie in the cell, and in another in the whole organism; and since it may be further proved that some cells, which do not differ from the rest in their mode of growth, are developed independently, we must ascribe to all cells an independent vitality, that is, such combinations of molecules as occur in any single cell, are capable of setting free the power by which it is enabled to take up fresh molecules. The cause of nutrition and growth resides not in the organism as a whole, but in the separate elementary parts—the cells. The failure of growth in the case of any particular cell, when separated from an organized body, is as slight an objection to this theory, as it is an objection against the independent vitality of a bee, that it cannot continue long in existence after being separated from its swarm. The manifestation of the power which resides in the cell depends upon conditions to which it is subject only when in connection with the whole (organism).

The question, then, as to the fundamental power of organized bodies resolves itself into that of the fundamental powers of the individual cells. We must now consider the general phenomena attending the formation of cells, in order to discover what powers may be presumed to exist in the cells to explain them. These phenomena may be arranged in two natural groups: first, those which relate to the combination of the molecules to form a cell, and which may be denominated the plastic phenomena of the cells; secondly, those which result from chemical changes either in the component particles of the cell itself, or in the surrounding cytoblastema, and which may be called metabolic phenomena (implying that which is liable to occasion or to suffer change) . . . .

These are the most important phenomena observed in the formation and development of cells. The unknown cause, presumed to be capable of explaining these processes in the cells, may be called the plastic power of the cells. We will, in the next place, proceed to determine how far a more accurate definition of this power may be deduced from these phenomena.

In the first place, there is a power of attraction exerted in the very commencement of the cell . . . .

The power of attraction may be uniform throughout the whole cell, but it may also be confined to single spots; the deposition of new molecules is then more vigorous at these spots, and the consequence of this uneven growth of the cell-membrane is a change in the form of the cell.

The attractive power of the cells manifests a certain form of election in its operation. It does not take up all the substances contained in the surrounding cytoblastema, but only particular ones, either those which are analogous with the substance already present in the cell (assimilation), or such as differ from it in chemical properties. The several layers grow by assimilation, but when a new layer is being formed, different material from that of the previously-formed layer is attracted: for the nucleolus, the nucleus and cell-membrane are composed of materials which differ in their chemical properties.

Such are the peculiarities of the plastic power of the cells, so far as they can as yet be drawn from observation. But the manifestations of this power presuppose another faculty of the cells. The cytoblastema, in which the cells are formed, contains the elements of the materials of which the cell is composed, but in other combinations: it is not a mere solution of cell-material, but it contains only certain organic substances in solution. The cells, therefore, not only attract materials from out of the cytoblastema, but they must have the faculty of producing chemical changes in its constituent particles. Besides which, all the parts of the cell itself may be chemically altered during the process of its vegetation. The unknown cause of all these phenomena, which we comprise under the term metabolic phenomena of the cells, we will denominate the *metabolic power* . . . .

I think therefore that, in order to explain the distinction between the cell-contents and the external cytoblastema, we must ascribe to the cell-membrane not only the power in general of chemically altering the substances which it is either in contact with, or has imbibed, but also of so separating them that certain substances appear on its inner, and others on its outer surface. The secretion of substances already present in the blood, as, for instance, of urea, by the cells with which the urinary tubes are lined, cannot be explained without such a faculty of the cells. There is, however, nothing so very hazardous in it, since it is a fact that different substances are separated in the decompositions produced by the galvanic pile. It might perhaps be conjectured from this peculiarity of the metabolic phenomena in the cells, that a particular position of the axes of the atoms composing the cell-membrane is essential for the production of these appearances.

Translated by Henry Smith

Reading and Discussion Questions

1. How does Schwann characterize the distinction between a teleological view of nature and a physical view?
2. What observations cause him to conclude that it is the cell (or lack of cells) that is the primary distinction between organic and inorganic forms of matter?

Natural History: The Theory of the Earth (1749)

Georges-Louis Leclerc, Comte de Buffon

82

> I have myself seen what once was solid land changed into sea; and again I have seen land made from the sea. Sea-shells have been seen lying far from the ocean, and an ancient anchor has been found on a mountain-top. What once was a level plain, down-flowing waters have made into a valley; and hills by the force of the floods have been washed into the sea.
>
> Ovid, *Metamorphoses*

Neither the figure of the earth, its motion, nor its external connections with the rest of the universe pertain to our present investigation. It is the internal structure of the globe, its composition, form, and manner of existence which we propose to examine. The general history of the earth should doubtless precede that of its productions, as a necessary study for those who wish to be acquainted with nature in her variety of shapes, and the detail of facts relative to the life and manners of animals, or to the culture and vegetation of plants belong not, perhaps, so much to natural history, as to the general deductions drawn from the observations that have been made upon the different materials which compose the terrestrial globe: as its heights, depth, and inequalities of its form; the motion of the sea, the direction of mountains, the situation of rocks and quarries, the rapidity and effects of currents in the ocean, etc. This is the history of nature in its most ample extent, and these are the operations by which every other effect is influenced and produced. The theory of these effects constitutes what may be termed a primary science, upon which the exact knowledge of particular appearances as well as terrestrial substances entirely depends. This description of science may fairly be considered as appertaining to physics; but does not all physical knowledge, in which no system is admitted, form part of the history of nature? In a subject of great magnitude, whose relative connections are difficult to trace, and where some facts are but partially known, and others uncertain and obscure, it is more easy to form a visionary system than to establish a rational theory; thus it is that the theory of the earth has only hitherto been treated in a vague and hypothetical manner; I shall therefore but slightly mention the singular notions of some authors who have written upon the subject.

What we shall say on this subject will doubtless be less extraordinary, and appear unimportant, if put in comparison with the grand systems just mentioned, but it should be remembered that it is an historian's business to describe, not invent; that no suppositions should be admitted upon subjects that depend upon facts and observation; that his imagination ought only to be exercised for the purpose of combining observations, rendering facts more general, and forming one connected whole, so as to present to the mind a distinct arrangement of clear ideas and probable conjectures; I say probable because we must not expect to give exact demonstration on this subject, that being confined to mathematical sciences, while our knowledge in physics and natural history depends solely upon experience, and is confined to reasoning upon inductions.

In the history of the earth, we shall therefore begin with those facts that have been obtained from the experience of time, together with what we have collected by our own observations.

This immense globe exhibits upon its surface heights, depths, plains, seas, lakes, marshes, rivers, caverns, gulfs, and volcanos; and upon the first view of these objects we cannot discover in their disposition either order or regularity. If we penetrate into its internal part, we shall there find metals, minerals, stones, bitumens, sands, earths, waters, and matters of every kind, placed as it were by chance, and without the smallest apparent design. Examining with a more strict attention we discover sunken mountains, caverns filled, rocks split and broken, countries swallowed up, and new islands rising from the ocean; we shall also perceive heavy substances placed above light ones, hard bodies surrounded with soft; in short we shall there find matter in every form, wet and dry, hot and cold, solid and brittle, mixed in such a sort of confusion as to leave room to compare them only to a mass of rubbish and the ruins of a wrecked world.

We inhabit these ruins, however, with a perfect security. The various generations of men, animals, and plants, succeed each other without interruption; the earth [abundantly supplies] their sustenance: the sea has its limits; its motions and the currents of air are regulated by fixed laws: the returns of the seasons are certain and regular; the severity of the winter being constantly succeeded by the beauties of the spring: everything appears in order, and the earth, formerly a chaos, is now a tranquil and delightful abode, where all is animated and regulated by such an amazing display of power and intelligence as fills us with admiration, and elevates our minds with the most sublime ideas of an all-potent and wonderful Creator.

Let us not then draw any hasty conclusions upon the irregularities of the surface of the earth, nor the apparent disorders in the interior parts, for we shall soon discover the utility, and even the necessity, of them; and, by considering them with a little attention, we shall perhaps find an order of which we had no conception, and a general connection that we could neither perceive nor comprehend by a slight examination: but in fact, our knowledge on this subject must always be confined. There are many parts of the surface of the globe with which we are entirely unacquainted, and have but partial ideas of the bottom of the sea, which in many places we have not been able to fathom. We can only penetrate into the coat of the earth; the greatest caverns and the deepest mines do not descend [beyond] the eighth-thousandth part of its diameter; we can therefore judge only of the external and mere superficial part; we know, indeed, that bulk for bulk the earth weighs four times heavier than the sun, and we also know the proportion its weight bears with other planets; but this is merely a relative estimation; we have no certain standard nor proportion; we are so entirely ignorant of the real weight of the materials, that the internal part of the globe may be a void space, or composed of matter a thousand times heavier than gold, nor is there any method to make further discoveries on this subject; and it is with the greatest difficulty any rational conjectures can be formed thereon.

We must therefore confine ourselves to a correct examination and description of the surface of the earth, and to those trifling depths to which we have been enabled to penetrate. The first object which presents itself is that immense quantity of water which covers the greatest part of the globe; this water always occupies the lowest ground, its surface always level, and constantly tending to equilibrium and rest; nevertheless it is kept in perpetual agitation by a powerful agent,[1] which opposing its natural tranquility impresses it with a regular periodical motion, alternately raising and depressing its waves, producing a vibration in the total mass, by disturbing the whole body to the greatest depths. This motion we know has existed from the commencement of time, and will continue as long as the sun and moon, which are the causes of it.

. . .

Let us now take a view of the earth. What prodigious differences do we find in different climates? What a variety of soils? What inequalities in the surface? But upon a minute and attentive observation we shall find the greatest chains of mountains are nearer the equator than the poles; that in the Old Continent their direction is more from the east to west than from the north to south, and that on the contrary in the new world they extend more from north to south than from east to west; but what is still more remarkable, the form and direction of those mountains, whose appearance is so very irregular, correspond so directly that the prominent angles of one mountain are always opposite to the concave angles of the neighboring mountain, and are of equal dimensions whether they are separated by a small valley or an extensive plain. I have also observed that opposite hills are nearly of the same height, and that in general mountains occupy

[1] See "Proofs . . .," art. 12. [Not included].

the middle of continents, islands, and promontories, which they divide by their greatest lengths.

In following the courses of the principal rivers I have likewise found that they are almost always perpendicular with those of the sea [coasts] into which they empty themselves; and that in the greatest part of their courses they proceed nearly in the direction of the mountains from which they derive their source.

. . .

Pursuing our examination in a more extensive view, we find that the upper strata, that surrounds the globe, is universally the same, [and] that this substance which serves for the growth and nourishment of animals and vegetables, is nothing but a composition of decayed animal and vegetable bodies reduced into such small particles that their former organization is not distinguishable. Penetrating a little further we find the real earth, beds of sand, limestone, argol, shells, marble, gravel chalk, etc. These beds are always parallel to each other and of the same thickness throughout their whole extent. In neighbouring hills beds of the same materials are invariably found upon the same levels, though the hills are separated by deep and extensive intervals. All beds of earth, even the most solid strata, as rocks, quarries of marble etc. are uniformly divided by perpendicular fissures; it is the same in the largest as well as smallest depths, and appears a rule which nature invariably pursues.

In the very bowels of the earth, on the tops of mountains, and even the most remote parts from the sea, shells, skeletons of fishes, marine plants, etc. are frequently found; and these shells, fishes, and plants, are exactly similar to those which exist in the ocean. There are a prodigious quantity of petrified shells to be met with in an infinity of places, not only enclosed in rocks, masses of marble, limestone, as well as in earths and clays, but are actually incorporated and filled with the very substance which surrounds them. In short, I find myself convinced, by repeated observations [...] that marbles, stones, chalks, marles, clay, sand, and almost all terrestrial substances, wherever they may be placed, are filled with shells, and other [debris from] the sea.

These facts being enumerated, let us now see what reasonable conclusions are to be drawn from them.

The changes and alterations which have happened to the earth in the space of the last two or three thousand years are very inconsiderable indeed when compared with those important revolutions which must have taken place in those ages which immediately followed the creation; for as all terrestrial substances could only acquire solidity by the continued action of gravity, it would be easy to demonstrate that the surface of the earth was much softer at first than it is a present, and consequently the same causes which now produce but slight and almost imperceptible changes during many ages, would then effect great revolutions in a very short space. It appears to be a certain fact, that the earth which we now inhabit, and even the tops of the highest mountains were formerly covered with the sea, for shells and other marine productions are frequently found in almost every part; it appears also that the water remained a considerable time on the surface of the earth, since in many places there have been discovered such prodigious banks of shells that it is impossible so great a multitude of animals could exist at the same time: this fact seems likewise to prove, that although the materials which composed the surface of the earth were then in a state of softness that rendered them easy to be disunited, moved and transported by the waters, yet that these removals were not made at once; they must indeed have been successive, gradual, and by degrees, because these kind of sea-productions are frequently met with more than a thousand feet below the surface, and such a considerable thickness of earth and stone could not have accumulated but by the length of time. If we were to suppose that at the deluge all the shell-fish were raised from the bottom of the sea, and transported over all the earth; besides the difficulty of establishing this supposition, is evident, that as we find shells incorporated in marble and in the rocks of the highest mountains, we must likewise suppose that all these marbles and rocks were formed at the same time, and that too at the very instant of the deluge; and besides, that previous to this great revolution there were neither mountains, marble, nor rocks, nor clays, nor matters of any kind similar to those we are at present acquainted with, as they almost all contain shells and other productions of the sea. Besides, at the time of the deluge the earth must have acquired a considerable degree of solidity, from the action of gravity, for more than sixteen centuries, and consequently it does not appear possible that the waters, during the short time the deluge lasted, should have overturned and dissolved its surface to the greatest depths we have since been enabled to penetrate.

But without dwelling longer on this point, which shall hereafter be more amply discussed, I shall confine myself to well-known observations and established facts. There is no doubt but that the waters of the sea at some period covered and remained for ages upon that part of the globe which is now known to be dry land; and consequently the whole continents of Asia, Europe, Africa, and America, were then the bottom of an ocean abounding with similar productions to those which the sea at present contains: it is equally certain that the different strata which compose the earth are parallel and horizontal, and it is evident their being in this situation is the operation of the waters which have collected and accumulated by degrees the different materials, and given them the same position as the water itself always assumes. We observe that the position of strata is almost universally horizontal: in plains it is exactly so, and it is only in the mountains that they are inclined to the horizon, from their having been originally formed by a sediment deposited upon an inclined base. Now I insist that these strata must have been formed by degrees, and not all at once, by any revolution whatever, because strata composed of heavy materials are very frequently found placed above light ones, which could not be, if, as some authors assert, the whole had been mixed with the waters at the time of the deluge, and afterwards precipitated; in that case everything must have had a very different appearance to that which now exists. The heaviest bodies would have descended first, and each particular stratum would have been arranged according to its weight and specific gravity, and we should not see solid rocks or metals placed above light sand any more than clay under coal.

We should also pay attention to another circumstance; it confirms what we have said on the formation of the strata; no other cause than the motions and sediments of water could possibly produce so regular a position of the strata; for the highest mountains are composed of parallel strata as well as the lowest plains, and therefore we cannot attribute the origin and formation of mountains to the shocks of earthquakes, or eruptions of volcanos. The small eminences which are sometimes raised by volcanos or convulsive motions of the earth are not by any means composed of parallel strata, they are a mere disordered heap of matters thrown confusedly together; but the horizontal and parallel position of the strata must necessarily proceed from the operations of a constant cause and motion always regulated and directed in the same uniform manner.

From repeated observations, and these incontrovertible facts, we are convinced that the dry part of the globe, which is now habitable, has remained for a long time under the waters of the sea, and consequently this earth underwent the same fluctuations and changes which the bottom of the ocean is at present actually undergoing. To discover therefore what formerly passed on the earth, let us examine what now passes at the bottom of the sea, and from thence we shall soon be enabled to draw rational conclusions with regard to the external form and internal composition of that which we inhabit.

From the creation the sea has constantly been subject to a regular flux and reflux: this motion, which raises and [lowers] the waters twice in every twenty-four hours, is principally occasioned by the action of the moon, and is much greater under the equator than in any other climates. The earth performs a rapid motion on its axis, and consequently has a centrifugal force, which is also the greatest at the equator; this latter, independent of actual observation, proves that the earth is not perfectly spherical, but that it must be more elevated under the equator than at the poles.

From these combined causes, the ebbing and flowing of the tides, and the motion of the earth, we may fairly conclude, that although the earth was a perfect sphere in its original form, yet its diurnal motion, together with the constant flux and reflux of the sea, must, by degrees, in the course of time, have raised the equatorial parts, by carrying mud, earth, sand, shells, etc. from other climes, and there depositing of them. Agreeable to this idea the greatest irregularities must be found, and, in fact, are found near the equator. Besides, as this motion of the tides is made by diurnal alternatives, and [has] been repeated, without interruption from the commencement of time, is it not natural to imagine, that each time the tide flows the water carries a small quantity of matter from one place to another, which may fall to the bottom like a sediment, and form those parallel and horizontal strata which are everywhere to be met with? For the whole motion of the water, in the flux and reflux being horizontal, the matters carried away with them will naturally be deposited in the same parallel direction.

It is therefore evident that the prodigious chains of mountains which run from the West to the East in the old continent, and from the North to the South in the new, must have been produced by the general motion of the tides; but the origin of all the inferior mountains must be attributed to the particular motions of currents, occasioned by the winds and other irregular agitations of the sea: they may probably have been produced by a combination of all those motions, which must be capable of infinite variations, since the winds and different positions of islands and coasts change the regular course of the tides, and compel them to flow in every possible direction: it is therefore not in the least astonishing that we should see considerable eminences whose courses have no determined direction. But it is sufficient for our present purpose to have demonstrated that mountains are not the produce of earthquakes, or other accidental causes, but that they are the effects resulting from the general order of nature, both as to their organization, and the position of the materials of which they are composed.

But how has it happened that this earth which we and our ancestors have inhabited for ages, which, from time immemorial, has been an immense continent, dry and removed from the reach of the waters, should, if formerly the bottom of the ocean, be actually larger than all the waters, and raised to such a height as to be distinctly separated from them? Having remained so long on the earth why have the waters now abandoned it? What accident, what cause could produce so great a change? Is it possible to conceive one possessed of sufficient power to produce such an amazing effect?

These questions are difficult to be resolved, but as the facts are certain and incontrovertible, the exact manner in which they happened may remain unknown, without prejudicing the conclusions that may be drawn from them; nevertheless by a little reflection we shall find at least plausible reasons for these changes. We daily observe the sea gaining ground on some coasts, and losing it on others; we know that the ocean has a continued regular motion from east to west; that it makes loud and violent efforts against the low lands and rocks which confine it; that there are whole provinces which human industry can hardly secure from the rage of the sea; but there are instances of islands rising above, and others being sunk under the waters. History speaks of much greater deluges and inundations. Ought not this to incline us to believe that the surface of the earth has undergone great revolutions, and that the sea may have quitted the greatest part of the earth which it formerly covered? Let us but suppose that the old and new worlds were formerly but one continent, and that the Atlantis of Plato, was sunk by a violent earthquake: the natural consequence would bc, that the sea would necessarily have flowed in from all sides, and formed what is now called the Atlantic Ocean, leaving vast continents dry, and possibly those which we now inhabit. This revolution therefore might be made [suddenly] by the opening of some vast cavern in the interior part of the globe, which an universal deluge must inevitably succeed: or possibly this change was not effected at once but required a length of time, which I am rather inclined to think: however these conjectures may be, it is certain the revolution has occurred, and in my opinion very naturally, for to judge of the future, as well as the past, we must carefully attend to what daily happens before our eyes. It is a fact clearly established by repeated observations of travellers that the ocean has a constant motion from the east to west; this motion, like the trade winds, is not only felt between the tropics but also throughout the temperate climates, and as near the poles as navigators have gone; of course the Pacific Ocean makes a continual effort against the coasts of Tartary, China, and India: the Indian Ocean acts against the east coast of Africa, and the Atlantic in like manner against all the eastern coasts of America; therefore the sea must have always [gained land on the east and lost it on the west], and still continues to do so; and this alone is sufficient to prove the possibility of the change of earth into sea, and sea into land. If in fact, such are the effects of the sea's motion from east to west, may we not very reasonably suppose that Asia and the eastern continent are the oldest countries in the world, and that Europe and part of Africa, especially the western coasts of these continents, as Great Britain, France, Spain, Mauritania, etc. are of a more modern date? Both history and physics agree in confirming this conjecture.

There are, however, many other causes which concur with the continual motion of the sea from east to west, in producing these effects.

In many places there are lands lower than the level of the sea, and which are only defended from it by an isthmus of rocks, or by banks and dikes of still weaker materials; these barriers must gradually be destroyed by the constant action of the sea, when the lands will be overflowed, and constantly make part of the ocean. Besides, are not mountains daily decreasing by the rains which loosen the earth, and carry it down into the valleys? It is also well known that floods wash the earth from the plains and high grounds into the small brooks and rivers, who in their turn convey into the sea. By these means the bottom of the sea is filling up by degrees, the surface of the earth lowering to a level, and nothing but time is necessary for the sea's successively changing places with the earth.

I speak not here of those remote causes which stand above our comprehension; of those convulsions of nature, whose least effects would be fatal to the world; the near approach of a comet, the absence of the moon, the introduction of a new planet, etc. are suppositions on which it is easy to give scope to the imagination. Such causes would produce any effect we chose, and from a single hypothesis of this nature, a thousand physical romances might be drawn, any of which the authors might term Theory of the Earth. As historians we reject these vain speculations; they are mere possibilities which suppose the destruction of the universe, in which our globe, like a particle of forsaken matter, escapes our observation and is no longer an object worthy [of] regard; but to preserve consistency, we must take the earth as it is, closely observing every part and by inductions judge of the future from what exists at present; in other respects we ought not to be affected by causes which seldom happen, and whose effects are always sudden and violent; they do not occur in the common course of nature; but effects which are daily repeated, motions which succeed each other without interruption, and operations that are constant, ought alone to be the ground of our reasoning.

Translated by J. S. Barr, with emendations by P. R. Sloan

Reading and Discussion Questions

1. Buffon says that we cannot expect "exact demonstrations" in our study of the internal structure of the globe or its history. But why not?
2. From observations of the surface of the earth, what conclusions can we draw about both its interior and its history?
3. What mechanisms are at work in changing the surface of the earth? What does Buffon's description of these mechanisms suggest about how old he takes the earth to be?

Natural History: History of Animals (1749)

Georges-Louis Leclerc, Comte de Buffon

83

Chapter II: Of Reproduction in General

We shall now make a more minute inspection into this common property of animal and vegetable nature; this power of producing its resemblance; this chain of successive individuals, which constitutes the real existence of the species; and without attaching ourselves to the generation of man, or to that of any particular kind of animal, let us inspect the phenomena of reproduction in general; let us collect facts, and enumerate the different methods nature makes use of to renew organized beings. The first, and as we think the most simple method, is to collect in one body an infinite number of similar organic bodies, and so to compose its substance, that there is not a part of it which does not contain a germ of the same species, and which cannot consequently of itself become a whole, resembling that of which it constitutes a part. This process seems to suppose a prodigious waste, and to carry with it profusion; yet it is a very common magnificence of nature, and one which manifests itself even in the most common and inferior kinds, such as worms, polyps, elms, willows, gooseberry-trees, and many other plants and insects, each part of which contains a whole, which by the single effect of expansion alone may become a plant or an insect. By considering organized beings, in this point of view, an individual is a whole, uniformly organized in all its parts; a compound of an infinity of resembling figures and similar parts, an assemblage of germs, of small individuals of the same kind, which can expand in the same mode according to circumstances, and form new bodies, composed like those from whence they proceed.

By examining this idea thoroughly, we shall discover a connection between animals, vegetables, and minerals, which we could not expect. Salts, and some other minerals, are composed of parts resembling each other, and to all that composes them; a grain of salt is a cube, composed of an infinity of smaller cubes, which we may easily perceive by a microscope: these are also composed of other cubes still smaller, as may be perceived with a better microscope; and we cannot doubt, but that the primitive and constituting particles of this salt are like-wise cubes so exceedingly minute as to escape our sight, and our imagination. Animals and plants which can multiply by all their parts, are organized bodies, of which the primitive and constituting parts are also organic and similar, and of which we discern the aggregate quantity, but cannot perceive the primitive parts except by reason and analogy.

This leads us to believe that there is an infinity of organic particles actually existing and living in nature, the substance of which is the same with that of organized bodies, just as there is, as we have recognized, an infinity of similar inert particles of inanimate bodies. And as it would perhaps be necessary for millions of small cubes of salt to accumulate in order to make the visible individual cube of sea-salt, so likewise millions of organic particles, like the whole, are required to form a single one out of that multiplicity of germs contained in an elm or a polyp; and as we must separate, bruise, and dissolve a cube of sea-salt to perceive, by means of crystallization, the small cubes of which it is composed; we must likewise separate the parts of an elm or polypus to discover, by means of vegetation and expansion, the small elms or polyps contained in those parts.

It therefore appears very probable, by the above reasons, that there really exists in nature an infinity of small organized beings, alike, in every respect, to the large organized bodies seen in the world; that these small organized beings are composed of living organic particles, which are common to animals and vegetables, and are their primitive and incorruptible particles; that the assemblage of these particles forms an animal or plant, and consequently that reproduction, or generation, is only a change of form made by the addition of these resembling parts alone, and that death or dissolution is nothing more than a separation of the same particles. Of the truth of this we apprehend there will not remain a doubt after reading the proofs we shall give in the following chapters. Besides, if we reflect on the manner in which trees grow, and consider how so considerable a volume can arise from so small an origin, we shall be convinced that it proceeds from the simple addition of small similar organized particles. A grain produces a young tree, which it contained in miniature. At the summit of this small tree a bud is formed, which contains the young tree for the succeeding year, and this bud is an organic part, resembling the young tree of the first year's growth. A similar bud appears the second year containing a tree for the third; and thus, successively, as long as the tree continues growing, at the extremity of each branch, new buds will form, which contain, in miniature, young trees like that of the first year. It is, therefore, evident, that trees are composed of small organized bodies, similar to themselves, and that the whole individual is formed by the assemblage of small resembling individuals.

But, it may be asked, were not all these organized bodies contained in the seed, and may not the order of their expansion be traced from that source? For the bud which first appeared was evidently surmounted by another similar bud, which was not expanded till the second year, and so on to the third: and consequently, the seed may be said really to contain all the buds, or young trees that would be produced for a hundred years, or till the dissolution of the tree itself. This seed, it is also plain, not only contained all the small organized bodies which one day must constitute the individual tree, but also every seed, every individual, and every succession of seeds and individuals, to the total destruction of the species.

This is the principal difficulty, and we shall examine it with the strictest attention. It is certain, that the seed produces, by the single expansion of the bud, or germ it contains, a young tree the first year, and that this tree existed in miniature in that bud, but it is not equally certain, that the bud of the second year, and those of succeeding, were all contained in the first seed, no more than that every organized body and seed, which must succeed to the end of the world, or to the destruction of the species would be. This opinion supposes a progress to infinity, and makes each existing individual, a source of eternal generations. The first seed, in that case, must have contained every plant of its kind which has existed or ever will exist; and the first man must actually and individually have contained in his loins every man which has or will appear on the face of the earth. Each seed, and each animal, agreeable to this opinion must have possessed within it an infinite posterity. But the more we suffer ourselves to wander into these kind of reasonings, the more we lose the sight of truth in the labyrinth of infinity; and, instead of clearing up and solving the question, we confuse and involve it in more obscurity; it is placing the object out of sight, and afterward saying it is impossible to see it.

Let us investigate a little these ideas of infinite progression and expansion. From whence do they arise? What do they represent? The ideas of infinity can only spring from an idea of that which is limited; for it is in that manner we have an idea of an infinity of succession, a geometrical infinity: each individual is a unit; many individuals compose a finite number, and the whole species is the infinite number. Thus in the same manner as a geometrical infinity may be demonstrated not to exist, so we may be assured, that an infinite progression or development does not exist; that it is only an abstract idea, a suppression of the idea of finity, of which take away the limits that necessarily terminate all size; and that, of course, we must reject from philosophy every opinion which leads to an idea of the actual existence of geometrical or arithmetical infinity.

When we ask, how creatures are reproduced? And it is answered that this multiplication was completely made in the first body, is it not acknowledging that they are ignorant how it is made, and renouncing the will of conceiving it? The question is asked, how one body produces its like? And it is answered, that the whole was created at once. Can we receive this as a solution? For whether one or a million of generations have passed the like difficulty remains, and so far from explaining the supposition of an indefinite number of germs, increases the obscurity, and renders it incomprehensible.

There are two kinds of questions, some pertaining to the first causes, the others only to particular effects; for example, if it is asked why matter is impenetrable? it must either remain unanswered, or be replied to by saying, matter is impenetrable, because it is impenetrable. It will be the same with respect to all the general qualities of matter, whether relative to gravity, extension, motion, or rest; no other reply can be given, and we shall not be surprised that such is the case, if we attentively consider, that in order to give a reason for a thing, we must have a different subject from which we may deduce a comparison, and therefore if the reason of a general cause is asked, that is, of a quality which belongs to all in general, and of which we have no subject to which it does not belong, we are consequently unable to reason upon it; from thence it is demonstrable that it would be useless to make such enquiries, since we should go against the supposition that quality is general and universal.

If, on the contrary, the reason of a particular effect depends immediately on one of the general causes above mentioned, and whether it partakes of the general effect immediately, or by a chain of other effects, the question will be equally solved, provided we distinctly perceive the dependence these effects have on each other, and the connections there are between them.

But if the particular effect, of which we seek the reason, does not appear to depend on these general effects, nor to have any analogy with other known effects; then, this effect being the only one of its kind, and having nothing in common with other effects at least known to us, the question is insolvable; because, not having, in this point, any known subject which has any connection with that we would explain, there is nothing from whence we can draw the reason sought after. When the reason of a general cause is demanded, it is unanswerable because it exists in every object and, on the other hand, the reason of a singular or isolated effect is not found, because not anything known has the same qualities. We cannot explain the reason of a general effect, without discovering one more general; whereas the reason of an isolated effect may be explained by the discovery of some other relative effect, which although we are ignorant of at present, chance or experience may bring to light.

Besides these, there is another kind of question, which may be called the question of fact. For example, why do trees, dogs, etc. exist? All these fact questions are totally insoluble, for those who answer them by final causes, do not consider that they take the effect for the cause; the connection particular objects have with us having no influence on their origin. Moral affinity can never become a physical reason.

We must carefully distinguish those questions where the *why* is used, from those where the *how* is employed, and more so from those where the *how many* is mentioned. *Why* is always relative to the cause of the effect, or to the effect itself.

How is relative to the mode from which the effect springs, and the *how many* has relation only to the proportionate quantity of the effect.

All these distinctions being explained, let us proceed to examine the question concerning the reproduction of bodies. If it is asked, why animals and vegetables reproduce? We shall clearly discover, that this being a question of fact, it is insolvable, and useless to endeavour at the solution of it. But if it is asked, *how* animals and vegetables reproduce? We reply by relating the history of the generation of every species of animals, and of the reproduction of each distinct vegetable; but, after having run over all the methods of an animal engendering its likeness, accompanied even with the most exact observations, we shall find it has only taught us facts without indicating causes; and that the apparent methods which Nature makes use of for reproduction, do not appear to have any connection with the effects resulting therefrom; we shall be still obliged to ask what is the secret mode by which she enables different bodies to propagate their own species.

The question is very different from the first and second; it gives liberty of enquiry and admits the employment of imagination, and therefore is not insolvable, for it does not immediately belong to a general cause; nor is it entirely a question of fact, for provided we can conceive a mode of reproduction dependent upon, or not repugnant to original causes, we shall have gained a satisfactory answer; and the more it shall have a connection with other effects of nature, the better foundation will it be raised upon.

By the question itself it is therefore permitted to form hypotheses, and to select that which shall appear to have the greatest analogy with the other phenomena of nature. But we must exclude from the number, all those which suppose the thing already done; for example, such as suppose that all the germs of the same species were contained in the first seed, or that every reproduction is a new creation, an immediate effect of the Almighty's will; because these hypotheses are questions of fact, and about which it is impossible to reason. We must also reject every hypothesis which might have final causes for its object; such as, we might say, that reproduction is made in order for the living to supply the place of the dead, that the earth may be always covered with vegetables, and peopled with animals; that man may find plenty for his subsistence, etc. because these hypotheses, instead of explaining the effects by physical causes, are founded only on arbitrary connections and moral conventions. At the same time we must not rely on these absolute axioms and physical problems, which so many people have improperly made use of, as principles; for example, that there is no fecundation made apart from the body, *nulla foecondatio extra corpus*; that every living thing is produced from an egg; that all generation supposes sexes, etc. We must not take these maxims in an absolute sense, but consider them only as signifying things generally performed in one particular mode rather than in any other.

Let us, therefore, search after an hypothesis which has not any of these defects, and by which we cannot fall into any of these inconveniences; if, then, we do not succeed in the explanation of the mechanism nature makes use of to effect the reproduction of beings, we shall, at least, arrive at something more probable than what has hitherto been advanced. As we can make moulds, by which we can give to the external parts of bodies whatever figure we please, let us suppose nature can form the same, by which she not only bestows on bodies the external figure but also the internal. Would not this be one mode by which reproduction may be performed?

Let us, then, consider on what foundation this supposition is raised; let us examine if it contains anything contradictory, and afterwards we shall discover what consequences maybe derived from it. Though our senses are only judges of the external parts of bodies, we perfectly comprehend external affection and different figures. We can also imitate nature, by representing external figures by different modes, as by painting, sculpture, and moulds; but although our senses are only judges of external qualities we know there are internal qualities, some of which are general, as gravity. This quality, or power, does not act relatively to surfaces but proportionably to the masses or quantities of matter; there are, therefore, very active qualities in nature, which even penetrate bodies to the most internal parts; but we shall never gain a perfect idea of these qualities, because, not being external, they cannot fall within the compass of our senses; but we can compare their effects, and deduce analogies therefrom, to give an account of the effects of similar qualities.

If our eyes, instead of representing to us the surface of objects only, were so formed as to show us the internal parts alone, we should then have clear ideas of the latter, without the smallest knowledge of the former. In this supposition the internal moulds, which I have supposed to be made use of by nature, might be as easily seen and conceived as the moulds for external figures. In that case we should have modes of imitating the internal parts of bodies as we now have for the external. These internal moulds, although we can never obtain them, can be possessed by nature, just as she has the property of gravity, which penetrates to the internal particles of matter. The supposition of these moulds being formed on good analogies it only remains for us to examine if it includes any contradiction.

It may be argued that the expression of an *internal mould* includes two contradictory ideas; that the idea of a mould can only be related to the surface, and that the internal, according to this, must have a connection with the whole mass, and, therefore, it might as well be called a massive surface, as an internal mould.

I admit, that when we are about to represent ideas which have not hitherto been expressed, we are obliged to make use of terms which seem contradictory; for this reason philosophers have often employed foreign terms on such occasions, instead of applying those in common use, and which have a received signification; but this artifice is useless, since we can show the opposition is only in the words, and that there is nothing contradictory in the idea. Now I affirm that a simple idea cannot contain a contradiction, that is, when we can form an idea of a thing, if this idea is simple it cannot be compounded, it cannot include any other idea, and, consequently, it will contain nothing opposite nor contrary.

Simple ideas are not only the primary apprehensions which strike us by the senses but also the primary comparisons which form from those apprehensions, for the first apprehension itself is always a comparison. The idea of the size of an object, or of its remoteness, necessarily includes a comparison with bulk or distance in general; therefore, when an idea only includes comparison it must be regarded as simple, and from that circumstance, as containing nothing contradictory. Such is the idea of the internal mould. There is a quality in nature, called *gravity,* which penetrates the internal parts of bodies. I take the idea of internal mould relatively to this quality, and, therefore, including only comparison, it contains nothing opposed or contrary.

Let us now see the consequences that may be deduced from this supposition; let us also search after facts corresponding therewith, as it will become so much the more probable, as the number of analogies shall be greater. Let us begin by unfolding this idea of internal moulds, and by explaining in what manner we understand it, we shall be brought to conceive the modes of reproduction.

...

Chapter III: Of Nutrition and Growth

The body of an animal is a kind of internal mould, in which the nutritive matter assimilates itself with the whole in such a manner that, without changing the order and proportion of the parts, each receives an augmentation, and it is this augmentation of bulk which some have called *expansion*, because they imagined every difficulty would be removed by the supposition that the animal was completely formed in the embryo, and that it would be easy to conceive that its parts would expand, or unfold, in proportion as it would increase by the addition of accessory matter.

But if we would have a clear idea of this augmentation and expansion, how can it be done otherwise than by considering the animal body, and each of its parts, as so many internal moulds which receive the accessory matter in the order that results from the position of all their parts? This expansion cannot be made by the addition to the surfaces alone, but, on the contrary, by an intimate susception which penetrates the mass, and thus increases the size of the parts, without changing the form, from whence it is necessary that the matter which serves for this expansion should penetrate the internal part in all its dimensions; it is also as necessary that this penetration be made in a certain order and proportion, so that no one point can receive more than another. Otherwise some parts would expand quicker than others, and the form would be entirely changed. Now what can prescribe this rule to accessory matter, and constrain it to arrive perpetually and proportionally to every point of the external parts, unless we conceive an internal mould? It therefore appears certain that the body of an animal or vegetable is an internal mould of a constant form, but one in which the mass and the volume may be increased proportionally, by the extension of this mould in all its external and internal dimensions, and that this extension also is made by the intussusception of any accessory or foreign matter which penetrates the internal part, and becomes similar to the form, and identical in substance with, the matter of the mould itself.

But of what nature is this matter which the animal or vegetable assimilates with its own substance? What can be the nature of that power which gives it the activity and necessary motion to penetrate the internal mould? And if such a power does exist must it not be similar to that by which the internal mould itself would be produced? . . .

These three questions include all that can be desired on this subject, and seem to depend on each other so much that I am persuaded the reproduction of an animal or vegetable cannot be explained in a satisfactory manner if a clear idea of the mode of the operation of nutrition is not obtained; we must, therefore, examine these three questions separately, in order to compare the consequences resulting therefrom.

The first, which relates to the nutritive nature of this matter, is in part resolved by the reasons we have already given, and will be fully demonstrated in the succeeding chapter. We shall show that there exists an infinity of living organic particles in nature; that their production is of little expense to nature, since their existence is constant and invariable; and that the causes of death only separate without destroying them. Therefore the matter which the animal or vegetable assimilates is an organic matter of the same nature as the animal or vegetable itself, and which consequently can augment the size without changing the form or quality of the matter of the mould, since it is, in fact, of the same form and quality as that which it is constituted with. Thus, in the quantity of aliments which the animal takes to support life, and to keep its organs in play, and in the sap, which the vegetable takes up by its roots and leaves, there is a great part thrown off by transpiration, secretion, and other excretory modes, and only a small portion retained for the nourishment of the parts and their expansion. It is very probable, that in the body of an animal or vegetable [a separation is made of the inert and organic parts of the food]; that the first are carried off by the causes just mentioned. Therefore only organic particles remain, and the distribution of them is made by means of some active power which conducts them to every part in an exact proportion, insomuch that neither receive more nor less than is needful for its equal nutrition, growth, or expansion.

The second question, is what can be the active power which causes this organic matter to penetrate and incorporate itself with this internal mould? By the preceding chapter it appears, that there exist in nature powers relative to the internal part of matter, and which have no relation with its external qualities. These powers, as already observed, will never come under our cognizance, because their action is made on the internal part of the body, whereas our senses cannot reach beyond what is external; it is therefore evident, that we shall never have a clear idea of the penetrating powers, nor of the manner by which they act; but it is not less certain that they exist, than that by their means most effects of nature are produced; we must attribute to them, the effects of nutrition and expansion, which cannot be effected by any other means than the penetration of the most intimate recesses of the original mould; in the same mode as gravity penetrates all parts of matter, so the power which impels or attracts the organic particles of food, penetrates into the internal parts of organized bodies, and as those bodies have a certain form, which we call the internal mould, the organic particles, impelled by the action of the penetrating force, cannot enter therein but in a certain order relative to this form, which consequently it cannot change, but only augment its dimensions, and thus produce the growth of organized bodies; and if in the organized body, expanded by this means, there are some particles, whose external and internal forms are like that of the whole body, from those reproduction will proceed.

The third question: is it not by a similar power that the internal mould itself is reproduced? It appears, that it is not only a similar but the same power which causes development and reproduction, for in an organized body which develops, if there is some particle like the whole, it is sufficient for that particle to become one day an organized body itself, perfectly similar to that of which it made a part. This particle will not at first present a figure striking enough for us to compare with the whole body; but when separated from that body, and receiving proper nourishment, it will begin to expand, and in a short time present a similar being, both externally and internally, as the body from which it had been separated: thus a willow or polyp, which contain more organic particles similar to the whole than most other substances, if cut into ever such a number of pieces, from each piece will spring a body similar to that from whence it was divided.

Now in a body in which every particle is like every other, the organization is the most simple, as we have observed in the first chapter; for it is only the repetition of the same form, and a composition of similar figures, all organized alike. It is for this reason, that the most simple bodies, or the most imperfect kinds, are reproduced with the greatest ease, and in the greatest plenty; whereas, if an organized body contains only some few particles like itself, then, as such alone can attain the second development. Consequently, the reproduction will be more difficult, and not so abundant in number; the organization of these bodies will also be more compounded, because the more the organized parts differ from the whole, the more the organization of this body will be perfect, and the more difficult the reproduction will be.

Nourishment, expansion, and propagation, then are the effects of one and the same and cause. The organized body is nourished by the particles of aliments analogous to it; it expands by the intimate susception of organical parts which agree with it, and it propagates because it contains some original particles which resemble itself. It only remains to examine, whether these similar organic particles come into the organized body by nutriment, or whether they were there before, and have an independent existence. If we suppose the latter, we shall fall in with the doctrine of the infinity of parts, or similar germs contained one in the other; the insufficiency and absurdity of which hypothesis we have already shown; we must therefore conclude that similar parts are extracted from the food; and after what has been said, we hope to explain the manner in which the organic molecules are formed, and how the minute particles unite.

There is, as we have said, a separation of the parts in the nutriment; the organic, from those analogous to the animal or vegetable, by transpiration and other excretory modes; the organical remain and serve for the expansion and nutriment of the body. But these organic parts must be of various kinds, and as each part of the body receives only those similar to itself, and that in a due proportion, it is very natural to imagine, that the superfluity of this organic matter will be sent back from every part of the body into one or more places, where all these organical molecules uniting, form small organized bodies like the first, and to which nothing is wanting but the mode of expansion for them to become individuals of the same species; for every part of the body sending back organized parts, like those of which they themselves are composed, it is necessary, that from the union of all these parts, there should result organized bodies like the first. This being admitted, may we not conclude this is the reason why, during the time of expansion and growth, organized bodies cannot produce, because the parts which expand, absorb the whole of the organic molecules which belong to them and not having any superfluous parts, consequently are incapable of reproduction.

This explanation of nutrition and reproduction will not probably be received by those who admit as the basis of their philosophy only a certain number of mechanical principles, and reject everything which does not depend on them; this is, they say, the great difference between the old philosophy and that of today: it is no longer permissible to postulate causes. It is necessary to give an account of everything by the laws of mechanics, and only those explanations are satisfactory which can be deduced from them.

And as that account which you give of nutrition and reproduction do not depend on these, we ought not to admit them. But I am quite of a different opinion from those philosophers; for it appears to me that, by admitting only a certain number of mechanical principles, they do not see how greatly they contract the bounds of philosophy, and that for one phenomenon that can be explained by a system so confined, a thousand would be found exceeding its limits.

The idea of explaining every phenomenon in nature by mechanical principles, was certainly a great and beautiful exertion, and one which Descartes first attempted. But this idea is only a project, and if properly founded have we the means of performing it? These mechanical principles are the extension of matter, its impenetrability, its motion, its external figure, its divisibility, and the communication of movement by impulsion, by elasticity, etc. The particular ideas of each of these qualities we have acquired by our senses, and regard them as principles, because they are general and belong to all matter. But are we certain these qualities are the only ones which matter possesses, or rather, must we not think these qualities, which we take for principles, are only modes of perception; and that if our senses were differently formed, we should discover in matter, qualities different from those which we have enumerated? To admit only those qualities in matter which are known to us, seems to be a vain and unfounded pretension. Matter may have many general qualities which we shall ever be ignorant of; she may also have others that human assiduity may discover, in the same manner as has recently been done with respect to gravity, which exists universally in all tangible matter. The cause of impulsion, and such other mechanical principles, will always be as impossible to find out as that of attraction, or such other general qualities. From hence is it not very reasonable to say, that mechanical principles are nothing but general effects, which experience has pointed out to us in matter, and that every time a new general effect is discovered, either by reflection, comparison, measure, or experience, a new mechanical principle will be gained, which may be used with as much certainty and advantage as any we are now acquainted with?

The defect of Aristotle's philosophy was making use of particular effects as common causes; and that of Descartes in making use of only a few general effects as causes, and excluding all the rest. The philosophy which appears to me would be the least deficient, is that where general effects only are made use of for causes, but seeking to augment the number of them, by endeavouring to generalize particular effects.

In my explanation of development and reproduction, I admit the received mechanical principles, the penetrating force of gravity, and by analogy I have strived to point out that there are other penetrating powers existing in organized bodies, which experience has confirmed. I have proved by facts, that matter inclines to organization, and that there exists an infinite number of organic particles. I have therefore only generalized some observations, without having advanced anything contrary to mechanical principles, when that term is used as it ought to be understood, as denoting the general effects of nature.

Translated by J.S. Barr, with emendations by P. R. Sloan

Reading and Discussion Questions

1. In this reading Buffon distinguishes between "why" questions, "how" questions, and "what is the secret mode by which" questions. Which of these types does he think should guide our biological investigations? What is wrong with the other two types of questions?
2. What is an "internal mould" and how does it function in Buffon's account of reproduction and nutrition?
3. Does Buffon believe that the laws of motion outlined by Descartes (and Newton) will be sufficient to describe the functioning of living creatures?

On the Increase of the Habitable Earth (1744)

Carl Linnaeus

84

Scripture and reason equally assures us that this astonishing machine, the Universe, was produced and created by an infinite Architect. For nothing is, without a cause; nor can man, endued with reason, admit an infinite series of secondary causes: we must stop therefore at some limit in this profession; a primary, infinite, and absolutely perfect cause.

Let us reflect upon our own natures, let us consider animals and insects, let us contemplate a single vegetable, we everywhere discover the most admirable wisdom, inimitable by any human of finite art: What genius, what art, can imitate one of those fibers whose various and infinite complications form the human body? In its most minute filament we see the finger of God, and the seal of the great Artificer of Universal Nature. If we turn out attention to the properties of the elements, the mind is lost in its own astonishment: If we survey the starts removed from us by such immense spaces (either by unassisted sight, or by those instruments with which art has improved it) consider their laws, their magnitudes, their courses in the depths of an infinite space; prescribed to hours, minutes, moments, our mind must be filled with the idea of an infinite power and wisdom, and their infinite Architect. The life of man would be insufficient to recapitulate in the briefest manner all the separate wonders of his production, much more thoroughly and fully to consider them.

Under the protection of the gracious being, I have determined on this occasion to enter into the reasons which have induced me to believe, *that at the beginning to the world, there was created only one single, sexual pair of every species of living things.*

To the proofs of this proposition, I request those who are my auditors to lend a favourable ear and willing attention.

Our holy Faith instructs us to believe that the Divinity created a single pair of the human kind, one individual Male, the other Female: The sacred writings of Moses acquaint us that they were placed in the Garden of Eden, and that Adam there gave names to every species of animal, God causing them to appear before him.

By a sexual pair I mean one male, and one female in every species where the individuals differ in sex: but there are certain classes of animals natural Hermaphrodites, and of these only a single individual was originally formed in each kind.

Experience teaches that in single and exclusive marriages both in men and animals, the offspring produced exceeds the number of the parents. These as they grow up become further multiplied; thus in the descending line, in every step which we stop to consider, we find the numbers always greater than in the next above which immediately precedes it; and the numbers of every species now exceed what they formerly were. If we trace this train back in the opposite order, and consider the ascending series, we shall find the number of individuals less and less at every step; so that many owe their origin to few, these to still fewer, and so on till the decreasing progression terminates in an individual pair; and the first link in this chain of secondary causes must be referred to an act of creation in the Deity.

To enter into the remainder of the subject with as much brevity as possible, I think myself not greatly in danger of error in laying down the following proposition: "*That the Continent in the first ages of the world lay immersed under the sea, except a single island in the midst of this immense ocean; where all animals lived commodiously, and all vegetables were produced the greatest luxuriance.*"

We had before learned from the declarations of revelation, and the testimony of reason, that of mankind a single individual only of each sex was created. By the Mosaic History we are informed that Paradise was given to Adam for an habitation, and that the various species of animals contributed there to his pleasure and conveniency.

Now if all the animals were in Paradise, which appears from Adam's giving them names, all the insects must have been inhabitants of Paradise; and therefore all the species of vegetables must have had their stations assigned them in this delightful garden. For every vegetable nourished its peculiar insect, and most insects are confined for their food to particular vegetables.

An infinite number of exams may be brought in proof of this:

The Silkwork cannot live or be propagated but where mulberry-tree is plentiful.
The Coccionella lives upon the Indian fig only.
Some species of fish subsist only upon particular worms, as the Greenland whale upon the Medusa. Some live only upon herbs, as the Cretan Scarus, or Char.
There are birds who feed only upon certain berries, as the Ficedula, upon figs and grapes.

Others who eat nothing but particular kinds of insects allotted to them, as the Pici: thus flies are the food of the Muscicapa, Shellfish to the Haematopos, Ants to the Myrmecophaga, Earthworms to the mole, the nocturnal Phalaenae to the Bat.

Fish and birds of prey live upon other species of their own tribe, they are the Hunters of their respective elements.

Thus one animal by the death of another supports its life, which it cannot prolong unless it finds a table spread, suited to its natural appetite.

Beside if that part of the globe occupied by land and the continent had been originally of the same extent which it is this day, it would have been difficult and almost impossible for Adam to have found every animal; their natural wildness would presently have dispersed them over the whole earth; to think the land was originally created of the same size that it is at present, equally full of herbs and trees, equally inhabited by animals, and only two of the human species placed in a single corner of it, is as absurd as to suppose the planet Jupiter similar in all respects to our globe, and abounding in plants and animals, but not to be inhabited by man or any other rational being, to consider this scene of beauty, and repay the debt of glory to his Creator.

Is it credible that the Deity should have replenished the whole earth with animals to destroy them all in a little time by a flood, except a pair of each species preserved in the ark?

He who has ordered all things with the most singular wisdom, and has regulated the number of the offspring of every kind of animal with a proportion so exact, employed certainly as accurate a calculation in creating them. He has done noting in vain, nothing inconsistent with the laws he has once laid down; if we should admit that many individuals of any species of living thing were created a dispersed over all the world, we at once assign limits to creation beyond which it could not extend; and what reason can plead for multiplying creations in one species, when by that of a few, a single pair, a single individual the same might be obtained.

Let us now proceed to the consideration of the earth itself, and we hope that truth of this proposition will clearly appear *a posteriori*.

It is evident from ocular inspection that the land increases from year to year, and that the bounds of our continent are extended.

We see the sea ports of East and West Bothnia every year decreasing, and becoming incapable of admitting vessels, by the sand and soil thrown up, which are always adding new increments to the shore; the inhabitants of the ports are obliged to change their seats, and sometimes remove a quarter of a mile nearer the sear; of this we have seen examples at Pithoa, Luloa and Hudwickval. On the Easter side of Gothland near Hoburg, the increase of the continent for the last 90 years is distinctly visible, being about 2 or 3 toises annually. New Slite and Kylle, in the same country, are enormous stones which rudely represent large temples, giants, and colossal statues in the magnitude, yet worked out of the most solid rock by the force of the water.

The two very tall mountains of Torfburg and Hoburg, in Gothland, are formed of calcareous rock, and were marked and hollowed out by the force of the water at the same time that all Gothland lay immersed in the sea, except these two mountains, which raised their heads out of the deep in the same manner, and with a similar appearance to the Carolinian islands in the present state.

It is impossible to see without admiration the immense masses of stone and rock which everywhere lie scatter without order upon the face of the earth; when broken they are found to consist of mica, quartz, and spat; an evident indication that they (as well as all other stones) were formed on earth, and produced in subterraneous places; but by the force of the waves they were freed from the superincumbent earth, and cast upon the shore; as we frequently see similar masses upon the coasts.

The inhabitants of West Bothnia have observed by marks upon stones that the sea decreases every ten years four inches five lines perpendicularly; which amounts in an age to four feet five inches. According to which calculation 6000 years ago the sea was 240 feet deeper than it is at this present.

To these proofs we may add the infinite number of shells of sea fish found in calcareous mountains. The calcareous Mountains at Raetvic and Dalicum are full of petrified Conchs and Orthocerotes. That Conchaceous earth so frequent in Helfingia is all composed of fragments of Mytuli, or conchs, of a dirty yellow color: no man can be ignorant that the sea, not the dry land, was the place of their production; that these Conchs and bivalves are found at a certain distance from land, and not in the deeps; who does not see that the sea cast forth upon the shore the testaceous covering of its dead shell fish? Hence we infer that the sea once penetrated 20 miles up into Dalekarlia, or further, wherever any vestiges of shells are to be discovered.

Whoever ascribes these circumstances to the Deluge, which came suddenly and was a suddenly past, must be totally ignorant in Natural Philosophy; and if he sees anything, sees with the eyes of others. The sea every year becomes deeper by casting up earth, sand, and stones upon the shores, hence the earth increases in breadth, the sea in depth, and occupies the void space made by the earthy matter it cast up, which it is straightened on every side in narrower limits.

About Hoburg in Gothland I have seen immense stones, moveable by no animals or human art, yet cast up by the sea upon the shore: they consisted of pure marble, of white and red particles; a species of stone not produced in this country but in the Carolinian islands, from whence without doubt they were transferred higher by the force of the waves, especially as they were upon the eastern shore. One stone of a particular size, of that kind called micaceous, lies at the distance of a quarter of a mile from Hoburg; it is too large for any application of human force to have moved, yet it does not lie in the place where it is generated; for the neighbouring matter is not proper for that purpose: from hence we may infer that it was brought hither by some violent commotion of the sea from Sweden or Muscovy, while Gothland was yet immersed under it.

Translated by F. J. Brand

Reading and Discussion Questions

1. Linnaeus takes for granted in the piece that creation occurred as described in Genesis. He is also an extremely accomplished observer of plants, animals, and their environments. What does this lead him to conclude about the surface of the earth and how it has changed over time? What evidence does he provide to support his views?
2. What mechanisms are at work in the alteration of the surface of the earth since its creation?

The Economy of Nature, a Dissertation Presided Over by Carl Linnaeus (1749)

Isaac Biberg

85

Section One

By the Economy of Nature we understand the all-wise disposition of the Creator in relation to natural things, by which they are fitted to produce general ends, and reciprocal uses.

All things contained in the compass of the universe declare, as it were, with one accord the infinite wisdom of the Creator. For whatever strikes our senses, whatever is the object of our thoughts, are so contrived, that they concur to make manifest the divine glory, i.e. the ultimate end which God proposed in all his works. Whoever duly turns his attention to the things on this our terraqueous globe, must necessarily confess, that they are so connected, so chained together, that they all aim at the same end, and to this end a vast number of intermediate ends are subservient. But as the intent of this treatise will not suffer me to consider them all, I shall at present only take notice of such as relate to the preservation of natural things. In order therefore to perpetuate the established course of nature in a continued series, the divine wisdom has thought fit, that all living creatures should constantly be employed in producing individuals; that all natural things should contribute and lend a helping hand to preserve every species; and lastly, that the death and destruction of one thing should always be subservient to the restitution of another. I seems to me that a greater subject than this cannot be found, nor one on which laborious men may more worthily employ their industry, or men of genius their penetration . . . .

Section 2

The world, or the terraqueous globe, which we inhabit, is everywhere surrounded with elements, and contains in its superficies the three kingdoms of nature, as they are called; the *fossil*, which constitutes the crust of the earth, the *vegetable*, which adorns the face of it, and draws the greatest part of its nourishment from the fossil kingdom, and the *animal*, which is sustained by the vegetable kingdom. Thus these three kingdoms cover, adorn and vary the superficies of our earth. It is not my design to make any inquiry concerning the center of the terraqueous globe. He, who likes hypotheses, may consult Descartes, Helmont, Kircher, and others. My business is to consider the external parts of it only, and whatever is obvious to the eye.

As to the strata of the earth and mountains, as far as we have hitherto been able to discover, the upper parts consist of *rag-stone*, the next of *slate*, the third of *marble* filled with petrifactions, the fourth again with *slate*, and lastly the lowest of *free-stone*. The habitable part of the earth, though it is scooped into various inequalities, yet is every where high in comparison with the water, and the farther it is from the sea, it is generally higher. Thus the waters in the lower places are not at rest, unless some obstacle confines them, and by that means form lakes, and marshes . . . .

We often see new meadows arise from marshes dryed up. This happens sooner when the *sphagnum* has laid a foundation; for this in process of time changes into a very porous mould, till almost the whole marsh is filled with it. After that the *rush* strikes root, and along with the *cotton grasses* constitutes a turf, raised in such a manner, that the roots get continually higher, and thus lay a more firm foundation for other plants, till the whole marsh is changed into a fine and delightful meadow; especially if the water happens to work itself a new passage . . . .

It is scarcely to be doubted, but that the rocks and stones dispersed over the globe were formed originally in, and form the earth; but when torrents of rain have softened, as they easily do, the soluble earth, and carried it down into the lower parts, we imagine it happens that these solid, and heavy bodies, being laid bare, stick out above the surface. We might also take notice of the wonderful effects of the tide, such as we see happen from time to time on the sea-shore, which being daily and nightly assaulted with repeated blows, at length gives way, and breaks off. Hence we see in most places the rubbish of the sea, and shores.

The winter by its frost prepares the earth, and mould, which thence are broken into very minute particles, and thus, being put into a mouldering state, become more fit for the nourishment of plants; nay by its snow it covers the seeds, and roots of plants, and thus by cold defends them from the force of cold. I must add also that the piercing frost of the winter purifies the atmosphere, and putrid waters, and makes them more wholesome for animals . . . .

The Vegetable Kingdom

Section 6

Anatomy abundantly proves, that all *plants* are *organic*, and living bodies; and that all *organic* bodies are propagated from an egg has been sufficiently demonstrated by the industry of the moderns. We therefore the rather, according to the opinion of the skillful, reject the equivocal generation of *plants*; and the more so, as it is certain that every living thing is produced from an egg. Now the seeds of *vegetables* are called eggs; these are different in every different *plant*, that the means being the same, each may multiply its species, and produce an offspring like its parent. We do not deny, that very many *plants* push forth from their roots fresh offsets for two or more years. Nay not a few *plants* may be propagated by branches, buds, suckers and leaves fixed in the ground, as likewise many trees. Hence their stems being divided into branches, may be looked on as roots above ground, for in the same way the roots creep underground; and divide into branches. And there is the more reason for thinking so, because we know that a tree will grow in an inverted situation, viz. the roots being placed upwards, and the head downwards, and buried in the ground; for then the branches will produce leaves, and flowers. The *lime-tree* will serve for an example, on which gardeners have chiefly made the experiment. Yet this by no means overturns the doctrine, that all *vegetables* are propagated by seeds; since it is clear that in each of the foregoing instances nothing vegetates but what was part of a plant, formerly produced from seed, so that, accurately speaking, without seed no new *plant* is produced.

Thus again *plants* produce seeds, but they are entirely unfit for propagation, unless fecundation precedes, which is performed by an intercourse between different sexes, as experience testifies. *Plants* therefore must be provided with *organs* of generation; in which respect they hold an analogy with *animals* . . . .

Moreover we cannot without admiration observe that most flowers expand themselves when the sun shines forth, whereas when clouds, rain, or the evening comes on, they close up, lest the genital dust should be coagulated, or rendered useless, so that it cannot be conveyed to the *stigmata*. But what is still more remarkable and wonderfull! When the fecundation is over, the flowers neither upon showers, nor evening coming on close themselves up. Hence when rain falls in the flowering time, the husbandman and gardener foretell a scarcity of fruits . . . I cannot help remarking one particular more, viz. that the organs of generation, which in the animal kingdom are by nature generally removed from sight, in the vegetable are exposed to the eyes of all, and that when their nuptials are celebrated, it is wonderfull what delight they afford to the spectator by their most beautiful colours and delicious odors. At this time bees, flies, and other insects suck honey out of their nectaries, not to mention the humming bird; and that from their effete dust the bees gather wax . . . .

Section 7

... *Berries* and other *pericarps*, are by nature allotted for aliment to animals, but with this condition, that while they eat the pulp they shall sow the *seeds*; for when they feed upon it they either disperse them at the same time, or, if they swallow them, they are returned with interest; for they always come out unhurt. It is not therefore surprising, that if a field be manured with recent mud or dung not quite rotten, various other plants, injurious to the farmer, should come up along with the grain, that is sowed. Many have believed that *barley*, or *rye* has been changes into *oats*, although all such kinds of metamorphoses are repugnant to the laws of generation, not considering that there is another cause of the phenomenon, viz. that the ground perhaps has been manured with horse-dung, in which the *seeds* of *oats*, coming entire from the horse, lye hid and produce that grain ....

Section 8

The great Author and Parent of all things, decreed, that the whole earth should be covered with plants, and that no place should be void, none barren. But since all countries have not the same changes of seasons, and every soil is not equally fit for every plants, He therefore, that no place should be without some, gave to every one of them such a nature, as might be chiefly adapted to the climate; so that some of them can bear and intense cold, others an equal degree of heat; some delight in dry ground, others in moist, &c. Hence the same plants grow only where there are the same seasons of the year, and the same soil ....

Section 10

Daily experience teaches us, that all *plants* as well as all other living things, must submit to death. They spring up, they grow, they flourish, they ripen their fruit, they wither, and at last, having finished their course, they die, and return to the dust again, from whence they first took their life. Thus all black mould, which every where covers the earth, for the greatest part is owing to dead *vegetables*. For all roots descend into the sand by their branches, and after a *plant* has lost its stem the root remains; but this too rots as last, and changes into mould. By this means this kind of earth is mixed with sand, by the contrivance of nature, nearly in the same way as dung thrown upon fields in wrought into the earth by the industry of the husbandman. The earth thus prepared offers again to *plants* from its bosom, what it has received from them. For when seeds are committed to the earth, they draw to themselves, accommodate to their nature, and turn into *plants*, the more subtle parts of this mould by the co-operation of the sun, air, clouds, rains, and winds; so that the tallest tree is, properly speaking, nothing but mould wonderfully compounded with air, and water, and modified by a virtue communicated to a small seed by the Creator. From these *plants*, when they die, just the same kind of mould is formed, as gave birth to them originally; but in such a manner, that it is in greater quantity than before. *Vegetables* therefore increase the black mould, whence fertility remains continually uninterrupted. Whereas the earth could not make good its annual consumption, unless it were constantly recruited by new supplies.

The Animal Kingdom

Section 11

The generation of animals holds the first place among all things, that raise our admiration, when we consider the works of the Creator; and that appointment particularly, by which he has regulated the conception of the *fetus*, and its exclusion, that it should be adapted to the disposition and way of each living animal, is most worthy of our attention.

We find no species of animals exempt from the stings of love, which is put into them to the end, that the Creator's mandate may be executed, *increase and multiply*; and that thus the egg, in which is contained the rudiment of the *fetus* may be fecundated; for without fecundation all eggs are unfit to produce an offspring ....

Section 12

... Most of the *insect* kind neither bear young nor hatch eggs; yet their tribes are the most numerous of all living creatures; insomuch that if the bulk of their bodies were proportionate to their quantity, they would scarce leave room for any other kinds of animals. Let us see therefore with what wisdom the Creator has managed about the propagation of these minute creatures. The females by natural instinct meet and copulate with the males; and afterwards lay their eggs, but not indiscriminately in every place; for they all know how to choose such places as may supply their offspring in its tender age with nourishment, and other things necessary to satisfy their natural wants; for the mother, soon after she has laid her eggs, dies, and were she to live she would not have it in her power to take care of her young.

Butterflies, moths, some *beetles, wevils, bugs, cuckow-spit insects, gall-insects, tree bugs*, &c. lay their eggs, on the leaves of plants, and every different tribe chooses its own species of plant. Nay there is scarce any plant, which does not afford nourishment to some insect; and still more, there is scarcely any part of a plant, which is not preferred by some of them. Thus one insect feeds upon the flower; another upon the trunk; another upon the root; and another upon the leaves. But we cannot help wondering particularly, when we see how the leaves of some trees, and plants, after eggs have been let into them, grow into galls; and form dwellings, as it were, for the young ones, where they may conveniently live . . . .

Section 15

As soon as animals come to maturity, and want no longer the care of their parents, they attend with the utmost labor, and industry, according to the law and economy appointed for every species, to the preservation of their lives. But that so great a number of them, which occur everywhere, may be supported, and a certain and fixed order may be kept up amongst them, behold the wonderful disposition of the Creator, in assigning to each species certain kinds of food, and in putting limits to their appetites. So that some live on particular species of plants, which particular regions, and soils only produce. Some on particular animalcula, and others on carcasses, and some even on mud and dung. For this reason Providence has ordained, that some should swim in certain regions of the watery element, others should fly; some should inhabit the torrid, the frigid, or the temperate zones, and others should frequent deserts, mountains, woods, pools or meadows, according as the food proper to their nature is found in sufficient quantity. By this means there is no terrestrial tract, no sea, no river, no country, but what contains, and nourishes various kinds of animals. Hence also an animal of one kind cannot rob those of another kind of its aliment; which, if it happened, would endanger their lives or health; and thus the world at all times affords nourishment to so many, and so large inhabitants, at the same time that nothing, which it produces, is useless or superfluous.

Section 18

. . . If we consider the end for which it pleased the Supreme Being to constitute such an order of nature, that some animals should be, as it were, created only to be miserably butchered by others, it seems that his Providence not only aimed at sustaining, but also keeping a just proportion amongst all the species; and so prevent any one of them increasing too much, to the detriment of men, and other animals. For if it be true, as it is most assuredly, that the surface of the earth can support only a certain number of inhabitants, they must all perish, if the same number were doubled, or tripled.

There are some viviparous *flies*, which bring forth 2000 young. These in a little time would fill the air, and like clouds intercept the rays of the sun, unless they were devoured by birds, spiders, and many other animals .... The *white fox* is of equal advantage in the Lapland alps; as he destroys the Norway *rats* which, are generated there in great abundance; and thus hinders them from increasing too much in proportion, which would be the destruction of vegetables.

It is sufficient for us, that nothing is made by Providence in vain, and that whatever is made, is made with supreme wisdom. For it does not become us to pry too boldly into all the designs of God. Let us not imagine, when these rapacious animals sometimes do us mischief, that the Creator planned the order of nature according to our private principles of economy; for the Laplanders have one way of living; the European husbandman another; the Hottentots and savages a third, where as the stupendous economy of the Deity is one throughout the globe, and if Providence does not always calculate exactly according to our way of reckoning, we ought to consider this affair in the same light, as when different seamen wait for a fair wind, every one, with respect to the part he is bound to, who we plainly see cannot all be satisfied.

Section 19

The whole earth would be overwhelmed with carcasses, and stinking bodies, if some animals did not delight to feed upon them. Therefore when an animal dyes, *bears, wolves, foxes, ravens,* &c. do not lose a moment till they have taken all away. But if a *horse* e.g. dyes near the public road, you will find him, after a few days, swoln, burst, and at last filled with innumerable *grubs* of carnivorous *flies,* by which he is entirely consumed, and removed out of the way, that he may not become a nuisance to passengers by his poisonous stench ....

Lice increase in a wonderful manner in the heads of children, that are scabby, nor are they without their use, for they consume the redundant humours.

The *beetle* kind in summer extract all moist and glutinous matter out of the dung of cattle, so that is becomes like dust, and is spread by the wind over the ground. Were it not for this, the vegetables that lye under the dung, would be so far from thriving, that all that spot would be rendered barren.

Section 20

Lastly; all these treasures of nature so artfully contrived, so wonderfully propagated, so providentially supported throughout her three kingdoms, seem intended by the Creator for the sake of man. Every thing may be made subservient to his use, if not immediately, yet mediately, not so to that of other animals. By the help of reason man tames the fiercest animals, pursues and catches the swiftest, nay he is able to reach even those, which lye hid in the bottom of the sea.

By the help of reason he increases the number of vegetables immensely, and does that by art, which nature, left to herself, could scarcely effect. By ingenuity he obtains from vegetables whatever is convenient or necessary for food, drink, clothing, medicine, navigation, and a thousand other purposes.

He has found the means of going down into the abyss of the earth, and almost searching its very bowels. With what artifice has he learned to get fragments from the most rocky mountains, to make the hardest stones fluid like water; to separate the usefull metal from the useless dross, and to turn the finest sand to some use! In short when we follow the series of created things, and consider how providentially one is made for the sake of another, the matter comes to this, that all things are made for the sake of man; and for this end more especially, that he by admiring the works of the Creator should extoll his glory, and at once enjoy all those things, of which he stands in need, in order to pass his life conveniently and pleasantly.

Section 21

This subject concerning the economy of nature, a very small part of which I have lightly touched upon, is of such importance and dignity, that if it were to be properly treated in all its parts, men would find wherewithal to employ almost all the powers of the mind. Nay time itself would fail before even the most acute human sagacity would be able to discover the amazing economy, laws, and exquisite structure of the least insect, since as Pliny observes, nature no where appears more herself, than in her most minute works. Every species of created beings deserves to engross one examiner.

If according to gross calculation we reckon in the world 200 species of *vegetables*, 300 of *worms*, 12000 of *insects*, 200 of *amphibious animals*, 2600 of *fishes*, 2000 of *birds*, 200 of *quadrupeds*; the whole sum of the species of living creatures will amount to 40000. Out of these our country has scarcely 3000, for we have discovered only about 1200 native plants, and about 1400 species of animals. We of the human race, who were created to praise and adore our Creator, unless we choose to be mere idle spectators, should and in duty ought to be affected with nothing so much as the pious consideration of this glorious palace. Most certainly if we were to improve and polish our minds by the knowledge of these things, we should beside the great use which would accrue to our economy, discover the more excellent economy of nature, and more strongly admire it when discovered.

Omnium elementorum alterni recurfi sunt, Quicquid alteri perit in alterum transit.—Seneca[1]

Translated by Benjamin Stillingfleet

[1] "Reciprocal changes occur in the case of all elements. Whatever is lost to one thing moves on to another."

Reading and Discussion Questions

1. In what sense is this text about "economy" in the natural world? Under what modern scientific discipline might this work be classified?
2. Does Biberg's Christian faith seem at all in conflict with his scientific goals here? Is there anywhere that it seems to have clouded his observational abilities, or does it merely inspire his devotion to the field?
3. How do lice and fleas and beetles all serve as evidence for the goodness of the creator?

An Essay on the Principle of Population (1798)

Thomas Malthus

86

Chapter 1

Question stated—Little prospect of a determination of it, from the enmity of the opposing parties—The principal argument against the perfectibility of man and of society has never been fairly answered—Nature of the difficulty arising from population—Outline of the principal argument of the Essay

The great and unlooked for discoveries that have taken place of late years in natural philosophy, the increasing diffusion of general knowledge from the extension of the art of printing, the ardent and unshackled spirit of inquiry that prevails throughout the lettered and even unlettered world, the new and extraordinary lights that have been thrown on political subjects which dazzle and astonish the understanding, and particularly that tremendous phenomenon in the political horizon, the French Revolution, which, like a blazing comet, seems destined either to inspire with fresh life and vigor, or to scorch up and destroy the shrinking inhabitants of the earth, have all concurred to lead many able men into the opinion that we were touching on a period big with the most important changes, changes that would in some measure be decisive of the future fate of mankind.

It has been said that the great question is now at issue, whether man shall henceforth start forwards with accelerated velocity towards illimitable, and hitherto unconceived improvement, or be condemned to a perpetual oscillation between happiness and misery, and after every effort remain still at an immeasurable distance from the wished-for goal.

Yet, anxiously as every friend of mankind must look forwards to the termination of this painful suspense, and eagerly as the inquiring mind would hail every ray of light that might assist its view into futurity, it is much to be lamented that the writers on each side of this momentous question still keep far aloof from each other. Their mutual arguments do not meet with a candid examination. The question is not brought to rest on fewer points, and even in theory scarcely seems to be approaching to a decision.

The advocate for the present order of things is apt to treat the sect of speculative philosophers either as a set of artful and designing knaves who preach up ardent benevolence and draw captivating pictures of a happier state of society only the better to enable them to destroy the present establishments and to forward their own deep-laid schemes of ambition, or as wild and mad-headed enthusiasts whose silly speculations and absurd paradoxes are not worthy the attention of any reasonable man.

The advocate for the perfectibility of man, and of society, retorts on the defender of establishments a more than equal contempt. He brands him as the slave of the most miserable and narrow prejudices; or as the defender of the abuses of civil society only because he profits by them. He paints him either as a character who prostitutes his understanding to his interest, or as one whose powers of mind are not of a size to grasp anything great and noble, who cannot see above five yards before him, and who must therefore be utterly unable to take in the views of the enlightened benefactor of mankind.

In this unamicable contest the cause of truth cannot but suffer. The really good arguments on each side of the question are not allowed to have their proper weight. Each pursues his own theory, little solicitous to correct or improve it by an attention to what is advanced by his opponents.

The friend of the present order of things condemns all political speculations in the gross. He will not even condescend to examine the grounds from which the perfectibility of society is inferred. Much less will he give himself the trouble in a fair and candid manner to attempt an exposition of their fallacy.

The speculative philosopher equally offends against the cause of truth. With eyes fixed on a happier state of society, the blessings of which he paints in the most captivating colors, he allows himself to indulge in the most bitter invectives against every present establishment, without applying his talents to consider the best and safest means of removing abuses and without seeming to be aware of the tremendous obstacles that threaten, even in theory, to oppose the progress of man towards perfection.

It is an acknowledged truth in philosophy that a just theory will always be confirmed by experiment. Yet so much friction, and so many minute circumstances occur in practice, which it is next to impossible for the most enlarged and penetrating mind to foresee, that on few subjects can any theory be pronounced just, till all the arguments against it have been maturely weighed and clearly and consistently refuted.

I have read some of the speculations on the perfectibility of man and of society with great pleasure. I have been warmed and delighted with the enchanting picture which they hold forth. I ardently wish for such happy improvements. But

I see great, and, to my understanding, unconquerable difficulties in the way to them. These difficulties it is my present purpose to state, declaring, at the same time, that so far from exulting in them, as a cause of triumph over the friends of innovation, nothing would give me greater pleasure than to see them completely removed.

. . .

In entering upon the argument I must premise that I put out of the question, at present, all mere conjectures, that is, all suppositions, the probable realization of which cannot be inferred upon any just philosophical grounds. A writer may tell me that he thinks man will ultimately become an ostrich. I cannot properly contradict him. But before he can expect to bring any reasonable person over to his opinion, he ought to shew that the necks of mankind have been gradually elongating, that the lips have grown harder and more prominent, that the legs and feet are daily altering their shape, and that the hair is beginning to change into stubs of feathers. And till the probability of so wonderful a conversion can be shewn, it is surely lost time and lost eloquence to expatiate on the happiness of man in such a state; to describe his powers, both of running and flying, to paint him in a condition where all narrow luxuries would be contemned, where he would be employed only in collecting the necessaries of life, and where, consequently, each man's share of labor would be light, and his portion of leisure ample.

I think I may fairly make two postulates:

First, That food is necessary to the existence of man.
Secondly, That the passion between the sexes is necessary and will remain nearly in its present state.

These two laws, ever since we have had any knowledge of mankind, appear to have been fixed laws of our nature, and, as we have not hitherto seen any alteration in them, we have no right to conclude that they will ever cease to be what they now are, without an immediate act of power in that Being who first arranged the system of the universe, and for the advantage of his creatures, still executes, according to fixed laws, all its various operations.

I do not know that any writer has supposed that on this earth man will ultimately be able to live without food. But Mr Godwin has conjectured that the passion between the sexes may in time be extinguished. As, however, he calls this part of his work a deviation into the land of conjecture, I will not dwell longer upon it at present than to say that the best arguments for the perfectibility of man are drawn from a contemplation of the great progress that he has already made from the savage state and the difficulty of saying where he is to stop. But towards the extinction of the passion between the sexes, no progress whatever has hitherto been made. It appears to exist in as much force at present as it did two thousand or four thousand years ago. There are individual exceptions now as there always have been. But, as these exceptions do not appear to increase in number, it would surely be a very unphilosophical mode of arguing to infer, merely from the existence of an exception, that the exception would, in time, become the rule, and the rule the exception.

Assuming then my postulata as granted, I say, that the power of population is indefinitely greater than the power in the earth to produce subsistence for man.

Population, when unchecked, increases in a geometrical ratio. Subsistence increases only in an arithmetical ratio. A slight acquaintance with numbers will shew the immensity of the first power in comparison of the second.

By that law of our nature which makes food necessary to the life of man, the effects of these two unequal powers must be kept equal.

This implies a strong and constantly operating check on population from the difficulty of subsistence.

This difficulty must fall somewhere and must necessarily be severely felt by a large portion of mankind.

Through the animal and vegetable kingdoms, nature has scattered the seeds of life abroad with the most profuse and liberal hand. She has been comparatively sparing in the room and the nourishment necessary to rear them. The germs of existence contained in this spot of earth, with ample food, and ample room to expand in, would fill millions of worlds in the course of a few thousand years. Necessity, that imperious all-pervading law of nature, restrains them within the prescribed bounds. The race of plants and the race of animals shrink under this great restrictive law. And the race of man cannot, by any efforts of reason, escape from it. Among plants and animals its effects are waste of seed, sickness, and premature death. Among mankind, misery and vice. The former, misery, is an absolutely necessary consequence of it. Vice is a highly probable consequence, and we therefore see it abundantly prevail, but it ought not, perhaps, to be called an absolutely necessary consequence. The ordeal of virtue is to resist all temptation to evil.

This natural inequality of the two powers of population and of production in the earth, and that great law of our nature which must constantly keep their effects equal, form the great difficulty that to me appears insurmountable in the way to the perfectibility of society. All other arguments are of slight and subordinate consideration in comparison of this. I see no way by which man can escape from the weight of this law which pervades all animated nature. No fancied equality, no agrarian regulations in their utmost extent, could remove the pressure of it even for a single century. And it appears, therefore, to be decisive against the possible existence of a society, all the members of which should live in ease, happiness, and comparative leisure; and feel no anxiety about providing the means of subsistence for themselves and families.

Consequently, if the premises are just, the argument is conclusive against the perfectibility of the mass of mankind.

I have thus sketched the general outline of the argument, but I will examine it more particularly, and I think it will be found that experience, the true source and foundation of all knowledge, invariably confirms its truth.

Reading and Discussion Question

1. What are Malthus' two postulates? Why does he believe that, taken together, they make the moral perfectibility of humans impossible?

Natural Theology: Or, Evidences of the Existence and Attributes of the Deity, Collected from the Appearances of Nature (1802)

William Paley

Chapter 1: State of the Argument

In crossing a heath, suppose I pitched my foot against a *stone*, and were asked how the stone came to be there; I might possibly answer, that, for any thing I knew to the contrary, it had lain there forever: nor would it perhaps be very easy to show the absurdity of this answer. But suppose I had found a *watch* upon the ground, and it should be inquired how the watch happened to be in that place; I should hardly think of the answer which I had before given, that, for any thing I knew, the watch might have always been there. Yet why should not this answer serve for the watch as well as for the stone? Why is it not as admissible in the second case, as in the first? For this reason, and for no other, viz. that, when we come to inspect the watch, we perceive (what we could not discover in the stone) that its several parts are framed and put together for a purpose, *e.g.* that they are so formed and adjusted as to produce motion, and that motion so regulated as to point out the hour of the day; that, if the different parts had been differently shaped from what they are, of a different size from what they are, or placed after any other manner, or in any other order, than that in which they are placed, either no motion at all would have been carried on in the machine, or none which would have answered the use that is now served by it. To reckon up a few of the plainest of these parts, and of their offices, all tending to one result:—We see a cylindrical box containing a coiled elastic spring, which, by its endeavour to relax itself, turns round the box. We next observe a flexible chain (artificially wrought for the sake of flexure), communicating the action of the spring from the box to the fusee. We then find a series of wheels, the teeth of which catch in, and apply to, each other, conducting the motion from the fusee to the balance, and from the balance to the pointer; and at the same time, by the size and shape of those wheels, so regulating that motion, as to terminate in causing an index, by an equable and measured progression, to pass over a given space in a given time. We take notice that the wheels are made of brass in order to keep them from rust; the springs of steel, no other metal being so elastic; that over the face of the watch there is placed a glass, a material employed in no other part of the work, but in the room of which, if there had been any other than a transparent substance, the hour could not be seen without opening the case. This mechanism being observed (it requires indeed an examination of the instrument, and perhaps some previous knowledge of the subject, to perceive and understand it; but being once, as we have said, observed and understood), the inference, we think, is inevitable, that the watch must have had a maker: that there must have existed, at some time, and at some place or other, an artificer or artificers who formed it for the purpose which we find it actually to answer; who comprehended its construction, and designed its use.

I. Nor would it, I apprehend, weaken the conclusion, that we had never seen a watch made; that we had never known an artist capable of making one; that we were altogether incapable of executing such a piece of workmanship ourselves, or of understanding in what manner it was performed; all this being no more than what is true of some exquisite remains of ancient art, of some lost arts, and, to the generality of mankind, of the more curious productions of modern manufacture. Does one man in a million know how oval frames are turned? Ignorance of this kind exalts our opinion of the unseen and unknown artist's skill, if he be unseen and unknown, but raises no doubt in our minds of the existence and agency of such an artist, at some former time, and in some place or other. Nor can I perceive that it varies at all the inference, whether the question arise concerning a human agent, or concerning an agent of a different species, or an agent possessing, in some respects, a different nature.

II. Neither, secondly, would it invalidate our conclusion, that the watch sometimes went wrong, or that it seldom went exactly right. The purpose of the machinery, the design, and the designer, might be evident, and in the case supposed would be evident, in whatever way we accounted for the irregularity of the movement, or whether we could account for it or not. It is not necessary that a machine be perfect, in order to show with what design it was made: still less necessary, where the only question is, whether it were made with any design at all.

III. Nor, thirdly, would it bring any uncertainty into the argument, if there were a few parts of the watch, concerning which we could not discover, or had not yet discovered, in what manner they conduced to the general effect; or even some parts, concerning which we could not ascertain, whether they conduced to that effect in any manner whatever. For, as to the first branch of the case;

if by the loss, or disorder, or decay of the parts in question, the movement of the watch were found in fact to be stopped, or disturbed, or retarded, no doubt would remain in our minds as to the utility or intention of these parts, although we should be unable to investigate the manner according to which, or the connexion by which, the ultimate effect depended upon their action or assistance; and the more complex is the machine, the more likely is this obscurity to arise. Then, as to the second thing supposed, namely, that there were parts which might be spared, without prejudice to the movement of the watch, and that we had proved this by experiment—these superfluous parts, even if we were completely assured that they were such, would not vacate the reasoning which we had instituted concerning other parts. The indication of contrivance remained, with respect to them, nearly as it was before.

IV. Nor, fourthly, would any man in his senses think the existence of the watch, with its various machinery, accounted for, by being told that it was one out of possible combinations of material forms; that whatever he had found in the place where he found the watch, must have contained some internal configuration or other; and that this configuration might be the structure now exhibited, viz. of the works of a watch, as well as a different structure.

V. Nor, fifthly, would it yield his inquiry more satisfaction to be answered, that there existed in things a principle of order, which had disposed the parts of the watch into their present form and situation. He never knew a watch made by the principle of order; nor can he even form to himself an idea of what is meant by a principle of order, distinct from the intelligence of the watch-maker.

VI. Sixthly, he would be surprised to hear that the mechanism of the watch was no proof of contrivance, only a motive to induce the mind to think so:

VII. And not less surprised to be informed, that the watch in his hand was nothing more than the result of the laws of *metallic* nature. It is a perversion of language to assign any law, as the efficient, operative cause of anything. A law presupposes an agent; for it is only the mode, according to which an agent proceeds: it implies a power; for it is the order, according to which that power acts. Without this agent, without this power, which are both distinct from itself, the *law* does nothing; is nothing. The expression, "the law of metallic nature," may sound strange and harsh to a philosophic ear; but it seems quite as justifiable as some others which are more familiar to him, such as "the law of vegetable nature," "the law of animal nature," or indeed as "the law of nature" in general, when assigned as the cause of phænomena, in exclusion of agency and power; or when it is substituted into the place of these.

VIII. Neither, lastly, would our observer be driven out of his conclusion, or from his confidence in its truth, by being told that he knew nothing at all about the matter. He knows enough for his argument: he knows the utility of the end: he knows the subserviency and adaptation of the means to the end. These points being known, his ignorance of other points, his doubts concerning other points, affect not the certainty of his reasoning. The consciousness of knowing little, need not beget a distrust of that which he does know.

Chapter 2: State of the Argument Continued

Suppose, in the next place, that the person who found the watch, should, after some time, discover that, in addition to all the properties which he had hitherto observed in it, it possessed the unexpected property of producing, in the course of its movement, another watch like itself (the thing is conceivable); that it contained within it a mechanism, a system of parts, a mould for instance, or a complex adjustment of lathes, files, and other tools, evidently and separately calculated for this purpose; let us inquire, what effect ought such a discovery to have upon his former conclusion.

I. The first effect would be to increase his admiration of the contrivance, and his conviction of the consummate skill of the contriver. Whether he regarded the object of the contrivance, the distinct apparatus, the intricate, yet in many parts intelligible mechanism, by which it was carried on, he would perceive, in this new observation, nothing but an additional reason for doing what he had already done—for referring the construction of the watch to design, and to supreme art. If that construction *without* this property, or which is the same thing, before this property had been noticed, proved intention and art to have been employed about it; still more strong would the proof appear, when he came to the knowledge of this further property, the crown and perfection of all the rest.

II. He would reflect, that though the watch before him were, *in some sense*, the maker of the watch, which was fabricated in the course of its movements, yet it was in a very different sense from that, in which a carpenter, for instance, is the maker of a chair; the author of its contrivance, the cause of the relation of its parts to their use. With respect to these, the first watch was no cause at all to the second: in no such sense as this was it the author of the constitution and order, either of the parts which the new watch contained, or of the parts by the aid and instrumentality of which it was produced. We might possibly say, but with great latitude of expression, that a stream of water ground corn: but no latitude of expression would allow us to say, no stretch of conjecture could lead us to think, that the stream of water built the mill, though it were too ancient for us to know who the builder was. What the stream of water does in the affair, is neither more nor less than this; by the application of an unintelligent impulse to a mechanism previously arranged, arranged independently of it, and arranged by intelligence, an effect is produced, viz. the corn is ground. But the effect results from the arrangement. The force of the stream cannot be said to be the cause or author of the effect, still less of the arrangement. Understanding and plan in the formation of the mill were not the less necessary, for any share which the water has in grinding the corn: yet is this share the same, as that which the watch would have contributed to the production of the new watch, upon the supposition assumed in the last section. Therefore,

III. Though it be now no longer probable, that the individual watch, which our observer had found, was made immediately by the hand of an artificer, yet doth not this alteration in anywise affect the inference, that an artificer had been originally employed and concerned in the production. The argument from design remains as it was. Marks of design and contrivance are no more accounted for now, than they were before. In the same thing, we may ask for the cause of different properties. We may ask for the cause of the colour of a body, of its hardness, of its head; and these causes may be all different. We are now asking for the cause of that subserviency to a use, that relation to an end, which we have remarked in the watch before us. No answer is given to this question, by telling us that a preceding watch produced it. There cannot be design without a designer; contrivance without a contriver; order without choice; arrangement, without anything capable of arranging; subserviency and relation to a purpose, without that which could intend a purpose; means suitable to an end, and executing their office, in accomplishing that end, without the end ever having been contemplated, or the means accommodated to it. Arrangement, disposition of parts, subserviency of means to an end, relation of instruments to a use, imply the presence of intelligence and mind. No one, therefore, can rationally believe, that the insensible, inanimate watch, from which the watch before us issued, was the proper cause of the mechanism we so much admire in it—could be truly said to have constructed the instrument, disposed its parts, assigned their office, determined their order, action, and mutual dependency, combined their several motions into one result, and that also a result connected with the utilities of other beings. All these properties, therefore, are as much unaccounted for, as they were before.

IV. Nor is anything gained by running the difficulty farther back, *i.e.* by supposing the watch before us to have been produced from another watch, that from a former, and so on indefinitely. Our going back ever so far, brings us no nearer to the least degree of satisfaction upon the subject. Contrivance is still unaccounted for . . . .

. . .

Here is contrivance, but no contriver; proofs of design, but no designer.

V. Our observer would further also reflect, that the maker of the watch before him, was, in truth and reality, the maker of every watch produced from it; there being no difference (except that the latter manifests a more exquisite skill) between the making of another watch with his own hands, by the mediation of files, lathes, chisels, &c. and the disposing, fixing, and inserting of these instruments, or of others equivalent to them, in the body of the watch already made in such a manner, as to form a new watch in the course of the movements which he had given to the old one. It is only working by one set of tools, instead of another.

The conclusion of which the *first* examination of the watch, of its works, construction, and movement, suggested, was, that it must have had, for the cause and author of that construction, an artificer, who understood its mechanism, and designed its use. This conclusion is invincible. A *second* examination presents us with a new discovery. The watch is found, in the course of its movement, to produce another watch, similar to itself; and not only so, but we perceive in it a system or organization, separately calculated for that purpose. What effect would this discovery have, or ought it to have, upon our former inference? What, as hath already been said, but to increase, beyond measure, our admiration of the skill, which had been employed in the formation of such a machine? Or shall it, instead of this, all at once turn us round to an opposite conclusion, viz. that no art or skill whatever has been concerned in the business, although all other evidences of art and skill remain as they were, and this last and supreme piece of art be now added to the rest? Can this be maintained without absurdity? Yet this is atheism.

Chapter 3: Application of the Argument

THIS is atheism: for every indication of contrivance, every manifestation of design, which existed in the watch, exists in the works of nature; with the difference, on the side of nature, of being greater and more, and that in a degree which exceeds all computation. I mean that the contrivances of nature surpass the contrivances of art, in the complexity, subtlety, and curiosity of the mechanism; and still more, if possible, do they go beyond them in number and variety; yet, in a multitude of cases, are not less evidently mechanical, not less evidently contrivances, not less evidently accommodated to their end, or suited to their office, than are the most perfect productions of human ingenuity.

I know no better method of introducing so large a subject, than that of comparing a single thing with a single thing; an eye, for example, with a telescope. As far as the examination of the instrument goes, there is precisely the same proof that the eye was made for vision, as there is that the telescope was made for assisting it. They are made upon the same principles; both being adjusted to the laws by which the transmission and refraction of rays of light are regulated. I speak not of the origin of the laws themselves; but such laws being fixed, the construction, in both cases, is adapted to them. For instance; these laws require, in order to produce the same effect, that the rays of light, in passing from water into the eye, should be refracted by a more convex surface, than when it passes out of air into the eye. Accordingly we find that the eye of a fish, in that part of it called the crystalline lens, is much rounder than the eye of terrestrial animals. What plainer manifestation of design can there be than this difference? What could a mathematical-instrument-maker have done more, to show his knowledge of his principle, his application of that knowledge, his suiting of his means to his end; I will not say to display the compass or excellence of his skill and art, for in these all comparison is indecorous, but to testify counsel, choice, consideration, purpose?

...

... Can anything be more decisive of contrivance than this is? The most secret laws of optics must have been known to the author of a structure endowed with such a capacity of change. It is as though an optician, when he had a nearer object to view, should *rectify* his instrument by putting in another glass, at the same time drawing out also his tube to a different length ....

But this, though much, is not the whole; by different species of animals the faculty we are describing is possessed, in degrees suited to the different range of vision which their mode of life, and of procuring their food, requires. *Birds*, for instance, in general, procure their food by means of their beak; and, the distance between the eye and the point of the beak being small, it becomes necessary that they should have the power of seeing very near objects distinctly. On the other hand, from being often elevated much above the ground, living in air, and moving through it with great velocity, they require, for their safety, as well as for assisting them in descrying their prey, a power of seeing at a great distance; a power of which, in birds of rapine, surprising examples are given. The fact accordingly is, that two peculiarities are found in the eyes of birds, both tending to *facilitate* the change upon which the adjustment of the eye to different distances depends. The one is a bony, yet, in most species, a flexible rim or hoop, surrounding the broadest part of the eye; which, confining the action of the muscles to that part, increases the effect of their lateral pressure upon the orb, by which pressure its axis is elongated for the purpose of looking at very near objects. The other is an additional muscle, called the marsupium, to draw, on occasion, the crystalline lens *back*, and to fit the same eye for the viewing of very distant objects. By these means, the eyes of birds can pass from one extreme to another of their scale of adjustment, with more ease and readiness than the eyes of other animals.

The eyes of *fishes* also, compared with those of terrestrial animals, exhibit certain distinctions of structure, adapted to their state and element. We have already observed upon the figure of the crystalline compensating by its roundness the density of the medium through which their light passes. To which we have to add, that the eyes of fish, in their natural and indolent state, appear to be adjusted to near objects, in this respect differing from the human eye, as well as those of quadrupeds and birds. The ordinary shape of the fish's eye being in a much higher degree convex than that of land-animals, a corresponding difference attends its muscular conformation, viz. that it is throughout calculated for *flattening* the eye. The *iris* also in the eyes of fish does not admit of contraction. This is a great difference, of which the probable reason is, that the diminished light in water is never too strong for the retina. In the *eel*, which has to work its head through sand and gravel, the roughest and harshest substances, there is placed before the eye, and at some distance from it, a transparent, horny, convex case or covering, which, without obstructing the sight, defends the organ. To such an animal, could anything be more wanted, or more useful?

Thus, in comparing the eyes of different kinds of animals, we see, in their resemblances and distinctions, one general plan laid down, and that plan varied with the varying exigences to which it is to be applied.

...

Chapter 14: Prospective Contrivances

I can hardly imagine to myself a more distinguishing mark, and, consequently, a more certain proof of design, than *preparation*, i.e. the providing of things beforehand, which are not to be used until a considerable time afterwards; for this implies a contemplation of the future, which belongs only to intelligence.

Of these *prospective* contrivances, the bodies of animals furnish various examples.

I. The human teeth afford an instance, not only of prospective contrivance, but of the completion of the contrivance being designedly suspended. They are formed within the gums, and there they stop: the fact being, that their further advance to maturity would not only be useless to the new-born animal, but extremely in its way; as it is evident that the act of *sucking*, by which it is for some time to be nourished, will be performed with more ease both to the nurse and to the infant, whilst the inside of the mouth and edges of the gums are smooth and soft than if set with hard pointed bones. By the time they are wanted the teeth are ready. They have been lodged within the gums for some months past but detained as it were, in their sockets so long as their further protrusion would interfere with the office to which the mouth is destined. Nature, namely, that intelligence which was employed in creation, looked beyond the first year of the infant's life; yet, whilst, she was providing for functions which were after that term to become necessary, was careful not to incommode those which preceded them. What renders it more probable that this is the effect of design, is, that the teeth are imperfect, whilst all other parts of the mouth are perfect. The lips are perfect, the tongue is perfect; the cheeks, the jaws, the palate, the pharynx, the larynx, are all perfect: the teeth alone are not so. This is the fact with respect to the human mouth: the fact also is, that the parts above enumerated, are called into use from the beginning; whereas the teeth would be only so many obstacles and annoyances, if they were there. When a contrary order is necessary, a contrary order prevails. In the worm of the beetle, as hatched from the egg, the teeth are the first things which arrive at perfection. The insect begins to gnaw as soon as it escapes from the shell, though its other parts be only gradually advancing to their maturity.

What has been observed of the teeth, is true of the *horns* of animals; and for the same reason. The horn of a calf or a lamb does not bud, or at least does not sprout to any considerable length, until the animal be capable of browsing upon its pasture: because such a substance upon the forehead of the young animal, would very much incommode the teat of the dam in the office of giving suck.

But in the case of the *teeth*—of the human teeth at least, the prospective contrivance looks still further. A succession of crops is provided, and provided from the beginning; a second tier being originally formed beneath the first, which do not come into use till several years afterwards. And this double or suppletory provision meets a difficulty in the mechanism of the mouth, which would have appeared almost unsurmountable. The expansion of the jaw (the consequence of the proportionable growth of the animal, and of its skull), necessarily separates the teeth of the first set, however compactly disposed, to a distance from one another, which would be very inconvenient. In due time, therefore, *i.e.* when the jaw has attained a great part of its dimensions, a new set of teeth springs up (loosening and pushing out the old ones before them), more exactly fitted to the space which they are to occupy, and rising also in such close ranks, as to allow for any extension of line which the subsequent enlargement of the head may occasion.

Compensation

Compensation is a species of relation. It is relation when the *defects* of one part, or of one organ, are supplied by the structure of another part or of another organ. Thus,

I. The short unbending neck of the *elephant*, is compensated by the length and flexibility of his *proboscis*. He could not have reached the ground without it; or, if it be supposed that he might have fed upon the fruit, leaves, or branches of trees, how was he to drink? Should it be asked, why is the elephant's neck so short? It may be answered, that the weight of a head so heavy could not have been supported at the end of a longer lever. To a form, therefore, in some respects necessary, but in some respects also inadequate to the occasion of the animal, a supplement is added, which exactly makes up the deficiency under which he laboured.

If it be suggested that this proboscis may have been produced, in a long course of generations, by the constant endeavour of the elephant to thrust out his nose (which is the general hypothesis by which it has lately been attempted to account for the forms of animated nature), I would ask, How was the animal to subsist in the meantime; during the process; *until* this prolongation of snout were completed? What was to become of the individual, whilst the species was perfecting?

Our business at present is, simply to point out the relation which this organ bears to the peculiar figure of the animal to which it belongs. And herein all things correspond. The necessity of the elephant's proboscis arises from the shortness of his neck; the shortness of the neck is rendered necessary by the weight of the head. Were we to enter into an examination of the structure and anatomy of the proboscis itself, we should see in it one of the most curious of all examples of animal mechanism. The disposition of the ringlets and fibres, for the purpose, first, of forming a long cartilaginous pipe; secondly, of contracting and lengthening that pipe; thirdly, of turning it in every direction at the will of the animal; with the superaddition at the end, of a fleshy production, of about the length and thickness of a finger, and performing the office of a finger, so as to pick up a straw from the ground; these properties of the same organ, taken together, exhibit a specimen, not only of design (which is attested by the advantage), but of consummate art, and, as I may say, of elaborate preparation, in accomplishing that design.

Chapter 20: Of Plants

I think a designed and studied mechanism to be, in general, more evident in animals than in *plants:* and it is unnecessary to dwell upon a weaker argument, where a stronger is at hand. There are, however, a few observations upon the vegetable kingdom, which lie so directly in our way, that it would be improper to pass by them without notice.

The one great intention of nature in the structure of plants seems to be the perfecting of the *seed;* and, what is part of the same intention, the preserving of it until it be *perfected.* This intention shows itself, in the first place, by the care which appears to be taken, to protect and ripen, by every advantage which can be given to them of situation in the plant, those parts which most immediately contribute to fructification, viz. the antheræ, the stamina, and the stigmata. These parts are usually lodged in the centre, the recesses, or the labyrinths of the flower; during their tender and immature state, are shut up in the stalk, or sheltered in the bud; as soon as they have acquired firmness of texture sufficient to bear exposure, and are ready to perform the important office which is assigned to them, they are disclosed to the light and air, by the bursting of the stem, or the expansion of the petals; after which they have, in many cases, by the very form of the flower during its blow, the light and warmth reflected upon them from the concave side of the cup. . . .

From the conformation of fruits alone, one might be led, even without experience, to suppose, that part of this provision was destined for the utilities of animals. As limited to the plant, the provision itself seems to go beyond its object. The flesh of an apple, the pulp of an orange, the meat of a plum, the fatness of the olive, appear to be more than sufficient for the nourishing of the seed or kernel. The event shows, that this redundancy, if it be one, ministers to the support and gratification of animal natures; and when we observe a provision to be more than sufficient for one purpose, yet wanted for another purpose, it is not unfair to conclude that both purposes were contemplated together. . . .

Chapter 26: The Goodness of the Deity

The proof of the *divine goodness* rests upon two propositions; each, as we contend, capable of being made out by observations drawn from the appearances of nature. The first is, "that, in a vast plurality of instances in which contrivance is perceived, the design of the contrivance is *beneficial.*" The second, "that the Deity has superadded *pleasure* to animal sensations, beyond what was necessary for any other purpose, or when the purpose, so far as it was necessary," might have been effected by the operation of pain.

First, "in a vast plurality of instances in which contrivance is perceived, the design of the contrivance is *beneficial.*"

No production of nature display contrivance so manifestly as the parts of animals; and the parts of animals have all of them, I believe, a real, and, with very few exceptions, all of them a known and intelligible subserviency to the use of the animal. Now, when the multitude of animals is considered, the number of parts in each, their figure and fitness, the faculties depending upon them, the variety of species, the complexity of structure, the success, in so many cases, and felicity of the result, we can never reflect, without the profoundest adoration, upon the character of that Being from whom all these things have proceeded: we cannot help acknowledging, what an exertion of benevolence creation was; of a benevolence how minute in its care, how vast in its comprehension! ...

...

... It is a happy world after all. The air, the earth, the water, teem with delighted existence. In a spring noon, or a summer evening, on whichever side I turn my eyes, myriads of happy beings crowd upon my view. "The insect youth are on the wing." Swarms of newborn *flies* are trying their pinions in the air. Their sportive motions, their wanton mazes, their gratuitous activity, their continual change of place without use or purpose, testify their joy, and the exultation which they feel in their lately discovered faculties. A *bee* amongst the flowers in spring, is one of the most cheerful objects that can be looked upon. Its life appears to be all enjoyment; so busy, and so pleased: yet it is only a specimen of insect life, with which, by reason of the animal being half domesticated, we happen to be better acquainted than we are with that of others. The *whole winged* insect tribe, it is probable, are equally intent upon their proper employments, and, under every variety of constitution, gratified, and perhaps equally gratified, by the offices which the Author of their nature has assigned to them. But the atmosphere is not the only scene or enjoyment for the insect race. Plants are covered with aphides, greedily sucking their juices, and constantly, as it should seem, in the act of sucking. It cannot be doubted but that this is a state of gratification. What else should fix them so close to the operation, and so long? Other species are *running about*, with an alacrity in their motions, which carries with it every mark of pleasure. Large patches of ground are sometimes half covered with these brisk and sprightly natures. If we look to what the *waters* produce, shoals of the fry of fish frequent the margins of rivers, of lakes, and of the sea itself. These are so happy, that they know not what to do with themselves. Their attitudes, their vivacity, their leaps out of the water, their frolics in it (which I have noticed a thousand times with equal attention and amusement), all conduce to show their excess of spirits, and are simply the effects of that excess. Walking by the sea-side, in a calm evening, upon a sandy shore, and with an ebbing tide, I have frequently remarked the appearance of a dark cloud, or, rather, very thick mist, hanging over the edge of the water, to the height, perhaps, of half a yard, and of the breadth of two or three yards, stretching along the coast as far as the eye could reach, and always retiring with the water. When this cloud came to be examined, it proved to be nothing else than so much space, filled with young *shrimps*, in the act of bounding into the air from the shallow margin of the water, or from the wet sand. If any motion of a mute animal could express delight, it was this: if they had meant to make signs of their happiness, they could not have done it more intelligibly. Suppose then, what I have no doubt of, each individual of this number to be in a state of positive enjoyment; what a sum, collectively, of gratification and pleasure have we here before our view!

. . .

But it will be said, that the instances which we have here brought forward, whether of vivacity or repose, or of apparent enjoyment derived from either, are picked and favourable instances. We answer, first, that they are instances, nevertheless, which comprise large provinces of sensitive existence; that every case which we have described, is the case of millions. At this moment, in every given moment of time, how many myriads of animals are eating their food, gratifying their appetites, ruminating in their holes, accomplishing their wishes, pursuing their pleasures, taking their pastimes! In each individual, how many things must go right for it to be at ease; yet how large a proportion out of every species is so in every assignable instant! Secondly, we contend, in the terms of our original proposition, that throughout the whole of life, as it is diffused in nature, and as far as we are acquainted with it, looking to the average of sensations, the plurality and the preponderancy is in favour of happiness by a vast excess. In our own species, in which perhaps the assertion may be more questionable than in any other, the prepollency of good over evil, of health, for example, and ease, over pain and distress, is evinced by the very notice which calamities excite. What inquiries does the sickness of our friends produce! What conversation their misfortunes! This shows that the common course of things is in favour of happiness: that happiness is the rule, misery the exception. Were the order reversed, our attention would be called to examples of health and competency, instead of disease and want . . . .

. . .

A single instance will make all this clear. Assuming the necessity of food for the support of animal life; it is requisite, that the animal be provided with organs, fitted for the procuring, receiving, and digesting of its food. It may be also necessary, that the animal be impelled by its sensations to exert its organs. But the pain of hunger would do all this. Why add pleasure to the act of eating; sweetness and relish to food? Why a new and appropriate sense for the perception of the pleasure? Why should the juice of a peach, applied to the palate, affect the part so differently from what it does when rubbed upon the palm of the hand? This is a constitution which, so far as appears to me, can be resolved into nothing but the pure benevolence of the Creator. Eating is necessary; but the pleasure attending it is not necessary: and that this pleasure depends, not only upon our being in possession of the sense of taste, which is different from every other, but upon a particular state of the organ in which it resides, a felicitous adaptation of the organ to the object, will be confessed by any one, who may happen to have experienced that vitiation of taste which frequently occurs in fevers, when every taste is irregular, and every one bad . . . .

. . .

In which-ever way we regard the senses, they appear to be specific gifts, ministering, not only to preservation, but to pleasure. But what we usually call the *senses*, are probably themselves far from being the only vehicles of enjoyment, or the whole of our constitution which is calculated for the same purpose. We have many internal sensations of the most agreeable kind, hardly referable to any of the five senses. Some physiologists have holden, that all secretion is pleasurable; and that the complacency which in health, without any external assignable object to excite it, we derive from life itself, is the effect of our secretions going on well within us. All this may be true: but if true, what reason can be assigned for it, except the will of the Creator?

Reading and Discussion Questions

1. What is the example of a watch found in a heath meant to stand as an analogy for? What feature of the watch does Paley take to cry out for explanation?
2. How does Paley's argument for the existence of an intelligent designer go? What are some of the observations that he cites as evidences of design?
3. What evidence does he see that the designer is not only in existence, but desires the good for each creation?

Zoological Philosophy: Exposition with Regard to the Natural History of Animals (1809)

J. B. Lamarck

88

First Part

Considerations of the Natural History of Animals, Their Characteristics, Their Interrelationships, Their Organic Structure, Their Distribution, Their Classification and Their Species

Chapter Three

Concerning Speciation among Living Things and the Idea Which We Should Attach to This Word

It is not a futile pursuit firmly to establish the idea which we should form about what are called species among living creatures and to investigate whether it is true that species have an absolute constancy, are as old as nature, and have all existed originally just as we see them today, or whether, subject to changes which could have taken place in the circumstances relevant to them, they have not changed their characteristics and shape with the passage of time (although extremely slowly).

The illumination of this question is not only of interest to our zoological and botanical knowledge but also is essential to the history of the earth.

I will show in one of the chapters which follow that each species has received from the influence of the circumstances which it encounters over a long period the habits which we know about and that these habits have themselves exerted influences on the parts of each individual of the species, to the point where they have modified these parts and have made them appropriate to the acquired habits. Let us first examine the idea which has developed about what is called a species.

We call species every collection of similar individuals produced by other individuals just like themselves.

This definition is exact, for every individual enjoying life always resembles very closely the one or those from which it came. But we add to this definition the assumption that the individuals who make up a species never vary in their specific characteristics and that therefore the species has an absolute constancy in nature.

It is precisely this assumption that I propose to contest, because clear proofs obtained through observation establish that it is not well founded.

The assumption almost universally admitted that living things make up eternally distinct species on account of their invariable characteristics and that the existence of these species is as ancient as nature herself was established at a time when people had not observed nature sufficiently and when the natural sciences were still almost nothing. The assumption is contradicted every day in the eyes of those who have looked at a great deal and have followed nature for a long time, and who have reaped the benefits of the large and rich collections in our museum.

Moreover, all those who are very busy studying natural history know that nowadays naturalists are extremely embarrassed in their attempts to define the objects which they have to consider species. In fact, not knowing that species have a constancy only relative to the duration of the circumstances in which all the individuals composing them are found and that some of these individuals, having undergone variations, make up races which modulate into some other neighbouring species, naturalists make decisions arbitrarily, by describing some individuals observed in different countries and in various environments as varieties and others as species. As a result, that section of work concerning the determination of species is becoming day by day increasingly defective, that is, more embarrassing and confusing.

In truth, it has been observed for a long time that there exist collections of individuals who so resemble each other in their organic structure, as well as by the totality of their parts, and who remain in the same condition generation after generation for as long as we have known about them that people have believed themselves justified in regarding these collections of similar individuals as making up just as many invariable species.

Now, not having attended to the fact that the individuals of a species must perpetuate themselves without variation, as long as the circumstances which influence their manner of life do not essentially vary and the existing prejudices agreeing well enough with the successive regeneration of similar individuals, people have assumed that each species did not vary, was also as old as nature, and was uniquely created by the work of the Supreme Author of everything which exists.

There is no doubt that nothing exists except by the will of the sublime Author of everything. But can we assign some rules to Him in the execution of His will and establish the method which He followed in this matter? Could not His infinite power have been capable of creating an order of things which gave life successively to everything which we see, as well as to everything existing which we do not know about?

To be sure, whatever His will, the immensity of his power is still the same and whatever the manner in which the Supreme Will carried out His work, nothing can diminish His grandeur.

Therefore, respecting the decrees of this infinite wisdom, I confine myself within the limits of a simple observer of nature. Then, if I manage to unravel something of the progress which nature has followed to bring about its productions, I will say, without fear of being wrong, that it has pleased her Author that nature has had this faculty and this power.

The notion of species among living creatures which people formed was very simple, easy to grasp, and seemed confirmed by the constancy in the apparent form of individuals which reproduction or generation perpetuated. Such individuals create for us a great number of those alleged species which we see every day.

However, the more we advance our knowledge of the different organic bodies which cover the surface of the earth almost everywhere, the greater becomes our embarrassment about determining what ought to be regarded as a species and, for even more compelling reasons, about limiting and distinguishing genera.

The more we collect the productions of nature and our collections grow richer, the more we see almost all the gaps being filled and our lines of separation being erased. We find ourselves reduced to an arbitrary determination, which sometimes leads us to seize upon the least differences among the varieties to form the characteristic of what we call species. Sometimes this makes us call certain individuals with slight differences a variety of some species. Other people consider these individuals constitute a separate species.

Let me repeat myself: the more our collections increase, the more we encounter proofs that everything is more or less nuanced, the remarkable differences disappear, and as often as not nature makes available to us for the creation of distinctions only minute and, so to speak, puerile particularities . . . .

When we formed these genera, we knew only a small number of their species; thus, they were easy to distinguish. But now that almost all the gaps between them have been filled, our specific differences are necessarily minute and very frequently insufficient.

Having well established this state of affairs, let us examine the causes which can have given rise to it. Let us see if nature possess means for that and if observation could have given us insight into this question.

A number of facts teach us that, to the extent that the individuals of one of our species change their situation, climate, manner of life, or habits, they obtain from that change influences which little by little alter the constancy and the proportions of their parts, shape, faculties, even their organic structure, with the result that everything in them participates, over time, in the mutations which they have experienced.

In the same climate, significantly different situations and exposures at first simply induce changes in the individuals who find themselves confronted with them. But as time passes, the continual difference in the situation of the individuals I'm talking about, who live and reproduce successively in the same circumstances, leads to changes in them which become, in some way, essential to their being, so that after many generations, following one after the other, these individuals, belonging originally to another species, find themselves at last transformed into a new species, distinct from the other.

For example, if the seeds of a grass or of any other plant common to a humid prairie are transported, by some circumstance or other, at first to the slope of a neighbouring hill, where the soil, although at a higher altitude, is still sufficiently damp to allow the plant to continue living, if then, after living there and reproducing many times in that spot, the plant little by little reaches the almost arid soil of the mountain slope and succeeds in subsisting there and perpetuates itself through a sequence of generations, it will then be so changed that botanists who come across it there will create a special species for it.

The same thing happens to animals which circumstances have forced to change their climate, manner of life, and habits. But for these, the causal influences which I have just mentioned require even more time than is the case with plants in order to effect notable changes in the individuals.

The idea of including, in the name species, a collection of similar individuals who perpetuate creatures like themselves through reproduction and who have thus existed in the same form for as long as nature necessarily requires that the individuals of the same species, in their reproductive acts, cannot mate with the individuals of another species.

Unfortunately, observation has demonstrated and still establishes every day that this idea has no foundation whatsoever. For hybrids, very well known among plants, and the matings which we often see between individuals of very different species among animals attest to the fact that the limits between these species, supposedly constant, were not as firm as people have imagined.

To be sure, often nothing results from these odd matings, above all when they involve very different types, and then the individuals produced are, in general, infertile. But then again, when the disparity is less great, we know that the flaws in question do not occur. Now, this method by itself is sufficient to create varieties gradually which then become races, and which, in time, make up what we call species.

In order to evaluate whether the traditional idea of species has some real foundation, let us look again at points which I have already established. They enable us to see the following:

1. All organic bodies of our earth are true products of nature, which she has brought forth successively over a long period of time;
2. In her progress, nature began, and begins again every day, by creating the simplest organic bodies, and she does not directly create anything except by this process, that is to say, by these first beginnings of organic structure which are designated by the expression spontaneous generation.
3. The first beginnings of animals and plants were formed in appropriate places and circumstances. Once the faculties of a commencing life and of organic movement were established, these animals and plants of necessity gradually developed organs, and, in time, they diversified these organs, as well as their parts.
4. The faculty of growth in each portion of an organic body is inherent in the first effects of life; it gave rise to different ways of multiplication and reproduction of individuals. In this process, the progress acquired in the composition of the organic structure and in the shape and diversity of parts was maintained.
5. With the help of a sufficient lapse of time, of circumstances which were necessarily favourable, of changes which every point on the surface of the earth has successively undergone, in a word, with the assistance of the power which new situations and habits have for modifying the organs of a body endowed with life, all those which exist now have been imperceptibly shaped just as we see them.
6. Finally, after a sequence of events like the above, living bodies have each experienced greater or lesser changes in the condition of their organic structure and their parts. What we call species have been created in this way imperceptibly and successively among them; they have a constancy which is only relative to their condition and cannot be as old as nature . . . .

Chapter Seven

Concerning the Influence of Circumstances on the Actions and Habits of Animals, and the Influence of the Actions and Habits of These Living Bodies as Causes Which Modify Their Organic Structure and Their Parts

What we are now concerned with is not a rational speculation but the examination of a reliable fact, a more universal one than people think and something to which we have neglected to pay the attention it deserves. Undoubtedly this is the case because on most occasions it is very difficult to recognize. This fact consists of the influence which circumstances exert on the different living things subject to them.

In truth, for quite a long time now we have noticed the influence of the different states of our organic structure on our characteristics, inclinations, actions, and even our ideas. But it seems to me that no one has yet made known the influence of our actions and habits on our own organic structure. Now, as these actions and habits are entirely dependent on the circumstances in which we usually find ourselves, I am going to try to show how great the influence is which these circumstances exert on the general form, the condition of the parts, and even on the organic structure of living things. Thus, this chapter is going to explore this very well established fact.

If we had not had numerous occasions to recognize quite clearly the effects of this influence on certain living bodies which we have transported into entirely new environments, very different from the ones where they used to live, and if we had not seen these effects and the changes resulting from them come to light in some way under our very eyes, the important fact under discussion would have always remained unknown to us.

The influence of circumstances is truly working always and everywhere on living bodies. But what makes this influence difficult for us to perceive is that its effects become perceptible or recognizable (especially in animals) only after a long passage of time.

Before laying out and examining the proofs for this noteworthy fact (something extremely important for *Zoological Philosophy*), let us summarize the thread of the ideas with which we started our analysis.

In the preceding paragraphs, we have seen that it is now an incontestable fact that, when we consider the animal scale in a sense opposite to the natural direction, we find that in the groups which form this scale there exists a sustained but irregular degradation in the organic structure of animals making up the groups, an increasing simplicity in the organization of living bodies and finally a corresponding diminution in the number of faculties in these beings.

This well known fact can provide us the greatest insights into the very order which nature followed in the production of all animals which she has brought into existence. But it does not show us why animals' organic structure, with its increasing complexity from the most imperfect right to the most improved, only displays an irregular gradation in which the range manifests a number of anomalies or gaps which have no apparent order amid their variety.

Now, in seeking out the reason for this peculiar irregularity in the growing complexity of animals' organic structure, if we consider the results of the influences which the infinitely various circumstances in all the regions of the earth exert on the general shape, parts, and even the organic structure of these animals, then everything will be clearly explained.

In fact, it will be quite clear that the condition in which we see every animal is, on the one hand, the product of the increasing complexity in organic structure which tends to create a regular gradation and, on the other hand, the product of influences of a multitude of very different circumstances which continuously tend to work against the regularity in the gradations of the growing complexity in organization.

Here it becomes necessary that I explain what I mean by the following expression: *Circumstances have an influence on the form and the organic structure of animals.* What this means is that by undergoing significant change, the circumstances proportionally alter, over time, both the form and the organic structure itself

True, if someone takes these expressions literally, he would say I was making a mistake. For no matter what the circumstances can be, they do not work to bring about directly any modification whatsoever in the shape and organic structure in animals.

But significant changes in the circumstances lead, for animals, to great changes in their needs. Such changes in the needs necessarily lead to changes in their actions. Now, if the new needs become constant or last a long time, the animals then acquire new habits which are just as long lasting as the needs which brought them about. That is what is easy to demonstrate and, indeed, requires no detailed explanation to be understood.

Thus, it is clear that a significant change in circumstances, once it becomes constant for a race of animals, leads these animals to new habits . . . .

In the plants where there are no actions and consequently, strictly speaking, no habits, significant changes in circumstances nonetheless lead to significant differences in the development of their parts. As a result, these different circumstances give rise to and develop certain parts, while they weaken several other parts and lead to their disappearance. But here everything exerts its effect through changes undergone in what the plant uses for nourishment, in what it absorbs and breathes, in the quality of heat, light, air, and humidity which the plant customarily then receives, and, finally, through the superiority which some of these various vital movements can gain over others.

Among individuals of the same species, if some are continually well nourished in circumstances favourable to their total development, while others find themselves in opposite circumstances, then there is produced a difference between the conditions of these individuals which gradually becomes very noticeable. How many examples I could cite concerning animals and plants which confirm the basis for this idea! Now, if circumstances remain the same, making the condition of the poorly nourished individuals habitual and constant, with suffering and malnourishment, their interior organic structure is finally changed. Reproduction among the individuals in question preserves the acquired modifications and ends up by giving rise to a race very different from the one made up of individuals who find themselves constantly in circumstances favourable to their development.

A very dry spring causes prairie grasses to grow very little, to remain thin and scrawny, to flower and bear fruit, although they have grown very little.

A spring mixed with hot days and rainy days brings about in these same grasses a generous growth, and the harvest of hay is then excellent.

But if with these plants some causes perpetuate unfavourable circumstances, they will vary proportionally, at first in their bearing or their general condition, and later in several specific characteristics.

For example, if a grain of some prairie grass or other is carried into a high place, onto a dry, arid, and rocky patch of land very exposed to the wind and can germinate there, the plant which can live in this place will always find itself malnourished, and if the individuals which it produces continue to exist in these poor circumstances, there will result a race truly different from the one which lives in the prairie (which is, however, the origin of the second race). The individuals of this new race will be small, scrawny in their parts, and some of their organs, having undergone more development than others, will then manifest strange proportions.

Those who have observed a great deal and consulted large collections have been convinced that as the conditions in the environment, exposure, climate, nourishment, way of life, and so on undergo changes, the characteristics of height, shape, proportions among the parts, colour, consistency, agility, and industry (for the animals) correspondingly change.

What nature does with a great deal of time, we do every day, when on our own we suddenly change the conditions in which a living plant and all the individuals of its species are found. . . .

The following fact proves (with respect to plants) how much a change in some important circumstance has an influence on changing the parts of living organisms . . . .

There is no doubt that, so far as animals are concerned, important changes in the circumstances where they usually live produce similar changes in their parts. But here the changes are much slower manifesting themselves than in the plants. Consequently, they are less perceptible to us and their cause less recognizable.

Now, the true order of things relevant to consider in all this consists in recognizing the following:

1. All slightly remarkable changes later maintained in circumstances where each race of animals is located works to create in that race a real change in its needs.
2. All changes in animals' needs require of them alternative actions to satisfy the new needs and, consequently, alternative habits.
3. Since the satisfaction of every new need demands new actions, it requires from the animal experiencing that need either the more frequent use of some of its parts which previously it used less often (something which develops and makes that part grow), or the use of new parts which the needs imperceptibly bring forth in the animal by the efforts of its interior feeling. This I will establish very soon by known facts.

Thus, to reach an understanding of the true causes of so many diverse forms and so many different habits, examples of which the known animals manifest to us, we must take into account the fact that the infinitely diversified and slowly changing conditions in which the animals of each race are successively located have led, in each of them, to new needs and necessarily to changes in their habits. Now, once this truth, which one cannot contest, is recognized, it will be easy to see how animals have been able to satisfy the new needs and to acquire new habits, if we give some attention to the two following laws of nature, which observation has always confirmed.

First Law

In every animal which has not exceeded the limit of its development, the more frequent and sustained use of any organ gradually strengthens this organ, develops it, makes it larger, and gives it a power proportional to the duration of this use; whereas, the constant lack of use of such an organ imperceptibly weakens it, makes it deteriorate, progressively diminishes it faculties, and ends by making it disappear.

Second Law

Everything which nature has made individuals acquire or lose through the influence of conditions to which their race has been exposed for a long time and, consequently, through the influence of the predominant use of some organ or by the influence of the constant disuse of this organ, nature preserves by reproduction in the new individuals arising from them, provided that the acquired changes are common to the two sexes or to those who have produced these new individuals.

These are the two constant truths which cannot be overlooked except by those who have never observed nor followed nature in her work or by those who have let themselves be led into the error which I am going to contest.

Once naturalists noticed that the forms of animals' parts are always linked to the use of these parts, they thought that the forms and the condition of the parts had led to the usage. Now, there is the mistake. For it is easy to demonstrate through observation that, by contrast, it is the needs and the use of the parts which have developed them, factors which even produced the parts at a time when they did not exist and which, consequently, gave rise to the condition in which we see them in each animal.

In order for that not to be the case, it would have been necessary for nature to create for the animal parts as many forms as required by the diversity of circumstances in which they have to live and that these forms, as well as the circumstances, never change.

That is certainly not the natural order which exists. If it had ever really been like that, we would not have race horses in the form of those in England; we would not have our large draught horses, so heavy and different from these race horses, for nature on her own did not produce anything like them. For the same reason we would not have basset hounds with crooked limbs, such swift-running greyhounds, water spaniels, and so on. We would not have tailless hens, fantail pigeons, and so on. Finally, we would be able to cultivate wild plants as much as we liked in the rich fertile soil of our gardens, without fear of seeing them change through long cultivation.

In this matter, for a long time we have had a feeling for what is really the case, because we developed the following sentence, which has become proverbial and universally known: *habits form a second nature*.

To be sure, if habits and the nature of every animal were incapable of ever changing, the proverb would be false, would not have arisen, and would not have been able to be preserved in the event someone had proposed it.

If one considers seriously everything which I have just revealed, one will sense that I grounded my views rationally when in my work entitled *Research Into Living Bodies* (p. 50), I laid down the following proposition:

"It is not the organs, that is, the nature and the form of the animal's body parts, which have given rise to its habits and special faculties, but, by contrast, its habits, manner of life, and circumstances of the individuals from which the animal comes to possess, over time, the form of its body, the number and condition of its organs, and finally the faculties which it enjoys."

Let people consider well this proposition and bring to it all the observations which nature and the state of things enable us to make all the time. Then its importance and reliability will become for us the most significant evidence.

Favourable times and circumstances are, as I have already said, the two main means employed by nature to bring into existence all her productions. We know that time has no limits for her and that, as a result, she always has time to spare.

As to the circumstances which she needed and which she still uses every day to vary everything which she continues to produce, we can say that circumstances are, in some way, for her inexhaustible.

The main circumstances arise from the influence of climates, various temperatures in the atmosphere and all the environmental surroundings, the variety of places and their exposure, habits, the most ordinary movements, the most frequent actions, finally the means of self-preservation, reproduction, and so on.

Now, as a result of these various influences, the faculties expand and grow stronger through use. With new habits preserved over a long time they diversify. Imperceptibly the arrangement, consistency, in a word, the nature and the condition of the parts, as well as the organs, undergo the consequences of all these influences, preserving and propagating themselves in reproduction.

These truths, which are only the consequences of the two natural laws set forth above, are, in every case, amply confirmed by the facts. They indicate clearly the march of nature in the variety of her productions.

But instead of contenting ourselves with generalities which we could consider hypothetical, let us examine the facts directly. Let us consider in animals what is produced by the use or lack of use of their organs on these very organs, according to the habits which each race has been compelled to acquire.

Now, I am going to prove that the constant lack of exercise with respect to an organ at first reduces its faculties, then gradually shrinks it, and ends up by making it disappear or even destroying it, if this lack of use continues for a long time in a sequence of successive generations of animals of the same race.

Then I will reveal how, by contrast, the habit of exercising an organ, in every animal which has not reached the limit in the diminution of its faculties, not only improves this organ's faculties and makes it grow, but also makes it develop and acquire dimensions which imperceptibly change it, so that in time it makes it quite different from the same organ examined in another animal which exercises it much less.

The lack of use of an organ, once it has become constant because of the habits which one has taken up, gradually diminishes that organ and ends up by making it disappear and even destroying it.

Many insects which, according to the natural characteristics of their order and even their genus, should have wings, lack them more less completely, because they do not use them. A number of coleoptera, orthopetera, hymenoptera, and hemiptera, and so on give us examples of this fact. The habits of these animals never put them in situations where they used their wings.

But it is not sufficient to provide an explanation for the cause which has led to the state of the organs of different animals, a condition which we observe is always the same in those of the same species. In addition, it is necessary to make known the alterations brought about in the organs of a single individual during its lifetime, solely as the product of a great mutation in the habits unique to the individuals of its species. The following extremely remarkable fact will complete the proof of the influence of habits on the condition of the organs and establish how much sustained changes in the habits of an individual lead to changes in the condition of the organs which are brought into action during the exercise of these habits.

Mr. Tenon, member of the Institute, has made known to the Class of Sciences, that he examined the intestinal canal of several men who had been passionate drinkers for a large part of their lives. He constantly found the organ shortened by an extraordinary amount in comparison with the same organ in all those who had not picked up the same habit.

We know that great drinkers, or those who have been addicted to drinking, eat very little solid food, that they eat almost nothing, and that the drink which they consume in abundance and frequently is sufficient to nourish them.

Now, since the alimentary fluids, especially spirit drinks, do not stay for long either in the stomach or in the intestine, among drinkers the stomach and the rest of the intestinal canal lose the habit of being distended, just as the stomachs of sedentary persons constantly busy with intellectual work who are accustomed to eating only a little gradually over time contract, and their intestines grow shorter.

This matter is not at all a question of a shrinking and a contraction brought about by a gathering in of the parts which would allow for an ordinary extension if these internal organs were filled, rather than undergoing a sustained emptiness. It is rather a question of a real and considerable shrinkage and contraction such that these organs would break rather than yield suddenly to causes which demand an ordinary extension.

Compare, at entirely similar ages, a man who, in order to free himself for studies and habitual intellectual work, has acquired the habit of eating very little with another who habitually takes plenty of exercise, frequently goes out of his house, and eats well. The stomach of the first will have very little capacity and will be filled by a very small quantity of nourishment, while the stomach of the second will have preserved and even increased its capacity.

There we have an organ strongly modified in its dimensions and capacity by the single cause of a change in habits over the lifetime of an individual.

The frequent use of an organ, once it becomes constant and habitually, increases the capacities of this organ, develops it, and makes it acquire dimensions and an active power which the organ does not possess in the animals which exercise it less ....

Translated by Ian Johnston

Reading and Discussion Questions

1. What assumption about species is Lamarck contesting in this reading? What evidence does he provide in support of his own view?
2. By what mechanisms does Lamarck believe living things change over time?
3. Lamarck believes that organisms are constantly (if gradually) increasing in complexity, and that the earth is very old. How then, does he explain the continued existence of simple organisms on the earth?

On the Law of the Correlation of Parts (1800)

Georges Cuvier

89

View of the Relations Which Exist amongst the Variations of the Several Organs

The preceding Article has pointed out the principal differences of which the organs, belonging to each animal function, are susceptible in their structure and operations. The number of these differences would have been much greater had we entered into details, and descended to the less important circumstances.

It is obvious, however, from the manner in which we have described them, that by supposing each of the differences of one organ united successively with those of every other, there would be produced a very considerable number of combinations, which would correspond with as many classes of animals. But these combinations which appear possible, when we consider them abstractedly, do not all exist in nature; because, in a state of life, the organs do not simply join their effects but act on each other, and concur altogether to one common object. Hence the modifications of any one of them exercise an influence on those of every other. Such of these modifications as cannot exist together, reciprocally exclude one another, while others are, as it were, called into the system; and this takes place, not only in the organs which have an immediate connexion, but in those which at the first view appear the most separate and independent.

In fact there is not one function which does not stand in need of the concurrence of almost all the others, and which is not more or less affected by their degree of energy.

Respiration, for example, cannot take place without the aid of the motion of the blood, since it consists in bringing that fluid in contact with the surrounding element; but as it is circulation that gives motion to the blood, it therefore is a necessary mean in producing respiration.

Circulation itself has its cause in the muscular action of the heart and arteries: it is produced, therefore, by the aid of irritability. That faculty, in its turn, derives its origin from the nervous fluid, and, consequently, from the function of sensibility which returns, by a kind of circle to the circulation of the blood, which is the cause of all the secretions, and of that of the nervous fluid as well as others.

Of what value would sensibility be, were it not aided by the muscular force, even in the most trifling circumstances? What would be the utility of the sense of feeling, were we not able to turn our hands towards palpable objects? And what would be the advantage of seeing, if we could not turn the head or eyes in every direction?

It is on this mutual dependence of the functions, and the aid they reciprocally yield to one another, that the laws which determine the relations of their organs are founded laws which have their origin in a necessity equal to that of metaphysical or mathematical laws: for it is evident that a suitable harmony between organs which act on one another, is a necessary condition of the existence of the being to which they belong; and that *if* any one of the functions were modified in a manner incompatible with the regulations of the others, that being could not exist.

Thus, among the animals that have blood-vessels, and enjoy a double circulation, those which respire the air by receiving it immediately into the cellular lungs, have always the two trunks of their arteries approximated, and furnished with muscular ventricles, but joined together in one mass; while those which respire only through the medium of water that passes between the folds of their branchiae, have always two separate trunks, whether they be both provided with ventricles as the *Sepia* [Squid] or have a ventricle for one only, like fishes and mollusca. . . .

The nervous system has likewise its relations to respiration with respect to the varieties observed in both those functions. The external senses have much less energy, and the brain is considerably smaller in the animals that have cold blood, in which that organ occupies only a small part of the cranium, than in those of warm blood, in which the brain fills the whole cavity. Doubtless, the little irritability of the fibre in those animals requires but a small degree of activity in the organs that put it in motion: lively sensations and strong passions would have too much exhausted their muscular force. In this manner the organs of sensation are immediately connected with those of respiration.

But to what secret cause is it owing, that in all the animals which respire by distinct organs, the medullary masses form a small number, and are collected in the cranium, or, at least detached from the spinal marrow, while in those that respire by trachiae, nearly equal ganglions are distributed throughout the whole extent of their nervous cord? How does it happen, too, that there is no nervous system apparent in animals which have no organs particularly designed for respiration? These two relations must be included amongst those whose causes are unknown to us.

Digestion, also, has its connection with respiration: the latter being one of the functions which consume and expel, with the greatest rapidity, the substances of which the body is composed, the digestive power is generally the greater in proportion as respiration is more complete, in order that the quantity which is acquired may be equal to that which escapes.

In consequence of the connection, that subsists between the organs of respiration and the modifications of several other functions, some of the latter have relations to one another which at first sight did not appear necessary. This is the reason why birds have in general an exceedingly strong stomach, and a very quick digestion. This also is the reason why their repasts are so frequently repeated; while reptiles, which among the red-blooded animals seem to be contrasted to them in every respect, astonish us by the little aliment they take, and the length of time they abstain from food. These differences in the digestive powers do not depend upon the nature of the organs of motion which characterize these two classes, but upon that of the organs of respiration, the modifications of which have an immediate relation with those of motion.

It is easy to perceive that these two very different degrees of digestive powers depend on two dispositions equally different in the alimentary organs, and that each of these dispositions must be co-existent with a corresponding one in the respiratory organs. The latter also being always connected with a disposition equally determined in the organs of motion, in those of sensation, and in those of circulation, each of those five systems of organs may be said to regulate and govern the others.

The system of digestive organs has also immediate relations with those of motion and sensation. The disposition of the alimentary canal determines, in a manner perfectly absolute, the kind of food by which the animal is nourished; but if the animal did not possess, in its senses and organs of motion, the means of distinguishing the kinds of aliment suited to its nature, it is obvious it could not exist.

An animal, therefore, which can only digest flesh, must, to preserve its species, have the power of discovering its prey, of pursuing it, of seizing it, of overcoming it, and tearing it in pieces. It is necessary, then, that this animal should have a penetrating eye, a quick small, a swift motion, address, and strength in the claws and in the jaws. Agreeably to this necessity, a sharp tooth, fitted for cutting flesh, is never coexistent in the same species, with a foot covered with horn, which can only support the animal, but with which it cannot grasp any thing; hence the law by which all hoofed animals are herbivorous; and also those still more detailed laws which are but corollaries of the first, that hoofs indicate molar teeth, with flat crowns, a long alimentary canal, a capacious or multiplied stomach, and several other relations of the same kind.

Those laws which determine the relations of the organs belonging to the different functions, likewise exercise their powers on the different parts of the same system, and connect its variations with equal force. The application of these laws is particularly evident in the alimentary system, the parts of which are more numerous and distinct. The form of the teeth, the length, the convolutions, and the dilatations of the alimentary canal, and the number and abundance of the dissolving liquors poured into it, have always an admirable relation to each other, and to the nature, the hardness, and the solubility of the substances the animal eats. This connection is so evident, that the skillful anatomist, upon knowing one of those parts, may easily conjecture most of the others, and may, agreeably to the preceding laws, even guess the extent of the other functions.

The same harmony exists between all parts of the system of the organs of motion; as each of those organs acts upon the rest, and experiences their action in its turn, particularly when the animal is completely in motion, all their forms have relation to one another. Not a bone is varied in its surfaces, in its curvatures, or in its eminences, without subjecting the other bones to proportionate variations: we may, therefore, on the view of one of them, form, with a certain degree of accuracy, an idea of the whole skeleton.

These laws of co-existence, which we have thus far pointed out, may be said to be reduced by reasoning from the knowledge we have of the reciprocal uses and functions of each organ. Observation having confirmed these laws, we are authorized to follow an opposite course under other circumstances; when, therefore, we observe constant relations of form, between certain organs, we may conclude that they exercise some influence on one another, and we may even make pretty accurate conjectures as to the uses of both .... It is only by a profound study of those relations, and by the discovery of those which have hitherto escaped our observation, that we can hope to extend physiology. Comparative anatomy may, therefore, be regarded as one of the richest sources of observation for perfecting that important branch of knowledge.

Nature never oversteps the bounds which the necessary conditions of existence prescribe to her; but whenever she is unconfined by these conditions, she displays all her fertility and variety. Never departing from the small number of combinations that are possible, between the essential modifications of important organs, she seems to sport with infinite caprice in all the accessory parts. In these there appears no necessity for a particular form or disposition. It even frequently happens, that particular forms and dispositions are created without any apparent view to utility. It seems sufficient that they should be possible, that is to say, that they do not destroy the harmony of the whole. In proportion, therefore, as we turn our attention from the principal organs to those which are less important, we discover increasing variations; and when we arrive at the surface of bodies where the nature of things requires that the parts least essential, and the injury of which is least dangerous, should be placed, the number of varieties becomes so considerable, that all the labours of naturalists have not yet been able to give us an account of them.

Among these numerous combinations there are necessarily many which have common parts, and there is always a certain number which exhibits very few differences; by the comparison therefore of those which resemble each other, we may establish a kind of series, which will appear to descend gradually from a primitive type. These considerations are the foundations of the ideas from which certain naturalists have formed *A Scale of Being*, the object of which is to exhibit the whole in one series, commencing with the most simple kind of organization—with that which possesses the least numerous and most common properties; so that the mind passes from one link of the chain to the other, almost without perceiving any interval, and, as it were, by insensible shades.

Indeed, when we confine ourselves within certain limits, and particularly when we consider each organ separately, and follow it through all the species of one class, we observe that its progression in the scale is preserved with a singular regularity; we even perceive the organ partially, or some vestige of it in species, in which it is not longer of any use; so that Nature seems to have left it there only to show how strictly she adheres to the law of doing nothing by sudden transitions: but, on the one hand, the organs do not all follow the same order of gradation; one is found in its highest degree of perfection in one species, while another is most perfect in a species altogether different. If, therefore, we were to class the different species according to each organ considered separately, it would be necessary to form as many series as we should adopt regulating organs; and to make a general scale of perfection, it would be necessary to calculate the effect resulting from each combination. This, however, is far from being practicable.

On the other hand, the gentle and insensible shades of gradation prevail so long as we confine ourselves to the same combination of the principal organs, and so long as the great central springs remain the same. All the animals in which this takes place, seem to be formed upon one common plan, which serves as the basis of all the little external modifications: but the moment we turn our attention to those animals in which other principal combinations take place, there is no longer any resemblance, and an interval or marked transition is obvious to every one.

Whatever arrangement may be given to vertebral animals, and those which have no vertebrae, we never shall succeed in placing at the bottom of one of those great classes, and at the head of the other, two animals which sufficiently resemble each other to serve as a link between them.

Translated by William Ross

Reading and Discussion Questions

1. What does Cuvier mean in stating that the parts of animals are "correlated"? What limitations does this place on the degree to which an animal can vary? What kinds of things can change, and what cannot?
2. How can we use comparison of parts between animals to assist in our efforts at classification? What does he mean by saying that most animals are constructed on a "common plan"?

Discourse on the Revolutionary Upheavals on the Surface of the Globe (1826)

Georges Cuvier

90

Introduction

In my work on *Fossil Bones*, I set myself the task of identifying the animals whose fossilized remains fill the surface strata of the earth. This project meant I had to travel along a path where we had so far taken only a few tentative steps. As a new sort of antiquarian, I had to learn to restore these memorials to past upheavals and, at the same time, to decipher their meaning. I had to collect and put together in their original order the fragments which made up these animals, to reconstruct the ancient creatures to which these fragments belonged, to recreate their proportions and characteristics, and finally to compare them to those alive today on the surface of the earth. This was an almost unknown art, which assumed a science hardly touched upon up until now, that of the laws which govern the formal coexistence of the various parts in organic beings. Thus, I had to prepare myself for these studies through a much longer research into animals which presently exist. Only an almost universal review of present creation could provide some proof for my results concerning created life long ago. But at the same time such a study had to provide me with a large collection of equally demonstrable rules and interconnections. In the course of this exploration into a small part of the theory of the earth, I would have to be able to subject the entire animal kingdom in some way to new laws.

I was sustained in this double task by the constant interest which it promised to have and by service to the universal science of anatomy, the essential basis of all those sciences dealing with organic entities, and to the physical history of the earth, the foundation of mineralogy, geography, and, we can say, even of human history and everything really important for human beings to know about themselves.

If one finds it interesting to follow in the infancy of our species the almost eradicated traces of so many extinct nations, how could one not also find it interesting to search in the shadows of the earth's infancy for the traces of revolutionary upheavals which have preceded the existence of all nations? We admire the force with which the human spirit has measured the movements of planets which nature seemed to have concealed forever from our view; human genius and science have stepped beyond the limits of space; some observations developed by reasoning have unveiled the mechanical workings of the world. Would there not also be some glory for human beings to know how to step beyond the limits of time and to recover, through some observations, the history of this earth and a succession of events which have preceded the birth of the human genus? No doubt the astronomers have proceeded more rapidly than the naturalists. The theory of the earth at the present time is rather like the one in which some philosophers believed that the sky was made of freestone [fine-grained sandstone or limestone] and the moon was as big as the Peloponnese. But, following Anaxagoras, Copernicus and Kepler opened up the road to Newton. And why one day should natural history not also have its own Newton?

Exposition

In this discourse I propose above all to present the plan and result of my work on fossil bones. I will try also to sketch a rapid picture of the attempts made so far to reconstruct the history of the earth's upheavals. No doubt, the facts which I have discovered form only a really small part of those which must make up this ancient history; but several of these lead to significant consequences, and the rigorous way in which I have proceeded in determining them encourages me to believe that people will look on them as points definitely settled, things which will constitute a special age in science. Finally, I hope that their newness will excuse the fact that I focus the major attention of my readers on them.

My object will be, first, to show by what connections the history of the fossil bones of land animals is linked to the theory of the earth and why they have a particular importance in this respect. Then I will develop the principles on which rests the art of sorting out these bones, or, in other words, of recognizing a genus and distinguishing a species by a single bone fragment, an art on whose reliability depends the reliability of all my work. I will give a quick indication of new species, of genera previously unknown, which the application of these principles has led me to discover, as well as of the various sorts of formations which contain them. And since the difference between these species and those today does not exceed certain limits, I will show that these limits are considerably greater than those which today distinguish the varieties of a common species. I will thus reveal just where these varieties could go, whether by the influence of time, or climate, or finally domestication.

In this way, I will proceed to the conclusion (and I shall invite my readers to conclude with me), that there must have been great events to bring about the much greater differences which I have recognized. I will develop then the particular revisions which my research must introduce into the opinions accepted up to the present time about the earth's revolutions. Finally I will examine up to what point the civil and religious history of people agrees with the results of the observations dealing with the physical history of the earth and the probabilities which these observations set concerning the time when human societies could have established permanent homes and arable fields and when, consequently, societies could have taken on a lasting form.

The Geological Record of Ancient Upheavals

The First Appearance of the Earth

When the traveler goes through fertile plains where tranquil waters nourish with their regular flow an abundant vegetation and where the ground, trodden by numerous people and decorated with flourishing villages, rich cities, and superb monuments, is never troubled except by ravages of war or by the oppression of men in power, he is not tempted to believe that nature has also had its internal wars and that the surface of the earth has been overthrown by revolutions and catastrophes. But his ideas change as soon as he seeks to dig through this soil, today so calm, or when he takes himself up into the hills which border the plain; his ideas expand, so to speak, with what he is looking at. They begin to embrace the extent and the grandeur of the ancient events as soon as he climbs up the higher mountains of which these are the foothills, or when he follows the stream beds which descend from these mountains and moves into their interior.

The First Proofs of Upheavals

The lowest and most level land areas show us, especially when we dig there to very great depths, nothing but horizontal layers of material more or less varied, which almost all contain innumerable products of the sea. Similar layers, with similar products, form the hills up to quite high elevations. Sometimes the shells are so numerous that they make up the entire mass of soil by themselves. They occur at elevations higher than the level of all seas, where no sea could be carried today by present causes. Not only are these shells encased in loose sand, but the hardest rocks often encrust them and are penetrated by them throughout. All the parts of the world, both hemispheres, all continents, and all islands of any size provide evidence of the same phenomenon. The time is past when ignorance could continue to maintain that these remains of organic bodies were simple games of nature, products conceived in the bosom of the earth by its creative forces, and the renewed efforts of certain metaphysicians will probably not be enough to make these old opinions acceptable. A scrupulous comparison of the shapes of these deposits, of their makeup and often even their chemical composition shows not the slightest difference between these fossil shells and those which the sea nourishes. Their preservation is no less perfect. Very often one observes there neither shattering nor fractures, nothing which signifies a violent movement. The smallest of them keep their most delicate parts, their most subtle crests, their slenderest features. Thus, not only have they lived in the sea, but they have been deposited by the sea, which has left them in the places where we find them. Moreover, this sea has remained in these locations, with a sufficient calm and duration to form deposits so regular, so thick, so extensive, and in places so solid, that they are full of the remains of marine animals. The sea basin therefore has provided evidence of at least one change, whether in extent or location. See what results already from the first inspections and the most superficial observation.

The traces of upheavals become more impressive when one moves a little higher, when one gets even closer to the foot of the great mountain ranges. There are still plenty of shell layers. We notice them, even thicker and more solid ones. The shells there are just as numerous and just as well preserved. But they are no longer the same species. Also, the strata which contain them are no longer generally horizontal. They lie obliquely, sometimes almost vertically. In contrast to the plains and the low hills, where it was necessary to dig deep to recognize the succession of layers, here we see them on the mountain flank, as we follow the valleys produced by their tearing apart. At the foot of the escarpments, immense masses of debris form rounded hillocks, whose height is increased by each thawing and each storm.

And those upright layers which form the crests of secondary mountains do not rest on the horizontal layers of hills which serve as their lower stages. By contrast, they sink under these hills, which rest on the slopes of these oblique strata. When we bore into the horizontal strata near mountains with oblique layers, we find these oblique layers deep down. Sometimes when the oblique layers are not very high, their summits are even crowned with horizontal strata. The oblique layers are therefore older than the horizontal layers. Since it is impossible, at least for most of them, not to have been formed horizontally, evidently they have been lifted up again and were in existence before the others which rest on top of them.

Thus, before forming these horizontal layers, the sea had formed other strata. These were for some reason or other broken, raised up, and overturned in thousands of ways. As several of these oblique layers which the sea formed in a previous age rise higher than the horizontal layers which succeeded them and which surrounded them, the causes which gave these layers their oblique orientation also made them protrude above the level of the sea and turned them into islands or at least reefs and uneven structures, whether they were raised again by an extreme condition or whether a contrasting subsidence made the waters sink. The second result is no less clear or less proven than the first for anyone who will take the trouble to study the monuments which provide evidence for these results.

Proofs That These Revolutions Have Been Numerous

But the revolutions and changes which are responsible for the present state of the earth are not limited to the upsetting of the ancient strata and to the ebbing of the sea after the formations of new layers. When we compare together in greater detail the various layers and the products of life which they conceal, we soon realize that this ancient sea did not continuously deposit the same type of stones nor the remains of animals of the same species, and that each of its deposits did not extend over all the surface which the sea covered. Successive variations took place, of which only the first ones were almost universal; the others appear to have been considerably less. The older the layers, the more each of them is uniform over a great extent; the newer the layers, the more they are limited and subject to variation within small distances. Thus, the changes in the strata were accompanied and followed by changes in the nature of the liquid and of the materials which it held in solution. When certain layers, appearing above the sea, split the surface with islands and protruding ranges, different changes could have taken place in several particular ocean basins.

We know that in the midst of such variations in the nature of the liquid, the animals which it nourished could not have stayed the same. Their species, even their genera, changed with the layers; and although there are some returns of species within small distances, it is true to state, in general, that the shells of the ancient layers have forms unique to them, that they disappear gradually and do not show up in the recent layers, even less in the present sea, where we never discover species analogous to them. Even several of their genera are not found there. The shells of recent layers, by contrast, are generically similar to those which live in our seas. In the most recent and least solid of these layers and in certain recent and limited deposits there are some species which the most practiced eye would not be able to distinguish from those which the neighbouring coasts nourish.

Thus in animal nature a succession of variations has taken place, brought about by changes in the liquid where the animals lived or at least by variations which corresponded to those changes. And these variations brought by degrees the classes of aquatic animals to their present condition. At last, when the sea left our continents for the last time, its inhabitants did not differ much from those which the sea still feeds today.

Finally, we say that if we examine with even greater care the remains of these organic creatures, we come to discover in the middle of the marine strata, even the most ancient ones, layers full of animal or vegetable products from land and fresh water. In the most recent layers (i.e., the ones closest to the surface) there are somewhere land animals are buried under masses of marine creatures. Thus, not only did the different catastrophes which moved the layers gradually make the various parts of our continent rise up from the bottom of the sea and reduce the size of the sea basin; but this basin has been moved in several directions. Often the regions converted into dry land have been covered again by the seas, whether they have sunk or the waters have been carried above them. As for the particular matter of the soil which the sea uncovered in its last retreat, the part which human beings and terrestrial animals live on right now, it had already been dry land once and had nourished at that time quadrupeds, birds, plants, and land forms of all sorts. Thus, the sea which left that land had previously invaded. The changes in the heights of the oceans did not therefore consist only in one withdrawal more or less gradual, more or less universal. It was a matter of a succession of various eruptions and retreats. The result of these has definitely been, however, a general lowering of the sea level.

Proofs That These Revolutions Have Been Sudden

But it is also really important to note that these eruptions and repeated retreats were not at all slow and did not all take place gradually. On the contrary, most of the catastrophes which brought them on have been sudden. That is especially easy to demonstrate for the last of these catastrophes, which by a double movement inundated and later left dry our present continents or at least a great part of the land which forms them today. That catastrophe also left in the northern countries the cadavers of great quadrupeds locked in the ice, preserved right up to our time with their skin, hair, and flesh. If they had not been frozen as soon as they were killed, decay would have caused them to decompose. On the other hand, this permanent freezing was not a factor previously in the places where these animals were trapped, for they would not have been able to live in such a temperature. Hence the same instant which killed the animals froze the country where they lived. This event was sudden, instantaneous, without any gradual development. What is so clearly demonstrated for this most recent catastrophe is hardly less so for the earlier ones. The rending, rearranging, and overturning of more ancient layers leave no doubt that sudden and violent causes placed them in the state in which we see them. The very force of the movements which the bodies of water experienced is still attested to by the mountain of remains and rounded pebbles interposed in many places between the solid layers. Thus, life on this earth has often been disturbed by dreadful events. Innumerable living creatures have been victims of these catastrophes. Some inhabitants of dry land have seen themselves swallowed up by floods; others living in the ocean depths when the bottom of the sea was lifted up suddenly were placed on dry land. Their very races were extinguished forever, leaving behind nothing in the world but some hardly recognizable debris for the natural scientist.

Such are the conclusions to which we are necessarily led by the objects which we meet at every step and which we can verify at every instant in almost every country. These huge and terrible events are clearly printed everywhere for the eye which knows how to read the story in their monuments.

But what is even more astonishing and what is no less certain is that life has not always existed on the earth and that it is easy for the observer to recognize the point where life began to deposit her productions.

Proofs That There Were Revolutionary Upheavals Before the Existence of Living Things

Let us keep climbing. Let us move up towards the great mountain ridges, towards the terraced summits of the great ranges. Soon these remains of marine animals, those innumerable shells, will become increasingly rare and will disappear altogether. We will reach layers of a different sort, which will contain no vestiges of living things at all. However, they will show by their crystallization and by their very stratification that they were also formed in a liquid state. Their oblique orientation and their escarpments will indicate that they also have been overturned. The manner in which they slant under the layers with shells will reveal that they were formed before them. Finally the height of their bare and bristling peaks rising above all these layers with shells will show that these summits had already left the water when the layers with shells were formed.

Such are the famous primitive or primordial mountains which cross our continents in different directions, rising up above the clouds, separating river basins, holding in their perpetual snow the reservoirs which supply the rivers' sources, and forming something like the skeleton and rough framework of the earth.

From a long way away the eye perceives in the indentations which split up the crests, in the sharp peaks which bristle there, evidence of the violent manner in which they were uplifted, very different from those rounded mountains or hills with long flat surfaces where the recent mound always remains in the condition in which it was peacefully deposited by the most recent seas.

These signs become more evident as one approaches. The valleys do not have gentle slopes any more or those jutting angles facing indentations opposite, which seem to indicate the beds of some ancient water course. They grow bigger or smaller without any rule. Their waters sometimes extend into lakes; at other times they hurtle down in torrents. Sometimes the rocks, coming suddenly together, form transverse dams, from which these same waters fall in cataracts. The ripped apart strata, revealing on one side a sharp perpendicular edge, present on the other side large obliquely oriented sections of their surface. They do not correspond in height. Those which, on one side, form the summit of an escarpment, disappear on the other and do not reappear any more.

However, some great naturalists have managed to demonstrate that, in the middle of all this disorder, a certain order still reigns and that these immense ranges, as bristling and overturned as they all are, themselves follow a succession which is almost the same in all the large mountain ranges. The granite, they say, which forms the central crests of most of these ranges and which is the highest of all the rocks, is also the rock which disappears under all the others. It is the most ancient of those which we have been given to see in the place which nature put it, whether it owes its origin to a universal liquid which, in earlier times, held everything in solution, or whether it was the first rock established by the cooling of a large fused mass or even by evaporation. Foliated rocks lean on the flanks of the granite and form the lateral crests of these large mountain ranges. Schists, porphyries, sandstones, and talus are mixed together in the strata. Finally granular marbles and other calcareous rocks without shells resting on schists form the outer peaks, lower terraces, and foothills of these ranges, and are the last work by which this unknown liquid, this sea without inhabitants, seems to have prepared the materials for the mollusks and zoophytes which soon must have deposited on the bottom an immense quantity of their shells or their coral. We even see the first products of these mollusks, these zoophytes, showing up in small numbers here and there among the latest layers of these primitive formations or in the part of the earth's crust which geologists have called the transitional areas. In these places we meet here and there layers with shells interposed with some granites more recent than the others, among various schists and between some late beds of granular marble. The life which wished to seize hold of this earth seems in these early times to have fought with inert nature which had previously dominated. Only after a relatively long time did life clearly get the upper hand, so that to life alone belonged the right to continue to increase the solid outer layer of the earth.

Thus it cannot be denied that the masses which today form our highest mountains were originally in a liquid state; for a long time they were covered by waters which did not sustain any life. Changes did not take place in the nature of the materials deposited only after the appearance of life. The masses formed previously changed, as well as those which were formed later. They have similarly provided evidence of the violent alterations in their positions. Some of these transformations took place at the time when these masses existed by themselves and were not covered with layers of shells. We have the proof of that in the overthrusting, tearing apart, and fissures which can be observed in these strata, as well as in those of later land masses, which, indeed, are more numerous and more marked.

But these primitive structures have experienced still other upheavals since the creation of the secondary formations and have perhaps caused or at least shared some of those which these secondary formations have themselves undergone. There are, in fact, considerable sections of primitive rocks totally bare, although in a lower location than many of the secondary formations. How could these not have been covered over again unless they appeared since the creation of these secondary formations? We find many voluminous blocks of ancient materials scattered in certain countries on the surfaces of secondary formations, separated by deep valleys or even by the arms of the sea from the peaks and crests where these blocks could have originated. It must be the case either that some eruptions threw them there or that the low places which stopped their movement did not exist at the time of their transport, or finally perhaps that the motion of the waters which carried them surpassed in violence anything which we can imagine nowadays.

Here then is a collection of events, a series of periods earlier to the present times, whose sequence can be verified without doubt, although the lengths of the intervals cannot be defined with precision. There are so many items which indicate the measure and the direction of this ancient chronology.

Present Geological Processes

Examination of the Causes Which Are Still at Work Today on the Surface of the Earth

Let us now consider what happens today on the earth; let us analyze the causes which still disturb its surface and determine the possible extent of their effects. This part of earth's story is all the more important because for a long time we thought we could explain earlier revolutionary upheavals by present causes, just as we readily explain past events in political history, when we know well the passions and the intrigues of our own times. But we are going to see that unfortunately things are not the same in the history of physics. The thread of the processes is broken; nature's march has changed; and none of the agents which she uses today would have been sufficient to produce these ancient works.

There now exist four active causes which contribute to alterations on the surface of our continents: rains and thaws which erode the steep mountains and throw debris at their feet; the moving waters which carry away this debris and go on to deposit it in places where their current slows down; the sea which undermines the foot of high coasts to create cliffs there and which throws back mounds of sand onto coasts of low elevation; and finally volcanoes which break through solid strata and raise or scatter on the surface piles of the material which they emit.

...

Constant Astronomical Causes

The pole of the earth moves in a circle around the pole of the ecliptic; its axis inclines more or less on the plane of this same ecliptic. But these two movements, whose causes nowadays are understood, are carried out in known directions and within known limits, and they are not at all proportional to effects like those whose magnitude we have just established. In every case, their excessive slowness would prevent them from explaining the catastrophes which we have just shown to have been sudden.

This last rationale applies to all slow actions which people have imagined, without doubt in the hope that their existence could not be denied, because it would always be easy to maintain that their very slowness renders them imperceptible. Whether this is true or not is inconsequential. Such forces explain nothing, because no slow action could have produced these sudden effects. Thus, whether there was a gradual diminution of the waters, whether the sea carried solid material in all directions, whether the temperature of the earth decreased or increased, none of these has overturned the strata, enclosed in ice large quadrupeds with their flesh and pelt, put on dry land shell fish still as well preserved today as if they had been caught while still alive, or finally destroyed entire species and genera.

These arguments have forcibly impressed the great majority of natural scientists. And among those who have sought to explain the present state of the earth, hardly anyone has attributed it entirely to slow causes, even less to causes working before our very eyes. This need to seek causes different from those which we see at work now is the same need which has led scientists to dream up so many extraordinary conjectures and made them commit errors and lose themselves in contradictions, so that the very name of their science, as I have said elsewhere, has for a long time been a subject of mockery for some prejudiced people who looked only at the systems which this situation created and who forgot the long and important series of established facts which it has made known.

History of Geology and Geological Systems

Ancient Systems of Geologists

For a long time we have accepted only two events, two periods of changes on the earth: the Creation and the Flood. All the efforts of geologists have tended to explain the present state of the earth by imagining a certain original state, later modified by the Flood. Each of them has speculated also about the nature of the causes, the actions, and effects of these events.

...

More Recent Systems

In our time, freer spirits than ever before have also wished to busy themselves with this important subject. Certain writers have reproduced and enormously extended de Maillet's ideas. They claim that all was liquid at the beginning, that the liquid engendered at first very simple animals, like monads or other microscopic infusorian species, and that, with the passage of time and the development of different habits, the animal races became more complex and diversified to the point where we see them today.

...

Some other writers have preferred Kepler's ideas. Like this great astronomer, they give the earth itself vital faculties. According to them, a fluid circles in the earth, and an assimilation takes place just as in animated bodies. Each of its parts is alive. Every elementary molecule has instinct and will; they attract and repel each other according to antipathies and sympathies; each sort of mineral can change immense masses into its own nature, as we convert our food into flesh and blood. The mountains are the respiratory organs of the earth, and the schists are the secretary organs. Through them sea water is decomposed to create the volcanic eruptions. The seams finally are the decaying teeth, the abscesses of the mineral kingdom, and the metals a product of decay and illness. That is why almost all of them feel unpleasant.

...

Divergences of All Systems

...

We could cite still twenty other systems every bit as different as the above. And, just to make sure there is no mistake about it, our intention is not to criticize the authors of these systems. On the contrary, we recognize that these ideas have generally been conceived by men of wit and wisdom, who have not ignored the facts, several of whom have even traveled for a long time to examine them and have gathered a great deal of important scientific information.

. . .

The Nature and Conditions of the Problem

To abandon this mathematical language, we will say that virtually all the authors of these systems, having paid attention only to certain difficulties which struck them more than others, determined to resolve those difficulties by more or less plausible means and put aside numerous other equally important difficulties.

. . .

Are there animals and plants unique to certain layers which do not occur in others? What are the species which appear first or those which come later? Do these two types of species sometimes appear together? Is there an alternating pattern in their return or, in other words, do the first ones return for a second time and then do the second ones disappear? Have these animals and plants all lived in the areas where we find their remains, or are there any which were carried there from somewhere else? Are they still alive today somewhere, or have they been destroyed completely or in part? Is there a constant connection between the age of these layers and the similarity or dissimilarity of their fossils with living things? Is there a climatic connection between fossils and those living things which resemble them the most? Is it possible to conclude from this that the transport of these beings, if there was one, took place from north to south or from east to west, or by radiating out and mixing? And can we distinguish the times of these transports by the layers which carry the imprints of these living things?

What is there to say about the causes of the earth's present condition, if one cannot reply to these questions, if one has not yet sufficient reason for choosing between the affirmative and the negative? Now, it is only too true that for a long time none of these points has been resolved beyond doubt and that we have hardly even dreamed that it would be good to clarify them before making up a system.

Reasons for the Neglect of These Conditions

One will find the reason for this odd situation if one reflects that geologists have all been either museum naturalists, who hardly ever examined the structure of mountains on their own, or mineralogists who have not studied with sufficient detail the innumerable varieties of animals and the infinite complexity of their various parts. The first have only made systems; the latter have provided excellent observations: they have truly laid down the foundations of the science. But they have not been able to raise an edifice upon it.

Progress of Mineral Geology

...

Certainly other scholars have studied the fossil remains of these organic bodies. They have collected them and drawn copies of them by the thousands. Their works will be valuable collections of materials. But more occupied with animals or with plants, considered in themselves, than with the theory of the earth, or looking upon these petrified remains or fossils as curiosities rather than as historical documents or, finally, contenting themselves with partial explanations for the deposit of each piece, they have almost always neglected to seek out general laws concerning the position or the relationship of the fossils with the strata.

Translated by unnamed translator

Reading and Discussion Questions

1. How does Cuvier combine his studies of anatomy with research in geology? How is traditional geological research furthered by careful studies of fossils?
2. What evidence do we have that the surface of the earth and the creatures that live on it have changed over time?
3. Does he believe that the same types of mechanisms that we see altering the surface of the earth today produced the geological changes that we see in the fossil record?
4. Buffon (citing Ovid), Linnaeus, and Cuvier all point out that sea shells are often found embedded in rocks, even in the mountains. What explanation for this fact do they all reject?

Lectures on Comparative Anatomy and Natural History of Fishes: General Conclusion on the Organization of Fishes (1828)

Georges Cuvier

91

It follows from this general exposition of all observations on the particular organization [of different species of fishes], that the fishes form an animal class distinct from all the others, and totally determined by its structure to live, move and carry out its essential activities in an aqueous element. This is their place in the creation. They have been there since their origin. They will remain there until the destruction of the existing order of things. It is only by vain metaphysical speculations, or by very superficial comparisons, that their class has been considered a development, perfection or ennoblement of the molluscs, or as a first sketch, or embryonic state, of the other classes of vertebrates.

Without doubt, the molluscs, like the fishes, breathe by gills; they have in common with them and with the other vertebrates a nervous and circulatory system, an intestinal canal and liver, and no one knows this better than myself, since I had been the first to make known, with some small degree of completeness, their anatomy and their zoological relations. Since the animal level of existence has only received a limited number of organs, it is necessary that at least some of these organs should be common to several classes. But where, otherwise, is there a resemblance? Is the framework of these animals, their system of locomotion, comparable in the least of their parts? And in what way are the organs common to the molluscs and fishes? Can they be shown to have the relations and connections to each other in the molluscs that they have in the fishes and other vertebrates? What transition in nature leads us from one to the other?

One could, I am aware, propose a definition which only took account of the features these groups have in common, and disregarded their differences. But this definition would always remain a pure mental abstraction, a purely nominal definition, a vain assemblage of words, which could never be represented by a common plan unless it was stripped of all the details needed to conceive of it. Such a method could then be used to connect all the organisms one wished together, since, in the final analysis, any two creatures, however remote they might be from each other, would always show some resemblances, if in nothing else but existence itself.

The single heart in the molluscs is placed in a way contrary to that of the fishes; it is located at the conjunction of the branchial veins and the arteries of the body. In several of them, the appendages are on the head. In others, the organs of generation are on the side. Often the organs of respiration are located above those of digestion, or are spread out on all or part of the dorsal surface. In short, they and the fishes have gills. This is all the similarity they have. Every time one wishes to begin from these purely verbal and metaphysical formulas, they are led to make prefigured comparisons.

For example, by one author, the shells of bivalves are taken to represent the opercula of fishes. Another takes the bone of a cuttlefish for a true bone; yet a third interprets the great plates of the sturgeon or box fish for an external skeleton.

Others seek their analogical relations in the crustaceans. The segments of the thorax are taken to represent the opercular bones of the fishes, and to be sure, one does find the gills under these segments. But if we penetrate a little further, everything is reversed. The nerve cord is on the ventral side in the crustaceans, and the heart is on the back, and this heart, like that of the molluscs, receives, rather than sends, the blood from the gills. Other authors, without explanation of the causes involved, have wished to see the rays or spiny apophyses of the vertebrae of fishes in the feet of the crustaceans. But then, one no longer has a perfection of the fishes, but rather a manifest degradation.

The parallelism of the fishes and the other vertebrates is not as completely misconceived. Here, at least, some sensible relationships in the number of organ systems and in their mutual connections takes its rise. But this is still far not only from an identity of structure, but even from the appearance of a progressive sequence.

The head of fishes, or better their skull, is divided into a number of bones that are almost like those seen in the skull of birds and lizard-like reptiles. There is also some resemblance, although much less complete, between these bones and those of the mammalian fetus, just as the circulation in the reptiles has some relation with that of the mammalian fetus. Thus the oviparous classes, and especially the reptiles, have been considered to be mammals arrested at an early period in their development. Extending the comparison to the fishes, in which respiration and circulation, as far as it concerns the (arrangement of) the vessels, is almost the same as in the frog tadpole and the other batracian reptiles, it has been concluded that the fishes prefigure these tadpoles. They are, as a consequence some kind of fetus of the second degree, a fetus of a fetus.

But even if these relationships in the number of bones are as complete as they are rare, and even if one forgets that the reptiles most close to the fishes—the frogs and salamanders—have, in all their manifestations, fewer bones of the skull and face than the fishes and even the mammals, this manner of viewing them will be no less defective. It considers, we might say, only one or two points of similarity and ignores all the rest, or else connects them to this system of arrangement only by making suppositions contrary to intuition.

It is no advance in understanding to arrive by arbitrary concessions at the conclusion that entire organ systems are reversed; or that the bones which belong to one organ are to be seen interspersed between those of another; or that bones arranged in one group of organisms in series are found one upon the other in the succeeding class: or lastly, that a whole group of bones, which shows a constant diminution and simplification in one group, like the small tympanic bones, should suddenly regain its original number of segments and increase in volume so as to exercise a totally different function, the protection of the gills. For examination reveals that there is not the proper number of bone segments, nor are the parts in the proper connection. In short, nothing yet supports these claimed analogies that have allegedly been attained by such a laborious process.

Let us suppose, as an example, that the vertebral spine (*apophyse épineuse*) of a fish vertebra is detached, and that one of its two halves is raised above the other. Let us even allow that in these circumstances, nature shaped these parts differently, and made this articulation so complex that it could be termed a "chain-like" articulation. Does this then give us the radial (*interépineux*) and ray of the dorsal fin which articulate together? No, because the radial itself is composed of three parts, and the ray, if it is a simply spiny ray, is still divided vertically into two halves. Or take the case of a soft ray, divided at the end into numerous branches and hundreds of small articulations. With regard to the six separate muscles attached to each of these rays, the evidence for a lack of an analogy (with the serial body muscles) is such that no one would dare assign them one. The same would be true, although the contrary has been claimed, if one tries to compare the opercular muscles to those connected to the small bones of the ear.

Without doubt, the bony apparatus bearing the gills of fishes has some relationship, although a rather distant one, with that bearing the branchial fringes of the tadpole or salamander. But that itself would prove that it cannot be the analog of the larynx and bronchial tubes, since the larynx and bronchials exist in these animals simultaneously with the gill apparatus. Furthermore, is there the least comparison to be made between the muscles of the gill apparatus in the two classes?

Now, if nature had created one set of muscles expressly for the reptiles, and another for the fishes, why could it not also have created separate bones for them?

Some anatomists have wanted to see an analogy between the opercular bones [*pieces operculaires des ouies*] of fishes and the ear bones of the mammals. But then the former could not be rudiments of the latter, but rather would be an enormous development of them. How can this be reconciled with the fact that it is precisely these same bones in the (amphibious) reptiles, such as the salamanders and frogs, which in their first stages are almost true fishes, which are most similar to those of the fishes. Yet the amphibians are, of all the vertebrates, those which show the most weak and rudimentary development of the small bones of the ear.

We ought, therefore, conclude that if there are resemblances between the structures of the fishes and those of other classes of vertebrates, it is only insofar as there are similarities in their functions. Let us conclude that if it can be said that these animals are ennobled molluscs, molluscs elevated by a degree, or the fetuses of reptiles, reptiles in an embryonic stage, it is only to be understood in an abstract and metaphysical sense. Even then it is a necessary conclusion only insofar as this abstract expression organizes our legitimate ideas. Particularly, we should conclude that there are no links in that imaginary chain of successive forms, in which one form can serve as the source of the other, since none of these could exist by itself. Nor is there that other, no less fictitious, chain of simultaneous and graded forms, which has its existence only in the imagination of some naturalists, more poets than observers of nature. Rather fish belong to the real chain of coexistent beings of creatures necessary to each other and to the whole, which, by their mutual interaction, maintain the order and harmony of the universe, a chain in which no portion can exist without all the others, and in which the coils, ceaselessly united or dispersed, embrace the globe in their contours.

Translated by P. R. Sloan

Reading and Discussion Question

1. What anatomical differences does Cuvier enumerate between the mollusks and the fishes? In what way do these points serve as a critique of Lamarck?

Principles of Geology, or the Modern Changes of the Earth and Its Inhabitants Considered as Illustrative of Geology (1830)

Charles Lyell

92

Book I

Chapter 1

. . .

Geology is the science which investigates the successive changes that have taken place in the organic and inorganic kingdoms of nature; it inquires into the causes of these changes, and the influence which they have exerted in modifying the surface and external structure of our planet.

By these researches into the state of the earth and its inhabitants at former periods, we acquire a more perfect knowledge of its present condition, and more comprehensive views concerning the laws now governing its animate and inanimate productions. When we study history, we obtain a more profound insight into human nature, by instituting a comparison between the present and former states of society. We trace the long series of events which have gradually led to the actual posture of affairs; and by connecting effects with their causes, we are enabled to classify and retain in the memory a multitude of complicated relations—the various peculiarities of national character—the different degrees of moral and intellectual refinement, and numerous other circumstances, which, without historical associations, would be uninteresting or imperfectly understood. As the present condition of nations is the result of many antecedent changes, some extremely remote, and others recent, some gradual, others sudden and violent; so the state, of the natural world is the result of a long succession of events; and if we would enlarge our experience of the present economy of nature, we must investigate the effects of her operations in former epochs.

We often discover with surprise, on looking back into the chronicles of nations, how the fortune of some battle has influenced the fate of millions of our contemporaries, when it has long been forgotten by the mass of the population. With this remote event we may find inseparably connected the geographical boundaries of a great state, the language now spoken by the inhabitants, their peculiar manners, laws, and religious opinions. But far more astonishing and unexpected are the connections brought to light, when we carry back our researches into the history of nature. The form of a coast, the configuration of the interior of a country, the existence and extent of lakes, valleys, and mountains, can often be traced to the former prevalence of earthquakes and volcanoes in regions which have long been undisturbed. To these remote convulsions the present fertility of some districts, the sterile character of others, the elevation of land above the sea, the climate, and various peculiarities, may be distinctly referred. On the other hand, many distinguishing features of the surface may often be ascribed to the operation, at a remote era, of slow and tranquil causes—to the gradual deposition of sediment in a lake or in the ocean, or to the prolific increase of testacea and corals.

To select another example, we find in certain localities subterranean deposits of coal, consisting of vegetable matter, formerly drifted into seas and lakes. These seas and lakes have since been filled up, the lands whereon the forests grew have disappeared or changed their form, the rivers and currents which floated the vegetable masses can no longer be traced, and the plants belonged to species which for ages have passed away from the surface of our planet. Yet the commercial prosperity, and numerical strength of a nation, may now be mainly dependent on the local distribution of fuel determined by that ancient state of things.

Geology is intimately related to almost all the physical sciences, as history is to the moral. An historian should, if possible, be at once profoundly acquainted with ethics, politics, jurisprudence, the military art, theology; in a word, with all branches of knowledge by which any insight into human affairs, or into the moral and intellectual nature of man, can be obtained. It would be no less desirable that a geologist should be well versed in chemistry, natural philosophy, mineralogy, zoology, comparative anatomy, botany; in short, in every science relating to organic and inorganic nature. With these accomplishments, the historian and geologist would rarely fail to draw correct and philosophical conclusions from the various monuments transmitted to them of former occurrences. They would know to what combination of causes analogous effects were referable, and they would often be enabled to supply, by inference, information concerning many events unrecorded in the defective archives of former ages. But as such extensive acquisitions are scarcely within the reach of any individual, it is necessary that men who have devoted their lives to different departments should unite their efforts; and as the historian receives assistance from the antiquary, and from those who have cultivated different branches of moral and political science, so the geologist should avail himself of the aid of many naturalists, and particularly of those who have studied the fossil remains of lost species of animals and plants.

...

It was long before the distinct nature and legitimate objects of geology were fully recognized, and it was at first confounded with many other branches of inquiry, just as the limits of history, poetry, and mythology were ill-defined in the infancy of civilization. Even in Werner's time, or at the close of the eighteenth century, geology appears to have been regarded as little other than a subordinate department of mineralogy; and Desmarest included it under the head of Physical Geography. But the most common and serious source of confusion arose from the notion, that it was the business of geology to discover the mode in which the earth originated, or, as some imagined, to study the effects of those cosmological causes which were employed by the Author of Nature to bring this planet out of a nascent and chaotic state into a more perfect and habitable condition. Hutton was the first who endeavored to draw a strong line of demarcation between his favorite science and cosmogony, for he declared that geology was in nowise concerned "with questions as to the origin of things."

An attempt will be made in the sequel of this work to demonstrate that geology differs as widely from cosmogony, as speculations concerning the mode of the first creation of man differ from history. But, before entering more at large on this controverted question, it will be desirable to trace the progress of opinion on this topic, from the earliest ages to the commencement of the present century.

Chapter 13

Uniformity in the Series of Past Changes in the Animate and Inanimate World

...

Origin of the doctrine of alternate periods of repose and disorder.

It has been truly observed, that when we arrange the fossiliferous formations in chronological order, they constitute a broken and defective series of monuments: we pass without any intermediate gradations, from systems of strata which are horizontal to other systems which are highly inclined, from rocks of peculiar mineral composition to others which have a character wholly distinct,—from one assemblage of organic remains to another, in which frequently all the species, and most of the genera, are different. These violations of continuity are so common, as to constitute the rule rather than the exception, and they have been considered by many geologists as conclusive in favor of sudden revolutions in the inanimate and animate world. According to the speculations of some writers, there have been in the past history of the planet alternate periods of tranquility and convulsion, the former enduring for ages, and resembling that state of things now experienced by man: the other brief, transient, and paroxysmal, giving rise to new mountains, seas, and valleys, annihilating one set of organic beings, and ushering in the creation of another.

It will be the object of the present chapter to demonstrate, that these theoretical views are not borne out by a fair interpretation of geological monuments. It is true that in the solid framework of the globe, we have a chronological chain of natural records, and that many links in this chain are wanting; but a careful consideration of all the phenomena will lead to the opinion that the series was originally defective,—that it has been rendered still more so by time—that a great part of what remains is inaccessible to man, and even of that fraction which is accessible, nine-tenths are to this day unexplored.

How the facts may be explained by assuming a uniform series of changes.

The readiest way, perhaps, of persuading the reader that we may dispense with great and sudden revolutions in the geological order of events, is by showing him how a regular and uninterrupted series of changes in the animate and inanimate world may give rise to such breaks in the sequence, and such unconformability of stratified rocks, as are usually thought to imply convulsions and catastrophes. It is scarcely necessary to state, that the order of events thus assumed to occur, for the sake of illustration, must be in harmony with all the conclusions legitimately drawn by geologists from the structure of the earth, and must be equally in accordance with the changes observed by man to be now going on in the living as well as in the inorganic creation. It may be necessary in the present state of science to supply some part of the assumed course of nature hypothetically; but if so, this must be done without any violation of probability, and always consistently with the analogy of what is known both of the past and present economy of our system. Although the discussion of so comprehensive a subject must carry the beginner far beyond his depth, it will also, it is hoped, stimulate his curiosity, and prepare him to read some elementary treatises on geology with advantage, and teach him the bearing on that science of the changes now in progress on the earth. At the same time it may enable him the better to understand the intimate connection between the second and third books of this work, the former of which is occupied with the changes in the inorganic, the latter with those of the organic creation.

In pursuance, then, of the plan above proposed, I shall consider in this chapter, first, what may be the course of fluctuation in the animate world; secondly, the mode in which contemporaneous subterranean movements affect the earth's crust; and, thirdly, the laws which regulate the deposition of sediment.

Uniformity of Change Considered First in Reference to the Living Creation

First, in regard to the vicissitudes of the living creation, all are agreed that the sedimentary strata found in the earth's crust are divisible into a variety of groups, more or less dissimilar in their organic remains and mineral composition. The conclusion universally drawn from the study and comparison of these fossiliferous groups is this, that at successive periods distinct tribes of animals and plants have inhabited the land and waters, and that the organic types of the newer formations are more analogous to species now existing, than those of more ancient rocks. If we then turn to the present state of the animate creation, and inquire whether it has now become fixed and stationary, we discover that, on the contrary, it is in a state of continual flux—that there are many causes in action which tend to the extinction of species, and which are conclusive against the doctrine of their unlimited durability. But natural history has been successfully cultivated for so short a period, that a few examples only of local, and perhaps but one or two of absolute, extirpation can as yet be proved, and these only where the interference of man has been conspicuous. It will nevertheless appear evident, from the facts and arguments detailed in the third book (from the thirty-seventh to the forty-second chapters, inclusive) that man is not the only exterminating agent; and that, independently of his intervention, the annihilation of species is promoted by the multiplication and gradual diffusion of every animal or plant. It will also appear, that every alteration in the physical geography and climate of the globe cannot fail to have the same tendency. If we proceed still farther, and inquire whether new species are substituted from time to time for those which die out, and whether there are certain laws appointed by the Author of Nature to regulate such new creations, we find that the period of human observation is as yet too short to afford data for determining so weighty a question. All that can be done is to show that the successive introduction of new species may be a constant part of the economy of the terrestrial system, without our having any right to expect that we should be in possession of direct proof of the fact. The appearance again and again of new species may easily have escaped detection, since the numbers of known animals and plants have augmented so rapidly within the memory of persons now living, as to have doubled in some classes, and quadrupled in others. It will also be remarked in the sequel (book iii. chap. 43), that it must always be more easy if species proceeded originally from single stocks, to prove that one which formerly abounded in a given district has ceased to be, than that another has been called into being for the first time. If, therefore, there be as yet only one or two unequivocal instances of extinction, namely, those of the dodo and solitaire, it is scarcely reasonable as yet to hope that we should be cognizant of a single instance of the first appearance of a new species.

Recent origin of man, and gradual approach in the tertiary fossils of successive periods from an extinct to the recent fauna.

The geologist, however, if required to advance some fact which may lend countenance to the opinion that in the most modern times, that is to say, after the greater part of the existing fauna and flora were established on the earth, there has still been a new species superadded, may point to man himself as furnishing the required illustration—for man must be regarded by the geologist as a creature of yesterday, not merely in reference to the past history of the organic world, but also in relation to that particular state of the animate creation of which he forms a part. The comparatively modern introduction of the human race is proved by the absence of the remains of man and his works, not only from all strata containing a certain proportion of fossil shells of extinct species, but even from a large part of the newest strata, in which all the fossil individuals are referable to species still living.

To enable the reader to appreciate the full force of this evidence, I shall give a slight sketch of the information obtained from the newer strata, respecting fluctuations in the animate world, in times immediately antecedent to the appearance of man.

In tracing the series of fossiliferous formations from the more ancient to the more modern, the first deposits in which we meet with assemblages of organic remains, having a near analogy to the fauna of certain parts of the globe in our own time, are those commonly called tertiary. Even in the Eocene, or oldest subdivision of these tertiary formations, some few of the testacea belong to existing species, although almost all of them, and apparently all the associated vertebrata, are now extinct. These Eocene strata are succeeded by a great number of more modern deposits, which depart gradually in the character of their fossils from the Eocene type, and approach more and more to that of the living creation. In the present state of science, it is chiefly by the aid of shells that we are enabled to arrive at these results, for of all classes the testacea are the most generally diffused in a fossil state, and may be called the medals principally employed by nature, in recording the chronology of past events. In the Miocene deposits, which are next in succession to the Eocene, we begin to find a considerable number, although still a minority, of recent species, intermixed with some fossils common to the preceding epoch. We then arrive at the Pliocene strata, in which species now contemporary with man begin to preponderate, and in the newest of which nine-tenths of the fossils agree with species still inhabiting the neighboring sea.

In this passing from the older to the newer members of the tertiary system we meet with many chasms, but none which separate entirely, by a broad line of demarcation, one state of the organic world from another. There are no signs of an abrupt termination of one fauna and flora, and the starting into life of new and wholly distinct forms. Although we are far from being able to demonstrate geologically an insensible transition from the Eocene to the Miocene, or even from the latter to the recent fauna, yet the more we enlarge and perfect our general survey, the more nearly do we approximate to such a continuous series, and the more gradually are we conducted from times when many of the genera and nearly all the species were extinct, to those in which scarcely a single species flourished which we do not know to exist at present . . . .

It had often been objected that the evidence of fossil species occurring in two consecutive formations, was confined to the testacea or zoophytes, the characters of which are less marked and decisive than those afforded by the vertebrate animals. But Mr. Owen has lately insisted on the important fact, that not a few of the quadrupeds which now inhabit our island, and among others the horse, the ass, the hog, the smaller wild ox, the goat, the red deer, the roe, the beaver, and many of the diminutive rodents, are the same as those which once coexisted with the mammoth, the great northern hippopotamus, two kinds of rhinoceros, and other mammalia long since extinct. "A part," he observes, "and not the whole of the modern tertiary fauna has perished, and hence we may conclude that the cause of their destruction has not been a violent and universal catastrophe from which none could escape."

Had we discovered evidence that man had come into the earth at a period as early as that when a large number of the fossil quadrupeds now living, and almost all the recent species of land, freshwater, and marine shells were in existence, we should have been compelled to ascribe a much higher antiquity to our species, than even the boldest speculations of the ethnologist require, for no small part of the great physical revolution depicted on the map of Europe (Pl. 3), before described, took place very gradually after the recent testacea abounded almost to the exclusion of the extinct. Thus, for example, in the deposits called the "northern drift," or the glacial formation of Europe and North America, the fossil marine shells can easily be identified with species either now inhabiting the neighboring sea, or living in the seas of higher latitudes. Yet they exhibit no memorials of the human race, or of articles fabricated by the hand of man . . . .

To conclude, it appears that, in going back from the recent to the Eocene period, we are carried by many successive steps from the fauna now contemporary with man to an assemblage of fossil species wholly different from those now living. In this retrospect we have not yet succeeded in tracing back a perfect transition from the recent to an extinct fauna; but there are usually so many species in common to the groups which stand next in succession as to show that there is no great chasm, no signs of a crisis when one class of organic beings was annihilated to give place suddenly to another. This analogy, therefore, derived from a period of the earth's history which can best be compared with the present state of things, and more thoroughly investigated than any other, leads to the conclusion that the extinction and creation of species, has been and is the result of a slow and gradual change in the organic world.

Uniformity of Change Considered, Secondly, in Reference to Subterranean Movements

To pass on to another of the three topics before proposed for discussion, the reader will find, in the account given in the second book of the earthquakes recorded in history, that certain countries have, from time immemorial, been rudely shaken again and again, while others, comprising by far the largest part of the globe, have remained to all appearance motionless. In the regions of convulsion rocks have been rent asunder, the surface has been forced up into ridges, chasms have opened, or the ground throughout large spaces has been permanently lifted up above or let down below its former level. In the regions of tranquility some areas have remained at rest, but others have been ascertained by a comparison of measurements, made at different periods, to have risen by an insensible motion, as in Sweden, or to have subsided very slowly, as in Greenland. That these same movements, whether ascending or descending, have continued for ages in the same direction has been established by geological evidence . . . .

To detect proofs of slow and gradual subsidence must in general be more difficult; but the theory which accounts for the form of circular coral reefs and lagoon islands, and which will be explained in the last chapter of the third book, will satisfy the reader that there are spaces on the globe, several thousand miles in circumference, throughout which the downward movement has predominated for ages, and yet the land has never, in a single instance, gone down suddenly for several hundred feet at once. Yet geology demonstrates that the persistency of subterranean movements in one direction has not been perpetual throughout all past time. There have been great oscillations of level by which a surface of dry land has been submerged to a depth of several thousand feet, and then at a period long subsequent raised again and made to emerge. Nor have the regions now motionless been always at rest; and some of those which are at present the theatres of reiterated earthquakes have formerly enjoyed a long continuance of tranquility. But although disturbances have ceased after having long prevailed, or have recommenced after a suspension for ages, there has been no universal disruption of the earth's crust or desolation of the surface since times the most remote. The non-occurrence of such a general convulsion is proved by the perfect horizontality now retained by some of the most ancient fossiliferous strata throughout wide areas . . . .

Uniformity of Change Considered, Thirdly, in Reference to Sedimentary Deposition

It now remains to speak of the laws governing the deposition of new strata. If we survey the surface of the globe we immediately perceive that it is divisible into areas of deposition and non-deposition, or, in other words, at any given time there are spaces which are the recipients, others which are not the recipients of sedimentary matter. No new strata, for example, are thrown down on dry land, which remains the same from year to year; whereas, in many parts of the bottom of seas and lakes, mud, sand, and pebbles are annually spread out by rivers and currents. There are also great masses of limestone growing in some seas, or in mid-ocean, chiefly composed of corals and shells.

No sediment deposited on dry land.

As to the dry land, so far from being the receptacle of fresh accessions of matter, it is exposed almost everywhere to waste away. Forests may be as dense and lofty as those of Brazil, and may swarm with quadrupeds, birds, and insects, yet at the end of ten thousand years one layer of black mould, a few inches thick, may be the sole representative of those myriads of trees, leaves, flowers, and fruits, those innumerable bones and skeletons of birds, quadrupeds, and reptiles, which tenanted the fertile region. Should this land be at length submerged, the waves of the sea may wash away in a few hours the scanty covering of mould, and it may merely impart a darker shade of color to the next stratum of marl, sand, or other matter newly thrown down. So also at the bottom of the ocean where no sediment is accumulating, sea-weed, zoophytes, fish, and even shells, may multiply for ages and decompose, leaving no vestige of their form or substance behind. Their decay, in water, although more slow, is as certain and eventually as complete as in the open air. Nor can they be perpetuated for indefinite periods in a fossil state, unless imbedded in some matrix which is impervious to water, or which at least does not allow a free percolation of that fluid, impregnated as it usually is, with a slight quantity of carbonic or other acid. Such a free percolation may be prevented either by the mineral nature of the matrix itself, or by the superposition of an impermeable stratum: but if unimpeded, the fossil shell or bone will be dissolved and removed, particle after particle, and thus entirely effaced, unless petrifaction or the substitution of mineral for organic matter happen to take place.

That there has been land as well as sea at all former geological periods, we know from the fact, that fossil trees and terrestrial plants are imbedded in rocks of every age. Occasionally lacustrine and fluviatile shells, insects, or the bones of amphibious or land reptiles, point to the same conclusion. The existence of dry land at all periods of the past implies, as before mentioned, the partial deposition of sediment, or its limitation to certain areas; and the next point to which I shall call the reader's attention, is the shifting of these areas from one region to another.

. . .

Why successive sedimentary groups contain distinct fossils.

If, in the next place, we assume, for reasons before stated, a continual extinction of species and introduction of others into the globe, it will then follow that the fossils of strata formed at two distant periods on the same spot, will differ even more certainly than the mineral composition of the same. For rocks of the same kind have sometimes been reproduced in the same district after a long interval of time, whereas there are no facts leading to the opinion that species which have once died out have ever been reproduced. The submergence then of land must be often attended by the commencement of a new class of sedimentary deposits, characterized by a new set of fossil animals and plants, while the reconversion of the bed of the sea into land may arrest at once and for an indefinite time the formation of geological monuments. Should the land again sink, strata will again be formed; but one or many entire revolutions in animal or vegetable life may have been completed in the interval.

Conditions requisite for the original completeness of a fossiliferous series.

If we infer, for reasons before explained, that fluctuations in the animate world are brought about by the slow and successive removal and creation of species, we shall be convinced that a rare combination of circumstances alone can give rise to such a series of strata as will bear testimony to a gradual passage from one state of organic life to another. To produce such strata nothing less will be requisite than the fortunate coincidence of the following conditions: first, a never-failing supply of sediment in the same region throughout a period of vast duration; secondly, the fitness of the deposit in every part for the permanent preservation of imbedded fossils; and, thirdly, a gradual subsidence to prevent the sea or lake from being filled up and converted into land.

It will appear in the chapter on coral reefs, that, in certain parts of the Pacific and Indian Oceans, most of these conditions, if not all, are complied with, and the constant growth of coral, keeping pace with the sinking of the bottom of the sea, seems to have gone on so slowly, for such indefinite periods, that the signs of a gradual change in organic life might probably be detected in that quarter of the globe, if we could explore its submarine geology. Instead of the growth of coralline limestone, let us suppose, in some other place, the continuous deposition of fluviatile mud and sand, such as the Ganges and Brahmapootra have poured for thousands of years into the Bay of Bengal. Part of this bay, although of considerable depth, might at length be filled up before an appreciable amount of change was effected in the fish, mollusca, and other inhabitants of the sea and neighboring land. But, if the bottom be lowered by sinking at the same rate that it is raised by fluviatile mud, the bay can never be turned into dry land. In that case one new layer of matter may be superimposed upon another for a thickness of many thousand feet, and the fossils of the inferior beds may differ greatly from those entombed in the uppermost, yet every intermediate gradation may be indicated in the passage from an older to a newer assemblage of species. Granting, however, that such an unbroken sequence of monuments may thus be elaborated in certain parts of the sea, and that the strata happen to be all of them well adapted to preserve the included fossils from decomposition, how many accidents must still concur before these submarine formations will be laid open to our investigation! The whole deposit must first be raised several thousand feet, in order to bring into view the very foundation; and during the process of exposure the superior beds must not be entirely swept away by denudation.

In the first place, the chances are as three to one against the mere emergence of the mass above the waters, because three-fourths of the globe are covered by the ocean. But if it be upheaved and made to constitute part of the dry land, it must also, before it can be available for our instruction, become part of that area already surveyed by geologists; and this area comprehends perhaps less than a tenth of the whole earth. In this small fraction of land already explored, and still very imperfectly known, we are required to find a set of strata, originally of limited extent, and probably much lessened by subsequent denudation.

Yet it is precisely because we do not encounter at every step the evidence of such gradations from one state of the organic world to another, that so many geologists embrace the doctrine of great and sudden revolutions in the history of the animate world. Not content with simply availing themselves, for the convenience of classification, of those gaps and chasms which here and there interrupt the continuity of the chronological series, as at present known, they deduce, from the frequency of these breaks in the chain of records, an irregular mode of succession in the events themselves both in the organic and inorganic world. But, besides that some links of the chain which once existed are now clearly lost and others concealed from view, we have good reason to suspect that it was never complete originally. It may undoubtedly be said, that strata have been always forming somewhere, and therefore at every moment of past time nature has added a page to her archives; but, in reference to this subject, it should be remembered that we can never hope to compile a consecutive history by gathering together monuments which were originally detached and scattered over the globe. For as the species of organic beings contemporaneously inhabiting remote regions are distinct, the fossils of the first of several periods which may be preserved in any one country, as in America, for example, will have no connection with those of a second period found in India, and will therefore no more enable us to trace the signs of a gradual change in the living creation, than a fragment of Chinese history will fill up a blank in the political annals of Europe.

The absence of any deposits of importance containing recent shells in Chili, or anywhere on the western coast of South America, naturally led Mr. Darwin to the conclusion that "where the bed of the sea is either stationary or rising, circumstances are far less favorable than where the level is sinking to the accumulation of conchiferous strata of sufficient thickness and extension to resist the average vast amount of denudation." An examination of the superficial clay, sand, and gravel of the most modern date in Norway and Sweden, where the land is also rising, would incline us to admit a similar proposition. Yet in these cases there has been a supply of sediment from the waste of the coast and the interior, especially in Patagonia and Chili . . . .

Consistency of the theory of gradual change with the existence of great breaks in the series.

To return to the general argument pursued in this chapter, it is assumed, for reasons above explained, that a slow change of species is in simultaneous operation everywhere throughout the habitable surface of sea and land; whereas the fossilization of plants and animals is confined to those areas where new strata are produced. These areas, as we have seen, are always shifting their position; so that the fossilizing process, by means of which the commemoration of the particular state of the organic world, at any given time, is affected, may be said to move about, visiting and revisiting different tracts in succession.

To make still more clear the supposed working of this machinery, I shall compare it to a somewhat analogous case that might be imagined to occur in the history of human affairs ....

Suppose we had discovered two buried cities at the foot of Vesuvius, immediately superimposed upon each other, with a great mass of tuff and lava intervening, just as Portici and Resina, if now covered with ashes, would overlie Herculaneum. An antiquary might possibly be entitled to infer, from the inscriptions on public edifices, that the inhabitants of the inferior and older city were Greeks, and those of the modern towns Italians. But he would reason very hastily if he also concluded from these data that there had been a sudden change from the Greek to the Italian language in Campania. But if he afterwards found *three* buried cities, one above the other, the intermediate one being Roman, while, as in the former example, the lowest was Greek and the uppermost Italian, he would then perceive the fallacy of his former opinion, and would begin to suspect that the catastrophes by which the cities were inhumed might have no relation whatever to the fluctuations in the language of the inhabitants; and that, as the Roman tongue had evidently intervened between the Greek and Italian, so many other dialects may have been spoken in succession, and the passage from the Greek to the Italian may have been very gradual; some terms growing obsolete, while others were introduced from time to time.

If this antiquary could have shown that the volcanic paroxysms of Vesuvius were so governed as that cities should be buried one above the other, just as often as any variation occurred in the language of the inhabitants, then, indeed, the abrupt passage from a Greek to a Roman, and from a Roman to an Italian city, would afford proof of fluctuations no less sudden in the language of the people.

So, in Geology, if we could assume that it is part of the plan of Nature to preserve, in every region of the globe, an unbroken series of monuments to commemorate the vicissitudes of the organic creation, we might infer the sudden extirpation of species, and the simultaneous introduction of others, as often as two formations in contact are found to include dissimilar organic fossils. But we must shut our eyes to the whole economy of the existing causes, aqueous, igneous, and organic, if we fail to perceive *that such is not the plan of Nature*.

Concluding remarks on the identity of the ancient and present system of terrestrial changes.

I shall now conclude the discussion of a question with which we have been occupied since the beginning of the fifth chapter; namely, whether there has been any interruption, from the remotest periods, of one uniform system of change in the animate and inanimate world. We were induced to enter into that inquiry by reflecting how much the progress of opinion in Geology had been influenced by the assumption that the analogy was slight in kind, and still more slight in degree, between the causes which produced the former revolutions of the globe, and those now in every-day operation. It appeared clear that the earlier geologists had not only a scanty acquaintance with existing changes, but were singularly unconscious of the amount of their ignorance. With the presumption naturally inspired by this unconsciousness, they had no hesitation in deciding at once that time could never enable the existing powers of nature to work out changes of great magnitude, still less such important revolutions as those which are brought to light by Geology. They, therefore, felt themselves at liberty to indulge their imaginations in guessing at what *might be*, rather than inquiring *what is*; in other words, they employed themselves in conjecturing what might have been the course of nature at a remote period, rather than in the investigation of what was the course of nature in their own times.

It appeared to them more philosophical to speculate on the possibilities of the past, than patiently to explore the realities of the present; and having invented theories under the influence of such maxims, they were consistently unwilling to test their validity by the criterion of their accordance with the ordinary operations of nature . . . .

The course directly opposed to this method of philosophizing consists in an earnest and patient inquiry, how far geological appearances are reconcilable with the effect of changes now in progress, or which may be in progress in regions inaccessible to us, and of which the reality is attested by volcanoes and subterranean movements. It also endeavors to estimate the aggregate result of ordinary operations multiplied by time, and cherishes a sanguine hope that the resources to be derived from observation and experiment, or from the study of nature such as she now is, are very far from being exhausted. For this reason all theories are rejected which involve the assumption of sudden and violent catastrophes and revolutions of the whole earth, and its inhabitants,—theories which are restrained by no reference to existing analogies, and in which a desire is manifested to cut, rather than patiently to untie, the Gordian knot.

We have now, at least, the advantage of knowing, from experience, that an opposite method has always put geologists on the road that leads to truth,—suggesting views which, although imperfect at first, have been found capable of improvement, until at last adopted by universal consent; while the method of speculating on a former distinct state of things and causes, has led invariably to a multitude of contradictory systems, which have been overthrown one after the other,—have been found incapable of modification,—and which have often required to be precisely reversed.

The remainder of this work will be devoted to an investigation of the changes now going on in the crust of the earth and its inhabitants. The importance which the student will attach to such researches will mainly depend in the degree of confidence which he feels in the principles above expounded.

. . .

Reading and Discussion Questions

1. What does Lyell mean in saying that change has been uniform in the inorganic world? Compare and contrast Lyell's methodical approach in geology to that of Cuvier or others we have read.
2. Does Lyell believe that all of the animals that currently exist on the earth have been there since its inception? Does he believe in extinction? What evidence does he provide for his beliefs?
3. How does his emphasis on uniform change relate to what he says about slow and gradual processes? What are some of the many pieces of evidence that suggest to Lyell that the earth is almost unimaginably old?
4. What does Lyell say about the difficulty of forming fossils?

Additional Resources

Appleman, Philip, ed. *Darwin: A Norton Critical Edition*. 3rd ed. W. W. Norton and Co., 2001.

Gregory, Frederick. *Natural Science in Western History*. 1st ed. Cengage Learning, 2007. A solid secondary source textbook, by a past president of the History of Science Society, for the whole period covered by the readings in this volume.

Hodge, Jonathan and Radick, Gregory, eds. *The Cambridge Companion to Darwin*. 2nd ed. Cambridge: Cambridge University Press, 2009.

Laudan, Rachel. "The History of Geology, 1780–1840." In *Companion to the History of Modern Science*. M. J. S. Hodge, R. C. Olby, N. Cantor, and J. R. R. Christie (eds). London: Routledge, 1990.

Maienschein, Jane. "Cell Theory and Development." In *Companion to the History of Modern Science*. M. J. S. Hodge, R. C. Olby, N. Cantor, and J. R. R. Christie (eds). London: Routledge, 1990.

Richards, Robert J. *The Romantic Conception of Life: Science and Philosophy in the Age of Goethe*. Chicago: University of Chicago Press, 2002.

Sloan, Phillip R. "The Buffon-Linnaeus Controversy." *Isis*, 67(3) (September 1, 1976), 356–75.

Sloan, Phillip R. "Natural History, 1670–1802." In *Companion to the History of Modern Science*. M. J. S. Hodge, R. C. Olby, N. Cantor, and J. R. R. Christie (eds). London: Routledge, 1990.

Wear, Andrew. "The Heart and Blood from Vesalius to Harvey." In *Companion to the History of Modern Science*. M. J. S. Hodge, R. C. Olby, N. Cantor, and J. R. R. Christie (eds). London: Routledge, 1990.

van Wyhe, John. *Darwin Online.* http://darwin-online.org.uk/. A vast collection of Darwin's works, manuscripts, letters, reviewers, and texts that Darwin read and referenced in his published works.

Part VII

The Emergence of Evolution: Darwin and His Interlocutors

On the Tendency of Species to Form Varieties; and on the Perpetuation of Varieties and Species by Natural Means of Selection (1858)

Charles Darwin and
Alfred Wallace

93

...

Communicated by Sir Charles Lyell, F.R.S., F.L.S., and J. D. Hooker, Esq., M.D., V.P.R.S., F.L.S

London, June 30th, 1858.

My Dear Sir—The accompanying papers, which we have the honour of communicating to the Linnean Society, and which all relate to the same subject, viz. the Laws which affect the Production of Varieties, Races, and Species, contain the results of the investigations of two indefatigable naturalists, Mr. Charles Darwin and Mr. Alfred Wallace.

These gentlemen having, independently and unknown to one another, conceived the same very ingenious theory to account for the appearance and perpetuation of varieties and of specific forms on our planet, may both fairly claim the merit of being original thinkers in this important line of inquiry; but neither of them having published his views, though Mr. Darwin has for many years past been repeatedly urged by us to do so, and both authors having now unreservedly placed their papers in our hands, we think it would best promote the interests of science that a selection from them should be laid before the Linnean Society.

Taken in the order of their dates, they consist of:—

1. Extracts from a MS. work on Species, by Mr. Darwin, which was sketched in 1839, and copied in 1844, when the copy was read by Dr. Hooker, and its contents afterwards communicated to Sir Charles Lyell. The first Part is devoted to "The Variation of Organic Beings under Domestication and in their Natural State;" and the second chapter of that Part, from which we propose to read to the Society the extracts referred to, is headed, "On the Variation of Organic Beings in a state of Nature; on the Natural Means of Selection; on the Comparison of Domestic Races and true Species."
2. An abstract of a private letter addressed to Professor Asa Gray, of Boston, U.S., in October 1857, by Mr. Darwin, in which he repeats his views, and which shows that these remained unaltered from 1839 to 1857.
3. An Essay by Mr. Wallace, entitled "On the Tendency of Varieties to depart indefinitely from the Original Type." This was written at Ternate in February 1858, for the perusal of his friend and correspondent Mr. Darwin, and sent to him with the expressed wish that it should be forwarded to Sir Charles Lyell, if Mr. Darwin thought it sufficiently novel and interesting. So highly did Mr. Darwin appreciate the value of the views therein set forth, that he proposed, in a letter to Sir Charles Lyell, to obtain Mr. Wallace's consent to allow the Essay to be published as soon as possible. Of this step we highly approved, provided Mr. Darwin did not withhold from the public, as he was strongly inclined to do (in favour of Mr. Wallace), the memoir which he had himself written on the same subject, and which, as before stated, one of us had perused in 1844, and the contents of which we had both of us been privy to for many years. On representing this to Mr. Darwin, he gave us permission to make what use we thought proper of his memoir, &c.; and in adopting our present course, of presenting it to the Linnean Society, we have explained to him that we are not solely considering the relative claims to priority of himself and his friend, but the interests of science generally; for we feel it to be desirable that views founded on a wide deduction from facts, and matured by years of reflection, should constitute at once a goal from which others may start, and that, while the scientific world is waiting for the appearance of Mr. Darwin's complete work, some of the leading results of his labours, as well as those of his able correspondent, should together be laid before the public.

We have the honour to be yours very obediently,
Charles Lyell.
Jos. D. Hooker.
J. J. Bennett, Esq.,
Secretary of the Linnean Society.

. . .

III. On the Tendency of Varieties to Depart Indefinitely from the Original Type. By Alfred Russel Wallace

One of the strongest arguments which have been adduced to prove the original and permanent distinctness of species is, that varieties produced in a state of domesticity are more or less unstable, and often have a tendency, if left to themselves, to return to the normal form of the parent species; and this instability is considered to be a distinctive peculiarity of all varieties, even of those occurring among wild animals in a state of nature, and to constitute a provision for preserving unchanged the originally created distinct species.

In the absence or scarcity of facts and observations as to varieties occurring among wild animals, this argument has had great weight with naturalists, and has led to a very general and somewhat prejudiced belief in the stability of species. Equally general, however, is the belief in what are called "permanent or true varieties,"—races of animals which continually propagate their like, but which differ so slightly (although constantly) from some other race, that the one is considered to be a variety of the other. Which is the variety and which the original species, there is generally no means of determining, except in those rare cases in which the one race has been known to produce an offspring unlike itself and resembling the other. This, however, would seem quite incompatible with the "permanent invariability of species," but the difficulty is overcome by assuming that such varieties have strict limits, and can never again vary further from the original type, although they may return to it, which, from the analogy of the domesticated animals, is considered to be highly probable, if not certainly proved.

It will be observed that this argument rests entirely on the assumption, that varieties occurring in a state of nature are in all respects analogous to or even identical with those of domestic animals, and are governed by the same laws as regards their permanence or further variation. But it is the object of the present paper to show that this assumption is altogether false, that there is a general principle in nature which will cause many varieties to survive the parent species, and to give rise to successive variations departing further and further from the original type, and which also produces, in domesticated animals, the tendency of varieties to return to the parent form.

The life of wild animals is a struggle for existence. The full exertion of all their faculties and all their energies is required to preserve their own existence and provide for that of their infant offspring. The possibility of procuring food during the least favourable seasons, and of escaping the attacks of their most dangerous enemies, are the primary conditions which determine the existence both of individuals and of entire species. These conditions will also determine the population of a species; and by a careful consideration of all the circumstances we may be enabled to comprehend, and in some degree to explain, what at first sight appears so inexplicable—the excessive abundance of some species, while others closely allied to them are very rare.

The general proportion that must obtain between certain groups of animals is readily seen. Large animals cannot be so abundant as small ones; the carnivora must be less numerous than the herbivora; eagles and lions can never be so plentiful as pigeons and antelopes; the wild asses of the Tartarian deserts cannot equal in numbers the horses of the more luxuriant prairies and pampas of America. The greater or less fecundity of an animal is often considered to be one of the chief causes of its abundance or scarcity; but a consideration of the facts will show us that it really has little or nothing to do with the matter. Even the least prolific of animals would increase rapidly if unchecked, whereas it is evident that the animal population of the globe must be stationary, or perhaps, through the influence of man, decreasing. Fluctuations there may be; but permanent increase, except in restricted localities, is almost impossible. For example, our own observation must convince us that birds do not go on increasing every year in a geometrical ratio, as they would do, were there not some powerful check to their natural increase. Very few birds produce less than two young ones each year, while many have six, eight, or ten; four will certainly be below the average; and if we suppose that each pair produce young only four times in their life, that will also be below the average, supposing them not to die either by violence or want of food. Yet at this rate how tremendous would be the increase in a few years from a single pair! A simple calculation will show that in fifteen years each pair of birds would have increased to nearly ten millions! Whereas we have no reason to believe that the number of the birds of any country increases at all in fifteen or in one hundred and fifty years. With such powers of increase the population must have reached its limits, and have become stationary, in a very few years after the origin of each species. It is evident, therefore, that each year an immense number of birds must perish—as many in fact as are born; and as on the lowest calculation the progeny are each year twice as numerous as their parents, it follows that, whatever be the average number of individuals existing in any given country, twice that number must perish annually,—a striking result, but one which seems at least highly probable, and is perhaps under rather than over the truth. It would therefore appear that, as far as the continuance of the species and the keeping up the average number of individuals are concerned, large broods are superfluous. On the average all above one become food for hawks and kites, wild cats and weasels, or perish of cold and hunger as winter comes on. This is strikingly proved by the case of particular species; for we find that their abundance in individuals bears no relation whatever to their fertility in producing offspring. Perhaps the most remarkable instance of an immense bird population is that of the passenger pigeon of the United States, which lays only one, or at most two eggs, and is said to rear generally but one young one. Why is this bird so extraordinarily abundant, while others producing two or three times as many young are much less plentiful? The explanation is not difficult. The food most congenial to this species, and on which it thrives best, is abundantly distributed over a very extensive region, offering such differences of soil and climate, that in one part or another of the area the supply never fails. The bird is capable of a very rapid and long-continued flight, so that it can pass without fatigue over the whole of the district it inhabits, and as soon as the supply of food begins to fail in one place is able to discover a fresh feeding-ground. This example strikingly shows us that the procuring a constant supply of wholesome food is almost the sole condition requisite for ensuring the rapid increase of a given species, since neither the limited fecundity, nor the unrestrained attacks of birds of prey and of man are here sufficient to check it. In no other birds are these peculiar circumstances so strikingly combined. Either their food is more liable to failure, or they have not sufficient power of wing to search for it over an extensive area, or during some season of the year it becomes very scarce, and less wholesome substitutes have to be found; and thus, though more fertile in offspring, they can never increase beyond the supply of food in the least favourable seasons. Many birds can only exist by migrating, when their food becomes scarce, to regions possessing a milder, or at least a different climate, though, as these migrating birds are seldom excessively abundant, it is evident that the countries they visit are still deficient in a constant and abundant supply of wholesome food. Those whose organization does not permit them to migrate when their food becomes periodically scarce, can never attain a large population. This is probably the reason why woodpeckers are scarce with us, while in the tropics they are among the most abundant of solitary birds. Thus the house sparrow is more abundant than the redbreast, because its food is more constant and plentiful,—seeds of grasses being preserved during the winter, and our farm-yards and stubble-fields furnishing an almost inexhaustible supply. Why, as a general rule, are aquatic, and especially sea birds, very numerous in individuals? Not because they are more prolific than others, generally the contrary; but because their food never fails, the sea-shores and river-banks daily swarming with a fresh supply of small mollusca and crustacea. Exactly the same laws will apply to mammals. Wild cats are prolific and have few enemies; why then are they never as abundant as rabbits? The only intelligible answer is, that their supply of food is more precarious. It appears evident, therefore, that so long as a country remains physically unchanged, the numbers of its animal population cannot materially increase. If one species does so, some others requiring the same kind of food must diminish in proportion. The numbers that die annually must be immense; and as the individual existence of each animal depends upon itself, those that die must be the weakest—the very young, the aged, and the diseased,—while those that prolong their existence can only be the most perfect in health and vigour—those who are best able to obtain food regularly, and avoid their numerous enemies. It is, as we commenced by remarking, "a struggle for existence," in which the weakest and least perfectly organized must always succumb.

Now it is clear that what takes place among the individuals of a species must also occur among the several allied species of a group,—viz. that those which are best adapted to obtain a regular supply of food, and to defend themselves against the attacks of their enemies and the vicissitudes of the seasons, must necessarily obtain and preserve a superiority in population; while those species which from some defect of power or organization are the least capable of counteracting the vicissitudes of food, supply, &c., must diminish in numbers, and, in extreme cases, become altogether extinct. Between these extremes the species will present various degrees of capacity for ensuring the means of preserving life; and it is thus we account for the abundance or rarity of species. Our ignorance will generally prevent us from accurately tracing the effects to their causes; but could we become perfectly acquainted with the organization and habits of the various species of animals, and could we measure the capacity of each for performing the different acts necessary to its safety and existence under all the varying circumstances by which it is surrounded, we might be able even to calculate the proportionate abundance of individuals which is the necessary result.

If now we have succeeded in establishing these two points—1st, that the animal population of a country is generally stationary, being kept down by a periodical deficiency of food, and other checks; and, 2nd, that the comparative abundance or scarcity of the individuals of the several species is entirely due to their organization and resulting habits, which, rendering it more difficult to procure a regular supply of food and to provide for their personal safety in some cases than in others, can only be balanced by a difference in the population which have to exist in a given area—we shall be in a condition to proceed to the consideration of varieties, to which the preceding remarks have a direct and very important application.

Most or perhaps all the variations from the typical form of a species must have some definite effect, however slight, on the habits or capacities of the individuals. Even a change of colour might, by rendering them more or less distinguishable, affect their safety; a greater or less development of hair might modify their habits. More important changes, such as an increase in the power or dimensions of the limbs or any of the external organs, would more or less affect their mode of procuring food or the range of country which they inhabit. It is also evident that most changes would affect, either favourably or adversely, the powers of prolonging existence. An antelope with shorter or weaker legs must necessarily suffer more from the attacks of the feline carnivora; the passenger pigeon with less powerful wings would sooner or later be affected in its powers of procuring a regular supply of food; and in both cases the result must necessarily be a diminution of the population of the modified species. If, on the other hand, any species should produce a variety having slightly increased powers of preserving existence, that variety must inevitably in time acquire a superiority in numbers. These results must follow as surely as old age, intemperance, or scarcity of food produce an increased mortality. In both cases there may be many individual exceptions; but on the average the rule will invariably be found to hold good. All varieties will therefore fall into two classes—those which under the same conditions would never reach the population of the parent species, and those which would in time obtain and keep a numerical superiority. Now, let some alteration of physical conditions occur in the district—a long period of drought, a destruction of vegetation by locusts, the irruption of some new carnivorous animal seeking "pastures new"—any change in fact tending to render existence more difficult to the species in question, and tasking its utmost powers to avoid complete extermination; it is evident that, of all the individuals composing the species, those forming the least numerous and most feebly organized variety would suffer first, and, were the pressure severe, must soon become extinct. The same causes continuing in action, the parent species would next suffer, would gradually diminish in numbers, and with a recurrence of similar unfavourable conditions might also become extinct. The superior variety would then alone remain, and on a return to favourable circumstances would rapidly increase in numbers and occupy the place of the extinct species and variety.

The variety would now have replaced the species, of which it would be a more perfectly developed and more highly organized form. It would be in all respects better adapted to secure its safety, and to prolong its individual existence and that of the race. Such a variety could not return to the original form; for that form is an inferior one, and could never compete with it for existence. Granted, therefore, a "tendency" to reproduce the original type of the species, still the variety must ever remain preponderant in numbers, and under adverse physical conditions again alone survive.

But this new, improved, and populous race might itself, in course of time, give rise to new varieties, exhibiting several diverging modifications of form, any of which, tending to increase the facilities for preserving existence, must, by the same general law, in their turn become predominant. Here, then, we have progression and continued divergence deduced from the general laws which regulate the existence of animals in a state of nature, and from the undisputed fact that varieties do frequently occur. It is not, however, contended that this result would be invariable; a change of physical conditions in the district might at times materially modify it, rendering the race which had been the most capable of supporting existence under the former conditions now the least so, and even causing the extinction of the newer and, for a time, superior race, while the old or parent species and its first inferior varieties continued to flourish. Variations in unimportant parts might also occur, having no perceptible effect on the life-preserving powers; and the varieties so furnished might run a course parallel with the parent species, either giving rise to further variations or returning to the former type. All we argue for is, that certain varieties have a tendency to maintain their existence longer than the original species, and this tendency must make itself felt; for though the doctrine of chances or averages can never be trusted to on a limited scale, yet, if applied to high numbers, the results come nearer to what theory demands, and, as we approach to an infinity of examples, become strictly accurate. Now the scale on which nature works is so vast—the numbers of individuals and periods of time with which she deals approach so near to infinity, that any cause, however slight, and however liable to be veiled and counteracted by accidental circumstances, must in the end produce its full legitimate results.

Let us now turn to domesticated animals, and inquire how varieties produced among them are affected by the principles here enunciated. The essential difference in the condition of wild and domestic animals is this,—that among the former, their well-being and very existence depend upon the full exercise and healthy condition of all their senses and physical powers, whereas, among the latter, these are only partially exercised, and in some cases are absolutely unused. A wild animal has to search, and often to labour, for every mouthful of food—to exercise sight, hearing, and smell in seeking it, and in avoiding dangers, in procuring shelter from the inclemency of the seasons, and in providing for the subsistence and safety of its offspring. There is no muscle of its body that is not called into daily and hourly activity; there is no sense or faculty that is not strengthened by continual exercise. The domestic animal, on the other hand, has food provided for it, is sheltered, and often confined, to guard it against the vicissitudes of the seasons, is carefully secured from the attacks of its natural enemies, and seldom even rears its young without human assistance. Half of its senses and faculties are quite useless; and the other half are but occasionally called into feeble exercise, while even its muscular system is only irregularly called into action.

Now when a variety of such an animal occurs, having increased power or capacity in any organ or sense, such increase is totally useless, is never called into action, and may even exist without the animal ever becoming aware of it. In the wild animal, on the contrary, all its faculties and powers being brought into full action for the necessities of existence, any increase becomes immediately available, is strengthened by exercise, and must even slightly modify the food, the habits, and the whole economy of the race. It creates as it were a new animal, one of superior powers, and which will necessarily increase in numbers and outlive those inferior to it.

Again, in the domesticated animal all variations have an equal chance of continuance; and those which would decidedly render a wild animal unable to compete with its fellows and continue its existence are no disadvantage whatever in a state of domesticity. Our quickly fattening pigs, short-legged sheep, pouter pigeons, and poodle dogs could never have come into existence in a state of nature, because the very first step towards such inferior forms would have led to the rapid extinction of the race; still less could they now exist in competition with their wild allies. The great speed but slight endurance of the race horse, the unwieldy strength of the ploughman's team, would both be useless in a state of nature. If turned wild on the pampas, such animals would probably soon become extinct, or under favourable circumstances might each lose those extreme qualities which would never be called into action, and in a few generations would revert to a common type, which must be that in which the various powers and faculties are so proportioned to each other as to be best adapted to procure food and secure safety,—that in which by the full exercise of every part of his organization the animal can alone continue to live. Domestic varieties, when turned wild, must return to something near the type of the original wild stock, or become altogether extinct.

We see, then, that no inferences as to varieties in a state of nature can be deduced from the observation of those occurring among domestic animals. The two are so much opposed to each other in every circumstance of their existence, that what applies to the one is almost sure not to apply to the other. Domestic animals are abnormal, irregular, artificial; they are subject to varieties which never occur and never can occur in a state of nature: their very existence depends altogether on human care; so far are many of them removed from that just proportion of faculties, that true balance of organization, by means of which alone an animal left to its own resources can preserve its existence and continue its race.

The hypothesis of Lamarck—that progressive changes in species have been produced by the attempts of animals to increase the development of their own organs, and thus modify their structure and habits—has been repeatedly and easily refuted by all writers on the subject of varieties and species, and it seems to have been considered that when this was done the whole question has been finally settled; but the view here developed renders such an hypothesis quite unnecessary, by showing that similar results must be produced by the action of principles constantly at work in nature. The powerful retractile talons of the falcon- and the cat-tribes have not been produced or increased by the volition of those animals; but among the different varieties which occurred in the earlier and less highly organized forms of these groups, those always survived longest which had the greatest facilities for seizing their prey. Neither did the giraffe acquire its long neck by desiring to reach the foliage of the more lofty shrubs, and constantly stretching its neck for the purpose, but because any varieties which occurred among its antitypes with a longer neck than usual at once secured a fresh range of pasture over the same ground as their shorter-necked companions, and on the first scarcity of food were thereby enabled to outlive them. Even the peculiar colours of many animals, especially insects, so closely resembling the soil or the leaves or the trunks on which they habitually reside, are explained on the same principle; for though in the course of ages varieties of many tints may have occurred, yet those races having colours best adapted to concealment from their enemies would inevitably survive the longest. We have also here an acting cause to account for that balance so often observed in nature,—a deficiency in one set of organs always being compensated by an increased development of some others—powerful wings accompanying weak feet, or great velocity making up for the absence of defensive weapons; for it has been shown that all varieties in which an unbalanced deficiency occurred could not long continue their existence. The action of this principle is exactly like that of the centrifugal governor of the steam engine, which checks and corrects any irregularities almost before they become evident; and in like manner no unbalanced deficiency in the animal kingdom can ever reach any conspicuous magnitude, because it would make itself felt at the very first step, by rendering existence difficult and extinction almost sure soon to follow. An origin such as is here advocated will also agree with the peculiar character of the modifications of form and structure which obtain in organized beings—the many lines of divergence from a central type, the increasing efficiency and power of a particular organ through a succession of allied species, and the remarkable persistence of unimportant parts such as colour, texture of plumage and hair, form of horns or crests, through a series of species differing considerably in more essential characters. It also furnishes us with a reason for that "more specialized structure" which Professor Owen states to be a characteristic of recent compared with extinct forms, and which would evidently be the result of the progressive modification of any organ applied to a special purpose in the animal economy.

We believe we have now shown that there is a tendency in nature to the continued progression of certain classes of varieties further and further from the original type—a progression to which there appears no reason to assign any definite limits—and that the same principle which produces this result in a state of nature will also explain why domestic varieties have a tendency to revert to the original type. This progression, by minute steps, in various directions, but always checked and balanced by the necessary conditions, subject to which alone existence can be preserved, may, it is believed, be followed out so as to agree with all the phenomena presented by organized beings, their extinction and succession in past ages, and all the extraordinary modifications of form, instinct, and habits which they exhibit.

Reading and Discussion Questions

1. After Wallace sent Darwin a manuscript describing his ideas in 1858, arrangements were made by Charles Lyell and Joseph Hooker for it to be presented jointly with a paper by Darwin at the Linnaean Society of London. Carefully explain the theory that Alfred Russel Wallace sets out in this essay.
2. In what important respects, if any, does Wallace's theory differ from the one we will see Darwin lay out in the next reading, *On the Origin of Species* (1859) and what does Darwin say about Wallace in the preface to that great work?

On the Origin of Species: *By Means of Natural Selection, or the Preservation of Favoured Races in the Struggle for Life* (1859)

Charles Darwin

94

Introduction

When on board H.M.S. 'Beagle,' as naturalist, I was much struck with certain facts in the distribution of the inhabitants of South America, and in the geological relations of the present to the past inhabitants of that continent. These facts seemed to me to throw some light on the origin of species—that mystery of mysteries, as it has been called by one of our greatest philosophers. On my return home, it occurred to me, in 1837, that something might perhaps be made out on this question by patiently accumulating and reflecting on all sorts of facts which could possibly have any bearing on it. After five years' work I allowed myself to speculate on the subject, and drew up some short notes; these I enlarged in 1844 into a sketch of the conclusions, which then seemed to me probable: from that period to the present day I have steadily pursued the same object. I hope that I may be excused for entering on these personal details, as I give them to show that I have not been hasty in coming to a decision.

My work is now nearly finished; but as it will take me two or three more years to complete it, and as my health is far from strong, I have been urged to publish this Abstract. I have more especially been induced to do this, as Mr. Wallace, who is now studying the natural history of the Malay archipelago, has arrived at almost exactly the same general conclusions that I have on the origin of species. Last year he sent to me a memoir on this subject, with a request that I would forward it to Sir Charles Lyell, who sent it to the Linnean Society, and it is published in the third volume of the Journal of that Society. Sir C. Lyell and Dr. Hooker, who both knew of my work—the latter having read my sketch of 1844—honoured me by thinking it advisable to publish, with Mr. Wallace's excellent memoir, some brief extracts from my manuscripts.

This Abstract, which I now publish, must necessarily be imperfect. I cannot here give references and authorities for my several statements; and I must trust to the reader reposing some confidence in my accuracy. No doubt errors will have crept in, though I hope I have always been cautious in trusting to good authorities alone. I can here give only the general conclusions at which I have arrived, with a few facts in illustration, but which, I hope, in most cases will suffice. No one can feel more sensible than I do of the necessity of hereafter publishing in detail all the facts, with references, on which my conclusions have been grounded; and I hope in a future work to do this. For I am well aware that scarcely a single point is discussed in this volume on which facts cannot be adduced, often apparently leading to conclusions directly opposite to those at which I have arrived. A fair result can be obtained only by fully stating and balancing the facts and arguments on both sides of each question; and this cannot possibly be here done.

I much regret that want of space prevents my having the satisfaction of acknowledging the generous assistance which I have received from very many naturalists, some of them personally unknown to me. I cannot, however, let this opportunity pass without expressing my deep obligations to Dr. Hooker, who for the last fifteen years has aided me in every possible way by his large stores of knowledge and his excellent judgment.

In considering the Origin of Species, it is quite conceivable that a naturalist, reflecting on the mutual affinities of organic beings, on their embryological relations, their geographical distribution, geological succession, and other such facts, might come to the conclusion that each species had not been independently created, but had descended, like varieties, from other species. Nevertheless, such a conclusion, even if well founded, would be unsatisfactory, until it could be shown how the innumerable species inhabiting this world have been modified, so as to acquire that perfection of structure and coadaptation which most justly excites our admiration. Naturalists continually refer to external conditions, such as climate, food, etc., as the only possible cause of variation. In one very limited sense, as we shall hereafter see, this may be true; but it is preposterous to attribute to mere external conditions, the structure, for instance, of the woodpecker, with its feet, tail, beak, and tongue, so admirably adapted to catch insects under the bark of trees. In the case of the mistletoe, which draws its nourishment from certain trees, which has seeds that must be transported by certain birds, and which has flowers with separate sexes absolutely requiring the agency of certain insects to bring pollen from one flower to the other, it is equally preposterous to account for the structure of this parasite, with its relations to several distinct organic beings, by the effects of external conditions, or of habit, or of the volition of the plant itself.

The author of the 'Vestiges of Creation' would, I presume, say that, after a certain unknown number of generations, some bird had given birth to a woodpecker, and some plant to the mistletoe, and that these had been produced perfect as we now see them; but this assumption seems to me to be no explanation, for it leaves the case of the coadaptations of organic beings to each other and to their physical conditions of life, untouched and unexplained.

It is, therefore, of the highest importance to gain a clear insight into the means of modification and coadaptation. At the commencement of my observations it seemed to me probable that a careful study of domesticated animals and of cultivated plants would offer the best chance of making out this obscure problem. Nor have I been disappointed; in this and in all other perplexing cases I have invariably found that our knowledge, imperfect though it be, of variation under domestication, afforded the best and safest clue. I may venture to express my conviction of the high value of such studies, although they have been very commonly neglected by naturalists.

From these considerations, I shall devote the first chapter of this Abstract to Variation under Domestication. We shall thus see that a large amount of hereditary modification is at least possible, and, what is equally or more important, we shall see how great is the power of man in accumulating by his Selection successive slight variations. I will then pass on to the variability of species in a state of nature; but I shall, unfortunately, be compelled to treat this subject far too briefly, as it can be treated properly only by giving long catalogues of facts. We shall, however, be enabled to discuss what circumstances are most favourable to variation. In the next chapter the Struggle for Existence amongst all organic beings throughout the world, which inevitably follows from their high geometrical powers of increase, will be treated of. This is the doctrine of Malthus, applied to the whole animal and vegetable kingdoms. As many more individuals of each species are born than can possibly survive; and as, consequently, there is a frequently recurring struggle for existence, it follows that any being, if it vary however slightly in any manner profitable to itself, under the complex and sometimes varying conditions of life, will have a better chance of surviving, and thus be NATURALLY SELECTED. From the strong principle of inheritance, any selected variety will tend to propagate its new and modified form.

This fundamental subject of Natural Selection will be treated at some length in the fourth chapter; and we shall then see how Natural Selection almost inevitably causes much Extinction of the less improved forms of life and induces what I have called Divergence of Character. In the next chapter I shall discuss the complex and little known laws of variation and of correlation of growth. In the four succeeding chapters, the most apparent and gravest difficulties on the theory will be given: namely, first, the difficulties of transitions, or in understanding how a simple being or a simple organ can be changed and perfected into a highly developed being or elaborately constructed organ; secondly the subject of Instinct, or the mental powers of animals, thirdly, Hybridism, or the infertility of species and the fertility of varieties when intercrossed; and fourthly, the imperfection of the Geological Record. In the next chapter I shall consider the geological succession of organic beings throughout time; in the eleventh and twelfth, their geographical distribution throughout space; in the thirteenth, their classification or mutual affinities, both when mature and in an embryonic condition. In the last chapter I shall give a brief recapitulation of the whole work, and a few concluding remarks.

No one ought to feel surprise at much remaining as yet unexplained in regard to the origin of species and varieties, if he makes due allowance for our profound ignorance in regard to the mutual relations of all the beings which live around us. Who can explain why one species ranges widely and is very numerous, and why another allied species has a narrow range and is rare? Yet these relations are of the highest importance, for they determine the present welfare, and, as I believe, the future success and modification of every inhabitant of this world. Still less do we know of the mutual relations of the innumerable inhabitants of the world during the many past geological epochs in its history. Although much remains obscure, and will long remain obscure, I can entertain no doubt, after the most deliberate study and dispassionate judgment of which I am capable, that the view which most naturalists entertain, and which I formerly entertained—namely, that each species has been independently created—is erroneous. I am fully convinced that species are not immutable; but that those belonging to what are called the same genera are lineal descendants of some other and generally extinct species, in the same manner as the acknowledged varieties of any one species are the descendants of that species. Furthermore, I am convinced that Natural Selection has been the main but not exclusive means of modification.

1. Variation under Domestication

Causes of Variability. Effects of Habit. Correlation of Growth. Inheritance. Character of Domestic Varieties. Difficulty of distinguishing between Varieties and Species. Origin of Domestic Varieties from one or more Species. Domestic Pigeons, their Differences and Origin. Principle of Selection anciently followed, its Effects. Methodical and Unconscious Selection. Unknown Origin of our Domestic Productions. Circumstances favourable to Man's power of Selection.

When we look to the individuals of the same variety or sub-variety of our older cultivated plants and animals, one of the first points which strikes us, is, that they generally differ much more from each other, than do the individuals of any one species or variety in a state of nature. When we reflect on the vast diversity of the plants and animals which have been cultivated, and which have varied during all ages under the most different climates and treatment, I think we are driven to conclude that this greater variability is simply due to our domestic productions having been raised under conditions of life not so uniform as, and somewhat different from, those to which the parent-species have been exposed under nature. There is, also, I think, some probability in the view propounded by Andrew Knight, that this variability may be partly connected with excess of food. It seems pretty clear that organic beings must be exposed during several generations to the new conditions of life to cause any appreciable amount of variation; and that when the organisation has once begun to vary, it generally continues to vary for many generations. No case is on record of a variable being ceasing to be variable under cultivation. Our oldest cultivated plants, such as wheat, still often yield new varieties: our oldest domesticated animals are still capable of rapid improvement or modification.

It has been disputed at what period of life the causes of variability, whatever they may be, generally act; whether during the early or late period of development of the embryo, or at the instant of conception. Geoffroy St. Hilaire's experiments show that unnatural treatment of the embryo causes monstrosities; and monstrosities cannot be separated by any clear line of distinction from mere variations. But I am strongly inclined to suspect that the most frequent cause of variability may be attributed to the male and female reproductive elements having been affected prior to the act of conception . . . .

Sterility has been said to be the bane of horticulture; but on this view we owe variability to the same cause which produces sterility; and variability is the source of all the choicest productions of the garden. I may add, that as some organisms will breed most freely under the most unnatural conditions (for instance, the rabbit and ferret kept in hutches), showing that their reproductive system has not been thus affected; so will some animals and plants withstand domestication or cultivation, and vary very slightly—perhaps hardly more than in a state of nature.

A long list could easily be given of "sporting plants"; by this term gardeners mean a single bud or offset, which suddenly assumes a new and sometimes very different character from that of the rest of the plant. Such buds can be propagated by grafting, etc., and sometimes by seed. These "sports" are extremely rare under nature, but far from rare under cultivation; and in this case we see that the treatment of the parent has affected a bud or offset, and not the ovules or pollen. But it is the opinion of most physiologists that there is no essential difference between a bud and an ovule in their earliest stages of formation; so that, in fact, "sports" support my view, that variability may be largely attributed to the ovules or pollen, or to both, having been affected by the treatment of the parent prior to the act of conception. These cases anyhow show that variation is not necessarily connected, as some authors have supposed, with the act of generation.

Seedlings from the same fruit, and the young of the same litter, sometimes differ considerably from each other, though both the young and the parents, as Muller has remarked, have apparently been exposed to exactly the same conditions of life; and this shows how unimportant the direct effects of the conditions of life are in comparison with the laws of reproduction, and of growth, and of inheritance; for had the action of the conditions been direct, if any of the young had varied, all would probably have varied in the same manner . . . . Nevertheless some slight amount of change may, I think, be attributed to the direct action of the conditions of life—as, in some cases, increased size from amount of food, colour from particular kinds of food and from light, and perhaps the thickness of fur from climate.

Habit also has a decided influence, as in the period of flowering with plants when transported from one climate to another. In animals it has a more marked effect; for instance, I find in the domestic duck that the bones of the wing weigh less and the bones of the leg more, in proportion to the whole skeleton, than do the same bones in the wild-duck; and I presume that this change may be safely attributed to the domestic duck flying much less, and walking more, than its wild parent. The great and inherited development of the udders in cows and goats in countries where they are habitually milked, in comparison with the state of these organs in other countries, is another instance of the effect of use. Not a single domestic animal can be named which has not in some country drooping ears; and the view suggested by some authors, that the drooping is due to the disuse of the muscles of the ear, from the animals not being much alarmed by danger, seems probable.

There are many laws regulating variation, some few of which can be dimly seen, and will be hereafter briefly mentioned. I will here only allude to what may be called correlation of growth. Any change in the embryo or larva will almost certainly entail changes in the mature animal. In monstrosities, the correlations between quite distinct parts are very curious; and many instances are given in Isidore Geoffroy St. Hilaire's great work on this subject. Breeders believe that long limbs are almost always accompanied by an elongated head. Some instances of correlation are quite whimsical; thus cats with blue eyes are invariably deaf; colour and constitutional peculiarities go together, of which many remarkable cases could be given amongst animals and plants. From the facts collected by Heusinger, it appears that white sheep and pigs are differently affected from coloured individuals by certain vegetable poisons. Hairless dogs have imperfect teeth; long-haired and coarse-haired animals are apt to have, as is asserted, long or many horns; pigeons with feathered feet have skin between their outer toes; pigeons with short beaks have small feet, and those with long beaks large feet. Hence, if man goes on selecting, and thus augmenting, any peculiarity, he will almost certainly unconsciously modify other parts of the structure, owing to the mysterious laws of the correlation of growth.

The result of the various, quite unknown, or dimly seen laws of variation is infinitely complex and diversified . . . .

Any variation which is not inherited is unimportant for us. But the number and diversity of inheritable deviations of structure, both those of slight and those of considerable physiological importance, is endless. Dr. Prosper Lucas's treatise, in two large volumes, is the fullest and the best on this subject. No breeder doubts how strong is the tendency to inheritance: like produces like is his fundamental belief: doubts have been thrown on this principle by theoretical writers alone. When a deviation appears not unfrequently, and we see it in the father and child, we cannot tell whether it may not be due to the same original cause acting on both; but when amongst individuals, apparently exposed to the same conditions, any very rare deviation, due to some extraordinary combination of circumstances, appears in the parent—say, once amongst several million individuals—and it reappears in the child, the mere doctrine of chances almost compels us to attribute its reappearance to inheritance. Everyone must have heard of cases of albinism, prickly skin, hairy bodies, etc., appearing in several members of the same family. If strange and rare deviations of structure are truly inherited, less strange and commoner deviations may be freely admitted to be inheritable. Perhaps the correct way of viewing the whole subject, would be, to look at the inheritance of every character whatever as the rule, and non-inheritance as the anomaly.

The laws governing inheritance are quite unknown; no one can say why the same peculiarity in different individuals of the same species, and in individuals of different species, is sometimes inherited and sometimes not so; why the child often reverts in certain characters to its grandfather or grandmother or other much more remote ancestor; why a peculiarity is often transmitted from one sex to both sexes or to one sex alone, more commonly but not exclusively to the like sex. It is a fact of some little importance to us, that peculiarities appearing in the males of our domestic breeds are often transmitted either exclusively, or in a much greater degree, to males alone. A much more important rule, which I think may be trusted, is that, at whatever period of life a peculiarity first appears, it tends to appear in the offspring at a corresponding age, though sometimes earlier. In many cases this could not be otherwise: thus the inherited peculiarities in the horns of cattle could appear only in the offspring when nearly mature; peculiarities in the silkworm are known to appear at the corresponding caterpillar or cocoon stage. But hereditary diseases and some other facts make me believe that the rule has a wider extension, and that when there is no apparent reason why a peculiarity should appear at any particular age, yet that it does tend to appear in the offspring at the same period at which it first appeared in the parent. I believe this rule to be of the highest importance in explaining the laws of embryology. These remarks are of course confined to the first APPEARANCE of the peculiarity, and not to its primary cause, which may have acted on the ovules or male element; in nearly the same manner as in the crossed offspring from a short-horned cow by a long-horned bull, the greater length of horn, though appearing late in life, is clearly due to the male element.

Having alluded to the subject of reversion, I may here refer to a statement often made by naturalists—namely, that our domestic varieties, when run wild, gradually but certainly revert in character to their aboriginal stocks. Hence it has been argued that no deductions can be drawn from domestic races to species in a state of nature. I have in vain endeavoured to discover on what decisive facts the above statement has so often and so boldly been made. There would be great difficulty in proving its truth: we may safely conclude that very many of the most strongly-marked domestic varieties could not possibly live in a wild state. In many cases we do not know what the aboriginal stock was, and so could not tell whether or not nearly perfect reversion had ensued. It would be quite necessary, in order to prevent the effects of intercrossing, that only a single variety should be turned loose in its new home. Nevertheless, as our varieties certainly do occasionally revert in some of their characters to ancestral forms, it seems to me not improbable, that if we could succeed in naturalising, or were to cultivate, during many generations, the several races, for instance, of the cabbage, in very poor soil (in which case, however, some effect would have to be attributed to the direct action of the poor soil), that they would to a large extent, or even wholly, revert to the wild aboriginal stock. Whether or not the experiment would succeed, is not of great importance for our line of argument; for by the experiment itself the conditions of life are changed. If it could be shown that our domestic varieties manifested a strong tendency to reversion,—that is, to lose their acquired characters, whilst kept under unchanged conditions, and whilst kept in a considerable body, so that free intercrossing might check, by blending together, any slight deviations of structure, in such case, I grant that we could deduce nothing from domestic varieties in regard to species. But there is not a shadow of evidence in favour of this view: to assert that we could not breed our cart and race-horses, long and short-horned cattle, and poultry of various breeds, and esculent vegetables, for an almost infinite number of generations, would be opposed to all experience. I may add, that when under nature the conditions of life do change, variations and reversions of character probably do occur; but natural selection, as will hereafter be explained, will determine how far the new characters thus arising shall be preserved.

...

When we attempt to estimate the amount of structural difference between the domestic races of the same species, we are soon involved in doubt, from not knowing whether they have descended from one or several parent-species. This point, if it could be cleared up, would be interesting; if, for instance, it could be shown that the greyhound, bloodhound, terrier, spaniel, and bull-dog, which we all know propagate their kind so truly, were the offspring of any single species, then such facts would have great weight in making us doubt about the immutability of the many very closely allied and natural species—for instance, of the many foxes—inhabiting different quarters of the world. I do not believe, as we shall presently see, that all our dogs have descended from any one wild species; but, in the case of some other domestic races, there is presumptive, or even strong, evidence in favour of this view.

. . .

Even in the case of the domestic dogs of the whole world, which I fully admit have probably descended from several wild species, I cannot doubt that there has been an immense amount of inherited variation. Who can believe that animals closely resembling the Italian greyhound, the bloodhound, the bull-dog, or Blenheim spaniel, etc.—so unlike all wild Canidae—ever existed freely in a state of nature? It has often been loosely said that all our races of dogs have been produced by the crossing of a few aboriginal species; but by crossing we can get only forms in some degree intermediate between their parents; and if we account for our several domestic races by this process, we must admit the former existence of the most extreme forms, as the Italian greyhound, bloodhound, bull-dog, etc., in the wild state. Moreover, the possibility of making distinct races by crossing has been greatly exaggerated. There can be no doubt that a race may be modified by occasional crosses, if aided by the careful selection of those individual mongrels, which present any desired character; but that a race could be obtained nearly intermediate between two extremely different races or species, I can hardly believe. Sir J. Sebright expressly experimentised for this object, and failed. The offspring from the first cross between two pure breeds is tolerably and sometimes (as I have found with pigeons) extremely uniform, and everything seems simple enough; but when these mongrels are crossed one with another for several generations, hardly two of them will be alike, and then the extreme difficulty, or rather utter hopelessness, of the task becomes apparent. Certainly, a breed intermediate between TWO VERY DISTINCT breeds could not be got without extreme care and long-continued selection; nor can I find a single case on record of a permanent race having been thus formed.

On the Breeds of the Domestic Pigeon

Believing that it is always best to study some special group, I have, after deliberation, taken up domestic pigeons. I have kept every breed which I could purchase or obtain, and have been most kindly favoured with skins from several quarters of the world, more especially by the Honourable W. Elliot from India, and by the Honourable C. Murray from Persia.

. . .

Great as the differences are between the breeds of pigeons, I am fully convinced that the common opinion of naturalists is correct, namely, that all have descended from the rock-pigeon (Columba livia), including under this term several geographical races or sub-species, which differ from each other in the most trifling respects.

. . .

From these several reasons, namely, the improbability of man having formerly got seven or eight supposed species of pigeons to breed freely under domestication;

these supposed species being quite unknown in a wild state, and their becoming nowhere feral; these species having very abnormal characters in certain respects, as compared with all other Columbidae, though so like in most other respects to the rock-pigeon; the blue colour and various marks occasionally appearing in all the breeds, both when kept pure and when crossed; the mongrel offspring being perfectly fertile;—from these several reasons, taken together, I can feel no doubt that all our domestic breeds have descended from the Columba livia with its geographical sub-species.

. . .

I have discussed the probable origin of domestic pigeons at some, yet quite insufficient, length; because when I first kept pigeons and watched the several kinds, knowing well how true they bred, I felt fully as much difficulty in believing that they could ever have descended from a common parent, as any naturalist could in coming to a similar conclusion in regard to the many species of finches, or other large groups of birds, in nature. One circumstance has struck me much; namely, that all the breeders of the various domestic animals and the cultivators of plants, with whom I have ever conversed, or whose treatises I have read, are firmly convinced that the several breeds to which each has attended, are descended from so many aboriginally distinct species. Ask, as I have asked, a celebrated raiser of Hereford cattle, whether his cattle might not have descended from long horns, and he will laugh you to scorn. I have never met a pigeon, or poultry, or duck, or rabbit fancier, who was not fully convinced that each main breed was descended from a distinct species. Van Mons, in his treatise on pears and apples, shows how utterly he disbelieves that the several sorts, for instance a Ribston-pippin or Codlin-apple, could ever have proceeded from the seeds of the same tree. Innumerable other examples could be given. The explanation, I think, is simple: from long-continued study they are strongly impressed with the differences between the several races; and though they well know that each race varies slightly, for they win their prizes by selecting such slight differences, yet they ignore all general arguments, and refuse to sum up in their minds slight differences accumulated during many successive generations. May not those naturalists who, knowing far less of the laws of inheritance than does the breeder, and knowing no more than he does of the intermediate links in the long lines of descent, yet admit that many of our domestic races have descended from the same parents—may they not learn a lesson of caution, when they deride the idea of species in a state of nature being lineal descendants of other species?

Selection

Let us now briefly consider the steps by which domestic races have been produced, either from one or from several allied species. Some little effect may, perhaps, be attributed to the direct action of the external conditions of life, and some little to habit; but he would be a bold man who would account by such agencies for the differences of a dray and race horse, a greyhound and bloodhound, a carrier and tumbler pigeon. One of the most remarkable features in our domesticated races is that we see in them adaptation, not indeed to the animal's or plant's own good, but to man's use or fancy. Some variations useful to him have probably arisen suddenly, or by one step; many botanists, for instance, believe that the fuller's teazle, with its hooks, which cannot be rivalled by any mechanical contrivance, is only a variety of the wild Dipsacus; and this amount of change may have suddenly arisen in a seedling. So it has probably been with the turnspit dog; and this is known to have been the case with the ancon sheep. But when we compare the dray-horse and race-horse, the dromedary and camel, the various breeds of sheep fitted either for cultivated land or mountain pasture, with the wool of one breed good for one purpose, and that of another breed for another purpose; when we compare the many breeds of dogs, each good for man in very different ways; when we compare the game-cock, so pertinacious in battle, with other breeds so little quarrelsome, with "everlasting layers" which never desire to sit, and with the bantam so small and elegant; when we compare the host of agricultural, culinary, orchard, and flower-garden races of plants, most useful to man at different seasons and for different purposes, or so beautiful in his eyes, we must, I think, look further than to mere variability. We cannot suppose that all the breeds were suddenly produced as perfect and as useful as we now see them; indeed, in several cases, we know that this has not been their history. The key is man's power of accumulative selection: nature gives successive variations; man adds them up in certain directions useful to him. In this sense he may be said to make for himself useful breeds.

. . .

At the present time, eminent breeders try by methodical selection, with a distinct object in view, to make a new strain or sub-breed, superior to anything existing in the country. But, for our purpose, a kind of Selection, which may be called Unconscious, and which results from every one trying to possess and breed from the best individual animals, is more important. Thus, a man who intends keeping pointers naturally tries to get as good dogs as he can, and afterwards breeds from his own best dogs, but he has no wish or expectation of permanently altering the breed. Nevertheless I cannot doubt that this process, continued during centuries, would improve and modify any breed, in the same way as Bakewell, Collins, etc., by this very same process, only carried on more methodically, did greatly modify, even during their own lifetimes, the forms and qualities of their cattle. Slow and insensible changes of this kind could never be recognised unless actual measurements or careful drawings of the breeds in question had been made long ago, which might serve for comparison. In some cases, however, unchanged or but little changed individuals of the same breed may be found in less civilised districts, where the breed has been less improved. There is reason to believe that King Charles's spaniel has been unconsciously modified to a large extent since the time of that monarch. Some highly competent authorities are convinced that the setter is directly derived from the spaniel, and has probably been slowly altered from it. It is known that the English pointer has been greatly changed within the last century, and in this case the change has, it is believed, been chiefly effected by crosses with the fox-hound; but what concerns us is, that the change has been effected unconsciously and gradually, and yet so effectually, that, though the old Spanish pointer certainly came from Spain, Mr. Borrow has not seen, as I am informed by him, any native dog in Spain like our pointer.

By a similar process of selection, and by careful training, the whole body of English racehorses have come to surpass in fleetness and size the parent Arab stock, so that the latter, by the regulations for the Goodwood Races, are favoured in the weights they carry. Lord Spencer and others have shown how the cattle of England have increased in weight and in early maturity, compared with the stock formerly kept in this country. By comparing the accounts given in old pigeon treatises of carriers and tumblers with these breeds as now existing in Britain, India, and Persia, we can, I think, clearly trace the stages through which they have insensibly passed, and come to differ so greatly from the rock-pigeon.

...

In regard to the domestic animals kept by uncivilised man, it should not be overlooked that they almost always have to struggle for their own food, at least during certain seasons. And in two countries very differently circumstanced, individuals of the same species, having slightly different constitutions or structure, would often succeed better in the one country than in the other, and thus by a process of "natural selection," as will hereafter be more fully explained, two sub-breeds might be formed. This, perhaps, partly explains what has been remarked by some authors, namely, that the varieties kept by savages have more of the character of species than the varieties kept in civilised countries.

On the view here given of the all-important part which selection by man has played, it becomes at once obvious, how it is that our domestic races show adaptation in their structure or in their habits to man's wants or fancies. We can, I think, further understand the frequently abnormal character of our domestic races, and likewise their differences being so great in external characters and relatively so slight in internal parts or organs. Man can hardly select, or only with much difficulty, any deviation of structure excepting such as is externally visible; and indeed he rarely cares for what is internal. He can never act by selection, excepting on variations which are first given to him in some slight degree by nature. No man would ever try to make a fantail, till he saw a pigeon with a tail developed in some slight degree in an unusual manner, or a pouter till he saw a pigeon with a crop of somewhat unusual size; and the more abnormal or unusual any character was when it first appeared, the more likely it would be to catch his attention. But to use such an expression as trying to make a fantail, is, I have no doubt, in most cases, utterly incorrect. The man who first selected a pigeon with a slightly larger tail, never dreamed what the descendants of that pigeon would become through long-continued, partly unconscious and partly methodical selection. Perhaps the parent bird of all fantails had only fourteen tail-feathers somewhat expanded, like the present Java fantail, or like individuals of other and distinct breeds, in which as many as seventeen tail-feathers have been counted. Perhaps the first pouter-pigeon did not inflate its crop much more than the turbit now does the upper part of its oesophagus,—a habit which is disregarded by all fanciers, as it is not one of the points of the breed.

Nor let it be thought that some great deviation of structure would be necessary to catch the fancier's eye: he perceives extremely small differences, and it is in human nature to value any novelty, however slight, in one's own possession.

. . .

I must now say a few words on the circumstances, favourable, or the reverse, to man's power of selection. A high degree of variability is obviously favourable, as freely giving the materials for selection to work on; not that mere individual differences are not amply sufficient, with extreme care, to allow of the accumulation of a large amount of modification in almost any desired direction. But as variations manifestly useful or pleasing to man appear only occasionally, the chance of their appearance will be much increased by a large number of individuals being kept; and hence this comes to be of the highest importance to success.

. . .

To sum up on the origin of our Domestic Races of animals and plants. I believe that the conditions of life, from their action on the reproductive system, are so far of the highest importance as causing variability. I do not believe that variability is an inherent and necessary contingency, under all circumstances, with all organic beings, as some authors have thought. The effects of variability are modified by various degrees of inheritance and of reversion. Variability is governed by many unknown laws, more especially by that of correlation of growth. Something may be attributed to the direct action of the conditions of life. Something must be attributed to use and disuse. The final result is thus rendered infinitely complex. In some cases, I do not doubt that the intercrossing of species, aboriginally distinct, has played an important part in the origin of our domestic productions. When in any country several domestic breeds have once been established, their occasional intercrossing, with the aid of selection, has, no doubt, largely aided in the formation of new sub-breeds; but the importance of the crossing of varieties has, I believe, been greatly exaggerated, both in regard to animals and to those plants which are propagated by seed. In plants which are temporarily propagated by cuttings, buds, etc., the importance of the crossing both of distinct species and of varieties is immense; for the cultivator here quite disregards the extreme variability both of hybrids and mongrels, and the frequent sterility of hybrids; but the cases of plants not propagated by seed are of little importance to us, for their endurance is only temporary. Over all these causes of Change I am convinced that the accumulative action of Selection, whether applied methodically and more quickly, or unconsciously and more slowly, but more efficiently, is by far the predominant Power.

2. Variation under Nature

Variability. Individual differences. Doubtful species. Wide ranging, much diffused, and common species vary most. Species of the larger genera in any country vary more than the species of the smaller genera. Many of the species of the larger genera resemble varieties in being very closely, but unequally, related to each other, and in having restricted ranges.

Before applying the principles arrived at in the last chapter to organic beings in a state of nature, we must briefly discuss whether these latter are subject to any variation. To treat this subject at all properly, a long catalogue of dry facts should be given; but these I shall reserve for my future work. Nor shall I here discuss the various definitions which have been given of the term species. No one definition has as yet satisfied all naturalists; yet every naturalist knows vaguely what he means when he speaks of a species. Generally the term includes the unknown element of a distinct act of creation. The term "variety" is almost equally difficult to define; but here community of descent is almost universally implied, though it can rarely be proved. We have also what are called monstrosities; but they graduate into varieties. By a monstrosity I presume is meant some considerable deviation of structure in one part, either injurious to or not useful to the species, and not generally propagated. Some authors use the term "variation" in a technical sense, as implying a modification directly due to the physical conditions of life; and "variations" in this sense are supposed not to be inherited: but who can say that the dwarfed condition of shells in the brackish waters of the Baltic, or dwarfed plants on Alpine summits, or the thicker fur of an animal from far northwards, would not in some cases be inherited for at least some few generations? and in this case I presume that the form would be called a variety.

Again, we have many slight differences which may be called individual differences, such as are known frequently to appear in the offspring from the same parents, or which may be presumed to have thus arisen, from being frequently observed in the individuals of the same species inhabiting the same confined locality. No one supposes that all the individuals of the same species are cast in the very same mould. These individual differences are highly important for us, as they afford materials for natural selection to accumulate, in the same manner as man can accumulate in any given direction individual differences in his domesticated productions . . . .

There is one point connected with individual differences, which seems to me extremely perplexing: I refer to those genera which have sometimes been called "protean" or "polymorphic," in which the species present an inordinate amount of variation; and hardly two naturalists can agree which forms to rank as species and which as varieties. We may instance Rubus, Rosa, and Hieracium amongst plants, several genera of insects, and several genera of Brachiopod shells. In most polymorphic genera some of the species have fixed and definite characters. Genera which are polymorphic in one country seem to be, with some few exceptions, polymorphic in other countries, and likewise, judging from Brachiopod shells, at former periods of time. These facts seem to be very perplexing, for they seem to show that this kind of variability is independent of the conditions of life. I am inclined to suspect that we see in these polymorphic genera variations in points of structure which are of no service or disservice to the species, and which consequently have not been seized on and rendered definite by natural selection, as hereafter will be explained.

Those forms which possess in some considerable degree the character of species, but which are so closely similar to some other forms, or are so closely linked to them by intermediate gradations, that naturalists do not like to rank them as distinct species, are in several respects the most important for us. We have every reason to believe that many of these doubtful and closely-allied forms have permanently retained their characters in their own country for a long time; for as long, as far as we know, as have good and true species. Practically, when a naturalist can unite two forms together by others having intermediate characters, he treats the one as a variety of the other, ranking the most common, but sometimes the one first described, as the species, and the other as the variety. But cases of great difficulty, which I will not here enumerate, sometimes occur in deciding whether or not to rank one form as a variety of another, even when they are closely connected by intermediate links; nor will the commonly-assumed hybrid nature of the intermediate links always remove the difficulty. In very many cases, however, one form is ranked as a variety of another, not because the intermediate links have actually been found, but because analogy leads the observer to suppose either that they do now somewhere exist, or may formerly have existed; and here a wide door for the entry of doubt and conjecture is opened.

Hence, in determining whether a form should be ranked as a species or a variety, the opinion of naturalists having sound judgment and wide experience seems the only guide to follow. We must, however, in many cases, decide by a majority of naturalists, for few well-marked and well-known varieties can be named which have not been ranked as species by at least some competent judges.

That varieties of this doubtful nature are far from uncommon cannot be disputed. Compare the several floras of Great Britain, of France or of the United States, drawn up by different botanists, and see what a surprising number of forms have been ranked by one botanist as good species, and by another as mere varieties. Mr. H. C. Watson, to whom I lie under deep obligation for assistance of all kinds, has marked for me 182 British plants, which are generally considered as varieties, but which have all been ranked by botanists as species; and in making this list he has omitted many trifling varieties, but which nevertheless have been ranked by some botanists as species, and he has entirely omitted several highly polymorphic genera. Under genera, including the most polymorphic forms, Mr. Babington gives 251 species, whereas Mr. Bentham gives only 112,—a difference of 139 doubtful forms! Amongst animals which unite for each birth, and which are highly locomotive, doubtful forms, ranked by one zoologist as a species and by another as a variety, can rarely be found within the same country, but are common in separated areas. How many of those birds and insects in North America and Europe, which differ very slightly from each other, have been ranked by one eminent naturalist as undoubted species, and by another as varieties, or, as they are often called, as geographical races! Many years ago, when comparing, and seeing others compare, the birds from the separate islands of the Galapagos Archipelago, both one with another, and with those from the American mainland, I was much struck how entirely vague and arbitrary is the distinction between species and varieties.

...

When a young naturalist commences the study of a group of organisms quite unknown to him, he is at first much perplexed to determine what differences to consider as specific, and what as varieties; for he knows nothing of the amount and kind of variation to which the group is subject; and this shows, at least, how very generally there is some variation. But if he confine his attention to one class within one country, he will soon make up his mind how to rank most of the doubtful forms. His general tendency will be to make many species, for he will become impressed, just like the pigeon or poultry-fancier before alluded to, with the amount of difference in the forms which he is continually studying; and he has little general knowledge of analogical variation in other groups and in other countries, by which to correct his first impressions. As he extends the range of his observations, he will meet with more cases of difficulty; for he will encounter a greater number of closely-allied forms. But if his observations be widely extended, he will in the end generally be enabled to make up his own mind which to call varieties and which species; but he will succeed in this at the expense of admitting much variation,—and the truth of this admission will often be disputed by other naturalists. When, moreover, he comes to study allied forms brought from countries not now continuous, in which case he can hardly hope to find the intermediate links between his doubtful forms, he will have to trust almost entirely to analogy, and his difficulties will rise to a climax.

Certainly no clear line of demarcation has as yet been drawn between species and sub-species—that is, the forms which in the opinion of some naturalists come very near to, but do not quite arrive at the rank of species; or, again, between sub-species and well-marked varieties, or between lesser varieties and individual differences. These differences blend into each other in an insensible series; and a series impresses the mind with the idea of an actual passage.

...

The passage from one stage of difference to another and higher stage may be, in some cases, due merely to the long-continued action of different physical conditions in two different regions; but I have not much faith in this view; and I attribute the passage of a variety, from a state in which it differs very slightly from its parent to one in which it differs more, to the action of natural selection in accumulating (as will hereafter be more fully explained) differences of structure in certain definite directions. Hence I believe a well-marked variety may be justly called an incipient species; but whether this belief be justifiable must be judged of by the general weight of the several facts and views given throughout this work.

. . .

From looking at species as only strongly-marked and well-defined varieties, I was led to anticipate that the species of the larger genera in each country would oftener present varieties, than the species of the smaller genera; for wherever many closely related species (i.e. species of the same genus) have been formed, many varieties or incipient species ought, as a general rule, to be now forming. Where many large trees grow, we expect to find saplings. Where many species of a genus have been formed through variation, circumstances have been favourable for variation; and hence we might expect that the circumstances would generally be still favourable to variation. On the other hand, if we look at each species as a special act of creation, there is no apparent reason why more varieties should occur in a group having many species, than in one having few.

. . .

Finally, then, varieties have the same general characters as species, for they cannot be distinguished from species,—except, firstly, by the discovery of intermediate linking forms, and the occurrence of such links cannot affect the actual characters of the forms which they connect; and except, secondly, by a certain amount of difference, for two forms, if differing very little, are generally ranked as varieties, notwithstanding that intermediate linking forms have not been discovered; but the amount of difference considered necessary to give to two forms the rank of species is quite indefinite. In genera having more than the average number of species in any country, the species of these genera have more than the average number of varieties. In large genera the species are apt to be closely, but unequally, allied together, forming little clusters round certain species. Species very closely allied to other species apparently have restricted ranges. In all these several respects the species of large genera present a strong analogy with varieties. And we can clearly understand these analogies, if species have once existed as varieties, and have thus originated: whereas, these analogies are utterly inexplicable if each species has been independently created.

We have, also, seen that it is the most flourishing and dominant species of the larger genera which on an average vary most; and varieties, as we shall hereafter see, tend to become converted into new and distinct species. The larger genera thus tend to become larger; and throughout nature the forms of life which are now dominant tend to become still more dominant by leaving many modified and dominant descendants. But by steps hereafter to be explained, the larger genera also tend to break up into smaller genera. And thus, the forms of life throughout the universe become divided into groups subordinate to groups.

3. Struggle for Existence

Bears on natural selection. The term used in a wide sense. Geometrical powers of increase. Rapid increase of naturalised animals and plants. Nature of the checks to increase. Competition universal. Effects of climate. Protection from the number of individuals. Complex relations of all animals and plants throughout nature. Struggle for life most severe between individuals and varieties of the same species; often severe between species of the same genus. The relation of organism to organism the most important of all relations.

Before entering on the subject of this chapter, I must make a few preliminary remarks, to show how the struggle for existence bears on Natural Selection. It has been seen in the last chapter that amongst organic beings in a state of nature there is some individual variability; indeed I am not aware that this has ever been disputed. It is immaterial for us whether a multitude of doubtful forms be called species or sub-species or varieties; what rank, for instance, the two or three hundred doubtful forms of British plants are entitled to hold, if the existence of any well-marked varieties be admitted. But the mere existence of individual variability and of some few well-marked varieties, though necessary as the foundation for the work, helps us but little in understanding how species arise in nature. How have all those exquisite adaptations of one part of the organisation to another part, and to the conditions of life, and of one distinct organic being to another being, been perfected? We see these beautiful co-adaptations most plainly in the woodpecker and mistletoe; and only a little less plainly in the humblest parasite which clings to the hairs of a quadruped or feathers of a bird; in the structure of the beetle which dives through the water; in the plumed seed which is wafted by the gentlest breeze; in short, we see beautiful adaptations everywhere and in every part of the organic world.

Again, it may be asked, how is it that varieties, which I have called incipient species, become ultimately converted into good and distinct species, which in most cases obviously differ from each other far more than do the varieties of the same species? How do those groups of species, which constitute what are called distinct genera, and which differ from each other more than do the species of the same genus, arise? All these results, as we shall more fully see in the next chapter, follow inevitably from the struggle for life. Owing to this struggle for life, any variation, however slight and from whatever cause proceeding, if it be in any degree profitable to an individual of any species, in its infinitely complex relations to other organic beings and to external nature, will tend to the preservation of that individual, and will generally be inherited by its offspring. The offspring, also, will thus have a better chance of surviving, for, of the many individuals of any species which are periodically born, but a small number can survive. I have called this principle, by which each slight variation, if useful, is preserved, by the term of Natural Selection, in order to mark its relation to man's power of selection. We have seen that man by selection can certainly produce great results, and can adapt organic beings to his own uses, through the accumulation of slight but useful variations, given to him by the hand of Nature. But Natural Selection, as we shall hereafter see, is a power incessantly ready for action, and is as immeasurably superior to man's feeble efforts, as the works of Nature are to those of Art.

We will now discuss in a little more detail the struggle for existence. In my future work this subject shall be treated, as it well deserves, at much greater length. The elder De Candolle and Lyell have largely and philosophically shown that all organic beings are exposed to severe competition. In regard to plants, no one has treated this subject with more spirit and ability than W. Herbert, Dean of Manchester, evidently the result of his great horticultural knowledge. Nothing is easier than to admit in words the truth of the universal struggle for life, or more difficult—at least I have found it so—than constantly to bear this conclusion in mind. Yet unless it be thoroughly engrained in the mind, I am convinced that the whole economy of nature, with every fact on distribution, rarity, abundance, extinction, and variation, will be dimly seen or quite misunderstood. We behold the face of nature bright with gladness, we often see superabundance of food; we do not see, or we forget, that the birds which are idly singing round us mostly live on insects or seeds, and are thus constantly destroying life; or we forget how largely these songsters, or their eggs, or their nestlings, are destroyed by birds and beasts of prey; we do not always bear in mind, that though food may be now superabundant, it is not so at all seasons of each recurring year.

I should premise that I use the term Struggle for Existence in a large and metaphorical sense, including dependence of one being on another, and including (which is more important) not only the life of the individual, but success in leaving progeny. Two canine animals in a time of dearth, may be truly said to struggle with each other which shall get food and live. But a plant on the edge of a desert is said to struggle for life against the drought, though more properly it should be said to be dependent on the moisture. A plant which annually produces a thousand seeds, of which on an average only one comes to maturity, may be more truly said to struggle with the plants of the same and other kinds which already clothe the ground. The mistletoe is dependent on the apple and a few other trees, but can only in a far-fetched sense be said to struggle with these trees, for if too many of these parasites grow on the same tree, it will languish and die. But several seedling mistletoes, growing close together on the same branch, may more truly be said to struggle with each other. As the mistletoe is disseminated by birds, its existence depends on birds; and it may metaphorically be said to struggle with other fruit-bearing plants, in order to tempt birds to devour and thus disseminate its seeds rather than those of other plants. In these several senses, which pass into each other, I use for convenience sake the general term of struggle for existence.

A struggle for existence inevitably follows from the high rate at which all organic beings tend to increase. Every being, which during its natural lifetime produces several eggs or seeds, must suffer destruction during some period of its life, and during some season or occasional year, otherwise, on the principle of geometrical increase, its numbers would quickly become so inordinately great that no country could support the product. Hence, as more individuals are produced than can possibly survive, there must in every case be a struggle for existence, either one individual with another of the same species, or with the individuals of distinct species, or with the physical conditions of life. It is the doctrine of Malthus applied with manifold force to the whole animal and vegetable kingdoms; for in this case there can be no artificial increase of food, and no prudential restraint from marriage. Although some species may be now increasing, more or less rapidly, in numbers, all cannot do so, for the world would not hold them.

There is no exception to the rule that every organic being naturally increases at so high a rate, that if not destroyed, the earth would soon be covered by the progeny of a single pair. Even slow-breeding man has doubled in twenty-five years, and at this rate, in a few thousand years, there would literally not be standing room for his progeny. Linnaeus has calculated that if an annual plant produced only two seeds—and there is no plant so unproductive as this—and their seedlings next year produced two, and so on, then in twenty years there would be a million plants. The elephant is reckoned to be the slowest breeder of all known animals, and I have taken some pains to estimate its probable minimum rate of natural increase: it will be under the mark to assume that it breeds when thirty years old, and goes on breeding till ninety years old, bringing forth three pair of young in this interval; if this be so, at the end of the fifth century there would be alive fifteen million elephants, descended from the first pair.

. . .

In a state of nature almost every plant produces seed, and amongst animals there are very few which do not annually pair. Hence we may confidently assert, that all plants and animals are tending to increase at a geometrical ratio, that all would most rapidly stock every station in which they could any how exist, and that the geometrical tendency to increase must be checked by destruction at some period of life . . . .

In looking at Nature, it is most necessary to keep the foregoing considerations always in mind—never to forget that every single organic being around us may be said to be striving to the utmost to increase in numbers; that each lives by a struggle at some period of its life; that heavy destruction inevitably falls either on the young or old, during each generation or at recurrent intervals. Lighten any check, mitigate the destruction ever so little, and the number of the species will almost instantaneously increase to any amount. The face of Nature may be compared to a yielding surface, with ten thousand sharp wedges packed close together and driven inwards by incessant blows, sometimes one wedge being struck, and then another with greater force.

What checks the natural tendency of each species to increase in number is most obscure. Look at the most vigorous species; by as much as it swarms in numbers, by so much will its tendency to increase be still further increased. We know not exactly what the checks are in even one single instance. Nor will this surprise anyone who reflects how ignorant we are on this head, even in regard to mankind, so incomparably better known than any other animal. This subject has been ably treated by several authors, and I shall, in my future work, discuss some of the checks at considerable length, more especially in regard to the feral animals of South America. Here I will make only a few remarks, just to recall to the reader's mind some of the chief points. Eggs or very young animals seem generally to suffer most, but this is not invariably the case. With plants there is a vast destruction of seeds, but, from some observations which I have made, I believe that it is the seedlings which suffer most from germinating in ground already thickly stocked with other plants. Seedlings, also, are destroyed in vast numbers by various enemies; for instance, on a piece of ground three feet long and two wide, dug and cleared, and where there could be no choking from other plants, I marked all the seedlings of our native weeds as they came up, and out of the 357 no less than 295 were destroyed, chiefly by slugs and insects. If turf which has long been mown, and the case would be the same with turf closely browsed by quadrupeds, be let to grow, the more vigorous plants gradually kill the less vigorous, though fully grown, plants: thus out of twenty species growing on a little plot of turf (three feet by four) nine species perished from the other species being allowed to grow up freely.

The amount of food for each species of course gives the extreme limit to which each can increase; but very frequently it is not the obtaining food, but the serving as prey to other animals, which determines the average numbers of a species. Thus, there seems to be little doubt that the stock of partridges, grouse, and hares on any large estate depends chiefly on the destruction of vermin. If not one head of game were shot during the next twenty years in England, and, at the same time, if no vermin were destroyed, there would, in all probability, be less game than at present, although hundreds of thousands of game animals are now annually killed. On the other hand, in some cases, as with the elephant and rhinoceros, none are destroyed by beasts of prey: even the tiger in India most rarely dares to attack a young elephant protected by its dam.

Climate plays an important part in determining the average numbers of a species, and periodical seasons of extreme cold or drought, I believe to be the most effective of all checks. I estimated that the winter of 1854–55 destroyed four-fifths of the birds in my own grounds; and this is a tremendous destruction, when we remember that ten percent is an extraordinarily severe mortality from epidemics with man. The action of climate seems at first sight to be quite independent of the struggle for existence; but in so far as climate chiefly acts in reducing food, it brings on the most severe struggle between the individuals, whether of the same or of distinct species, which subsist on the same kind of food. Even when climate, for instance extreme cold, acts directly, it will be the least vigorous, or those which have got least food through the advancing winter, which will suffer most. When we travel from south to north, or from a damp region to a dry, we invariably see some species gradually getting rarer and rarer, and finally disappearing; and the change of climate being conspicuous, we are tempted to attribute the whole effect to its direct action. But this is a very false view: we forget that each species, even where it most abounds, is constantly suffering enormous destruction at some period of its life, from enemies or from competitors for the same place and food; and if these enemies or competitors be in the least degree favoured by any slight change of climate, they will increase in numbers, and, as each area is already fully stocked with inhabitants, the other species will decrease. When we travel southward and see a species decreasing in numbers, we may feel sure that the cause lies quite as much in other species being favoured, as in this one being hurt. So it is when we travel northward, but in a somewhat lesser degree, for the number of species of all kinds, and therefore of competitors, decreases northwards; hence in going northward, or in ascending a mountain, we far oftener meet with stunted forms, due to the DIRECTLY injurious action of climate, than we do in proceeding southwards or in descending a mountain. When we reach the Arctic regions, or snow-capped summits, or absolute deserts, the struggle for life is almost exclusively with the elements.

That climate acts in main part indirectly by favouring other species, we may clearly see in the prodigious number of plants in our gardens which can perfectly well endure our climate, but which never become naturalised, for they cannot compete with our native plants, nor resist destruction by our native animals.

When a species, owing to highly favourable circumstances, increases inordinately in numbers in a small tract, epidemics—at least, this seems generally to occur with our game animals—often ensue: and here we have a limiting check independent of the struggle for life. But even some of these so-called epidemics appear to be due to parasitic worms, which have from some cause, possibly in part through facility of diffusion amongst the crowded animals, been disproportionably favoured: and here comes in a sort of struggle between the parasite and its prey.

On the other hand, in many cases, a large stock of individuals of the same species, relatively to the numbers of its enemies, is absolutely necessary for its preservation. Thus we can easily raise plenty of corn and rape-seed, etc., in our fields, because the seeds are in great excess compared with the number of birds which feed on them; nor can the birds, though having a superabundance of food at this one season, increase in number proportionally to the supply of seed, as their numbers are checked during winter: but anyone who has tried, knows how troublesome it is to get seed from a few wheat or other such plants in a garden; I have in this case lost every single seed. This view of the necessity of a large stock of the same species for its preservation, explains, I believe, some singular facts in nature, such as that of very rare plants being sometimes extremely abundant in the few spots where they do occur; and that of some social plants being social, that is, abounding in individuals, even on the extreme confines of their range. For in such cases, we may believe, that a plant could exist only where the conditions of its life were so favourable that many could exist together, and thus save each other from utter destruction. I should add that the good effects of frequent intercrossing, and the ill effects of close interbreeding, probably come into play in some of these cases; but on this intricate subject I will not here enlarge.

...

I am tempted to give one more instance showing how plants and animals, most remote in the scale of nature, are bound together by a web of complex relations. I shall hereafter have occasion to show that the exotic Lobelia fulgens, in this part of England, is never visited by insects, and consequently, from its peculiar structure, never can set a seed. Many of our orchidaceous plants absolutely require the visits of moths to remove their pollen-masses and thus to fertilise them. I have, also, reason to believe that humble-bees are indispensable to the fertilisation of the heartsease (Viola tricolor), for other bees do not visit this flower. From experiments which I have tried, I have found that the visits of bees, if not indispensable, are at least highly beneficial to the fertilisation of our clovers; but humble-bees alone visit the common red clover (Trifolium pratense), as other bees cannot reach the nectar. Hence I have very little doubt, that if the whole genus of humble-bees became extinct or very rare in England, the heartsease and red clover would become very rare, or wholly disappear. The number of humble-bees in any district depends in a great degree on the number of field-mice, which destroy their combs and nests; and Mr. H. Newman, who has long attended to the habits of humble-bees, believes that "more than two thirds of them are thus destroyed all over England." Now the number of mice is largely dependent, as everyone knows, on the number of cats; and Mr. Newman says, "Near villages and small towns I have found the nests of humble-bees more numerous than elsewhere, which I attribute to the number of cats that destroy the mice." Hence it is quite credible that the presence of a feline animal in large numbers in a district might determine, through the intervention first of mice and then of bees, the frequency of certain flowers in that district!

In the case of every species, many different checks, acting at different periods of life, and during different seasons or years, probably come into play; some one check or some few being generally the most potent, but all concurring in determining the average number or even the existence of the species. In some cases it can be shown that widely-different checks act on the same species in different districts. When we look at the plants and bushes clothing an entangled bank, we are tempted to attribute their proportional numbers and kinds to what we call chance. But how false a view is this! Everyone has heard that when an American forest is cut down, a very different vegetation springs up; but it has been observed that the trees now growing on the ancient Indian mounds, in the

Southern United States, display the same beautiful diversity and proportion of kinds as in the surrounding virgin forests. What a struggle between the several kinds of trees must here have gone on during long centuries, each annually scattering its seeds by the thousand; what war between insect and insect—between insects, snails, and other animals with birds and beasts of prey—all striving to increase, and all feeding on each other or on the trees or their seeds and seedlings, or on the other plants which first clothed the ground and thus checked the growth of the trees! Throw up a handful of feathers, and all must fall to the ground according to definite laws; but how simple is this problem compared to the action and reaction of the innumerable plants and animals which have determined, in the course of centuries, the proportional numbers and kinds of trees now growing on the old Indian ruins!

The dependency of one organic being on another, as of a parasite on its prey, lies generally between beings remote in the scale of nature. This is often the case with those which may strictly be said to struggle with each other for existence, as in the case of locusts and grass-feeding quadrupeds. But the struggle almost invariably will be most severe between the individuals of the same species, for they frequent the same districts, require the same food, and are exposed to the same dangers.

. . .

As species of the same genus have usually, though by no means invariably, some similarity in habits and constitution, and always in structure, the struggle will generally be more severe between species of the same genus, when they come into competition with each other, than between species of distinct genera. We see this in the recent extension over parts of the United States of one species of swallow having caused the decrease of another species. The recent increase of the missel-thrush in parts of Scotland has caused the decrease of the song-thrush. How frequently we hear of one species of rat taking the place of another species under the most different climates! In Russia the small Asiatic cockroach has everywhere driven before it its great congener. One species of charlock will supplant another, and so in other cases. We can dimly see why the competition should be most severe between allied forms, which fill nearly the same place in the economy of nature; but probably in no one case could we precisely say why one species has been victorious over another in the great battle of life.

A corollary of the highest importance may be deduced from the foregoing remarks, namely, that the structure of every organic being is related, in the most essential yet often hidden manner, to that of all other organic beings, with which it comes into competition for food or residence, or from which it has to escape, or on which it preys. This is obvious in the structure of the teeth and talons of the tiger; and in that of the legs and claws of the parasite which clings to the hair on the tiger's body. But in the beautifully plumed seed of the dandelion, and in the flattened and fringed legs of the water-beetle, the relation seems at first confined to the elements of air and water. Yet the advantage of plumed seeds no doubt stands in the closest relation to the land being already thickly clothed by other plants; so that the seeds may be widely distributed and fall on unoccupied ground. In the water-beetle, the structure of its legs, so well adapted for diving, allows it to compete with other aquatic insects, to hunt for its own prey, and to escape serving as prey to other animals.

...

It is good thus to try in our imagination to give any form some advantage over another. Probably in no single instance should we know what to do, so as to succeed. It will convince us of our ignorance on the mutual relations of all organic beings; a conviction as necessary, as it seems to be difficult to acquire. All that we can do, is to keep steadily in mind that each organic being is striving to increase at a geometrical ratio; that each at some period of its life, during some season of the year, during each generation or at intervals, has to struggle for life, and to suffer great destruction. When we reflect on this struggle, we may console ourselves with the full belief, that the war of nature is not incessant, that no fear is felt, that death is generally prompt, and that the vigorous, the healthy, and the happy survive and multiply.

4. Natural Selection

Natural Selection: its power compared with man's selection, its power on characters of trifling importance, its power at all ages and on both sexes. Sexual Selection. On the generality of intercrosses between individuals of the same species. Circumstances favourable and unfavourable to Natural Selection, namely, intercrossing, isolation, number of individuals. Slow action. Extinction caused by Natural Selection. Divergence of Character, related to the diversity of inhabitants of any small area, and to naturalisation. Action of Natural Selection, through Divergence of Character and Extinction, on the descendants from a common parent. Explains the Grouping of all organic beings.

How will the struggle for existence, discussed too briefly in the last chapter, act in regard to variation? Can the principle of selection, which we have seen is so potent in the hands of man, apply in nature? I think we shall see that it can act most effectually. Let it be borne in mind in what an endless number of strange peculiarities our domestic productions, and, in a lesser degree, those under nature, vary; and how strong the hereditary tendency is. Under domestication, it may be truly said that the whole organisation becomes in some degree plastic. Let it be borne in mind how infinitely complex and close-fitting are the mutual relations of all organic beings to each other and to their physical conditions of life. Can it, then, be thought improbable, seeing that variations useful to man have undoubtedly occurred, that other variations useful in some way to each being in the great and complex battle of life, should sometimes occur in the course of thousands of generations? If such do occur, can we doubt (remembering that many more individuals are born than can possibly survive) that individuals having any advantage, however slight, over others, would have the best chance of surviving and of procreating their kind? On the other hand, we may feel sure that any variation in the least degree injurious would be rigidly destroyed. This preservation of favourable variations and the rejection of injurious variations, I call Natural Selection. Variations neither useful nor injurious would not be affected by natural selection, and would be left a fluctuating element, as perhaps we see in the species called polymorphic.

We shall best understand the probable course of natural selection by taking the case of a country undergoing some physical change, for instance, of climate. The proportional numbers of its inhabitants would almost immediately undergo a change, and some species might become extinct. We may conclude, from what we have seen of the intimate and complex manner in which the inhabitants of each country are bound together, that any change in the numerical proportions of some of the inhabitants, independently of the change of climate itself, would most seriously affect many of the others. If the country were open on its borders, new forms would certainly immigrate, and this also would seriously disturb the relations of some of the former inhabitants. Let it be remembered how powerful the influence of a single introduced tree or mammal has been shown to be. But in the case of an island, or of a country partly surrounded by barriers, into which new and better adapted forms could not freely enter, we should then have places in the economy of nature which would assuredly be better filled up, if some of the original inhabitants were in some manner modified; for, had the area been open to immigration, these same places would have been seized on by intruders. In such case, every slight modification, which in the course of ages chanced to arise, and which in any way favoured the individuals of any of the species, by better adapting them to their altered conditions, would tend to be preserved; and natural selection would thus have free scope for the work of improvement.

We have reason to believe, as stated in the first chapter, that a change in the conditions of life, by specially acting on the reproductive system, causes or increases variability; and in the foregoing case the conditions of life are supposed to have undergone a change, and this would manifestly be favourable to natural selection, by giving a better chance of profitable variations occurring; and unless profitable variations do occur, natural selection can do nothing. Not that, as I believe, any extreme amount of variability is necessary; as man can certainly produce great results by adding up in any given direction mere individual differences, so could Nature, but far more easily, from having incomparably longer time at her disposal. Nor do I believe that any great physical change, as of climate, or any unusual degree of isolation to check immigration, is actually necessary to produce new and unoccupied places for natural selection to fill up by modifying and improving some of the varying inhabitants. For as all the inhabitants of each country are struggling together with nicely balanced forces, extremely slight modifications in the structure or habits of one inhabitant would often give it an advantage over others; and still further modifications of the same kind would often still further increase the advantage. No country can be named in which all the native inhabitants are now so perfectly adapted to each other and to the physical conditions under which they live, that none of them could anyhow be improved; for in all countries, the natives have been so far conquered by naturalised productions, that they have allowed foreigners to take firm possession of the land. And as foreigners have thus everywhere beaten some of the natives, we may safely conclude that the natives might have been modified with advantage, so as to have better resisted such intruders.

As man can produce and certainly has produced a great result by his methodical and unconscious means of selection, what may not nature effect? Man can act only on external and visible characters: nature cares nothing for appearances, except in so far as they may be useful to any being. She can act on every internal organ, on every shade of constitutional difference, on the whole machinery of life. Man selects only for his own good; Nature only for that of the being which she tends. Every selected character is fully exercised by her; and the being is placed under well-suited conditions of life. Man keeps the natives of many climates in the same country; he seldom exercises each selected character in some peculiar and fitting manner; he feeds a long and a short beaked pigeon on the same food; he does not exercise a long-backed or long-legged quadruped in any peculiar manner; he exposes sheep with long and short wool to the same climate. He does not allow the most vigorous males to struggle for the females. He does not rigidly destroy all inferior animals, but protects during each varying season, as far as lies in his power, all his productions. He often begins his selection by some half-monstrous form; or at least by some modification prominent enough to catch his eye, or to be plainly useful to him. Under nature, the slightest difference of structure or constitution may well turn the nicely-balanced scale in the struggle for life, and so be preserved. How fleeting are the wishes and efforts of man! how short his time! and consequently how poor will his products be, compared with those accumulated by nature during whole geological periods. Can we wonder, then, that nature's productions should be far "truer" in character than man's productions; that they should be infinitely better adapted to the most complex conditions of life, and should plainly bear the stamp of far higher workmanship?

It may be said that natural selection is daily and hourly scrutinising, throughout the world, every variation, even the slightest; rejecting that which is bad, preserving and adding up all that is good; silently and insensibly working, whenever and wherever opportunity offers, at the improvement of each organic being in relation to its organic and inorganic conditions of life. We see nothing of these slow changes in progress, until the hand of time has marked the long lapse of ages, and then so imperfect is our view into long past geological ages, that we only see that the forms of life are now different from what they formerly were.

Although natural selection can act only through and for the good of each being, yet characters and structures, which we are apt to consider as of very trifling importance, may thus be acted on. When we see leaf-eating insects green, and bark-feeders mottled-grey; the alpine ptarmigan white in winter, the red-grouse the colour of heather, and the black-grouse that of peaty earth, we must believe that these tints are of service to these birds and insects in preserving them from danger.

...

Natural selection will modify the structure of the young in relation to the parent, and of the parent in relation to the young. In social animals it will adapt the structure of each individual for the benefit of the community; if each in consequence profits by the selected change. What natural selection cannot do, is to modify the structure of one species, without giving it any advantage, for the good of another species; and though statements to this effect may be found in works of natural history, I cannot find one case which will bear investigation. A structure used only once in an animal's whole life, if of high importance to it, might be modified to any extent by natural selection; for instance, the great jaws possessed by certain insects, and used exclusively for opening the cocoon—or the hard tip to the beak of nestling birds, used for breaking the egg. It has been asserted, that of the best short-beaked tumbler-pigeons more perish in the egg than are able to get out of it; so that fanciers assist in the act of hatching. Now, if nature had to make the beak of a full-grown pigeon very short for the bird's own advantage, the process of modification would be very slow, and there would be simultaneously the most rigorous selection of the young birds within the egg, which had the most powerful and hardest beaks, for all with weak beaks would inevitably perish: or, more delicate and more easily broken shells might be selected, the thickness of the shell being known to vary like every other structure.

Sexual Selection

Inasmuch as peculiarities often appear under domestication in one sex and become hereditarily attached to that sex, the same fact probably occurs under nature, and if so, natural selection will be able to modify one sex in its functional relations to the other sex, or in relation to wholly different habits of life in the two sexes, as is sometimes the case with insects. And this leads me to say a few words on what I call Sexual Selection. This depends, not on a struggle for existence, but on a struggle between the males for possession of the females; the result is not death to the unsuccessful competitor, but few or no offspring. Sexual selection is, therefore, less rigorous than natural selection. Generally, the most vigorous males, those which are best fitted for their places in nature, will leave most progeny. But in many cases, victory will depend not on general vigour, but on having special weapons, confined to the male sex. A hornless stag or spurless cock would have a poor chance of leaving offspring. Sexual selection by always allowing the victor to breed might surely give indomitable courage, length to the spur, and strength to the wing to strike in the spurred leg, as well as the brutal cock-fighter, who knows well that he can improve his breed by careful selection of the best cocks. How low in the scale of nature this law of battle descends, I know not; male alligators have been described as fighting, bellowing, and whirling round, like Indians in a war-dance, for the possession of the females; male salmons have been seen fighting all day long; male stag-beetles often bear wounds from the huge mandibles of other males. The war is, perhaps, severest between the males of polygamous animals, and these seem oftenest provided with special weapons. The males of carnivorous animals are already well armed; though to them and to others, special means of defence may be given through means of sexual selection, as the mane to the lion, the shoulder-pad to the boar, and the hooked jaw to the male salmon; for the shield may be as important for victory, as the sword or spear.

Amongst birds, the contest is often of a more peaceful character. All those who have attended to the subject, believe that there is the severest rivalry between the males of many species to attract by singing the females. The rock-thrush of Guiana, birds of Paradise, and some others, congregate; and successive males display their gorgeous plumage and perform strange antics before the females, which standing by as spectators, at last choose the most attractive partner. Those who have closely attended to birds in confinement well know that they often take individual preferences and dislikes: thus Sir R. Heron has described how one pied peacock was eminently attractive to all his hen birds. It may appear childish to attribute any effect to such apparently weak means: I cannot here enter on the details necessary to support this view; but if man can in a short time give elegant carriage and beauty to his bantams, according to his standard of beauty, I can see no good reason to doubt that female birds, by selecting, during thousands of generations, the most melodious or beautiful males, according to their standard of beauty, might produce a marked effect. I strongly suspect that some well-known laws with respect to the plumage of male and female birds, in comparison with the plumage of the young, can be explained on the view of plumage having been chiefly modified by sexual selection, acting when the birds have come to the breeding age or during the breeding season; the modifications thus produced being inherited at corresponding ages or seasons, either by the males alone, or by the males and females; but I have not space here to enter on this subject.

Thus it is, as I believe, that when the males and females of any animal have the same general habits of life, but differ in structure, colour, or ornament, such differences have been mainly caused by sexual selection; that is, individual males have had, in successive generations, some slight advantage over other males, in their weapons, means of defence, or charms; and have transmitted these advantages to their male offspring. Yet, I would not wish to attribute all such sexual differences to this agency: for we see peculiarities arising and becoming attached to the male sex in our domestic animals (as the wattle in male carriers, horn-like protuberances in the cocks of certain fowls, etc.), which we cannot believe to be either useful to the males in battle, or attractive to the females. We see analogous cases under nature, for instance, the tuft of hair on the breast of the turkey-cock, which can hardly be either useful or ornamental to this bird;—indeed, had the tuft appeared under domestication, it would have been called a monstrosity.

Illustrations of the Action of Natural Selection

In order to make it clear how, as I believe, natural selection acts, I must beg permission to give one or two imaginary illustrations. Let us take the case of a wolf, which preys on various animals, securing some by craft, some by strength, and some by fleetness; and let us suppose that the fleetest prey, a deer for instance, had from any change in the country increased in numbers, or that other prey had decreased in numbers, during that season of the year when the wolf is hardest pressed for food. I can under such circumstances see no reason to doubt that the swiftest and slimmest wolves would have the best chance of surviving, and so be preserved or selected,—provided always that they retained strength to master their prey at this or at some other period of the year, when they might be compelled to prey on other animals. I can see no more reason to doubt this, than that man can improve the fleetness of his greyhounds by careful and methodical selection, or by that unconscious selection which results from each man trying to keep the best dogs without any thought of modifying the breed.

Even without any change in the proportional numbers of the animals on which our wolf preyed, a cub might be born with an innate tendency to pursue certain kinds of prey. Nor can this be thought very improbable; for we often observe great differences in the natural tendencies of our domestic animals; one cat, for instance, taking to catch rats, another mice; one cat, according to Mr. St. John, bringing home winged game, another hares or rabbits, and another hunting on marshy ground and almost nightly catching woodcocks or snipes. The tendency to catch rats rather than mice is known to be inherited. Now, if any slight innate change of habit or of structure benefited an individual wolf, it would have the best chance of surviving and of leaving offspring. Some of its young would probably inherit the same habits or structure, and by the repetition of this process, a new variety might be formed which would either supplant or coexist with the parent-form of wolf. Or, again, the wolves inhabiting a mountainous district, and those frequenting the lowlands, would naturally be forced to hunt different prey; and from the continued preservation of the individuals best fitted for the two sites, two varieties might slowly be formed. These varieties would cross and blend where they met; but to this subject of intercrossing we shall soon have to return. I may add, that, according to Mr. Pierce, there are two varieties of the wolf inhabiting the Catskill Mountains in the United States, one with a light greyhound-like form, which pursues deer, and the other more bulky, with shorter legs, which more frequently attacks the shepherd's flocks.

...

I am well aware that this doctrine of natural selection, exemplified in the above imaginary instances, is open to the same objections which were at first urged against Sir Charles Lyell's noble views on "the modern changes of the earth, as illustrative of geology"; but we now very seldom hear the action, for instance, of the coast-waves, called a trifling and insignificant cause, when applied to the excavation of gigantic valleys or to the formation of the longest lines of inland cliffs. Natural selection can act only by the preservation and accumulation of infinitesimally small inherited modifications, each profitable to the preserved being; and as modern geology has almost banished such views as the excavation of a great valley by a single diluvial wave, so will natural selection, if it be a true principle, banish the belief of the continued creation of new organic beings, or of any great and sudden modification in their structure.

...

To sum up the circumstances favourable and unfavourable to natural selection, as far as the extreme intricacy of the subject permits. I conclude, looking to the future, that for terrestrial productions a large continental area, which will probably undergo many oscillations of level, and which consequently will exist for long periods in a broken condition, will be the most favourable for the production of many new forms of life, likely to endure long and to spread widely. For the area will first have existed as a continent, and the inhabitants, at this period numerous in individuals and kinds, will have been subjected to very severe competition. When converted by subsidence into large separate islands, there will still exist many individuals of the same species on each island: intercrossing on the confines of the range of each species will thus be checked: after physical changes of any kind, immigration will be prevented, so that new places in the polity of each island will have to be filled up by modifications of the old inhabitants; and time will be allowed for the varieties in each to become well modified and perfected. When, by renewed elevation, the islands shall be re-converted into a continental area, there will again be severe competition: the most favoured or improved varieties will be enabled to spread: there will be much extinction of the less improved forms, and the relative proportional numbers of the various inhabitants of the renewed continent will again be changed; and again there will be a fair field for natural selection to improve still further the inhabitants, and thus produce new species.

That natural selection will always act with extreme slowness, I fully admit. Its action depends on there being places in the polity of nature, which can be better occupied by some of the inhabitants of the country undergoing modification of some kind. The existence of such places will often depend on physical changes, which are generally very slow, and on the immigration of better adapted forms having been checked. But the action of natural selection will probably still oftener depend on some of the inhabitants becoming slowly modified; the mutual relations of many of the other inhabitants being thus disturbed. Nothing can be effected, unless favourable variations occur, and variation itself is apparently always a very slow process. The process will often be greatly retarded by free intercrossing. Many will exclaim that these several causes are amply sufficient wholly to stop the action of natural selection. I do not believe so. On the other hand, I do believe that natural selection will always act very slowly, often only at long intervals of time, and generally on only a very few of the inhabitants of the same region at the same time. I further believe, that this very slow, intermittent action of natural selection accords perfectly well with what geology tells us of the rate and manner at which the inhabitants of this world have changed.

Slow though the process of selection may be, if feeble man can do much by his powers of artificial selection, I can see no limit to the amount of change, to the beauty and infinite complexity of the coadaptations between all organic beings, one with another and with their physical conditions of life, which may be effected in the long course of time by nature's power of selection.

Extinction

This subject will be more fully discussed in our chapter on Geology; but it must be here alluded to from being intimately connected with natural selection. Natural selection acts solely through the preservation of variations in some way advantageous, which consequently endure. But as from the high geometrical powers of increase of all organic beings, each area is already fully stocked with inhabitants, it follows that as each selected and favoured form increases in number, so will the less favoured forms decrease and become rare. Rarity, as geology tells us, is the precursor to extinction . . . .

Furthermore, the species which are most numerous in individuals will have the best chance of producing within any given period favourable variations. We have evidence of this, in the facts given in the second chapter, showing that it is the common species which afford the greatest number of recorded varieties, or incipient species. Hence, rare species will be less quickly modified or improved within any given period, and they will consequently be beaten in the race for life by the modified descendants of the commoner species.

From these several considerations I think it inevitably follows, that as new species in the course of time are formed through natural selection, others will become rarer and rarer, and finally extinct. The forms which stand in closest competition with those undergoing modification and improvement, will naturally suffer most. And we have seen in the chapter on the Struggle for Existence that it is the most closely-allied forms,—varieties of the same species, and species of the same genus or of related genera,—which, from having nearly the same structure, constitution, and habits, generally come into the severest competition with each other. Consequently, each new variety or species, during the progress of its formation, will generally press hardest on its nearest kindred, and tend to exterminate them . . . .

Divergence of Character

The principle, which I have designated by this term, is of high importance on my theory, and explains, as I believe, several important facts. In the first place, varieties, even strongly-marked ones, though having somewhat of the character of species—as is shown by the hopeless doubts in many cases how to rank them—yet certainly differ from each other far less than do good and distinct species. Nevertheless, according to my view, varieties are species in the process of formation, or are, as I have called them, incipient species. How, then, does the lesser difference between varieties become augmented into the greater difference between species? That this does habitually happen, we must infer from most of the innumerable species throughout nature presenting well-marked differences; whereas varieties, the supposed prototypes and parents of future well-marked species, present slight and ill-defined differences. Mere chance, as we may call it, might cause one variety to differ in some character from its parents, and the offspring of this variety again to differ from its parent in the very same character and in a greater degree; but this alone would never account for so habitual and large an amount of difference as that between varieties of the same species and species of the same genus.

As has always been my practice, let us seek light on this head from our domestic productions. We shall here find something analogous. A fancier is struck by a pigeon having a slightly shorter beak; another fancier is struck by a pigeon having a rather longer beak; and on the acknowledged principle that "fanciers do not and will not admire a medium standard, but like extremes," they both go on (as has actually occurred with tumbler-pigeons) choosing and breeding from birds with longer and longer beaks, or with shorter and shorter beaks. Again, we may suppose that at an early period one man preferred swifter horses; another stronger and more bulky horses. The early differences would be very slight; in the course of time, from the continued selection of swifter horses by some breeders, and of stronger ones by others, the differences would become greater, and would be noted as forming two sub-breeds; finally, after the lapse of centuries, the sub-breeds would become converted into two well-established and distinct breeds. As the differences slowly become greater, the inferior animals with intermediate characters, being neither very swift nor very strong, will have been neglected, and will have tended to disappear. Here, then, we see in man's productions the action of what may be called the principle of divergence, causing differences, at first barely appreciable, steadily to increase, and the breeds to diverge in character both from each other and from their common parent.

But how, it may be asked, can any analogous principle apply in nature? I believe it can and does apply most efficiently, from the simple circumstance that the more diversified the descendants from any one species become in structure, constitution, and habits, by so much will they be better enabled to seize on many and widely diversified places in the polity of nature, and so be enabled to increase in numbers.

We can clearly see this in the case of animals with simple habits. Take the case of a carnivorous quadruped, of which the number that can be supported in any country has long ago arrived at its full average. If its natural powers of increase be allowed to act, it can succeed in increasing (the country not undergoing any change in its conditions) only by its varying descendants seizing on places at present occupied by other animals: some of them, for instance, being enabled to feed on new kinds of prey, either dead or alive; some inhabiting new stations, climbing trees, frequenting water, and some perhaps becoming less carnivorous. The more diversified in habits and structure the descendants of our carnivorous animal became, the more places they would be enabled to occupy. What applies to one animal will apply throughout all time to all animals—that is, if they vary—for otherwise natural selection can do nothing. So it will be with plants. It has been experimentally proved, that if a plot of ground be sown with one species of grass, and a similar plot be sown with several distinct genera of grasses, a greater number of plants and a greater weight of dry herbage can thus be raised. The same has been found to hold good when first one variety and then several mixed varieties of wheat have been sown on equal spaces of ground. Hence, if any one species of grass were to go on varying, and those varieties were continually selected which differed from each other in at all the same manner as distinct species and genera of grasses differ from each other, a greater number of individual plants of this species of grass, including its modified descendants, would succeed in living on the same piece of ground. And we well know that each species and each variety of grass is annually sowing almost countless seeds; and thus, as it may be said, is striving its utmost to increase its numbers. Consequently, I cannot doubt that in the course of many thousands of generations, the most distinct varieties of any one species of grass would always have the best chance of succeeding and of increasing in numbers, and thus of supplanting the less distinct varieties; and varieties, when rendered very distinct from each other, take the rank of species.

The truth of the principle, that the greatest amount of life can be supported by great diversification of structure, is seen under many natural circumstances . . . .

After the foregoing discussion, which ought to have been much amplified, we may, I think, assume that the modified descendants of any one species will succeed by so much the better as they become more diversified in structure, and are thus enabled to encroach on places occupied by other beings. Now let us see how this principle of great benefit being derived from divergence of character, combined with the principles of natural selection and of extinction, will tend to act.

The accompanying diagram will aid us in understanding this rather perplexing subject. Let A to L represent the species of a genus large in its own country; these species are supposed to resemble each other in unequal degrees, as is so generally the case in nature, and as is represented in the diagram by the letters standing at unequal distances. I have said a large genus, because we have seen in the second chapter, that on an average more of the species of large genera vary than of small genera; and the varying species of the large genera present a greater number of varieties. We have, also, seen that the species, which are the commonest and the most widely-diffused, vary more than rare species with restricted ranges. Let (A) be a common, widely-diffused, and varying species, belonging to a genus large in its own country. The little fan of diverging dotted lines of unequal lengths proceeding from (A), may represent its varying offspring. The variations are supposed to be extremely slight, but of the most diversified nature; they are not supposed all to appear simultaneously, but often after long intervals of time; nor are they all supposed to endure for equal periods. Only those variations which are in some way profitable will be preserved or naturally selected. And here the importance of the principle of benefit being derived from divergence of character comes in; for this will generally lead to the most different or divergent variations (represented by the outer dotted lines) being preserved and accumulated by natural selection. When a dotted line reaches one of the horizontal lines, and is there marked by a small numbered letter, a sufficient amount of variation is supposed to have been accumulated to have formed

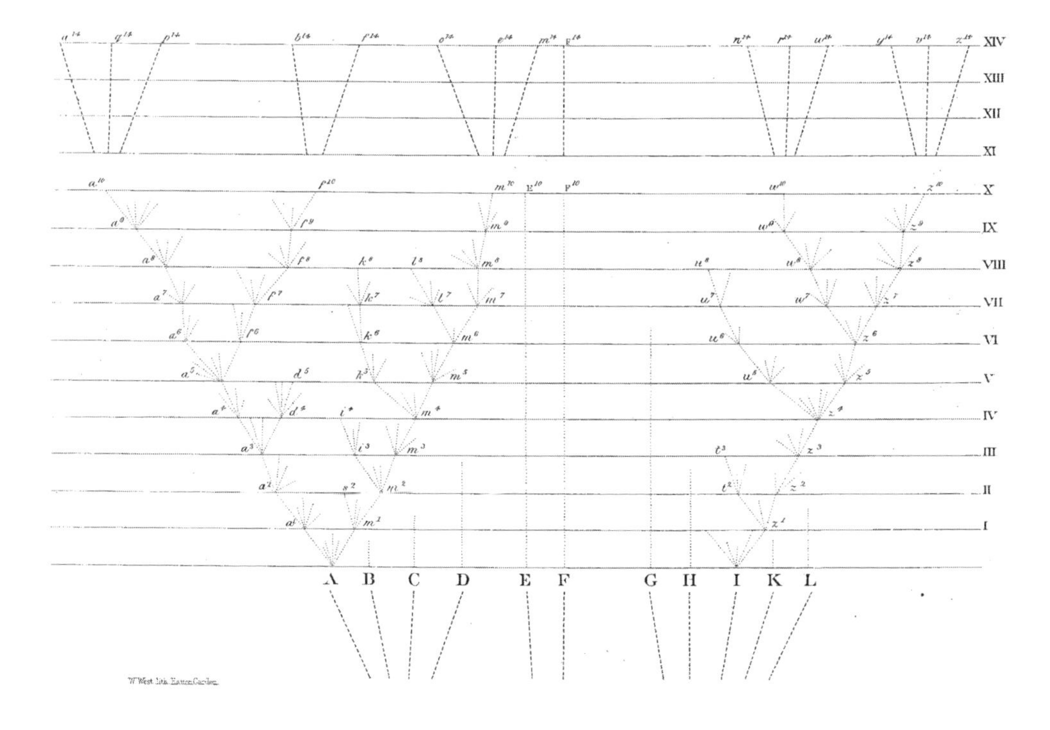

a fairly well-marked variety, such as would be thought worthy of record in a systematic work.

The intervals between the horizontal lines in the diagram, may represent each a thousand generations; but it would have been better if each had represented ten thousand generations. After a thousand generations, species (A) is supposed to have produced two fairly well-marked varieties, namely a1 and m1. These two varieties will generally continue to be exposed to the same conditions which made their parents variable, and the tendency to variability is in itself hereditary, consequently they will tend to vary, and generally to vary in nearly the same manner as their parents varied. Moreover, these two varieties, being only slightly modified forms, will tend to inherit those advantages which made their common parent (A) more numerous than most of the other inhabitants of the same country; they will likewise partake of those more general advantages which made the genus to which the parent-species belonged, a large genus in its own country. And these circumstances we know to be favourable to the production of new varieties.

If, then, these two varieties be variable, the most divergent of their variations will generally be preserved during the next thousand generations. And after this interval, variety a1 is supposed in the diagram to have produced variety a2, which will, owing to the principle of divergence, differ more from (A) than did variety a1. Variety m1 is supposed to have produced two varieties, namely m2 and s2, differing from each other, and more considerably from their common parent (A). We may continue the process by similar steps for any length of time; some of the varieties, after each thousand generations, producing only a single variety, but in a more and more modified condition, some producing two or three varieties, and some failing to produce any. Thus the varieties or modified descendants, proceeding from the common parent (A), will generally go on increasing in number and diverging in character. In the diagram the process is represented up to the ten-thousandth generation, and under a condensed and simplified form up to the fourteen-thousandth generation.

But I must here remark that I do not suppose that the process ever goes on so regularly as is represented in the diagram, though in itself made somewhat irregular. I am far from thinking that the most divergent varieties will invariably prevail and multiply: a medium form may often long endure, and may or may not produce more than one modified descendant; for natural selection will always act according to the nature of the places which are either unoccupied or not perfectly occupied by other beings; and this will depend on infinitely complex relations. But as a general rule, the more diversified in structure the descendants from any one species can be rendered, the more places they will be enabled to seize on, and the more their modified progeny will be increased. In our diagram the line of succession is broken at regular intervals by small numbered letters marking the successive forms which have become sufficiently distinct to be recorded as varieties. But these breaks are imaginary, and might have been inserted anywhere, after intervals long enough to have allowed the accumulation of a considerable amount of divergent variation.

...

But during the process of modification, represented in the diagram, another of our principles, namely that of extinction, will have played an important part. As in each fully stocked country natural selection necessarily acts by the selected form having some advantage in the struggle for life over other forms, there will be a constant tendency in the improved descendants of any one species to supplant and exterminate in each stage of descent their predecessors and their original parent. For it should be remembered that the competition will generally be most severe between those forms which are most nearly related to each other in habits, constitution, and structure. Hence all the intermediate forms between the earlier and later states, that is between the less and more improved state of a species, as well as the original parent-species itself, will generally tend to become extinct. So it probably will be with many whole collateral lines of descent, which will be conquered by later and improved lines of descent. If, however, the modified offspring of a species get into some distinct country, or become quickly adapted to some quite new station, in which child and parent do not come into competition, both may continue to exist.

...

It is worthwhile to reflect for a moment on the character of the new species F14, which is supposed not to have diverged much in character, but to have retained the form of (F), either unaltered or altered only in a slight degree. In this case, its affinities to the other fourteen new species will be of a curious and circuitous nature. Having descended from a form which stood between the two parent-species (A) and (I), now supposed to be extinct and unknown, it will be in some degree intermediate in character between the two groups descended from these species. But as these two groups have gone on diverging in character from the type of their parents, the new species (F14) will not be directly intermediate between them, but rather between types of the two groups; and every naturalist will be able to bring some such case before his mind.

In the diagram, each horizontal line has hitherto been supposed to represent a thousand generations, but each may represent a million or hundred million generations, and likewise a section of the successive strata of the earth's crust including extinct remains. We shall, when we come to our chapter on Geology, have to refer again to this subject, and I think we shall then see that the diagram throws light on the affinities of extinct beings, which, though generally belonging to the same orders, or families, or genera, with those now living, yet are often, in some degree, intermediate in character between existing groups; and we can understand this fact, for the extinct species lived at very ancient epochs when the branching lines of descent had diverged less.

I see no reason to limit the process of modification, as now explained, to the formation of genera alone. If, in our diagram, we suppose the amount of change represented by each successive group of diverging dotted lines to be very great, the forms marked a14 to p14, those marked b14 and f14, and those marked o14 to m14, will form three very distinct genera. We shall also have two very distinct genera descended from (I) and as these latter two genera, both from continued divergence of character and from inheritance from a different parent, will differ widely from the three genera descended from (A), the two little groups of genera will form two distinct families, or even orders, according to the amount of divergent modification supposed to be represented in the diagram. And the two new families, or orders, will have descended from two species of the original genus; and these two species are supposed to have descended from one species of a still more ancient and unknown genus.

We have seen that in each country it is the species of the larger genera which oftenest present varieties or incipient species. This, indeed, might have been expected; for as natural selection acts through one form having some advantage over other forms in the struggle for existence, it will chiefly act on those which already have some advantage; and the largeness of any group shows that its species have inherited from a common ancestor some advantage in common. Hence, the struggle for the production of new and modified descendants, will mainly lie between the larger groups, which are all trying to increase in number. One large group will slowly conquer another large group, reduce its numbers, and thus lessen its chance of further variation and improvement. Within the same large group, the later and more highly perfected sub-groups, from branching out and seizing on many new places in the polity of Nature, will constantly tend to supplant and destroy the earlier and less improved sub-groups. Small and broken groups and sub-groups will finally tend to disappear. Looking to the future, we can predict that the groups of organic beings which are now large and triumphant, and which are least broken up, that is, which as yet have suffered least extinction, will for a long period continue to increase. But which groups will ultimately prevail, no man can predict; for we well know that many groups, formerly most extensively developed, have now become extinct. Looking still more remotely to the future, we may predict that, owing to the continued and steady increase of the larger groups, a multitude of smaller groups will become utterly extinct, and leave no modified descendants; and consequently that of the species living at any one period, extremely few will transmit descendants to a remote futurity. I shall have to return to this subject in the chapter on Classification, but I may add that on this view of extremely few of the more ancient species having transmitted descendants, and on the view of all the descendants of the same species making a class, we can understand how it is that there exist but very few classes in each main division of the animal and vegetable kingdoms. Although extremely few of the most ancient species may now have living and modified descendants, yet at the most remote geological period, the earth may have been as well peopled with many species of many genera, families, orders, and classes, as at the present day.

Summary of Chapter

If during the long course of ages and under varying conditions of life, organic beings vary at all in the several parts of their organisation, and I think this cannot be disputed; if there be, owing to the high geometrical powers of increase of each species, at some age, season, or year, a severe struggle for life, and this certainly cannot be disputed; then, considering the infinite complexity of the relations of all organic beings to each other and to their conditions of existence, causing an infinite diversity in structure, constitution, and habits, to be advantageous to them, I think it would be a most extraordinary fact if no variation ever had occurred useful to each being's own welfare, in the same way as so many variations have occurred useful to man. But if variations useful to any organic being do occur, assuredly individuals thus characterised will have the best chance of being preserved in the struggle for life; and from the strong principle of inheritance they will tend to produce offspring similarly characterised. This principle of preservation, I have called, for the sake of brevity, Natural Selection. Natural selection, on the principle of qualities being inherited at corresponding ages, can modify the egg, seed, or young, as easily as the adult. Amongst many animals, sexual selection will give its aid to ordinary selection, by assuring to the most vigorous and best adapted males the greatest number of offspring. Sexual selection will also give characters useful to the males alone, in their struggles with other males.

Whether natural selection has really thus acted in nature, in modifying and adapting the various forms of life to their several conditions and stations, must be judged of by the general tenour and balance of evidence given in the following chapters. But we already see how it entails extinction; and how largely extinction has acted in the world's history, geology plainly declares. Natural selection, also, leads to divergence of character; for more living beings can be supported on the same area the more they diverge in structure, habits, and constitution, of which we see proof by looking at the inhabitants of any small spot or at naturalised productions. Therefore during the modification of the descendants of any one species, and during the incessant struggle of all species to increase in numbers, the more diversified these descendants become, the better will be their chance of succeeding in the battle of life. Thus the small differences distinguishing varieties of the same species, will steadily tend to increase till they come to equal the greater differences between species of the same genus, or even of distinct genera.

We have seen that it is the common, the widely-diffused, and widely-ranging species, belonging to the larger genera, which vary most; and these will tend to transmit to their modified offspring that superiority which now makes them dominant in their own countries. Natural selection, as has just been remarked, leads to divergence of character and to much extinction of the less improved and intermediate forms of life. On these principles, I believe, the nature of the affinities of all organic beings may be explained. It is a truly wonderful fact—the wonder of which we are apt to overlook from familiarity—that all animals and all plants throughout all time and space should be related to each other in group subordinate to group, in the manner which we everywhere behold—namely, varieties of the same species most closely related together, species of the same genus less closely and unequally related together, forming sections and sub-genera, species of distinct genera much less closely related, and genera related in different degrees, forming sub-families, families, orders, sub-classes, and classes. The several subordinate groups in any class cannot be ranked in a single file, but seem rather to be clustered round points, and these round other points, and so on in almost endless cycles. On the view that each species has been independently created, I can see no explanation of this great fact in the classification of all organic beings; but, to the best of my judgment, it is explained through inheritance and the complex action of natural selection, entailing extinction and divergence of character, as we have seen illustrated in the diagram.

The affinities of all the beings of the same class have sometimes been represented by a great tree. I believe this simile largely speaks the truth. The green and budding twigs may represent existing species; and those produced during each former year may represent the long succession of extinct species. At each period of growth all the growing twigs have tried to branch out on all sides, and to overtop and kill the surrounding twigs and branches, in the same manner as species and groups of species have tried to overmaster other species in the great battle for life. The limbs divided into great branches, and these into lesser and lesser branches, were themselves once, when the tree was small, budding twigs; and this connexion of the former and present buds by ramifying branches may well represent the classification of all extinct and living species in groups subordinate to groups. Of the many twigs which flourished when the tree was a mere bush, only two or three, now grown into great branches, yet survive and bear all the other branches; so with the species which lived during long-past geological periods, very few now have living and modified descendants. From the first growth of the tree, many a limb and branch has decayed and dropped off; and these lost branches of various sizes may represent those whole orders, families, and genera which have now no living representatives, and which are known to us only from having been found in a fossil state. As we here and there see a thin straggling branch springing from a fork low down in a tree, and which by some chance has been favoured and is still alive on its summit, so we occasionally see an animal like the Ornithorhynchus or Lepidosiren, which in some small degree connects by its affinities two large branches of life, and which has apparently been saved from fatal competition by having inhabited a protected station. As buds give rise by growth to fresh buds, and these, if vigorous, branch out and overtop on all sides many a feebler branch, so by generation I believe it has been with the great Tree of Life, which fills with its dead and broken branches the crust of the earth, and covers the surface with its ever branching and beautiful ramifications.

. . .

9. On the Imperfection of the Geological Record

On the absence of intermediate varieties at the present day. On the nature of extinct intermediate varieties; on their number. On the vast lapse of time, as inferred from the rate of deposition and of denudation. On the poorness of our palaeontological collections. On the intermittence of geological formations. On the absence of intermediate varieties in any one formation. On the sudden appearance of groups of species. On their sudden appearance in the lowest known fossiliferous strata.

In the sixth chapter I enumerated the chief objections which might be justly urged against the views maintained in this volume. Most of them have now been discussed. One, namely the distinctness of specific forms, and their not being blended together by innumerable transitional links, is a very obvious difficulty. I assigned reasons why such links do not commonly occur at the present day, under the circumstances apparently most favourable for their presence, namely on an extensive and continuous area with graduated physical conditions. I endeavoured to show, that the life of each species depends in a more important manner on the presence of other already defined organic forms, than on climate; and, therefore, that the really governing conditions of life do not graduate away quite insensibly like heat or moisture. I endeavoured, also, to show that intermediate varieties, from existing in lesser numbers than the forms which they connect, will generally be beaten out and exterminated during the course of further modification and improvement. The main cause, however, of innumerable intermediate links not now occurring everywhere throughout nature depends on the very process of natural selection, through which new varieties continually take the places of and exterminate their parent-forms. But just in proportion as this process of extermination has acted on an enormous scale, so must the number of intermediate varieties, which have formerly existed on the earth, be truly enormous. Why then is not every geological formation and every stratum full of such intermediate links? Geology assuredly does not reveal any such finely graduated organic chain; and this, perhaps, is the most obvious and gravest objection which can be urged against my theory. The explanation lies, as I believe, in the extreme imperfection of the geological record.

In the first place it should always be borne in mind what sort of intermediate forms must, on my theory, have formerly existed. I have found it difficult, when looking at any two species, to avoid picturing to myself, forms DIRECTLY intermediate between them. But this is a wholly false view; we should always look for forms intermediate between each species and a common but unknown progenitor; and the progenitor will generally have differed in some respects from all its modified descendants. To give a simple illustration: the fantail and pouter pigeons have both descended from the rock-pigeon; if we possessed all the intermediate varieties which have ever existed, we should have an extremely close series between both and the rock-pigeon; but we should have no varieties directly intermediate between the fantail and pouter; none, for instance, combining a tail somewhat expanded with a crop somewhat enlarged, the characteristic features of these two breeds. These two breeds, moreover, have become so much modified, that if we had no historical or indirect evidence regarding their origin, it would not have been possible to have determined from a mere comparison of their structure with that of the rock-pigeon, whether they had descended from this species or from some other allied species, such as C. oenas.

So with natural species, if we look to forms very distinct, for instance to the horse and tapir, we have no reason to suppose that links ever existed directly intermediate between them, but between each and an unknown common parent. The common parent will have had in its whole organisation much general resemblance to the tapir and to the horse; but in some points of structure may have differed considerably from both, even perhaps more than they differ from each other. Hence in all such cases, we should be unable to recognise the parent-form of any two or more species, even if we closely compared the structure of the parent with that of its modified descendants, unless at the same time we had a nearly perfect chain of the intermediate links.

It is just possible by my theory, that one of two living forms might have descended from the other; for instance, a horse from a tapir; and in this case DIRECT intermediate links will have existed between them. But such a case would imply that one form had remained for a very long period unaltered, whilst its descendants had undergone a vast amount of change; and the principle of competition between organism and organism, between child and parent, will render this a very rare event; for in all cases the new and improved forms of life will tend to supplant the old and unimproved forms.

By the theory of natural selection all living species have been connected with the parent-species of each genus, by differences not greater than we see between the varieties of the same species at the present day; and these parent-species, now generally extinct, have in their turn been similarly connected with more ancient species; and so on backwards, always converging to the common ancestor of each great class. So that the number of intermediate and transitional links, between all living and extinct species, must have been inconceivably great. But assuredly, if this theory be true, such have lived upon this earth.

On the Lapse of Time

Independently of our not finding fossil remains of such infinitely numerous connecting links, it may be objected, that time will not have sufficed for so great an amount of organic change, all changes having been effected very slowly through natural selection. It is hardly possible for me even to recall to the reader, who may not be a practical geologist, the facts leading the mind feebly to comprehend the lapse of time. He who can read Sir Charles Lyell's grand work on the Principles of Geology, which the future historian will recognise as having produced a revolution in natural science, yet does not admit how incomprehensibly vast have been the past periods of time, may at once close this volume. Not that it suffices to study the Principles of Geology, or to read special treatises by different observers on separate formations, and to mark how each author attempts to give an inadequate idea of the duration of each formation or even each stratum. A man must for years examine for himself great piles of superimposed strata, and watch the sea at work grinding down old rocks and making fresh sediment, before he can hope to comprehend anything of the lapse of time, the monuments of which we see around us.

It is good to wander along lines of sea-coast, when formed of moderately hard rocks, and mark the process of degradation. The tides in most cases reach the cliffs only for a short time twice a day, and the waves eat into them only when they are charged with sand or pebbles; for there is reason to believe that pure water can effect little or nothing in wearing away rock. At last the base of the cliff is undermined, huge fragments fall down, and these remaining fixed, have to be worn away, atom by atom, until reduced in size they can be rolled about by the waves, and then are more quickly ground into pebbles, sand, or mud. But how often do we see along the bases of retreating cliffs rounded boulders, all thickly clothed by marine productions, showing how little they are abraded and how seldom they are rolled about! Moreover, if we follow for a few miles any line of rocky cliff, which is undergoing degradation, we find that it is only here and there, along a short length or round a promontory, that the cliffs are at the present time suffering. The appearance of the surface and the vegetation show that elsewhere years have elapsed since the waters washed their base.

. . .

I am tempted to give one other case, the well-known one of the denudation of the Weald. Though it must be admitted that the denudation of the Weald has been a mere trifle, in comparison with that which has removed masses of our palaeozoic strata, in parts ten thousand feet in thickness, as shown in Professor Ramsay's masterly memoir on this subject. Yet it is an admirable lesson to stand on the North Downs and to look at the distant South Downs; for, remembering that at no great distance to the west the northern and southern escarpments meet and close, one can safely picture to oneself the great dome of rocks which must have covered up the Weald within so limited a period as since the latter part of the Chalk formation. The distance from the northern to the southern Downs is about 22 miles, and the thickness of the several formations is on an average about 1100 feet, as I am informed by Professor Ramsay. But if, as some geologists suppose, a range of older rocks underlies the Weald, on the flanks of which the overlying sedimentary deposits might have accumulated in thinner masses than elsewhere, the above estimate would be erroneous; but this source of doubt probably would not greatly affect the estimate as applied to the western extremity of the district. If, then, we knew the rate at which the sea commonly wears away a line of cliff of any given height, we could measure the time requisite to have denuded the Weald. This, of course, cannot be done; but we may, in order to form some crude notion on the subject, assume that the sea would eat into cliffs 500 feet in height at the rate of one inch in a century. This will at first appear much too small an allowance; but it is the same as if we were to assume a cliff one yard in height to be eaten back along a whole line of coast at the rate of one yard in nearly every twenty-two years. I doubt whether any rock, even as soft as chalk, would yield at this rate excepting on the most exposed coasts; though no doubt the degradation of a lofty cliff would be more rapid from the breakage of the fallen fragments. On the other hand, I do not believe that any line of coast, ten or twenty miles in length, ever suffers degradation at the same time along its whole indented length; and we must remember that almost all strata contain harder layers or nodules, which from long resisting attrition form a breakwater at the base. Hence, under ordinary circumstances, I conclude that for a cliff 500 feet in height, a denudation of one inch per century for the whole length would be an ample allowance. At this rate, on the above data, the denudation of the Weald must have required 306,662,400 years; or say three hundred million years.

The action of fresh water on the gently inclined Wealden district, when upraised, could hardly have been great, but it would somewhat reduce the above estimate. On the other hand, during oscillations of level, which we know this area has undergone, the surface may have existed for millions of years as land, and thus have escaped the action of the sea: when deeply submerged for perhaps equally long periods, it would, likewise, have escaped the action of the coast-waves. So that in all probability a far longer period than 300 million years has elapsed since the latter part of the Secondary period.

I have made these few remarks because it is highly important for us to gain some notion, however imperfect, of the lapse of years. During each of these years, over the whole world, the land and the water has been peopled by hosts of living forms. What an infinite number of generations, which the mind cannot grasp, must have succeeded each other in the long roll of years! Now turn to our richest geological museums, and what a paltry display we behold!

On the Poorness of Our Palaeontological Collections

That our palaeontological collections are very imperfect, is admitted by everyone. The remark of that admirable palaeontologist, the late Edward Forbes, should not be forgotten, namely, that numbers of our fossil species are known and named from single and often broken specimens, or from a few specimens collected on some one spot. Only a small portion of the surface of the earth has been geologically explored, and no part with sufficient care, as the important discoveries made every year in Europe prove. No organism wholly soft can be preserved. Shells and bones will decay and disappear when left on the bottom of the sea, where sediment is not accumulating .... I suspect that but few of the very many animals which live on the beach between high and low watermark are preserved.

...

If then, there be some degree of truth in these remarks, we have no right to expect to find in our geological formations, an infinite number of those fine transitional forms, which on my theory assuredly have connected all the past and present species of the same group into one long and branching chain of life. We ought only to look for a few links, some more closely, some more distantly related to each other; and these links, let them be ever so close, if found in different stages of the same formation, would, by most palaeontologists, be ranked as distinct species. But I do not pretend that I should ever have suspected how poor a record of the mutations of life, the best preserved geological section presented, had not the difficulty of our not discovering innumerable transitional links between the species which appeared at the commencement and close of each formation, pressed so hardly on my theory.

On the Sudden Appearance of Whole Groups of Allied Species

The abrupt manner in which whole groups of species suddenly appear in certain formations, has been urged by several palaeontologists, for instance, by Agassiz, Pictet, and by none more forcibly than by Professor Sedgwick, as a fatal objection to the belief in the transmutation of species. If numerous species, belonging to the same genera or families, have really started into life all at once, the fact would be fatal to the theory of descent with slow modification through natural selection. For the development of a group of forms, all of which have descended from some one progenitor, must have been an extremely slow process; and the progenitors must have lived long ages before their modified descendants. But we continually over-rate the perfection of the geological record, and falsely infer, because certain genera or families have not been found beneath a certain stage, that they did not exist before that stage. We continually forget how large the world is, compared with the area over which our geological formations have been carefully examined; we forget that groups of species may elsewhere have long existed and have slowly multiplied before they invaded the ancient archipelagoes of Europe and of the United States. We do not make due allowance for the enormous intervals of time, which have probably elapsed between our consecutive formations,—longer perhaps in some cases than the time required for the accumulation of each formation. These intervals will have given time for the multiplication of species from some one or some few parent-forms; and in the succeeding formation such species will appear as if suddenly created.

...

On the Sudden Appearance of Groups of Allied Species in the Lowest Known Fossiliferous Strata

There is another and allied difficulty, which is much graver. I allude to the manner in which numbers of species of the same group, suddenly appear in the lowest known fossiliferous rocks. Most of the arguments which have convinced me that all the existing species of the same group have descended from one progenitor, apply with nearly equal force to the earliest known species. For instance, I cannot doubt that all the Silurian trilobites have descended from some one crustacean, which must have lived long before the Silurian age, and which probably differed greatly from any known animal. Some of the most ancient Silurian animals, as the Nautilus, Lingula, etc., do not differ much from living species; and it cannot on my theory be supposed, that these old species were the progenitors of all the species of the orders to which they belong, for they do not present characters in any degree intermediate between them. If, moreover, they had been the progenitors of these orders, they would almost certainly have been long ago supplanted and exterminated by their numerous and improved descendants.

Consequently, if my theory be true, it is indisputable that before the lowest Silurian stratum was deposited, long periods elapsed, as long as, or probably far longer than, the whole interval from the Silurian age to the present day; and that during these vast, yet quite unknown, periods of time, the world swarmed with living creatures.

To the question why we do not find records of these vast primordial periods, I can give no satisfactory answer.

...

For my part, following out Lyell's metaphor, I look at the natural geological record, as a history of the world imperfectly kept, and written in a changing dialect; of this history we possess the last volume alone, relating only to two or three countries. Of this volume, only here and there a short chapter has been preserved; and of each page, only here and there a few lines. Each word of the slowly-changing language, in which the history is supposed to be written, being more or less different in the interrupted succession of chapters, may represent the apparently abruptly changed forms of life, entombed in our consecutive, but widely separated formations. On this view, the difficulties above discussed are greatly diminished, or even disappear.

13. Mutual Affinities of Organic Beings: Morphology: Embryology: Rudimentary Organs

CLASSIFICATION, groups subordinate to groups. Natural system. Rules and difficulties in classification, explained on the theory of descent with modification. Classification of varieties. Descent always used in classification. Analogical or adaptive characters. Affinities, general, complex and radiating. Extinction separates and defines groups. MORPHOLOGY, between members of the same class, between parts of the same individual. EMBRYOLOGY, laws of, explained by variations not supervening at an early age, and being inherited at a corresponding age. RUDIMENTARY ORGANS; their origin explained. Summary.

From the first dawn of life, all organic beings are found to resemble each other in descending degrees, so that they can be classed in groups under groups. This classification is evidently not arbitrary like the grouping of the stars in constellations. The existence of groups would have been of simple signification, if one group had been exclusively fitted to inhabit the land, and another the water; one to feed on flesh, another on vegetable matter, and so on; but the case is widely different in nature; for it is notorious how commonly members of even the same subgroup have different habits. In our second and fourth chapters, on Variation and on Natural Selection, I have attempted to show that it is the widely ranging, the much diffused and common, that is the dominant species belonging to the larger genera, which vary most. The varieties, or incipient species, thus produced ultimately become converted, as I believe, into new and distinct species; and these, on the principle of inheritance, tend to produce other new and dominant species. Consequently the groups which are now large, and which generally include many dominant species, tend to go on increasing indefinitely in size. I further attempted to show that from the varying descendants of each species trying to occupy as many and as different places as possible in the economy of nature, there is a constant tendency in their characters to diverge. This conclusion was supported by looking at the great diversity of the forms of life which, in any small area, come into the closest competition, and by looking to certain facts in naturalisation.

I attempted also to show that there is a constant tendency in the forms which are increasing in number and diverging in character, to supplant and exterminate the less divergent, the less improved, and preceding forms. I request the reader to turn to the diagram illustrating the action, as formerly explained, of these several principles; and he will see that the inevitable result is that the modified descendants proceeding from one progenitor become broken up into groups subordinate to groups. In the diagram each letter on the uppermost line may represent a genus including several species; and all the genera on this line form together one class, for all have descended from one ancient but unseen parent, and, consequently, have inherited something in common. But the three genera on the left hand have, on this same principle, much in common, and form a sub-family, distinct from that including the next two genera on the right hand, which diverged from a common parent at the fifth stage of descent. These five genera have also much, though less, in common; and they form a family distinct from that including the three genera still further to the right hand, which diverged at a still earlier period. And all these genera, descended from (A), form an order distinct from the genera descended from (I). So that we here have many species descended from a single progenitor grouped into genera; and the genera are included in, or subordinate to, sub-families, families, and orders, all united into one class. Thus, the grand fact in natural history of the subordination of group under group, which, from its familiarity, does not always sufficiently strike us, is in my judgment fully explained.

Naturalists try to arrange the species, genera, and families in each class, on what is called the Natural System. But what is meant by this system? Some authors look at it merely as a scheme for arranging together those living objects which are most alike, and for separating those which are most unlike; or as an artificial means for enunciating, as briefly as possible, general propositions,—that is, by one sentence to give the characters common, for instance, to all mammals, by another those common to all carnivora, by another those common to the dog-genus, and then by adding a single sentence, a full description is given of each kind of dog. The ingenuity and utility of this system are indisputable. But many naturalists think that something more is meant by the Natural System; they believe that it reveals the plan of the Creator; but unless it be specified whether order in time or space, or what else is meant by the plan of the Creator, it seems to me that nothing is thus added to our knowledge. Such expressions as that famous one of Linnaeus, and which we often meet with in a more or less concealed form, that the characters do not make the genus, but that the genus gives the characters, seem to imply that something more is included in our classification, than mere resemblance. I believe that something more is included; and that propinquity of descent,—the only known cause of the similarity of organic beings,—is the bond, hidden as it is by various degrees of modification, which is partially revealed to us by our classifications.

Let us now consider the rules followed in classification, and the difficulties which are encountered on the view that classification either gives some unknown plan of creation, or is simply a scheme for enunciating general propositions and of placing together the forms most like each other. It might have been thought (and was in ancient times thought) that those parts of the structure which determined the habits of life, and the general place of each being in the economy of nature, would be of very high importance in classification. Nothing can be more false. No one regards the external similarity of a mouse to a shrew, of a dugong to a whale, of a whale to a fish, as of any importance.

. . .

All the foregoing rules and aids and difficulties in classification are explained, if I do not greatly deceive myself, on the view that the natural system is founded on descent with modification; that the characters which naturalists consider as showing true affinity between any two or more species, are those which have been inherited from a common parent, and, in so far, all true classification is genealogical; that community of descent is the hidden bond which naturalists have been unconsciously seeking, and not some unknown plan of creation, or the enunciation of general propositions, and the mere putting together and separating objects more or less alike.

. . .

Thus, on the view which I hold, the natural system is genealogical in its arrangement, like a pedigree; but the degrees of modification which the different groups have undergone, have to be expressed by ranking them under different so-called genera, sub-families, families, sections, orders, and classes.

. . .

We can understand, on these views, the very important distinction between real affinities and analogical or adaptive resemblances. Lamarck first called attention to this distinction, and he has been ably followed by Macleay and others. The resemblance, in the shape of the body and in the fin-like anterior limbs, between the dugong, which is a pachydermatous animal, and the whale, and between both these mammals and fishes, is analogical. Amongst insects there are innumerable instances: thus Linnaeus, misled by external appearances, actually classed an homopterous insect as a moth. We see something of the same kind even in our domestic varieties, as in the thickened stems of the common and Swedish turnip. The resemblance of the greyhound and racehorse is hardly more fanciful than the analogies which have been drawn by some authors between very distinct animals. On my view of characters being of real importance for classification, only in so far as they reveal descent, we can clearly understand why analogical or adaptive character, although of the utmost importance to the welfare of the being, are almost valueless to the systematist. For animals, belonging to two most distinct lines of descent, may readily become adapted to similar conditions, and thus assume a close external resemblance; but such resemblances will not reveal—will rather tend to conceal their blood-relationship to their proper lines of descent. We can also understand the apparent paradox, that the very same characters are analogical when one class or order is compared with another, but give true affinities when the members of the same class or order are compared one with another: thus the shape of the body and fin-like limbs are only analogical when whales are compared with fishes, being adaptations in both classes for swimming through the water; but the shape of the body and fin-like limbs serve as characters exhibiting true affinity between the several members of the whale family; for these cetaceans agree in so many characters, great and small, that we cannot doubt that they have inherited their general shape of body and structure of limbs from a common ancestor. So it is with fishes.

...

Morphology

We have seen that the members of the same class, independently of their habits of life, resemble each other in the general plan of their organisation. This resemblance is often expressed by the term "unity of type"; or by saying that the several parts and organs in the different species of the class are homologous. The whole subject is included under the general name of Morphology. This is the most interesting department of natural history, and may be said to be its very soul. What can be more curious than that the hand of a man, formed for grasping, that of a mole for digging, the leg of the horse, the paddle of the porpoise, and the wing of the bat, should all be constructed on the same pattern, and should include the same bones, in the same relative positions? Geoffroy St. Hilaire has insisted strongly on the high importance of relative connexion in homologous organs: the parts may change to almost any extent in form and size, and yet they always remain connected together in the same order. We never find, for instance, the bones of the arm and forearm, or of the thigh and leg, transposed. Hence the same names can be given to the homologous bones in widely different animals. We see the same great law in the construction of the mouths of insects: what can be more different than the immensely long spiral proboscis of a sphinx-moth, the curious folded one of a bee or bug, and the great jaws of a beetle?—yet all these organs, serving for such different purposes, are formed by infinitely numerous modifications of an upper lip, mandibles, and two pairs of maxillae. Analogous laws govern the construction of the mouths and limbs of crustaceans. So it is with the flowers of plants.

Nothing can be more hopeless than to attempt to explain this similarity of pattern in members of the same class, by utility or by the doctrine of final causes. The hopelessness of the attempt has been expressly admitted by Owen in his most interesting work on the 'Nature of Limbs.' On the ordinary view of the independent creation of each being, we can only say that so it is;—that it has so pleased the Creator to construct each animal and plant.

The explanation is manifest on the theory of the natural selection of successive slight modifications,—each modification being profitable in some way to the modified form, but often affecting by correlation of growth other parts of the organisation. In changes of this nature, there will be little or no tendency to modify the original pattern, or to transpose parts. The bones of a limb might be shortened and widened to any extent, and become gradually enveloped in thick membrane, so as to serve as a fin; or a webbed foot might have all its bones, or certain bones, lengthened to any extent, and the membrane connecting them increased to any extent, so as to serve as a wing: yet in all this great amount of modification there will be no tendency to alter the framework of bones or the relative connexion of the several parts. If we suppose that the ancient progenitor, the archetype as it may be called, of all mammals, had its limbs constructed on the existing general pattern, for whatever purpose they served, we can at once perceive the plain signification of the homologous construction of the limbs throughout the whole class.

. . .

How inexplicable are these facts on the ordinary view of creation! Why should the brain be enclosed in a box composed of such numerous and such extraordinarily shaped pieces of bone? As Owen has remarked, the benefit derived from the yielding of the separate pieces in the act of parturition of mammals, will by no means explain the same construction in the skulls of birds. Why should similar bones have been created in the formation of the wing and leg of a bat, used as they are for such totally different purposes? Why should one crustacean, which has an extremely complex mouth formed of many parts, consequently always have fewer legs; or conversely, those with many legs have simpler mouths? Why should the sepals, petals, stamens, and pistils in any individual flower, though fitted for such widely different purposes, be all constructed on the same pattern?

On the theory of natural selection, we can satisfactorily answer these questions. In the vertebrata, we see a series of internal vertebrae bearing certain processes and appendages; in the articulata, we see the body divided into a series of segments, bearing external appendages; and in flowering plants, we see a series of successive spiral whorls of leaves. An indefinite repetition of the same part or organ is the common characteristic (as Owen has observed) of all low or little-modified forms; therefore we may readily believe that the unknown progenitor of the vertebrata possessed many vertebrae; the unknown progenitor of the articulata, many segments; and the unknown progenitor of flowering plants, many spiral whorls of leaves. We have formerly seen that parts many times repeated are eminently liable to vary in number and structure; consequently it is quite probable that natural selection, during a long-continued course of modification, should have seized on a certain number of the primordially similar elements, many times repeated, and have adapted them to the most diverse purposes. And as the whole amount of modification will have been effected by slight successive steps, we need not wonder at discovering in such parts or organs, a certain degree of fundamental resemblance, retained by the strong principle of inheritance.

. . .

Embryology

It has already been casually remarked that certain organs in the individual, which when mature become widely different and serve for different purposes, are in the embryo exactly alike. The embryos, also, of distinct animals within the same class are often strikingly similar: a better proof of this cannot be given, than a circumstance mentioned by Agassiz, namely, that having forgotten to ticket the embryo of some vertebrate animal, he cannot now tell whether it be that of a mammal, bird, or reptile. The vermiform larvae of moths, flies, beetles, etc., resemble each other much more closely than do the mature insects; but in the case of larvae, the embryos are active, and have been adapted for special lines of life. A trace of the law of embryonic resemblance, sometimes lasts till a rather late age: thus birds of the same genus, and of closely allied genera, often resemble each other in their first and second plumage; as we see in the spotted feathers in the thrush group. In the cat tribe, most of the species are striped or spotted in lines; and stripes can be plainly distinguished in the whelp of the lion. We occasionally though rarely see something of this kind in plants: thus the embryonic leaves of the ulex or furze, and the first leaves of the phyllodineous acaceas, are pinnate or divided like the ordinary leaves of the leguminosae.

The points of structure, in which the embryos of widely different animals of the same class resemble each other, often have no direct relation to their conditions of existence. We cannot, for instance, suppose that in the embryos of the vertebrata the peculiar loop-like course of the arteries near the branchial slits are related to similar conditions,—in the young mammal which is nourished in the womb of its mother, in the egg of the bird which is hatched in a nest, and in the spawn of a frog under water. We have no more reason to believe in such a relation, than we have to believe that the same bones in the hand of a man, wing of a bat, and fin of a porpoise, are related to similar conditions of life. No one will suppose that the stripes on the whelp of a lion, or the spots on the young blackbird, are of any use to these animals, or are related to the conditions to which they are exposed.

. . .

But there is no obvious reason why, for instance, the wing of a bat, or the fin of a porpoise, should not have been sketched out with all the parts in proper proportion, as soon as any structure became visible in the embryo. And in some whole groups of animals and in certain members of other groups, the embryo does not at any period differ widely from the adult: thus Owen has remarked in regard to cuttle-fish, "there is no metamorphosis; the cephalopodic character is manifested long before the parts of the embryo are completed"; and again in spiders, "there is nothing worthy to be called a metamorphosis." The larvae of insects, whether adapted to the most diverse and active habits, or quite inactive, being fed by their parents or placed in the midst of proper nutriment, yet nearly all pass through a similar worm-like stage of development; but in some few cases, as in that of Aphis, if we look to the admirable drawings by Professor Huxley of the development of this insect, we see no trace of the vermiform stage.

How, then, can we explain these several facts in embryology,—namely the very general, but not universal difference in structure between the embryo and the adult;—of parts in the same individual embryo, which ultimately become very unlike and serve for diverse purposes, being at this early period of growth alike;—of embryos of different species within the same class, generally, but not universally, resembling each other;—of the structure of the embryo not being closely related to its conditions of existence, except when the embryo becomes at any period of life active and has to provide for itself;—of the embryo apparently having sometimes a higher organisation than the mature animal, into which it is developed. I believe that all these facts can be explained, as follows, on the view of descent with modification.

...

The fore-limbs, for instance, which served as legs in the parent-species, may become, by a long course of modification, adapted in one descendant to act as hands, in another as paddles, in another as wings; and on the above two principles—namely of each successive modification supervening at a rather late age, and being inherited at a corresponding late age—the fore-limbs in the embryos of the several descendants of the parent-species will still resemble each other closely, for they will not have been modified. But in each individual new species, the embryonic fore-limbs will differ greatly from the fore-limbs in the mature animal; the limbs in the latter having undergone much modification at a rather late period of life, and having thus been converted into hands, or paddles, or wings. Whatever influence long-continued exercise or use on the one hand, and disuse on the other, may have in modifying an organ, such influence will mainly affect the mature animal, which has come to its full powers of activity and has to gain its own living; and the effects thus produced will be inherited at a corresponding mature age. Whereas the young will remain unmodified, or be modified in a lesser degree, by the effects of use and disuse.

...

As all the organic beings, extinct and recent, which have ever lived on this earth have to be classed together, and as all have been connected by the finest gradations, the best, or indeed, if our collections were nearly perfect, the only possible arrangement, would be genealogical. Descent being on my view the hidden bond of connexion which naturalists have been seeking under the term of the natural system. On this view we can understand how it is that, in the eyes of most naturalists, the structure of the embryo is even more important for classification than that of the adult. For the embryo is the animal in its less modified state; and in so far it reveals the structure of its progenitor.

...

Thus, as it seems to me, the leading facts in embryology, which are second in importance to none in natural history, are explained on the principle of slight modifications not appearing, in the many descendants from someone ancient progenitor, at a very early period in the life of each, though perhaps caused at the earliest, and being inherited at a corresponding not early period. Embryology rises greatly in interest, when we thus look at the embryo as a picture, more or less obscured, of the common parent-form of each great class of animals.

Rudimentary, Atrophied, or Aborted Organs

Organs or parts in this strange condition, bearing the stamp of inutility, are extremely common throughout nature. For instance, rudimentary mammae are very general in the males of mammals: I presume that the "bastard-wing" in birds may be safely considered as a digit in a rudimentary state: in very many snakes one lobe of the lungs is rudimentary; in other snakes there are rudiments of the pelvis and hind limbs. Some of the cases of rudimentary organs are extremely curious; for instance, the presence of teeth in foetal whales, which when grown up have not a tooth in their heads; and the presence of teeth, which never cut through the gums, in the upper jaws of our unborn calves. It has even been stated on good authority that rudiments of teeth can be detected in the beaks of certain embryonic birds. Nothing can be plainer than that wings are formed for flight, yet in how many insects do we see wings so reduced in size as to be utterly incapable of flight, and not rarely lying under wing-cases, firmly soldered together!

...

On my view of descent with modification, the origin of rudimentary organs is simple. We have plenty of cases of rudimentary organs in our domestic productions,—as the stump of a tail in tailless breeds,—the vestige of an ear in earless breeds,—the reappearance of minute dangling horns in hornless breeds of cattle, more especially, according to Youatt, in young animals,—and the state of the whole flower in the cauliflower. We often see rudiments of various parts in monsters. But I doubt whether any of these cases throw light on the origin of rudimentary organs in a state of nature, further than by showing that rudiments can be produced; for I doubt whether species under nature ever undergo abrupt changes. I believe that disuse has been the main agency; that it has led in successive generations to the gradual reduction of various organs, until they have become rudimentary,—as in the case of the eyes of animals inhabiting dark caverns, and of the wings of birds inhabiting oceanic islands, which have seldom been forced to take flight, and have ultimately lost the power of flying. Again, an organ useful under certain conditions, might become injurious under others, as with the wings of beetles living on small and exposed islands; and in this case natural selection would continue slowly to reduce the organ, until it was rendered harmless and rudimentary.

. . .

Summary

In this chapter I have attempted to show, that the subordination of group to group in all organisms throughout all time; that the nature of the relationship, by which all living and extinct beings are united by complex, radiating, and circuitous lines of affinities into one grand system; the rules followed and the difficulties encountered by naturalists in their classifications; the value set upon characters, if constant and prevalent, whether of high vital importance, or of the most trifling importance, or, as in rudimentary organs, of no importance; the wide opposition in value between analogical or adaptive characters, and characters of true affinity; and other such rules;—all naturally follow on the view of the common parentage of those forms which are considered by naturalists as allied, together with their modification through natural selection, with its contingencies of extinction and divergence of character. In considering this view of classification, it should be borne in mind that the element of descent has been universally used in ranking together the sexes, ages, and acknowledged varieties of the same species, however different they may be in structure. If we extend the use of this element of descent,—the only certainly known cause of similarity in organic beings,—we shall understand what is meant by the natural system: it is genealogical in its attempted arrangement, with the grades of acquired difference marked by the terms varieties, species, genera, families, orders, and classes.

On this same view of descent with modification, all the great facts in Morphology become intelligible,—whether we look to the same pattern displayed in the homologous organs, to whatever purpose applied, of the different species of a class; or to the homologous parts constructed on the same pattern in each individual animal and plant.

On the principle of successive slight variations, not necessarily or generally supervening at a very early period of life, and being inherited at a corresponding period, we can understand the great leading facts in Embryology; namely, the resemblance in an individual embryo of the homologous parts, which when matured will become widely different from each other in structure and function; and the resemblance in different species of a class of the homologous parts or organs, though fitted in the adult members for purposes as different as possible. Larvae are active embryos, which have become specially modified in relation to their habits of life, through the principle of modifications being inherited at corresponding ages. On this same principle—and bearing in mind, that when organs are reduced in size, either from disuse or selection, it will generally be at that period of life when the being has to provide for its own wants, and bearing in mind how strong is the principle of inheritance—the occurrence of rudimentary organs and their final abortion, present to us no inexplicable difficulties; on the contrary, their presence might have been even anticipated. The importance of embryological characters and of rudimentary organs in classification is intelligible, on the view that an arrangement is only so far natural as it is genealogical.

Finally, the several classes of facts which have been considered in this chapter, seem to me to proclaim so plainly, that the innumerable species, genera, and families of organic beings, with which this world is peopled, have all descended, each within its own class or group, from common parents, and have all been modified in the course of descent, that I should without hesitation adopt this view, even if it were unsupported by other facts or arguments.

14. Recapitulation and Conclusion

Recapitulation of the difficulties on the theory of Natural Selection. Recapitulation of the general and special circumstances in its favour. Causes of the general belief in the immutability of species. How far the theory of natural selection may be extended. Effects of its adoption on the study of Natural history. Concluding remarks.

As this whole volume is one long argument, it may be convenient to the reader to have the leading facts and inferences briefly recapitulated.

That many and grave objections may be advanced against the theory of descent with modification through natural selection, I do not deny. I have endeavoured to give to them their full force. Nothing at first can appear more difficult to believe than that the more complex organs and instincts should have been perfected, not by means superior to, though analogous with, human reason, but by the accumulation of innumerable slight variations, each good for the individual possessor. Nevertheless, this difficulty, though appearing to our imagination insuperably great, cannot be considered real if we admit the following propositions, namely,—that gradations in the perfection of any organ or instinct, which we may consider, either do now exist or could have existed, each good of its kind,—that all organs and instincts are, in ever so slight a degree, variable,—and, lastly, that there is a struggle for existence leading to the preservation of each profitable deviation of structure or instinct. The truth of these propositions cannot, I think, be disputed.

It is, no doubt, extremely difficult even to conjecture by what gradations many structures have been perfected, more especially amongst broken and failing groups of organic beings; but we see so many strange gradations in nature, as is proclaimed by the canon, "Natura non facit saltum," that we ought to be extremely cautious in saying that any organ or instinct, or any whole being, could not have arrived at its present state by many graduated steps. There are, it must be admitted, cases of special difficulty on the theory of natural selection; and one of the most curious of these is the existence of two or three defined castes of workers or sterile females in the same community of ants; but I have attempted to show how this difficulty can be mastered.

With respect to the almost universal sterility of species when first crossed, which forms so remarkable a contrast with the almost universal fertility of varieties when crossed, I must refer the reader to the recapitulation of the facts given at the end of the eighth chapter, which seem to me conclusively to show that this sterility is no more a special endowment than is the incapacity of two trees to be grafted together, but that it is incidental on constitutional differences in the reproductive systems of the intercrossed species. We see the truth of this conclusion in the vast difference in the result, when the same two species are crossed reciprocally; that is, when one species is first used as the father and then as the mother.

The fertility of varieties when intercrossed and of their mongrel offspring cannot be considered as universal; nor is their very general fertility surprising when we remember that it is not likely that either their constitutions or their reproductive systems should have been profoundly modified. Moreover, most of the varieties which have been experimentised on have been produced under domestication; and as domestication apparently tends to eliminate sterility, we ought not to expect it also to produce sterility.

The sterility of hybrids is a very different case from that of first crosses, for their reproductive organs are more or less functionally impotent; whereas in first crosses the organs on both sides are in a perfect condition. As we continually see that organisms of all kinds are rendered in some degree sterile from their constitutions having been disturbed by slightly different and new conditions of life, we need not feel surprise at hybrids being in some degree sterile, for their constitutions can hardly fail to have been disturbed from being compounded of two distinct organisations. This parallelism is supported by another parallel, but directly opposite, class of facts; namely, that the vigour and fertility of all organic beings are increased by slight changes in their conditions of life, and that the offspring of slightly modified forms or varieties acquire from being crossed increased vigour and fertility. So that, on the one hand, considerable changes in the conditions of life and crosses between greatly modified forms, lessen fertility; and on the other hand, lesser changes in the conditions of life and crosses between less modified forms, increase fertility.

Turning to geographical distribution, the difficulties encountered on the theory of descent with modification are grave enough. All the individuals of the same species, and all the species of the same genus, or even higher group, must have descended from common parents; and therefore, in however distant and isolated parts of the world they are now found, they must in the course of successive generations have passed from some one part to the others . . . .

As on the theory of natural selection an interminable number of intermediate forms must have existed, linking together all the species in each group by gradations as fine as our present varieties, it may be asked, Why do we not see these linking forms all around us? Why are not all organic beings blended together in an inextricable chaos? With respect to existing forms, we should remember that we have no right to expect (excepting in rare cases) to discover DIRECTLY connecting links between them, but only between each and some extinct and supplanted form. Even on a wide area, which has during a long period remained continuous, and of which the climate and other conditions of life change insensibly in going from a district occupied by one species into another district occupied by a closely allied species, we have no just right to expect often to find intermediate varieties in the intermediate zone. For we have reason to believe that only a few species are undergoing change at any one period; and all changes are slowly effected. I have also shown that the intermediate varieties which will at first probably exist in the intermediate zones, will be liable to be supplanted by the allied forms on either hand; and the latter, from existing in greater numbers, will generally be modified and improved at a quicker rate than the intermediate varieties, which exist in lesser numbers; so that the intermediate varieties will, in the long run, be supplanted and exterminated.

On this doctrine of the extermination of an infinitude of connecting links, between the living and extinct inhabitants of the world, and at each successive period between the extinct and still older species, why is not every geological formation charged with such links? Why does not every collection of fossil remains afford plain evidence of the gradation and mutation of the forms of life? We meet with no such evidence, and this is the most obvious and forcible of the many objections which may be urged against my theory. Why, again, do whole groups of allied species appear, though certainly they often falsely appear, to have come in suddenly on the several geological stages? Why do we not find great piles of strata beneath the Silurian system, stored with the remains of the progenitors of the Silurian groups of fossils? For certainly on my theory such strata must somewhere have been deposited at these ancient and utterly unknown epochs in the world's history.

I can answer these questions and grave objections only on the supposition that the geological record is far more imperfect than most geologists believe. It cannot be objected that there has not been time sufficient for any amount of organic change; for the lapse of time has been so great as to be utterly inappreciable by the human intellect. The number of specimens in all our museums is absolutely as nothing compared with the countless generations of countless species which certainly have existed. We should not be able to recognise a species as the parent of any one or more species if we were to examine them ever so closely, unless we likewise possessed many of the intermediate links between their past or parent and present states; and these many links we could hardly ever expect to discover, owing to the imperfection of the geological record. Numerous existing doubtful forms could be named which are probably varieties; but who will pretend that in future ages so many fossil links will be discovered, that naturalists will be able to decide, on the common view, whether or not these doubtful forms are varieties? As long as most of the links between any two species are unknown, if any one link or intermediate variety be discovered, it will simply be classed as another and distinct species. Only a small portion of the world has been geologically explored. Only organic beings of certain classes can be preserved in a fossil condition, at least in any great number . . . .

With respect to the absence of fossiliferous formations beneath the lowest Silurian strata, I can only recur to the hypothesis given in the ninth chapter. That the geological record is imperfect all will admit; but that it is imperfect to the degree which I require, few will be inclined to admit. If we look to long enough intervals of time, geology plainly declares that all species have changed; and they have changed in the manner which my theory requires, for they have changed slowly and in a graduated manner. We clearly see this in the fossil remains from consecutive formations invariably being much more closely related to each other, than are the fossils from formations distant from each other in time.

Such is the sum of the several chief objections and difficulties which may justly be urged against my theory; and I have now briefly recapitulated the answers and explanations which can be given to them. I have felt these difficulties far too heavily during many years to doubt their weight. But it deserves especial notice that the more important objections relate to questions on which we are confessedly ignorant; nor do we know how ignorant we are. We do not know all the possible transitional gradations between the simplest and the most perfect organs; it cannot be pretended that we know all the varied means of Distribution during the long lapse of years, or that we know how imperfect the Geological Record is. Grave as these several difficulties are, in my judgment they do not overthrow the theory of descent with modification.

Now let us turn to the other side of the argument. Under domestication we see much variability. This seems to be mainly due to the reproductive system being eminently susceptible to changes in the conditions of life; so that this system, when not rendered impotent, fails to reproduce offspring exactly like the parent-form. Variability is governed by many complex laws,—by correlation of growth, by use and disuse, and by the direct action of the physical conditions of life. There is much difficulty in ascertaining how much modification our domestic productions have undergone; but we may safely infer that the amount has been large, and that modifications can be inherited for long periods. As long as the conditions of life remain the same, we have reason to believe that a modification, which has already been inherited for many generations, may continue to be inherited for an almost infinite number of generations. On the other hand we have evidence that variability, when it has once come into play, does not wholly cease; for new varieties are still occasionally produced by our most anciently domesticated productions.

Man does not actually produce variability; he only unintentionally exposes organic beings to new conditions of life, and then nature acts on the organisation, and causes variability. But man can and does select the variations given to him by nature, and thus accumulate them in any desired manner. He thus adapts animals and plants for his own benefit or pleasure. He may do this methodically, or he may do it unconsciously by preserving the individuals most useful to him at the time, without any thought of altering the breed. It is certain that he can largely influence the character of a breed by selecting, in each successive generation, individual differences so slight as to be quite inappreciable by an uneducated eye. This process of selection has been the great agency in the production of the most distinct and useful domestic breeds. That many of the breeds produced by man have to a large extent the character of natural species, is shown by the inextricable doubts whether very many of them are varieties or aboriginal species.

There is no obvious reason why the principles which have acted so efficiently under domestication should not have acted under nature. In the preservation of favoured individuals and races, during the constantly-recurrent Struggle for Existence, we see the most powerful and ever-acting means of selection. The struggle for existence inevitably follows from the high geometrical ratio of increase which is common to all organic beings. This high rate of increase is proved by calculation, by the effects of a succession of peculiar seasons, and by the results of naturalisation, as explained in the third chapter. More individuals are born than can possibly survive. A grain in the balance will determine which individual shall live and which shall die,—which variety or species shall increase in number, and which shall decrease, or finally become extinct. As the individuals of the same species come in all respects into the closest competition with each other, the struggle will generally be most severe between them; it will be almost equally severe between the varieties of the same species, and next in severity between the species of the same genus. But the struggle will often be very severe between beings most remote in the scale of nature. The slightest advantage in one being, at any age or during any season, over those with which it comes into competition, or better adaptation in however slight a degree to the surrounding physical conditions, will turn the balance.

With animals having separated sexes there will in most cases be a struggle between the males for possession of the females. The most vigorous individuals, or those which have most successfully struggled with their conditions of life, will generally leave most progeny. But success will often depend on having special weapons or means of defence, or on the charms of the males; and the slightest advantage will lead to victory.

As geology plainly proclaims that each land has undergone great physical changes, we might have expected that organic beings would have varied under nature, in the same way as they generally have varied under the changed conditions of domestication. And if there be any variability under nature, it would be an unaccountable fact if natural selection had not come into play. It has often been asserted, but the assertion is quite incapable of proof, that the amount of variation under nature is a strictly limited quantity. Man, though acting on external characters alone and often capriciously, can produce within a short period a great result by adding up mere individual differences in his domestic productions; and every one admits that there are at least individual differences in species under nature. But, besides such differences, all naturalists have admitted the existence of varieties, which they think sufficiently distinct to be worthy of record in systematic works. No one can draw any clear distinction between individual differences and slight varieties; or between more plainly marked varieties and sub-species, and species. Let it be observed how naturalists differ in the rank which they assign to the many representative forms in Europe and North America.

If then we have under nature variability and a powerful agent always ready to act and select, why should we doubt that variations in any way useful to beings, under their excessively complex relations of life, would be preserved, accumulated, and inherited? Why, if man can by patience select variations most useful to himself, should nature fail in selecting variations useful, under changing conditions of life, to her living products? What limit can be put to this power, acting during long ages and rigidly scrutinising the whole constitution, structure, and habits of each creature,—favouring the good and rejecting the bad? I can see no limit to this power, in slowly and beautifully adapting each form to the most complex relations of life. The theory of natural selection, even if we looked no further than this, seems to me to be in itself probable. I have already recapitulated, as fairly as I could, the opposed difficulties and objections: now let us turn to the special facts and arguments in favour of the theory.

On the view that species are only strongly marked and permanent varieties, and that each species first existed as a variety, we can see why it is that no line of demarcation can be drawn between species, commonly supposed to have been produced by special acts of creation, and varieties which are acknowledged to have been produced by secondary laws. On this same view we can understand how it is that in each region where many species of a genus have been produced, and where they now flourish, these same species should present many varieties; for where the manufactory of species has been active, we might expect, as a general rule, to find it still in action; and this is the case if varieties be incipient species. Moreover, the species of the larger genera, which afford the greater number of varieties or incipient species, retain to a certain degree the character of varieties; for they differ from each other by a less amount of difference than do the species of smaller genera. The closely allied species also of the larger genera apparently have restricted ranges, and they are clustered in little groups round other species—in which respects they resemble varieties. These are strange relations on the view of each species having been independently created, but are intelligible if all species first existed as varieties.

As each species tends by its geometrical ratio of reproduction to increase inordinately in number; and as the modified descendants of each species will be enabled to increase by so much the more as they become more diversified in habits and structure, so as to be enabled to seize on many and widely different places in the economy of nature, there will be a constant tendency in natural selection to preserve the most divergent offspring of any one species. Hence during a long-continued course of modification, the slight differences, characteristic of varieties of the same species, tend to be augmented into the greater differences characteristic of species of the same genus. New and improved varieties will inevitably supplant and exterminate the older, less improved and intermediate varieties; and thus species are rendered to a large extent defined and distinct objects. Dominant species belonging to the larger groups tend to give birth to new and dominant forms; so that each large group tends to become still larger, and at the same time more divergent in character. But as all groups cannot thus succeed in increasing in size, for the world would not hold them, the more dominant groups beat the less dominant. This tendency in the large groups to go on increasing in size and diverging in character, together with the almost inevitable contingency of much extinction, explains the arrangement of all the forms of life, in groups subordinate to groups, all within a few great classes, which we now see everywhere around us, and which has prevailed throughout all time. This grand fact of the grouping of all organic beings seems to me utterly inexplicable on the theory of creation.

As natural selection acts solely by accumulating slight, successive, favourable variations, it can produce no great or sudden modification; it can act only by very short and slow steps. Hence the canon of "Natura non facit saltum," which every fresh addition to our knowledge tends to make more strictly correct, is on this theory simply intelligible. We can plainly see why nature is prodigal in variety, though niggard in innovation. But why this should be a law of nature if each species has been independently created, no man can explain.

Many other facts are, as it seems to me, explicable on this theory. How strange it is that a bird, under the form of woodpecker, should have been created to prey on insects on the ground; that upland geese, which never or rarely swim, should have been created with webbed feet; that a thrush should have been created to dive and feed on sub-aquatic insects; and that a petrel should have been created with habits and structure fitting it for the life of an auk or grebe! and so on in endless other cases. But on the view of each species constantly trying to increase in number, with natural selection always ready to adapt the slowly varying descendants of each to any unoccupied or ill-occupied place in nature, these facts cease to be strange, or perhaps might even have been anticipated.

As natural selection acts by competition, it adapts the inhabitants of each country only in relation to the degree of perfection of their associates; so that we need feel no surprise at the inhabitants of any one country, although on the ordinary view supposed to have been specially created and adapted for that country, being beaten and supplanted by the naturalised productions from another land. Nor ought we to marvel if all the contrivances in nature be not, as far as we can judge, absolutely perfect; and if some of them be abhorrent to our ideas of fitness. We need not marvel at the sting of the bee causing the bee's own death; at drones being produced in such vast numbers for one single act, and being then slaughtered by their sterile sisters; at the astonishing waste of pollen by our fir-trees; at the instinctive hatred of the queen bee for her own fertile daughters; at ichneumonidae feeding within the live bodies of caterpillars; and at other such cases. The wonder indeed is, on the theory of natural selection, that more cases of the want of absolute perfection have not been observed.

. . .

Looking to geographical distribution, if we admit that there has been during the long course of ages much migration from one part of the world to another, owing to former climatal and geographical changes and to the many occasional and unknown means of dispersal, then we can understand, on the theory of descent with modification, most of the great leading facts in Distribution.

. . .

The fact, as we have seen, that all past and present organic beings constitute one grand natural system, with group subordinate to group, and with extinct groups often falling in between recent groups, is intelligible on the theory of natural selection with its contingencies of extinction and divergence of character. On these same principles we see how it is, that the mutual affinities of the species and genera within each class are so complex and circuitous. We see why certain characters are far more serviceable than others for classification;—why adaptive characters, though of paramount importance to the being, are of hardly any importance in classification; why characters derived from rudimentary parts, though of no service to the being, are often of high classificatory value; and why embryological characters are the most valuable of all. The real affinities of all organic beings are due to inheritance or community of descent. The natural system is a genealogical arrangement, in which we have to discover the lines of descent by the most permanent characters, however slight their vital importance may be.

The framework of bones being the same in the hand of a man, wing of a bat, fin of the porpoise, and leg of the horse,—the same number of vertebrae forming the neck of the giraffe and of the elephant,—and innumerable other such facts, at once explain themselves on the theory of descent with slow and slight successive modifications. The similarity of pattern in the wing and leg of a bat, though used for such different purpose,—in the jaws and legs of a crab,—in the petals, stamens, and pistils of a flower, is likewise intelligible on the view of the gradual modification of parts or organs, which were alike in the early progenitor of each class. On the principle of successive variations not always supervening at an early age, and being inherited at a corresponding not early period of life, we can clearly see why the embryos of mammals, birds, reptiles, and fishes should be so closely alike, and should be so unlike the adult forms. We may cease marvelling at the embryo of an air-breathing mammal or bird having branchial slits and arteries running in loops, like those in a fish which has to breathe the air dissolved in water, by the aid of well-developed branchiae.

...

I have now recapitulated the chief facts and considerations which have thoroughly convinced me that species have changed, and are still slowly changing by the preservation and accumulation of successive slight favourable variations. Why, it may be asked, have all the most eminent living naturalists and geologists rejected this view of the mutability of species? It cannot be asserted that organic beings in a state of nature are subject to no variation; it cannot be proved that the amount of variation in the course of long ages is a limited quantity; no clear distinction has been, or can be, drawn between species and well-marked varieties. It cannot be maintained that species when intercrossed are invariably sterile, and varieties invariably fertile; or that sterility is a special endowment and sign of creation. The belief that species were immutable productions was almost unavoidable as long as the history of the world was thought to be of short duration; and now that we have acquired some idea of the lapse of time, we are too apt to assume, without proof, that the geological record is so perfect that it would have afforded us plain evidence of the mutation of species, if they had undergone mutation.

But the chief cause of our natural unwillingness to admit that one species has given birth to other and distinct species, is that we are always slow in admitting any great change of which we do not see the intermediate steps. The difficulty is the same as that felt by so many geologists, when Lyell first insisted that long lines of inland cliffs had been formed, and great valleys excavated, by the slow action of the coast-waves. The mind cannot possibly grasp the full meaning of the term of a hundred million years; it cannot add up and perceive the full effects of many slight variations, accumulated during an almost infinite number of generations.

Although I am fully convinced of the truth of the views given in this volume under the form of an abstract, I by no means expect to convince experienced naturalists whose minds are stocked with a multitude of facts all viewed, during a long course of years, from a point of view directly opposite to mine. It is so easy to hide our ignorance under such expressions as the "plan of creation," "unity of design," etc., and to think that we give an explanation when we only restate a fact. Anyone whose disposition leads him to attach more weight to unexplained difficulties than to the explanation of a certain number of facts will certainly reject my theory. A few naturalists, endowed with much flexibility of mind, and who have already begun to doubt on the immutability of species, may be influenced by this volume; but I look with confidence to the future, to young and rising naturalists, who will be able to view both sides of the question with impartiality. Whoever is led to believe that species are mutable will do good service by conscientiously expressing his conviction; for only thus can the load of prejudice by which this subject is overwhelmed be removed.

. . .

It may be asked how far I extend the doctrine of the modification of species. The question is difficult to answer, because the more distinct the forms are which we may consider, by so much the arguments fall away in force. But some arguments of the greatest weight extend very far. All the members of whole classes can be connected together by chains of affinities, and all can be classified on the same principle, in groups subordinate to groups. Fossil remains sometimes tend to fill up very wide intervals between existing orders. Organs in a rudimentary condition plainly show that an early progenitor had the organ in a fully developed state; and this in some instances necessarily implies an enormous amount of modification in the descendants. Throughout whole classes various structures are formed on the same pattern, and at an embryonic age the species closely resemble each other. Therefore I cannot doubt that the theory of descent with modification embraces all the members of the same class. I believe that animals have descended from at most only four or five progenitors, and plants from an equal or lesser number.

Analogy would lead me one step further, namely, to the belief that all animals and plants have descended from some one prototype. But analogy may be a deceitful guide. Nevertheless all living things have much in common, in their chemical composition, their germinal vesicles, their cellular structure, and their laws of growth and reproduction. We see this even in so trifling a circumstance as that the same poison often similarly affects plants and animals; or that the poison secreted by the gall-fly produces monstrous growths on the wild rose or oak-tree. Therefore I should infer from analogy that probably all the organic beings which have ever lived on this earth have descended from some one primordial form, into which life was first breathed.

When the views entertained in this volume on the origin of species, or when analogous views are generally admitted, we can dimly foresee that there will be a considerable revolution in natural history.

. . .

The other and more general departments of natural history will rise greatly in interest. The terms used by naturalists of affinity, relationship, community of type, paternity, morphology, adaptive characters, rudimentary and aborted organs, etc., will cease to be metaphorical, and will have a plain signification. When we no longer look at an organic being as a savage looks at a ship, as at something wholly beyond his comprehension; when we regard every production of nature as one which has had a history; when we contemplate every complex structure and instinct as the summing up of many contrivances, each useful to the possessor, nearly in the same way as when we look at any great mechanical invention as the summing up of the labour, the experience, the reason, and even the blunders of numerous workmen; when we thus view each organic being, how far more interesting, I speak from experience, will the study of natural history become!

A grand and almost untrodden field of inquiry will be opened, on the causes and laws of variation, on correlation of growth, on the effects of use and disuse, on the direct action of external conditions, and so forth. The study of domestic productions will rise immensely in value. A new variety raised by man will be a far more important and interesting subject for study than one more species added to the infinitude of already recorded species. Our classifications will come to be, as far as they can be so made, genealogies; and will then truly give what may be called the plan of creation. The rules for classifying will no doubt become simpler when we have a definite object in view.

...

During early periods of the earth's history, when the forms of life were probably fewer and simpler, the rate of change was probably slower; and at the first dawn of life, when very few forms of the simplest structure existed, the rate of change may have been slow in an extreme degree. The whole history of the world, as at present known, although of a length quite incomprehensible by us, will hereafter be recognised as a mere fragment of time, compared with the ages which have elapsed since the first creature, the progenitor of innumerable extinct and living descendants, was created.

In the distant future I see open fields for far more important researches. Psychology will be based on a new foundation, that of the necessary acquirement of each mental power and capacity by gradation. Light will be thrown on the origin of man and his history.

Authors of the highest eminence seem to be fully satisfied with the view that each species has been independently created. To my mind it accords better with what we know of the laws impressed on matter by the Creator, that the production and extinction of the past and present inhabitants of the world should have been due to secondary causes, like those determining the birth and death of the individual. When I view all beings not as special creations, but as the lineal descendants of some few beings which lived long before the first bed of the Silurian system was deposited, they seem to me to become ennobled. Judging from the past, we may safely infer that not one living species will transmit its unaltered likeness to a distant futurity. And of the species now living very few will transmit progeny of any kind to a far distant futurity; for the manner in which all organic beings are grouped, shows that the greater number of species of each genus, and all the species of many genera, have left no descendants, but have become utterly extinct. We can so far take a prophetic glance into futurity as to foretell that it will be the common and widely-spread species, belonging to the larger and dominant groups, which will ultimately prevail and procreate new and dominant species. As all the living forms of life are the lineal descendants of those which lived long before the Silurian epoch, we may feel certain that the ordinary succession by generation has never once been broken, and that no cataclysm has desolated the whole world. Hence we may look with some confidence to a secure future of equally inappreciable length. And as natural selection works solely by and for the good of each being, all corporeal and mental endowments will tend to progress towards perfection.

It is interesting to contemplate an entangled bank, clothed with many plants of many kinds, with birds singing on the bushes, with various insects flitting about, and with worms crawling through the damp earth, and to reflect that these elaborately constructed forms, so different from each other, and dependent on each other in so complex a manner, have all been produced by laws acting around us. These laws, taken in the largest sense, being Growth with Reproduction; Inheritance which is almost implied by reproduction; Variability from the indirect and direct action of the external conditions of life, and from use and disuse; a Ratio of Increase so high as to lead to a Struggle for Life, and as a consequence to Natural Selection, entailing Divergence of Character and the Extinction of less-improved forms. Thus, from the war of nature, from famine and death, the most exalted object which we are capable of conceiving, namely, the production of the higher animals, directly follows. There is grandeur in this view of life, with its several powers, having been originally breathed into a few forms or into one; and that, whilst this planet has gone cycling on according to the fixed law of gravity, from so simple a beginning endless forms most beautiful and most wonderful have been, and are being, evolved.

Reading and Discussion Questions

1. What influences on Darwin's thinking in the *Origin* are evident from previous works that we have read? In particular, can you identify specific ways in which Darwin engages with ideas we encountered in Lamarck's *Zoological Philosophy,* Cuvier's *Law of the Correlation of the Parts,* Thomas Malthus' *An Essay on the Principle of Population,* Paley's *Natural Theology,* and Lyell's *Principles of Geology*?
2. It will be important to Darwin's theory both that there are slight individual differences among progeny and that some variation is heritable. How would you characterize the main points that Darwin makes about these ideas in Chapter 1 and what importance do they have? What does Darwin mean by artificial selection?
3. In what ways does Darwin work to erode the reader's confidence that a clear distinction can be drawn between varieties and species in Chapter 2? What does Darwin take such difficulties in classification to suggest? How does the way that Darwin proposes to think about species differ from or compare to views we have encountered in other readings?
4. Chapter 3 of the *Origin* is entitled "Struggle for Existence." What does Darwin mean by that and what role does it play in his theory? What sorts of factors might hold the tendency for populations to increase at high rates in check?
5. What does Darwin mean by "natural selection" and how does it relate to the struggle for existence?

6. Darwin articulates a theory of descent with modification by means of natural selection. Explain the central theory put forward Darwin's *Origin of Species* (1859), focusing particularly on the first four chapters.
7. Toward the end of the *Origin*, in Chapter 14, Darwin describes the entire book as "one long argument" in support of his theory. How would you reconstruct that argument, the case that he makes for his theory?
8. In what sense could such a theory account for how diverse species arise? What does Darwin mean by divergence of character? Why does Darwin think that populations will tend to diversify?
9. Consider the diagram from the *Origin of Species*. Which of the following does Darwin intend to be inferred from the diagram? That species u^8 goes extinct in time era VIII? That the more diversified descendants from any one species become, the more their modified progeny will tend to be increased? That species F^{10} goes extinct in time era X? That by time era II, the descendants of species A have successfully outcompeted the descendants of B & C?
10. In Chapter 9, Darwin considers the objection that: if there really has been a gradual transition between forms we should expect to see more intermediates. How does Darwin respond to imperfection of the geological record and what other objections does he anticipate and respond to? What criticisms does he think are most worthy of consideration? Can you think of other objections?
11. Darwin attempts to support his theory by arguing that it provides coherent explanations for a broad scope of diverse phenomena. Think of at least three distinct kinds of biological observations that Darwin thinks that his theory can explain. Discuss how Darwin uses such arguments, particularly in chapters 13 and 14 of the *Origin*.

The Descent of Man, and Selection in Relation to Sex (1871)

Charles Darwin

95

Introduction

The nature of the following work will be best understood by a brief account of how it came to be written. During many years I collected notes on the origin or descent of man, without any intention of publishing on the subject, but rather with the determination not to publish, as I thought that I should thus only add to the prejudices against my views. It seemed to me sufficient to indicate, in the first edition of my 'Origin of Species,' that by this work "light would be thrown on the origin of man and his history;" and this implies that man must be included with other organic beings in any general conclusion respecting his manner of appearance on this earth. Now the case wears a wholly different aspect. When a naturalist like Carl Vogt ventures to say in his address as President of the National Institution of Geneva (1869), "*personne, en Europe au moins, n'ose plus soutenir la création indépendante et de toutes pièces, des espèces*," it is manifest that at least a large number of naturalists must admit that species are the modified descendants of other species; and this especially holds good with the younger and rising naturalists. The greater number accept the agency of natural selection; though some urge, whether with justice the future must decide, that I have greatly overrated its importance. Of the older and honoured chiefs in natural science, many unfortunately are still opposed to evolution in every form.

In consequence of the views now adopted by most naturalists, and which will ultimately, as in every other case, be followed by other men, I have been led to put together my notes, so as to see how far the general conclusions arrived at in my former works were applicable to man. This seemed all the more desirable as I had never deliberately applied these views to a species taken singly. When we confine our attention to any one form, we are deprived of the weighty arguments derived from the nature of the affinities which connect together whole groups of organisms—their geographical distribution in past and present times, and their geological succession. The homological structure, embryological development, and rudimentary organs of a species, whether it be man or any other animal, to which our attention may be directed, remain to be considered; but these great classes of facts afford, as it appears to me, ample and conclusive evidence in favour of the principle of gradual evolution. The strong support derived from the other arguments should, however, always be kept before the mind.

The sole object of this work is to consider, firstly, whether man, like every other species, is descended from some pre-existing form; secondly, the manner of his development; and thirdly, the value of the differences between the so-called races of man. As I shall confine myself to these points, it will not be necessary to describe in detail the differences between the several races—an enormous subject which has been fully discussed in many valuable works. The high antiquity of man has recently been demonstrated by the labours of a host of eminent men, beginning with M. Boucher de Perthes; and this is the indispensable basis for understanding his origin. I shall, therefore, take this conclusion for granted, and may refer my readers to the admirable treatises of Sir Charles Lyell, Sir John Lubbock, and others. Nor shall I have occasion to do more than to allude to the amount of difference between man and the anthropomorphous apes; for Prof. Huxley, in the opinion of most competent judges, has conclusively shewn that in every single visible character man differs less from the higher apes than these do from the lower members of the same order of Primates.

This work contains hardly any original facts in regard to man; but as the conclusions at which I arrived, after drawing up a rough draft, appeared to me interesting, I thought that they might interest others. It has often and confidently been asserted, that man's origin can never be known: but ignorance more frequently begets confidence than does knowledge: it is those who know little, and not those who know much, who so positively assert that this or that problem will never be solved by science. The conclusion that man is the co-descendant with other species of some ancient, lower, and extinct form, is not in any degree new. Lamarck long ago came to this conclusion, which has lately been maintained by several eminent naturalists and philosophers; for instance by Wallace, Huxley, Lyell, Vogt, Lubbock, Büchner, Rolle, &c., and especially by Häckel. This last naturalist, besides his great work, '*Generelle Morphologie*' (1866), has recently (1868, with a second edit. in 1870), published his '*Natürliche Schöpfungsgeschichte*,' in which he fully discusses the genealogy of man. If this work had appeared before my essay had been written, I should probably never have completed it. Almost all the conclusions at which I have arrived I find confirmed by this naturalist, whose knowledge on many points is much fuller than mine. Wherever I have added any fact or view from Prof. Häckel's writings, I give his authority in the text, other statements I leave as they originally stood in my manuscript, occasionally giving in the foot-notes references to his works, as a confirmation of the more doubtful or interesting points.

During many years it has seemed to me highly probable that sexual selection has played an important part in differentiating the races of man; but in my 'Origin of Species' (first edition, p. 199) I contented myself by merely alluding to this belief. When I came to apply this view to man, I found it indispensable to treat the whole subject in full detail. Consequently the second part of the present work, treating of sexual selection, has extended to an inordinate length, compared with the first part; but this could not be avoided.

I had intended adding to the present volumes an essay on the expression of the various emotions by man and the lower animals. My attention was called to this subject many years ago by Sir Charles Bell's admirable work. This illustrious anatomist maintains that man is endowed with certain muscles solely for the sake of expressing his emotions. As this view is obviously opposed to the belief that man is descended from some other and lower form, it was necessary for me to consider it. I likewise wished to ascertain how far the emotions are expressed in the same manner by the different races of man. But owing to the length of the present work, I have thought it better to reserve my essay, which is partially completed, for separate publication.

Part I.—The Descent of Man

Chapter I. The Evidence of the Descent of Man from Some Lower Form

Nature of the evidence bearing on the origin of man—Homologous structures in man and the lower animals—Miscellaneous points of correspondence—Development—Rudimentary structures, muscles, sense-organs, hair, bones, reproductive organs, &c.—The bearing of these three great classes of facts on the origin of man.

He who wishes to decide whether man is the modified descendant of some pre-existing form, would probably first enquire whether man varies, however slightly, in bodily structure and in mental faculties; and if so, whether the variations are transmitted to his offspring in accordance with the laws which prevail with the lower animals; such as that of the transmission of characters to the same age or sex. Again, are the variations the result, as far as our ignorance permits us to judge, of the same general causes, and are they governed by the same general laws, as in the case of other organisms; for instance by correlation, the inherited effects of use and disuse, &c.? Is man subject to similar malconformations, the result of arrested development, of reduplication of parts, &c., and does he display in any of his anomalies reversion to some former and ancient type of structure? It might also naturally be enquired whether man, like so many other animals, has given rise to varieties and subraces, differing but slightly from each other, or to races differing so much that they must be classed as doubtful species? How are such races distributed over the world; and how, when crossed, do they react on each other, both in the first and succeeding generations? And so with many other points.

The enquirer would next come to the important point, whether man tends to increase at so rapid a rate, as to lead to occasional severe struggles for existence, and consequently to beneficial variations, whether in body or mind, being preserved, and injurious ones eliminated. Do the races or species of men, whichever term may be applied, encroach on and replace each other, so that some finally become extinct? We shall see that all these questions, as indeed is obvious in respect to most of them, must be answered in the affirmative, in the same manner as with the lower animals. But the several considerations just referred to may be conveniently deferred for a time; and we will first see how far the bodily structure of man shows traces, more or less plain, of his descent from some lower form. In the two succeeding chapters the mental powers of man, in comparison with those of the lower animals, will be considered.

The Bodily Structure of Man.—It is notorious that man is constructed on the same general type or model with other mammals. All the bones in his skeleton can be compared with corresponding bones in a monkey, bat, or seal. So it is with his muscles, nerves, blood-vessels and internal viscera. The brain, the most important of all the organs, follows the same law, as shewn by Huxley and other anatomists. Bischoff, who is a hostile witness, admits that every chief fissure and fold in the brain of man has its analogy in that of the orang; but he adds that at no period of development do their brains perfectly agree; nor could this be expected, for otherwise their mental powers would have been the same . . . .

Man is liable to receive from the lower animals, and to communicate to them, certain diseases as hydrophobia, variola, the glanders, &c.; and this fact proves the close similarity of their tissues and blood, both in minute structure and composition, far more plainly than does their comparison under the best microscope, or by the aid of the best chemical analysis. Monkeys are liable to many of the same non-contagious diseases as we are; thus Rengger, who carefully observed for a long time the *Cebus Azaræ* in its native land, found it liable to catarrh, with the usual symptoms, and which when often recurrent led to consumption. These monkeys suffered also from apoplexy, inflammation of the bowels, and cataract in the eye. The younger ones when shedding their milk-teeth often died from fever. Medicines produced the same effect on them as on us. Many kinds of monkeys have a strong taste for tea, coffee, and spirituous liquors: they will also, as I have myself seen, smoke tobacco with pleasure. Brehm asserts that the natives of north-eastern Africa catch the wild baboons by exposing vessels with strong beer, by which they are made drunk. He has seen some of these animals, which he kept in confinement, in this state; and he gives a laughable account of their behaviour and strange grimaces. On the following morning they were very cross and dismal; they held their aching heads with both hands and wore a most pitiable expression: when beer or wine was offered them, they turned away with disgust, but relished the juice of lemons. An American monkey, an Ateles, after getting drunk on brandy, would never touch it again, and thus was wiser than many men. These trifling facts prove how similar the nerves of taste must be in monkeys and man, and how similarly their whole nervous system is affected.

. . .

The whole process of that most important function, the reproduction of the species, is strikingly the same in all mammals, from the first act of courtship by the male to the birth and nurturing of the young. Monkeys are born in almost as helpless a condition as our own infants; and in certain genera the young differ fully as much in appearance from the adults, as do our children from their full-grown parents. It has been urged by some writers as an important distinction, that with man the young arrive at maturity at a much later age than with any other animal; but if we look to the races of mankind which inhabit tropical countries the difference is not great, for the orang is believed not to be adult till the age of from ten to fifteen years.[1] Man differs from woman in size, bodily strength, hairyness, &c., as well as in mind, in the same manner as do the two sexes of many mammals. It is, in short, scarcely possible to exaggerate the close correspondence in general structure, in the minute structure of the tissues, in chemical composition and in constitution, between man and the higher animals, especially the anthropomorphous apes.

Embryonic Development.—Man is developed from an ovule, about the 125th of an inch in diameter, which differs in no respect from the ovules of other animals. The embryo itself at a very early period can hardly be distinguished from that of other members of the vertebrate kingdom. At this period the arteries run in arch-like branches, as if to carry the blood to branchiæ which are not present in the higher vertebrata, though the slits on the sides of the neck still remain (*f*, *g*, fig. 1), marking their former position. At a somewhat later period, when the extremities are developed, "the feet of lizards and mammals," as the illustrious Von Baer remarks, "the wings and feet of birds, no less than the hands and feet of man, all arise from the same fundamental form." It is, says Prof. Huxley, "quite in the later stages of development that the young human being presents marked differences from the young ape, while the latter departs as much from the dog in its developments, as the man does. Startling as this last assertion may appcar to bc, it is demonstrably true."

As some of my readers may never have seen a drawing of an embryo, I have given one of man and another of a dog, at about the same early stage of development, carefully copied from two works of undoubted accuracy.[2]

[1] Huxley, 'Man's Place in Nature,' 1863, p. 34.

[2] The human embryo (upper fig.) is from Ecker, 'Icones Phys.,' 1851–1859, tab. xxx. fig. 2. This embryo was ten lines in length, so that the drawing is much magnified. The embryo of the dog is from

After the foregoing statements made by such high authorities, it would be superfluous on my part to give a number of borrowed details, shewing that the embryo of man closely resembles that of other mammals .... I will conclude with a quotation from Huxley, who after asking, does man originate in a different way from a dog, bird, frog or fish? says, "the reply is not doubtful for a moment; without question, the mode of origin and the early stages of the development of man are identical with those of the animals immediately below him in the scale: without a doubt in these respects, he is far nearer to apes, than the apes are to the dog."

Rudiments.—This subject, though not intrinsically more important than the two last, will for several reasons be here treated with more fullness. Not one of the higher animals can be named which does not bear some part in a rudimentary condition; and man forms no exception to the rule. Rudimentary organs must be distinguished from those that are nascent; though in some cases the distinction is not easy. The former are either absolutely useless, such as the *mammæ* of male quadrupeds, or the incisor teeth of ruminants which never cut through the gums; or they are of such slight service to their present possessors, that we cannot suppose that they were developed under the conditions which now exist. Organs in this latter state are not strictly rudimentary, but they are tending in this direction. Nascent organs, on the other hand, though not fully developed, are of high service to their possessors, and are capable of further development. Rudimentary organs are eminently variable; and this is partly intelligible, as they are useless or nearly useless, and consequently are no longer subjected to natural selection. They often become wholly suppressed. When this occurs, they are nevertheless liable to occasional reappearance through reversion; and this is a circumstance well worthy of attention.

...

Some few persons have the power of contracting the superficial muscles on their scalps; and these muscles are in a variable and partially rudimentary condition. M. A. de Candolle has communicated to me a curious instance of the long-continued persistence or inheritance of this power, as well as of its unusual development. He knows a family, in which one member, the present head of a family, could, when a youth, pitch several heavy books from his head by the movement of the scalp alone; and he won wagers by performing this feat. His father, uncle, grandfather, and all his three children possess the same power to the same unusual degree. This family became divided eight generations ago into two branches; so that the head of the above-mentioned branch is cousin in the seventh degree to the head of the other branch. This distant cousin resides in

Bischoff, 'Entwicklungsgeschichte des Hunde-Eies,' 1845, tab. xi. fig. 42 B. This drawing is five times magnified, the embryo being 25 days old. The internal viscera have been omitted, and the uterine appendages in both drawings removed. I was directed to these figures by Prof. Huxley, from whose work, 'Man's Place in Nature.' the idea of giving them was taken.

Fig. 1. Upper figure human embryo, from Ecker. Lower figure that of a dog, from Bischoff.

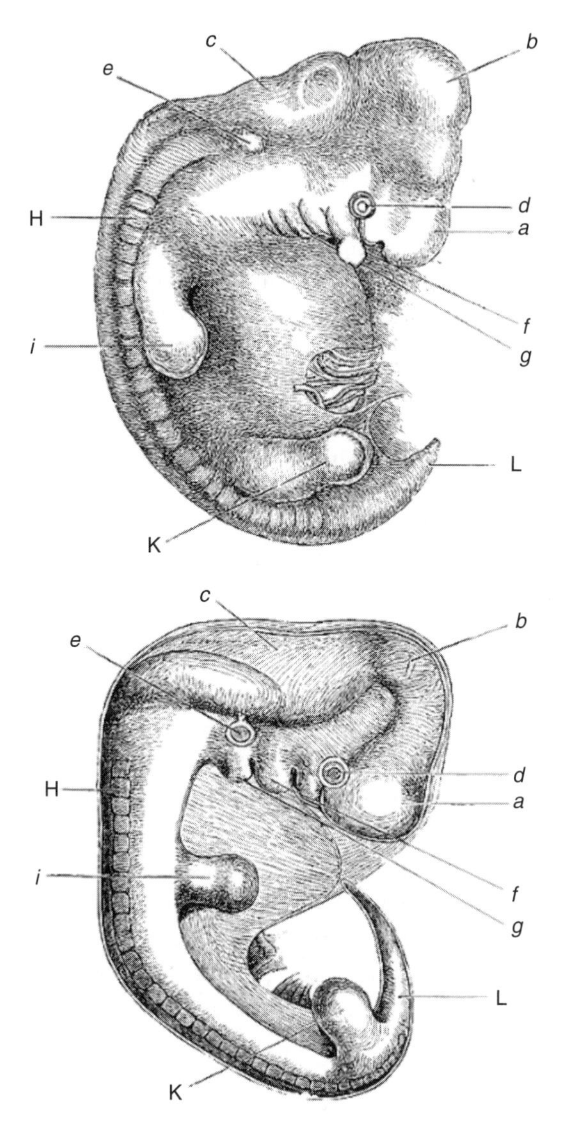

a. Fore-brain, cerebral hemispheres, &c.

b. Mid-brain, corpora quadrigemina.

c. Hind-brain, cerebellum, medulla oblongata.

d. Eye.

e. Ear.

f. First visceral arch.

g. Second visceral arch.

H. Vertebral columns and muscles in process of development.

i. Anterior } extremities

K. Posterior } extremities

L. Tail or os coccyx.

another part of France, and on being asked whether he possessed the same faculty, immediately exhibited his power. This case offers a good illustration how persistently an absolutely useless faculty may be transmitted.

The extrinsic muscles which serve to move the whole external ear, and the intrinsic muscles which move the different parts, all of which belong to the system of the panniculus, are in a rudimentary condition in man; they are also variable in development, or at least in function. I have seen one man who could draw his ears forwards, and another who could draw them backwards; and from what one of these persons told me, it is probable that most of us by often touching our ears and thus directing our attention towards them, could by repeated trials recover some power of movement. The faculty of erecting the ears and of directing them to different points of the compass, is no doubt of the highest service to many animals, as they thus perceive the point of danger; but I have never heard of a man who possessed the least power of erecting his ears,—the one movement which might be of use to him. The whole external shell of the ear may be considered a rudiment, together with the various folds and prominences (helix and anti-helix, tragus and anti-tragus, &c.) which in the lower animals strengthen and support the ear when erect, without adding much to its weight. Some authors, however, suppose that the cartilage of the shell serves to transmit vibrations to the acoustic nerve; but Mr. Toynbee, after collecting all the known evidence on this head, concludes that the external shell is of no distinct use. The ears of the chimpanzee and orang are curiously like those of man, and I am assured by the keepers in the Zoological Gardens that these animals never move or erect them; so that they are in an equally rudimentary condition, as far as function is concerned, as in man. Why these animals, as well as the progenitors of man, should have lost the power of erecting their ears we cannot say. It may be, though I am not quite satisfied with this view, that owing to their arboreal habits and great strength they were but little exposed to danger, and so during a lengthened period moved their ears but little, and thus gradually lost the power of moving them. This would be a parallel case with that of those large and heavy birds, which from inhabiting oceanic islands have not been exposed to the attacks of beasts of prey, and have consequently lost the power of using their wings for flight.

The celebrated sculptor, Mr. Woolner, informs me of one little peculiarity in the external ear, which he has often observed both in men and women, and of which he perceived the full signification. His attention was first called to the subject whilst at work on his figure of Puck, to which he had given pointed ears. He was thus led to examine the ears of various monkeys, and subsequently more carefully those of man. The peculiarity consists in a little blunt point, projecting from the inwardly folded margin, or helix. Mr. Woolner made an exact model of one such case, and has sent me the accompanying drawing (Fig. 2). These points not only project inwards, but often a little outwards, so that they are visible when the head is viewed from directly in front or behind. They are variable in size and somewhat in position, standing either a little higher or lower; and

Fig. 2. Human Ear, modelled and drawn by Mr. Woolner.

a. **The projecting point.**

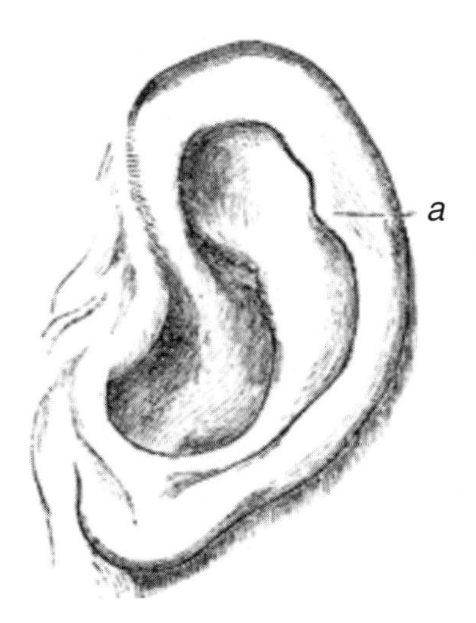

they sometimes occur on one ear and not on the other. Now the meaning of these projections is not, I think, doubtful; but it may be thought that they offer too trifling a character to be worth notice. This thought, however, is as false as it is natural. Every character, however slight, must be the result of some definite cause; and if it occurs in many individuals deserves consideration. The helix obviously consists of the extreme margin of the ear folded inwards; and this folding appears to be in some manner connected with the whole external ear being permanently pressed backwards. In many monkeys, which do not stand high in the order, as baboons and some species of macacus, the upper portion of the ear is slightly pointed, and the margin is not at all folded inwards; but if the margin were to be thus folded, a slight point would necessarily project inwards and probably a little outwards. This could actually be observed in a specimen of the *Ateles beelzebuth* in the Zoological Gardens; and we may safely conclude that it is a similar structure—a vestige of formerly pointed ears—which occasionally reappears in man . . . .

Man differs conspicuously from all the other Primates in being almost naked. But a few short straggling hairs are found over the greater part of the body in the male sex, and fine down on that of the female sex. In individuals belonging to the same race these hairs are highly variable, not only in abundance, but likewise in position: thus the shoulders in some Europeans are quite naked, whilst in others they bear thick tufts of hair. There can be little doubt that the hairs thus scattered over the body are the rudiments of the uniform hairy coat of the lower animals. This view is rendered all the more probable, as it is known that fine, short, and pale-coloured hairs on the limbs and other parts of the body occasionally become developed into "thickset, long, and rather coarse dark hairs," when abnormally nourished near old-standing inflamed surfaces.

...

It appears as if the posterior molar or wisdom-teeth were tending to become rudimentary in the more civilised races of man. These teeth are rather smaller than the other molars, as is likewise the case with the corresponding teeth in the chimpanzee and orang; and they have only two separate fangs. They do not cut through the gums till about the seventeenth year, and I am assured by dentists that they are much more liable to decay, and are earlier lost, than the other teeth. It is also remarkable that they are much more liable to vary both in structure and in the period of their development than the other teeth. In the Melanian races, on the other hand, the wisdom-teeth are usually furnished with three separate fangs, and are generally sound: they also differ from the other molars in size less than in the Caucasian races. Prof. Schaaffhausen accounts for this difference between the races by "the posterior dental portion of the jaw being always shortened" in those that are civilised, and this shortening may, I presume, be safely attributed to civilised men habitually feeding on soft, cooked food, and thus using their jaws less. I am informed by Mr. Brace that it is becoming quite a common practice in the United States to remove some of the molar teeth of children, as the jaw does not grow large enough for the perfect development of the normal number.

...

The bearing of the three great classes of facts now given is unmistakeable. But it would be superfluous here fully to recapitulate the line of argument given in detail in my 'Origin of Species.' The homological construction of the whole frame in the members of the same class is intelligible, if we admit their descent from a common progenitor, together with their subsequent adaptation to diversified conditions. On any other view the similarity of pattern between the hand of a man or monkey, the foot of a horse, the flipper of a seal, the wing of a bat, &c., is utterly inexplicable. It is no scientific explanation to assert that they have all been formed on the same ideal plan. With respect to development, we can clearly understand, on the principle of variations supervening at a rather late embryonic period, and being inherited at a corresponding period, how it is that the embryos of wonderfully different forms should still retain, more or less perfectly, the structure of their common progenitor. No other explanation has ever been given of the marvellous fact that the embryo of a man, dog, seal, bat, reptile, &c., can at first hardly be distinguished from each other. In order to understand the existence of rudimentary organs, we have only to suppose that a former progenitor possessed the parts in question in a perfect state, and that under changed habits of life they became greatly reduced, either from simple disuse, or through the natural selection of those individuals which were least encumbered with a superfluous part, aided by the other means previously indicated.

Thus we can understand how it has come to pass that man and all other vertebrate animals have been constructed on the same general model, why they pass through the same early stages of development, and why they retain certain rudiments in common. Consequently we ought frankly to admit their community of descent: to take any other view, is to admit that our own structure and that of all the animals around us, is a mere snare laid to entrap our judgment. This conclusion is greatly strengthened, if we look to the members of the whole animal series, and consider the evidence derived from their affinities or classification, their geographical distribution and geological succession. It is only our natural prejudice, and that arrogance which made our forefathers declare that they were descended from demi-gods, which leads us to demur to this conclusion. But the time will before long come when it will be thought wonderful, that naturalists, who were well acquainted with the comparative structure and development of man and other mammals, should have believed that each was the work of a separate act of creation.

Chapter II. Comparison of the Mental Powers of Man and the Lower Animals

The difference in mental power between the highest ape and the lowest savage, immense—Certain instincts in common—The emotions—Curiosity—Imitation—Attention—Memory—Imagination—Reason—Progressive improvement—Tools and weapons used by animals—Language—Self-consciousness—Sense of beauty—Belief in God, spiritual agencies, superstitions.

We have seen in the last chapter that man bears in his bodily structure clear traces of his descent from some lower form; but it may be urged that, as man differs so greatly in his mental power from all other animals, there must be some error in this conclusion. No doubt the difference in this respect is enormous, even if we compare the mind of one of the lowest savages, who has no words to express any number higher than four, and who uses no abstract terms for the commonest objects or affections, with that of the most highly organised ape. The difference would, no doubt, still remain immense, even if one of the higher apes had been improved or civilised as much as a dog has been in comparison with its parent-form, the wolf or jackal. The Fuegians rank amongst the lowest barbarians; but I was continually struck with surprise how closely the three natives on board H.M.S. "Beagle," who had lived some years in England and could talk a little English, resembled us in disposition and in most of our mental faculties. If no organic being excepting man had possessed any mental power, or if his powers had been of a wholly different nature from those of the lower animals, then we should never have been able to convince ourselves that our high faculties had been gradually developed. But it can be clearly shewn that there is no fundamental difference of this kind. We must also admit that there is a much wider interval in mental power between one of the lowest fishes, as a lamprey or lancelet, and one of the higher apes, than between an ape and man; yet this immense interval is filled up by numberless gradations.

Nor is the difference slight in moral disposition between a barbarian, such as the man described by the old navigator Byron, who dashed his child on the rocks for dropping a basket of sea-urchins, and a Howard or Clarkson; and in intellect, between a savage who does not use any abstract terms, and a Newton or Shakespeare. Differences of this kind between the highest men of the highest races and the lowest savages, are connected by the finest gradations. Therefore it is possible that they might pass and be developed into each other.

My object in this chapter is solely to shew that there is no fundamental difference between man and the higher mammals in their mental faculties. Each division of the subject might have been extended into a separate essay, but must here be treated briefly. As no classification of the mental powers has been universally accepted, I shall arrange my remarks in the order most convenient for my purpose; and will select those facts which have most struck me, with the hope that they may produce some effect on the reader.

. . .

To return to our immediate subject: the lower animals, like man, manifestly feel pleasure and pain, happiness and misery. Happiness is never better exhibited than by young animals, such as puppies, kittens, lambs, &c., when playing together, like our own children. Even insects play together, as has been described by that excellent observer, P. Huber, who saw ants chasing and pretending to bite each other, like so many puppies.

The fact that the lower animals are excited by the same emotions as ourselves is so well established, that it will not be necessary to weary the reader by many details. Terror acts in the same manner on them as on us, causing the muscles to tremble, the heart to palpitate, the sphincters to be relaxed, and the hair to stand on end. Suspicion, the offspring of fear, is eminently characteristic of most wild animals. Courage and timidity are extremely variable qualities in the individuals of the same species, as is plainly seen in our dogs. Some dogs and horses are ill-tempered and easily turn sulky; others are good-tempered; and these qualities are certainly inherited. Everyone knows how liable animals are to furious rage, and how plainly they show it. Many anecdotes, probably true, have been published on the long-delayed and artful revenge of various animals. The accurate Rengger and Brehm state that the American and African monkeys which they kept tame, certainly revenged themselves. The love of a dog for his master is notorious; in the agony of death he has been known to caress his master, and everyone has heard of the dog suffering under vivisection, who licked the hand of the operator; this man, unless he had a heart of stone, must have felt remorse to the last hour of his life. As Whewell has remarked, "who that reads the touching instances of maternal affection, related so often of the women of all nations, and of the females of all animals, can doubt that the principle of action is the same in the two cases?"

...

Most of the more complex emotions are common to the higher animals and ourselves. Everyone has seen how jealous a dog is of his master's affection, if lavished on any other creature; and I have observed the same fact with monkeys. This shews that animals not only love, but have the desire to be loved. Animals manifestly feel emulation. They love approbation or praise; and a dog carrying a basket for his master exhibits in a high degree self-complacency or pride. There can, I think, be no doubt that a dog feels shame, as distinct from fear, and something very like modesty when begging too often for food. A great dog scorns the snarling of a little dog, and this may be called magnanimity. Several observers have stated that monkeys certainly dislike being laughed at; and they sometimes invent imaginary offences. In the Zoological Gardens I saw a baboon who always got into a furious rage when his keeper took out a letter or book and read it aloud to him; and his rage was so violent that, as I witnessed on one occasion, he bit his own leg till the blood flowed.

We will now turn to the more intellectual emotions and faculties, which are very important, as forming the basis for the development of the higher mental powers. Animals manifestly enjoy excitement and suffer from ennui, as may be seen with dogs, and, according to Rengger, with monkeys. All animals feel Wonder, and many exhibit Curiosity. They sometimes suffer from this latter quality, as when the hunter plays antics and thus attracts them; I have witnessed this with deer, and so it is with the wary chamois, and with some kinds of wild-ducks. Brehm gives a curious account of the instinctive dread which his monkeys exhibited towards snakes; but their curiosity was so great that they could not desist from occasionally satiating their horror in a most human fashion, by lifting up the lid of the box in which the snakes were kept. I was so much surprised at his account, that I took a stuffed and coiled-up snake into the monkey-house at the Zoological Gardens, and the excitement thus caused was one of the most curious spectacles which I ever beheld. Three species of Cercopithecus were the most alarmed; they dashed about their cages and uttered sharp signal-cries of danger, which were understood by the other monkeys. A few young monkeys and one old Anubis baboon alone took no notice of the snake. I then placed the stuffed specimen on the ground in one of the larger compartments. After a time all the monkeys collected round it in a large circle, and staring intently, presented a most ludicrous appearance. They became extremely nervous; so that when a wooden ball, with which they were familiar as a plaything, was accidently moved in the straw, under which it was partly hidden, they all instantly started away. These monkeys behaved very differently when a dead fish, a mouse, and some other new objects were placed in their cages; for though at first frightened, they soon approached, handled and examined them. I then placed a live snake in a paper bag, with the mouth loosely closed, in one of the larger compartments. One of the monkeys immediately approached, cautiously opened the bag a little, peeped in, and instantly dashed away. Then I witnessed what Brehm has described, for monkey after monkey, with head raised high and turned on one side, could not resist taking momentary peeps into the upright bag, at the dreadful object lying quiet at the bottom. It would almost appear as if monkeys had some notion of zoological affinities, for those kept by Brehm exhibited a strange, though mistaken, instinctive dread of innocent lizards and frogs. An orang, also, has been known to be much alarmed at the first sight of a turtle.

The principle of *Imitation* is strong in man, and especially in man in a barbarous state. Desor has remarked that no animal voluntarily imitates an action performed by man, until in the ascending scale we come to monkeys, which are well-known to be ridiculous mockers. Animals, however, sometimes imitate each others' actions: thus two species of wolves, which had been reared by dogs, learned to bark, as does sometimes the jackal, but whether this can be called voluntary imitation is another question. From one account which I have read, there is reason to believe that puppies nursed by cats sometimes learn to lick their feet and thus to clean their faces: it is at least certain, as I hear from a perfectly trustworthy friend, that some dogs behave in this manner. Birds imitate the songs of their parents, and sometimes those of other birds; and parrots are notorious imitators of any sound which they often hear.

Hardly any faculty is more important for the intellectual progress of man than the power of *Attention*. Animals clearly manifest this power, as when a cat watches by a hole and prepares to spring on its prey. Wild animals sometimes become so absorbed when thus engaged, that they may be easily approached. Mr. Bartlett has given me a curious proof how variable this faculty is in monkeys. A man who trains monkeys to act used to purchase common kinds from the Zoological Society at the price of five pounds for each; but he offered to give double the price, if he might keep three or four of them for a few days, in order to select one. When asked how he could possibly so soon learn whether a particular monkey would turn out a good actor, he answered that it all depended on their power of attention. If when he was talking and explaining anything to a monkey, its attention was easily distracted, as by a fly on the wall or other trifling object, the case was hopeless. If he tried by punishment to make an inattentive monkey act, it turned sulky. On the other hand, a monkey which carefully attended to him could always be trained.

It is almost superfluous to state that animals have excellent *Memories* for persons and places. A baboon at the Cape of Good Hope, as I have been informed by Sir Andrew Smith, recognised him with joy after an absence of nine months.

I had a dog who was savage and averse to all strangers, and I purposely tried his memory after an absence of five years and two days. I went near the stable where he lived, and shouted to him in my old manner; he showed no joy, but instantly followed me out walking and obeyed me, exactly as if I had parted with him only half-an-hour before. A train of old associations, dormant during five years, had thus been instantaneously awakened in his mind. Even ants, as P. Huber has clearly shewn, recognised their fellow-ants belonging to the same community after a separation of four months. Animals can certainly by some means judge of the intervals of time between recurrent events.

The *Imagination* is one of the highest prerogatives of man. By this faculty he unites, independently of the will, former images and ideas, and thus creates brilliant and novel results. A poet, as Jean Paul Richter remarks,[3] "who must reflect whether he shall make a character say yes or no—to the devil with him; he is only a stupid corpse." Dreaming gives us the best notion of this power; as Jean Paul again says, "The dream is an involuntary art of poetry." The value of the products of our imagination depends of course on the number, accuracy, and clearness of our impressions; on our judgment and taste in selecting or rejecting the involuntary combinations, and to a certain extent on our power of voluntarily combining them. As dogs, cats, horses, and probably all the higher animals, even birds, as is stated on good authority,[4] have vivid dreams, and this is shewn by their movements and voice, we must admit that they possess some power of imagination.

Of all the faculties of the human mind, it will, I presume, be admitted that *Reason* stands at the summit. Few persons any longer dispute that animals possess some power of reasoning. Animals may constantly be seen to pause, deliberate, and resolve. It is a significant fact, that the more the habits of any particular animal are studied by a naturalist, the more he attributes to reason and the less to unlearnt instincts.[5] In future chapters we shall see that some animals extremely low in the scale apparently display a certain amount of reason. No doubt it is often difficult to distinguish between the power of reason and that of instinct. Thus Dr. Hayes, in his work on 'The Open Polar Sea,' repeatedly remarks that his dogs, instead of continuing to draw the sledges in a compact body, diverged and separated when they came to thin ice, so that their weight might be more evenly distributed. This was often the first warning and notice which the travellers received that the ice was becoming thin and dangerous. Now, did the dogs act thus from the experience of each individual, or from the example of the older and wiser dogs, or from an inherited habit, that is from an instinct? This instinct might possibly have arisen since the time, long ago, when dogs were

[3] Quoted in Dr. Maudsley's 'Physiology and Pathology of Mind,' 1868, pp. 19, 220.

[4] Dr. Jerdon, 'Birds of India,' vol. i. 1862, p. xxi.

[5] Mr. L. H. Morgan's work on 'The American Beaver,' 1868, offers a good illustration of this remark. I cannot, however, avoid thinking that he goes too far in underrating the power of Instinct.

first employed by the natives in drawing their sledges; or the Arctic wolves, the parent-stock of the Esquimaux dog, may have acquired this instinct, impelling them not to attack their prey in a close pack when on thin ice. Questions of this kind are most difficult to answer.

So many facts have been recorded in various works shewing that animals possess some degree of reason, that I will here give only two or three instances, authenticated by Rengger, and relating to American monkeys, which stand low in their order. He states that when he first gave eggs to his monkeys, they smashed them and thus lost much of their contents; afterwards they gently hit one end against some hard body, and picked off the bits of shell with their fingers. After cutting themselves only once with any sharp tool, they would not touch it again, or would handle it with the greatest care. Lumps of sugar were often given them wrapped up in paper; and Rengger sometimes put a live wasp in the paper, so that in hastily unfolding it they got stung; after this had once happened, they always first held the packet to their ears to detect any movement within. Any one who is not convinced by such facts as these, and by what he may observe with his own dogs, that animals can reason, would not be convinced by anything that I could add. Nevertheless I will give one case with respect to dogs, as it rests on two distinct observers, and can hardly depend on the modification of any instinct.

Mr. Colquhoun[6] winged two wild-ducks, which fell on the opposite side of a stream; his retriever tried to bring over both at once, but could not succeed; she then, though never before known to ruffle a feather, deliberately killed one, brought over the other, and returned for the dead bird. Col. Hutchinson relates that two partridges were shot at once, one being killed, the other wounded; the latter ran away, and was caught by the retriever, who on her return came across the dead bird; "she stopped, evidently greatly puzzled, and after one or two trials, finding she could not take it up without permitting the escape of the winged bird, she considered a moment, then deliberately murdered it by giving it a severe crunch, and afterwards brought away both together. This was the only known instance of her ever having wilfully injured any game." Here we have reason, though not quite perfect, for the retriever might have brought the wounded bird first and then returned for the dead one, as in the case of the two wild-ducks.

The muleteers in S. America say, "I will not give you the mule whose step is easiest, but *la mas rational,*—the one that reasons best;" and Humboldt[7] adds, "this popular expression, dictated by long experience, combats the system of animated machines, better perhaps than all the arguments of speculative philosophy."

. . .

[6] 'The Moor and the Loch,' p. 45. Col. Hutchinson on 'Dog Breaking,' 1850, p. 46.
[7] 'Personal Narrative,' *Eng. translat.*, vol. iii. p. 106.

Chapter III. Comparison of the Mental Powers of Man and the Lower Animals —*Continued*

The moral sense—Fundamental proposition—The qualities of social animals—Origin of sociability—Struggle between opposed instincts—Man a social animal—The more enduring social instincts conquer other less persistent instincts—The social virtues alone regarded by savages—The self-regarding virtues acquired at a later stage of development—The importance of the judgment of the members of the same community on conduct—Transmission of moral tendencies—Summary.

...

Concluding Remarks.—Philosophers of the derivative[8] school of morals formerly assumed that the foundation of morality lay in a form of Selfishness; but more recently in the "Greatest Happiness principle." According to the view given above, the moral sense is fundamentally identical with the social instincts; and in the case of the lower animals it would be absurd to speak of these instincts as having been developed from selfishness, or for the happiness of the community. They have, however, certainly been developed for the general good of the community. The term, general good, may be defined as the means by which the greatest possible number of individuals can be reared in full vigour and health, with all their faculties perfect, under the conditions to which they are exposed. As the social instincts both of man and the lower animals have no doubt been developed by the same steps, it would be advisable, if found practicable, to use the same definition in both cases, and to take as the test of morality, the general good or welfare of the community, rather than the general happiness; but this definition would perhaps require some limitation on account of political ethics.

When a man risks his life to save that of a fellow-creature, it seems more appropriate to say that he acts for the general good or welfare, rather than for the general happiness of mankind. No doubt the welfare and the happiness of the individual usually coincide; and a contented, happy tribe will flourish better than one that is discontented and unhappy. We have seen that at an early period in the history of man, the expressed wishes of the community will have naturally influenced to a large extent the conduct of each member; and as all wish for

[8] This term is used in an able article in the 'Westminster Review,' Oct. 1869, p. 498. For the Greatest Happiness principle, see J. S. Mill, 'Utilitarianism,' p. 17.

happiness, the "greatest happiness principle" will have become a most important secondary guide and object; the social instincts, including sympathy, always serving as the primary impulse and guide. Thus the reproach of laying the foundation of the most noble part of our nature in the base principle of selfishness is removed; unless indeed the satisfaction which every animal feels when it follows its proper instincts, and the dissatisfaction felt when prevented, be called selfish.

The expression of the wishes and judgment of the members of the same community, at first by oral and afterwards by written language, serves, as just remarked, as a most important secondary guide of conduct, in aid of the social instincts, but sometimes in opposition to them. This latter fact is well exemplified by the *Law of Honour*, that is the law of the opinion of our equals, and not of all our countrymen. The breach of this law, even when the breach is known to be strictly accordant with true morality, has caused many a man more agony than a real crime. We recognise the same influence in the burning sense of shame which most of us have felt even after the interval of years, when calling to mind some accidental breach of a trifling though fixed rule of etiquette. The judgment of the community will generally be guided by some rude experience of what is best in the long run for all the members; but this judgment will not rarely err from ignorance and from weak powers of reasoning. Hence the strangest customs and superstitions, in complete opposition to the true welfare and happiness of mankind, have become all-powerful throughout the world. We see this in the horror felt by a Hindoo who breaks his caste, in the shame of a Mahometan woman who exposes her face, and in innumerable other instances. It would be difficult to distinguish between the remorse felt by a Hindoo who has eaten unclean food, from that felt after committing a theft; but the former would probably be the more severe.

How so many absurd rules of conduct, as well as so many absurd religious beliefs, have originated we do not know; nor how it is that they have become, in all quarters of the world, so deeply impressed on the mind of men; but it is worthy of remark that a belief constantly inculcated during the early years of life, whilst the brain is impressible, appears to acquire almost the nature of an instinct; and the very essence of an instinct is that it is followed independently of reason. Neither can we say why certain admirable virtues, such as the love of truth, are much more highly appreciated by some savage tribes than by others[9]; nor, again, why similar differences prevail even amongst civilised nations. Knowing how firmly fixed many strange customs and superstitions have become, we need feel no surprise that the self-regarding virtues should now appear to us so natural, supported as they are by reason, as to be thought innate, although they were not valued by man in his early condition.

Notwithstanding many sources of doubt, man can generally and readily distinguish between the higher and lower moral rules. The higher are founded on

[9] Good instances are given by Mr. Wallace in 'Scientific Opinion,' Sept. 15, 1869; and more fully in his 'Contributions to the Theory of Natural Selection,' 1870, p. 353.

the social instincts, and relate to the welfare of others. They are supported by the approbation of our fellow-men and by reason. The lower rules, though some of them when implying self-sacrifice hardly deserve to be called lower, relate chiefly to self, and owe their origin to public opinion, when matured by experience and cultivated; for they are not practised by rude tribes.

As man advances in civilisation, and small tribes are united into larger communities, the simplest reason would tell each individual that he ought to extend his social instincts and sympathies to all the members of the same nation, though personally unknown to him. This point being once reached, there is only an artificial barrier to prevent his sympathies extending to the men of all nations and races. If, indeed, such men are separated from him by great differences in appearance or habits, experience unfortunately shews us how long it is before we look at them as our fellow-creatures. Sympathy beyond the confines of man, that is humanity to the lower animals, seems to be one of the latest moral acquisitions. It is apparently unfelt by savages, except towards their pets. How little the old Romans knew of it is shewn by their abhorrent gladiatorial exhibitions. The very idea of humanity, as far as I could observe, was new to most of the Gauchos of the Pampas. This virtue, one of the noblest with which man is endowed, seems to arise incidentally from our sympathies becoming more tender and more widely diffused, until they are extended to all sentient beings. As soon as this virtue is honoured and practised by some few men, it spreads through instruction and example to the young, and eventually through public opinion.

The highest stage in moral culture at which we can arrive, is when we recognise that we ought to control our thoughts, and "not even in inmost thought to think again the sins that made the past so pleasant to us."[10] Whatever makes any bad action familiar to the mind, renders its performance by so much the easier. As Marcus Aurelius long ago said, "Such as are thy habitual thoughts, such also will be the character of thy mind; for the soul is dyed by the thoughts."[11]

Our great philosopher, Herbert Spencer, has recently explained his views on the moral sense. He says,[12] "I believe that the experiences of utility organised and consolidated through all past generations of the human race, have been producing corresponding modifications, which, by continued transmission and accumulation, have become in us certain faculties of moral intuition—certain emotions responding to right and wrong conduct, which have no apparent basis in the individual experiences of utility." There is not the least inherent improbability, as it seems to me, in virtuous tendencies being more or less strongly inherited; for, not to mention the various dispositions and habits transmitted by many of our domestic animals, I have heard of cases in which a desire to steal and a tendency to lie appeared to run in families of the upper ranks; and as stealing is so rare a crime

[10] Tennyson, 'Idylls of the King,' p. 244.

[11] 'The Thoughts of the Emperor M. Aurelius Antoninus,' *Eng. translat.*, 2nd edit., 1869, p. 112. Marcus Aurelius was born A.D. 121.

[12] Letter to Mr. Mill in Bain's 'Mental and Moral Science,' 1868, p. 722.

in the wealthy classes, we can hardly account by accidental coincidence for the tendency occurring in two or three members of the same family. If bad tendencies are transmitted, it is probable that good ones are likewise transmitted. Excepting through the principle of the transmission of moral tendencies, we cannot understand the differences believed to exist in this respect between the various races of mankind. We have, however, as yet, hardly sufficient evidence on this head.

Even the partial transmission of virtuous tendencies would be an immense assistance to the primary impulse derived directly from the social instincts, and indirectly from the approbation of our fellow-men. Admitting for the moment that virtuous tendencies are inherited, it appears probable, at least in such cases as chastity, temperance, humanity to animals, &c., that they become first impressed on the mental organisation through habit, instruction, and example, continued during several generations in the same family, and in a quite subordinate degree, or not at all, by the individuals possessing such virtues, having succeeded best in the struggle for life. My chief source of doubt with respect to any such inheritance, is that senseless customs, superstitions, and tastes, such as the horror of a Hindoo for unclean food, ought on the same principle to be transmitted. Although this in itself is perhaps not less probable than that animals should acquire inherited tastes for certain kinds of food or fear of certain foes, I have not met with any evidence in support of the transmission of superstitious customs or senseless habits.

Finally, the social instincts which no doubt were acquired by man, as by the lower animals, for the good of the community, will from the first have given to him some wish to aid his fellows, and some feeling of sympathy. Such impulses will have served him at a very early period as a rude rule of right and wrong. But as man gradually advanced in intellectual power and was enabled to trace the more remote consequences of his actions; as he acquired sufficient knowledge to reject baneful customs and superstitions; as he regarded more and more not only the welfare but the happiness of his fellow-men; as from habit, following on beneficial experience, instruction, and example, his sympathies became more tender and widely diffused, so as to extend to the men of all races, to the imbecile, the maimed, and other useless members of society, and finally to the lower animals,—so would the standard of his morality rise higher and higher. And it is admitted by moralists of the derivative school and by some intuitionists, that the standard of morality has risen since an early period in the history of man.[13]

As a struggle may sometimes be seen going on between the various instincts of the lower animals, it is not surprising that there should be a struggle in man between his social instincts, with their derived virtues, and his lower, though at the moment, stronger impulses or desires. This, as Mr. Galton[14] has remarked,

[13] A writer in the 'North British Review' (July, 1869, p. 531), well capable of forming a sound judgment, expresses himself strongly to this effect. Mr. Lecky ('Hist. of Morals,' vol. i. p. 143) seems to a certain extent to coincide.

[14] See his remarkable work on 'Hereditary Genius,' 1869, p. 349. The Duke of Argyll ('Primeval Man,' 1869, p. 188) has some good remarks on the contest in man's nature between right and wrong.

is all the less surprising, as man has emerged from a state of barbarism within a comparatively recent period. After having yielded to some temptation we feel a sense of dissatisfaction, analogous to that felt from other unsatisfied instincts, called in this case conscience; for we cannot prevent past images and impressions continually passing through our minds, and these in their weakened state we compare with the ever-present social instincts, or with habits gained in early youth and strengthened during our whole lives, perhaps inherited, so that they are at last rendered almost as strong as instincts. Looking to future generations, there is no cause to fear that the social instincts will grow weaker, and we may expect that virtuous habits will grow stronger, becoming perhaps fixed by inheritance. In this case the struggle between our higher and lower impulses will be less severe, and virtue will be triumphant.

Summary of the two last Chapters.—There can be no doubt that the difference between the mind of the lowest man and that of the highest animal is immense. An anthropomorphous ape, if he could take a dispassionate view of his own case, would admit that though he could form an artful plan to plunder a garden—though he could use stones for fighting or for breaking open nuts, yet that the thought of fashioning a stone into a tool was quite beyond his scope. Still less, as he would admit, could he follow out a train of metaphysical reasoning, or solve a mathematical problem, or reflect on God, or admire a grand natural scene. Some apes, however, would probably declare that they could and did admire the beauty of the coloured skin and fur of their partners in marriage. They would admit, that though they could make other apes understand by cries some of their perceptions and simpler wants, the notion of expressing definite ideas by definite sounds had never crossed their minds. They might insist that they were ready to aid their fellow-apes of the same troop in many ways, to risk their lives for them, and to take charge of their orphans; but they would be forced to acknowledge that disinterested love for all living creatures, the most noble attribute of man, was quite beyond their comprehension.

Nevertheless the difference in mind between man and the higher animals, great as it is, is certainly one of degree and not of kind. We have seen that the senses and intuitions, the various emotions and faculties, such as love, memory, attention, curiosity, imitation, reason, &c., of which man boasts, may be found in an incipient, or even sometimes in a well-developed condition, in the lower animals. They are also capable of some inherited improvement, as we see in the domestic dog compared with the wolf or jackal. If it be maintained that certain powers, such as self-consciousness, abstraction, &c., are peculiar to man, it may well be that these are the incidental results of other highly-advanced intellectual faculties; and these again are mainly the result of the continued use of a highly developed language. At what age does the new-born infant possess the power of abstraction, or become selfconscious and reflect on its own existence? We cannot answer; nor can we answer in regard to the ascending organic scale. The half-art and half-instinct of language still bears the stamp of its gradual evolution. The ennobling belief in God is not universal with man; and the belief in active spiritual agencies naturally follows from his other mental powers. The moral sense perhaps affords the best and highest distinction between man and the lower animals; but I need not say anything on this head, as I have so lately endeavoured to shew that the social instincts,—the prime principle of man's moral constitution[15]—with the aid of active intellectual powers and the effects of habit, naturally lead to the golden rule, "As ye would that men should do to you, do ye to them likewise;" and this lies at the foundation of morality.

In a future chapter I shall make some few remarks on the probable steps and means by which the several mental and moral faculties of man have been gradually evolved. That this at least is possible ought not to be denied, when we daily see their development in every infant; and when we may trace a perfect gradation from the mind of an utter idiot, lower than that of the lowest animal, to the mind of a Newton.

Chapter IV. On the Manner of Development of Man from Some Lower Form

Variability of body and mind in man—Inheritance—Causes of variability—Laws of variation the same in man as in the lower animals—Direct action of the conditions of life—Effects of the increased use and disuse of parts—Arrested development—Reversion—Correlated variation—Rate of increase—Checks to increase—Natural selection—Man the most dominant animal in the world—Importance of his corporeal structure—The causes which have led to his becoming erect—Consequent changes of structure—Decrease in size of the canine teeth—Increased size and altered shape of the skull—Nakedness—Absence of a tail—Defenceless condition of man.

We have seen in the first chapter that the homological structure of man, his embryological development and the rudiments which he still retains, all declare in the plainest manner that he is descended from some lower form. The possession of exalted mental powers is no insuperable objection to this conclusion. In order that an ape-like creature should have been transformed into man, it is necessary that this early form, as well as many successive links, should all have varied in mind and body. It is impossible to obtain direct evidence on this head; but if it can be shewn that man now varies—that his variations are induced by the same general causes, and obey the same general laws, as in the case of the lower animals—there can be little doubt that the preceding intermediate links

[15] 'The Thoughts of Marcus Aurelius,' &c., p. 139.

varied in a like manner. The variations at each successive stage of descent must, also, have been in some manner accumulated and fixed.

The facts and conclusions to be given in this chapter relate almost exclusively to the probable means by which the transformation of man has been effected, as far as his bodily structure is concerned. The following chapter will be devoted to the development of his intellectual and moral faculties. But the present discussion likewise bears on the origin of the different races or species of mankind, whichever term may be preferred.

It is manifest that man is now subject to much variability. No two individuals of the same race are quite alike. We may compare millions of faces, and each will be distinct. There is an equally great amount of diversity in the proportions and dimensions of the various parts of the body; the length of the legs being one of the most variable points. Although in some quarters of the world an elongated skull, and in other quarters a short skull prevails, yet there is great diversity of shape even within the limits of the same race, as with the aborigines of America and South Australia,—the latter a race "probably as pure and homogeneous in blood, customs, and language as any in existence"—and even with the inhabitants of so confined an area as the Sandwich Islands.

. . .

If we consider all the races of man, as forming a single species, his range is enormous; but some separate races, as the Americans and Polynesians, have very wide ranges. It is a well-known law that widely-ranging species are much more variable than species with restricted ranges; and the variability of man may with more truth be compared with that of widely-ranging species, than with that of domesticated animals.

. . .

Natural Selection.—We have now seen that man is variable in body and mind; and that the variations are induced, either directly or indirectly, by the same general causes, and obey the same general laws, as with the lower animals . . . . I cannot, therefore, understand how it is that Mr. Wallace maintains, that "natural selection could only have endowed the savage with a brain a little superior to that of an ape."

. . .

As the various mental faculties were gradually developed, the brain would almost certainly have become larger. No one, I presume, doubts that the large size of the brain in man, relatively to his body, in comparison with that of the gorilla or orang, is closely connected with his higher mental powers. We meet with closely analogous facts with insects, in which the cerebral ganglia are of extraordinary dimensions in ants; these ganglia in all the Hymenoptera being many times larger than in the less intelligent orders, such as beetles. On the other hand, no one supposes that the intellect of any two animals or of any two men can be accurately gauged by the cubic contents of their skulls. It is certain that there may be extraordinary mental activity with an extremely small absolute mass of nervous matter: thus the wonderfully diversified instincts, mental powers, and affections of ants are generally known, yet their cerebral ganglia are not so large as the quarter of a small pin's head. Under this latter point of view, the brain of an ant is one of the most marvellous atoms of matter in the world, perhaps more marvellous than the brain of man.

The belief that there exists in man some close relation between the size of the brain and the development of the intellectual faculties is supported by the comparison of the skulls of savage and civilised races, of ancient and modern people, and by the analogy of the whole vertebrate series. Dr. J. Barnard Davis has proved by many careful measurements, that the mean internal capacity of the skull in Europeans is 92·3 cubic inches; in Americans 87·5; in Asiatics 87·1; and in Australians only 81·9 inches. Professor Broca found that skulls from graves in Paris of the nineteenth century, were larger than those from vaults of the twelfth century, in the proportion of 1484 to 1426; and Prichard is persuaded that the present inhabitants of Britain have "much more capacious brain-cases" than the ancient inhabitants. Nevertheless it must be admitted that some skulls of very high antiquity, such as the famous one of Neanderthal, are well developed and capacious. With respect to the lower animals, M. E. Lartet, by comparing the crania of tertiary and recent mammals, belonging to the same groups, has come to the remarkable conclusion that the brain is generally larger and the convolutions more complex in the more recent form. On the other hand I have shewn that the brains of domestic rabbits are considerably reduced in bulk, in comparison with those of the wild rabbit or hare; and this may be attributed to their having been closely confined during many generations, so that they have exerted but little their intellect, instincts, senses, and voluntary movements.

. . .

Conclusion.—In this chapter we have seen that as man at the present day is liable, like every other animal, to multiform individual differences or slight variations, so no doubt were the early progenitors of man; the variations being then as now induced by the same general causes, and governed by the same general and complex laws. As all animals tend to multiply beyond their means of subsistence, so it must have been with the progenitors of man; and this will inevitably have led to a struggle for existence and to natural selection. This latter process will have been greatly aided by the inherited effects of the increased use of parts; these two processes incessantly reacting on each other. It appears, also, as we shall hereafter see, that various unimportant characters have been acquired by man through sexual selection. An unexplained residuum of change, perhaps a large one, must be left to the assumed uniform action of those unknown agencies, which occasionally induce strongly-marked and abrupt deviations of structure in our domestic productions.

Judging from the habits of savages and of the greater number of the Quadrumana, primeval men, and even the ape-like progenitors of man, probably lived in society. With strictly social animals, natural selection sometimes acts indirectly on the individual, through the preservation of variations which are beneficial only to the community. A community including a large number of well-endowed individuals increases in number and is victorious over other and less well-endowed communities; although each separate member may gain no advantage over the other members of the same community. With associated insects many remarkable structures, which are of little or no service to the individual or its own offspring, such as the pollen-collecting apparatus, or the sting of the worker-bee, or the great jaws of soldier-ants, have been thus acquired. With the higher social animals, I am not aware that any structure has been modified solely for the good of the community, though some are of secondary service to it. For instance, the horns of ruminants and the great canine teeth of baboons appear to have been acquired by the males as weapons for sexual strife, but they are used in defence of the herd or troop. In regard to certain mental faculties the case, as we shall see in the following chapter, is wholly different; for these faculties have been chiefly, or even exclusively, gained for the benefit of the community; the individuals composing the community being at the same time indirectly benefited.

It has often been objected to such views as the foregoing, that man is one of the most helpless and defenceless creatures in the world; and that during his early and less well-developed condition he would have been still more helpless. The Duke of Argyll, for instance, insists that "the human frame has diverged from the structure of brutes, in the direction of greater physical helplessness and weakness. That is to say, it is a divergence which of all others it is most impossible to ascribe to mere natural selection." He adduces the naked and unprotected state of the body, the absence of great teeth or claws for defence, the little strength of man, his small speed in running, and his slight power of smell, by which to discover food or to avoid danger. To these deficiencies there might have been added the still more serious loss of the power of quickly climbing trees, so as to escape from enemies. Seeing that the unclothed Fuegians can exist under their wretched climate, the loss of hair would not have been a great injury to primeval man, if he inhabited a warm country. When we compare defenceless man with the apes, many of which are provided with formidable canine teeth, we must remember that these in their fully-developed condition are possessed by the males alone, being chiefly used by them for fighting with their rivals; yet the females which are not thus provided, are able to survive.

In regard to bodily size or strength, we do not know whether man is descended from some comparatively small species, like the chimpanzee, or from one as powerful as the gorilla; and, therefore, we cannot say whether man has become larger and stronger, or smaller and weaker, in comparison with his progenitors. We should, however, bear in mind that an animal possessing great size, strength, and ferocity, and which, like the gorilla, could defend itself from all enemies, would probably, though not necessarily, have failed to become social; and this would most effectually have checked the acquirement by man of his higher mental qualities, such as sympathy and the love of his fellow-creatures. Hence it might have been an immense advantage to man to have sprung from some comparatively weak creature.

The slight corporeal strength of man, his little speed, his want of natural weapons, &c., are more than counterbalanced, firstly by his intellectual powers, through which he has, whilst still remaining in a barbarous state, formed for himself weapons, tools, &c., and secondly by his social qualities which lead him to give aid to his fellow-men and to receive it in return. No country in the world abounds in a greater degree with dangerous beasts than Southern Africa; no country presents more fearful physical hardships than the Arctic regions; yet one of the puniest races, namely, the Bushmen, maintain themselves in Southern Africa, as do the dwarfed Esquimaux in the Arctic regions. The early progenitors of man were, no doubt, inferior in intellect, and probably in social disposition, to the lowest existing savages; but it is quite conceivable that they might have existed, or even flourished, if, whilst they gradually lost their brute-like powers, such as climbing trees, &c., they at the same time advanced in intellect. But granting that the progenitors of man were far more helpless and defenceless than any existing savages, if they had inhabited some warm continent or large island, such as Australia or New Guinea, or Borneo (the latter island being now tenanted by the orang), they would not have been exposed to any special danger. In an area as large as one of these islands, the competition between tribe and tribe would have been sufficient, under favourable conditions, to have raised man, through the survival of the fittest, combined with the inherited effects of habit, to his present high position in the organic scale.

Chapter VI. On the Affinities and Genealogy of Man

Position of man in the animal series—The natural system genealogical—Adaptive characters of slight value—Various small points of resemblance between man and the Quadrumana—Rank of man in the natural system—Birthplace and antiquity of man—Absence of fossil connecting-links—Lower stages in the genealogy of man, as inferred, firstly from his affinities and secondly from his structure—Early androgynous condition of the Vertebrata—Conclusion.

Even if it be granted that the difference between man and his nearest allies is as great in corporeal structure as some naturalists maintain, and although we must grant that the difference between them is immense in mental power, yet the facts given in the previous chapters declare, as it appears to me, in the plainest manner, that man is descended from some lower form, notwithstanding that connecting-links have not hitherto been discovered.

Man is liable to numerous, slight, and diversified variations, which are induced by the same general causes, are governed and transmitted in accordance with the same general laws, as in the lower animals. Man tends to multiply at so rapid a rate that his offspring are necessarily exposed to a struggle for existence, and consequently to natural selection. He has given rise to many races, some of which are so different that they have often been ranked by naturalists as distinct species. His body is constructed on the same homological plan as that of other mammals, independently of the uses to which the several parts may be put. He passes through the same phases of embryological development. He retains many rudimentary and useless structures, which no doubt were once serviceable. Characters occasionally make their reappearance in him, which we have every reason to believe were possessed by his early progenitors. If the origin of man had been wholly different from that of all other animals, these various appearances would be mere empty deceptions; but such an admission is incredible. These appearances, on the other hand, are intelligible, at least to a large extent, if man is the co-descendant with other mammals of some unknown and lower form.

Some naturalists, from being deeply impressed with the mental and spiritual powers of man, have divided the whole organic world into three kingdoms, the Human, the Animal, and the Vegetable, thus giving to man a separate kingdom. Spiritual powers cannot be compared or classed by the naturalist; but he may endeavour to shew, as I have done, that the mental faculties of man and the lower animals do not differ in kind, although immensely in degree. A difference in degree, however great, does not justify us in placing man in a distinct kingdom, as will perhaps be best illustrated by comparing the mental powers of two insects, namely, a coccus or scale-insect and an ant, which undoubtedly belong to the same class. The difference is here greater, though of a somewhat different kind, than that between man and the highest mammal. The female coccus, whilst young, attaches itself by its proboscis to a plant; sucks the sap but never moves again; is fertilised and lays eggs; and this is its whole history. On the other hand, to describe the habits and mental powers of a female ant, would require, as Pierre Huber has shewn, a large volume; I may, however, briefly specify a few points. Ants communicate information to each other, and several unite for the same work, or games of play. They recognise their fellow-ants after months of absence. They build great edifices, keep them clean, close the doors in the evening, and post sentries. They make roads, and even tunnels under rivers. They collect food for the community, and when an object, too large for entrance, is brought to the nest, they enlarge the door, and afterwards build it up again. They go out to battle in regular bands, and freely sacrifice their lives for the common weal. They emigrate in accordance with a preconcerted plan. They capture slaves. They keep Aphides as milch-cows. They move the eggs of their aphides, as well as their own eggs and cocoons, into warm parts of the nest, in order that they may be quickly hatched; and endless similar facts could be given. On the whole, the difference in mental power between an ant and a coccus is immense; yet no one has ever dreamed of placing them in distinct classes, much less in distinct kingdoms. No doubt this interval is bridged over by the intermediate mental powers of many other insects; and this is not the case with man and the higher apes. But we have every reason to believe that breaks in the series are simply the result of many forms having become extinct.

Professor Owen, relying chiefly on the structure of the brain, has divided the mammalian series into four sub-classes. One of these he devotes to man; in another he places both the marsupials and the monotremata; so that he makes man as distinct from all other mammals as are these two latter groups conjoined. This view has not been accepted, as far as I am aware, by any naturalist capable of forming an independent judgment, and therefore need not here be further considered.

We can understand why a classification founded on any single character or organ—even an organ so wonderfully complex and important as the brain—or on the high development of the mental faculties, is almost sure to prove unsatisfactory. This principle has indeed been tried with hymenopterous insects; but when thus classed by their habits or instincts, the arrangement proved thoroughly artificial. (3. Westwood, 'Modern Classification of Insects,' vol. ii. 1840, p. 87.) Classifications may, of course, be based on any character whatever, as on size, colour, or the element inhabited; but naturalists have long felt a profound conviction that there is a natural system. This system, it is now generally admitted, must be, as far as possible, genealogical in arrangement,—that is, the co-descendants of the same form must be kept together in one group, apart from the co-descendants of any other form; but if the parent-forms are related, so will be their descendants, and the two groups together will form a larger group. The amount of difference between the several groups—that is the amount of modification which each has undergone—is expressed by such terms as genera, families, orders, and classes. As we have no record of the lines of descent, the pedigree can be discovered only by observing the degrees of resemblance between the beings which are to be classed. For this object numerous points of resemblance are of much more importance than the amount of similarity or dissimilarity in a few points. If two languages were found to resemble each other in a multitude of words and points of construction, they would be universally recognised as having sprung from a common source, notwithstanding that they differed greatly in some few words or points of construction. But with organic beings the points of resemblance must not consist of adaptations to similar habits of life: two animals may, for instance, have had their whole frames modified for living in the water, and yet they will not be brought any nearer to each other in the natural system. Hence we can see how it is that resemblances in several unimportant structures, in useless and rudimentary organs, or not now functionally active, or in an embryological condition, are by far the most serviceable for classification; for they can hardly be due to adaptations within a late period; and thus they reveal the old lines of descent or of true affinity.

We can further see why a great amount of modification in some one character ought not to lead us to separate widely any two organisms. A part which already differs much from the same part in other allied forms has already, according to the theory of evolution, varied much; consequently it would (as long as the organism remained exposed to the same exciting conditions) be liable to further variations of the same kind; and these, if beneficial, would be preserved, and thus be continually augmented. In many cases the continued development of a part, for instance, of the beak of a bird, or of the teeth of a mammal, would not aid the species in gaining its food, or for any other object; but with man we can see no definite limit to the continued development of the brain and mental faculties, as far as advantage is concerned. Therefore in determining the position of man in the natural or genealogical system, the extreme development of his brain ought not to outweigh a multitude of resemblances in other less important or quite unimportant points.

The greater number of naturalists who have taken into consideration the whole structure of man, including his mental faculties, have followed Blumenbach and Cuvier, and have placed man in a separate Order, under the title of the Bimana, and therefore on an equality with the orders of the Quadrumana, Carnivora, etc. Recently many of our best naturalists have recurred to the view first propounded by Linnaeus, so remarkable for his sagacity, and have placed man in the same Order with the Quadrumana, under the title of the Primates. The justice of this conclusion will be admitted: for in the first place, we must bear in mind the comparative insignificance for classification of the great development of the brain in man, and that the strongly-marked differences between the skulls of man and the Quadrumana (lately insisted upon by Bischoff, Aeby, and others) apparently follow from their differently developed brains. In the second place, we must remember that nearly all the other and more important differences between man and the Quadrumana are manifestly adaptive in their nature, and relate chiefly to the erect position of man; such as the structure of his hand, foot, and pelvis, the curvature of his spine, and the position of his head. The family of Seals offers a good illustration of the small importance of adaptive characters for classification. These animals differ from all other Carnivora in the form of their bodies and in the structure of their limbs, far more than does man from the higher apes; yet in most systems, from that of Cuvier to the most recent one by Mr. Flower (4. 'Proceedings Zoological Society,' 1863, p. 4.), seals are ranked as a mere family in the Order of the Carnivora. If man had not been his own classifier, he would never have thought of founding a separate order for his own reception . . . .

Chapter VII. On the Races of Man

The nature and value of specific characters—Application to the races of man—Arguments in favour of, and opposed to, ranking the so-called races of man as distinct species—Sub-species—Monogenists and polygenists—Convergence of character—Numerous points of resemblance in body and mind between the most distinct races of man—The state of man when he first spread over the earth—Each race not descended from a single pair—The extinction of races—The formation of races—The effects of crossing—Slight influence of the direct action of the conditions of life—Slight or no influence of natural selection—Sexual selection.

. . .

The question whether mankind consists of one or several species has of late years been much agitated by anthropologists, who are divided into two schools of monogenists and polygenists. Those who do not admit the principle of evolution, must look at species either as separate creations or as in some manner distinct entities; and they must decide what forms to rank as species by the analogy of other organic beings which are commonly thus received. But it is a hopeless endeavour to decide this point on sound grounds, until some definition of the term "species" is generally accepted; and the definition must not include an element which cannot possibly be ascertained, such as an act of creation. We might as well attempt without any definition to decide whether a certain number of houses should be called a village, or town, or city. We have a practical illustration of the difficulty in the never-ending doubts whether many closely-allied mammals, birds, insects, and plants, which represent each other in North America and Europe, should be ranked species or geographical races; and so it is with the productions of many islands situated at some little distance from the nearest continent.

Those naturalists, on the other hand, who admit the principle of evolution, and this is now admitted by the greater number of rising men, will feel no doubt that all the races of man are descended from a single primitive stock; whether or not they think fit to designate them as distinct species, for the sake of expressing their amount of difference. With our domestic animals the question whether the various races have arisen from one or more species is different. Although all such races, as well as all the natural species within the same genus, have undoubtedly sprung from the same primitive stock, yet it is a fit subject for discussion, whether, for instance, all the domestic races of the dog have acquired their present differences since some one species was first domesticated and bred by man; or whether they owe some of their characters to inheritance from distinct species, which had already been modified in a state of nature. With mankind no such question can arise, for he cannot be said to have been domesticated at any particular period.

When the races of man diverged at an extremely remote epoch from their common progenitor, they will have differed but little from each other, and been few in number; consequently they will then, as far as their distinguishing characters are concerned, have had less claim to rank as distinct species, than the existing so-called races. Nevertheless such early races would perhaps have been ranked by some naturalists as distinct species, so arbitrary is the term, if their differences, although extremely slight, had been more constant than at present, and had not graduated into each other.

It is, however, possible, though far from probable, that the early progenitors of man might at first have diverged much in character, until they became more unlike each other than are any existing races; but that subsequently, as suggested by Vogt, they converged in character. When man selects for the same object the offspring of two distinct species, he sometimes induces, as far as general appearance is concerned, a considerable amount of convergence. This is the case, as shewn by Von Nathusius, with the improved breeds of pigs, which are descended from two distinct species; and in a less well-marked manner with the improved breeds of cattle. A great anatomist, Gratiolet, maintains that the anthropomorphous apes do not form a natural sub-group; but that the orang is a highly developed gibbon or semnopithecus; the chimpanzee a highly developed macacus; and the gorilla a highly developed mandrill. If this conclusion, which rests almost exclusively on brain-characters, be admitted, we should have a case of convergence at least in external characters, for the anthropomorphous apes are certainly more like each other in many points than they are to other apes. All analogical resemblances, as of a whale to a fish, may indeed be said to be cases of convergence; but this term has never been applied to superficial and adaptive resemblances. It would be extremely rash in most cases to attribute to convergence close similarity in many points of structure in beings which had once been widely different. The form of a crystal is determined solely by the molecular forces, and it is not surprising that dissimilar substances should sometimes assume the same form; but with organic beings we should bear in mind that the form of each depends on an infinitude of complex relations, namely on the variations which have arisen, these being due to causes far too intricate to be followed out,—on the nature of the variations which have been preserved, and this depends on the surrounding physical conditions, and in a still higher degree on the surrounding organisms with which each has come into competition,—and lastly, on inheritance (in itself a fluctuating element) from innumerable progenitors, all of which have had their forms determined through equally complex relations.

It appears utterly incredible that two organisms, if differing in a marked manner, should ever afterwards converge so closely as to lead to a near approach to identity throughout their whole organisation. In the case of the convergent pigs above referred to, evidence of their descent from two primitive stocks is still plainly retained, according to Von Nathusius, in certain bones of their skulls. If the races of man were descended, as supposed by some naturalists, from two or more distinct species, which had differed as much, or nearly as much, from each other, as the orang differs from the gorilla, it can hardly be doubted that marked differences in the structure of certain bones would still have been discoverable in man as he now exists.

Although the existing races of man differ in many respects, as in colour, hair, shape of skull, proportions of the body, &c., yet if their whole organisation be taken into consideration they are found to resemble each other closely in a multitude of points. Many of these points are of so unimportant or of so singular a nature, that it is extremely improbable that they should have been independently acquired by aboriginally distinct species or races. The same remark holds good with equal or greater force with respect to the numerous points of mental similarity between the most distinct races of man. The American aborigines, Negroes and Europeans differ as much from each other in mind as any three races that can be named; yet I was incessantly struck, whilst living with the Fuegians on board the "Beagle," with the many little traits of character, shewing how similar their minds were to ours; and so it was with a full-blooded negro with whom I happened once to be intimate.

He who will carefully read Mr. Tylor's and Sir J. Lubbock's interesting works can hardly fail to be deeply impressed with the close similarity between the men of all races in tastes, dispositions and habits. This is shewn by the pleasure which they all take in dancing, rude music, acting, painting, tattooing, and otherwise decorating themselves,—in their mutual comprehension of gesture-language—and, as I shall be able to shew in a future essay, by the same expression in their features, and by the same inarticulate cries, when they are excited by various emotions. This similarity, or rather identity, is striking, when contrasted with the different expressions which may be observed in distinct species of monkeys. There is good evidence that the art of shooting with bows and arrows has not been handed down from any common progenitor of mankind, yet the stone arrow-heads, brought from the most distant parts of the world and manufactured at the most remote periods, are, as Nilsson has shewn, almost identical; and this fact can only be accounted for by the various races having similar inventive or mental powers. The same observation has been made by archæologists with respect to certain widely-prevalent ornaments, such as zigzags, &c.; and with respect to various simple beliefs and customs, such as the burying of the dead under megalithic structures. I remember observing in South America, that there, as in so many other parts of the world, man has generally chosen the summits of lofty hills, on which to throw up piles of stones, either for the sake of recording some remarkable event, or for burying his dead.

Now when naturalists observe a close agreement in numerous small details of habits, tastes and dispositions between two or more domestic races, or between nearly-allied natural forms, they use this fact as an argument that all are descended from a common progenitor who was thus endowed; and consequently that all should be classed under the same species. The same argument may be applied with much force to the races of man.

As it is improbable that the numerous and unimportant points of resemblance between the several races of man in bodily structure and mental faculties (I do not here refer to similar customs) should all have been independently acquired, they must have been inherited from progenitors who were thus characterised. We thus gain some insight into the early state of man, before he had spread step by step over the face of the earth. The spreading of man to regions widely separated by the sea, no doubt, preceded any considerable amount of divergence of character in the several races; for otherwise we should sometimes meet with the same race in distinct continents; and this is never the case. Sir J. Lubbock, after comparing the arts now practised by savages in all parts of the world, specifies those which man could not have known, when he first wandered from his original birthplace; for if once learnt they would never have been forgotten. He thus shews that "the spear, which is but a development of the knife-point, and the club, which is but a long hammer, are the only things left." He admits, however, that the art of making fire probably had already been discovered, for it is common to all the races now existing, and was known to the ancient cave-inhabitants of Europe. Perhaps the art of making rude canoes or rafts was likewise known; but as man existed at a remote epoch, when the land in many places stood at a very different level, he would have been able, without the aid of canoes, to have spread widely. Sir J. Lubbock further remarks how improbable it is that our earliest ancestors could have "counted as high as ten, considering that so many races now in existence cannot get beyond four." Nevertheless, at this early period, the intellectual and social faculties of man could hardly have been inferior in any extreme degree to those now possessed by the lowest savages; otherwise primeval man could not have been so eminently successful in the struggle for life, as proved by his early and wide diffusion.

From the fundamental differences between certain languages, some philologists have inferred that when man first became widely diffused he was not a speaking animal; but it may be suspected that languages, far less perfect than any now spoken, aided by gestures, might have been used, and yet have left no traces on subsequent and more highly-developed tongues. Without the use of some language, however imperfect, it appears doubtful whether man's intellect could have risen to the standard implied by his dominant position at an early period.

Whether primeval man, when he possessed very few arts of the rudest kind, and when his power of language was extremely imperfect, would have deserved to be called man, must depend on the definition which we employ. In a series of forms graduating insensibly from some ape-like creature to man as he now exists, it would be impossible to fix on any definite point when the term "man" ought to be used. But this is a matter of very little importance. So again it is almost a matter of indifference whether the so-called races of man are thus designated, or are ranked as species or sub-species; but the latter term appears the most appropriate. Finally, we may conclude that when the principles of evolution are generally accepted, as they surely will be before long, the dispute between the monogenists and the polygenists will die a silent and unobserved death.

One other question ought not to be passed over without notice, namely, whether, as is sometimes assumed, each sub-species or race of man has sprung from a single pair of progenitors. With our domestic animals a new race can readily be formed from a single pair possessing some new character, or even from a single individual thus characterised, by carefully matching the varying offspring; but most of our races have been formed, not intentionally from a selected pair, but unconsciously by the preservation of many individuals which have varied, however slightly, in some useful or desired manner. If in one country stronger and heavier horses, and in another country lighter and fleeter horses, were habitually preferred, we may feel sure that two distinct sub-breeds would, in the course of time, be produced, without any particular pairs or individuals having been separated and bred from in either country. Many races have been thus formed, and their manner of formation is closely analogous with that of natural species. We know, also, that the horses which have been brought to the Falkland Islands have become, during successive generations, smaller and weaker, whilst those which have run wild on the Pampas have acquired larger and coarser heads; and such changes are manifestly due, not to any one pair, but to all the individuals having been subjected to the same conditions, aided, perhaps, by the principle of reversion. The new sub-breeds in none of these cases are descended from any single pair, but from many individuals which have varied in different degrees, but in the same general manner; and we may conclude that the races of man have been similarly produced, the modifications being either the direct result of exposure to different conditions, or the indirect result of some form of selection. But to this latter subject we shall presently return.

On the Extinction of the Races of Man.—The partial and complete extinction of many races and sub-races of man are historically known events. Humboldt saw in South America a parrot which was the sole living creature that could speak the language of a lost tribe. Ancient monuments and stone implements found in all parts of the world, of which no tradition is preserved by the present inhabitants, indicate much extinction. Some small and broken tribes, remnants of former races, still survive in isolated and generally mountainous districts. In Europe the ancient races were all, according to Schaaffhausen, "lower in the scale than the rudest living savages;" they must therefore have differed, to a certain extent, from any existing race . . . .

On the Formation of the Races of Man.—It may be premised that when we find the same race, though broken up into distinct tribes, ranging over a great area, as over America, we may attribute their general resemblance to descent from a common stock. In some cases the crossing of races already, distinct has led to the formation of new races. The singular fact that Europeans and Hindoos, who belong to the same Aryan stock and speak a language fundamentally the same, differ widely in appearance, whilst Europeans differ but little from Jews, who belong to the Semitic stock and speak quite another language, has been accounted for by Broca through the Aryan branches having been largely crossed during their wide diffusion by various indigenous tribes. When two races in close contact cross, the first result is a heterogeneous mixture: thus Mr. Hunter, in describing the Santali or hill-tribes of India, says that hundreds of imperceptible gradations may be traced "from the black, squat tribes of the mountains to the tall olive-coloured Brahman, with his intellectual brow, calm eyes, and high but narrow head;" so that it is necessary in courts of justice to ask the witnesses whether they are Santalis or Hindoos. Whether a heterogeneous people, such as the inhabitants of some of the Polynesian islands, formed by the crossing of two distinct races, with few or no pure members left, would ever become homogeneous, is not known from direct evidence. But as with our domesticated animals, a crossed breed can certainly, in the course of a few generations, be fixed and made uniform by careful selection, we may infer that the free and prolonged intercrossing during many generations of a heterogeneous mixture would supply the place of selection, and overcome any tendency to reversion, so that a crossed race would ultimately become homogeneous, though it might not partake in an equal degree of the characters of the two parent-races.

Of all the differences between the races of man, the colour of the skin is the most conspicuous and one of the best marked. Differences of this kind, it was formerly thought, could be accounted for by long exposure under different climates; but Pallas first shewed that this view is not tenable, and he has been followed by almost all anthropologists. The view has been rejected chiefly because the distribution of the variously coloured races, most of whom must have long inhabited their present homes, does not coincide with corresponding differences of climate. Weight must also be given to such cases as that of the Dutch families, who, as we hear on excellent authority, have not undergone the least change of colour, after residing for three centuries in South Africa. The uniform appearance in various parts of the world of gypsies and Jews, though the uniformity of the latter has been somewhat exaggerated, is likewise an argument on the same side. A very damp or a very dry atmosphere has been supposed to be more influential in modifying the colour of the skin than mere heat; but as D'Orbigny in South America, and Livingstone in Africa, arrived at diametrically opposite conclusions with respect to dampness and dryness, any conclusion on this head must be considered as very doubtful.

Various facts, which I have elsewhere given, prove that the colour of the skin and hair is sometimes correlated in a surprising manner with a complete immunity from the action of certain vegetable poisons and from the attacks of certain parasites. Hence it occurred to me, that negroes and other dark races might have acquired their dark tints by the darker individuals escaping during a long series of generations from the deadly influence of the miasmas of their native countries.

I afterwards found that the same idea had long ago occurred to Dr. Wells. That negroes, and even mulattoes, are almost completely exempt from the yellow-fever, which is so destructive in tropical America, has long been known. They likewise escape to a large extent the fatal intermittent fevers that prevail along, at least, 2600 miles of the shores of Africa, and which annually cause one-fifth of the white settlers to die, and another fifth to return home invalided . . . .

That the immunity of the negro is in any degree correlated with the colour of his skin is a mere conjecture: it may be correlated with some difference in his blood, nervous system, or other tissues . . . .

Although with our present knowledge we cannot account for the strongly-marked differences in colour between the races of man, either through correlation with constitutional peculiarities, or through the direct action of climate; yet we must not quite ignore the latter agency, for there is good reason to believe that some inherited effect is thus produced.

. . .

If, however, we look to the races of man, as distributed over the world, we must infer that their characteristic differences cannot be accounted for by the direct action of different conditions of life, even after exposure to them for an enormous period of time. The Esquimaux live exclusively on animal food; they are clothed in thick fur, and are exposed to intense cold and to prolonged darkness; yet they do not differ in any extreme degree from the inhabitants of Southern China, who live entirely on vegetable food and are exposed almost naked to a hot, glaring climate. The unclothed Fuegians live on the marine productions of their inhospitable shores; the Botocudos of Brazil wander about the hot forests of the interior and live chiefly on vegetable productions; yet these tribes resemble each other so closely that the Fuegians on board the "Beagle" were mistaken by some Brazilians for Botocudos. The Botocudos again, as well as the other inhabitants of tropical America, are wholly different from the Negroes who inhabit the opposite shores of the Atlantic, are exposed to a nearly similar climate, and follow nearly the same habits of life.

Nor can the differences between the races of man be accounted for, except to a quite insignificant degree, by the inherited effects of the increased or decreased use of parts. Men who habitually live in canoes, may have their legs somewhat stunted; those who inhabit lofty regions have their chests enlarged; and those who constantly use certain sense-organs have the cavities in which they are lodged somewhat increased in size, and their features consequently a little modified. With civilised nations, the reduced size of the jaws from lessened use, the habitual play of different muscles serving to express different emotions, and the increased size of the brain from greater intellectual activity, have together produced a considerable effect on their general appearance in comparison with savages. It is also possible that increased bodily stature, with no corresponding increase in the size of the brain, may have given to some races (judging from the previously adduced cases of the rabbits) an elongated skull of the dolichocephalic type.

. . .

We have now seen that the characteristic differences between the races of man cannot be accounted for in a satisfactory manner by the direct action of the conditions of life, nor by the effects of the continued use of parts, nor through the principle of correlation. We are therefore led to inquire whether slight individual differences, to which man is eminently liable, may not have been preserved and augmented during a long series of generations through natural selection. But here we are at once met by the objection that beneficial variations alone can be thus preserved; and as far as we are enabled to judge (although always liable to error on this head) not one of the external differences between the races of man are of any direct or special service to him. The intellectual and moral or social faculties must of course be excepted from this remark; but differences in these faculties can have had little or no influence on external characters. The variability of all the characteristic differences between the races, before referred to, likewise indicates that these differences cannot be of much importance; for, had they been important, they would long ago have been either fixed and preserved, or eliminated. In this respect man resembles those forms, called by naturalists protean or polymorphic, which have remained extremely variable, owing, as it seems, to their variations being of an indifferent nature, and consequently to their having escaped the action of natural selection.

We have thus far been baffled in all our attempts to account for the differences between the races of man; but there remains one important agency, namely Sexual Selection, which appears to have acted as powerfully on man, as on many other animals. I do not intend to assert that sexual selection will account for all the differences between the races. An unexplained residuum is left, about which we can in our ignorance only say, that as individuals are continually born with, for instance, heads a little rounder or narrower, and with noses a little longer or shorter, such slight differences might become fixed and uniform, if the unknown agencies which induced them were to act in a more constant manner, aided by long-continued intercrossing. Such modifications come under the provisional class, alluded to in our fourth chapter, which for the want of a better term have been called spontaneous variations. Nor do I pretend that the effects of sexual selection can be indicated with scientific precision; but it can be shewn that it would be an inexplicable fact if man had not been modified by this agency, which has acted so powerfully on innumerable animals, both high and low in the scale. It can further be shewn that the differences between the races of man, as in colour, hairyness, form of features, &c., are of the nature which it might have been expected would have been acted on by sexual selection. But in order to treat this subject in a fitting manner, I have found it necessary to pass the whole animal kingdom in review; I have therefore devoted to it the Second Part of this work.

At the close I shall return to man, and, after attempting to shew how far he has been modified through sexual selection, will give a brief summary of the chapters in this First Part.

Part II.—Sexual Selection

Chapter XIX. Secondary Sexual Characters of Man

Differences between man and woman—Causes of such differences and of certain characters common to both sexes—Law of battle—Differences in mental powers—and voice—On the influence of beauty in determining the marriages of mankind—Attention paid by savages to ornaments—Their ideas of beauty in woman—The tendency to exaggerate each natural peculiarity.

With mankind the differences between the sexes are greater than in most species of Quadrumana, but not so great as in some, for instance, the mandrill. Man on an average is considerably taller, heavier, and stronger than woman, with squarer shoulders and more plainly-pronounced muscles. Owing to the relation which exists between muscular development and the projection of the brows, the superciliary ridge is generally more strongly marked in man than in woman. His body, and especially his face, is more hairy, and his voice has a different and more powerful tone. In certain tribes the women are said, whether truly I know not, to differ slightly in tint from the men; and with Europeans, the women are perhaps the more brightly coloured of the two, as may be seen when both sexes have been equally exposed to the weather.

Man is more courageous, pugnacious, and energetic than woman, and has a more inventive genius. His brain is absolutely larger, but whether relatively to the larger size of his body, in comparison with that of woman, has not, I believe been fully ascertained. In woman the face is rounder; the jaws and the base of the skull smaller; the outlines of her body rounder, in parts more prominent; and her pelvis is broader than in man; but this latter character may perhaps be considered rather as a primary than a secondary sexual character. She comes to maturity at an earlier age than man.

As with animals of all classes, so with man, the distinctive characters of the male sex are not fully developed until he is nearly mature; and if emasculated they never appear. The beard, for instance, is a secondary sexual character, and male children are beardless, though at an early age they have abundant hair on their heads. It is probably due to the rather late appearance in life of the successive variations, by which man acquired his masculine characters, that they are transmitted to the male sex alone. Male and female children resemble each other closely, like the young of so many other animals in which the adult sexes differ; they likewise resemble the mature female much more closely, than the mature male. The female, however, ultimately assumes certain distinctive characters, and in the formation of her skull, is said to be intermediate between the child and the man. Again, as the young of closely allied though distinct species do not differ nearly so much from each other as do the adults, so it is with the children of the different races of man. Some have even maintained that race-differences cannot be detected in the infantile skull. In regard to colour, the new-born negro child is reddish nut-brown, which soon becomes slaty-grey; the black colour being fully developed within a year in the Sudan, but not until three years in Egypt. The eyes of the negro are at first blue, and the hair chestnut-brown rather than black, being curled only at the ends. The children of the Australians immediately after birth are yellowish-brown, and become dark at a later age. Those of the Guaranys of Paraguay are whitish-yellow, but they acquire in the course of a few weeks the yellowish-brown tint of their parents. Similar observations have been made in other parts of America.

. . .

There can be little doubt that the greater size and strength of man, in comparison with woman, together with his broader shoulders, more developed muscles, rugged outline of body, his greater courage and pugnacity, are all due in chief part to inheritance from some early male progenitor, who, like the existing anthropoid apes, was thus characterised. These characters will, however, have been preserved or even augmented during the long ages whilst man was still in a barbarous condition, by the strongest and boldest men having succeeded best in the general struggle for life, as well as in securing wives, and thus having left a large number of offspring. It is not probable that the greater strength of man was primarily acquired through the inherited effects of his having worked harder than woman for his own subsistence and that of his family; for the women in all barbarous nations are compelled to work at least as hard as the men. With civilised people the arbitrament of battle for the possession of the women has long ceased; on the other hand, the men, as a general rule, have to work harder than the women for their mutual subsistence; and thus their greater strength will have been kept up.

Difference in the Mental Powers of the two Sexes.—With respect to differences of this nature between man and woman, it is probable that sexual selection has played a very important part. I am aware that some writers doubt whether there is any inherent difference; but this is at least probable from the analogy of the lower animals which present other secondary sexual characters. No one will dispute that the bull differs in disposition from the cow, the wild-boar from the sow, the stallion from the mare, and, as is well known to the keepers of menageries, the males of the larger apes from the females. Woman seems to differ from man in mental disposition, chiefly in her greater tenderness and less selfishness; and this holds good even with savages, as shewn by a well-known passage in Mungo Park's Travels, and by statements made by many other travellers. Woman, owing to her maternal instincts, displays these qualities towards her infants in an eminent degree; therefore it is likely that she should often extend them towards her fellow-creatures. Man is the rival of other men; he delights in competition, and this leads to ambition which passes too easily into selfishness. These latter qualities seem to be his natural and unfortunate birthright. It is generally admitted that with woman the powers of intuition, of rapid perception, and perhaps of imitation, are more strongly marked than in man; but some, at least, of these faculties are characteristic of the lower races, and therefore of a past and lower state of civilisation.

The chief distinction in the intellectual powers of the two sexes is shewn by man attaining to a higher eminence, in whatever he takes up, than woman can attain—whether requiring deep thought, reason, or imagination, or merely the use of the senses and hands. If two lists were made of the most eminent men and women in poetry, painting, sculpture, music,—comprising composition and performance, history, science, and philosophy, with half-a-dozen names under each subject, the two lists would not bear comparison. We may also infer, from the law of the deviation of averages, so well illustrated by Mr. Galton, in his work on 'Hereditary Genius,' that if men are capable of decided eminence over women in many subjects, the average standard of mental power in man must be above that of woman.

The half-human male progenitors of man, and men in a savage state, have struggled together during many generations for the possession of the females. But mere bodily strength and size would do little for victory, unless associated with courage, perseverance, and determined energy. With social animals, the young males have to pass through many a contest before they win a female, and the older males have to retain their females by renewed battles. They have, also, in the case of man, to defend their females, as well as their young, from enemies of all kinds, and to hunt for their joint subsistence. But to avoid enemies, or to attack them with success, to capture wild animals, and to invent and fashion weapons, requires the aid of the higher mental faculties, namely, observation, reason, invention, or imagination. These various faculties will thus have been continually put to the test, and selected during manhood; they will, moreover, have been strengthened by use during this same period of life. Consequently, in accordance with the principle often alluded to, we might expect that they would at least tend to be transmitted chiefly to the male offspring at the corresponding period of manhood.

Now, when two men are put into competition, or a man with a woman, who possess every mental quality in the same perfection, with the exception that the one has higher energy, perseverance, and courage, this one will generally become more eminent, whatever the object may be, and will gain the victory. He may be said to possess genius—for genius has been declared by a great authority to be patience; and patience, in this sense, means unflinching, undaunted perseverance. But this view of genius is perhaps deficient; for without the higher powers of the imagination and reason, no eminent success in many subjects can be gained. But these latter as well as the former faculties will have been developed in man, partly through sexual selection,—that is, through the contest of rival males, and partly through natural selection,—that is, from success in the general struggle for life; and as in both cases the struggle will have been during maturity, the characters thus gained will have been transmitted more fully to the male than to the female offspring. Thus man has ultimately become superior to woman. It is, indeed, fortunate that the law of the equal transmission of characters to both sexes has commonly prevailed throughout the whole class of mammals; otherwise it is probable that man would have become as superior in mental endowment to woman, as the peacock is in ornamental plumage to the peahen.

. . .

Chapter XXI. General Summary and Conclusion

Main conclusion that man is descended from some lower form—Manner of development—Genealogy of man—Intellectual and moral faculties—Sexual selection—Concluding remarks.

A brief summary will here be sufficient to recall to the reader's mind the more salient points in this work. Many of the views which have been advanced are highly speculative, and some no doubt will prove erroneous; but I have in every case given the reasons which have led me to one view rather than to another. It seemed worthwhile to try how far the principle of evolution would throw light on some of the more complex problems in the natural history of man. False facts are highly injurious to the progress of science, for they often long endure; but false views, if supported by some evidence, do little harm, as everyone takes a salutary pleasure in proving their falseness; and when this is done, one path towards error is closed and the road to truth is often at the same time opened.

The main conclusion arrived at in this work, and now held by many naturalists who are well competent to form a sound judgment, is that man is descended from some less highly organised form. The grounds upon which this conclusion rests will never be shaken, for the close similarity between man and the lower animals in embryonic development, as well as in innumerable points of structure and constitution, both of high and of the most trifling importance,—the rudiments which he retains, and the abnormal reversions to which he is occasionally liable,—are facts which cannot be disputed. They have long been known, but until recently they told us nothing with respect to the origin of man. Now when viewed by the light of our knowledge of the whole organic world, their meaning is unmistakeable. The great principle of evolution stands up clear and firm, when these groups of facts are considered in connection with others, such as the mutual affinities of the members of the same group, their geographical distribution in past and present times, and their geological succession. It is incredible that all these facts should speak falsely. He who is not content to look, like a savage, at the phenomena of nature as disconnected, cannot any longer believe that man is the work of a separate act of creation. He will be forced to admit that the close resemblance of the embryo of man to that, for instance, of a dog—the construction of his skull, limbs, and whole frame, independently of the uses to which the parts may be put, on the same plan with that of other mammals—the occasional reappearance of various structures, for instance of several distinct muscles, which man does not normally possess, but which are common to the Quadrumana—and a crowd of analogous facts—all point in the plainest manner to the conclusion that man is the co-descendant with other mammals of a common progenitor.

We have seen that man incessantly presents individual differences in all parts of his body and in his mental faculties. These differences or variations seem to be induced by the same general causes, and to obey the same laws as with the lower animals. In both cases similar laws of inheritance prevail. Man tends to increase at a greater rate than his means of subsistence; consequently he is occasionally subjected to a severe struggle for existence, and natural selection will have effected whatever lies within its scope. A succession of strongly-marked variations of a similar nature are by no means requisite; slight fluctuating differences in the individual suffice for the work of natural selection. We may feel assured that the inherited effects of the long-continued use or disuse of parts will have done much in the same direction with natural selection. Modifications formerly of importance, though no longer of any special use, will be long inherited. When one part is modified, other parts will change through the principle of correlation, of which we have instances in many curious cases of correlated monstrosities. Something may be attributed to the direct and definite action of the surrounding conditions of life, such as abundant food, heat, or moisture; and lastly, many characters of slight physiological importance, some indeed of considerable importance, have been gained through sexual selection.

No doubt man, as well as every other animal, presents structures, which as far as we can judge with our little knowledge, are not now of any service to him, nor have been so during any former period of his existence, either in relation to his general conditions of life, or of one sex to the other. Such structures cannot be accounted for by any form of selection, or by the inherited effects of the use and disuse of parts. We know, however, that many strange and strongly-marked peculiarities of structure occasionally appear in our domesticated productions, and if the unknown causes which produce them were to act more uniformly, they would probably become common to all the individuals of the species. We may hope hereafter to understand something about the causes of such occasional modifications, especially through the study of monstrosities: hence the labours of experimentalists, such as those of M. Camille Dareste, are full of promise for the future. In the greater number of cases we can only say that the cause of each slight variation and of each monstrosity lies much more in the nature or constitution of the organism, than in the nature of the surrounding conditions; though new and changed conditions certainly play an important part in exciting organic changes of all kinds.

Through the means just specified, aided perhaps by others as yet undiscovered, man has been raised to his present state. But since he attained to the rank of manhood, he has diverged into distinct races, or as they may be more appropriately called sub-species. Some of these, for instance the Negro and European, are so distinct that, if specimens had been brought to a naturalist without any further information, they would undoubtedly have been considered by him as good and true species. Nevertheless all the races agree in so many unimportant details of structure and in so many mental peculiarities, that these can be accounted for only through inheritance from a common progenitor; and a progenitor thus characterised would probably have deserved to rank as man.

It must not be supposed that the divergence of each race from the other races, and of all the races from a common stock, can be traced back to any one pair of progenitors. On the contrary, at every stage in the process of modification, all the individuals which were in any way best fitted for their conditions of life, though in different degrees, would have survived in greater numbers than the less well fitted. The process would have been like that followed by man, when he does not intentionally select particular individuals, but breeds from all the superior and neglects all the inferior individuals. He thus slowly but surely modifies his stock, and unconsciously forms a new strain. So with respect to modifications, acquired independently of selection, and due to variations arising from the nature of the organism and the action of the surrounding conditions, or from changed habits of life, no single pair will have been modified in a much greater degree than the other pairs which inhabit the same country, for all will have been continually blended through free intercrossing.

By considering the embryological structure of man,—the homologies which he presents with the lower animals,—the rudiments which he retains,—and the reversions to which he is liable, we can partly recall in imagination the former condition of our early progenitors; and can approximately place them in their proper position in the zoological series. We thus learn that man is descended from a hairy quadruped, furnished with a tail and pointed ears, probably arboreal in its habits, and an inhabitant of the Old World. This creature, if its whole structure had been examined by a naturalist, would have been classed amongst the Quadrumana, as surely as would the common and still more ancient progenitor of the Old and New World monkeys. The Quadrumana and all the higher mammals are probably derived from an ancient marsupial animal, and this through a long line of diversified forms, either from some reptile-like or some amphibian-like creature, and this again from some fish-like animal. In the dim obscurity of the past we can see that the early progenitor of all the Vertebrata must have been an aquatic animal, provided with branchiæ, with the two sexes united in the same individual, and with the most important organs of the body (such as the brain and heart) imperfectly developed. This animal seems to have been more like the larvæ of our existing marine Ascidians than any other known form.

The greatest difficulty which presents itself, when we are driven to the above conclusion on the origin of man, is the high standard of intellectual power and of moral disposition which he has attained. But everyone who admits the general principle of evolution, must see that the mental powers of the higher animals, which are the same in kind with those of mankind, though so different in degree, are capable of advancement. Thus the interval between the mental powers of one of the higher apes and of a fish, or between those of an ant and scale-insect, is immense. The development of these powers in animals does not offer any special difficulty; for with our domesticated animals, the mental faculties are certainly variable, and the variations are inherited. No one doubts that these faculties are of the utmost importance to animals in a state of nature. Therefore the conditions are favourable for their development through natural selection. The same conclusion may be extended to man; the intellect must have been all-important to him, even at a very remote period, enabling him to use language, to invent and make weapons, tools, traps, &c.; by which means, in combination with his social habits, he long ago became the most dominant of all living creatures.

A great stride in the development of the intellect will have followed, as soon as, through a previous considerable advance, the half-art and half-instinct of language came into use; for the continued use of language will have reacted on the brain, and produced an inherited effect; and this again will have reacted on the improvement of language. The large size of the brain in man, in comparison with that of the lower animals, relatively to the size of their bodies, may be attributed in chief part, as Mr. Chauncey Wright has well remarked, to the early use of some simple form of language,—that wonderful engine which affixes signs to all sorts of objects and qualities, and excites trains of thought which would never arise from the mere impression of the senses, and if they did arise could not be followed out. The higher intellectual powers of man, such as those of ratiocination, abstraction, self-consciousness, &c., will have followed from the continued improvement of other mental faculties; but without considerable culture of the mind, both in the race and in the individual, it is doubtful whether these high powers would be exercised, and thus fully attained.

The development of the moral qualities is a more interesting and difficult problem. Their foundation lies in the social instincts, including in this term the family ties. These instincts are of a highly complex nature, and in the case of the lower animals give special tendencies towards certain definite actions; but the more important elements for us are love, and the distinct emotion of sympathy. Animals endowed with the social instincts take pleasure in each other's company, warn each other of danger, defend and aid each other in many ways. These instincts are not extended to all the individuals of the species, but only to those of the same community. As they are highly beneficial to the species, they have in all probability been acquired through natural selection.

A moral being is one who is capable of comparing his past and future actions and motives,—of approving of some and disapproving of others; and the fact that man is the one being who with certainty can be thus designated makes the greatest of all distinctions between him and the lower animals. But in our third chapter I have endeavoured to shew that the moral sense follows, firstly, from the enduring and always present nature of the social instincts, in which respect man agrees with the lower animals; and secondly, from his mental faculties being highly active and his impressions of past events extremely vivid, in which respects he differs from the lower animals. Owing to this condition of mind, man cannot avoid looking backwards and comparing the impressions of past events and actions. He also continually looks forward. Hence after some temporary desire or passion has mastered his social instincts, he will reflect and compare the now weakened impression of such past impulses, with the ever present social instinct; and he will then feel that sense of dissatisfaction which all unsatisfied instincts leave behind them. Consequently he resolves to act differently for the future—and this is conscience. Any instinct which is permanently stronger or more enduring than another, gives rise to a feeling which we express by saying that it ought to be obeyed. A pointer dog, if able to reflect on his past conduct, would say to himself, I ought (as indeed we say of him) to have pointed at that hare and not have yielded to the passing temptation of hunting it.

Social animals are partly impelled by a wish to aid the members of the same community in a general manner, but more commonly to perform certain definite actions. Man is impelled by the same general wish to aid his fellows, but has few or no special instincts. He differs also from the lower animals in being able to express his desires by words, which thus become the guide to the aid required and bestowed. The motive to give aid is likewise somewhat modified in man: it no longer consists solely of a blind instinctive impulse, but is largely influenced by the praise or blame of his fellow men. Both the appreciation and the bestowal of praise and blame rest on sympathy; and this emotion, as we have seen, is one of the most important elements of the social instincts. Sympathy, though gained as an instinct, is also much strengthened by exercise or habit. As all men desire their own happiness, praise or blame is bestowed on actions and motives, according as they lead to this end; and as happiness is an essential part of the general good, the greatest-happiness principle indirectly serves as a nearly safe standard of right and wrong. As the reasoning powers advance and experience is gained, the more remote effects of certain lines of conduct on the character of the individual, and on the general good, are perceived; and then the self-regarding virtues, from coming within the scope of public opinion, receive praise, and their opposites receive blame. But with the less civilised nations reason often errs, and many bad customs and base superstitions come within the same scope, and consequently are esteemed as high virtues, and their breach as heavy crimes.

The moral faculties are generally esteemed, and with justice, as of higher value than the intellectual powers. But we should always bear in mind that the activity of the mind in vividly recalling past impressions is one of the fundamental though secondary bases of conscience. This fact affords the strongest argument for educating and stimulating in all possible ways the intellectual faculties of every human being. No doubt a man with a torpid mind, if his social affections and sympathies are well developed, will be led to good actions, and may have a fairly sensitive conscience. But whatever renders the imagination of men more vivid and strengthens the habit of recalling and comparing past impressions, will make the conscience more sensitive, and may even compensate to a certain extent for weak social affections and sympathies.

The moral nature of man has reached the highest standard as yet attained, partly through the advancement of the reasoning powers and consequently of a just public opinion, but especially through the sympathies being rendered more tender and widely diffused through the effects of habit, example, instruction, and reflection. It is not improbable that virtuous tendencies may through long practice be inherited. With the more civilised races, the conviction of the existence of an all-seeing Deity has had a potent influence on the advancement of morality. Ultimately man no longer accepts the praise or blame of his fellows as his chief guide, though few escape this influence, but his habitual convictions controlled by reason afford him the safest rule. His conscience then becomes his supreme judge and monitor. Nevertheless the first foundation or origin of the moral sense lies in the social instincts, including sympathy; and these instincts no doubt were primarily gained, as in the case of the lower animals, through natural selection.

The belief in God has often been advanced as not only the greatest, but the most complete of all the distinctions between man and the lower animals. It is however impossible, as we have seen, to maintain that this belief is innate or instinctive in man. On the other hand a belief in all-pervading spiritual agencies seems to be universal; and apparently follows from a considerable advance in the reasoning powers of man, and from a still greater advance in his faculties of imagination, curiosity and wonder. I am aware that the assumed instinctive belief in God has been used by many persons as an argument for His existence. But this is a rash argument, as we should thus be compelled to believe in the existence of many cruel and malignant spirits, possessing only a little more power than man; for the belief in them is far more general than of a beneficent Deity. The idea of a universal and beneficent Creator of the universe does not seem to arise in the mind of man, until he has been elevated by long-continued culture.

He who believes in the advancement of man from some lowly-organised form, will naturally ask how does this bear on the belief in the immortality of the soul. The barbarous races of man, as Sir J. Lubbock has shewn, possess no clear belief of this kind; but arguments derived from the primeval beliefs of savages are, as we have just seen, of little or no avail. Few persons feel any anxiety from the impossibility of determining at what precise period in the development of the individual, from the first trace of the minute germinal vesicle to the child either before or after birth, man becomes an immortal being; and there is no greater cause for anxiety because the period in the gradually ascending organic scale cannot possibly be determined.

I am aware that the conclusions arrived at in this work will be denounced by some as highly irreligious; but he who thus denounces them is bound to shew why it is more irreligious to explain the origin of man as a distinct species by descent from some lower form, through the laws of variation and natural selection, than to explain the birth of the individual through the laws of ordinary reproduction. The birth both of the species and of the individual are equally parts of that grand sequence of events, which our minds refuse to accept as the result of blind chance. The understanding revolts at such a conclusion, whether or not we are able to believe that every slight variation of structure,—the union of each pair in marriage,—the dissemination of each seed,—and other such events, have all been ordained for some special purpose.

Sexual selection has been treated at great length in these volumes; for, as I have attempted to shew, it has played an important part in the history of the organic world. As summaries have been given to each chapter, it would be superfluous here to add a detailed summary. I am aware that much remains doubtful, but I have endeavoured to give a fair view of the whole case. In the lower divisions of the animal kingdom, sexual selection seems to have done nothing: such animals are often affixed for life to the same spot, or have the two sexes combined in the same individual, or what is still more important, their perceptive and intellectual faculties are not sufficiently advanced to allow of the feelings of love and jealousy, or of the exertion of choice. When, however, we come to the Arthropoda and Vertebrata, even to the lowest classes in these two great Sub-Kingdoms, sexual selection has effected much; and it deserves notice that we here find the intellectual faculties developed, but in two very distinct lines, to the highest standard, namely in the Hymenoptera (ants, bees, &c.) amongst the Arthropoda, and in the Mammalia, including man, amongst the Vertebrata.

In the most distinct classes of the animal kingdom, with mammals, birds, reptiles, fishes, insects, and even crustaceans, the differences between the sexes follow almost exactly the same rules. The males are almost always the wooers; and they alone are armed with special weapons for fighting with their rivals. They are generally stronger and larger than the females, and are endowed with the requisite qualities of courage and pugnacity. They are provided, either exclusively or in a much higher degree than the females, with organs for producing vocal or instrumental music, and with odoriferous glands. They are ornamented with infinitely diversified appendages, and with the most brilliant or conspicuous colours, often arranged in elegant patterns, whilst the females are left unadorned.

When the sexes differ in more important structures, it is the male which is provided with special sense-organs for discovering the female, with locomotive organs for reaching her, and often with prehensile organs for holding her. These various structures for securing or charming the female are often developed in the male during only part of the year, namely the breeding season. They have in many cases been transferred in a greater or less degree to the females; and in the latter case they appear in her as mere rudiments. They are lost by the males after emasculation. Generally they are not developed in the male during early youth, but appear a short time before the age for reproduction. Hence in most cases the young of both sexes resemble each other; and the female resembles her young offspring throughout life. In almost every great class a few anomalous cases occur in which there has been an almost complete transposition of the characters proper to the two sexes; the females assuming characters which properly belong to the males. This surprising uniformity in the laws regulating the differences between the sexes in so many and such widely separated classes, is intelligible if we admit the action throughout all the higher divisions of the animal kingdom of one common cause, namely sexual selection.

Sexual selection depends on the success of certain individuals over others of the same sex in relation to the propagation of the species; whilst natural selection depends on the success of both sexes, at all ages, in relation to the general conditions of life. The sexual struggle is of two kinds; in the one it is between the individuals of the same sex, generally the male sex, in order to drive away or kill their rivals, the females remaining passive; whilst in the other, the struggle is likewise between the individuals of the same sex, in order to excite or charm those of the opposite sex, generally the females, which no longer remain passive, but select the more agreeable partners. This latter kind of selection is closely analogous to that which man unintentionally, yet effectually, brings to bear on his domesticated productions, when he continues for a long time choosing the most pleasing or useful individuals, without any wish to modify the breed.

The laws of inheritance determine whether characters gained through sexual selection by either sex shall be transmitted to the same sex, or to both sexes; as well as the age at which they shall be developed. It appears that variations which arise late in life are commonly transmitted to one and the same sex.

Variability is the necessary basis for the action of selection, and is wholly independent of it. It follows from this, that variations of the same general nature have often been taken advantage of and accumulated through sexual selection in relation to the propagation of the species, and through natural selection in relation to the general purposes of life. Hence secondary sexual characters, when equally transmitted to both sexes can be distinguished from ordinary specific characters only by the light of analogy. The modifications acquired through sexual selection are often so strongly pronounced that the two sexes have frequently been ranked as distinct species, or even as distinct genera. Such strongly-marked differences must be in some manner highly important; and we know that they have been acquired in some instances at the cost not only of inconvenience, but of exposure to actual danger.

The belief in the power of sexual selection rests chiefly on the following considerations. The characters which we have the best reason for supposing to have been thus acquired are confined to one sex; and this alone renders it probable that they are in some way connected with the act of reproduction. These characters in innumerable instances are fully developed only at maturity; and often during only a part of the year, which is always the breeding-season. The males (passing over a few exceptional cases) are the most active in courtship; they are the best armed, and are rendered the most attractive in various ways. It is to be especially observed that the males display their attractions with elaborate care in the presence of the females; and that they rarely or never display them excepting during the season of love. It is incredible that all this display should be purposeless. Lastly we have distinct evidence with some quadrupeds and birds that the individuals of the one sex are capable of feeling a strong antipathy or preference for certain individuals of the opposite sex.

Bearing these facts in mind, and not forgetting the marked results of man's unconscious selection, it seems to me almost certain that if the individuals of one sex were during a long series of generations to prefer pairing with certain individuals of the other sex, characterised in some peculiar manner, the offspring would slowly but surely become modified in this same manner. I have not attempted to conceal that, excepting when the males are more numerous than the females, or when polygamy prevails, it is doubtful how the more attractive males succeed in leaving a larger number of offspring to inherit their superiority in ornaments or other charms than the less attractive males; but I have shewn that this would probably follow from the females,—especially the more vigorous females which would be the first to breed, preferring not only the more attractive but at the same time the more vigorous and victorious males.

Although we have some positive evidence that birds appreciate bright and beautiful objects, as with the Bower-birds of Australia, and although they certainly appreciate the power of song, yet I fully admit that it is an astonishing fact that the females of many birds and some mammals should be endowed with sufficient taste for what has apparently been effected through sexual selection; and this is even more astonishing in the case of reptiles, fish, and insects. But we really know very little about the minds of the lower animals. It cannot be supposed that male Birds of Paradise or Peacocks, for instance, should take so much pains in erecting, spreading, and vibrating their beautiful plumes before the females for no purpose. We should remember the fact given on excellent authority in a former chapter, namely that several peahens, when debarred from an admired male, remained widows during a whole season rather than pair with another bird.

. . .

Everyone who admits the principle of evolution, and yet feels great difficulty in admitting that female mammals, birds, reptiles, and fish, could have acquired the high standard of taste which is implied by the beauty of the males, and which generally coincides with our own standard, should reflect that in each member of the vertebrate series the nerve-cells of the brain are the direct offshoots of those possessed by the common progenitor of the whole group. It thus becomes intelligible that the brain and mental faculties should be capable under similar conditions of nearly the same course of development, and consequently of performing nearly the same functions.

The reader who has taken the trouble to go through the several chapters devoted to sexual selection, will be able to judge how far the conclusions at which I have arrived are supported by sufficient evidence. If he accepts these conclusions, he may, I think, safely extend them to mankind; but it would be superfluous here to repeat what I have so lately said on the manner in which sexual selection has apparently acted on both the male and female side, causing the two sexes of man to differ in body and mind, and the several races to differ from each other in various characters, as well as from their ancient and lowly-organised progenitors.

He who admits the principle of sexual selection will be led to the remarkable conclusion that the cerebral system not only regulates most of the existing functions of the body, but has indirectly influenced the progressive development of various bodily structures and of certain mental qualities. Courage, pugnacity, perseverance, strength and size of body, weapons of all kinds, musical organs, both vocal and instrumental, bright colours, stripes and marks, and ornamental appendages, have all been indirectly gained by the one sex or the other, through the influence of love and jealousy, through the appreciation of the beautiful in sound, colour or form, and through the exertion of a choice; and these powers of the mind manifestly depend on the development of the cerebral system.

Man scans with scrupulous care the character and pedigree of his horses, cattle, and dogs before he matches them; but when he comes to his own marriage he rarely, or never, takes any such care. He is impelled by nearly the same motives as are the lower animals when left to their own free choice, though he is in so far superior to them that he highly values mental charms and virtues. On the other hand he is strongly attracted by mere wealth or rank. Yet he might by selection do something not only for the bodily constitution and frame of his offspring, but for their intellectual and moral qualities. Both sexes ought to refrain from marriage if in any marked degree inferior in body or mind; but such hopes are Utopian and will never be even partially realised until the laws of inheritance are thoroughly known. All do good service who aid towards this end. When the principles of breeding and of inheritance are better understood, we shall not hear ignorant members of our legislature rejecting with scorn a plan for ascertaining by an easy method whether or not consanguineous marriages are injurious to man.

The advancement of the welfare of mankind is a most intricate problem: all ought to refrain from marriage who cannot avoid abject poverty for their children; for poverty is not only a great evil, but tends to its own increase by leading to recklessness in marriage. On the other hand, as Mr. Galton has remarked, if the prudent avoid marriage, whilst the reckless marry, the inferior members will tend to supplant the better members of society. Man, like every other animal, has no doubt advanced to his present high condition through a struggle for existence consequent on his rapid multiplication; and if he is to advance still higher he must remain subject to a severe struggle. Otherwise he would soon sink into indolence, and the more highly-gifted men would not be more successful in the battle of life than the less gifted. Hence our natural rate of increase, though leading to many and obvious evils, must not be greatly diminished by any means. There should be open competition for all men; and the most able should not be prevented by laws or customs from succeeding best and rearing the largest number of offspring. Important as the struggle for existence has been and even still is, yet as far as the highest part of man's nature is concerned there are other agencies more important. For the moral qualities are advanced, either directly or indirectly, much more through the effects of habit, the reasoning powers, instruction, religion, &c., than through natural selection; though to this latter agency the social instincts, which afforded the basis for the development of the moral sense, may be safely attributed.

The main conclusion arrived at in this work, namely that man is descended from some lowly-organised form, will, I regret to think, be highly distasteful to many persons. But there can hardly be a doubt that we are descended from barbarians. The astonishment which I felt on first seeing a party of Fuegians on a wild and broken shore will never be forgotten by me, for the reflection at once rushed into my mind—such were our ancestors. These men were absolutely naked and bedaubed with paint, their long hair was tangled, their mouths frothed with excitement, and their expression was wild, startled, and distrustful. They possessed hardly any arts, and like wild animals lived on what they could catch; they had no government, and were merciless to everyone not of their own small tribe. He who has seen a savage in his native land will not feel much shame, if forced to acknowledge that the blood of some more humble creature flows in his veins. For my own part I would as soon be descended from that heroic little monkey, who braved his dreaded enemy in order to save the life of his keeper; or from that old baboon, who, descending from the mountains, carried away in triumph his young comrade from a crowd of astonished dogs—as from a savage who delights to torture his enemies, offers up bloody sacrifices, practises infanticide without remorse, treats his wives like slaves, knows no decency, and is haunted by the grossest superstitions.

Man may be excused for feeling some pride at having risen, though not through his own exertions, to the very summit of the organic scale; and the fact of his having thus risen, instead of having been aboriginally placed there, may give him hopes for a still higher destiny in the distant future. But we are not here concerned with hopes or fears, only with the truth as far as our reason allows us to discover it. I have given the evidence to the best of my ability; and we must acknowledge, as it seems to me, that man with all his noble qualities, with sympathy which feels for the most debased, with benevolence which extends not only to other men but to the humblest living creature, with his god-like intellect which has penetrated into the movements and constitution of the solar system—with all these exalted powers—Man still bears in his bodily frame the indelible stamp of his lowly origin.

Reading and Discussion Questions

1. Why do you think that Darwin includes the bit about monkeys getting drunk and smoking tobacco? Is this just a humorous anecdote or is there a deeper point here?
2. While Darwin avoided the question of human origins in the *Origin*, in the *Descent* it is clear that he intends to fully integrate human beings into his evolutionary perspective. How, if at all, might this issue change how people think about our places in nature or bear on issues of race, gender, religion, and morality?
3. How would you characterize Darwin's attitudes toward issues of race and gender, as expressed in this selection? To what extent would you say that his views on these topics are grounded in biology and to what extent are they simply reflective of his cultural context—Victorian era values characteristic of the time and place in which he lived? On the topics of both race and of gender, what is Darwin suggesting, what differences is he trying to explain, what evidence does he provide, and what role might social values be playing in the discussion?

The "Doctrine of Uniformity" in Geology Briefly Refuted (1866)

Kelvin (William Thomson)

96

The "Doctrine of Uniformity" in Geology, as held by many of the most eminent of British geologists, assumes that the earth's surface and upper crust have been nearly as they are at present in temperature and other physical qualities during millions of millions of years. But the heat which we know, by observation, to be now conducted out of the earth yearly is so great, that if this action had been going on with any approach to uniformity for 20,000 million years, the amount of heat lost out of the earth would have been about as much as would heat, by 100° Cent., a quantity of ordinary surface rock of 100 times the earth's bulk. (See calculation appended.) This would be more than enough to melt a mass of surface rock equal in bulk to the whole earth. No hypothesis as to chemical action, internal fluidity, effects of pressure at great depth, or possible character of substances in the interior of the earth, possessing the smallest vestige of probability, can justify the supposition that the earth's crust has remained nearly as it is, while from the whole, or from any part, of the earth, so great a quantity of heat has been lost.

Appendix: Estimate of Present Annual Loss of Heat from the Earth

Let A be the area of the earth's surface, D the increase of depth in any locality for which the temperature increases by 1° Cent., and k the conductivity per annum of the strata in the same locality. The heat conducted out per annum per square foot of surface in that locality is k/D. Hence, if we give k and D proper average values for the whole upper crust of the earth, the quantity conducted out across the whole earth's surface per annum will be (kA)/D. The bulk of a sphere being its surface multiplied by 1/3 of its radius, the thermal capacity of a mass of rock equal in bulk to the earth, and of specific heat *s* per unit of bulk is (1/3) *Ars*. Hence (3k)/(D*rs*) is the elevation of temperature which a quantity of heat equal to that lost from the earth in a year, would produce in a mass of rock equal in bulk to the whole earth. The laboratory experiments of Peclet; Observations on Underground Temperature in three kinds of rock in and near Edinburgh, by Forbes; in two Swedish strata, by Ångström, and at the Royal Observatory, Greenwich, give values of the conductivity in gramme-water units of heat per square centimetre, per 1° per centimetre of variation of temperature, per second, from .002 (marble, Peclet) to .0107 (sandstone of Craigleith quarry, Forbes); and .005 may be taken as a rough average. Hence, as there are 31,557,000 seconds in a year, we have $k = .005 \times 31{,}557{,}000$, or approximately 16×10^4. The thermal capacity of surface rock is somewhere about half that of equal bulk of water; so that we may take $s = .5$. And the increase of temperature downwards may be taken as roughly averaging 1° Cent. per 30 metres; so, that, D = 3000 centimetres. Lastly, the earth's quadrant being according to the first foundation of the French metrical system, about 10^9 centimetres, we may take, in a rough estimate such as the present, $r = 6 \times 10^8$ centimetres. Hence,

$$\frac{3k}{Drs} = \frac{3\times16\times10^4}{3000\times6\times10^8\times.5} = \frac{8}{15\times10^6}$$

This, multiplied by 20,000 × 106, amounts to 10,000, or to 100 times as much heat as would warm 10 times the earth's bulk of surface rock by 1° Cent.

Reading and Discussion Questions

1. William Thomson (Lord Kelvin) draws on thermodynamics to argue that there simply had not been enough time for evolutionary process of the type that Darwin describes to account for the diversity of life on earth. How old does he think that the earth is and what calculations related to the Earth's cooling serve as the basis for his argument?
2. If we are faced with a conflict between our best science in physics and best work in biology and geology at a given time, to which should we defer?

Review of "The Origin of Species" (1867)

Fleeming Jenkin

The theory proposed by Mr. Darwin as sufficient to account for the origin of species has been received as probably, and even as certainly true, by many who from their knowledge of physiology, natural history, and geology, are competent to form an intelligent opinion. The facts, they think, are consistent with the theory. Small differences are observed between animals and their offspring. Greater differences are observed between varieties known to be sprung form a common stock. The differences between what have been termed species are sometimes hardly greater in appearance than those between varieties owning a common origin. Even when species differ more widely, the difference they say, is one of degree only, not of kind. They can see no clear, definite distinction by which to decide in all cases, whether two animals have sprung from a common ancestor or not. They feel warranted in concluding, that for aught the structure of animals shows to the contrary, they may be descended from a few ancestors only—nay, even from a single pair . . . .

Some persons seem to have thought his theory dangerous to religion, morality, and what not. Others have tried to laugh it out of court. We can share neither the fears of the former nor the merriment of the latter; and, on the contrary, own to feeling the greatest admiration both for the ingenuity of the doctrine and for the temper in which it was broached, although, from a consideration of the following arguments, our opinion is adverse to its truth.

Variability. Darwin's theory requires that there shall be no limit to the possible differences between descendants and their progenitors, or, at least, that if there be limits, they shall be at so great a distance as to comprehend the utmost differences between any known forms of life. The variability required, if not infinite, is indefinite. Experience with domestic animals and cultivated plants shows that great variability exists. Darwin calls special attention to the differences between the various fancy pigeons, which, he says, are descended from one stock; between various breeds of cattle and horses, and some other domestic animals. He states that these differences are greater than those which induce some naturalists to class many specimens as distinct species. These differences are infinitely small as compared with the range required by his theory, but he assumes that by accumulation of successive difference any degree of variation may be produced . . . .

We all believe that a breeder, starting business with a considerable stock of average horses, could, by selection, in a very few generations, obtain horses able to run much faster than any of their sires or dams; in time perhaps he would obtain descendants running twice as fast as their ancestors and possibly equal to our race-horses. But would not the difference in speed between each successive generation be less and less? Hundreds of skilful men are yearly breeding thousands of racers. Wealth and honour await the main who can breed one horse to run one part in five thousand faster than his fellows. As a matter of experience, have our racers improved in speed by one part in a thousand during the last twenty generations? Could we not double the speed of a cart-horse in twenty generations? Here is the analogy with our cannon-ball; the rate of variation in a given direction is not constant, is not erratic; it is a constantly diminishing rate, tending therefore to a limit . . . .

We are thus led to believe that whatever new point in the variable beast, bird, or flower, be chosen as desirable by a fancier, this point can be rapidly approached at first, but that the rate of approach quickly diminishes, tending to a limit never to be attained. Darwin says that our oldest cultivated plants still yield new varieties. Granted; but the new variations are not successive variations in one direction. Horses could be produced with very long or with very short ears, very long or short hair, with large or small hooves, with peculiar colour, eyes, teeth, perhaps. In short, whatever variation we perceive of ordinary occurrence might by selection be carried to an extravagant excess. If a large annual prize were offered for any of these novel peculiarities, probably the variation in the first few years would be remarkable, but in twenty years' time the judges would be much puzzled to which breeder the prize should fall, and the maximum excellence would be known and expressed in figures, so that an eighth of an inch more or less would determine success or failure.

A given animal or plant appears to be contained, as it were, within a sphere of variation; one individual lies near one portion of the surface; another individual, of the same species, near another part of the surface; the average animal at the centre. Any individual may produce descendants varying in any direction, but is more likely to produce descendants varying towards the centre of the sphere, and the variations in that direction will be greater in amount than the variations towards the surface. Thus, a set of racers of equal merit indiscriminately breeding will produce more colts and foals of inferior than of superior speed, and the falling off of the degenerate will be greater than the improvement of the select. A set of Clydesdale prize horses would produce more colts and foals of inferior than superior strength. More seedlings of 'Senateur Vaisse' will be inferior to him in size and colour than superior. The tendency to revert, admitted by Darwin, is generalized in the simile of the sphere here suggested. On the other hand, Darwin insists very sufficiently on the rapidity with which new peculiarities are produced; and this rapidity is quite as essential to the argument now urged as subsequent slowness.

We hope this argument is now plain. However slow the rate of variation might be, even though it were only one part in a thousand per twenty or two thousand generations, yet if it were constant or erratic we might believe that, in untold time, it would lead to untold distance; but if in every case we find that deviation from an average individual can be rapidly effected at first, and that the rate of deviation steadily diminishes till it reaches an almost imperceptible amount, then we are as much entitled to assume a limit to the possible deviation as we are to the progress of a cannon-ball from a knowledge of the law of diminution in its speed. This limit to the variation of species seems to be established for all cases of man's selection. What argument does Darwin offer showing that the law of variation will be different when the variation occurs slowly, not rapidly? The law may be different, but is there any experimental ground for believing that it *is* different? Darwin says (p. 153), 'The struggle between natural selection, on the one hand, and the tendency to reversion and variability on the other hand, will in the course of time cease, and that the most abnormally developed organs may be made constant, I can see no reason to doubt.' But what reason have we to believe this? Darwin says the variability will disappear by the continued rejection of the individuals tending to revert to a former condition; but is there any experimental ground for believing that the variability *will* disappear; and, secondly, if the variety can become fixed, that it will in time become ready to vary still more in the original direction, passing that limit which we think has just been shown to exist in the case of man's selection? It is peculiarly difficult to see how natural selection could reject individuals having a tendency to produce offspring reverting to an original stock. The tendency to produce offspring more like their superior parents than their inferior grandfathers can surely be of no advantage to any individual in the struggle for life . . . .

Although many domestic animals and plants are highly variable, there appears to be a limit to their variation in any one direction. This limit is shown by the fact that new points are at first rapidly gained, but afterwards more slowly, while finally no further perceptible change can be effected. Great, therefore, as the variability is, we are not free to assume that successive variations of the same kind can be accumulated. There is no experimental reason for believing that the limit would be removed to a great distance, or passed, simply because it was approached by very slow degrees, instead of by more rapid steps. There is no reason to believe that a fresh variability is acquired by long selection of one form; on the contrary, we know that with the oldest breeds it is easier to bring about a diminution than an increase in the points of excellence. The sphere of variation is a simile embodying this view;—each point of the sphere corresponding to a different individual of the same race, the centre to the average animal, the surface to the limit in various directions. The individual near the centre may have offspring varying in all directions with nearly equal rapidity. A variety near the surface may be made to approach it still nearer, but has a greater tendency to vary in every other direction. The sphere may be conceived as large for some species and small for others.

Efficiency of Natural Selection. Those individuals of any species which are most adapted to the life they lead, live on an average longer than those which are less adapted to the circumstances in which the species is placed. The individuals which live the longest will have the most numerous offspring, and as the offspring on the whole resemble their parents, the descendants from any given generation will on the whole resemble the more favoured rather than the less favoured individuals of the species. So much of the theory of natural selection will hardly be denied; but it will be worth while to consider how far this process can tend to cause a variation in some one direction. It is clear that it will frequently, and indeed generally, tend to prevent any deviation from the common type. The mere existence of a species is a proof that it is tolerably well adapted to the life it must lead; many of the variations which may occur will be variations for the worse, and natural selection will assuredly stamp these out. A white grouse in the heather, or a white hare on a fallow would be sooner detected by its enemies than one of the usual plumage or colour. Even so, any favourable deviation must, according to the very terms of the statement, give its fortunate possessor a better chance of life; but this conclusion differs widely from the supposed consequence that a whole species may or will gradually acquire some one new quality, or wholly change in one direction and in the same manner. In arguing this point, two distinct kinds of possible variation must be separately considered: *first*, that kind of common variation which must be conceived as not only possible, but inevitable, in each individual of the species, such as longer and shorter legs, better or worse hearing, etc.; and, *secondly*, that kind of variation which only occurs rarely, and may be called a sport of nature, or more briefly a 'sport,' as when a child is born with six fingers on each hand. The common variation is not limited to one part of any animal, but occurs in all; and when we say that on the whole the stronger live longer than the weaker, we mean that in some cases long life will have been due to good lungs, in others to good ears, in others to good legs. There are few cases in which one faculty is pre-eminently useful to an animal beyond all other faculties, and where that is not so, the effect of natural selection will simply be to kill the weakly, and insure a sound, healthy, well-developed breed. If we could admit the principle of a gradual accumulation of improvements, natural selection would gradually improve the breed of everything, making the hare of the present generation run faster, hear better, digest better, than his ancestors; his enemies, the weasels, greyhounds, etc., would have improved likewise, so that perhaps the hare would not be really better off; but at any rate the direction of the change would be from a war of pigmies to a war of Titans. Opinions may differ as to the evidence of this gradual perfectibility of all things, but it is beside the question to argue this point, as the origin of species requires not the gradual improvement of animals retaining the same habits and structure, but such modification of those habits and structure as will actually lead to the appearance of new organs. We freely admit, that if an accumulation of slight improvements be possible, natural selection might improve hares as hares, and weasels as weasels, that is to say, it might produce animals having every useful faculty and every useful organ of their ancestors developed to a higher degree; more than this, it may obliterate some once useful organs when circumstances have so changed that they are no longer useful, for since that organ will weigh for nothing in the struggle of life, the average animal must be calculated as though it did not exist . . . .

The vague use of an imperfectly understood doctrine of chance has led Darwinian supporters, first, to confuse the two cases above distinguished; and, secondly to imagine that a very slight balance in favour of some individual sport must lead to its perpetuation. All that can be said, is that in the above example the favoured sport would be preserved once in fifty times. Let us consider what will be its influence on the main stock when preserved. It will breed and have a progeny of say 100; now this progeny will, on the whole, be intermediate between the average individual and the sport. The odds in favour of one of this generation of the new breed will be, say 1 to 1, as compared with the average individual; the odds in their favour will therefore be less than that of their parent; but owing to their greater number, the chances are that about 1 of them would survive. Unless these breed together, a most improbable event, their progeny would again approach the average individual; there would be 150 of them, and their superiority would be say in the ratio of 1 to 1; the probability would now be that nearly two of them would survive, and have 200 children, with an eighth superiority. Rather more than two of these would survive; but the superiority would again dwindle, until after a few generations it would no longer be observed and would count for no more in the struggle for life, than any of the hundred trifling advantages which occur in the ordinary organs. An illustration will bring this conception home. Suppose a white man to have been wrecked on an island inhabited by negroes, and to have established himself in friendly relations with a powerful tribe, whose customs he has learnt. Suppose him to possess the physical strength, energy, and ability of a dominant white race, and let the food and climate of the island suit his constitution; grant him every advantage which we can conceive a white to possess over the native; concede that in the struggle for existence his chance of a long life will be much superior to that of the native chiefs; yet from all these admissions, there does not follow the conclusion that, after a limited or unlimited number of generations, the inhabitants of the island will be white. Our shipwrecked hero would probably become king; he would kill a great many blacks in the struggle for existence; he would have a great many wives and children, while many of his subjects would live and die as bachelors; an insurance company would accept his life at perhaps one-tenth of the premium which they would exact from the most favoured of the negroes. Our white's qualities would certainly tend very much to preserve him to good old age, and yet he would not suffice in any number of generations to turn his subjects' descendants white. It may be said that the white colour is not the cause of the superiority. True, but it may be used simply to bring before the senses the way in which qualities belonging to one individual in a large number must be gradually obliterated. In the first generation there will be some dozens of intelligent young mulattoes, much superior in average intelligence to the negroes. We might expect the throne for some generations to be occupied by a more or less yellow king; but can any one believe that the whole island will gradually acquire a white, or even a yellow population, or that the islanders would acquire the energy, courage, ingenuity, patience, self-control, endurance, in virtue of which qualities our hero killed so many of their ancestors, and begot so many children; those qualities, in fact, which the struggle for existence would select, if it could select anything?

Here is a case in which a variety was introduced, with far greater advantages than any sport every heard of, advantages tending to its preservation, and yet powerless to perpetuate the new variety.

Darwin says that in the struggle for life a grain may turn the balance in favour of a given structure, which will then be preserved. But one of the weights in the scale of nature is due to the number of a given tribe. Let there be 7000 A's and 7000 B's, representing two varieties of a given animal, and let all the B's, in virtue of a slight difference of structure, have the better chance of life by 1/7000th part. We must allow that there is a slight probability that the descendants of B will supplant the descendants of A; but let there be only 7001 A's against 7000 B's at first, and the chances are once more equal, while if there be 7002 A's to start, the odds would be laid on the A's. True, they stand a greater chance of being killed; but then they can better afford to be killed. The grain will only turn the scales when these are very nicely balanced, and an advantage in numbers counts for weight, even as an advantage in structure. As the numbers of the favoured variety diminish, so must its relative advantage increase, if the chance of its existence is to surpass the chance of its extinction, until hardly any conceivable advantage would enable the descendants of a single pair to exterminate the descendants of many thousands if they and their descendants are supposed to breed freely with the inferior variety, and so gradually lose their ascendancy. If it is impossible that any sport or accidental variation in a single individual, however favourable to life, should be preserved and transmitted by natural selection, still less can slight and imperceptible variations, occurring in single individuals be garnered up and transmitted to continually increasing numbers; for if a very highly-favoured white cannot blanch a nation of negroes, it will hardly be contended that a comparatively very dull mulatto has a good chance of producing a tawny tribe; the idea, which seems almost absurd when presented in connexion with a practical case, rests on a fallacy of exceedingly common occurrence in mechanics and physics generally. When a man shows that a tendency to produce a given effect exists he often thinks he has proved that the effect must follow. He does not take into account the opposing tendencies, much less does he measure the various forces, with a view to calculate the result. For instance, there is a tendency on the part of a submarine cable to assume a catenary curve, and very high authorities once said it would; but, in fact, forces neglected by them utterly alter the curve from the catenary. There is a tendency on the part of the same cables, as usually made, to untwist entirely; luckily there are opposing forces, and they untwist very little. These cases will hardly seem obvious; but what should we say to a man who asserted that the centrifugal tendency of the earth must send it off in a tangent? One tendency is balanced or outbalanced by others; the advantage of structure possessed by an isolated specimen is enormously outbalanced by the advantage of numbers possessed by the others.

Reading and Discussion Questions

1. What does Jenkin suggest we observe about the limits of variability in breeding domesticated animals? How does this observation serve as a critique of Darwin's theory of evolution by natural selection? How might Darwin respond to this critique?
2. Jenkin objects to Darwin's theory on the ground that even enormously advantageous variations that might arise in a population will be lost or swamped out in just a few subsequent generations. Individual variations, such as an unusually tall person or an unusually fast cheetah, will tend to get lost because of the blending that results with all of the other short people or fast cheetahs. There will be a regression to the mean. Why is this objection particularly difficult for Darwin to respond to, given the assumptions he shares about heredity with Jenkin?

Letter from Adam Sedgwick to Charles Darwin (November 24, 1859)

Adam Sedgwick

98

My dear Darwin,

I write to thank you for your work on the origin of Species. It came, I think, in the latter part of last week; but it may have come a few days sooner, & been overlooked among my bookparcels, which often remain unopened when I am lazy, or busy with any work before me. So soon as I opened it I began to read it, & I finished it, after many interruptions, on Tuesday. Yesterday I was employed 1st. in preparing for my lecture—2dly. In attending a meeting of my brother Fellows to discuss the final propositions of the Parliamentary Commissions. 3d. In lecturing. 4thly In hearing the conclusion of the discussion & the College reply whereby in conformity with my own wishes we accepted the scheme of the Commission 5th. In dining with an old friend at Clare College—6thly In adjourning to the weekly meeting of the Ray Club, from which I returned at 10. P.M.—dog-tired & hardly able to climb my staircase—Lastly in looking thro' the Times to see what was going on in the busy world—

I do not state this to fill space (tho' I believe that Nature does abhor a vacuum); but to prove that my reply & my thanks are sent to you by the earliest leisure I have; tho' this is but a very contracted opportunity.—If I did not think you a good tempered & truth loving man I should not tell you that, (spite of the great knowledge; store of facts; capital views of the correlations of the various parts of organic nature; admirable hints about the diffusions, thro' wide regions, of nearly related organic beings; &c &c) I have read your book with more pain than pleasure. Parts of it I admired greatly; parts I laughed at till my sides were almost sore; other parts I read with absolute sorrow; because I think them utterly false & grievously mischievous—You have deserted—after a start in that tram-road of all solid physical truth—the true method of induction—& started up a machinery as wild I think as Bishop Wilkin's locomotive that was to sail with us to the Moon. Many of your wide conclusions are based upon assumptions which can neither be proved nor disproved. Why then express them in the language & arrangements of philosophical induction?—

As to your grand principle—natural selection—what is it but a secondary consequence of supposed, or known, primary facts. Development is a better word because more close to the cause of the fact. For you do not deny causation. I call (in the abstract) causation the will of God: & I can prove that He acts for the good of His creatures. He also acts by laws which we can study & comprehend—Acting by law, & under what is called final cause, comprehends, I think, your whole principle. You write of "natural selection" as if it were done consciously by the selecting agent. 'Tis but a consequence of the presupposed development, & the subsequent battle for life.—

This view of nature you have stated admirably; tho' admitted by all naturalists & denied by no one of common sense. We all admit development as a fact of history; but how came it about? Here, in language, & still more in logic, we are point blank at issue—There is a moral or metaphysical part of nature as well as a physical. A man who denies this is deep in the mire of folly. Tis the crown & glory of organic science that it does thro' final cause, link material to moral; & yet does not allow us to mingle them in our first conception of laws, & our classification of such laws whether we consider one side of nature or the other— You have ignored this link; &, if I do not mistake your meaning, you have done your best in one or two pregnant cases to break it. Were it possible (which thank God it is not) to break it, humanity in my mind, would suffer a damage that might brutalize it—& sink the human race into a lower grade of degradation than any into which it has fallen since its written records tell us of its history. Take the case of the bee cells. If your development produced the successive modification of the bee & its cells (which no mortal can prove) final cause would stand good as the directing cause under which the successive generations acted & gradually improved— Passages in your book, like that to which I have alluded (& there are others almost as bad) greatly shocked my moral taste. I think in speculating upon organic descent, you over state the evidence of geology; & that you under state it while you are talking of the broken links of your natural pedigree: but my paper is nearly done, & I must go to my lecture room—

Lastly then, I greatly dislike the concluding chapter—not as a summary—for in that light it appears good—but I dislike it from the tone of triumphant confidence in which you appeal to the rising generation (in a tone I condemned in the author of the Vestiges), & prophesy of things not yet in the womb of time; nor, (if we are to trust the accumulated experience of human sense & the inferences of its logic) ever likely to be found anywhere but in the fertile womb of man's imagination.—

And now to say a word about a son of a monkey & an old friend of yours. I am better, far better than I was last year. I have been lecturing three days a week (formerly I gave six a week) without much fatigue but I find, by the loss of activity & memory, & of all productive powers, that my bodily frame is sinking slowly towards the earth. But I have visions of the future. They are as much a part of myself as my stomach & my heart; & tho visions are to have their antitype in solid fruition of what is best & greatest. But on one condition only—that I humbly accept God's revelation of himself both in His works & in His word; & do my best to act in conformity with that knowledge which He only can give me, & He only can sustain me in doing. If you & I do all this we shall meet in heaven.

I have written in a hurry & in a spirit of brotherly love. Therefore forgive any sentence you happen to dislike; & believe me, spite of our disagreement in some points of the deepest moral interest, your true-hearted old friend.

A. Sedgwick.
Cambridge
November 24, 1859

Reading and Discussion Questions

1. Sedgwick was a very accomplished geologist, and taught Darwin the techniques of field geological research while Darwin was a student at Cambridge. What does he mean that Darwin "has deserted ... the true method of induction"? Is this a fair critique of Darwin's work?
2. Sedgwick is deeply distressed that Darwin has abandoned discussion of final cause, which, he thinks, links the "material to moral." Why is he concerned that Darwin's view of evolution will be ethically damaging to humanity?

Review of "Darwin on the Origin of Species" (1860)

Richard Owen

99

Mr. Darwin refers to the multitude of the individual of every species, which, from one cause or another, perish either before, or soon after attaining maturity.

'Owing to this struggle for life, any variation, however slight and from whatever cause proceeding, if it be in any degree profitable to an individual of any species, in its infinitely complex relations to other organic beings and to external nature, will tend to the preservation of that individual, and will generally be inherited by its offspring. The offspring, also, will thus have a better chance of surviving, for, of the many individuals of any species which are periodically born, but a small number can survive. I have called this principle, by which each slight variation, if useful, is preserved, by the term of Natural Selection, in order to mark its relation to man's power of selection. We have seen that man by selection can certainly produce great results, and can adapt organic beings to his own uses, through the accumulation of slight but useful variations, given to him by the hand of Nature. But Natural Selection, as we shall hereafter see, is a power incessantly ready for action, and is as immeasurably superior to man's feeble efforts, as the works of Nature are to those of Art.' (p. 61)

The scientific world has looked forward with great interest to the facts which Mr. Darwin might finally deem adequate to the support of his theory on this supreme question in biology, and to the course of inductive original research which might issue in throwing light on 'that mystery of mysteries.' But having now cited the chief, if not the whole, of the original observations adduced by its author in the volume now before us, our disappointment may be conceived. Failing the adequacy of such observations, not merely to carry conviction, but to give a colour to the hypothesis, we were then left to confide in the superior grasp of mind, strength of intellect, clearness and precision of thought and expression, which raise one man so far above his contemporaries, as to enable him to discern in the common stock of facts, of coincidences, correlations and analogies in Natural History, deeper and truer conclusions than his fellow-labourers had been able to reach.

These expectations, we must confess, received a check on perusing the first sentence in the book.

'When on board H.M.S. "Beagle," as naturalist, I was much struck with certain facts in the distribution of the inhabitants of South America, and in the geological relations of the present to the past inhabitants of that continent. These facts seemed to me to throw some light on the origin of species that mystery of mysteries, as it has been called by some of our greatest philosophers.' (p.1)

What is there, we asked ourselves, as we closed the volume to ponder on this paragraph, what can there possibly be in the inhabitants, we suppose he means aboriginal inhabitants, of South America, or in their distribution on that continent, to have suggested to any mind that man might be a transmuted ape, or to throw any light on the origin of the human or other species? Mr. Darwin must be aware of what is commonly understood by an 'uninhabited island;' he may, however, mean by the inhabitants of South America, not the human kind only, whether aboriginal or otherwise, but all the lower animals. Yet again, why are the fresh-water polypes or sponges to be called 'inhabitants' more than the plants? Perhaps what was meant might be, that the distribution and geological relations of the organised beings generally in South America, had suggested transmutational views. They have commonly suggested ideas as to the independent origin of such localized kinds of plants and animals. But what the 'certain facts' were, and what may be the nature of the light which they threw upon the mysterious beginning of species, is not mentioned or further alluded to in the present work . . . .

'Isolation, also,' says Mr. Darwin, is an important element in the process of natural selection.' But how can one select if a thing be 'isolated'? Even using the word in the sense of a confined area, Mr. Darwin admits that the conditions of life 'throughout such area, will tend to modify all the individuals of a species in the same manner, in relation to the same conditions.' (P. 104.) No evidence, however, is given of a species having ever been created in that way; but granting the hypothetical influence and transmutation, there is no selection here. The author adds, 'Although I do not doubt that isolation is of considerable importance in the production of new species, on the whole, I am inclined to believe, that largeness of area is of more importance in the production of species capable of spreading widely.' (p. 105.)

Now, on such a question as the origin of species, and in an express, formal, scientific treatise on the subject, the expression of a belief, where one looks for a demonstration, is simply provoking. We are not concerned in the author's beliefs or inclinations to believe. Belief is a state of mind short of actual knowledge. It is a state which may govern action, when based upon a tacit admission of the mind's incompetency to prove a proposition, coupled with submissive acceptance of an authoritative dogma, or worship of a favourite idol of the mind. We readily concede, and it needs, indeed, no ghost to reveal the fact, that the wider the area in which a species may be produced, the more widely it will spread. But we fail to discern its import in respect of the great question at issue.

We have read and studied with care most of the monographs conveying the results of close investigations of particular groups of animals, but have not found, what Darwin asserts to be the fact, at least as regards all those investigators of particular groups of animals and plants whose treatises he has read, viz., that their authors 'are one and all firmly convinced that each of the well-marked forms or species was at the first independently created.' Our experience has been that the monographers referred to have rarely committed themselves to any conjectural hypothesis whatever, upon the origin of the species which they have closely studied.

Darwin appeals from the 'experienced naturalists whose minds are stocked with a multitude of facts' which he assumes to have been 'viewed from a point of view opposite to his own,' to the 'few naturalists endowed with much flexibility of mind,' for a favourable reception of his hypothesis. We must confess that the minds to whose conclusions we incline to bow belong to that truth-loving, truth-seeking, truth-imparting class which Robert Brown, Bojanus, Rudolphi, Cuvier, Ehrenberg, Herold, Kölliker, and Siebold, worthily exemplify. The rightly and sagaciously generalising intellect is associated with the power of endurance of continuous and laborious research, exemplarily manifested in such monographs as we have quoted below. Their authors are the men who trouble the intellectual world little with their beliefs, but enrich it greatly with their proofs. If close and long-continued research, sustained by the determination to get accurate results, blunted, as Mr. Darwin seems to imply, the far-seeing discovering faculty, then are we driven to this paradox, viz., that the elucidation of the higher problems, nay the highest, in Biology, is to be sought for or expected in the lucubrations of those naturalists whose minds are not weighted or troubled with more than a discursive and superficial knowledge of nature.

Lasting and fruitful conclusions have, indeed, hitherto been based only on the possession of knowledge; now we are called upon to accept an hypothesis on the plea of want of knowledge. The geological record, it is averred, is so imperfect! But what human record is not? Especially must the record of past organisms be much less perfect than of present ones. We freely admit it. But when Mr. Darwin, in reference to the absence of the intermediate fossil forms required by his hypothesis and only the zootomical zoologist can approximatively appreciate their immense numbers the countless hosts of transitional links which, on 'natural selection,' must certainly have existed at one period or another of the world's history when Mr. Darwin exclaims what may be, or what may not be, the forms yet forthcoming out of the graveyards of strata, we would reply, that our only ground for prophesying of what may come, is by the analogy of what has come to light. We may expect, e.g., a chambered-shell from a secondary rock; but not the evidence of a creature linking on the cuttle-fish to the lump-fish.

Mr. Darwin asks, 'How is it that varieties, which I have called incipient species, become ultimately good and distinct species?' To which we rejoin with the question: Do they become good and distinct species? Is there any one instance proved by observed facts of such transmutation? We have searched the volume in vain for such. When we see the intervals that divide most species from their nearest congeners, in the recent and especially the fossil series, we either doubt the fact of progressive conversion, or, as Mr. Darwin remarks in his letter to Dr. Asa Gray, one's 'imagination must fill up very wide blanks.'

The last ichthyosaurus, by which the genus disappears in the chalk, is hardly distinguishable specifically from the first ichthyosaurus, which abruptly introduces that strange form of sea-lizard in the lias. The oldest Pterodactyle is as thorough and complete a one as the latest. No contrast can be more remarkable, nor, we believe, more instructive, than the abundance of evidence of the various species of ichthyosaurus throughout the marine strata of the oolite and cretaceous periods, and the utter blank in reference to any form calculated to enlighten us as to whence the ichthyosaurus came, or what it graduated into, before or after these periods. The Enaliosauria of the secondary seas were superseded by the Cetacea of the tertiary ones.

Professor Agassiz affirms:—

> 'Between two successive geological periods, changes have taken place among plants and animals. But none of those primordial forms of life which naturalists call species, are known to have changed during any of these periods. It cannot be denied that the species of different successive periods are supposed by some naturalists to derive their distinguishing features from changes which have taken place in those of preceding ages, but this is a mere supposition, supported neither by physiological nor by geological evidence; and the assumption that animals and plants may change in a similar manner during one and the same manner is equally gratuitous.'

Cuvier adduced the evidence of the birds and beasts which had been preserved in the tombs of Egypt, to prove that no change in their specific characters had taken place during the thousands of years two, three, or five which had elapsed, according to the monumental evidence, since the individuals of those species were the subjects of the mummifier's skill.

Professor Agassiz adduces evidence to show that there are animals of species now living which have been for a much longer period inhabitants of our globe.

'It has been possible' he writes, 'to trace the formation and growth of our coral reefs, especially in Florida, with sufficient precision to ascertain that it must take about eight thousand years for one of those coral walls to rise from its foundation to the level of the surface of the ocean. There are around the southernmost extremity of Florida alone, four such reefs, concentric with one another, which can be shown to have grown up one after the other. This gives for the beginning of the first of these reefs an age of over thirty thousand years; and yet the corals by which they were all built up are the same identical species in all of them. These facts, then, furnish as direct evidence as we can obtain in any branch of physical inquiry, that some, at least, of the species of animals now existing, have been in existence over thirty thousand years, and have not undergone the slightest change during the whole of that period.'

To this, of course, the transmutationists reply that a still longer period of time might do what thirty thousand years have not done.

Professor Baden Powell, for example, affirms; 'Though each species may have possessed its peculiarities unchanged for a lapse of time, the fact that when long periods are considered, all those of our earlier period are replaced by new ones at a later period, proves that species change in the end, provided a sufficiently long time is granted.' But here lies the fallacy: it merely proves that species are changed, it gives us no evidence as to the mode of change; transmutation, gradual or abrupt, is in this case mere assumption. We have no objection on any score to the change; we have the greatest desire to know how it is brought about. Owen has long stated his belief that some pre-ordained law or secondary cause is operative in bringing about the change; but our knowledge of such law, if such exists, can only be acquired on the prescribed terms. We, therefore, regard the painstaking and minute comparisons by Cuvier of the osteological and every other character that could be tested in the mummified ibis, cat, or crocodile, with those of the species living in his time; and the equally philosophical investigations of the polypes operating at an interval of 30,000 years in the building up of coral reefs, by the profound palæontologist of Neuchatel, as of far higher value in reference to the inductive determination of the question of the origin of species that the speculations of de Maillet, Buffon, Lamarck, 'Vestiges,' Baden Powell, or Darwin.

The essential element in the complex idea of species, as it has been variously framed and defined by naturalists, viz., the blood-relationship between all the individuals of such species, is annihilated on the hypothesis of 'natural selection.' According to this view a genus, a family, an order, a class, a sub-kingdom, the individuals severally representing these grades of difference or relationship, now differ from individuals of the same species only in degree: the species, like every other group, is a mere creature of the brain; it is no longer from nature. With the present evidence from form, structure, and procreative phenomena, of the truth of the opposite proposition, that 'classification is the task of science, but species the work of nature,' we believe that this aphorism will endure; we are certain that it has not yet been refuted; and we repeat in the words of Linnæus, *'Classis et Ordo est sapientiæ, Species naturæ opus'*.

Reading and Discussion Questions

1. Owen regarded himself as a student and inheritor of Cuvier's work. What about this piece reflects his commitment to Cuvier's principles?
2. Some of the critiques of Darwin in this piece are deeply unfair, and some are well founded. Find an example of each.

Darwin and His Reviewers (1860)

Asa Gray

100

The origin of species, like all origination, like the institution of any other natural state or order, is beyond our immediate ken. We see or may learn how things go on; we can only frame hypotheses as to how they began.

Two hypotheses divide the scientific world, very unequally, upon the origin of the existing diversity of the plants and animals which surround us. One assumes that the actual kinds are primordial; the other, that they are derivative. One, that all kinds originated supernaturally and directly as such, and have continued unchanged in the order of Nature; the other, that the present kinds appeared in some sort of genealogical connection with other and earlier kinds, that they became what they now are in the course of time and in the order of Nature.

Or, bringing in the word species, which is well defined as "the perennial succession of individuals," commonly of very like individuals—as a close corporation of individuals perpetuated by generation, instead of election—and reducing the question to mathematical simplicity of statement: species are lines of individuals coming down from the past and running on to the future; lines receding, therefore, from our view in either direction. Within our limited observation they appear to be parallel lines, as a general thing neither approaching to nor diverging from each other.

The first hypothesis assumes that they were parallel from the unknown beginning and will be to the unknown end. The second hypothesis assumes that the apparent parallelism is not real and complete, at least aboriginally, but approximate or temporary; that we should find the lines convergent in the past, if we could trace them far enough; that some of them, if produced back, would fall into certain fragments of lines, which have left traces in the past, lying not exactly in the same direction, and these farther back into others to which they are equally unparallel. It will also claim that the present lines, whether on the whole really or only approximately parallel, sometimes fork or send off branches on one side or the other, producing new lines (varieties), which run for a while, and for aught we know indefinitely when not interfered with, near and approximately parallel to the parent line. This claim it can establish; and it may also show that these close subsidiary lines may branch or vary again, and that those branches or varieties which are best adapted to the existing conditions may be continued, while others stop or die out. And so we may have the basis of a real theory of the diversification of species and here indeed, there is a real, though a narrow, established ground to build upon But as systems of organic Nature, both doctrines are equally hypotheses, are suppositions of what there is no proof of from experience, assumed in order to account for the observed phenomena, and supported by such indirect evidence as can be had.

Even when the upholders of the former and more popular system mix up revelation with scientific discussion—which we decline to do—they by no means thereby render their view other than hypothetical. Agreeing that plants and animals were produced by Omnipotent fiat does not exclude the idea of natural order and what we call secondary causes. The record of the fiat—"Let the earth bring forth grass, the herb yielding seed," etc., "and it was so;" "let the earth bring forth the living creature after his kind, cattle and creeping thing and beast of the earth after his kind, and it was so"—seems even to imply them. Agreeing that they were formed of "the dust of the ground," and of thin air, only leads to the conclusion that the pristine individuals were corporeally constituted like existing individuals, produced through natural agencies. To agree that they were created "after their kinds" determines nothing as to what were the original kinds, nor in what mode, during what time, and in what connections it pleased the Almighty to introduce the first individuals of each sort upon the earth. Scientifically considered, the two opposing doctrines are equally hypothetical.

The two views very unequally divide the scientific world; so that believers in "the divine right of majorities" need not hesitate which side to take, at least for the present. Up to a time quite within the memory of a generation still on the stage, two hypotheses about the nature of light very unequally divided the scientific world. But the small minority has already prevailed: the emission theory has gone out; the undulatory or wave theory, after some fluctuation, has reached high tide, and is now the pervading, the fully-established system. There was an intervening time during which most physicists held their opinions in suspense.

The adoption of the undulatory theory of light called for the extension of the same theory to heat, and this promptly suggested the hypothesis of a correlation, material connection, and transmutability of heat, light, electricity, magnetism, etc.; which hypothesis the physicists held in absolute suspense until very lately, but are now generally adopting. If not already established as a system, it promises soon to become so. At least, it is generally received as a tenable and probably true hypothesis.

Parallel to this, however less cogent the reasons, Darwin and others, having shown it likely that some varieties of plants or animals have diverged in time into cognate species, or into forms as different as species, are led to infer that all species of a genus may have thus diverged from a common stock, and thence to suppose a higher community of origin in ages still farther back, and so on. Following the safe example of the physicists, and acknowledging the fact of the diversification of a once homogeneous species into varieties, we may receive the theory of the evolution of these into species, even while for the present we hold the hypothesis of a further evolution in cool suspense or in grave suspicion. In respect to very many questions a wise man's mind rests long in a state neither of belief nor unbelief. But your intellectually short-sighted people are apt to be preternaturally clear-sighted, and to find their way very plain to positive conclusions upon one side or the other of every mooted question.

In fact, most people, and some philosophers, refuse to hold questions in abeyance, however incompetent they may be to decide them. And, curiously enough, the more difficult, recondite, and perplexing the questions or hypotheses are—such, for instance, as those about organic Nature—the more impatient they are of suspense. Sometimes, and evidently in the present case, this impatience grows out of a fear that a new hypothesis may endanger cherished and most important beliefs. Impatience under such circumstances is not unnatural, though perhaps needless, and, if so, unwise.

To us the present revival of the derivative hypothesis, in a more winning shape than it ever before had, was not unexpected. We wonder that any thoughtful observer of the course of investigation and of speculation in science should not have foreseen it, and have learned at length to take its inevitable coming patiently; the more so, as in Darwin's treatise it comes in a purely scientific form, addressed only to scientific men. The notoriety and wide popular perusal of this treatise appear to have astonished the author even more than the book itself has astonished the reading world. Coming as the new presentation does from a naturalist of acknowledged character and ability and marked by a conscientiousness and candor which have not always been reciprocated we have thought it simply right to set forth the doctrine as fairly and as favorably as we could. There are plenty to decry it and the whole theory is widely exposed to attack. For the arguments on the other side we may look to the numerous adverse publications which Darwin's volume has already called out and especially to those reviews which propose directly to refute it. Taking various lines and reflecting very diverse modes of thought, these hostile critics may be expected to concentrate and enforce the principal objections which can be brought to bear against the derivative hypothesis in general, and Darwin's new exposition of it in particular.

Upon the opposing side of the question we have read with attention—1. An article in the North American Review for April last; 2. One in the Christian Examiner, Boston, for May; 3. M. Pictet's article in the *Bibliotheque Universelle*, which we have already made considerable use of, which seems throughout most able and correct, and which in tone and fairness is admirably in contrast with—4. The article in the Edinburgh Review for May, attributed—although against a large amount of internal presumptive evidence—to the most distinguished

British comparative anatomist; 5. An article in the North British Review for May; 6. Prof. Agassiz has afforded an early opportunity to peruse the criticisms he makes in the forthcoming third volume of his great work, by a publication of them in advance in the American Journal of Science for July.

In our survey of the lively discussion which has been raised, it matters little how our own particular opinions may incline. But we may confess to an impression, thus far, that the doctrine of the permanent and complete immutability of species has not been established, and may fairly be doubted. We believe that species vary, and that "Natural Selection" works; but we suspect that its operation, like every analogous natural operation, may be limited by something else. Just as every species by its natural rate of reproduction would soon completely fill any country it could live in, but does not, being checked by some other species or some other condition—so it may be surmised that variation and natural selection have their struggle and consequent check, or are limited by something inherent in the constitution of organic beings . . . .

Proposing now to criticise the critics, so far as to see what their most general and comprehensive objections amount to, we must needs begin with the American reviewers, and with their arguments adduced to prove that a derivative hypothesis ought not to be true, or is not possible, philosophical, or theistic.

It must not be forgotten that on former occasions very confident judgments have been pronounced by very competent persons, which have not been finally ratified. Of the two great minds of the seventeenth century, Newton and Leibnitz, both profoundly religious as well as philosophical, one produced the theory of gravitation, the other objected to that theory that it was subversive of natural religion. The nebular hypothesis—a natural consequence of the theory of gravitation and of the subsequent progress of physical and astronomical discovery—has been denounced as atheistical even down to our own day. But it is now largely adopted by the most theistical natural philosophers as a tenable and perhaps sufficient hypothesis, and where not accepted is no longer objected to, so far as we know, on philosophical or religious grounds.

The gist of the philosophical objections urged by the two Boston reviewers against an hypothesis of the derivation of species—or at least against Darwin's particular hypothesis—is, that it is incompatible with the idea of any manifestation of design in the universe, that it denies final causes. A serious objection this, and one that demands very serious attention . . . .

In their scientific objections the two reviewers take somewhat different lines; but their philosophical and theological arguments strikingly coincide. They agree in emphatically asserting that Darwin's hypothesis of the origination of species through variation and natural selection "repudiates the whole doctrine of final causes," and "all indication of design or purpose in the organic world . . . is neither more nor less than a formal denial of any agency beyond that of a blind chance in the developing or perfecting of the organs or instincts of created beings . . . . It is in vain that the apologists of this hypothesis might say that it merely attributes a different mode and time to the Divine agency—that all the qualities subsequently appearing in their descendants must have been implanted, and have remained latent in the original pair." Such a view, the Examiner declares, "is nowhere stated in this book, and would be, we are sure, disclaimed by the author."

We should like to be informed of the grounds of this sureness. The marked rejection of spontaneous generation—the statement of a belief that all animals have descended from four or five progenitors, and plants from an equal or lesser number, or, perhaps, if constrained to it by analogy, "from some one primordial form into which life was first breathed"—coupled with the expression, "To my mind it accords better with what we know of the laws impressed on matter by the Creator, that the production and extinction of the past and present inhabitants of the world should have been due to secondary causes," than "that each species has been independently created"—these and similar expressions lead us to suppose that the author probably does accept the kind of view which the Examiner is sure he would disclaim. At least, we charitably see nothing in his scientific theory to hinder his adoption of Lord Bacon's "Confession of Faith" in this regard—"That, notwithstanding God hath rested and ceased from creating, yet, nevertheless, he doth accomplish and fulfill his divine will in all things, great and small, singular and general, as fully and exactly by providence as he could by miracle and new creation, though his working be not immediate and direct, but by compass; not violating Nature, which is his own law upon the creature."

However that may be, it is undeniable that Mr. Darwin has purposely been silent upon the philosophical and theological applications of his theory. This reticence, under the circumstances, argues design, and raises inquiry as to the final cause or reason why. Here, as in higher instances, confident as we are that there is a final cause, we must not be overconfident that we can infer the particular or true one. Perhaps the author is more familiar with natural-historical than with philosophical inquiries, and, not having decided which particular theory about efficient cause is best founded, he meanwhile argues the scientific questions concerned—all that relates to secondary causes—upon purely scientific grounds, as he must do in any case. Perhaps, confident, as he evidently is, that his view will finally be adopted, he may enjoy a sort of satisfaction in hearing it denounced as sheer atheism by the inconsiderate, and afterward, when it takes its place with the nebular hypothesis and the like, see this judgment reversed, as we suppose it would be in such event.

Whatever Mr. Darwin's philosophy may be, or whether he has any, is a matter of no consequence at all, compared with the important questions, whether a theory to account for the origination and diversification of animal and vegetable forms through the operation of secondary causes does or does not exclude design; and whether the establishment by adequate evidence of Darwin's particular theory of diversification through variation and natural selection would essentially alter the present scientific and philosophical grounds for theistic views of Nature. The unqualified affirmative judgment rendered by the two Boston reviewers, evidently able and practised reasoners, "must give us pause."

We hesitate to advance our conclusions in opposition to theirs. But, after full and serious consideration, we are constrained to say that, in our opinion, the adoption of a derivative hypothesis, and of Darwin's particular hypothesis, if we understand it, would leave the doctrines of final causes, utility, and special design, just where they were before. We do not pretend that the subject is not environed with difficulties. Every view is so environed; and every shifting of the view is likely, if it removes some difficulties, to bring others into prominence. But we cannot perceive that Darwin's theory brings in any new kind of scientific difficulty, that is, any with which philosophical naturalists were not already familiar.

Since natural science deals only with secondary or natural causes, the scientific terms of a theory of derivation of species—no less than of a theory of dynamics—must needs be the same to the theist as to the atheist. The difference appears only when the inquiry is carried up to the question of primary cause—a question which belongs to philosophy. Wherefore, Darwin's reticence about efficient cause does not disturb us. He considers only the scientific questions. As already stated, we think that a theistic view of Nature is implied in his book, and we must charitably refrain from suggesting the contrary until the contrary is logically deduced from his premises. If, however, he anywhere maintains that the natural causes through which species are diversified operate without an ordaining and directing intelligence, and that the orderly arrangements and admirable adaptations we see all around us are fortuitous or blind, undesigned results—that the eye, though it came to see, was not designed for seeing, nor the hand for handling—then, we suppose, he is justly chargeable with denying, and very needlessly denying, all design in organic Nature; otherwise, we suppose not. Why, if Darwin's well-known passage about the eye [III-10] equivocal though some of the language be—does not imply ordaining and directing intelligence, then he refutes his own theory as effectually as any of his opponents are likely to do. He asks:

> May we not believe that [under variation proceeding long enough, generation multiplying the better variations times enough, and natural selection securing the improvements] a living optical instrument might be thus formed as superior to one of glass as the works of the Creator are to those of man?

This must mean one of two things: either that the living instrument was made and perfected under (which is the same thing as by) an intelligent First Cause, or that it was not. If it was, then theism is asserted; and as to the mode of operation, how do we know, and why must we believe, that, fitting precedent forms being in existence, a living instrument (so different from a lifeless manufacture) would be originated and perfected in any other way, or that this is not the fitting way? If it means that it was not, if he so misuses words that by the Creator he intends an unintelligent power, undirected force, or necessity, then he has put his case so as to invite disbelief in it. For then blind forces have produced not only manifest adaptions of means to specific ends—which is absurd enough—but better adjusted and more perfect instruments or machines than intellect (that is, human intellect) can contrive and human skill execute—which no sane person will believe.

On the other hand, if Darwin even admits—we will not say adopts—the theistic view, he may save himself much needless trouble in the endeavor to account for the absence of every sort of intermediate form. Those in the line between one species and another supposed to be derived from it he may be bound to provide; but as to "an infinite number of other varieties not intermediate, gross, rude, and purposeless, the unmeaning creations of an unconscious cause," born only to perish, which a relentless reviewer has imposed upon his theory—rightly enough upon the atheistic alternative—the theistic view rids him at once of this "scum of creation." For, as species do not now vary at all times and places and in all directions, nor produce crude, vague, imperfect, and useless forms, there is no reason for supposing that they ever did. Good-for-nothing monstrosities, failures of purpose rather than purposeless, indeed, sometimes occur; but these are just as anomalous and unlikely upon Darwin's theory as upon any other. For his particular theory is based, and even over-strictly insists, upon the most universal of physiological laws, namely, that successive generations shall differ only slightly, if at all, from their parents; and this effectively excludes crude and impotent forms. Wherefore, if we believe that the species were designed, and that natural propagation was designed, how can we say that the actual varieties of the species were not equally designed? Have we not similar grounds for inferring design in the supposed varieties of species, that we have in the case of the supposed species of a genus? When a naturalist comes to regard as three closely related species what he before took to be so many varieties of one species how has he thereby strengthened our conviction that the three forms are designed to have the differences which they actually exhibit? Wherefore so long as gradatory, orderly, and adapted forms in Nature argue design, and at least while the physical cause of variation is utterly unknown and mysterious, we should advise Mr. Darwin to assume in the philosophy of his hypothesis that variation has been led along certain beneficial lines. Streams flowing over a sloping plain by gravitation (here the counterpart of natural selection) may have worn their actual channels as they flowed; yet their particular courses may have been assigned; and where we see them forming definite and useful lines of irrigation, after a manner unaccountable on the laws of gravitation and dynamics, we should believe that the distribution was designed.

To insist, therefore, that the new hypothesis of the derivative origin of the actual species is incompatible with final causes and design, is to take a position which we must consider philosophically untenable. We must also regard it as highly unwise and dangerous, in the present state and present prospects of physical and physiological science. We should expect the philosophical atheist or skeptic to take this ground; also, until better informed, the unlearned and unphilosophical believer; but we should think that the thoughtful theistic philosopher would take the other side. Not to do so seems to concede that only supernatural events can be shown to be designed, which no theist can admit—seems also to misconceive the scope and meaning of all ordinary arguments for design in Nature. This misconception is shared both by the reviewers and the reviewed. At least, Mr. Darwin uses expressions which imply that the natural forms which surround us, because they have a history or natural sequence, could have been only generally, but not particularly designed—a view at once superficial and contradictory; whereas his true line should be, that his hypothesis concerns the order and not the cause, the how and not the why of the phenomena, and so leaves the question of design just where it was before . . . .

It is very easy to assume that, because events in Nature are in one sense accidental, and the operative forces which bring them to pass are themselves blind and unintelligent (physically considered, all forces are), therefore they are undirected, or that he who describes these events as the results of such forces thereby assumes that they are undirected. This is the assumption of the Boston reviewers, and of Mr. Agassiz, who insists that the only alternative to the doctrine, that all organized beings were supernaturally created just as they are, is, that they have arisen spontaneously through the omnipotence of matter. [III-11]

As to all this, nothing is easier than to bring out in the conclusion what you introduce in the premises. If you import atheism into your conception of variation and natural selection, you can readily exhibit it in the result. If you do not put it in, perhaps there need be none to come out. While the mechanician is considering a steamboat or locomotive-engine as a material organism, and contemplating the fuel, water, and steam, the source of the mechanical forces, and how they operate, he may not have occasion to mention the engineer. But, the orderly and special results accomplished, the why the movements are in this or that particular direction, etc., is inexplicable without him. If Mr. Darwin believes that the events which he supposes to have occurred and the results we behold were undirected and undesigned, or if the physicist believes that the natural forces to which he refers phenomena are uncaused and undirected, no argument is needed to show that such belief is atheism. But the admission of the phenomena and of these natural processes and forces does not necessitate any such belief, nor even render it one whit less improbable than before . . . .

Finally, it is worth noticing that, though natural selection is scientifically explicable, variation is not. Thus far the cause of variation, or the reason why the offspring is sometimes unlike the parents, is just as mysterious as the reason why it is generally like the parents. It is now as inexplicable as any other origination; and, if ever explained, the explanation will only carry up the sequence of secondary causes one step farther, and bring us in face of a somewhat different problem, but which will have the same element of mystery that the problem of variation has now. Circumstances may preserve or may destroy the variations man may use or direct them but selection whether artificial or natural no more originates them than man originates the power which turns a wheel when he dams a stream and lets the water fall upon it. The origination of this power is a question about efficient cause. The tendency of science in respect to this obviously is not toward the omnipotence of matter, as some suppose, but toward the omnipotence of spirit.

So the real question we come to is as to the way in which we are to conceive intelligent and efficient cause to be exerted, and upon what exerted. Are we bound to suppose efficient cause in all cases exerted upon nothing to evoke something into existence—and this thousands of times repeated, when a slight change in the details would make all the difference between successive species? Why may not the new species, or some of them, be designed diversifications of the old?

There are, perhaps, only three views of efficient cause which may claim to be both philosophical and theistic:

1. The view of its exertion at the beginning of time, endowing matter and created things with forces which do the work and produce the phenomena.
2. This same view, with the theory of insulated interpositions, or occasional direct action, engrafted upon it—the view that events and operations in general go on in virtue simply of forces communicated at the first, but that now and then, and only now and then, the Deity puts his hand directly to the work.
3. The theory of the immediate, orderly, and constant, however infinitely diversified, action of the intelligent efficient Cause.

It must be allowed that, while the third is preeminently the Christian view, all three are philosophically compatible with design in Nature. The second is probably the popular conception. Perhaps most thoughtful people oscillate from the middle view toward the first or the third—adopting the first on some occasions, the third on others. Those philosophers who like and expect to settle all mooted questions will take one or the other extreme. The Examiner inclines toward, the North American reviewer fully adopts, the third view, to the logical extent of maintaining that "the origin of an individual, as well as the origin of a species or a genus, can be explained only by the direct action of an intelligent creative cause." To silence his critics, this is the line for Mr. Darwin to take; for it at once and completely relieves his scientific theory from every theological objection which his reviewers have urged against it . . . .

Among the unanswerable, perhaps the weightiest of the objections, is that of the absence, in geological deposits, of vestiges of the intermediate forms which the theory requires to have existed. Here all that Mr. Darwin can do is to insist upon the extreme imperfection of the geological record and the uncertainty of negative evidence. But, withal, he allows the force of the objection almost as much as his opponents urge it—so much so, indeed, that two of his English critics turn the concession unfairly upon him, and charge him with actually basing his hypothesis upon these and similar difficulties—as if he held it because of the difficulties, and not in spite of them; a handsome return for his candor!

As to this imperfection of the geological record, perhaps we should get a fair and intelligible illustration of it by imagining the existing animals and plants of New England, with all their remains and products since the arrival of the Mayflower, to be annihilated; and that, in the coming time, the geologists of a new colony, dropped by the New Zealand fleet on its way to explore the ruins of London, undertake, after fifty years of examination, to reconstruct in a catalogue the flora and fauna of our day, that is, from the close of the glacial period to the present time. With all the advantages of a surface exploration, what a beggarly account it would be! How many of the land animals and plants which are enumerated in the Massachusetts official reports would it be likely to contain? . . .

It has been attempted to destroy the very foundation of Darwin's hypothesis by denying that there are any wild varieties, to speak of, for natural selection to operate upon. We cannot gravely sit down to prove that wild varieties abound. We should think it just as necessary to prove that snow falls in winter. That variation among plants cannot be largely due to hybridism, and that their variation in Nature is not essentially different from much that occurs in domestication, and, in the long-run, probably hardly less in amount, we could show if our space permitted.

As to the sterility of hybrids, that can no longer be insisted upon as absolutely true, nor be practically used as a test between species and varieties, unless we allow that hares and rabbits are of one species. That such sterility, whether total or partial, subserves a purpose in keeping species apart, and was so designed, we do not doubt. But the critics fail to perceive that this sterility proves nothing whatever against the derivative origin of the actual species; for it may as well have been intended to keep separate those forms which have reached a certain amount of divergence, as those which were always thus distinct.

The argument for the permanence of species, drawn from the identity with those now living of cats, birds, and other animals preserved in Egyptian catacombs, was good enough as used by Cuvier against St. Hilaire, that is, against the supposition that time brings about a gradual alteration of whole species; but it goes for little against Darwin, unless it be proved that species never vary, or that the perpetuation of a variety necessitates the extinction of the parent breed. For Darwin clearly maintains—what the facts warrant—that the mass of a species remains fixed so long as it exists at all, though it may set off a variety now and then. The variety may finally supersede the parent form, or it may coexist with it; yet it does not in the least hinder the unvaried stock from continuing true to the breed, unless it crosses with it. The common law of inheritance may be expected to keep both the original and the variety mainly true as long as they last, and none the less so because they have given rise to occasional varieties. The tailless Manx cats, like the curtailed fox in the fable, have not induced the normal breeds to dispense with their tails, nor have the Dorkings (apparently known to Pliny) affected the permanence of the common sort of fowl.

As to the objection that the lower forms of life ought, on Darwin's theory, to have been long ago improved out of existence, and replaced by higher forms, the objectors forget what a vacuum that would leave below, and what a vast field there is to which a simple organization is best adapted, and where an advance would be no improvement, but the contrary. To accumulate the greatest amount of being upon a given space, and to provide as much enjoyment of life as can be under the conditions, is what Nature seems to aim at; and this is effected by diversification.

Finally, we advise nobody to accept Darwin's or any other derivative theory as true. The time has not come for that, and perhaps never will. We also advise against a similar credulity on the other side, in a blind faith that species—that the manifold sorts and forms of existing animals and vegetables—"have no secondary cause." The contrary is already not unlikely, and we suppose will hereafter become more and more probable. But we are confident that, if a derivative hypothesis ever is established, it will be so on a solid theistic ground.

Meanwhile an inevitable and legitimate hypothesis is on trial—an hypothesis thus far not untenable—a trial just now very useful to science, and, we conclude, not harmful to religion, unless injudicious assailants temporarily make it so.

One good effect is already manifest; its enabling the advocates of the hypothesis of a multiplicity of human species to perceive the double insecurity of their ground. When the races of men are admitted to be of one species, the corollary, that they are of one origin, may be expected to follow. Those who allow them to be of one species must admit an actual diversification into strongly-marked and persistent varieties, and so admit the basis of fact upon which the Darwinian hypothesis is built; while those, on the other hand, who recognize several or numerous human species, will hardly be able to maintain that such species were primordial and supernatural in the ordinary sense of the word.

The English mind is prone to positivism and kindred forms of materialistic philosophy, and we must expect the derivative theory to be taken up in that interest. We have no predilection for that school, but the contrary. If we had, we might have looked complacently upon a line of criticism which would indirectly, but effectively, play into the hands of positivists and materialistic atheists generally. The wiser and stronger ground to take is, that the derivative hypothesis leaves the argument for design, and therefore for a designer, as valid as it ever was; that to do any work by an instrument must require, and therefore presuppose, the exertion rather of more than of less power than to do it directly; that whoever would be a consistent theist should believe that Design in the natural world is coextensive with Providence, and hold as firmly to the one as he does to the other, in spite of the wholly similar and apparently insuperable difficulties which the mind encounters whenever it endeavors to develop the idea into a system, either in the material and organic, or in the moral world. It is enough, in the way of obviating objections, to show that the philosophical difficulties of the one are the same, and only the same, as of the other.

Reading and Discussion Questions

1. What are the "two hypotheses" that Gray suggests currently divide the scientific world? What is the significance of calling them both hypotheses?
2. Unlike Sedgwick, Gray firmly defends Darwin for not addressing final cause. Why does he think Darwin is right to avoid discussions of final cause? How does he defend Darwin against the charges of atheism?
3. Gray was an abolitionist, living in Boston in 1860. Consider the significance of the second-to-last paragraph in this light.

The Coming of Age of "The Origin of Species" (1880)

Thomas Henry Huxley

VII. The Coming of Age of "The Origin Of Species"

Many of you will be familiar with the aspect of this small green-covered book. It is a copy of the first edition of the "Origin of Species," and bears the date of its production—the 1st of October 1859. Only a few months, therefore, are needed to complete the full tale of twenty-one years since its birthday.

Those whose memories carry them back to this time will remember that the infant was remarkably lively, and that a great number of excellent persons mistook its manifestations of a vigorous individuality for mere naughtiness; in fact there was a very pretty turmoil about its cradle. My recollections of the period are particularly vivid, for, having conceived a tender affection for a child of what appeared to me to be such remarkable promise, I acted for some time in the capacity of a sort of under-nurse, and thus came in for my share of the storms which threatened the very life of the young creature. For some years it was undoubtedly warm work; but considering how exceedingly unpleasant the apparition of the newcomer must have been to those who did not fall in love with him at first sight, I think it is to the credit of our age that the war was not fiercer, and that the more bitter and unscrupulous forms of opposition died away as soon as they did.

I speak of this period as of something past and gone, possessing merely an historical, I had almost said an antiquarian interest. For, during the second decade of the existence of the "Origin of Species," opposition, though by no means dead, assumed a different aspect. On the part of all those who had any reason to respect themselves, it assumed a thoroughly respectful character. By this time, the dullest began to perceive that the child was not likely to perish of any congenital weakness or infantile disorder, but was growing into a stalwart personage, upon whom mere goody scoldings and threatenings with the birch-rod were quite thrown away.

In fact, those who have watched the progress of science within the last ten years will bear me out to the full, when I assert that there is no field of biological inquiry in which the influence of the "Origin of Species" is not traceable; the foremost men of science in every country are either avowed champions of its leading doctrines, or at any rate abstain from opposing them; a host of young and ardent investigators seek for and find inspiration and guidance in Mr. Darwin's great work; and the general doctrine of evolution, to one side of which it gives expression, obtains, in the phenomena of biology, a firm base of operations whence it may conduct its conquest of the whole realm of Nature.

History warns us, however, that it is the customary fate of new truths to begin as heresies and to end as superstitions; and, as matters now stand, it is hardly rash to anticipate that, in another twenty years, the new generation, educated under the influences of the present day, will be in danger of accepting the main doctrines of the "Origin of Species," with as little reflection, and it may be with as little justification, as so many of our contemporaries, twenty years ago, rejected them.

Against any such a consummation let us all devoutly pray; for the scientific spirit is of more value than its products, and irrationally held truths may be more harmful than reasoned errors. Now the essence of the scientific spirit is criticism. It tells us that whenever a doctrine claims our assent we should reply, Take it if you can compel it. The struggle for existence holds as much in the intellectual as in the physical world. A theory is a species of thinking, and its right to exist is coextensive with its power of resisting extinction by its rivals.

From this point of view, it appears to me that it would be but a poor way of celebrating the Coming of Age of the "Origin of Species," were I merely to dwell upon the facts, undoubted and remarkable as they are, of its far-reaching influence and of the great following of ardent disciples who are occupied in spreading and developing its doctrines. Mere insanities and inanities have before now swollen to portentous size in the course of twenty years. Let us rather ask this prodigious change in opinion to justify itself: let us inquire whether anything has happened since 1859, which will explain, on rational grounds, why so many are worshipping that which they burned, and burning that which they worshipped. It is only in this way that we shall acquire the means of judging whether the movement we have witnessed is a mere eddy of fashion, or truly one with the irreversible current of intellectual progress, and, like it, safe from retrogressive reaction.

Every belief is the product of two factors: the first is the state of the mind to which the evidence in favour of that belief is presented; and the second is the logical cogency of the evidence itself. In both these respects, the history of biological science during the last twenty years appears to me to afford an ample explanation of the change which has taken place; and a brief consideration of the salient events of that history will enable us to understand why, if the "Origin of Species" appeared now, it would meet with a very different reception from that which greeted it in 1859.

One-and-twenty years ago, in spite of the work commenced by Hutton and continued with rare skill and patience by Lyell, the dominant view of the past history of the earth was catastrophic. Great and sudden physical revolutions, wholesale creations and extinctions of living beings, were the ordinary machinery of the geological epic brought into fashion by the misapplied genius of Cuvier. It was gravely maintained and taught that the end of every geological epoch was signalised by a cataclysm, by which every living being on the globe was swept away, to be replaced by a brand-new creation when the world returned to quiescence. A scheme of nature which appeared to be modelled on the likeness of a succession of rubbers of whist, at the end of each of which the players upset the table and called for a new pack, did not seem to shock anybody.

I may be wrong, but I doubt if, at the present time, there is a single responsible representative of these opinions left. The progress of scientific geology has elevated the fundamental principle of uniformitarianism, that the explanation of the past is to be sought in the study of the present, into the position of an axiom; and the wild speculations of the catastrophists, to which we all listened with respect a quarter of a century ago, would hardly find a single patient hearer at the present day. No physical geologist now dreams of seeking, outside the range of known natural causes, for the explanation of anything that happened millions of years ago, any more than he would be guilty of the like absurdity in regard to current events.

The effect of this change of opinion upon biological speculation is obvious. For, if there have been no periodical general physical catastrophes, what brought about the assumed general extinctions and re-creations of life which are the corresponding biological catastrophes? And, if no such interruptions of the ordinary course of nature have taken place in the organic, any more than in the inorganic, world, what alternative is there to the admission of evolution?

The doctrine of evolution in biology is the necessary result of the logical application of the principles of uniformitarianism to the phenomena of life. Darwin is the natural successor of Hutton and Lyell, and the "Origin of Species" the logical sequence of the "Principles of Geology."

The fundamental doctrine of the "Origin of Species," as of all forms of the theory of evolution applied to biology, is "that the innumerable species, genera, and families of organic beings with which the world is peopled have all descended, each within its own class or group, from common parents, and have all been modified in the course of descent." [Origin of Species, ed. I, p. 457]

And, in view of the facts of geology, it follows that all living animals and plants "are the lineal descendants of those which lived long before the Silurian epoch." [Origin of Species, p. 458]

It is an obvious consequence of this theory of descent with modification, as it is sometimes called, that all plants and animals, however different they may now be, must, at one time or other, have been connected by direct or indirect intermediate gradations, and that the appearance of isolation presented by various groups of organic beings must be unreal.

No part of Mr. Darwin's work ran more directly counter to the prepossessions of naturalists twenty years ago than this. And such prepossessions were very excusable, for there was undoubtedly a great deal to be said, at that time, in favour of the fixity of species and of the existence of great breaks, which there was no obvious or probable means of filling up, between various groups of organic beings.

For various reasons, scientific and unscientific, much had been made of the hiatus between man and the rest of the higher mammalia, and it is no wonder that issue was first joined on this part of the controversy. I have no wish to revive past and happily forgotten controversies; but I must state the simple fact that the distinctions in the cerebral and other characters, which were so hotly affirmed to separate man from all other animals in 1860, have all been demonstrated to be non-existent, and that the contrary doctrine is now universally accepted and taught.

But there were other cases in which the wide structural gaps asserted to exist between one group of animals and another were by no means fictitious; and, when such structural breaks were real, Mr. Darwin could account for them only by supposing that the intermediate forms which once existed had become extinct. In a remarkable passage he says—

> "We may thus account even for the distinctness of whole classes from each other—for instance, of birds from all other vertebrate animals—by the belief that many animal forms of life have been utterly lost, through which the early progenitors of birds were formerly connected with the early progenitors of the other vertebrate classes" [Origin of Species, p. 431].

Adverse criticism made merry over such suggestions as these. Of course it was easy to get out of the difficulty by supposing extinction; but where was the slightest evidence that such intermediate forms between birds and reptiles as the hypothesis required ever existed? And then probably followed a tirade upon this terrible forsaking of the paths of "Baconian induction."

But the progress of knowledge has justified Mr. Darwin to an extent which could hardly have been anticipated. In 1862, the specimen of Archæopteryx, which, until the last two or three years, has remained unique, was discovered; and it is an animal which, in its feathers and the greater part of its organisation, is a veritable bird, while, in other parts, it is as distinctly reptilian.

In 1868, I had the honour of bringing under your notice, in this theatre, the results of investigations made, up to that time, into the anatomical characters of certain ancient reptiles, which showed the nature of the modifications in virtue of which the type of the quadrupedal reptile passed into that of a bipedal bird; and abundant confirmatory evidence of the justice of the conclusions which I then laid before you has since come to light.

In 1875, the discovery of the toothed birds of the cretaceous formation in North America by Professor Marsh completed the series of transitional forms between birds and reptiles, and removed Mr. Darwin's proposition that "many animal forms of life have been utterly lost, through which the early progenitors of birds were formerly connected with the early progenitors of the other vertebrate classes," from the region of hypothesis to that of demonstrable fact.

In 1859, there appeared to be a very sharp and clear hiatus between vertebrated and invertebrated animals, not only in their structure, but, what was more important, in their development. I do not think that we even yet know the precise links of connection between the two; but the investigations of Kowalewsky and others upon the development of Amphioxus and of the Tunicataprove, beyond a doubt, that the differences which were supposed to constitute a barrier between the two are non-existent. There is no longer any difficulty in understanding how the vertebrate type may have arisen from the invertebrate, though the full proof of the manner in which the transition was actually effected may still be lacking.

Again, in 1859, there appeared to be a no less sharp separation between the two great groups of flowering and flowerless plants. It is only subsequently that the series of remarkable investigations inaugurated by Hofmeister has brought to light the extraordinary and altogether unexpected modifications of the reproductive apparatus in the Lycopodiaceæ, the Rhizocarpeæ, and the Gymnospermeæ, by which the ferns and the mosses are gradually connected with the Phanerogamic division of the vegetable world.

So, again, it is only since 1859 that we have acquired that wealth of knowledge of the lowest forms of life which demonstrates the futility of any attempt to separate the lowest plants from the lowest animals, and shows that the two kingdoms of living nature have a common borderland which belongs to both, or to neither.

Thus it will be observed that the whole tendency of biological investigation, since 1859, has been in the direction of removing the difficulties which the apparent breaks in the series created at that time; and the recognition of gradation is the first step towards the acceptance of evolution.

As another great factor in bringing about the change of opinion which has taken place among naturalists, I count the astonishing progress which has been made in the study of embryology. Twenty years ago, not only were we devoid of any accurate knowledge of the mode of development of many groups of animals and plants, but the methods of investigation were rude and imperfect. At the present time, there is no important group of organic beings the development of which has not been carefully studied; and the modern methods of hardening and section-making enable the embryologist to determine the nature of the process, in each case, with a degree of minuteness and accuracy which is truly astonishing to those whose memories carry them back to the beginnings of modern histology. And the results of these embryological investigations are in complete harmony with the requirements of the doctrine of evolution. The first beginnings of all the higher forms of animal life are similar, and however diverse their adult conditions, they start from a common foundation. Moreover, the process of development of the animal or the plant from its primary egg, or germ, is a true process of evolution–a progress from almost formless to more or less highly organised matter, in virtue of the properties inherent in that matter.

To those who are familiar with the process of development, all a priori objections to the doctrine of biological evolution appear childish. Anyone who has watched the gradual formation of a complicated animal from the protoplasmic mass, which constitutes the essential element of a frog's or a hen's egg, has had under his eyes sufficient evidence that a similar evolution of the whole animal world from the like foundation is, at any rate, possible.

Yet another product of investigation has largely contributed to the removal of the objections to the doctrine of evolution current in 1859. It is the proof afforded by successive discoveries that Mr. Darwin did not over-estimate the imperfection of the geological record. No more striking illustration of this is needed than a comparison of our knowledge of the mammalian fauna of the Tertiary epoch in 1859 with its present condition. M. Gaudry's researches on the fossils of Pikermi were published in 1868, those of Messrs. Leidy, Marsh, and Cope, on the fossils of the Western Territories of America, have appeared almost wholly since 1870, those of M. Filhol on the phosphorites of Quercy in 1878. The general effect of these investigations has been to introduce to us a multitude of extinct animals, the existence of which was previously hardly suspected; just as if zoologists were to become acquainted with a country, hitherto unknown, as rich in novel forms of life as Brazil or South Africa once were to Europeans. Indeed, the fossil fauna of the Western Territories of America bid fair to exceed in interest and importance all other known Tertiary deposits put together; and yet, with the exception of the case of the American tertiaries, these investigations have extended over very limited areas; and, at Pikermi, were confined to an extremely small space.

Such appear to me to be the chief events in the history of the progress of knowledge during the last twenty years, which account for the changed feeling with which the doctrine of evolution is at present regarded by those who have followed the advance of biological science, in respect of those problems which bear indirectly upon that doctrine.

But all this remains mere secondary evidence. It may remove dissent, but it does not compel assent. Primary and direct evidence in favour of evolution can be furnished only by palæontology. The geological record, so soon as it approaches completeness, must, when properly questioned, yield either an affirmative or a negative answer: if evolution has taken place, there will its mark be left; if it has not taken place, there will lie its refutation.

What was the state of matters in 1859? Let us hear Mr. Darwin, who may be trusted always to state the case against himself as strongly as possible.

"On this doctrine of the extermination of an infinitude of connecting links between the living and extinct inhabitants of the world, and at each successive period between the extinct and still older species, why is not every geological formation charged with such links? Why does not every collection of fossil remains afford plain evidence of the gradation and mutation of the forms of life? We meet with no such evidence, and this is the most obvious and plausible of the many objections which may be urged against my theory." [Origin of Species, ed. 1, p. 463.]

Nothing could have been more useful to the opposition than this characteristically candid avowal, twisted as it immediately was into an admission that the writer's views were contradicted by the facts of palæontology. But, in fact, Mr. Darwin made no such admission. What he says in effect is, not that palæontological evidence is against him, but that it is not distinctly in his favour; and, without attempting to attenuate the fact, he accounts for it by the scantiness and the imperfection of that evidence.

What is the state of the case now, when, as we have seen, the amount of our knowledge respecting the mammalia of the Tertiary epoch is increased fifty-fold, and in some directions even approaches completeness?

Simply this, that, if the doctrine of evolution had not existed, palaeontologists must have invented it, so irresistibly is it forced upon the mind by the study of the remains of the Tertiary mammalia which have been brought to light since 1859.

Among the fossils of Pikermi, Gaudry found the successive stages by which the ancient civets passed into the more modern hyænas; through the Tertiary deposits of Western America, Marsh tracked the successive forms by which the ancient stock of the horse has passed into its present form; and innumerable less complete indications of the mode of evolution of other groups of the higher mammalia have been obtained. In the remarkable memoir on the phosphorites of Quercy, to which I have referred, M. Filhol describes no fewer than seventeen varieties of the genus Cynodictis, which fill up all the interval between the viverrine animals and the bear-like dog Amphicyon; nor do I know any solid ground of objection to the supposition that, in this Cynodictis-Amphicyon group, we have the stock whence all the Viveridæ, Felidæ, Hyænidæ, Canidæ, and perhaps the Procyonidæ and Ursidæ, of the present fauna have been evolved. On the contrary, there is a great deal to be said in favour.

In the course of summing up his results, M. Filhol observes:—

> During the epoch of the phosphorites, great changes took place in animal forms, and almost the same types as those which now exist became defined from one another.

> Under the influence of natural conditions of which we have no exact knowledge, though traces of them are discoverable, species have been modified in a thousand ways: races have arisen which, becoming fixed, have thus produced a corresponding number of secondary species.

In 1859, language of which this is an unintentional paraphrase, occurring in the "Origin of Species," was scouted as wild speculation; at present, it is a sober statement of the conclusions to which an acute and critically-minded investigator is led by large and patient study of the facts of palæontology. I venture to repeat what I have said before, that so far as the animal world is concerned, evolution is no longer a speculation, but a statement of historical fact. It takes its place alongside of those accepted truths which must be reckoned with by philosophers of all schools.

Thus when, on the first day of October next, "The Origin of Species" comes of age, the promise of its youth will be amply fulfilled; and we shall be prepared to congratulate the venerated author of the book, not only that the greatness of his achievement and its enduring influence upon the progress of knowledge have won him a place beside our Harvey; but, still more, that, like Harvey, he has lived long enough to outlast detraction and opposition, and to see the stone that the builders rejected become the head-stone of the corner.

Reading and Discussion Questions

1. In the 21 years between the publication of the *Origin* and Huxley's essay, what new evidence has been found that Huxley suggests support Darwin's theory?
2. Huxley says that "The doctrine of evolution in biology is the necessary result of the logical application of the principles of uniformitarianism to the phenomena of life." Explain what he means by this. What does this suggest about the importance of Lyell's geological work for Darwin's theory?

The Variation of Animals and Plants under Domestication (1868)

Charles Darwin

102

Volume II

Chapter 2.XXVII Provisional Hypothesis of Pangenesis

...

Every one would wish to explain to himself, even in an imperfect manner, how it is possible for a character possessed by some remote ancestor suddenly to reappear in the offspring; how the effects of increased or decreased use of a limb can be transmitted to the child; how the male sexual element can act not solely on the ovules, but occasionally on the mother-form; how a hybrid can be produced by the union of the cellular tissue of two plants independently of the organs of generation; how a limb can be reproduced on the exact line of amputation, with neither too much nor too little added; how the same organism may be produced by such widely different processes, as budding and true seminal generation; and, lastly, how of two allied forms, one passes in the course of its development through the most complex metamorphoses, and the other does not do so, though when mature both are alike in every detail of structure. I am aware that my view is merely a provisional hypothesis or speculation; but until a better one be advanced, it will serve to bring together a multitude of facts which are at present left disconnected by any efficient cause. As Whewell, the historian of the inductive sciences, remarks:—"Hypotheses may often be of service to science, when they involve a certain portion of incompleteness, and even of error."

Under this point of view I venture to advance the hypothesis of Pangenesis, which implies that every separate part of the whole organisation reproduces itself. So that ovules, spermatozoa, and pollen-grains,—the fertilised egg or seed, as well as buds,—include and consist of a multitude of germs thrown off from each separate part or unit.

. . .

In the First Part I will enumerate as briefly as I can the groups of facts which seem to demand connection; but certain subjects, not hitherto discussed, must be treated at disproportionate length. In the Second Part the hypothesis will be given; and after considering how far the necessary assumptions are in themselves improbable, we shall see whether it serves to bring under a single point of view the various facts.

Part I

Reproduction may be divided into two main classes, namely, sexual and asexual. The latter is effected in many ways—by the formation of buds of various kinds, and by fissiparous generation, that is by spontaneous or artificial division. It is notorious that some of the lower animals, when cut into many pieces, reproduce so many perfect individuals: Lyonnet cut a Nais or freshwater worm into nearly forty pieces, and these all reproduced perfect animals. (27/2. Quoted by Paget 'Lectures on Pathology' 1853 page 159.)

. . .

Regrowth of Amputated Parts

This subject deserves a little further discussion. A multitude of the lower animals and some vertebrates possess this wonderful power. For instance, Spallanzani cut off the legs and tail of the same salamander six times successively, and Bonnet (27/18. Spallanzani 'An Essay on Animal Reproduction' translated by Dr. Maty 1769 page 79. Bonnet 'Oeuvres d'Hist. Nat.' tome 5 part 1 4to. edition 1781 pages 343, 350.) did so eight times; and on each occasion the limbs were reproduced on the exact line of amputation, with no part deficient or in excess. An allied animal, the axolotl, had a limb bitten off, which was reproduced in an abnormal condition, but when this was amputated it was replaced by a perfect limb. (27/19. Vulpian as quoted by Prof. Faivre 'La Variabilite des Especes' 1868 page 112.) The new limbs in these cases bud forth, and are developed in the same manner as during the regular development of a young animal. For instance, with the Amblystoma lurida, three toes are first developed, then the fourth, and on the hind-feet the fifth, and so it is with a reproduced limb. (27/20. Dr. P. Hoy 'The American Naturalist' September 1871 page 579.)

...

In the case of those animals which may be bisected or chopped into pieces, and of which every fragment will reproduce the whole, the power of regrowth must be diffused throughout the whole body. Nevertheless there seems to be much truth in the view maintained by Prof. Lessona (27/23. 'Atti della Soc. Ital. di Sc. Nat.' volume 11 1869 page 493.), that this capacity is generally a localised and special one, serving to replace parts which are eminently liable to be lost in each particular animal. The most striking case in favour of this view, is that the terrestrial salamander, according to Lessona, cannot reproduce lost parts, whilst another species of the same genus, the aquatic salamander, has extraordinary powers of regrowth, as we have just seen; and this animal is eminently liable to have its limbs, tail, eyes and jaws bitten off by other tritons.

...

Direct Action of the Male Element on the Female

In the eleventh chapter, abundant proofs were given that foreign pollen occasionally affects in a direct manner the mother-plant. Thus, when Gallesio fertilised an orange-flower with pollen from the lemon, the fruit bore stripes of perfectly characterised lemon-peel. With peas, several observers have seen the colour of the seed-coats and even of the pod directly affected by the pollen of a distinct variety. So it has been with the fruit of the apple, which consists of the modified calyx and upper part of the flower-stalk. In ordinary cases these parts are wholly formed by the mother-plant. We here see that the formative elements included within the male element or pollen of one variety can affect and hybridise, not the part which they are properly adapted to affect, namely, the ovules, but the partially-developed tissues of a distinct variety or species. We are thus brought halfway towards a graft-hybrid, in which the formative elements included within the tissues of one individual combine with those included in the tissues of a distinct variety or species, thus giving rise to a new and intermediate form, independently of the male or female sexual organs.

...

Variability and Inheritance

...

The principle of reversion, recently alluded to, is one of the most wonderful of the attributes of Inheritance. It proves to us that the transmission of a character and its development, which ordinarily go together and thus escape discrimination, are distinct powers; and these powers in some cases are even antagonistic, for each acts alternately in successive generations. Reversion is not a rare event, depending on some unusual or favourable combination of circumstances, but occurs so regularly with crossed animals and plants, and so frequently with uncrossed breeds, that it is evidently an essential part of the principle of inheritance. We know that changed conditions have the power of evoking long-lost characters, as in the case of animals becoming feral. The act of crossing in itself possesses this power in a high degree. What can be more wonderful than that characters, which have disappeared during scores, or hundreds, or even thousands of generations, should suddenly reappear perfectly developed, as in the case of pigeons and fowls, both when purely bred and especially when crossed; or as with the zebrine stripes on dun-coloured horses, and other such cases? Many monstrosities come under this same head, as when rudimentary organs are redeveloped, or when an organ which we must believe was possessed by an early progenitor of the species, but of which not even a rudiment is left, suddenly reappears, as with the fifth stamen in some Scrophulariaceae. We have already seen that reversion acts in bud-reproduction; and we know that it occasionally acts during the growth of the same individual animal, especially, but not exclusively, if of crossed parentage,—as in the rare cases described of fowls, pigeons, cattle, and rabbits, which have reverted to the colours of one of their parents or ancestors as they advanced in years.

We are led to believe, as formerly explained, that every character which occasionally reappears is present in a latent form in each generation, in nearly the same manner as in male and female animals the secondary characters of the opposite sex lie latent and ready to be evolved when the reproductive organs are injured. This comparison of the secondary sexual characters which lie latent in both sexes, with other latent characters, is the more appropriate from the case recorded of a Hen, which assumed some of the masculine characters, not of her own race, but of an early progenitor; she thus exhibited at the same time the redevelopment of latent characters of both kinds. In every living creature we may feel assured that a host of long-lost characters lie ready to be evolved under proper conditions. How can we make intelligible and connect with other facts, this wonderful and common capacity of reversion,—this power of calling back to life long-lost characters?

Part II

I have now enumerated the chief facts which every one would desire to see connected by some intelligible bond. This can be done, if we make the following assumptions, and much may be advanced in favour of the chief one. The secondary assumptions can likewise be supported by various physiological considerations. It is universally admitted that the cells or units of the body increase by self-division or proliferation, retaining the same nature, and that they ultimately become converted into the various tissues and substances of the body. But besides this means of increase I assume that the units throw off minute granules which are dispersed throughout the whole system; that these, when supplied with proper nutriment, multiply by self-division, and are ultimately developed into units like those from which they were originally derived. These granules may be called gemmules. They are collected from all parts of the system to constitute the sexual elements, and their development in the next generation forms a new being; but they are likewise capable of transmission in a dormant state to future generations and may then be developed. Their development depends on their union with other partially developed or nascent cells which precede them in the regular course of growth. Why I use the term union, will be seen when we discuss the direct action of pollen on the tissues of the mother-plant. Gemmules are supposed to be thrown off by every unit, not only during the adult state, but during each stage of development of every organism; but not necessarily during the continued existence of the same unit. Lastly, I assume that the gemmules in their dormant state have a mutual affinity for each other, leading to their aggregation into buds or into the sexual elements. Hence, it is not the reproductive organs or buds which generate new organisms, but the units of which each individual is composed. These assumptions constitute the provisional hypothesis which I have called Pangenesis.

...

Before proceeding to show, firstly, how far these assumptions are in themselves probable, and secondly, how far they connect and explain the various groups of facts with which we are concerned, it may be useful to give an illustration, as simple as possible, of the hypothesis. If one of the Protozoa be formed, as it appears under the microscope, of a small mass of homogeneous gelatinous matter, a minute particle or gemmule thrown off from any part and nourished under favourable circumstances would reproduce the whole; but if the upper and lower surfaces were to differ in texture from each other and from the central portion, then all three parts would have to throw off gemmules, which when aggregated by mutual affinity would form either buds or the sexual elements, and would ultimately be developed into a similar organism. Precisely the same view may be extended to one of the higher animals; although in this case many thousand gemmules must be thrown off from the various parts of the body at each stage of development; these gemmules being developed in union with pre-existing nascent cells in due order of succession.

Physiologists maintain, as we have seen, that each unit of the body, though to a large extent dependent on others, is likewise to a certain extent independent or autonomous, and has the power of increasing by self-division. I go one step further, and assume that each unit casts off free gemmules which are dispersed throughout the system, and are capable under proper conditions of being developed into similar units.

...

The retention of free and undeveloped gemmules in the same body from early youth to old age will appear improbable, but we should remember how long seeds lie dormant in the earth and buds in the bark of a tree. Their transmission from generation to generation will appear still more improbable; but here again we should remember that many rudimentary and useless organs have been transmitted during an indefinite number of generations. We shall presently see how well the long-continued transmission of undeveloped gemmules explains many facts.

As each unit, or group of similar units, throughout the body, casts off its gemmules, and as all are contained within the smallest ovule, and within each spermatozoon or pollen-grain, and as some animals and plants produce an astonishing number of pollen-grains and ovules (27/48. Mr. F. Buckland found 6,867,840 eggs in a cod-fish ('Land and Water' 1868 page 62) .... From the data arrived at by Sir W. Thomson, my son George finds that a cube of 1/10000 of an inch of glass or water must consist of between 16 million millions, and 131 thousand million million molecules. No doubt the molecules of which an organism is formed are larger, from being more complex, than those of an inorganic substance, and probably many molecules go to the formation of a gemmule; but when we bear in mind that a cube of 1/10000 of an inch is much smaller than any pollen-grain, ovule or bud, we can see what a vast number of gemmules one of these bodies might contain.

The gemmules derived from each part or organ must be thoroughly dispersed throughout the whole system. We know, for instance, that even a minute fragment of a leaf of a Begonia will reproduce the whole plant; and that if a freshwater worm is chopped into small pieces, each will reproduce the whole animal. Considering also the minuteness of the gemmules and the permeability of all organic tissues, the thorough dispersion of the gemmules is not surprising.

...

The assumed elective affinity of each gemmule for that particular cell which precedes it in due order of development is supported by many analogies.

...

Thus far we have been able by the aid of our hypothesis to throw some obscure light on the problems which have come before us; but it must be confessed that many points remain altogether doubtful. Thus it is useless to speculate at what period of development each unit of the body casts off its gemmules, as the whole subject of the development of the various tissues is as yet far from clear. We do not know whether the gemmules are merely collected by some unknown means at certain seasons within the reproductive organs, or whether after being thus collected they rapidly multiply there, as the flow of blood to these organs at each breeding season seems to render probable.

. . .

Having now endeavoured to show that the several foregoing assumptions are to a certain extent supported by analogous facts, and having alluded to some of the most doubtful points, we will consider how far the hypothesis brings under a single point of view the various cases enumerated in the First Part. All the forms of reproduction graduate into one another and agree in their product; for it is impossible to distinguish between organisms produced from buds, from self-division, or from fertilised germs; such organisms are liable to variations of the same nature and to reversions of the same kind; and as, according to our hypothesis, all the forms of reproduction depend on the aggregation of gemmules derived from the whole body, we can understand this remarkable agreement. Parthenogenesis is no longer wonderful, and if we did not know that great good followed from the union of the sexual elements derived from two distinct individuals, the wonder would be that parthenogenesis did not occur much oftener than it does. On any ordinary theory of reproduction the formation of graft-hybrids, and the action of the male element on the tissues of the mother-plant, as well as on the future progeny of female animals, are great anomalies; but they are intelligible on our hypothesis. The reproductive organs do not actually create the sexual elements; they merely determine the aggregation and perhaps the multiplication of the gemmules in a special manner. These organs, however, together with their accessory parts, have high functions to perform. They adapt one or both elements for independent temporary existence, and for mutual union. The stigmatic secretion acts on the pollen of a plant of the same species in a wholly different manner to what it does on the pollen of one belonging to a distinct genus or family. The spermatophores of the Cephalopoda are wonderfully complex structures, which were formerly mistaken for parasitic worms; and the spermatozoa of some animals possess attributes which, if observed in an independent animal, would be put down to instinct guided by sense-organs,—as when the spermatozoa of an insect find their way into the minute micropyle of the egg.

. . .

Hardly any fact in physiology is more wonderful than the power of regrowth; for instance, that a snail should be able to reproduce its head, or a salamander its eyes, tail, and legs, exactly at the points where they have been cut off. Such cases are explained by the presence of gemmules derived from each part, and disseminated throughout the body.

...

With respect to hybridism, pangenesis agrees well with most of the ascertained facts. We must believe, as previously shown, that several gemmules are requisite for the development of each cell or unit. But from the occurrence of parthenogenesis, more especially from those cases in which an embryo is only partially formed, we may infer that the female element generally includes gemmules in nearly sufficient number for independent development, so that when united with the male element the gemmules are superabundant. Now, when two species or races are crossed reciprocally, the offspring do not commonly differ, and this shows that the sexual elements agree in power, in accordance with the view that both include the same gemmules. Hybrids and mongrels are also generally intermediate in character between the two parent-forms, yet occasionally they closely resemble one parent in one part and the other parent in another part, or even in their whole structure: nor is this difficult to understand on the admission that the gemmules in the fertilised germ are superabundant in number, and that those derived from one parent may have some advantage in number, affinity, or vigour over those derived from the other parent. Crossed forms sometimes exhibit the colour or other characters of either parent in stripes or blotches; and this occurs in the first generation, or through reversion in succeeding bud and seminal generations, of which fact several instances were given in the eleventh chapter.

...

Finally, we see that on the hypothesis of pangenesis variability depends on at least two distinct groups of causes. Firstly, the deficiency, superabundance, and transposition of gemmules, and the redevelopment of those which have long been dormant; the gemmules themselves not having undergone any modification; and such changes will amply account for much fluctuating variability. Secondly, the direct action of changed conditions on the organisation, and of the increased use or disuse of parts; and in this case the gemmules from the modified units will be themselves modified, and, when sufficiently multiplied, will supplant the old gemmules and be developed into new structures.

...

The last subject that need be discussed, namely, Reversion, rests on the principle that transmission and development, though generally acting in conjunction, are distinct powers; and the transmission of gemmules with their subsequent development shows us how this is possible. We plainly see the distinction in the many cases in which a grandfather transmits to his grandson, through his daughter, characters which she does not, or cannot, possess. But before proceeding, it will be advisable to say a few words about latent or dormant characters. Most, or perhaps all, of the secondary characters, which appertain to one sex, lie dormant in the other sex; that is, gemmules capable of development into the secondary male sexual characters are included within the female; and conversely female characters in the male: we have evidence of this in certain masculine characters, both corporeal and mental, appearing in the female, when her ovaria are diseased or when they fail to act from old age. In like manner female characters appear in castrated males, as in the shape of the horns of the ox, and in the absence of horns in castrated stags.

...

The tendency to reversion is often induced by a change of conditions, and in the plainest manner by crossing. Crossed forms of the first generation are generally nearly intermediate in character between their two parents; but in the next generation the offspring commonly revert to one or both of their grandparents, and occasionally to more remote ancestors. How can we account for these facts?

...

Conclusion

The hypothesis of Pangenesis, as applied to the several great classes of facts just discussed, no doubt is extremely complex, but so are the facts. The chief assumption is that all the units of the body, besides having the universally admitted power of growing by self-division, throw off minute gemmules which are dispersed through the system. Nor can this assumption be considered as too bold, for we know from the cases of graft-hybridisation that formative matter of some kind is present in the tissues of plants, which is capable of combining with that included in another individual, and of reproducing every unit of the whole organism. But we have further to assume that the gemmules grow, multiply, and aggregate themselves into buds and the sexual elements; their development depending on their union with other nascent cells or units. They are also believed to be capable of transmission in a dormant state, like seeds in the ground, to successive generations.

In a highly-organised animal, the gemmules thrown off from each different unit throughout the body must be inconceivably numerous and minute. Each unit of each part, as it changes during development, and we know that some insects undergo at least twenty metamorphoses, must throw off its gemmules. But the same cells may long continue to increase by self-division, and even become modified by absorbing peculiar nutriment, without necessarily throwing off modified gemmules. All organic beings, moreover, include many dormant gemmules derived from their grandparents and more remote progenitors, but not from all their progenitors. These almost infinitely numerous and minute gemmules are contained within each bud, ovule, spermatozoon, and pollen-grain. Such an admission will be declared impossible; but number and size are only relative difficulties. Independent organisms exist which are barely visible under the highest powers of the microscope, and their germs must be excessively minute. Particles of infectious matter, so small as to be wafted by the wind or to adhere to smooth paper, will multiply so rapidly as to infect within a short time the whole body of a large animal. We should also reflect on the admitted number and minuteness of the molecules composing a particle of ordinary matter. The difficulty, therefore, which at first appears insurmountable, of believing in the existence of gemmules so numerous and small as they must be according to our hypothesis, has no great weight.

The units of the body are generally admitted by physiologists to be autonomous. I go one step further and assume that they throw off reproductive gemmules. Thus an organism does not generate its kind as a whole, but each separate unit generates its kind. It has often been said by naturalists that each cell of a plant has the potential capacity of reproducing the whole plant; but it has this power only in virtue of containing gemmules derived from every part. When a cell or unit is from some cause modified, the gemmules derived from it will be in like manner modified. If our hypothesis be provisionally accepted, we must look at all the forms of asexual reproduction, whether occurring at maturity or during youth, as fundamentally the same, and dependent on the mutual aggregation and multiplication of the gemmules. The regrowth of an amputated limb and the healing of a wound is the same process partially carried out. Buds apparently include nascent cells, belonging to that stage of development at which the budding occurs, and these cells are ready to unite with the gemmules derived from the next succeeding cells. The sexual elements, on the other hand, do not include such nascent cells; and the male and female elements taken separately do not contain a sufficient number of gemmules for independent development, except in the cases of parthenogenesis. The development of each being, including all the forms of metamorphosis and metagenesis, depends on the presence of gemmules thrown off at each period of life, and on their development, at a corresponding period, in union with preceding cells. Such cells may be said to be fertilised by the gemmules which come next in due order of development. Thus the act of ordinary impregnation and the development of each part in each being are closely analogous processes. The child, strictly speaking, does not grow into the man, but includes germs which slowly and successively become developed and form the man. In the child, as well as in the adult, each part generates the same part. Inheritance must be looked at as merely a form of growth, like the self-division of a lowly-organised unicellular organism. Reversion depends on the transmission from the forefather to his descendants of dormant gemmules, which occasionally become developed under certain known or unknown conditions. Each animal and plant may be compared with a bed of soil full of seeds, some of which soon germinate, some lie dormant for a period, whilst others perish. When we hear it said that a man carries in his constitution the seeds of an inherited disease, there is much truth in the expression. No other attempt, as far as I am aware, has been made, imperfect as this confessedly is, to connect under one point of view these several grand classes of facts. An organic being is a microcosm—a little universe, formed of a host of self-propagating organisms, inconceivably minute and numerous as the stars in heaven.

Reading and Discussion Questions

1. Darwin's hypothesis of pangenesis represents his attempt to account for the reproduction and development of organisms. What is the theory? In particular, what are gemmules, what size are they, where might they be expected to be found in the body, and what role does Darwin take them to play in inheritance?
2. Darwin argues that a wide variety of different sorts of observations might be explained from the theoretical point of view that he presents for our consideration. How might the theory explain how a chopped up earthworm can reproduce the whole organism? What other biological observations does Darwin think that his theory might account for?

On Certain Correlated Variations in Carcinus maenas (1893)

W. F. R. Weldon

103

In previous communications I have discussed the variations in size occurring in one or two organs of the common shrimp (*Orangon vulgaris*). In these papers it has been shown (1) that the observed deviations from the average size of every organ measured are grouped symmetrically about the average, and occur with a frequency corresponding closely to that indicated by the probability integral; and (2) that the "degree of correlation" between a given pair of organs is approximately the same in each of five local races of the species (*Roy. Soc. Proc.* vol. 47, p. 445, and vol. 51, p. 2). In what follows I shall describe the results obtained by measuring certain parts of the shore crab (*Carcinus moenas*) in two samples, one from the Bay of Naples, and one from Plymouth Sound, each sample consisting of 1,000 adult females.

The measurements made were as follows:

1. *The total length of the carapace* (fig. 1, AB), in a straight line from the tip of the median inter-orbital tooth to the middle of the posterior margin.
2. *The total breadth* of the carapace, in a straight line from tip to tip of the posterior lateral teeth (fig. 1, EF).
3. *The frontal breadth,* from tip to tip of the anterior lateral teeth (fig. 1, CD).
4. *The right antero-lateral margin,* from the tip of the median inter-orbital tooth to the tip of the postero-lateral tooth (fig. 1, AF).
5. *The right dentary margin*, measured in a straight line from the tip of the antero-lateral to the tip of the postero-lateral tooth (fig 1, DF).
6. *The left antero-lateral margin*, measured in the same way as the right.

Fig. 1. Diagram to show the parts of the carapace measured. The diagram is drawn to scale, the right half representing a perfectly average Plymouth crab, the left an average crab from Naples.

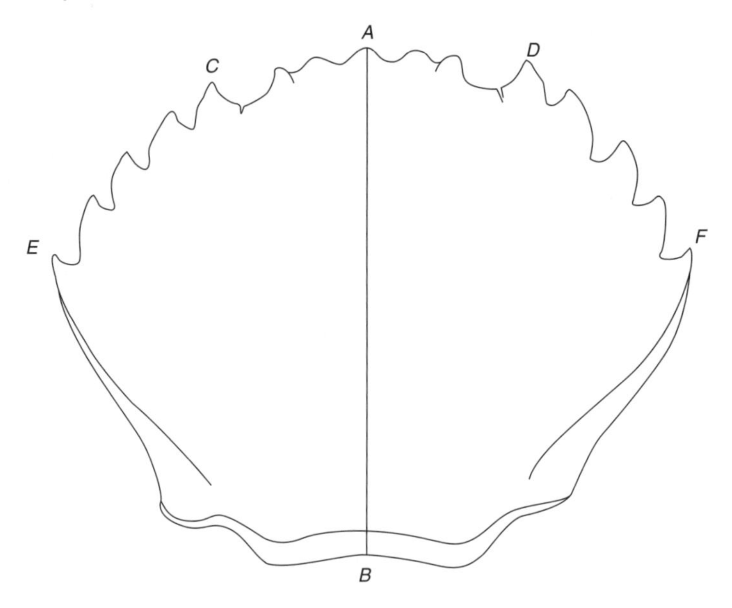

7. *The left dentary margin.*
8. *The sternal breadth,* measured between the articulations of the great chelas.
9. *The meropodite of the right chela*, measured, in a straight line between the inner articulations.
10. *The carpopodite of the right chela*, from the inner articulation, in a straight line to the tip.
11. *The proximal portion of the same carpopodite*, in a straight line from the inner articulation to the tip of the inner spine, at the base of the dactylopodite.

The dimensions 2–11 were expressed in terms of the total length of the carapace taken as 1000; and, in order to reduce the effect of possible errors of measurement, the values so obtained were grouped together in fours, the groups being so selected that no two individuals in anyone of them differed by more than 0.004 of the carapace length.

As an example of the way in which the values thus obtained were distributed, the measurements of the right antero-lateral margin in Naples and in Plymouth may be examined. The results of these measurements are shown in Tables I and II. The frequency with which every observed magnitude of this portion of the carapace occurred in the Naples specimens is given in the second column of Table I. The arithmetic mean of all these values is 752.22 thousandths of the carapace length; and the observations will be seen to cluster with a fair degree

Fig-. 2. Diagram showing the frequency of occurrence of all observed lengths of the antero-lateral margin of the carapace in 999 female crabs from Naples. The abscissa scale represents thousandths of the total carapace length. The vertical scale represents numbers of individuals.

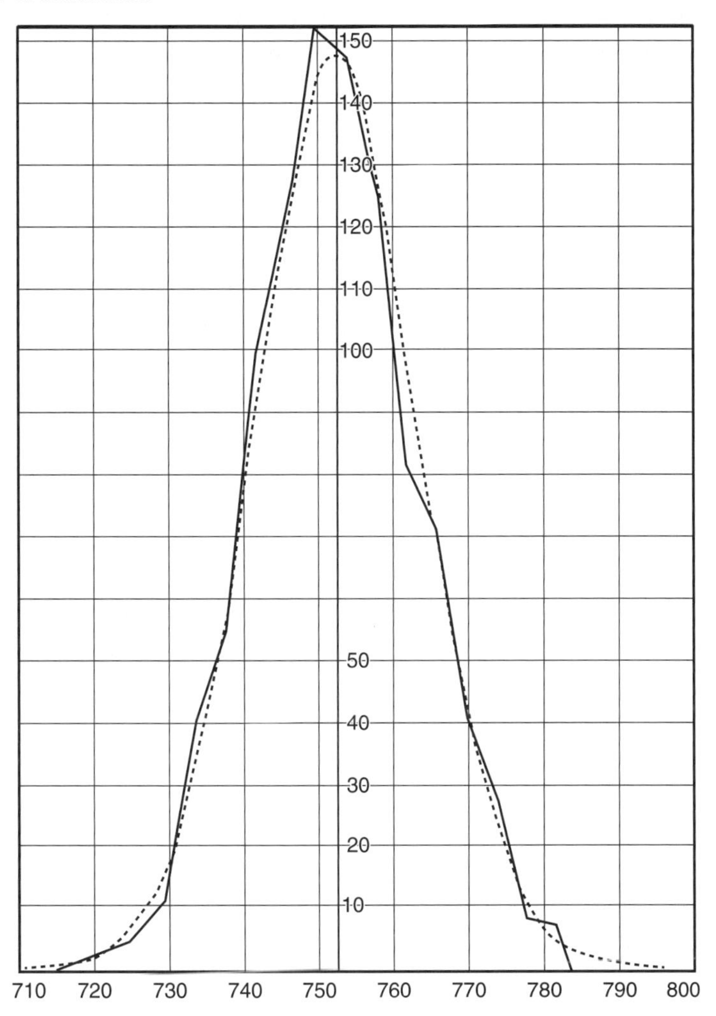

of symmetry around this value, the symmetry of distribution being, perhaps, more readily seen by the eye in the diagram, fig. 2. The total number of individuals in the sample was 999, and of these 513 had the antero-posterior margin greater than the average size, 486 having this portion of the carapace below the average. The arithmetic mean of all the deviations from the average, or "mean error" of distribution, was found to be 8.71 units; and the modulus is therefore $8.71 \times 1.77 = 15.42$ units. A probability curve, with modulus = 15.42 units, has been drawn by a dotted line in fig. 2; and the close agreement between this curve and the observed curve of distribution, which is indicated by a thick, line, is very striking. In order to make a more accurate comparison possible, the number of individuals corresponding to each observed magnitude, on the hypothesis that this probability curve represents the real distribution about the mean, has been calculated from the tables of the probability integral, and is given in the third column of Table I. In spite of some considerable discrepancies, the general agreement between the second and third columns of the table is undeniable.

The right antero-lateral margin of the Plymouth individuals, when treated in a similar way, gave the following results:—

Table I. Distribution of Lengths of Antero-lateral Margin of Carapace in 999 Female *Carcinus moenas* from Naples.

Dimension in thousandths of carapace length.	Number of individuals observed.	Number calculated from probability integral (c = 15.42
792–795	2	
788–791	0	16.4
784–787	0	16.4
780–783	7	16.4
776–779	8	16.4
772–775	28	22.2
768–771	41	42.1
764–767	72	70.2
760–763	82	102.0
756–759	126	130.0
752–755	147	144.8
748–751	152	141.1
744–747	121	120.3
740–743	98	87.0
736–739	55	59.6
732–735	40	34.3
728–731	11	16.6
724–727	5	11.8
720–723	3	11.8
716–719	1	11.8

Arithmetic mean 762.70 thousandths.
Mean error 9.77 thousandths.
Modulus 17.29 thousandths.

The frequency with which individual deviations from the average occur is compared with that indicated by a probability equation of the appropriate modulus in Table II.

These two examples will give a fair idea of the extent to which the distribution of the observed magnitudes of each organ about the mean of all of them

Table II. Distribution of lengths of Anterolateral Margin of Carapace in 999 Female *Carcinus* from Plymouth.

Length in thousandths of carapace length.	Number of individuals observed.	Number calculated from probability integral (c = 15.42
796–799	6	21.5
792–795	3	21.5
788–791	7	21.5
784–787	19	23.2
780–783	44	45.2
776–779	70	61.6
772–775	94	87.5
768–771	101	110.5
764–767	140	125.6
760–763	128	134.0
756–759	105	112.7
752–755	100	97.2
748–751	79	62.2
744–747	47	53.4
740–743	25	29.2
736–739	20	15.8
732–735	4	13.2
728–731	6	13.2
692–695	1	13.2

corresponds to that indicated by the probability equation. A similar treatment of every other set of measures would serve no useful purpose; it will be sufficient to give, in the following table, the mean value, and the probable error of distribution about that value, of every organ measured. The probable error is given below, instead of the mean error, because it is the constant which has the smallest numerical value of any in general use. This property renders the probable error more convenient than either the mean error, the modulus, or the error of mean squares, in the determination of the degree of correlation which will be described below.

...

The only case in which an undoubtedly asymmetrical result was obtained is that of the frontal breadth of the Naples specimens. From an inspection of the curve of distribution of these magnitudes, I was led to hope that the result obtained might arise from the presence, in the sample measured, of two races of individuals, clustered symmetrically about separate mean magnitudes. Professor Karl Pearson has been kind enough to test this supposition for me: he finds that the observed distribution corresponds fairly well with that resulting from the grouping of two series of individuals, one with a mean frontal breadth of 630.62 thousandths, and a probable error of 12.06 thousandths; the other with a mean breadth of 654.66 thousandths, and a probable error of 8.41 thousandths. Of the first race, Professor Pearson's calculation gives 414.5 individuals, of the second, 585.5.

...

We may, therefore, assume that the female *Carcinus moenas* is slightly dimorphic in Naples with respect to its frontal breadth; and that the individuals belonging to the two types are distributed in the proportion of nearly two to three.

2. — The Correlation of Pairs of Organs: Galton's Function

The method adopted to determine the degree of correlation between two organs was that proposed by Mr. Galton ('Roy. Soc. Proc.,' vol. 40, p. 63). The measures obtained were sorted into groups, such that in each group the deviation X of an organ A from its average was constant. The mean deviation from its average of a second organ B was determined in each of these groups.

...

It may, therefore, be asserted that the investigation which has been described does not demonstrate a difference between the value of Galton's function for a given pair of organs in Naples and the corresponding value in Plymouth. The values obtained are not in all cases shown to be identical, but the differences between them are within the limits of error of the method employed; and in the worst case it has been shown that the errors arising from a neglect of the observed discrepancy between two corresponding values are not of a very serious kind. So that in any discussion of the variation of the twenty-three pairs of organs discussed in the present paper, or of the pairs of shrimp organs discussed in my previous communication, it may be assumed as at least an empirical working rule that Galton's function has the same value in all local races. The question whether this empirical rule is rigidly true will have to be determined by fuller investigation, based on larger samples: but the value of a merely empirical expression for the relation between abnormality of one organ and that of another is very great. It cannot be too strongly urged that the problem of animal evolution is essentially a statistical problem: that before we can properly estimate the changes at present going on in a race or species we must know accurately (a) the percentage of animals which exhibit a given amount of abnormality with regard to a particular character; (b) the degree of abnormality of other organs which accompanies a given abnormality of one; (c) the difference between the death rate per cent, in animals of different degrees of abnormality with respect to any organ; (d) the abnormality of offspring in terms of the abnormality of parents, and vice versa. These are all questions of arithmetic; and when we know the numerical answers to these questions for a number of species we shall know the direction and the rate of change in these species at the present day a knowledge which is the only legitimate basis for speculations as to their past history and future fate.

Reading and Discussion Questions

1. Darwin tended to talk about traits in *qualitative* terms, noticing deaf cats with blue eyes, pigeons with small feathered feet that have skin between the outer toes, hairless dogs with imperfect teeth and droopy ears, and the complex social interactions of a beehive. What steps does Weldon take to *quantify* the traits that are under investigation?
2. How does statistical analysis of the study of crab populations in Weldon's paper provide empirical evidence for gradual changes in a population by means of natural selection? Do you think that work such as Weldon's could establish that natural selection is taking place, independently of hypotheses about their underlying causes?

Experiments in Plant Hybridization (1866)[1]

Gregor Mendel

104

Introductory Remarks

Experience of artificial fertilization, such as is effected with ornamental plants in order to obtain new variations in color, has led to the experiments which will here be discussed. The striking regularity with which the same hybrid forms always reappeared whenever fertilization took place between the same species induced further experiments to be undertaken, the object of which was to follow up the developments of the hybrids in their progeny.

To this object numerous careful observers, such as Kölreuter, Gärtner, Herbert, Lecoq, Wichura and others, have devoted a part of their lives with inexhaustible perseverance. Gärtner especially in his work *Die Bastarderzeugung im Pflanzenreiche* [The Production of Hybrids in the Vegetable Kingdom], has recorded very valuable observations; and quite recently Wichura published the results of some profound investigations into the hybrids of the Willow. That, so far, no generally applicable law governing the formation and development of hybrids has been successfully formulated can hardly be wondered at by anyone who is acquainted with the extent of the task, and can appreciate the difficulties with which experiments of this class have to contend. A final decision can only be arrived at when we shall have before us the results of *detailed experiments* made on plants belonging to the most diverse orders.

Those who survey the work done in this department will arrive at the conviction that among all the numerous experiments made, not one has been carried out to such an extent and in such a way as to make it possible to determine the

[1] Read at the February 8th, and March 8th, 1865, meetings of the Brünn Natural History Society, published in 1866.

number of different forms under which the offspring of the hybrids appear, or to arrange these forms with certainty according to their separate generations, or definitely to ascertain their statistical relations.

It requires indeed some courage to undertake a labor of such far-reaching extent; this appears, however, to be the only right way by which we can finally reach the solution of a question the importance of which cannot be overestimated in connection with the history of the evolution of organic forms.

The paper now presented records the results of such a detailed experiment. This experiment was practically confined to a small plant group, and is now, after eight years' pursuit, concluded in all essentials. Whether the plan upon which the separate experiments were conducted and carried out was the best suited to attain the desired end is left to the friendly decision of the reader.

Selection of the Experimental Plants

The value and utility of any experiment are determined by the fitness of the material to the purpose for which it is used, and thus in the case before us it cannot be immaterial what plants are subjected to experiment and in what manner such experiment is conducted.

The selection of the plant group which shall serve for experiments of this kind must be made with all possible care if it be desired to avoid from the outset every risk of questionable results.

The experimental plants must necessarily:

Possess constant differentiating characteristics

The hybrids of such plants must, during the flowering period, be protected from the influence of all foreign pollen, or be easily capable of such protection.

The hybrids and their offspring should suffer no marked disturbance in their fertility in the successive generations.

Accidental impregnation by foreign pollen, if it occurred during the experiments and were not recognized, would lead to entirely erroneous conclusions. Reduced fertility or entire sterility of certain forms, such as occurs in the offspring of many hybrids, would render the experiments very difficult or entirely frustrate them. In order to discover the relations in which the hybrid forms stand towards each other and also towards their progenitors it appears to be necessary that all members of the series developed in each successive generations should be, *without exception*, subjected to observation.

At the very outset special attention was devoted to the *Leguminosae* on account of their peculiar floral structure. Experiments which were made with several members of this family led to the result that the genus *Pisum* was found to possess the necessary qualifications.

Some thoroughly distinct forms of this genus possess characters which are constant, and easily and certainly recognizable, and when their hybrids are mutually crossed they yield perfectly fertile progeny. Furthermore, a disturbance through foreign pollen cannot easily occur, since the fertilizing organs are closely packed inside the keel and the anthers burst within the bud, so that the stigma becomes covered with pollen even before the flower opens. This circumstance is especially important. As additional advantages worth mentioning, there may be cited the easy culture of these plants in the open ground and in pots, and also their relatively short period of growth. Artificial fertilization is certainly a somewhat elaborate process, but nearly always succeeds. For this purpose the bud is opened before it is perfectly developed, the keel is removed, and each stamen carefully extracted by means of forceps, after which the stigma can at once be dusted over with the foreign pollen.

In all, thirty-four more or less distinct varieties of Peas were obtained from several seedsmen and subjected to a two year's trial. In the case of one variety there were noticed, among a larger number of plants all alike, a few forms which were markedly different. These, however, did not vary in the following year, and agreed entirely with another variety obtained from the same seedsman; the seeds were therefore doubtless merely accidentally mixed. All the other varieties yielded perfectly constant and similar offspring; at any rate, no essential difference was observed during two trial years. For fertilization twenty-two of these were selected and cultivated during the whole period of the experiments. They remained constant without any exception.

Their systematic classification is difficult and uncertain. If we adopt the strictest definition of a species, according to which only those individuals belong to a species which under precisely the same circumstances display precisely similar characters, no two of these varieties could be referred to one species. According to the opinion of experts, however, the majority belong to the species *Pisum sativum*; while the rest are regarded and classed, some as sub-species of *P. sativum*, and some as independent species, such as *P. quadratum*, *P. saccharatum*, and *P. umbellatum*. The positions, however, which may be assigned to them in a classificatory system are quite immaterial for the purposes of the experiments in question. It has so far been found to be just as impossible to draw a sharp line between the hybrids of species and varieties as between species and varieties themselves.

Division and Arrangement of the Experiments

If two plants which differ constantly in one or several characters be crossed, numerous experiments have demonstrated that the common characters are transmitted unchanged to the hybrids and their progeny; but each pair of differentiating characters, on the other hand, unite in the hybrid to form a new character, which in the progeny of the hybrid is usually variable. The object of the experiment was to observe these variations in the case of each pair of differentiating characters, and to deduce the law according to which they appear in successive generations. The experiment resolves itself therefore into just as many separate experiments as there are constantly differentiating characters presented in the experimental plants.

The various forms of Peas selected for crossing showed differences in length and color of the stem; in the size and form of the leaves; in the position, color, size of the flowers; in the length of the flower stalk; in the color, form, and size of the pods; in the form and size of the seeds; and in the color of the seed-coats and of the albumen (endosperm). Some of the characters noted do not permit of a sharp and certain separation, since the difference is of a "more or less" nature, which is often difficult to define. Such characters could not be utilized for the separate experiments; these could only be applied to characters which stand out clearly and definitely in the plants. Lastly, the result must show whether they, in their entirety, observe a regular behavior in their hybrid unions, and whether from these facts any conclusion can be reached regarding those characters which possess a subordinate significance in the type.

The characters which were selected for experiment relate:

To the *difference in the form of the ripe seeds*. These are either round or roundish, the depressions, if any, occur on the surface, being always only shallow; or they are irregularly angular and deeply wrinkled (*P. quadratum*).

To the *difference in the color of the seed albumen* (endosperm). The albumen of the ripe seeds is either pale yellow, bright yellow and orange colored, or it possesses a more or less intense green tint. This difference of color is easily seen in the seeds as their coats are transparent.

To the *difference in the color of the seed-coat*. This is either white, with which character white flowers are constantly correlated; or it is gray, gray-brown, leather-brown, with or without violet spotting, in which case the color of the standards is violet, that of the wings purple, and the stem in the axils of the leaves is of a reddish tint. The gray seed-coats become dark brown in boiling water.

To the *difference in the form of the ripe pods*. These are either simply inflated, not contracted in places; or they are deeply constricted between the seeds and more or less wrinkled (*P. saccharatum*).

To the *difference in the color of the unripe pods*. They are either light to dark green, or vividly yellow, in which coloring the stalks, leaf-veins, and calyx participate.[2]

To the *difference in the position of the flowers*. They are either axial, that is, distributed along the main stem; or they are terminal, that is, bunched at the top

[2] One species possesses a beautifully brownish-red colored pod, which when ripening turns to violet and blue. Trials with this character were only begun last year.

of the stem and arranged almost in a false umbel; in this case the upper part of the stem is more or less widened in section (*P. umbellatum*).

To the *difference in the length of the stem*. The length of the stem is very various in some forms; it is, however, a constant character for each, in so far that healthy plants, grown in the same soil, are only subject to unimportant variations in this character. In experiments with this character, in order to be able to discriminate with certainty, the long axis of 6 to 7 ft. was always crossed with the short one of ¾ ft. to 1½ ft.

Each two of the differentiating characters enumerated above were united by cross-fertilization. There were made for the 1st experiment 60 fertilizations on 15 plants. 2nd experiment 58 fertilizations on 10 plants. 3rd experiment 35 fertilizations on 10 plants. 4th experiment 40 fertilizations on 10 plants. 5th experiment 23 fertilizations on 5 plants. 6th experiment 34 fertilizations on 10 plants. 7th experiment 37 fertilizations on 10 plants.

From a larger number of plants of the same variety only the most vigorous were chosen for fertilization. Weakly plants always afford uncertain results, because even in the first generation of hybrids, and still more so in the subsequent ones, many of the offspring either entirely fail to flower or only form a few and inferior seeds.

Furthermore, in all the experiments reciprocal crossings were effected in such a way that each of the two varieties which in one set of fertilizations served as seed-bearer in the other set was used as the pollen plant.

The plants were grown in garden beds, a few also in pots, and were maintained in their natural upright position by means of sticks, branches of trees, and strings stretched between. For each experiment a number of pot plants were placed during the blooming period in a greenhouse, to serve as control plants for the main experiment in the open as regards possible disturbance by insects. Among the insects which visit Peas the beetle *Buchus pisi* might be detrimental to the experiments should it appear in numbers. The female of this species is known to lay the eggs in the flower, and in so doing opens the keel; upon the tarsi of one specimen, which was caught in a flower, some pollen grains could clearly be seen under a lens. Mention must also be made of a circumstance which possibly might lead to the introduction of foreign pollen. It occurs, for instance, in some rare cases that certain parts of an otherwise normally developed flower wither, resulting in a partial exposure of the fertilizing organs. A defective development of the keel has also been observed, owing to which the stigma and anthers remained partially covered. It also sometimes happens that the pollen does not reach full perfection. In this event there occurs a gradual lengthening of the pistil during the blooming period, until the stigmatic tip protrudes at the point of the keel. This remarkable appearance has also been observed in hybrids of *Phaseolus* and *Lathyrus*.

The risk of false impregnation by foreign pollen is, however, a very slight one with *Pisum*, and is quite incapable of disturbing the general result. Among more than 10,000 plants which were carefully examined there were only a very few cases where an indubitable false impregnation had occurred. Since in the greenhouse such a case was never remarked, it may well be supposed that *Brucus pisi*, and possibly also the described abnormalities in the floral structure, were to blame.

The Forms of the Hybrids

Experiments which in previous years were made with ornamental plants have already afforded evidence that the hybrids, as a rule, are not exactly intermediate between the parental species. With some of the more striking characters, those, for instance, which relate to the form and size of the leaves, the pubescence of the several parts, etc., the intermediate, indeed, is nearly always to be seen; in other cases, however, one of the two parental characters is so preponderant that it is difficult, or quite impossible, to detect the other in the hybrid.

This is precisely the case with the Pea hybrids. In the case of each of the seven crosses the hybrid-character resembles that of one of the parental forms so closely that the other either escapes observation completely or cannot be detected with certainty. This circumstance is of great importance in the determination and classification of the forms under which the offspring of the hybrids appear. Henceforth in this paper those characters which are transmitted entire, or almost unchanged in the hybridization, and therefore in themselves constitute the characters of the hybrid, are termed the *dominant*, and those which become latent in the process *recessive*. The expression *recessive* has been chosen because the characters thereby designated withdraw or entirely disappear in the hybrids, but nevertheless reappear unchanged in their progeny, as will be demonstrated later on.

It was furthermore shown by the whole of the experiments that it is perfectly immaterial whether the dominant character belongs to the seed plant or to the pollen plant; the form of the hybrid remains identical in both cases. This interesting fact was also emphasized by Gärtner, with the remark that even the most practiced expert is not in a position to determine in a hybrid which of the two parental species was the seed or the pollen plant.

Of the differentiating characters which were used in the experiments the following are dominant:

The round or roundish form of the seed with or without shallow depressions.
The yellow coloring of the seed albumen.
The gray, gray-brown, or leather brown color of the seed-coat, in association with violet-red blossoms and reddish spots in the leaf axils.
The simply inflated form of the pod.
The green coloring of the unripe pod in association with the same color of the stems, the leaf-veins and the calyx.
The distribution of the flowers along the stem.

The greater length of stem.

With regard to this last character it must be stated that the longer of the two parental stems is usually exceeded by the hybrid, a fact which is possibly only attributable to the greater luxuriance which appears in all parts of plants when stems of very different lengths are crossed. Thus, for instance, in repeated experiments, stems of 1 ft. and 6 ft. in length yielded without exception hybrids which varied in length between 6 ft. and 7½ ft.

The hybrid seeds in the experiments with seed-coat are often more spotted, and the spots sometimes coalesce into small bluish-violet patches. The spotting also frequently appears even when it is absent as a parental character.

The hybrid forms of the seed-shape and of the [color of the] albumen are developed immediately after the artificial fertilization by the mere influence of the foreign pollen. They can, therefore, be observed even in the first year of experiment, whilst all the other characters naturally only appear in the following year in such plants as have been raised from the crossed seed.

The First Generation from the Hybrids

In this generation there reappear, together with the dominant characters, also the recessive ones with their peculiarities fully developed, and this occurs in the definitely expressed average proportion of three to one, so that among each four plants of this generation three display the dominant character and one the recessive.

This relates without exception to all the characters which were investigated in the experiments. The angular wrinkled form of the seed, the green color of the albumen, the white color of the seed-coats and the flowers, the constrictions of the pods, the yellow color of the unripe pod, of the stalk, of the calyx, and of the leaf venation, the umbel-like form of the inflorescence, and the dwarfed stem, all reappear in the numerical proportion given, without any essential alteration. *Transitional forms were not observed in any experiment.*

Since the hybrids resulting from reciprocal crosses are formed alike and present no appreciable difference in their subsequent development, consequently these results can be reckoned together in each experiment. The relative numbers which were obtained for each pair of differentiating characters are as follows:

Expt 1: Form of seed. From 253 hybrids 7,324 seeds were obtained in the second trial year. Among them were 5,474 round or roundish ones and 1,850 angular wrinkled ones. Therefrom the ratio 2.96:1 is deduced.
Expt 2: Color of albumen. 258 plants yielded 8,023 seeds, 6,022 yellow, and 2,001 green; their ratio, therefore, is as 3.01:1.

In these two experiments each pod yielded usually both kinds of seed. In well-developed pods which contained on the average six to nine seeds, it often happened that all the seeds were round (Expt. 1) or all yellow (Expt. 2); on the other hand there were never observed more than five wrinkled or five green ones on one pod. It appears to make no difference whether the pods are developed early or later in the hybrid or whether they spring from the main axis or from a lateral one. In some few plants only a few seeds developed in the first formed pods, and these possessed exclusively one of the two characters, but in the subsequently developed pods the normal proportions were maintained nevertheless.

As in separate pods, so did the distribution of the characters vary in separate plants. By way of illustration the first ten individuals from both series of experiments may serve.

	Experiment 1 Form of the Seed		Experiment 2 Color of the Albumen	
Plants	*round*	*wrinkled*	*yellow*	*green*
1	45	12	25	11
2	27	8	32	7
3	24	7	14	5
4	19	10	70	27
5	32	11	24	13
6	26	6	20	6
7	88	24	32	13
8	22	10	44	9
9	28	6	50	14
10	25	7	44	18

As extremes in the distribution of the two seed characters in one plant, there were observed in Expt. 1 an instance of 43 round and only two angular, and another of 14 round and 15 angular seeds. In Expt. 2 there was a case of 32 yellow and only one green seed, but also one of 20 yellow and 19 green.

These two experiments are important for the determination of the average ratios, because with a smaller number of experimental plants they show that very considerable fluctuations may occur. In counting the seeds, also, especially in Expt. 2, some care is requisite, since in some of the seeds of many plants the green color of the albumen is less developed, and at first may be easily overlooked. The cause of this partial disappearance of the green coloring has no connection with the hybrid-character of the plants, as it likewise occurs in the parental variety. This peculiarity is also confined to the individual and is not inherited by the offspring. In luxuriant plants this appearance was frequently noted. Seeds which are damaged by insects during their development often vary in color and form, but with a little practice in sorting, errors are easily avoided. It is almost superfluous to mention that the pods must remain on the plants until they are thoroughly ripened and have become dried, since it is only then that the shape and color of the seed are fully developed.

Expt. 3: Color of the seed-coats. Among 929 plants, 705 bore violet-red flowers and gray-brown seed-coats; 224 had white flowers and white seed-coats, giving the proportion 3.15:1.
Expt. 4: Form of pods. Of 1,181 plants, 882 had them simply inflated, and in 299 they were constricted. Resulting ratio, 2.95:1.
Expt. 5: Color of the unripe pods. The number of trial plants was 580, of which 428 had green pods and 152 yellow ones. Consequently these stand in the ratio of 2.82:1.
Expt. 6: Position of flowers. Among 858 cases 651 had inflorescences axial and 207 terminal. Ratio, 3.14:1.
Expt. 7: Length of stem. Out of 1,064 plants, in 787 cases the stem was long, and in 277 short. Hence a mutual ratio of 2.84:1. In this experiment the dwarfed plants were carefully lifted and transferred to a special bed. This precaution was necessary, as otherwise they would have perished through being overgrown by their tall relatives. Even in their quite young state they can be easily picked out by their compact growth and thick dark-green foliage.

If now the results of the whole of the experiments be brought together, there is found, as between the number of forms with the dominant and recessive characters, an average ratio of 2.98:1, or 3:1.

The dominant character can have here a *double signification*—viz. that of a parental character or a hybrid–character. In which of the two significations it appears in each separate case can only be determined by the following generation. As a parental character it must pass over unchanged to the whole of the offspring; as a hybrid-character, on the other hand, it must maintain the same behavior as in the first generation.

The Second Generation from the Hybrids

Those forms which in the first generation exhibit the recessive character do not further vary in the second generation as regards this character; they remain constant in their offspring.

It is otherwise with those which possess the dominant character in the first generation [bred from the hybrids i.e., the F_2 in modern terminology]. Of these *two*-thirds yield offspring which display the dominant and recessive characters in the proportion of three to one, and thereby show exactly the same ratio as the hybrid forms, while only *one*-third remains with the dominant character constant.

The separate experiments yielded the following results:

Expt. 1: Among 565 plants which were raised from round seeds of the first generation, 193 yielded round seeds only, and remained therefore constant in this character; 372, however, gave both round and wrinkled seeds, in the proportion of 3:1. The number of the hybrids, therefore, as compared with the constants is 1.93:1.

Expt. 2: Of 519 plants which were raised from seeds whose albumen was of yellow color in the first generation, 166 yielded exclusively yellow, while 353 yielded yellow and green seeds in the proportion of 3:1. There resulted, therefore, a division into hybrid and constant forms in the proportion of 2.13:1.

For each separate trial in the following experiments 100 plants were selected which displayed the dominant character in the first generation, and in order to ascertain the significance of this, ten seeds of each were cultivated.

Expt. 3: The offspring of 36 plants yielded exclusively gray-brown seed-coats, while of the offspring of 64 plants some had gray-brown and some had white.

Expt. 4: The offspring of 29 plants had only simply inflated pods; of the offspring of 71, on the other hand, some had inflated and some constricted.

Expt. 5: The offspring of 40 plants had only green pods; of the offspring of 60 plants some had green, some yellow ones.

Expt. 6: The offspring of 33 plants had only axial flowers; of the offspring of 67, on the other hand, some had axial and some terminal flowers.

Expt. 7: The offspring of 28 plants inherited the long axis, of those of 72 plants some the long and some the short axis.

In each of these experiments a certain number of the plants came constant with the dominant character. For the determination of the proportion in which the separation of the forms with the constantly persistent character results, the two first experiments are especially important, since in these a larger number of plants can be compared. The ratios 1.93:1 and 2.13:1 gave together almost exactly the average ratio of 2:1. Experiment 6 gave a quite concordant result; in the others the ratio varies more or less, as was only to be expected in view of the smaller number of 100 trial plants. Experiment 5, which shows the greatest departure, was repeated, and then in lieu of the ratio of 60:40, that of 65:35 resulted. *The average ratio of 2 to 1 appears, therefore, as fixed with certainty*. It is therefore demonstrated that, of those forms which possess the dominant character in the first generation, two-thirds have the hybrid-character, while one-third remains constant with the dominant character.

The ratio 3:1, in accordance with which the distribution of the dominant and recessive characters results in the first generation, resolves itself therefore in all experiments into the ratio of 2:1:1, if the dominant character be differentiated according to its significance as a hybrid-character or as a parental one. Since the members of the first generation spring directly from the seed of the hybrids, it is now clear that the hybrids form seeds having one or other of the two differentiating characters, and of these one-half develop again the hybrid form, while the other half yield plants which remain constant and receive the dominant or the recessive characters in equal numbers.

The Subsequent Generations from the Hybrids

The proportions in which the descendants of the hybrids develop and split up in the first and second generations presumably hold good for all subsequent progeny. Experiments 1 and 2 have already been carried through six generations; 3 and 7 through five; and 4, 5, and 6 through four; these experiments being continued from the third generation with a small number of plants, and no departure from the rule has been perceptible. The offspring of the hybrids separated in each generation in the ratio of 2:1:1 into hybrids and constant forms.

If ***A*** be taken as denoting one of the two constant characters, for instance the dominant, ***a***, the recessive, and ***Aa*** the hybrid form in which both are conjoined, the expression ***A*** + 2***Aa*** + ***a*** shows the terms in the series for the progeny of the hybrids of two differentiating characters.

The observation made by Gärtner, Kölreuter, and others, that hybrids are inclined to revert to the parental forms, is also confirmed by the experiments described. It is seen that the number of the hybrids which arise from one fertilization, as compared with the number of forms which become constant, and their progeny from generation to generation, is continually diminishing, but that nevertheless they could not entirely disappear. If an average equality of

Generation				Ratios				
	A	*Aa*	*a*	*A*	:	*Aa*	:	*a*
1	1	2	1	1	:	2	:	1
2	6	4	6	3	:	2	:	3
3	28	8	28	7	:	2	:	7
4	120	16	120	15	:	2	:	15
5	496	32	496	31	:	2	:	31
n				$2^n - 1$	:	2	:	$2^n - 1$

fertility in all plants in all generations be assumed, and if, furthermore, each hybrid forms seed of which one-half yields hybrids again, while the other half is constant to both characters in equal proportions, the ratio of numbers for the offspring in each generation is seen by the following summary, in which ***A*** and ***a*** denote again the two parental characters, and ***Aa*** the hybrid forms. For brevity's sake it may be assumed that each plant in each generation furnishes only four seeds.

In the tenth generation, for instance, $2^n - 1 = 1{,}023$. There result, therefore, in each 2,048 plants which arise in this generation 1,023 with the constant dominant character, 1,023 with the recessive character, and only two hybrids.

The Offspring of the Hybrids in Which Several Differentiating Characters Are Associated

In the experiments above described plants were used which differed only on one essential character. The next task consisted in ascertaining whether the law of development discovered in these applied to each pair of differentiating characters when several diverse characters are united in the hybrid by crossing.

As regards the form of the hybrids in these cases, the experiments showed throughout that this invariably more nearly approaches to that one of the two parental plants which possesses the greater number of dominant characters. If, for instance, the seed plant has a short stem, terminal white flowers, and simply inflated pods; the pollen plant, on the other hand, a long stem, violet-red flowers distributed along the stem, and constricted pods; the hybrid resembles the seed parent only in the form of the pod; in the other characters it agrees with the pollen parent. Should one of the two parental types possess only dominant characters, then the hybrid is scarcely or not at all distinguishable from it.

Two experiments were made with a considerable number of plants. In the first experiment the parental plants differed in the form of the seed and in the color of the albumen; in the second in the form of the seed, in the color of the albumen, and in the color of the seed-coats. Experiments with seed characters give the result in the simplest and most certain way.

Expt. 1.—	***AB***, seed parents	***ab***, pollen parents
	A, form round	***a***, form wrinkled
	B, albumen yellow	***b***, albumen green

In order to facilitate study of the data in these experiments, the different characters of the seed plant will be indicated by ***A***, ***B***, ***C***, those of the pollen plant by ***a***, ***b***, ***c***, and the hybrid forms of the characters by ***Aa***, ***Bb***, and ***Cc***.

315	round and yellow,
101	wrinkled and yellow,
108	round and green,
32	wrinkled and green.

The fertilized seeds appeared round and yellow like those of the seed parents. The plants raised therefrom yielded seeds of four sorts, which frequently presented themselves in one pod. In all, 556 seeds were yielded by 15 plants, and of these there were:

All were sown the following year. Eleven of the round yellow seeds did not yield plants, and three plants did not form seeds. Among the rest:

38	had round yellow seeds	***AB***
65	round yellow and green seeds	***ABb***
60	round yellow and wrinkled yellow seeds	***AaB***
138	round yellow and green, wrinkled yellow and green seeds	***AaBb***

From the wrinkled yellow seeds 96 resulting plants bore seed, of which:

28 had only wrinkled yellow seeds	***aB***
68 wrinkled yellow and green seeds	***aBb***

From 108 round green seeds 102 resulting plants fruited, of which:

35 had only round green seeds	***Ab***
67 round and wrinkled green seeds	***Aab***

The wrinkled green seeds yielded 30 plants which bore seeds all of like character; they remained constant ***ab***.

The offspring of the hybrids appeared therefore under nine different forms, some of them in very unequal numbers. When these are collected and coordinated we find:

38	plants with the sign			***AB***
35	"	"	"	***Ab***
28	"	"	"	***aB***
30	"	"	"	***ab***
65	"	"	"	***ABb***
68	"	"	"	***aBb***
60	"	"	"	***AaB***
67	"	"	"	***Aab***
138	"	"	"	***AaBb***

The whole of the forms may be classed into three essentially different groups. The first includes those with the signs ***AB***, ***Ab***, ***aB***, and ***ab***: they possess only constant characters and do not vary again in the next generation. Each of these forms is represented on the average 33 times. The second group includes the signs ***ABb***, ***aBb***, ***AaB***, ***Aab***: these are constant in one character and hybrid in another, and vary in the next generation only as regards the hybrid-character. Each of these appears on any average 65 times. The form ***AaBb*** occurs 138 times: it is hybrid in both characters, and behaves exactly as do the hybrids from which it is derived.

If the numbers in which the forms belonging to these classes appear be compared, the ratios of 1:2:4 are unmistakably evident. The numbers 33, 65, 138 present very fair approximations to the ratio numbers of 33, 66, 132.

The development series consists, therefore, of nine classes, of which four appear therein always once and are constant in both characters; the forms ***AB***, ***ab***, resemble the parental forms, the two others present combinations between the conjoined characters ***A***, ***a***, ***B***, ***b***, which combinations are likewise possibly constant. Four classes appear always twice, and are constant in one character and hybrid in the other. One class appears four times, and is hybrid in both characters. Consequently, the offspring of the hybrids, if two kinds of differentiating characters are combined therein, are represented by the expression ***AB*** + ***Ab*** + ***aB*** + ***ab*** + 2***ABb*** + 2***aBb*** + 2***AaB*** + 2***Aab*** + 4***AaBb***.

This expression is indisputably a combination series in which the two expressions for the characters ***A*** and ***a***, ***B*** and ***b*** are combined. We arrive at the full number of the classes of the series by the combination of the expressions: ***A*** + 2***Aa*** + ***a*** and ***B*** + 2***Bb*** + ***b***

Expt. 2.—	***ABC***, seed parents	***abc***, pollen parents
	A, form round	***a***, form wrinkled
	B, albumen yellow	***b***, albumen green
	C, seed coat grey-brown	***c***, seed coat white

This experiment was made in precisely the same way as the previous one. Among all the experiments it demanded the most time and trouble. From 24 hybrids 687 seeds were obtained in all: these were all either spotted, gray-brown or gray-green, round or wrinkled. From these in the following year 639 plants fruited, and as further investigation showed, there were among them:

8	plants	***ABC***	22	plants	***ABCc***	45	plants	***ABbCc***
14	"	***Abc***	17	"	***AbCc***	36	"	***aBbCc***
9	"	***AbC***	25	"	***aBCc***	38	"	***AaBCc***
11	"	***Abc***	20	"	***abCc***	40	"	***AabCc***
8	"	***aBC***	15	"	***ABbC***	49	"	***AaBbC***
10	"	***aBc***	18	"	***Abbc***	48	"	***AaBbc***
10	"	***abC***	19	"	***aBbC***			
7	"	***abc***	24	"	***aBbc***			
			14	"	***AaBC***	78	"	***AaBbCc***
			18	"	***AaBc***			
			20	"	***AabC***			
			16	"	***Aabc***			

The whole expression contains 27 terms. Of these eight are constant in all characters, and each appears on the average ten times; twelve are constant in two characters, and hybrid in the third; each appears on the average 19 times; six are constant in one character and hybrid in the other two; each appears on the average 43 times. One form appears 78 times and is hybrid in all of the characters. The ratios 10:19:43:78 agree so closely with the ratios 10:20:40:80, or 1:2:4:8 that this last undoubtedly represents the true value.

The development of the hybrids when the original parents differ in three characters results therefore according to the following expression: ***ABC*** + ***ABc*** + ***AbC*** + ***Abc*** + ***aBC*** + ***aBc*** + ***abC*** + ***abc*** + 2***ABCc*** + 2***AbCc*** + 2***aBCc*** + 2***abCc*** + 2***ABbC*** + 2***ABbc*** + 2***aBbC*** + 2***aBbc*** + 2***AaBC*** + 2***AaBc*** + 2***AabC*** + 2***Aabc*** + 4***ABbCc*** + 4***aBbCc*** + 4***AaBCc*** + 4***AabCc*** + 4***AaBbC*** + 4***AaBbc*** + 8***AaBbCc***.

Here also is involved a combination series in which the expressions for the characters ***A*** and ***a***, ***B*** and ***b***, ***C*** and ***c***, are united. The expressions: ***A*** + 2***Aa*** + ***a*** ***B*** + 2***Bb*** + ***b*** and ***C*** + 2***Cc*** + ***c*** give all the classes of the series. The constant combinations which occur therein agree with all combinations which are possible between the characters ***A***, ***B***, ***C***, ***a***, ***b***, ***c***; two thereof, ***ABC*** and ***abc***, resemble the two original parental stocks.

In addition, further experiments were made with a smaller number of experimental plants in which the remaining characters by twos and threes were united as hybrids: all yielded approximately the same results. There is therefore no doubt that for the whole of the characters involved in the experiments the principle applies *that the offspring of the hybrids in which several essentially different characters are combined exhibit the terms of a series of combinations, in which the developmental series for each pair of differentiating characters are united.* It is demonstrated at the same time that *the relation of each pair of different characters in hybrid union is independent of the other differences in the two original parental stocks.*

If *n* represents the number of the differentiating characters in the two original stocks, 3^n gives the number of terms of the combination series, 4^n the number of individuals which belong to the series, and 2^n the number of unions which remain constant. The series therefore contains, if the original stocks differ in four characters, $3^4 = 81$ classes, $4^4 = 256$ individuals, and $2^4 = 16$ constant forms: or, which is the same, among each 256 offspring of the hybrids are 81 different combinations, 16 of which are constant.

All constant combinations which in Peas are possible by the combination of the said seven differentiating characters were actually obtained by repeated crossing. Their number is given by $2^7 = 128$. Thereby is simultaneously given the practical proof *that the constant characters which appear in the several varieties of a group of plants may be obtained in all the associations which are possible according to the laws of combination, by means of repeated artificial fertilization.*

As regards the flowering time of the hybrids, the experiments are not yet concluded. It can, however, already be stated that the time stands almost exactly between those of the seed and pollen parents, and that the constitution of the hybrids with respect to this character probably follows the rule ascertained in the case of the other characters. The forms which are selected for experiments of this class must have a difference of at least 20 days from the middle flowering period of one to that of the other; furthermore, the seeds when sown must all be placed at the same depth in the earth, so that they may germinate simultaneously. Also, during the whole flowering period, the more important variations in temperature must be taken into account, and the partial hastening or delaying of the flowering which may result there from. It is clear that this experiment presents many difficulties to be overcome and necessitates great attention.

If we endeavor to collate in a brief form the results arrived at, we find that those differentiating characters, which admit of easy and certain recognition in the experimental plants, *all behave exactly alike in their hybrid associations.* The offspring of the hybrids of each pair of differentiating characters are, one-half, and hybrid again, while the other half are constant in equal proportions having the characters of the seed and pollen parents respectively. If several differentiating characters are combined by cross-fertilization in a hybrid, the resulting offspring form the terms of a combination series in which the combination series for each pair of differentiating characters are united.

The uniformity of behavior shown by the whole of the characters submitted to experiment permits, and fully justifies, the acceptance of the principle that a similar relation exists in the other characters which appear less sharply defined in plants, and therefore could not be included in the separate experiments. An experiment with peduncles of different lengths gave on the whole a fairly satisfactory results, although the differentiation and serial arrangement of the forms could not be effected with that certainty which is indispensable for correct experiment.

Reading and Discussion Questions

1. Mendel presented his research to the Nature Research Society of Brünn in 1865. However, his results were not widely known until his paper was "rediscovered" around 1900. What do you take to be the point of his paper? What are his main conclusions? What is he trying to establish with these studies? Who is his intended audience?
2. Mendel, like Harvey, was trying to study very complex phenomena that he could only observe indirectly. What choices does he make that help to simplify the system under study? What data does he eliminate in the course of arriving at his quantitative results? Are these choices a reflection of good scientific methodology—of isolating, idealization, and looking for mathematical patterns in large amounts of data? Or do they (as Mendel's few early readers insisted) oversimplify, over-generalize, and lack statistical significance?

Additional Resources

Bowler, Peter J. *Evolution: The History of an Idea*. 3rd ed. Berkeley: University of California Press, 2003.

Darwin, Charles. *On the Origin of Species: A Facsimile of the First Edition*. Cambridge: Harvard University Press, 2003.

Gray, Asa. *Darwiniana: Essays and Reviews Pertaining to Darwinism*. Teddington: The Echo Library, 2006.

Gregory, Frederick. *Natural Science in Western History*. 1st ed. Cengage Learning, 2007. A solid secondary source textbook, by a past president of the History of Science Society, for the whole period covered by the readings in this volume.

Hodge, Jonathan, and Radick, Gregory, eds. *The Cambridge Companion to Darwin*. 2nd ed. Cambridge: Cambridge University Press, 2009.

Kelly, Ned, and Ralling, Christopher, producers. Friend, Martyn, director. *The Voyage of Charles Darwin*. UK: BBC, 1978. Old, and never available for commercial release, this is now available (albeit in mediocre video quality) on YouTube. It remains the best researched and produced film of Darwin's life and work—in seven hour-long episodes. The BBC outfitted a ship to look like the Beagle, sailed it around the world filled with sailors from the Royal Navy and a handful of actors, tracing the actual voyages Darwin took as a young man. Nearly every word spoken by the actor playing Darwin is actually taken from his journals, published writings, or letters. http://darwin-online.org.uk/contents.html

Robbins, Robert J. *Electronic Scholarly Publishing*. http://www.esp.org. This website specializes in the history of biology, going all the way back to Aristotle, with a particular focus on genetics.

Secord, James A., et al. *Darwin Correspondence Project*. https://www.darwinproject.ac.uk/.

van Wyhe, John. *Darwin Online*. http://darwin-online.org.uk/. A vast collection of Darwin's works, manuscripts, letters, reviewers, and texts that he read and referenced in his published works.

Bibliography

Sources (in order of appearance)

Part I The Birth of Natural Philosophy: Science and Mathematics in the Ancient Hellenistic World

Hippocrates, *Nature of Man*. In: *Hippocrates, Heracleitus. Nature of Man. Regimen in Health. Humours. Aphorisms. Regimen 1–3. Dreams. Heracleitus: On the Universe*; translated by W. H. S. Jones. Loeb Classical Library 150. Cambridge: Harvard University Press, 1931.

Lucretius, *De Rerum Natura*, translated by William Ellery Leonard. New York: E. P. Dutton, 1916.

Plato, *Timaeus, Philebus, and Republic*. In: *Plato, the Dialogues of Plato Translated into English with Analyses and Introductions*, 3rd edition revised and corrected; translated by Benjamin Jowett. Oxford: Oxford University Press, 1892. *Philebus* 55c–59e; *Republic* VI, 507d –VII, 537b; *Timaeus* 27a–34b, 37d–43b, 47a–48e, 52d–57d.

Aristotle, *The Categories*, translated by E. M. Edghill. 3–5, 1b10–4b19; 14–15, 15a14–15b32.

Aristotle, *Posterior Analytics*, translated by G. R. G. Mure. I.1–2, 71a1–72b4; I.13, 78a22–79a16.

Aristotle, *On the Heavens*, translated by J. L. Stocks. II.10–14, 291a35–298b20.

Aristotle, *Meteorology*, translated by E. W. Webster. I: 1–4, 9, 13–14; VI: 1–2 (lines 338a20–342a15, 346b16–347a12, 349a12–349b2, 351a19–352a29, 378b10–26, 379b10–25, 370b33–380a5).

Aristotle, *Physics*, translated by R. P. Hardie and R. K. Gaye. II.1–3, 192b9–195b30; IV.8, 214b12–216b20; VII.1–2, 241b34–245b2; VII.4, 248a10–249b26.

Aristotle, *On the Soul*, translated by J. A. Smith. II.1–4 412a1–416b31.

Aristotle, *On the Parts of Animals*, translated by William Ogle. I.1–4, 639a1–644b20.

Aristotle, *The History of Animals*, translated by D'Arcy Wentworth Thompson. I.1–3, 486a5–489a19; II.1 497b3–501b4.

Aristotle, *On the Generation of Animals*, translated by Arthur Platt. I.1–2, 715a1–716b13; II.1, 731b19–734b28.

Archimedes, *On the Equilibrium of Planes or the Centres of Gravity of Planes*. In: *The Works of Archimedes*, edited in Modern Notation, with introductory chapters by Thomas L. Heath; translated by Thomas L. Heath. Cambridge: Cambridge University Press, 1897.

Euclid, *The Thirteen Books of Euclid's Elements. Volume 1. Introduction and Books I, II,* 2nd edition, translated by Thomas L. Heath from the Text of Heiberg, with introduction and commentary by Thomas L. Heath. Cambridge: Cambridge University Press, 1926.

Apollonius of Perga, *Treatise on Conic Sections: Edited in Modern Notation with Introductions, Including an Essay on the Earlier History of the Subject,* translated by Thomas L. Heath. Cambridge: Cambridge University Press, 1896.

Aristarchus of Samos, *On the Sizes and Distances of the Sun and the Moon*. In: *Aristarchus of Samos, the Ancient Copernicus: A History of Greek Astronomy to Aristarchus, Together with Aristarchus's Treatise on the Sizes and Distances of the Sun and Moon, A New Greek Text with Translation and Notes; translated by Thomas L. Heath.* Oxford: Clarendon Press, 1913.

Eratosthenes, *Measurement of the Earth* as described by Cleomedes in *On the Circular Motion of the Heavenly Bodies*. In: *Selections Illustrating the History of Greek Mathematics: From Aristarchus to Pappus,* Vol. 2; edited by T. E. Page, E. Capps, W. H. D. Rouse, L. A. Post, and E. H. Warmington; translated by I. Thomas. London; Cambridge, MA: William Heinemann Ltd; Harvard University Press, 1941, reprinted in 1951, 1957.

Ptolemy, *Almagest*. In: *Ptolemy, Copernicus, Kepler: Great Books of the Western World,* Vol. 16; edited by Robert Maynard Hutchins; translated by R. Catesby Taliaferro. Chicago: Encyclopaedia Britannica, 1952.

Part II Translation, Appropriation, and Critical Engagement: Science in the Roman and Medieval Islamic and European Worlds

Galen, *On the Natural Faculties*, translated by Arthur John Brock. Loeb Classical Library 71. London; Cambridge, MA: William Heinemann Ltd; Harvard University Press, 1916.

Philoponus, *Commentary on Aristotle's Physics*. In: *A Source Book in Greek Science* (Source Books in the History of the Sciences); edited by Morris R. Cohen and I. E. Drabkin; translated by I. E. Drabkin. Cambridge: Harvard University Press, 1948.

Ibn Sina (Avicenna), *On Medicine*. In: *The Sacred Books and Early Literature of the East*, Vol. VI: *Medieval Arabia*; edited by Charles F. Horne. New York: Parke, Austin, & Lipscomb, 1917.

Ibn Sina (Avicenna), *The Book of the Remedy*. In Avicennae, "De congelatione et conglutinatione lapidum" being sections of the Kitab-al-Shifa, the Latin and Arabic text edited, with an English translation of the latter and with critical notes by E. J. Holmyard and D. C. Mandeville. Paris: Librairie Orientaliste Paul Geuthner, 1927.

Al-Bīrūnī, *The Book of Instruction in the Elements of the Arts of Astrology by Abū'l Rayḥān Muḥammad ibn Aḥmad Al-Bīrūnī*. Reproduced from *Brit. Mus. MS. Or. 8349*; translated by R. Ramsay Wright. London: Luzac & Co., 1934.

Ibn al-Haytham (Alhazen), *Perspectiva (Book of Optics)*. In: *A Source Book in Medieval Science*; edited by Edward Grant; translated by David C. Lindberg. Cambridge, MA: Harvard University Press, 1974.

Albertus Magnus (Albert the Great), *On Plants*. In: *A Source Book in Medieval Science*; edited by Edward Grant; translated by David C. Lindberg. Cambridge, MA: Harvard University Press, 1974.

Aquinas, *On the Motion of the Heart (De Motu Cordis)*, translated by Gregory Froelich. Available online at: http://www.theologywebsite.com/etext/aquinas/motionofheart.shtml..

Roger Bacon, *Opus Majus: On Experimental Science*. In: *The Library of Original Sources*, Vol. V: *The Early Medieval World*; edited by Oliver J. Thatcher. Milwaukee: University Research Extension Co., 1901, 369–76.

Oresme, Nicole (c. 1360). *Tractatus de configurationibus qualitatum et motuum (A Treatise on the Configuration of Qualities and Motions)*. Paris. In: *The Science of Mechanics in the Middle Ages*; translated by Marshall Clagett. Madison: The University of Wisconsin Press, 1959.

Buridan, John (1509). *Subtilissimae Quaestiones super octo Physicorum libros Aristotelis (Question on the Eight Books of the Physics of Aristotle)*. Paris. In: *The Science of Mechanics in the Middle Ages*; translated by Marshall Clagett. Madison: The University of Wisconsin Press, 1959.

Part III Revolutions in Astronomy and Mechanics: From Copernicus to Newton

Copernicus, Nicolaus (1515). *The Commentariolus*. In: *Three Copernican Treatises: The "Commentariolus of Copernicus," the "Letter against Werner," the "Narratio Prima of Rheticus"*; 3rd edition, *translated with introduction and notes by Edward Rosen*. New York: Dover, 1959.

Copernicus, Nicolaus (1543). *On the Revolutions of the Celestial Spheres*. In: *Prefaces and Prologues: To Famous Books*, edited by Charles William Eliot. The Harvard Classics. Vol. 39. New York: P.F. Collier & Son, 1909–14.

Osiander, Andreas (1543). *Preface to On the Revolutions*. In: *On the Revolutions of the Celestial Spheres; The First Translation into This Language of De Revolutionibus Orbium Coelestium* (from the Text of the Edition Published by the Societas Copernicana at Thorn, 1873), Nicolaus Copernicus; translated by Charles Glenn Wallis. Annapolis: St. John's Bookstore, 1939.

Brahe, Tycho (1598). *Astronomiæ instauratæ mechanica (Instruments for the Restoration of Astronomy)* Wandsbek. In: *Tycho Brahe's Description of His Instruments and Scientific Work as Given in Astronomiae Instauratae Mechanica*; translated and edited by H. Ræder, E. Strömgren, and B. Strömgren. Copenhagen: Munksgaard, 1946.

Kepler, Johannes (1618–21). *Epitome of Copernican Astronomy*, translated by Charles Glenn Wallis. In: Robert Maynard Hutchins (ed.), *Ptolemy, Copernicus, Kepler: Great Books of the Western World*, Vol. 16. Chicago: Encyclopaedia Britannica, 1952.

Galilei, Galileo (1610). *Starry Messenger*, with *Message to Cosimo de'Medici*. In: *The Sidereal Messenger of Galileo Galilei and a Part of the Preface to Kepler's Dioptrics Containing the Original Account of Galileo's Astronomical Discoveries*; translated 1880 by Edward Stafford Carlos. London. Reprinted, London: Dawson's of Pall Mall, 1960.

Galilei, Galileo (1632). *Dialogue concerning the Two Chief World Systems*. In: *Dialogue on the Great World Systems. Galileo Galilei*; translated by Thomas Salusbury; revised and annotated by Giorgio de Santillana. Chicago: University of Chicago Press, 1953.

Galilei, Galileo (1638). *Dialogue concerning the Two New Sciences*. In: *Dialogues concerning Two New Sciences by Galileo Galilei*; translated by Henry Crew and Alfonso de Salvio. New York: Macmillan, 1914.

Descartes, René (1664). *Le Monde, ou Traite de la lumiere (The World, or Treatise on the Light)*, translated by Michael Sean Mahoney. New York: Abaris Books, 1979.

Descartes, René (1637). *Discours de la methode pour bien conduire sa raison, & chercher la verité dans les sciences: plus la dioptrique, les meteores, et la geometrie, qui sont des essais de cete method (Discourse on the Method of Rightly Conducting the Reason and Seeking Truth in the Sciences, together with the Dioptrics, the Meteors, and the Geometry, which are essays in this method)*. Leiden: Jan Maire. In: *The Method, Meditations and Philosophy of Descartes*; translated from the original texts, with a new introductory essay, historical and critical by John Veitch and a special introduction by Frank Sewall. Washington: M. Walter Dunne, 1901.

Descartes, René (1644). *Principia philosophiae (Principles of Philosophy)*. Amsterdam: Elzevir. In: *The Method, Meditations and Philosophy of Descartes*; translated from the Original Texts, with a new introductory essay, historical and critical by John Veitch and a special introduction by Frank Sewall. Washington: M. Walter Dunne, 1901.

Bacon, Francis (1620). *Novum Organum Or True Suggestions for the Interpretation of Nature*, edited by Joseph Devey. New York: P.F. Collier & Son, 1902.

Bacon, Francis (1627). *The New Atlantis.* Vol. III, Part 2. The Harvard Classics. New York: P.F. Collier & Son, 1909–14.

Hooke, Robert (1674). *An Attempt to Prove the Motion of the Earth from Observations Made by Robert Hooke, Fellow of the Royal Society*. London.

Newton, Isaac (1687). *Philosophia naturalis principia mathematica or The Mathematical Principles of Natural Philosophy,* London; translated into English by Andrew Motte 1729 *To Which Is Added Newton's System of the World.* First American edition, Carefully Revised and Corrected, With a Life of the Author, by N. W. Chittenden. New York: Daniel Adee, 1846.

Buffon, Georges-Louis Leclerc, Comte de (1749). *Proof of The Theory of the Earth: On the Formation of the Planets*. In: *Natural History, Volume I (of X) Containing a Theory of the Earth, a General History of Man, of the Brute Creation, and of Vegetables, Mineral, &c. &c.*; translated by James Smith Barr. London: H.D. Symonds, 1797.

Part IV Investigating the Invisible: Light, Electricity, and Magnetism

Newton, Isaac (1671/2). "A Letter [to Henry Oldenburg] of Mr. Isaac Newton, Professor of the Mathematicks in the University of Cambridge; Containing His New Theory About Light and Colors." *Philosophical Transactions of the Royal Society of London* 6 (1671/2).

Hooke, Robert (1671/2). *Critique of Newton's Theory of Light and Colors*. In: *The History of the Royal Society*, vol. 3. London: Thomas Birch, 1757.

Newton, Isaac (1704). *Opticks, Or A Treatise of the Reflections, Refractions, Inflections and Colours of Life*, 4th edition, corrected. London: Printed for William Innys at the West-End of St. Paul's, 1730.

Huygens, Christiaan (1690). *Treatise on Light, in which are explained the causes of that which occurs in Reflexion, and in Refraction, and particularly in the strange Refraction of Iceland Crystal*, translated by Silvanus P. Thompson. Chicago: University of Chicago Press, 1945.

Young, Thomas (1802). "The Bakerian Lecture: On the Theory of Light and Colours." *Philosophical Transactions of the Royal Society of London* 92, 12–48.

Franklin, Benjamin (1747). *Experiments on Electricity, Written to Peter Collinson, Philadelphia, September 1, 1747*. In: Benjamin Franklin, *The Works of Benjamin Franklin, including the Private as well as the Official and Scientific Correspondence, together with the Unmutilated and Correct Version of the Autobiography*, compiled and edited by John Bigelow. New York: G.P. Putnam's Sons, 1904. The Federal Edition in 12 volumes. Vol. II (Letters and Misc. Writings 1735–53).

Oersted, John Christian (1820). "Experiments on the Effect of a Current of Electricity on the Magnetic Needle." *Annals of Philosophy* 16(4), 273–6.

Faraday, Michael (1859–60). *The Forces of Matter, Delivered before a Juvenile Auditory at the Royal Institution of Great Britain during the Christmas Holidays of 1859–60*. Reprinted in *Scientific Papers: Physics, Chemistry, Astronomy, Geology, Volume XXX* (The Harvard Classics), edited by Charles W. Eliot. New York: P.F. Collier & Son, 1909–14.

Faraday, Michael (1849). *Experimental Researches in Electricity*, 2nd edition, Vol. 1. London: Richard and John Edward Taylor, printers and publishers to the University of London, Red Lion Court, Fleet Street.

Maxwell, James Clerk (1865). "A Dynamical Theory of the Electromagnetic Field." *Philosophical Transactions of the Royal Society of London* 155, 459–512.

Part V Elements in Transition: Chemistry, Air, Atoms, and Heat

Paracelsus, *Of the Nature of Things* (1537). In: *The Alchemy Reader: From Hermes Trismegistus to Isaac Newton*; edited by Stanton J. Linden. Cambridge: Cambridge University Press, 2003.

Newton, Isaac (George Starkey). *The Key* (c. 1650s to 1670s). In: *The Alchemy Reader: From Hermes Trismegistus to Isaac Newton*; edited by Stanton J. Linden; translated by Betty Jo Teeter Dobbs. Cambridge: Cambridge University Press, 2003.

Boyle, Robert (1672). *Tracts Written by the Honourable Robert Boyle, Containing New Experiments, Touching the Relation between Flame and Air* and *About Explosions*. In: *A Source Book in Chemistry: 1400–1900*; edited by Henry M. Leicester and Herbert S. Klickstein. Cambridge: Harvard University Press, 1952.

Boyle, Robert (1674). *Of the Excellency and Grounds of the Corpuscular or Mechanical Philosophy*. In: *The excellency of theology compar'd with natural philosophy (as both are objects of men's study)/discours'd of in a letter to a friend by T.H.R.B.E. Fellow of the Royal Society; To which are annex'd some Occasional Thoughts about the Excellency and Grounds of the Mechanical Hypothesis*. London.

Becher, Johann Joachim (1669). *Concerning the First Principle of Metals and Stones, Which Is Called Vitreous Stone, or Stony Earth* [terra lapidia] *and Improperly Salt* from "*Acta Laboratorii Chymici Monacensis, Seu Physicae Subterranae.*" In: *A Source Book in Chemistry: 1400–1900*; edited by Henry M. Leicester and

Herbert S. Klickstein; translated by Henry M. Leicester and Herbert S. Klickstein. Cambridge: Harvard University Press, 1952.

Stahl, Georg Ernst (1697). *Foundation of the Fermentative Art* (*Zymotechnia Fundamentalis seu Fermentationis Theoria generalis*). In: *A Source Book in Chemistry: 1400–1900*; edited by Henry M. Leicester and Herbert S. Klickstein; translated by Henry M. Leicester and Herbert S. Klickstein. Cambridge: Harvard University Press, 1952.

Stahl, Georg Ernst (1718). *Random Thoughts and Useful Concerns* (*Zufällige Gedanken und nützliche Bedencken*). In: *A Source Book in Chemistry: 1400–1900*; edited by Henry M. Leicester and Herbert S. Klickstein; translated by Henry M. Leicester and Herbert S. Klickstein. Cambridge: Harvard University Press, 1952.

Stahl, Georg Ernst (1723). *Dogmatic and Experiential Foundations of Chemistry* (*Fundamenta chymiae dogmaticae et experimentalis*). In: *A Source Book in Chemistry: 1400–1900*; edited by Henry M. Leicester and Herbert S. Klickstein; translated by Peter Shaw. Cambridge: Harvard University Press, 1952.

Priestley, Joseph (1775). *Experiments and Observations on Different Kinds of Air*, Vol. 2. London. Section III. As translated in Alembic Club Reprint No. 7, Edinburgh, 1893.

Lavoisier, Antoine-Laurent (1789). *Traité élémentaire de chimie (Elements of Chemistry, In a New Systematic Order, Containing all the Modern Discoveries*), translated by Robert Kerr. Edinburgh: W. Creech, 1790.

Lavoisier, Antoine-Laurent and Laplace, Pierre-Simon marquis de (1780). *Memoir on Heat*. In: *Great Experiments in Biology*; edited by Mordecai L. Gabriel and Seymour Fogel; translated by Mordecai L. Gabriel. Englewood Cliffs: Prentice-Hall, 1955.

Dalton, John (1808). *A New System of Chemical Philosophy, Part I.* Manchester.

Gay-Lussac, Joseph Louis (1809). *Memoir on the Combination of Gaseous Substances with Each Other, Mémoires de la Société d'Arcueil* 2, 207–34. From Henry A. Boorse and Lloyd Motz, eds, *The World of the Atom*, Vol. 1. New York: Basic Books, 1966, as translated in Alembic Club Reprint No. 4, Edinburgh, 1890.

Amedeo Avogadro (1811). "Essay on a Manner of Determining the Relative Masses of the Elementary Molecules of Bodies, and the Proportions in Which They Enter into These Compounds." *Journal de Physique* 73, 58–76. As translated in Alembic Club Reprint No. 4, Edinburgh, 1890.

Carnot, Sadi (1824). *Réflexions sur la puissance motrice du feu et sur les machines propres à développer cette puissance (Reflection on the Motive Power of Heat and on Engines Suitable for Developing this Power)*. Paris: Bachelier. In: *The Second Law of Thermodynamics: Memoirs by Carnot, Clausius, and Thomson*; edited by William Francis Magie; translated by William Francis Magie. New York, London: Harper & Brothers, 1899.

Clausius, Rudolf (1857). "Ueber die Art der Bewegung, welche wir Winne nennen." *Annalen der Physik* 100, 353–80 (1857); translated into English by John Tyndall as "On the Nature of the Motion We Call Heat," *Philosophical Magazine* 14, 108–27 (1857).

Cannizzaro, Stanislao (1858). "Letter of Professor Stanislao Cannizzaro to Professor S. de Luca: Sketch of a Course of Chemical Philosophy." *Il Nuovo Cimento* 7, 321–66.

Mendeleev, Dmitri (1889). "The Periodic Law of the Chemical Elements." *Journal of the Chemical Society* 55 (1889), 634–56.

Kelvin, Lord (William Thomson) (1851–52). "On the Dynamical Theory of Heat, with numerical results deduced from Mr. Joule's equivalent of a Thermal Unit, and M. Regnault's Observations on Steam." *Transactions of the Royal Society of*

Edinburgh, March 1851, and *Philosophical Magazine* IV, 1852. Reprinted in Kelvin, William Thomson, *Mathematical and Physical Papers*, Vol. 1, Art. XLVIII, 174–332.
Helmholtz, Hermann von (1854). "On the Interaction of Natural Forces" (Lecture delivered February 7, 1854 at Königsberg), translated by John Tyndall. *The London, Edinburgh, and Dublin Philosophical Magazine and Journal of Science* 11, ser. 4 (1856), 489–518. Also printed in: *American Journal of Science and Arts* (1857) 24, 189–216.

Part VI The Earth and All Its Creatures: Developments in Geology and Biology

Harvey, William (1628). *On The Motion Of The Heart And Blood In Animals.* In: *The Works of William Harvey, M.D., Physician to the King, Professor of Anatomy and Surgery to the College of Physicians,* translated from the Latin, with a Life of the Author, by Robert Willis, MD. London: Printed for the Sydenham Society, 1847.
Descartes, René (1664). *L'homme, et un Traitté de la formation du foetus (Treatise on Man and On the Formation of the Fetus [Description of the Human Body])*, ed. Claude Clerselier. Paris: Charles Angot. Translated from original French by P. R. Sloan. Used with permission of Professor P. R. Sloan.
Hooke, Robert (1665). *Micrographia, or some Physiological Descriptions of Minute Bodies made by Magnifying Glasses with Observations and Inquiries thereupon.* London: John Martyn and James Allestry.
Schwann, Theodor (1839). *Microscopical Researches into the Accordance in the Structure and Growth of Animals and Plants*, translated into English by Henry Smith, for the Sydenham Society, 1847.
Buffon, Georges-Louis Leclerc, Comte de (1749). *The Theory of the Earth.* In: *Natural History, Volume I (of X) Containing a Theory of the Earth, a General History of Man, of the Brute Creation, and of Vegetables, Mineral, &c. &c;* translated into English from the original 1749 French *Imprimerie Royale* edition of the *Histoire naturelle* by James Smith Barr. London: H.D. Symonds, 1797. With emendations from the original French by P. R. Sloan.
Buffon, Georges-Louis Leclerc, Comte de (1749). *History of Animals.* In: *Natural History, Volume II (of X) Containing a Theory of the Earth, a General History of Man, of the Brute Creation, and of Vegetables, Mineral, &c. &c*; translated into English from the original 1749 French *Imprimerie Royale* edition of the *Histoire naturelle* by James Smith Barr. London: H.D. Symonds, 1797. With emendations from the original French by P. R. Sloan. Used with permission of Professor P. R. Sloan.
Linnaeus, Carl (1744). *On the Increase of the Habitable Earth.* In: Carl Linnaeus, *Selected Dissertations from the Amoenitates Academicae (1749–1790), a Supplement to Mr. Stillingfleet's Tracts Relating to Natural History*; translated by F. J. Brand. London: Robinson, et. Robson, 1781, Dissertation II.
Biberg, Isaac J. (1749). "*Specimen Academicum de OEconomia Naturae Quam Praeside Dn. D. Carlolo Linnaeo submittet I. J. Berg d. IV Martii MDCCXLIX*" (*The Economy of Nature*, a dissertation presided over by Carl Linnaeus), Uppsala. In: Carl Linnaeus (ed.), *Miscellaneous Tracts Relating to Natural History, Husbandry, and Physick. To Which Is Added The Calendar of Flora*, 2nd edition, Corrected and

Augmented, etc.; translated and edited by Benjamin Stillingfleet. London: R. and J. Dodsley, S. Baker, and T. Payne, 1762.
Malthus, Thomas (1798). *An Essay on the Principle of Population.* London: Printed for J. Johnson.
Paley, William (1802). *Natural Theology: or, Evidences of the Existence and Attributes of the Deity*, 12th edition. London: Printed for J. Faulder, 1809.
Lamarck, Jean-Baptiste (1809). *Philosophie Zoologique ou exposition des considérations relatives à l'histoire naturelle des animaux* (*Zoological Philosophy: Exposition with Regard to the Natural History of Animals*). Paris: Museum d'Histoire Naturelle (Jardin des Plantes); translated by Ian Johnston.
Cuvier, Georges (1800). *First Lecture, Article 4: On the Law of the Correlation of Parts.* In: Georges Cuvier, *Lecons d'anatomie comparée* (*Lectures on Comparative Anatomy)*, Vol. 1, Paris; translated by William Ross with James Macartney. London: Wilson and Co. for T. N. Longman and O. Rees, 1802, 46–61.
Cuvier, Georges (1826). *Discours sur les révolutions de la surface du globe, et sur les changemens qu'elles ont produits dans le règne animal (Discourse on the Revolutionary Upheavals on the Surface of the Globe, and on the changes which they have produced in the animal kingdom)*; translated anonymously from the French with Illustrations and a Glossary. Philadelphia: Carey & Lea, 1831.
Cuvier, Georges (1828). "General Conclusion on the Organization of the Fishes." In: Georges Cuvier and A. Valenciennes, *Histoire naturelle des Poissons* (Paris: Levrault, 1828), I, 543–51; translated into English by P. R. Sloan.
Lyell, Charles (1830). *Principles of Geology, or, The Modern Changes of the Earth and its Inhabitants Considered as Illustrative of Geology.* New York: D. Appleton & Co.

Part VII The Emergence of Evolution: Darwin and His Interlocutors

Wallace, Alfred Russel (1858). "On the Tendency of Species to form Varieties," published together with Charles Darwin and A. R. Wallace, "On the tendency of species to form varieties; and on the perpetuation of varieties and species by natural means of selection." *Journal of the Proceedings of the Linnean Society of London, Zoology* 3 (August 20, 1858), 45–50.
Darwin, Charles (1859). *On the Origin of Species, Or the Preservation of Favoured Races in the Struggle for Life,* 1st edition. London: John Murray, 1859.
Darwin, Charles (1871). *Descent of Man, and Selection in Relation to Sex*, 2nd edition, revised and augmented. London: John Murray, 1874.
Kelvin, Lord (William Thomson) (1866). "The 'Doctrine of Uniformity' in Geology Briefly Refuted." *Proceedings of the Royal Society of Edinburgh* 5, 512–13.
Fleeming Jenkin (1867). "Review of '*The Origin of Species.*'" *The North British Review* 46, 277–318.
Sedgwick, Adam (1859). Letter from Adam Sedgewick to Charles Darwin (November 24, 1859).
Owen, Richard (1860). "Darwin on the Origin of Species." *Edinburgh Review* 3, 487–532.
Gray, Asa (1860). "Darwin and His Reviewers." *Atlantic Monthly* 6, 406–25.
Huxley, T. H. (1880, July 3). "The Coming of Age of the *Origin of Species.*" *Science* 1, 15–20.

Darwin, Charles (1868). *The Variation of Animals and Plants under Domestication*, Vol. 2 2nd edition 1875. London: John Murray.

Weldon, W. F. R. (1893). "On Certain Correlated Variations in *Carcinus maenas*." *Proceedings of the Royal Society of London* 54, 318–29.

Mendel, Gregor (1866). *Versuche über Plflanzenhybriden. Verhand- lungen des naturforschenden Vereines in Brünn, Bd. IV für das Jahr* 1865 ("Experiments in Plant Hybridization," Read at the February 8 and March 8, 1865, meetings of the Brünn Natural History Society), *Abhandlungen*, 3–47; translated into English by William Bateson in 1901, with some minor corrections and changes provided by Roger Blumberg.

Index